Periodic Table of the Elements with the Gmelin System Numbers

1 H 2																	2 He 1
3 Li 20	4 Be 26											5 B 13	6 C 14	7 N 4	8 O 3	9 F 5	10 Ne 1
11 Na 21	12 Mg 27											13 Al 35	14 Si 15	15 P 16	16 S 9	17 Cl 6	18 Ar 1
19 * K 22	20 Ca 28	21 Sc 39	22 Ti 41	23 V 48	24 Cr 52	25 Mn 56	26 Fe 59	27 Co 58	28 Ni 57	29 Cu 60	30 Zn 32	31 Ga 36	32 Ge 45	33 As 17	34 Se 10	35 Br 7	36 Kr 1
37 Rb 24	38 Sr 29	39 Y 39	40 Zr 42	41 Nb 49	42 Mo 53	43 Tc 69	44 Ru 63	45 Rh 64	46 Pd 65	47 Ag 61	48 Cd 33	49 In 37	50 Sn 46	51 Sb 18	52 Te 11	53 I 8	54 Xe 1
55 Cs 25	56 Ba 30	57** La 39	72 Hf 43	73 Ta 50	74 W 54	75 Re 70	76 Os 66	77 Ir 67	78 Pt 68	79 Au 62	80 Hg 34	81 Tl 38	82 Pb 47	83 Bi 19	84 Po 12	85 At 8a	86 Rn 1
87 Fr 25a	88 Ra 31	89*** Ac 40	104 71	105 71													

* NH_4 23

**Lanthanides 39	58 Ce	59 Pr	60 Nd	61 Pm	62 Sm	63 Eu	64 Gd	65 Tb	66 Dy	67 Ho	68 Er	69 Tm	70 Yb	71 Lu
***Actinides	90 Th 44	91 Pa 51	92 U 55	93 Np 71	94 Pu 71	95 Am 71	96 Cm 71	97 Bk 71	98 Cf 71	99 Es 71	100 Fm 71	101 Md 71	102 No 71	103 Lr 71

A Key to the Gmelin System is given on the Inside Back Cover

Gmelin Handbook of Inorganic and Organometallic Chemistry

8th Edition

Gmelin Handbook of Inorganic and Organometallic Chemistry

8th Edition

Gmelin Handbuch der Anorganischen Chemie

Achte, völlig neu bearbeitete Auflage

PREPARED AND ISSUED BY

Gmelin-Institut für Anorganische Chemie
der Max-Planck-Gesellschaft
zur Förderung der Wissenschaften
Director: Ekkehard Fluck

FOUNDED BY

Leopold Gmelin

8TH EDITION

8th Edition begun under the auspices of the
Deutsche Chemische Gesellschaft by R. J. Meyer

CONTINUED BY

E. H. E. Pietsch and A. Kotowski, and by
Margot Becke-Goehring

Springer-Verlag Berlin Heidelberg GmbH 1991

GMELIN HANDBOOK

Dr. J. von Jouanne

Dr. L. Berg, Dr. H. Bergmann, Dr. J. Faust, J. Füssel, Dr. H. Katscher, Dr. R. Keim, Dr. E. Koch, Dipl.-Phys. D. Koschel, Dr. A. Kubny, Dr. P. Merlet, Dr. M. Mirbach, Prof. Dr. W. Petz, Dr. F.A. Schröder, Dr. A. Slawisch, Dr. W. Töpper

Dr. R. Albrecht, Dr. G. Bär, D. Barthel, Dr. N. Baumann, Dr. K. Behrends, Dr. W. Behrendt, D. Benzaid, Dr. R. Bohrer, K.D. Bonn, Dr. U. Busch, Dipl.-Ing. V.A. Chavizon, E. Cloos, Dipl.-Phys. G. Czack, A. Dittmar, Dipl.-Geol. R. Ditz, R. Dowideit, Dipl.-Chem. M. Drößmar, P. Dürr, Dr. H.-J. Fachmann, B. Fischer, Dipl.-Ing. N. Gagel, Dipl.-Phys. D. Gras, Dr. K. Greiner, Dipl.-Bibl. W. Grieser, Dr. I. Haas, Dr. R. Haubold, Dipl.-Min. H. Hein, Dipl.-Phys. C. Heinrich-Sterzel, H.-P. Hente, H.W. Herold, U. Hettwer, Dr. G. Hönes, Dr. W. Hoffmann, G. Horndasch, Dr. W. Huisl, Dr. M. Irmler, B. Jaeger, Dr. R. Jotter, Dipl.-Chem. P. Kämpf, Dr. B. Kalbskopf, Dipl.-Chem. W. Karl, H.-G. Karrenberg, Dipl.-Phys. H. Keller-Rudek, B. Kirchner, Dipl.-Chem. C. Koeppel, R. Kolb, Dr. M. Kotowski, E. Kranz, Dipl.-Chem. I. Kreuzbichler, Dr. V. Kruppa, Dr. W. Kurtz, M. Langer, Dr. B. Ledüc, Dr. A. Leonard, H. Mathis, E. Meinhard, M. Meßer, C. Metz, K. Meyer, E. Mlitzke, Dipl.-Chem. B. Mohsin, Dr. U. Neu-Becker, K. Nöring, Dipl.-Min. U. Nohl, Dr. U. Ohms-Bredemann, I. Rangnow, Dipl.-Phys. H.-J. Richter-Ditten, E. Rudolph, G. Rudolph, Dipl.-Chem. S. Ruprecht, Dr. B. Sarbas, Dr. H. Schäfer, Dr. R. Schemm, Dr. D. Schiöberg, V. Schlicht, Dipl.-Chem. D. Schneider, A. Schwärzel, Dr. B. Schwager, Dr. C. Strametz, G. Strauss, Dr. G. Swoboda, Dr. D. Tille, A. Tuttas, H.-M. Wagner, Dipl.-Phys. J. Wagner, R. Wagner, Dr. E. Warkentin, Dr. C. Weber, Dr. A. Wietelmann, Dr. B. Wöbke, K. Wolff

GMELIN ONLINE

Dr. R. Deplanque

Dr. P. Kuhn, Dr. G. Olbrich

Dr. R. Baier, Dr. B. Becker, Dipl.-Chem. E. Best, Dr. H.-U. Böhmer, Dipl.-Phys. R. Bost, Dr. A. Brandl, Dr. R. Braun, Dipl.-Chem. R. Durban, R. Hanz, Dr. A. Kirchhoff, Dipl.-Chem. H. Köttelwesch, Dipl.-Ing. W. Korba, Dr. M. Kunz, Dr. L. Leichner, Dipl.-Chem. R. Maass, Dr. A. Nebel, Dipl.-Chem. R. Nohl, Dr. B. Rempfer, Dr. U. Tölle, Dipl.-Ing. H. Vanecek

Organometallic Compounds in the Gmelin Handbook

The following listing indicates in which volumes these compounds are discussed or are referred to:

Ag	Silber B 5 (1975)
Au	Organogold Compounds (1980)
Be	Organoberyllium Compounds 1 (1987)
Bi	Bismut-Organische Verbindungen (1977)
Co	Kobalt-Organische Verbindungen 1 (1973), 2 (1973), Kobalt Erg.-Bd. A (1961), B 1 (1963), B 2 (1964)
Cr	Chrom-Organische Verbindungen (1971)
Cu	Organocopper Compounds 1 (1985), 2 (1983), 3 (1986), 4 (1987), Index (1987)
Fe	Eisen-Organische Verbindungen A 1 (1974), A 2 (1977), A 3 (1978), A 4 (1980), A 5 (1981), A 6 (1977), A 7 (1980), Organoiron Compounds A 8 (1985), A 9 (1989), A 10, Eisen-Organische Verbindungen B 1 (partly in English; 1976), Organoiron Compounds B 2 (1978), Eisen-Organische Verbindungen B 3 (partly in English; 1979), B 4 (1978), B 5 (1978), Organoiron Compounds B 6 (1981), B 7 (1981), B 8 to B 10 (1985), B 11 (1983), B 12 (1984), B 13 (1988), B 14 (1989), B 15 (1989), B 16a (1990), B 16b (1990), B 17 (1990), B 18 (1991), Eisen-Organische Verbindungen C 1 (1979), C 2 (1979), Organoiron Compounds C 3 (1980), C 4 (1981), C 5 (1981), C 7 (1985), and Eisen B (1929–1932)
Ga	Organogallium Compounds 1 (1986)
Ge	Organogermanium Compounds 1 (1988), 2 (1989), 3 (1990)
Hf	Organohafnium Compounds (1973)
In	Organoindium Compounds 1 (1991) **present volume**
Mo	Organomolybdenum Compounds 6 (1990), 7 (1991)
Nb	Niob B 4 (1973)
Ni	Nickel-Organische Verbindungen 1 (1975), 2 (1974), Register (1975), Nickel B 3 (1966), and C 1 (1968), C 2 (1969)
Np, Pu	Transurane C (partly in English; 1972)
Pb	Organolead Compounds 1 (1987), 2 (1990)
Pt	Platin C (1939) and D (1957)
Re	Organorhenium 1 (1989), 2 (1989)
Ru	Ruthenium Erg.-Bd. (1970)
Sb	Organoantimony Compounds 1 (1981), 2 (1981), 3 (1982), 4 (1986), 5 (1990)
Sc, Y, La to Lu	D 6 (1983)
Sn	Zinn-Organische Verbindungen 1 (1975), 2 (1975), 3 (1976), 4 (1976), 5 (1978), 6 (1979), Organotin Compounds 7 (1980), 8 (1981), 9 (1982), 10 (1983), 11 (1984), 12 (1985), 13 (1986), 14 (1987), 15 (1988), 16 (1988), 17 (1989), 18 (1990), 19 (1991)
Ta	Tantal B 2 (1971)
Ti	Titan-Organische Verbindungen 1 (1977), 2 (1980), Organotitanium Compounds 3 (1984), 4 and Register (1984), 5 (1990)
U	Uranium Suppl. Vol. E 2 (1980)
V	Vanadium-Organische Verbindungen (1971), Vanadium B (1967)
Zr	Organozirconium Compounds (1973)

Gmelin Handbook of Inorganic and Organometallic Chemistry

8th Edition

In

Organoindium Compounds 1

With 101 illustrations

AUTHOR Johann Weidlein

EDITOR Wolfgang Petz

FORMULA INDEX Bernd Kalbskopf, Hans-Jürgen Richter-Ditten
Edgar Rudolph

CHIEF EDITOR Wolfgang Petz

Springer-Verlag Berlin Heidelberg GmbH 1991

LITERATURE CLOSING DATE: SPRING 1991

Library of Congress Catalog Card Number: Agr 25-1383

ISBN 978-3-662-09146-3 ISBN 978-3-662-09144-9 (eBook)
DOI 10.1007/978-3-662-09144-9

Originally published by Springer-Verlag Berlin Heidelberg New York London Paris Tokyo Hongkong Barcelona in 1991
Softcover reprint of the hardcover 8th edition 1991

Typesetting

Preface

The present volume contains all compounds in which at least one indium-carbon bonding interaction can be assumed. The compilation starts with the simplest compound of trivalent indium, $In(CH_3)_3$, and ends with studies about the interaction of indium with carbon monoxide in an argon matrix. Literature coverage is intended to be complete to spring 1991 with various examples up to September 1991.

The arrangement is closely related to that of the organogallium volume and documents the similarities between the two elements. Following the indium triorganyls and their adducts with Lewis bases in Section 1, the broad field of compounds of the general type R_nInX_{3-n} (n = 1, 2) is treated in sections 2 to 9; X represents a ligand bonded with a non-carbon atom to the indium atom. The arrangement of the various ligands follows the order group 17, 16, 15, etc. elements, with few compounds having direct indium-transition metal bonds. Ionic species, predominantly $[R_nInX_{4-n}]^-$ compounds (n = 1 to 4), close the series of trivalent organoindium compounds and are collected in Section 11. Compounds of formally low valent indium (In^{II}, In^{I}, and In^{0}), with one R_2InInR_2 species having an In-In bond, form Section 12; an extended chapter therein is dedicated to the young area of Cp*In compounds in which formal In^{I} is coordinated in an η^5 manner.

About 100 X-ray studies, mainly from recent papers, reveal the structural diversity of the compounds as a result of the balance between Lewis acidity originating from the natural electron deficiency of group 13 compounds, the basicity of the ligands or parts of the ligands, and the steric requirements of the ligands. These properties are also responsible for the various degrees of association (dimers, trimers, or polymers) of these compounds and the high affinity to (mainly) hard Lewis bases D. However, for better classification only the monomeric formula unit is used in the Formula Index.

Up to the last two decades preparative, spectroscopic, and structural results in organoindium chemistry have been no more than accompaniments to the well studied and more extensive chemistry of organogallium, or even organoaluminum and organothallium compounds. The discovery of indium organic chemistry in recent time as an independent branch is due to the fact that volatile compounds from this family were studied as potential sources for gas phase epitaxy (MOVPE) production of semiconducting layers. However, we have not intended this volume to cover completely this area of the literature, but some recent literature is added in the chapter concerning the application of $In(CH_3)_3$.

The similarity of organoindium chemistry to organogallium chemistry has at best resulted in reviews of the organo group 13 elements describing the few known details of the corresponding indium chemistry in the appendices. Thus, we have renounced a general literature list; the reader is referred to the Gmelin organogallium volume and to the many annual reports in the series "Specialist Periodical Report" Organometallic Chemistry **1** [1972] to **20** [1991] and to the very useful reviews about spectroscopic data from various authors in "Specialist Periodical Report" Spectrosc. Prop. Inorg. Organomet. Compounds **1** [1968] to **24** [1991].

We hope that this timely and comprehensive review supports the rise of organoindium chemistry and can give new impetus to active researchers all over the world.

Stuttgart and Frankfurt
November 1991

Johann Weidlein
Wolfgang Petz

Explanations, Abbreviations, and Units

Abbreviations are used in the text and in the tables; units are omitted in some tables for the sake of conciseness. This necessitates the following clarification:

Temperatures are given in °C, otherwise K stands for Kelvin. Abbreviations used with temperatures are m.p. for melting point, b.p. for boiling point, dec. for decomposition, and subl. for sublimation. Terms like 80 °C/0.1 Torr mean the boiling or sublimation point at a pressure of 0.1 Torr. **Densities** D are given in g/cm^3. D_c and D_m distinguish calculated and measured values, respectively.

NMR represents **nuclear magnetic resonance**. Chemical shifts are given as δ values in ppm and positive to low field from the following reference substances: $Si(CH_3)_4$ for ^1H and ^{13}C, $BF_3 \cdot O(C_2H_5)_2$ for ^{11}B, $CFCl_3$ for ^{19}F, H_3PO_4 for ^{31}P, $Sn(CH_3)_4$ for ^{119}Sn, and $Pb(CH_3)_4$ for ^{207}Pb. Multiplicities of the signals are abbreviated as s, d, t, q (singlet to quartet), quint, sext, sept (quintet to septet), and m (multiplet); terms like dd (double doublet) and t's (triplets) are also used. Assignments referring to labeled structural formulas are given in the form C-4, H-3.5. Coupling constants J in Hz appear usually in parentheses behind the δ values, along with the multiplicity and the assignment, and refer to the respective nucleus. If a more precise designation is necessary, they are given as, e.g., $^nJ(C,H)$ or J(1,3) referring to labeled formulas.

NQR represents **nuclear quadrupole resonance** and the values for ^{115}In, ^{79}Br, etc. are given in MHz (ν_n, e^2Qq/h); η in %.

Optical spectra are labeled as IR (infrared), Raman, and UV (electronic spectrum including the visible region). IR bands and Raman lines are given in cm^{-1}; the assigned bands are usually labeled with the symbols ν for stretching vibration and δ for deformation vibration. Intensities occur either in the common qualitative terms (s, m, w, vs, etc.) or as numerical relative intensities; p and dp mean polarized and depolarized, respectively. The UV absorption maxima, λ_{max}, are given in nm followed by the extinction coefficient ε (L·cm^{-1}·mol^{-1}) or log ε in parentheses; sh means shoulder.

Photoelectron spectra are abbreviated PE, e.g., PE/He(I), with the ionization energies in eV.

Solvents or the **physical state** of the sample and the temperature (in °C or K) are given in parentheses immediately after the spectral symbol, e.g., IR (solid), ^{13}C NMR (C_6D_6, 50 °C), or at the end of the data if spectra for various media are reported. Common solvents are given by their formula (C_6H_{12}=cyclohexane) except THF, DMF, and HMPT, which represent tetrahydrofuran, dimethylformamide, and hexamethylphosphoric acid triamide, respectively.

The data of **mass spectra** are given as m/e or m/z, relative intensity in parentheses, and fragment ions in brackets; $[M]^+$ is the molecular ion and m* represents a metastable peak.

Electron spin resonance is abbreviated as ESR. Radicals, e.g., $InR_n^{\cdot}$, are characterized by their g-factors; hyperfine splittings a are given in G values.

Figures give only selected parameters. Barred bond lengths (in pm) or angles (°) are mean values for parameters of the same type.

Table of Contents

Organoindium Compounds

1 Indium Triorganyls

1.1 Compounds of the InR_3 Type

1.1.1 Trimethylindium and Its Adducts

1.1.1.1 $In(CH_3)_3$

1.1.1.1.1 Preparation

Almost all the preparative procedures applicable to $Ga(CH_3)_3$ [30] can, with slight alteration, be used to prepare $In(CH_3)_3$.

$In(CH_3)_3$ was first prepared by transmetalation of electrolytically purified indium metal and excess $Hg(CH_3)_2$ in the presence of a small amount of $HgCl_2$. It was made in a special glass apparatus at a temperature of about 100 °C, using CO_2 at atmospheric pressure to exclude air [1]. Detailed description of the apparatus is given in reference [1]. The extremely long reaction time of 8 to 10 days [1, 4] (30 days was also reported [8]) can be diminished to 2 to 3 days, if a simple glass pressure bottle and temperatures of 130 °C (starting at 165 °C) are used [16]. $In(CH_3)_3$ can be obtained in 2 to 3 h if activated In metal (obtained by reduction of $InCl_3$ with Na or K) is allowed to react with $Hg(CH_3)_2$ at 100 °C under N_2 [17]. The yield was also increased from 30% [4] or 50% [1, 8] to close to 100% [16, 17]. The unreacted $Hg(CH_3)_2$ was removed by briefly evacuating the system (0 °C/10^{-4} Torr), and the $In(CH_3)_3$ purified by repeated vacuum sublimation [1, 16, 17]. The advantage of this procedure lies in the formation of solvent-free trimethylindium.

Transalkylation of an indium trihalogenide with a solvent-free trialkylaluminium (1:3 mole ratio), first used for the preparation of $In(C_2H_5)_3$ [7], can also be used to prepare $In(CH_3)_3$. Usually, the isolation of InR_3 succeeds only after the accumulated R_2AlX (X=Cl, Br) is complexed by adding KX to form $K[R_2AlX_2]$ [7]. KF as a complexing agent, added after the reaction of $InCl_3$ and $Al(CH_3)_3$ (3:1:3 mole ratio) was complete, has proved to be especially favorable for preparing $In(CH_3)_3$, with yields of about 70% being achieved [32].

Trimethylaluminium, in the form of $K[Al(CH_3)_3Cl]$, is also suitable for making $In(CH_3)_3$ [15]. In this method final purification is accomplished by repeated sublimation of the crude product, after addition of KF or CsF [7, 32]. The reaction is also appropriate for higher trialkylindium compounds [7]. The yield and purity of the trialkyls can be increased by adding to the reaction mixture an alkane whose boiling point lies between that of the alkylaluminium used and that of the expected alkylindium [18].

All other methods produce only adducts of $In(CH_3)_3$. For the production of $(CH_3)_3In \cdot O(C_2H_5)_2$ from various starting materials and ether as the solvent, see Section 1.1.1.2.1 [2, 3, 9, 11, 23 to 25, 29].

An elegant variation on alkylation by ethereal $LiCH_3$ is represented by the reaction of $InCl_3$ and $LiCH_3$ in 1:4 ratio, which first forms the indate $[Li(O(C_2H_5)_2)_n][In(CH_3)_4]$ [14]. The coordinated ether can be completely removed in vacuum under mild conditions. The ether-free indate is then reacted with $InCl_3$ (3:1 mole ratio) in an inert solvent; $In(CH_3)_3$ forms in 80% yield [33, 37]; for the preparation from $Li[In(CH_3)_4]$ with $InCl_3$ in refluxing benzene, see [38].

The ether adduct $(CH_3)_3In \cdot O(C_2H_5)_2$ has been used for the preparation of $In(CH_3)_3$. The adduct has been obtained from the reaction of InX_3 compounds with $LiCH_3$ [11, 29], with CH_3MgBr [23, 25], and with $Mg(CH_3)_2$ [24, 40], as well as from In and CH_3I [9]. All of these preparations are done in the presence of ether. See Sections 1.1.1.2 and 1.1.1.2.1 for preparation information.

The principal problem in preparing $In(CH_3)_3$ from its etherate is the quantitative separation of the ether. This can be done by distillation of the adduct after addition of benzene [9] or by cracking the etherate at normal pressure and with temperatures increasing from 100 to 150 °C [22, 32].

Trimethylindium containing ≦1 ppm of diisopentyl ether was obtained in 64% yield when a mixture of powdered In–Mg alloy and CH_3I was treated dropwise with diisopentyl ether over 5 h and refluxed for 2 h [39].

Another way to obtain solvent-free $In(CH_3)_3$ having high purity, is the thermal decomposition of the solid diadduct, $\{(CH_3)_3In\}_2\{(C_6H_5)_2PCH_2CH_2P(C_6H_5)_2\}$ (p. 52) at 100 to 120 °C [26, 28, 35]. Similar purifications via adducts with N-donor ligands are described in [31]. The formation of $In(CH_3)_3$ on heating of $(CH_3)_3In \cdot MBDA$ (MBDA = 4,4′-methylenebis(N,N-dimethylaniline)) at 80 to 130 °C/10^{-2} Torr is described in a patent [34].

$In(CH_3)_3$ formed by disproportionation of $In(CH_3)_2(C_2H_5)$ when this compound was distilled just above room temperature in vacuum [36].

$In(CH_3)_3$ has been evaluated for contamination from solvents and metal alkyls used during the preparation by gas chromatography and atomic absorption spectroscopy in addition to simple elementary analysis (C and H as CH_4, In as In_2O_3 [1]) [19 to 21]. The sensitivity of these methods lies between 5 and 8×10^{-6} wt% using solid supports based on silicon. Trace impurities of Si, Zn, and Mg alkyls on the order of ppm or less were determined by inductively coupled plasma emission spectroscopy (ICP). As a check on the results, electrical and optical measurements were made on InP semiconductor layers (obtained by gas phase epitaxy of analyzed $In(CH_3)_3$) [27].

The standard enthalpy of formation, ΔH_f°, for solid $In(CH_3)_3$ was calculated to be 29.5 kcal/mol based upon experimental data from the reaction of solid $In(CH_3)_3$ with excess bromine in $CHCl_3$ at 25 °C. The calculation made use of $\Delta H_f^\circ = -102.5$ kcal/mol for solid $InBr_3$, and $\Delta H_f^\circ = -9.0$ kcal/mol for gaseous CH_3Br. Using the sublimation enthalpy, $\Delta H_s^\circ = 11.6$ kcal/mol, the standard enthalpy of formation of gaseous $In(CH_3)_3$ was calculated to be $\Delta H_f^\circ = 41.4$ kcal/mol [12]. This corresponds closely to the value of 41 ± 3 kcal/mol calculated from the results of the thermal decomposition of $In(CH_3)_3(g)$ (p. 11) [10]. For comparisons with other $M(CH_3)_n$ compounds, see [5, 6, 13].

References:

[1] Dennis, L.M.; Work, R.W.; Rochow, E.G. (J. Am. Chem. Soc. **56** [1934] 1047/9).

[2] Runge, F.; Zimmermann, W.; Pfeiffer, H.; Pfeiffer, I. (Z. Anorg. Allgem. Chem. **267** [1951] 39/48).

[3] Coates, G.E.; Whitcombe, R.A. (J. Chem. Soc. **1956** 3351/4).

[4] van der Kelen, G.P. (Bull. Soc. Chim. Belges **65** [1956] 343/9).

[5] Long, L.H. (Pure Appl. Chem. **2** [1961] 61/6).

[6] Skinner, R.A.; Bennett, J.E.; Pedley, J.B. (Pure Appl. Chem. **2** [1961] 17/24).

[7] Eisch, J.J. (J. Am. Chem. Soc. **84** [1962] 3605/10).

[8] Muller, N.; Otermat, A.L. (Inorg. Chem. **2** [1963] 1075/6).

[9] Todt, E.; Dötzer, R. (Z. Anorg. Allgem. Chem. **321** [1963] 120/3).

[10] Jacko, M.G.; Price, S.J.W. (Can. J. Chem. **42** [1964] 1198/1205).

[11] Clark, H.C.; Pickard, A.L. (J. Organometal. Chem. **8** [1967] 427/34).
[12] Clark, W.D.; Price, S.J.W. (Can. J. Chem. **46** [1968] 1633/4).
[13] Cox, J.D.; Pilcher, G. (Thermochemistry of Organic and Organometallic Compounds, Academic, London 1970, pp. 466/8).
[14] Hoffmann, K.; Weiss, E. (J. Organometal. Chem. **37** [1972] 1/8).
[15] Golubinskaya, L.M.; Bregadze, V.I.; Okhlobystin, O.Yu. (U.S.S.R. 375293 [1973] from C.A. **79** [1973] No. 53528).
[16] Krommes, P.; Lorberth, J. (Inorg. Nucl. Chem. Letters **9** [1973] 587/9).
[17] Chao, L.-C.; Rieke, R.D. (Synth. React. Inorg. Metal.-Org. Chem. **4** [1974] 373/8).
[18] Efremov, E.A.; Falaleev, V.A.; Quinberg, E.E.; Strel'chenko, S.S.; Lebedev, V.V.; Fedorov, V.A.; Sel'vernskii, Ya.D. (U.S.S.R. 417429 [1974] from C.A. **81** [1974] No. 49813).
[19] Zorin, A.D.; Umilin, V.A. (Zavodsk. Lab. **40** [1974] 502/3 from C.A. **81** [1974] No. 99028).
[20] Zorin, A.D.; Sorokina, N.M.; Lokhov, N.S.; Rachkova, O.F.; Umilin, V.A. (Zavodsk. Lab. **44** [1978] 157/9 from C.A. **89** [1978] No. 16326).

[21] Shushunova, A.F.; Demarin, V.T.; Makin, G.I.; Sklemina, L.V.; Rudnevskii, N.K.; Aleksandrov, Yu.A. (Zh. Analit. Khim. **35** [1980] 349/52; J. Anal. Chem. [USSR] **35** [1980] 240/3 from C.A. **92** [1980] No. 226120).
[22] Efremov, E.A.; Fedorov, V.A.; Ferapontov, A.P.; Dorogova, L.A.; Kozyrkin, B.I.; Filippov, E.P.; Sredinskaya, I.A. (U.S.S.R. 810700 [1981] from C.A. **95** [1981] No. 62396).
[23] Cole-Hamilton, D.J.; Gerrard, N.D.; Jones, A.C.; Holliday, A.K.; Mullin, J.B.; U.K. Secretary of State for Defence (Brit. Appl. 2125795 [1984] from C.A. **101** [1984] No. 91212).
[24] Jones, A.C.; Gerrard, N.D.; Cole-Hamilton, D.J.; Holliday, A.K. (J. Organometal. Chem. **265** [1984] 9/15).
[25] Jones, A.C.; Holliday, A.K.; Cole-Hamilton, D.J.; Ahmad, M.M.; Gerrard, N.D. (J. Cryst. Growth **68** [1984] 1/9).
[26] Bradley, D.C.; Faktor, M.M.; Scott, M.; White, E.A.D. (J. Cryst. Growth **75** [1986] 101/6).
[27] Jones, A.C.; Jacobs, P.R.; Cafferty, R.; Scott, M.D.; Moore, A.H.; Wright, P.J. (J. Cryst. Growth **77** [1986] 47/54).
[28] Moore, A.H.; Scott, M.D.; Davies, J.I.; Bradley, D.C.; Faktor, M.M.; Chudzynska, H. (J. Cryst. Growth **77** [1986] 19/22).
[29] Reier, F.-W.; Wolfram, P.; Schumann, H. (J. Cryst. Growth **77** [1986] 23/6).
[30] Gmelin Handbook "Organogallium Compounds" 1, 1987.

[31] Forster, D.F.; Rushworth, S.A.; Cole-Hamilton, D.J.; Jones, A.C.; Stagg, J.P. (Chemtronics **3** [1988] 38/43).
[32] Laube, G.; Kohler, U.; Weidlein, J.; Scholz, F.; Streubel, K.; Dieter, R.J.; Karl, N.; Gerdon, M. (J. Cryst. Growth **93** [1988] 45/51).
[33] Reier, F.W.; Wolfram, P.; Schumann, H. (J. Cryst. Growth **93** [1988] 41/4).
[34] Cole-Hamilton, D.J.; Forster, D.F.; Rushworth, S.; Plessey Co. PLC (Brit. Appl. 2201418 [1987/88] from C.A. **110** [1989] No. 57844).
[35] Bradley, D.C.; Faktor, M.M.; Frigo, D.M.; Zheng, D.H. (Chemtronics **3** [1988] 53/5).
[36] Bradley, D.C.; Chudzynska, H; Frigo, D.M. (Chemtronics **3** [1988] 159/61).
[37] Reier, F.W.; Wolfram, P.; Schumann, H. (Ger. Offen. 3742525 [1989] from C.A. **111** [1989] No. 106317).
[38] Hallock, R.B.; Manzik, S.J.; Mitchell, T.; Hui, B.C.; Morton Thiokol. Inc. (U.S. 4847399 [1989] from C.A. **112** [1990] No. 56270).
[39] Ninomiya, T.; Nakamura, K.; Imori, T; Nippon Mining Co., Ltd. (Japan. Kokai Tokkyo Koho 02-167 290 [1990] from C.A. **113** [1990] No. 191628).
[40] Forster, D.F.; Rushworth, S.A.; Cole-Hamilton, D.J.; Cafferty, R.; Harrison, J.; Parkes, P. (J. Chem. Soc. Dalton Trans. **1988** 7/11).

1.1.1.1.2 Physical and Structural Properties

Appearance, Melting Point, Boiling Point, Density. Trimethylindium is a colorless solid which crystallizes readily and refracts light strongly. Rapid sublimation produces needles; slow sublimation gives large prisms or rectangular solids [1]. When $In(CH_3)_3$ is condensed from the gas phase at about −78 °C, mostly needles are formed, and these show a substantially higher volatility than material condensed at room temperature. However, after about one hour a transformation seems to occur. A precise characterization of this unstable form of $In(CH_3)_3$ has not yet been reported [42]; the existence of a second phase is denied in [44]. The melting point of the "normal" form is 89 to 89.5 °C [1, 14, 16]; other values reported are 83 °C [6] and 88 to 88.5 °C [2, 8, 19]. The boiling point of the compound, extrapolated to normal pressure, is reported as 135.8 to 136 °C [2, 19].

$In(CH_3)_3$ can be distilled without decomposition only at reduced pressure; reported boiling temperatures are: 82 to 93 °C/0.1 to 0.2 Torr [29] and 66 to 67 °C/12 Torr [12]. The last value doubtless represents a vacuum sublimation. Vacuum sublimation is the preferred purification method. At about 10^{-3} Torr the sublimation occurs at room temperature at a preparatively useful speed (comment by the author; see also [1, 14, 42]).

The density of crystalline $In(CH_3)_3$ as measured by pycnometer is 1.568 g/cm³ [1].

Vapor Pressure. Vapor pressure as a function of temperature, $\log p = A - B/T$, has been determined by various authors; values for constants A and B, and the enthalpy of vaporization, ΔH_v (in kcal/mol), are given below:

A	9.2197	10.520	8.238	11.14	9.86
B	2535	3014	2190	3240	2830
range in °C	25 to 127	50 to 88	88 to 135	−7.7 to 70.4	−16 to 20
ΔH_v	11.6	13.8 ± 0.4	10.0 ± 0.1		
Ref.	[1]	[2]	[2]	[42]	[41]

In [42] the vapor pressure measurement was made under flowing hydrogen (25 cm³/min), while in [41] the static method with a specially developed apparatus was used. The values reported in [1] are confirmed in [14]. Corresponding vapor pressures for the group 13 MR_3 compounds ($R = CH_3$, C_2H_5, C_3H_7, C_4H_9) are collected in [26]. Comparative studies of vapor pressure data for the trimethyl and triethyl compounds of Al, Ga, In, and As are reported in [41]. The "low temperature form" of $In(CH_3)_3$ shows (in preliminary measurements) a vapor pressure about 10 times higher than the "normal" form [42]; for a contrary report, see [44].

ΔH_v of $In(CH_3)_3$ was determined by the chromatographic method to be 9.07 kcal/mol [26].

A few vapor pressures, measured on solid $In(CH_3)_3$ at various temperatures, are listed below:

t in °C	0	17	18	30	30	50	70
p in Torr	0.2 to 0.21	1.0	1.93	5.0*)	7.2	23.8	72.1
Ref.	[14, 30]	[33]	[37]	[26]	[1]	[1]	[1]

*) Extrapolated value.

Thermodynamic Functions. Thermochemical data for $In(CH_3)_3$ gas (calculated by statistical thermodynamic functions) are: $H^\circ_{298.15} - H^\circ_0 = 5.738$ kcal/mol, $C_{p(298)} = 28.59$, and $S_{(298)} = 88.72$ cal · mol⁻¹ · K⁻¹ for the temperature range 298 to 1200 K [25].

Partial Charges. The partial charges on the central atom and the methyl residues of the $In(CH_3)_3$ molecule were estimated, leading to an insignificant change in the electronegativity of the H atom (+0.03, similar to $Ga(CH_3)_3$) and to a net charge of −0.065 on the CH_3 ligand [5]; see also [4].

NMR Spectra. Chemical shift values (δ in ppm) of the ^1H and ^{13}C NMR spectra are presented below for various solvents and temperatures. All ^1H δ values with exception of that from [28, 39] had been initially given relative to internal cyclopentane (δ = 1.51 ppm):

solvent		t	δ (^1H)	δ (^{13}C)	Ref.
C_5H_{10}-c	(sat. sol.)	38 °C	−0.05		[24]
C_5H_{10}-c	(dil. sol.)	38 °C	−0.01		[10]
C_5H_{10}-c	(dil. sol.)	−60 °C	−0.23		[10]
CH_2Cl_2	(5 mol %)	38 °C	−0.04		[10]
CH_2Cl_2	(5 mol %)	−90 °C	−0.32		[10]
C_6H_6	(1 to 15 mol %)	30 °C	−0.19		[10, 40]
C_6D_6	(10 mol %)	35 °C	−0.18	0.3	[28, 39]
C_7H_8	(1 to 15 mol %)	−90 °C	−0.21		[10]
$CH_3OCH_2CH_2OCH_3$		35 °C	−0.42		[24]

Only sharp singlets, showing no concentration dependence, are observed. On cooling no broadening or splitting occurs. It is therefore concluded that $In(CH_3)_3$ in solution is monomeric [10]. The relatively large shift in 1,2-dimethoxyethane is attributed to the formation of an adduct. The coupling constant, $J(^1H,^{13}C) = 126$ Hz, is listed along with corresponding data for other organometallic compounds in [9].

The ^{115}In nuclear quadrupole resonance spectra indicate a planar structure for $In(CH_3)_3$ even in the solid state; occupation of the p_z orbital (perpendicular to the InC_3 plane) is on the order of zero and, therefore, either no, or only very loose associations are present [17, 18, 20]. The nuclear spin, $I = {}^9/_2$ for ^{115}In, would produce the four transition frequencies, $\nu_1(\pm 1/2 \leftrightarrow \pm 3/2)$, $\nu_2(\pm 3/2 \leftrightarrow \pm 5/2)$, $\nu_3(\pm 5/2 \leftrightarrow \pm 7/2)$, and $\nu_4(\pm 7/2 \leftrightarrow \pm 9/2)$, but generally only the first three can be observed for $In(CH_3)_3$. The measured data are:

T (in K)	e^2Qq/h (MHz)	η (%)	transition frequencies (MHz) ν_1	ν_2	ν_3	Ref.
77	1115.12	14.0	53.90	90.90	138.91	[17, 20]
296	1101.7 ± 1.0	11.3 ± 0.1	50.93	90.39	137.35	[18]

IR and Raman Spectroscopy. From the standpoint of molecular spectroscopy $In(CH_3)_3$ has received little attention. The first Raman spectroscopic investigations proposed that trimethylindium, in both the solid and liquid states, is a pyramidal molecule with C_{3v} symmetry [7]. The very complex spectra, and the resulting misinterpretations, are undoubtedly caused by the use of impure alkyls. More careful investigations of gaseous, liquid, and dissolved $In(CH_3)_3$ are described in [13, 28]; interpretation of these spectra indicates a trigonal planar structure of D_{3h} symmetry.

By far the most intense Raman line is at 467 [13] or 471 cm^{-1} [28], which is associated with the symmetrical InC_3 vibration of A'_1 mode. The value of 524 cm^{-1} [23] is too high for this vibration and implies to [7] an impurity or a decomposition. The associated asymmetric vibration of E′ mode is (unlike the symmetrical mode) IR-active as well as Raman-active, and is observed as intense absorption bands in the IR at 500 cm^{-1} (gas) [13], or about 495 cm^{-1} (liquid or solution) [11, 13, 28]. Only the IR spectrum is illustrated in [11].

Another very characteristic absorption of the symmetrical CH_3 deformation [13, 23, 28, 34] occurs around 1155 cm^{-1}. Concerning the deformations of the InC_3 framework, however, up to now only the E′ mode at 125 [28] or 132 cm^{-1} [13] has been located; see Table 1.

Comparative studies of the IR spectra of Ga, In, Sb trialkyls [11], and $M(CH_3)_3$ series for M = B, Ga, In [13] have been conducted. Simple force constant calculations (with CH_3 as a point of mass 15) give the stretching force constant for $In(CH_3)_3$, f(In–C) = 1.92 [11] or 1.93 N/cm [13].

IR diode laser spectroscopy was used to clarify the mechanisms of the reactions which occur during gas phase epitaxy of $In(CH_3)_3$; important criteria are the change in intensity of the CH_3 rocking motion at 730 cm^{-1} and the absorption by unstable methyl radicals (formed by thermal decomposition) near 610 cm^{-1} (δ_sCH_3). The minimum detection limit was 8.1×10^{13} molecules/cm^3 [38].

The frequency values of the IR and Raman spectra of gaseous and fluid (melted and dissolved) $In(CH_3)_3$ along with data of the deuterated compound, $In(CD_3)_3$, [13, 28] are summarized in Table 1. Table 2 contains the Raman frequencies of solid trimethylindium; the marked splitting of the important In–C framework bonds is accounted for by nearest-neighbor interactions in the crystal [27, 44]. According to [44] the difference between the

Table 1
Vibrational Spectra and Assignments for $In(CH_3)_3$ and $In(CD_3)_3$ in Various States [13, 28]. Wavenumbers in cm^{-1}.

$In(CH_3)_3$ gas IR [13]	liquid Raman [13]	liquid IR [28]	liquid Raman [28]	$In(CD_3)_3$ C_6H_6-sol. IR [28]	liquid Raman [28]	assignment D_{3h} [28]
3000 s	2974 mw, dp	2960 s	2970 w, dp	–	–	$\nu_{as}(CH_3)$ (A''_2, E′, E″)
2920 s	2910 ms, p	2905 s	2915 ms, p	–	–	$\nu_s(CH_3)$ (A'_1, E′)
2860 m	2861 w, sh	2840 mw	2850 w, p	–	–	$2 \times \delta_{as}(CH_3)$
–	–	–	–	2215 m	2227 w, dp	$\nu_{as}(CD_3)$ (A''_2, E′, E″)
–	–	–	–	2075 m, br	2120 ms, p	$\nu_s(CD_3)$ (A'_1, E′)
–	–	–	–	–	2105 s, p	$2 \times \delta_{as}(CD_3)$
–	–	1425 vw	–	–	–	$\delta_{as}(CH_3)$
–	–	–	–	1010 m, br	–	$\delta_{as}(CD_3)$
1155 vw	1157 s, dp	1154 m	1157 ms, p	–	–	
–	–	–	1152 w, sh, dp	–	–	$\delta_s(CH_3)$ (A'_1, E′)
–	1116 m, br, dp	1106 ms	1118 vw	–	–	
–	–	–	–	–	894 ms, p	
–	–	–	–	895 m	901 w, sh, dp	$\delta_s(CD_3)$ (A'_1, E′)
–	–	–	–	850 m	862 w, dp	
725 s	725 w, br, dp	720 s, br	720 vw	–	–	$\varrho(CH_3)$ (A''_2, E′)
687 ms	635 w, br, dp	–	633 vw	–	–	
–	–	–	–	540 s, br	–	$\varrho(CD_3)$ (A''_2, E′)
500 s	495 s, dp	490 s	492 ms, dp	450 s	455 m, dp	$\nu_{as}(InC_3)$ (E′)
–	467 vs, p	–	471 vvs, p	–	431 vvs, p	$\nu_s(InC_3)$ (A'_1)
–	132 s, dp	–	125 w, dp	–	–	$\delta_{as}(InC_3)$ (E′)

Table 2
Raman Spectra and Assignments for Solid $In(CH_3)_3$.
Wavenumbers in cm^{-1}

crystal [27]	crystal [44]	frost [44]	assignments [27]
2990 w	2975 mw, br	–	$\nu_{as}(CH_3)$
2918 m	2915 m	–	$\nu_s(CH_3)$
1450 vw, br	–	–	$\delta_{as}(CH_3)$
1190 w	–	–	
1158 s	1151 s	1153 s	$\delta_s(CH_3$, terminal)
1117 s	1123 m	–	
1099 s	1094 m	1105 m, br	$\delta_s(CH_3$, bridge)
750 w, br	–	–	$\varrho(CH_3)$
625 w, br	–	–	$\varrho(CH_3)$
513 m	–	–	
499 s	498 s	497 s	ν(InC, terminal)
494 vs	488 s	490 s	
–	480 m	–	
467 vs	462 vs	463 vs	ν(InC, bridge)
145 sh	155	–	
132 s	125	–	δ(InC)
–	86	–	

Raman spectra of a crystalline and an amorphous sample, obtained from condensation of $In(CH_3)_3$ gas at −46 °C, is not evidence for another, more stable crystalline phase. No significant orientation effects were found in the spectrum of a single crystal when the sample position in the beam of a laser Raman spectrometer was altered; the spectrum is identical to that of a polycrystalline sample [44].

UV Spectroscopy. The UV spectrum of $In(CH_3)_3$ gas at 3 Torr in H_2 (15 °C and 760 Torr total pressure) shows a broad absorption at $\lambda_{max}=213$ nm with $\varepsilon_{max}\approx 2500$ $L \cdot cm^{-1} \cdot mol^{-1}$. This molar absorption coefficient is sufficiently large to enable interaction with radiation by photolysis. The spectrum is illustrated in [32]. UV spectroscopic investigations of the pyrolysis of $In(CH_3)_3$ under H_2 or N_2 as carrier gas at temperatures between about 200 and 275 °C are described in [35] and [36]. Photoabsorption cross sections of $In(CH_3)_3$ and $Ga(CH_3)_3$ were measured in the 106 to 270 nm range at a sample pressure of 20 mTorr and Rydberg assignments were given using PES data (see below); for $In(CH_3)_3$ band maxima were recorded at 211, 163, and 125 nm with term values of 26000, 11900, and 27500 cm^{-1}, respectively [43].

Photoelectronic Spectrum. The He(I) photoelectron spectrum (PES) of $In(CH_3)_3$ is depicted in a figure and compared with that of $Ga(CH_3)_3$. A weak band at about 9 eV was assigned as ionization from the 3e′ MO, showing mainly metal-carbon bonding characteristics. The second broad band at about 13.5 eV represents the superimposed binding energies of the σ_{C-H} orbitals of 1e″, 3a′, and 2e′ MO's [43]. The photoelectron spectrum of $In(CH_3)_3$ and other metal alkyl compounds from He(II) irradiation is mentioned in [45].

Structural Properties. $In(CH_3)_3$ crystallizes in the tetragonal space group $C^4_{4h}-P4_2/n$ (No. 86) with eight molecules in the unit cell with the lattice constants a = b = 1321.66(11), c = 640.39(9) pm; $D_c = 1.899$ g/cm³. The data were collected at 273 K [44]. Earlier data collected at room temperature are given as a=b=1324±1 and c=644±1 pm; D_c=1.88 g/cm^3 [8]. It forms cyclic pseudotetramers with S_4-symmetry linked by intermolecular In ··· C(1′) contacts of 308.3(12) pm. The tetramer units, in which each In is 441 pm out of the plane of the three others, are loosely associated with one another in the crystal lattice by 4 longer In ··· C(2″) contacts of 355.8(15) pm, resulting in a three dimensional structure. In the individual $In(CH_3)_3$ unit, the C–In–C angles are significantly different from 120° but the sum of 359.94° indicates a planar InC_3 skeleton [44]. In contrast to [8] no significant difference in the In–C(2) and the In–C(3) distances is reported in [44]. The molecular structure is depicted in **Fig. 1A** [44].

Within each tetramer unit are four $In(CH_3)_3$ molecules linked together through nearly linear, but unsymmetrical, In–C ··· In electron-deficient bridge bonds (e.g., In–C(1′) ··· In′). Of the three methyl groups in each molecule, one functions as a bridge within the tetramer (e.g., C(1)), a second forms the connection to a neighboring tetramer (e.g., C(2)), and the third methyl group C(3) is not a bridge. The resulting fivefold coordination at the indium atom can be described as axially elongated trigonal bipyramidal as shown in **Fig 1B**. [8, 44]; for a further (incomplete) set of distances and angles see also [15].

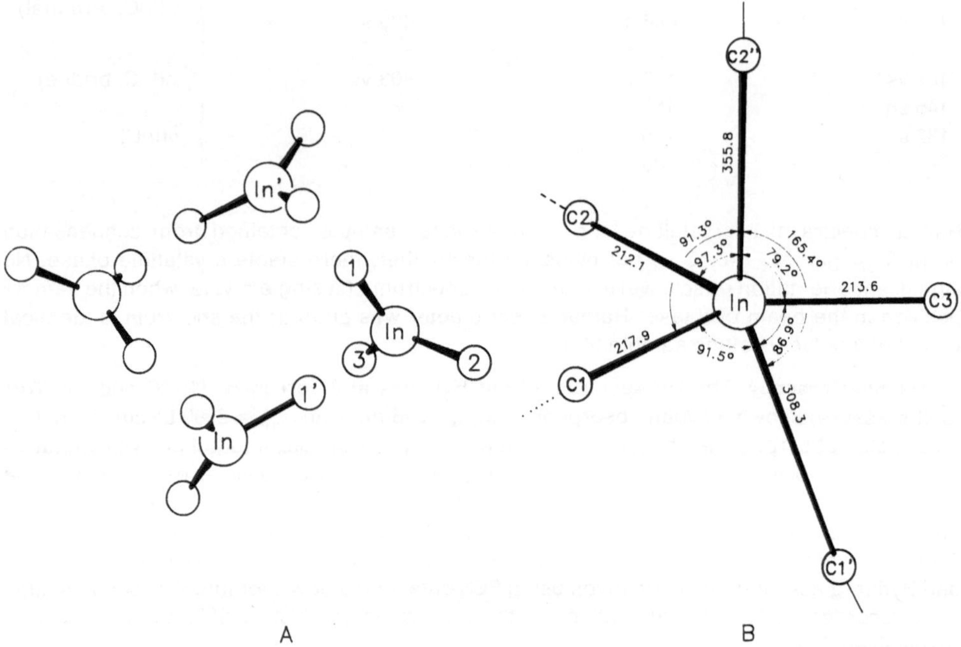

Fig. 1. View of one tetrameric unit of the crystal structure of $In(CH_3)_3$ (A); coordination at the In atom (B) [44].

Other bond angles (°)

In ··· C(1′)–In′	168.1(6)		
C(1′) ··· In–C(2)	94.3(5)	In ··· C(2″)–In″	166.2(6)

The molecular structure of $In(CH_3)_3$ has been determined several times by electron diffraction studies in the gas phase. The calculations in [3] are based on both the planar (D_{3h} symmetry) and a pyramidal (C_{3v} symmetry) configuration, but no decision could be made in favor of either form. An insignificant deviation of 3±2.5° from complete planarity of the molecule is reported in [21], while in [31] the trigonal-planar configuration was the fixed condition assumed for the calculation of distances and angles. The following tabulation contains structural parameters for $In(CH_3)_3$; comparisons with other $M(CH_3)_n$ (M = Be, Mg, Zn, Cd, Hg, B, Al, Ga, and Tl) are presented in [31]:

In-C (pm)	216(4)	209.3(6)	216.1(3)
C-H (pm)	109	114(3)	112.4(4)
C-In-C angle	120° + 109.5°	see text	120°
H-C-In angle	109.5°	112.7(8)°	109.8(4)°
Ref.	[3]	[21]	[31]

References:

[1] Dennis, L.M.; Work, R.W.; Rochow, E.G. (J. Am. Chem. Soc. **56** [1934] 1047/9).
[2] Laubengayer, A.W.; Gilliam, W.F. (J. Am. Chem. Soc. **63** [1941] 477/9).
[3] Pauling, L.; Laubengayer, A.W. (J. Am. Chem. Soc. **63** [1941] 480/1).
[4] Sanderson, R.T. (J. Chem. Educ. **32** [1955] 140/1).
[5] Sanderson, R.T. (J. Am. Chem. Soc. **77** [1955] 4531/2).
[6] van der Kelen, G.P. (Bull. Soc. Chim. Belges **65** [1956] 343/9).
[7] van der Kelen, G.P.; Herman, M.A. (Bull. Soc. Chim. Belges **65** [1956] 362/76).
[8] Amma, E.L.; Rundle, R.E. (J. Am. Chem. Soc. **80** [1958] 4141/5).
[9] Muller, N. (J. Chem. Phys. **36** [1962] 359/63).
[10] Muller, N.; Otermat, A.L. (Inorg. Chem. **2** [1963] 1075/6).

[11] Oswald, F. (Z. Anal. Chem. **197** [1963] 309/22).
[12] Todt, E.; Dötzer, R. (Z. Anorg. Allgem. Chem. **321** [1963] 120/3).
[13] Hall, J.R.; Woodward, L.A.; Ebsworth, E.A.V. (Spectrochim. Acta **20** [1964] 1249/56).
[14] Jacko, M.G.; Price, S.J.W. (Can. J. Chem. **42** [1964] 1198/205).
[15] Messmer, G.G.; Amma, E.L. (unpublished results from Vranka, R.G.; Amma, E.L.; J. Am. Chem. Soc. **89** [1967] 3121/6).
[16] Clark, W.D.; Price, S.J.W. (Can. J. Chem. **46** [1968] 1633/4).
[17] Svergun, V.I.; Bednova, L.M.; Okhlobystin, O.Yu.; Semin, G.K. (Izv. Akad. Nauk SSSR Ser. Khim. **1970** 1449; Bull. Acad. Sci. USSR Div. Chem. Sci. **1970** 1377).
[18] Golubinskaya, L.M.; Bregadze, V.I.; Bryuchova, E.V.; Svergun, V.I.; Semin, G.K.; Okhlobystin, O.Yu. (J. Organomet. Chem. **40** [1972] 275/9).
[19] Razuvaev, G.A.; Gribov, B.G.; Domrachev, G.A.; Salamatin, B.A. (Organometallic Compounds in Electronics [in Russian], Nauka, Moscow 1972 from [26]).
[20] Patterson, D.B.; Carnevale, A. (J. Chem. Phys. **59** [1973] 6464/7).

[21] Barbe, G.; Hencher, J.L.; Shen, Q.; Tuck, D.G. (Can. J. Chem. **52** [1974] 3936/40).
[22] Chao, L.-C.; Rieke, R.D. (Synth. React. Inorg. Metal-Org. Chem. **4** [1974] 373/8).
[23] Kurbakova, A.P.; Leimes, L.A.; Aleksanyan, V.T.; Golubinskaya, L.M.; Zorina, E.N.; Bregadze, V.I. (Zh. Strukt. Khim. **15** [1974] 1083/92; J. Struct. Chem. [USSR] **15** [1974] 961/9).
[24] Weibel, A.T.; Oliver, J.P. (J. Organometal. Chem. **74** [1974] 155/66).
[25] Malkova, A.S.; Pashinkin, A.S. (Zh. Fiz. Khim. **49** [1975] 2160; Russ. J. Phys. Chem. **49** [1975] 2170/1).
[26] Vanchagova, V.K.; Zorin, A.D.; Umilin, V.A. (Zh. Obshch. Khim. **46** [1976] 989/93; J. Gen. Chem. [USSR] **46** [1976] 985/8).
[27] von Dahlen, K.-H.; Dehnicke, K. (Chem. Ber. **110** [1977] 383/94).
[28] Karschin, J.; Mann, G.; Weidlein, J. (unpublished results 1979).

[29] Efremov, E.A.; Fedorov, V.A.; Ferapontov, A.P.; Dorogova, L.A.; Kozyrkin, B.I.; Filippov, E.P.; Sredinskaya, I.A. (U.S.S.R. 810700 [1981] from C.A. **95** [1981] No. 62396).

[30] Ludowise, M.J.; Cooper, C.B., III; Saxena, R.R. (J. Electron. Mater. **10** [1981] 1051/68).

[31] Fjeldberg, T.; Haaland, A.; Seip, R.; Shen, Q.; Weidlein, J. (Acta Chem. Scand. A **36** [1982] 495/9).

[32] Haigh, J. (J. Mater. Sci. **18** [1983] 1072/6).

[33] Hsu, C.C.; Cohen, R.M.; Stringfellow, G.B. (J. Cryst. Growth **63** [1983] 8/12).

[34] Cheng, C.H.; Jones, K.A.; Motyl, K.M. (J. Electron. Mater. **13** [1984] 703/26).

[35] Haigh, J.; O'Brien, S. (J. Cryst. Growth **68** [1984] 550/6).

[36] Karlicek, R.; Long, J.A.; Donnelly, V.M. (J. Cryst. Growth **68** [1984] 123/7).

[37] Alfa Products, Morton Thiokol Inc. (Alfa Organometallics for Vapor Phase Epitaxy, Literature and Product Review 1986, Danvers, Mass., 1986).

[38] Butler, J.E.; Bottka, N.; Sillmon, R.S.; Gaskill, D.K. (J. Cryst. Growth **77** [1986] 163/71).

[39] Bradley, D.C.; Dawes, H.; Frigo, D.M.; Hursthouse, M.B.; Hussain, B. (J. Organometal. Chem. **325** [1987] 55/67).

[40] Hodes, H.D.; Berry, A.D. (J. Organometal. Chem. **336** [1987] 299/305).

[41] Kayser, O.; Heinecke, H.; Brauers, A.; Lueth, H.; Balk, P. (Chemtronics **3** [1988] 90/3).

[42] Stagg, J.P. (Chemtronics **3** [1988] 44/9).

[43] Ibuki, T.; Hiraya, A.; Shobatake, K.; Matsumi, Y.; Kawasaki, M. (Chem. Phys. Letters **160** [1989] 152/6).

[44] Blake, A. J.; Cradock, S. (J. Chem. Soc. Dalton Trans. **1990** 2393/6).

[45] Bancroft, G.M.; Coatsworth, L.L.; Creber, D.K.; Tse, J. (Phys. Scr. **16** [1977] 217/20; CA. **88** [1977] No. 179867).

1.1.1.1.3 Molar Mass, Mass Spectrum, Thermal Decomposition, and Behavior toward Radiation

$In(CH_3)_3$ is monomeric in solution [2, 4], and not tetrameric as reported earlier [1]. Vapor pressure measurements on solutions of $In(CH_3)_3$ in various solvents provide the following values [2]:

solvent	C_5H_{10}-c	CH_2Cl_2	CH_2Cl_2	C_6H_6
mol % $In(CH_3)_3$	0.4	1.86	3	3 to 24
molar mass	178	188	168	175 to 181

Cryoscopic molecular weight determinations in benzene give values of 151 ± 10, and do not indicate any tendency toward association [4].

Mass spectral investigations are described in [6] and [20]. Almost exclusively indium-containing fragments are found and the results from [20] are collected in Table 3. Additional fragments given in [6] without intensities (source temperature 20 to 35 °C) are $[InH_2]^+$, $[InC]^+$, $[InC_3H_6]^+$, and $[InC_3H_8]^+$. The base peak of the spectrum is that of the methyl deficient molecular fragment, $[InC_2H_6]^+$, whose relative intensity is influenced neither by the source temperature nor by the ionization energy (in [6] the measurements were carried out from about 30 to 300 °C at 70, 50, 20, 12, and 10 eV). The molecular ion abundance for the $M(CH_3)_3$ series follows the order $Al > B > Ga > In > Tl$; the M^+ ion abundance increases in the sequence $B < Al < Ga < In < Tl$ from 0.5% for boron to 36% for thallium, and reflects the successive weakening of the M–C bond in these homologues [6]. Higher associations or their fragments are not observed. Metastable peaks for the transitions $[In(CH_3)_2]^+ \rightarrow In^+ + C_2H_6$ and $[InCH_3]^+ \rightarrow In^+ + CH_3$ have been identified [6].

Table 3.
Mass Spectrum of $In(CH_3)_3$ (natural isotops ^{115}In, 95.7%; ^{113}In, 4.3%).
Source temperature 100 °C, electron impact energy 70 eV [20].

fragment ion	rel. intensities in %	fragment ion	rel. intensities in %
$[^{113}In]^+$	2.6	$[^{115}InCH_4]^+$	0.8
$[^{115}In]^+$	56.6	$[^{113}InC_2H_5]^+$	0.1
$[^{115}InH]^+$	0.2	$[^{113}InC_2H_6]^+$	5.1
$[^{113}InCH_2]^+$	0.6	$[^{115}InC_2H_5]^+$	1.8
$[^{113}InCH_3]^+$	1.6	$[^{115}InC_2H_6]^+$	100.0
$[^{115}InCH_2]^+$	1.3	$[^{115}InC_2H_7]^+$	2.3
$[^{115}InCH_3]^+$	14.5	$[^{115}InC_3H_9]^+$	0.1

Under static conditions the thermal decomposition of $In(CH_3)_3$ (0.1 to 0.5 Torr) begins at 270 °C. A temperature increase of 20 °C causes the decomposition rate to increase more than 5-fold. Correlation of the change in intensity of the UV absorption band at 211 nm versus the temperature serves as the basis for calculating ca. 50 kcal/mol as the activation energy of decomposition [15]; for pyrolysis onset temperatures of $In(CH_3)_3$ and some other element alkyl compounds, see [23].

The decomposition products at 400 °C were determined to be CH_4 (84%), C_2H_6 (12%), and C_2H_4 (4%) [9].

Thermal decomposition under organometallic vapor phase epitaxy (MOVPE) conditions using absorption IR diode laser spectroscopy (Section 1.1.1.1.2) as monitoring technique showed the formation of CH_3 radicals. In the presence of H_2 (total of 7.6 Torr pressure) maximum methyl production was observed at 515 °C; the methane signal was first observed (8.1×10^{13} molecules/cm^3 $In(CH_3)_3$) at a concentration of 1.4×10^{13} molecules/cm^3 at 520 °C [17]; at 400 °C the methane concentration was too low to be registered on the IR spectrum [12].

Attempts using atomic absorption spectrometry (AAS) failed to demonstrate that pyrolysis forms elemental In; under the conditions of measurement (3 Torr $In(CH_3)_3$ in an H_2 atmosphere between 20 and 650 °C) the minimum detection limit of 8×10^{12} atoms/cm^3 was not attained [13].

Pyrolysis of $In(CH_3)_3$ at temperatures of 550 to 781 K (277 to 508 °C) with toluene as carrier gas and at total pressures of 6.0 to 33.5 Torr was studied. The gas mixture (methane, ethane, ethylene, ethylbenzene) formed in the decomposition was measured, and indium was analyzed directly [3]. The decomposition occurs in three steps:

(1) $In(CH_3)_3 \rightarrow In(CH_3)_2 + CH_3$

(2) $In(CH_3)_2 \rightarrow In(CH_3) + CH_3$

(3) $In(CH_3) \rightarrow In + CH_3$

Steps (1) and (2) take place in rapid sequence at all measured temperatures. Step (3) is only important above 400 °C; below this temperature the formation of a white film of polymeric $In(CH_3)_n$ is observed [3]; for the pyrolysis in the presence of $Ga(CH_3)_3$, see [22].

At a total pressure of 13 Torr and a toluene/trimethylindium ratio >150 the decomposition rate constants $\log k_1\ (s^{-1}) = 15.72 - (47.2/2.303\ RT)$ and $\log k_3\ (s^{-1}) = 10.91 - (38.7/2.303\ RT)$

are obtained. The dissociation energies are $D_1[(CH_3)_2In-CH_3]=47.2$ kcal/mol and $D_3[In-CH_3]=40.7$ kcal/mol. A value of 35 ± 5 kcal/mol was estimated for $D_2[(CH_3)In-CH_3]$ [3].

From D_1 and D_3 of [3] and the formation enthalpy of $In(CH_3)_3$ (p. 2), the mean $In-CH_3$ bond energy, $E(In-CH_3)$, was calculated to be 38.9 kcal/mol and $D_2[CH_3In-CH_3]=28.8$ kcal/mol was obtained [5].

Hydrogen and nitrogen were used in [14, 18] as carrier gases to investigate the first order decomposition reaction; the rate constants were found to be $\ln k_{H_2}$ $(s^{-1})=27.57-35.9$ kcal/mol · RT and $\ln k_{N_2}(s^{-1})=29.12-40.5$ kcal/mol · RT, respectively [18].

A mass spectral study of the pyrolysis of $In(CH_3)_3$ at normal pressure and temperatures from 200 to 400 °C (partly at 500 °C), using H_2, D_2, and He as carrier gases, gave the overall rate constants $\log k_{H_2}$ $(s^{-1})=15.0-42.6$ (kcal/mol)/2.303 RT, $\log k_{D_2}$ $(s^{-1})=13.4-39.8$ (kcal/mol)/2.303 RT, and $\log k_{He}$ $(s^{-1})=17.9-54.0$ (kcal/mol)/2.303 RT, respectively. Parameters for the formation of the pyrolysis products CH_4, C_2H_6, and CH_3D were also reported [19].

The catalytic influences of fused quartz and indium phosphide (important to thermal decomposition of $In(CH_3)_3$ in gas phase epitaxy) were described in detail in [13, 14]; detection of decomposition products was done by UV spectroscopy.

The 222 nm photochemical decomposition of $In(CH_3)_3$ adsorbed on quartz at 150 K was investigated and the time-of-flight profiles of the photofragments ejected into the gas phase were measured. A film of In metal was produced on the surface. The neutral species CH_3, In, $InCH_3$ and $In(CH_3)_2$, in addition to the cations $[CH_3]^+$, In^+, and $[InCH_3]^+$, were identified as photoproducts. A Maxwell-Boltzmann velocity distribution were found for the neutral photofragments with different translation temperatures for each species. The photoproduct translation temperatures for CH_3, $InCH_3$, and $In(CH_3)_2$ as a function of surface coverage are given in a table [21]; see also [24].

Excitation of $In(CH_3)_3$ at ~3 Torr with a pulsed laser (390 to 650 nm, 10^7 to 10^{10} W/cm²) leads to dissociation with formation of neutral In atoms, which undergo a one-, two-, or three-photon ionization within the photolysis laser pulse. The method was used to study the ground and excited states of In atoms and transitions between these states; comparisons with the multiphoton ionization spectra of the homologous elements Al and Ga are made in [11, 16].

Atoms of In are also formed by the spark discharge of a 0.5 μF condenser (23 kV) through $In(CH_3)_3$; their laser transitions (at 1.873 and 2.378 μ for In(I)) were determined [7], and compared with the laser wave lengths of Hg, Pb, Sb, Ga, Tl, and Bi, also generated from metal alkyls. UV photolytic decomposition of $In(CH_3)_3$ was used to separate In from InP substrate materials on borosilicate glasses [10]. Highly resolved indium 4d photoelectron spectra can be obtained if $In(CH_3)_3$ is irradiated with He(II) [8].

References:

[1] Dennis, L.M.; Work, R.W.; Rochow, E.G. (J. Am. Chem. Soc. **56** [1934] 1047/9).
[2] Muller, N.; Otermat, A.L. (Inorg. Chem. **2** [1963] 1075/6).
[3] Jacko, M.G.; Price, S.J.W. (Can. J. Chem. **42** [1964] 1198/1205).
[4] Muller, N.; Otermat, A.L. (Inorg. Chem. **4** [1965] 296/9).
[5] Clark, W.D.; Price, S.J.W. (Can. J. Chem. **46** [1968] 1633/4).
[6] Glockling, F.; Strafford, R.G. (J. Chem. Soc. A **1971** 1761/3).
[7] Chou, M.S.; Cool, T.A. (J. Appl. Phys. **47** [1976] 1055/61).

[8] Bancroft, G.M.; Coatsworth, L.L.; Creber, D.K.; Tse, J. (Phys. Scr. **16** [1977] 217/20; C.A. **88** [1977] No. 179867).
[9] Travkin, N.N.; Skachkov, B.K.; Tonoyan, I.G.; Kozyrkin, B.I. (Zh. Obshch. Khim. **48** [1978] 2678/81; J. Gen. Chem. [USSR] **48** [1978] 2428/31).
[10] Aylett, M.R.; Haigh, J. (Mater. Res. Soc. Symp. Proc. **17** [1983] 177/82; C.A. **99** [1983] No. 96685).

[11] Mitchell, S.A.; Hackett, P.A. (J. Chem. Phys. **79** [1983] 4815/22).
[12] Cheng, C.H.; Jones, K.A.; Motyl, K.M. (J. Electron. Mater. **13** [1984] 703/26).
[13] Haigh, J.; O'Brien, S. (J. Cryst. Growth **67** [1984] 75/8).
[14] Haigh, J.; O'Brien, S. (J. Cryst. Growth **68** [1984] 550/6).
[15] Karlicek, R.; Long, J.A.; Donnelly, V.M. (J. Cryst. Growth **68** [1984] 123/7).
[16] Mitchell, S.A.; Hackett, P.A. (J. Phys. Chem. **89** [1985] 1509/14).
[17] Butler, J.E.; Bottka, N.; Sillmon, R.S.; Gaskill, D.K. (J. Cryst. Growth **77** [1986] 163/71).
[18] Larsen, C.A.; Stringfellow, G.B. (J. Cryst. Growth **75** [1986] 247/54).
[19] Buchan, N.I.; Larsen, C.A.; Stringfellow, G.B. (J. Cryst. Growth **92** [1988] 591/604).
[20] Reier, F.W.; Wolfram, P.; Schumann, H. (J. Cryst. Growth **93** [1988] 41/4).

[21] Horwitz, J.S.; Villa, E.; Hsu, D.S.Y. (J. Phys. Chem. **94** [1990] 7214/9).
[22] Larsen, C.A.; Buchan, N.I.; Li, S.H.; Stringfellow, G.B. (J. Cryst. Growth **102** [1990] 103/16).
[23] Jackson, D.A., Jr. (J. Cryst. Growth **87** [1988] 205/12).
[24] Callender, C.N.; Rayner, D.M.; Hackett, P.A. (Appl. Phys. B **47** [1988] 7/15).

1.1.1.1.4 Reactions of $In(CH_3)_3$

The chemical reactions of $In(CH_3)_3$ are summarized in Table 4, which is subdivided according to reactions with elements, inorganic, organic, and organometallic reactants.

Table 4
Reactions of $In(CH_3)_3$.
An asterisk indicates further information at the end of the table. CH_3 groups on the In are abbreviated as R.

No.	reactant (reaction type)	products and remarks	Ref.	page
with elements				
*1	O_2 (II)	$R_2InOOCH_3$ (at −78 °C)	[63]	209
2	M^I	$M^I[InR_4]$ in $O(C_2H_5)_2$ at −60 °C; M = Na, K, Rb, Cs	[33]	337
with inorganic compounds				
3	HCl (III)	R_2InCl in $O(C_2H_5)_2$	[12]	118
		$RInCl_2$	[15]	
4	H_2O (III)	R_2InOH in C_6H_6; no pure compound isolated	[53]	
		$(RInO)_n$	[15]	180
5	SO_3 (II)	$R_2InOSO_2CH_3$ in CH_2Cl_2 (see also reaction No. 34)	[36]	211

Table 4 (continued)

No.	reactant (reaction type)	products and remarks	Ref.	page
6	$O_2S(OH)_2$ (III)	$[R_2In]_2SO_4$ in C_5H_{12} at -50 to -40 °C	[42]	222
7	SO_2 (II)	$R_2InOOSCH_3$ at < -50 °C $In(OOSCH_3)_3$ at > -50 °C	[25] [23, 26]	211
8	NH_3, PH_3 (III)	$(R_2In(N,P)H_2)$ unisolated intermediates in the decomposition reactions	[1, 2, 84]	312
9	X_2POOH (III)	X = H: $R_2InOOPH_2$ in C_6H_6 X = F: $R_2InOOPF_2$ in C_6H_6 X = Cl: $R_2InOOPCl_2$ in C_6H_6	[50, 52] [50, 52] [50, 52]	214 214 214
10	$F_2P(=NCH_3)OH$	R_2InF in C_5H_{12}, C_6H_{14}, or C_6H_6	[73]	117
11	HCN (III)	$(R_2InCN)_4$	[4]	176
12	$B_{10}H_{14}$ (III)	2:1 ratio gives $[R_2In][R_2InB_{10}H_{12}]$ 1:1 ratio gives $RInB_{10}H_{12}$	[61]	326
13	$B_4H_8C_2$-2,3 (III)	$RInC_2B_4H_6$	[32]	327
14	$BF_3 \cdot O(C_2H_5)_2$ (IV)	R_2InF	[13]	117
15	InX_3 (IV)	X = Cl gives R_2InCl X = Br gives R_2InBr X = I gives R_2InI	[13] [56] [13]	119 147 157
16	M^IH (I)	$M^I[R_3InH]$, M^I = Li, Na, K, no pure compounds isolated	[60, 66]	346
17	K_2S, K_2Se (I)	$K_2[S(InR_3)_4]$, $K_2[Se(InR_3)_4]$	[58]	364
with organic compounds (Type III, if not stated otherwise)				
*18	CH_3I (special)	eutectic mixture	[74]	–
19	$(CF_3)_2CO$ (II)	$R_2InOC(CF_3)_2CH_3$ in ether	[13]	182
20	$C_6H_5C{\equiv}CH$	$R_2InC{\equiv}CC_6H_5$	[16]	101
21	R′OH	R_2InOR'; R′ = CH_3, R′ = C_2H_5, C_6H_5 R′ = C_4H_9-t	[48] [26] [8, 10]	181 181 181
22	t-C_4H_9OOH	$R_2InOOC_4H_9$-t in C_7H_{16}	[63]	209
23	$R'COCH_2COR''$	$R_2InOC(R')CHOR''$; R′ = R″ = CH_3 R′ = CF_3, R″ = CH_3, C_4H_9-t, C_6H_5	[22] [44]	190 191

Table 4 (continued)

No.	reactant (reaction type)	products and remarks	Ref.	page
24	2-hydroxytropone (structure)	$R_2InOC_7H_5O$	[76]	191
25	o-nitrophenol (structure: NO_2, OH)	$R_2InOC_6H_4NO_2$	[13]	192
26	R′COOH	$R_2InOOCR'$; $R' = H$	[41]	197
		$R' = CH_3$	[40, 45, 46]	198
		$R' = CF_3$	[15]	198
27	HOOCCOOH	$(R_2In)_2O_4C_2$ in C_6H_6	[37]	206
28	$(CH_3)_2NCH_2CH_2OH$	$R_2InOCH_2CH_2N(CH_3)_2$	[34]	192
29	8-hydroxyquinoline (structure: N, OH)	$R_2InOC_9H_6N$	[13, 34]	192
30	$(CH_3)_2C{=}NOH$	$R_2InON{=}C(CH_3)_2$	[14]	214
31	CH_3SH	R_2InSCH_3	[47]	239
32	toluene-3,4-dithiol (structure: CH_3, SH, SH)	$RInS_2C_6H_3CH_3$	[29]	246
33	CH_3COSH	$R_2InOSCCH_3$	[39]	204
34	CH_3SO_2OH	$R_2InOSO_2CH_3$ (see also reaction No. 5)	[36, 49]	213
35	$(CH_3)_2SONH$	$R_2InONS(CH_3)_2$	[18]	283
36	$(CH_3)_2C{=}NCl$ (special)	$R_2InN{=}C(CH_3)_2$	[69]	265
37	$(CH_3)_3SiN{=}S{=}NSi(CH_3)_3$ (II)	$R_2In(NSi(CH_3)_3)_2SCH_3$	[87]	283
38	$NH(CH_3)_2$	$R_2InN(CH_3)_2$	[1, 7]	252
39	$NHC_6H_5(CH_3)$	$R_2InN(CH_3)C_6H_5$	[75]	252
40	2-(methylamino)pyridine (structure: N, N, CH_3, H)	$RIn(N(CH_3)C_5H_4N)_2$	[78]	302

Table 4 (continued)

No.	reactant (reaction type)	products and remarks	Ref.	page
41	R' NH	$R_2InNC_4H_3R'$; $R' = COH, COCH_3$	[55]	278
42	NH N CH_3	$R_2InN_2C_3CH_3$, also with 3,5-dimethylpyrazole	[68]	278
43	H N N	$R_2InN_2C_7H_5$	[68]	279
44	H N CH_3 N	$R_2InN_2C_3H_2CH_3$	[54]	279
45	H N N	$R_2InN_2C_7H_5$	[31]	279
46	$CH_3C(O)N(CH_3)H$	$R_2InN(CH_3)COCH_3$	[43]	267
47	$CH_3C(=NCH_3)N(CH_3)H$	$R_2InN(CH_3)C(=NCH_3)CH_3$	[64]	268
48	$CH_3C(O)N(CH_3)N(CH_3)H$	$R_2InN(CH_3)N(CH_3)COCH_3 \cdot InR_3$	[72]	271
49	$CH_3NHC(O)C(O)N(CH_3)H$	$(R_2InNCH_3CO)_2$	[57]	289
50	$CH_3C(O)NHNHC(O)CH_3$	$(R_2InNC(O)CH_3)_2$	[59]	267
51	$CH_3NHC(=NCH_3)$-$C(=NCH_3)N(CH_3)H$	$(R_2InNCH_3C(=NCH_3))_2$	[67]	291
52	$CH_3NHC(S)C(S)N(CH_3)H$	$(R_2InNCH_3C(S))_2$	[77]	293
53	R'_2PH	$R_2InPR'_2$; $R' = CH_3$; $R' = C_2H_5$, C_4H_9-t, C_6H_5	[6] [6, 86]	312
54	R'_2POOH	$R_2InOOPR'_2$; $R' = CH_3$; $R' = C_6H_5$	[5, 35] [81]	215
55	$(CH_3)_2POSH$	$R_2InOSP(CH_3)_2$	[50, 51]	215
56	$(CH_3)_2PSSH$	$R_2InSSP(CH_3)_2$	[5]	241
57	$(CH_3)_2P(=NCH_3)OH$	$R_2InOP(=NCH_3)(CH_3)_2$	[73]	215
58	$(CH_3)_2P(=NCH_3)SH$	$R_2InSP(=NCH_3)(CH_3)_2$	[73]	241
59	$(C_6H_5)_3P=NH$	$R_2InN=P(C_6H_5)_3$	[9, 17]	283

Table 4 (continued)

No.	reactant (reaction type)	products and remarks	Ref.	page
60	$(CH_3)_3P{=}NSn(CH_3)_3$	$R_2InN{=}P(CH_3)_3$	[21]	283
61	$(C_6H_5)_2P({=}NSiR_3)NHSiR_3$	$R_2InN(SiR_3)P({=}NSiR_3)(C_6H_5)_2$	[20]	286
62	R'_2AsH	$R_2InAsR'_2$; $R'=CH_3$ $R'=C_4H_9$-t, C_6H_5	[6] [3, 82]	312 –
63	$(CH_3)_2AsOOH$	$R_2InOOAs(CH_3)_2$	[18, 36]	215
64	$(CH_3)_3SiOH$	$R_2InOSi(CH_3)_3$, also for $(C_6H_5)_3SiOH$	[10]	188
65	$\{(CH_3)_2SiS\}_3$	$(RInS)_n$	[24]	245
66	$(CH_3)_2Si(NHC_4H_9\text{-t})_2$	$\{(R_2In)(t\text{-}C_4H_9)N\}_2Si(CH_3)_2$	[79]	288
67	$(CH_3)_2Si(NC_4H_9\text{-t})_2Sn$ (special)	$RIn(NC_4H_9\text{-t})_2Si(CH_3)_2$	[80]	310
68	$(CH_3)_3GeOH$	$R_2InOGe(CH_3)_3$	[12]	188
with metal organic compounds (mainly reactions of Type IV)				
69	$Li[Sn(CH_3)_3]$ (I)	$Li[R_3InSn(CH_3)_3]$; no pure compound isolated	[38]	367
70	$(CH_3)_2AsX_2$	$[As(CH_3)_4][R_2InX_2]$; X = Cl, Br, I	[62, 70]	349
71	$[As(CH_3)_4]X$	$[As(CH_3)_4][R_3InX]$; X = Cl, Br, I	[62, 70]	349
72	$As(CH_3)_5$	$[As(CH_3)_4][InR_4]$; also for $Sb(CH_3)_5$	[65]	339
*73	$Al(CH_3)_3$	exchange reactions studied by ^1H NMR	[11]	–
74	$Al(CH_3)_3$ + $(CH_3)_2Si(N{=}P(CH_3)_3)_2$	$[(CH_3)_2Si(N{=}P(CH_3)_3)_2Al(CH_3)_2][In(CH_3)_4]$	[28]	339
75	$In(C_2H_5)_3$	$InR_2C_2H_5$, $InR(C_2H_5)_2$	[83, 85]	112
76	$In(CH{=}CH_2)_3$	$InR_2CH{=}CH_2$	[71]	103
77	$LiCH_3$	$Li[InR_4]$	[33]	338
*78	$M(CH_3)_2$	M = Zn, Cd; exchange reactions studied by ^1H NMR	[19, 27]	–
79	$(C_5H_5)M(CO)_3H$	$In(M(CO)_3C_5H_5)_3$; M = Mo, W	[30]	–
80	$Re(CO)_5H$	$In(Re(CO)_5)_3$	[30]	–
81	$CH_3Ir(P(CH_3)_3)_4$	$Ir(CH_3)_2(In(CH_3)_2)(P(CH_3)_3)_3$	[88]	331

* Further information:

There are four different reaction types:

Type I: Lewis bases form adducts with the electron-poor $In(CH_3)_3$. Adduct formation appears to be the first step in every reaction of this compound. However, because the number of In–CH_3 bonds remains unchanged, most of the stable and characterized products of this kind are not listed in Table 4, but are described in Tables 5 and 6 and sections related to them.

Type II: Electrophiles X (such as SO_2, SO_3, RN=C=NR, etc.) can be inserted into one of the In–C bonds of $In(CH_3)_3$, producing monomeric or oligomeric compounds of the general composition $(CH_3)_2In\text{-}X\text{-}CH_3$.

Type III: Proton acids react with $In(CH_3)_3$ to eliminate CH_4; this is the most frequently used method for synthesizing dimethylindium compounds.

Type IV: Ligand exchange between EX_3 and $In(CH_3)_3$ to form either $(CH_3)_2InX$ or CH_3InX_2, depending on the mole ratio of the reactants.

The starting material for each of these reactions can be pure $In(CH_3)_3$ (dissolved in an inert aprotic solvent) or the (unstable) diethyl etherate, $(CH_3)_3In \cdot nO(C_2H_5)_2$. In reactions Nos. 28, 40, 48, and 66 pure $In(CH_3)_3$ has been used because of possible further reactions of the etherate.

Reaction 1 is quantitative in n-C_7H_{16} or n-C_9H_{20}. At room temperature pure $In(CH_3)_3$ ignites spontaneously in air; the peroxide is assumed to be an intermediate [63]. The etherate does not show this extreme sensitivity to oxidation.

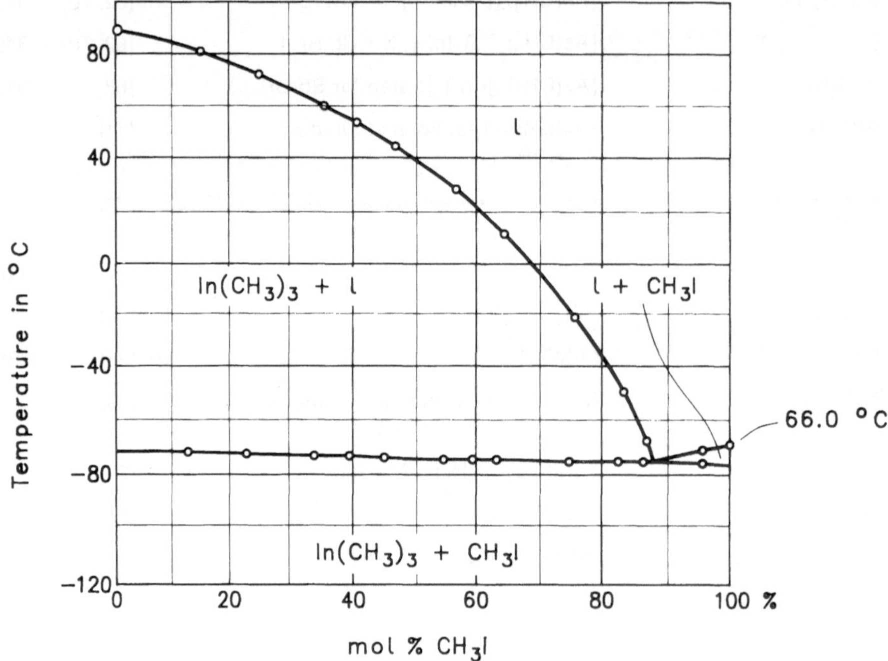

Fig. 2. Solid-liquid diagram of the $In(CH_3)_3$-CH_3I system [74].

Reaction 18. Liquid-solid phase equilibria in the $In(CH_3)_3$-CH_3I system have been studied by thermal analysis. The fusibility diagram shown in **Fig. 2** is of the eutectic type. The eutectic is 88 mol% CH_3I and melts at −72 °C [74].

Reaction 73. The exchange of CH_3 groups between $Al(CH_3)_3$ and $In(CH_3)_3$ in toluene at -50 ± 5 °C was investigated by ^1H NMR spectroscopy. The results are presented as a plot of log $1/\tau$ versus 1/T; the methyl exchange has an Arrhenius activation energy of 16.4 ± 1 kcal/mol [11].

Reaction 78. Exchange reactions between $Zn(CH_3)_2$ and $In(CH_3)_3$ in $O(C_2H_5)_2$ at −8.8 °C, in $N(C_2H_5)_3$ at 70.5 °C, and in CH_2Cl_2 at −90 °C were studied by ^1H NMR. The reactions are second order and the exchange rates decrease with increasing coordination ability of the solvent [27]. Similar studies were made on mixtures of $Cd(CH_3)_2$ and $In(CH_3)_3$ in CH_2Cl_2 at −41 °C [19]; however, the rate constants from [19] have been recalculated in [27] using a modified data treatment. Kinetic parameters are tabulated below; k_1 is the rate constant for the formation of the binuclear intermediate.

solvent	k_1 at 25 °C ($L \cdot mol^{-1} \cdot s^{-1}$)	E_a (kcal/mol)	$\Delta H^{\neq}$ (kcal/mol)	$\Delta S^{\neq}$ ($cal \cdot mol^{-1} \cdot K^{-1}$)
with $Zn(CH_3)_2$ [27]				
$O(C_2H_5)_2$	2900	9.1 ± 0.2	8.5	−14
$N(C_2H_5)_3$	17.4	16.3 ± 0.7	15.7	+ 2
CH_2Cl_2		<5		
with $Cd(CH_3)_2$ [19, 27]				
CH_2Cl_2 [19]	1300	8.4 ± 0.2		−18
CH_2Cl_2 [27]	7800	8.4 ± 0.2	7.8	−14

References:

[1] Coates, G. E.; Whitcombe, R. A. (J. Chem. Soc. **1956** 3351/4).
[2] Didchenko, R.; Alix, J. E.; Toeniskoetter, R. H. (J. Inorg. Nucl. Chem. **14** [1960] 35/7).
[3] Coates, G. E.; Graham, J. (J. Chem. Soc. **1963** 233/7).
[4] Coates, G. E.; Mukherjee, R. N. (J. Chem. Soc. **1963** 229/33).
[5] Coates, G. E.; Mukherjee, R. N. (J. Chem. Soc. **1964** 1295/303).
[6] Beachley, O. T.; Coates, G. E. (J. Chem. Soc. **1965** 3241/7).
[7] Beachley, O. T.; Coates, G. E.; Kohnstam, G. (J. Chem. Soc. **1965** 3248/52).
[8] Schmidbaur, H. (Angew. Chem. **77** [1965] 169/70; Angew. Chem. Intern. Ed. Engl. **4** [1965] 152/3).
[9] Schmidbaur, H.; Kuhr, G.; Krüger, W. (Angew. Chem. **77** [1965] 866; Angew. Chem. Intern. Ed. Engl. **4** [1965] 877).
[10] Schmidbaur, H.; Schindler, F. (Chem. Ber. **99** [1966] 2178/86).

[11] Williams, K. C.; Brown, T. L. (J. Am. Chem. Soc. **88** [1966] 5460/5).
[12] Armer, B.; Schmidbaur, H. (Chem. Ber. **100** [1967] 1521/35).
[13] Clark, H. C.; Pickard, A. L. (J. Organometal. Chem. **8** [1967] 427/34).
[14] Jennings, J. R.; Wade, K. (J. Chem. Soc. A **1967** 1333/9).
[15] Clark, H. C.; Pickard, A. L. (J. Organometal. Chem. **13** [1968] 61/71).
[16] Jeffery, E. A.; Mole, T. (J. Organometal. Chem. **11** [1968] 393/8).
[17] Schmidbaur, H.; Jonas, G. (Chem. Ber. **101** [1968] 1271/85).

[18] Schmidbaur, H.; Kammel, G. (J. Organometal. Chem. **14** [1968] P28/P29).
[19] Henold, K.; Soulati, J.; Oliver, J. P. (J. Am. Chem. Soc. **91** [1969] 3171/4).
[20] Schmidbaur, H.; Schwirten, K.; Pickel, H.-H. (Chem. Ber. **102** [1969] 564/7).

[21] Wolfsberger, W.; Schmidbaur, H. (J. Organometal. Chem. **17** [1969] 41/51).
[22] Hobbs, C. W.; Tobias, R. S. (Inorg. Chem. **9** [1970] 1998/2004).
[23] Hsieh, A. T. T. (Inorg. Nucl. Chem. Letters **6** [1970] 767/71).
[24] Okawara, R.; Yasuda, K. (Japan. 70-32692 [1970] from C.A. **74** [1971] No. 112621).
[25] Weidlein, J. (J. Organometal. Chem. **24** [1970] 63/75).
[26] Hsieh, A. T. T. (J. Organometal. Chem. **27** [1971] 293/301).
[27] Soulati, J.; Henold, K. L.; Oliver, J. P. (J. Am. Chem. Soc. **93** [1971] 5694/8).
[28] Wolfsberger, W.; Schmidbaur, H. (J. Organometal. Chem. **27** [1971] 181/4).
[29] Berniaz, A. F.; Tuck, D. G. (J. Organometal. Chem. **46** [1972] 243/50).
[30] Hsieh, A. T. T.; Mays, M. J. (J. Organometal. Chem. **37** [1972] 9/14).

[31] Garnovskii, A. D.; Okhlobystin, O. Yu.; Osipov, O. A.; Yunusov, K. M.; Kolodyazhnyi, Yu. V.; Golubinskaya, L. M.; Svergun, V. I. (Zh. Obshch. Khim. **42** [1972] 920/5; J. Gen. Chem. [USSR] **42** [1972] 910/4).
[32] Grimes, R. N.; Rademaker, W. J.; Denniston, M. L.; Bryan, R. F.; Greene, P. T. (J. Am. Chem. Soc. **94** [1972] 1865/9).
[33] Hoffmann, K.; Weiss, E. (J. Organometal. Chem. **37** [1972] 1/8).
[34] Maeda, T.; Okawara, R. (J. Organometal. Chem. **39** [1972] 87/91).
[35] Olapinski, H.; Schaible, B.; Weidlein, J. (J. Organometal. Chem. **43** [1972] 107/16).
[36] Olapinski, H.; Weidlein, J. (J. Organometal. Chem. **35** [1972] C53/C56).
[37] Schwering, H.-U.; Hausen, H.-D.; Weidlein, J. (Z. Anorg. Allgem. Chem. **391** [1972] 97/106).
[38] Weibel, A. T.; Oliver, J. P. (J. Am. Chem. Soc. **94** [1972] 8590/2).
[39] Hausen, H.-D.; Guder, H. J. (J. Organometal. Chem. **57** [1973] 243/53).
[40] Hausen, H.-D.; Schwering, H.-U. (Z. Anorg. Allgem. Chem. **398** [1973] 119/28).

[41] Lindel, W.; Huber, F. (Z. Naturforsch. **28b** [1973] 517/8).
[42] Olapinski, H.; Weidlein, J. (J. Organometal. Chem. **54** [1973] 87/93).
[43] Schwering, H.-U.; Weidlein, J. (Chimia [Switz.] **27** [1973] 535/8).
[44] Chung, H. L.; Tuck, D. G. (Can. J. Chem. **52** [1974] 3944/9).
[45] Habeeb, J. J.; Tuck, D. G. (Can. J. Chem. **52** [1974] 3950/4).
[46] Lindel, W.; Huber, F. (Z. Anorg. Allgem. Chem. **408** [1974] 167/74).
[47] Mann, W. G. (Diss. Univ. Stuttgart 1974).
[48] Mann, G.; Olapinski, H.; Ott, R.; Weidlein, J. (Z. Anorg. Allgem. Chem. **410** [1974] 195/205).
[49] Olapinski, H.; Weidlein, J.; Hausen, H.-D. (J. Organometal. Chem. **64** [1974] 193/204).
[50] Schaible, B.; Haubold, W.; Weidlein, J. (Z. Anorg. Allgem. Chem. **403** [1974] 289/300).

[51] Schaible, B.; Roessel, K.; Weidlein, J.; Hausen, H.-D. (Z. Anorg. Allgem. Chem. **409** [1974] 176/84).
[52] Schaible, B.; Weidlein, J. (Z. Anorg. Allgem. Chem. **403** [1974] 301/9).
[53] Schwering, H.-U.; Olapinski, H.; Jungk, E.; Weidlein, J. (J. Organometal. Chem. **76** [1974] 315/24).
[54] Breakell, K. R.; Rendle, D. F.; Storr, A.; Trotter, J. (J. Chem. Soc. Dalton Trans. **1975** 1584/9).
[55] Chung, H. L.; Tuck, D. G. (Can. J. Chem. **53** [1975] 3492/7).
[56] Hausen, H.-D.; Mertz, K.; Weidlein, J.; Schwarz, W. (J. Organometal. Chem. **93** [1975] 291/6).
[57] Schwering, H.-U.; Weidlein, J.; Fischer, P. (J. Organometal. Chem. **84** [1975] 17/37).

[58] von Dahlen, K.-H.; Dehnicke, K. (Chem. Ber. **110** [1977] 383/94).
[59] Eberwein, B.; Lieb, W.; Weidlein, J. (Z. Naturforsch. **32b** [1977] 32/6).
[60] Gavrilenko, V. V.; Kolesov, V. S.; Zakharkin, L. I. (Zh. Obshch. Khim. **47** [1977] 964; J. Gen. Chem. [USSR] **47** [1977] 881).

[61] Greenwood, N. N. (Pure Appl. Chem. **49** [1977] 791/801).
[62] Widler, H.-J.; Schwarz, W.; Hausen, H.-D.; Weidlein, J. (Z. Anorg. Allgem. Chem. **435** [1977] 179/90).
[63] Aleksandrov, Yu. A.; Chikinova, N. V.; Makin, G. I.; Kornilova, N. V.; Bregadze, V. I. (Zh. Obshch. Khim. **48** [1978] 467; J. Gen. Chem. [USSR] **48** [1978] 417).
[64] Gerstner, F.; Weidlein, J. (Z. Naturforsch. **33b** [1978] 24/9).
[65] Tatzel, G.; Schrem, H.; Weidlein, J. (Spectrochim. Acta A **34** [1978] 549/59).
[66] Gavrilenko, V. V.; Kolesov, V. S.; Zakharkin, L. I. (Zh. Obshch. Khim. **49** [1979] 1845/8; J. Gen. Chem. [USSR] **49** [1979] 1623/6).
[67] Gerstner, F.; Schwarz, W.; Hausen, H.-D.; Weidlein, J. (J. Organometal. Chem. **175** [1979] 33/47).
[68] Peterson, L. K.; Thé, K. I. (Can. J. Chem. **57** [1979] 2520/2).
[69] Weller, F.; Müller, U. (Chem. Ber. **112** [1979] 2039/44).
[70] Widler, H.-J.; Weidlein, J. (Z. Naturforsch. **34b** [1979] 18/22).

[71] Fries, W.; Sille, K.; Weidlein, J.; Haaland, A. (Spectrochim. Acta A **36** [1980] 611/9).
[72] Gerstner, F.; Hausen, H.-D.; Weidlein, J. (J. Organometal. Chem. **197** [1980] 135/46).
[73] Schrem, H.; Weidlein, J. (Z. Anorg. Allgem. Chem. **465** [1980] 109/19).
[74] Skachkov, B. K.; Kozyrkin, B. I.; Bogdanova, L. N.; Gribov, B. G. (Zh. Fiz. Khim. **54** [1980] 1851/2; Russ. J. Phys. Chem. **54** [1980] 1050).
[75] Beachley, O. T., Jr.; Bueno, C.; Churchill, M. R.; Hallock, R. B.; Simmons, R. G. (Inorg. Chem. **20** [1981] 2423/8).
[76] Waller, I.; Halder, T.; Schwarz, W.; Weidlein, J. (J. Organometal. Chem. **232** [1982] 99/112).
[77] Halder, T.; Schwarz, W.; Weidlein, J.; Fischer, P. (J. Organometal. Chem. **246** [1983] 29/48).
[78] Arif, A. M.; Bradley, D. C.; Frigo, D. M.; Hursthouse, M. B.; Hussain, B. (J. Chem. Soc. Chem. Commun. **1985** 783/4).
[79] Veith, M.; Lange, H.; Belo, A.; Recktenwald, O. (Chem. Ber. **118** [1985] 1600/15).
[80] Veith, M.; Lange, H.; Recktenwald, O.; Frank, W. (J. Organometal. Chem. **294** [1985] 273/94).

[81] Arif, A. M.; Barron, A. R. (Polyhedron **7** [1988] 2091/4).
[82] Arif, A. M.; Benac, B. L.; Cowley, A. H.; Jones, R. A.; Kidd, K. B.; Nunn, C. W. (New J. Chem. **12** [1988] 553/7).
[83] Bradley, D. C.; Chudzynska, H.; Frigo, D. M. (Chemtronics **3** [1988] 159/61).
[84] Buchan, N. I.; Larsen, C. A.; Stringfellow, G. B. (J. Cryst. Growth **92** [1988] 605/15).
[85] Hui, B. C.; Lorberth, J.; Melas, A. A. (US 4 720 560 [1988]; Eur. 181706 [1986]; C.A. **105** [1986] No. 79150).
[86] Aitchison, K. A.; Backer-Dirks, J. D. J.; Bradley, D. C.; Faktor, M. M.; Frigo, D. M.; Hurtshouse, M. B.; Hussain, B.; Short, R. L. (J. Organometal. Chem. **366** [1989] 11/23).
[87] Kottmair-Maieron, D.; Lechler, R.; Weidlein, J. (Z. Anorg. Allgem. Chem. **593** [1991] 111/23).
[88] Thorn, D. L.; Harlow, R. L. (J. Am. Chem. Soc. **111** [1989] 2575/80).

1.1.1.1.5 Applications of $In(CH_3)_3$

As do $Al(C_2H_5)_3$ and $Ga(CH_3)_3$, $In(CH_3)_3$ activates the dimerizations of propylene, butene, and heptene in continuous reactions at temperatures from 200 to 350 °C and pressures between 100 and 200 atm [2].

$In(CH_3)_3$, in association with halogenides of Ti, V, or Ta, is used as a cocatalyst for the polymerization of ethylene. The strongly exothermic reaction is carried out in the absence of solvent under N_2 at 300 lb/in^2 ($\sim$20 atm) [3].

Finely divided SiO_2 or C serves as carrier material for $In(CH_3)_3$ (and other alkyl derivatives of B, Ga, Tl) to polymerize mono and diolefins [7].

Butadiene can be polymerized in high yield to polybutadiene with the aid of $MoCl_5/InR_3$ catalysts [8].

$In(CH_3)_3$ catalyzes (as do other alkyl derivatives of B, Al, Ga, Zn, Cd, Hg) the formation of hexacarbonyls of Cr, Mo, and W, starting with the corresponding sulfides, oxides, or chlorides in THF solution and CO at temperatures between 0 and 200 °C and pressures up to 15000 lb/in^2 (about 1000 atm) [5]. On the other hand $In(CH_3)_3$ drastically and irreversibly diminishes the catalytic activity of platinum in hydrogenation reactions [1].

$In(CH_3)_3$ serves as starting material for the production of In atoms by UV or pulsed laser irradiation, or by spark discharge (see p. 12). In the presence of oxygen the decomposition of $In(CH_3)_3$ forms thin films of In_2O_3 [9].

By far the greatest importance of $In(CH_3)_3$ in the past 5 to 10 years is the production of single crystal III/V (13/15 according to the new notation) semiconductor layers. The work is based on the investigations of Didchenko [4], who obtained InP by decomposing a mixture of $In(CH_3)_3$ and PH_3 at nearly 300 °C. For the production of photodetectors, dye lasers, photodiodes, and other semiconductor components, development includes not only the III/V binary compounds (such as InP, InAs, or InSb [6]), but also ternary ($Ga_{1-x}In_xP$, $Ga_{1-x}In_xAs$, etc.) and quaternary (GaInAsP, GaInAsSb, etc.) products, formed by MOCVD (metal organic chemical vapor deposition), metal organic chemical vapor epitaxy (MOCVE), or the recent vacuum technique, molecular beam epitaxy (MBE). All the procedures are based on the thermal decomposition of a mixture of $In(CH_3)_3$ and volatile MX_3 compounds of the group 15 elements (M = P, As, Sb and X = H, CH_3, C_2H_5, Hal, and others), highly diluted with hydrogen (only rarely, helium or argon); the separation of the III/V components at the surface of solid substrates can be controlled by temperature changes and variations in the reactant ratio.

Insufficient space is available to present the complicated apparatus and preparative details of the different methods; the reader may wish to refer to review papers [10 to 18, 20 to 22] and a specialized literature collection [19]. A list of individual works in which $In(CH_3)_3$ is used in gas phase epitaxy is arranged at the end of this section; no claim to its completeness is made.

References:

[1] Maxted, E.B.; Moon, K.L. (J. Chem. Soc. **1949** 2171/4).
[2] Ziegler, K. (Brit. 775384 [1957] from C.A. **1958** 12893).
[3] Crawford, J.W.C. (Brit. 795971 [1958] from C.A. **1959** 1836).
[4] Didchenko, R.; Alix, J.E.; Toeniskoetter, R.H. (J. Inorg. Nucl. Chem. **14** [1960] 35/7).
[5] Podall, H.E.; Shapiro, H. (U.S. 2952517 [1960] from C.A. **1961** No. 4903).
[6] Harrison, B.C.; Tompkins, E.H. (Inorg. Chem. **1** [1962] 951/3).

[7] Cabot Corp. (Fr. 1375984 [1964] from C.A. **62** [1965] 16403).
[8] Naylor, F.E. (U.S. 3232920 [1966] from C.A. **64** [1966] 11423).
[9] Turner, P.; Howson, R.P.; Bishop, C.A. (Thin Solid Films **83** [1981] 253/8).
[10] Benz, K.W.; Haspeklo, H.; Bosch, R. (J. Phys. Colloq. [Paris] **43** [1982] C5-393/C5-399).

[11] Stringfellow, G.B. (J. Cryst. Growth **62** [1983] 225/9).
[12] Dupuis, R.D. (Science **226** [1984] 623/9).
[13] Karlicek, R.; Long, J.A.; Donnelly, V.M. (J. Cryst. Growth **68** [1984] 123/7).
[14] Razeghi, M.; Duchemin, J.P. (J. Cryst. Growth **70** [1984] 145/9).
[15] Stringfellow, G.B. (J. Cryst. Growth **68** [1984] 111/22).
[16] Griffiths, R.J.M. (Chem. Ind. [London] **1985** 247/51).
[17] Ludowise, M.J. (J. Appl. Phys. **58** [1985] R31/R55).
[18] Stringfellow, G.B. (Semicond. Semimetals **22** [1985] 209/59).
[19] Alfa Products, Morton Thiokol Inc. (Alfa Organometallics for Vapor Phase Epitaxy, Literature and Product Review, Danvers, Mass., 1980 [1983/86], References, 1960 to 1985).
[20] Wilmsen, C.W. (Physics and Chemistry of III/V Compound Semiconductor Interfaces, Plenum, New York 1985).

[21] Yoshida, M.; Watanabe, H. (J. Electrochem. Soc. **132** [1985] 1733/40).
[22] Stringfellow, G.B. (Conf. Ser. Inst. Phys. [London] No. 79 [1986] 115/20; C.A. **105** [1986] No. 124372).

List of individual works on MOCVD in which primarily $In(CH_3)_3$ is used in industrial research and application:

Pashinkin, A. S.; Malkova, A. S.; Fedorov, V. A.; Rodionov, A. V.; Sveshnikov, Yu. N. (Izv. Akad. Nauk SSSR Neorg. Mater. **14** [1978] 623/6; Inorg. Mater. [USSR] **14** [1978] 483/9; C.A. **88** [1978] No. 198620).
Moss, R. H.; Evans, J. S. (J. Cryst. Growth **55** [1981] 129/34).
Nippon Telegraph and Telephone Public Corp. (Japan. 81-50509 [1981] from C.A. **95** [1981] No. 715141).
Pearsall, T. P.; Hirtz, J. P. (J. Cryst. Growth **54** [1981] 127/31).
Yang, J. J.; Ruth, R. P.; Manasevit, H. M. (J. Appl. Phys. **52** [1981] 6729/34).
Guldner, Y.; Vieren, J. P.; Voisin, P.; Voos, M.; Razeghi, M.; Poisson, M. A. (Appl. Phys. Letters **40** [1982] 877/9).
Bass, S. J.; Pickering, C.; Young, M. L. (J. Cryst. Growth **64** [1983] 68/75).
Goetz, K. H.; Bimberg, D.; Jürgensen, H.; Selders, J.; Solomonov, A. V.; Gilinskii, G. F.; Razeghi, M. (J. Appl. Phys. **54** [1983] 4543/52).
Hino, I.; Gomyo, A.; Kobayashi, K.; Suzuki, T.; Nishida, K. (Appl. Phys. Letters **43** [1983] 987/9).
Hsu, C. C.; Cohen, R. M.; Stringfellow, G. B. (Proc. Electrochem. Soc. **83**-13 [1983] 193/200).

Hsu, C. C.; Cohen, R. M.; Stringfellow, G. B. (J. Cryst. Growth **63** [1983] 8/12).
Kuo, C. P.; Cohen, R. M.; Stringfellow, G. B. (J. Cryst. Growth **64** [1983] 461/70).
Ludowise, M. J.; Dietze, W. T.; Lewis, C. R. (Conf. Ser. Inst. Phys. [London] No. 65 [1983] 93/100).
Matsushita Electric Industrial Co. Ltd. (Japan. 83-132921 [1983]).
Sacilotti, M.; Mircea, A.; Azoulay, R. (J. Cryst. Growth **63** [1983] 111/5).
Ogura, M.; Mizuta, M.; Hase, N.; Kukimoto, H. (Japan. J. Appl. Phys. **22** Pt. 1 [1983] 658/62).
Sogou, S.; Kameyama, A.; Katsda, H.; Miyamoto, Y.; Furuya, K.; Suematsu, Y. (Electron. Letters **19** [1983] 1036/7).
Bass, S. J.; Young, M. L. (J. Cryst. Growth **68** [1984] 311/8).

Bedair, S. M.; Katsuyama, T.; Chiang, P. D.; El-Masry, N. A.; Tischler, M.; Timmons, M. (J. Cryst. Growth **68** [1984] 477/82).
Chiang, P. K.; Bedair, S. M. (J. Electrochem. Soc. **131** [1984] 2422/6).
Dean, P. J.; Skolnick, M. S.; Taylor, L. L. (J. Appl. Phys. **55** [1984] 957/63).
Fraas, L. M.; McLeod, P. S.; Cape, J. A.; Partain, L. D. (J. Cryst. Growth **68** [1984] 490/6).
Haigh, J.; O'Brien, S. (J. Cryst. Growth **67** [1984] 75/8).
Ikeda, M.; Honda, M.; Mori, Y.; Keneko, K.; Watanabe, N. (Appl. Phys. Letters **45** [1984] 964/6).
Kamijoh, T.; Takano, H.; Sakuta, M. (J. Cryst. Growth **67** [1984] 144/6).
Kuo, C. P.; Yuan, J. S.; Cohen, R. M.; Dunn, J.; Stringfellow, G. B. (Appl. Phys. Letters **44** [1984] 550/2).
Lewis, C. R.; Ford, C. W.; Werthen, J. G. (Appl. Phys. Letters **45** [1984] 895/7).
Long, J. A.; Riggs, V. G.; Johnston, W. D., Jr. (J. Cryst. Growth **69** [1984] 10/4).
Mircea, A.; Azoulay, R.; Dugrand, L.; Mellet, R.; Rao, K.; Sacilotti, M. (J. Electron. Mater. **13** [1984] 603/20).
NEC Corp. (Japan. 84-172718 [1984] from C.A. **102** [1985] No. 104065).

Nelson, A. W.; Westbrook, L. D. (J. Appl. Phys. **55** [1984] 3103/8).
Nippon Electric Co., Ltd. (Japan. 84-22321 [1984] from C.A. **101** [1984] No. 31549).
Oishi, M.; Nojima, S.; Kuroiwa, K. (Japan. J. Appl. Phys. **23** Pt. 2 [1984] 625/7).
Ogura, M.; Mizuta, M.; Hase, N.; Kukimoto, H. (Japan. J. Appl. Phys. **23** Pt. 1 [1984] 79/83).
Scott, M. D.; Normann, A. G.; Bradley, R. R. (J. Cryst. Growth **68** [1984] 319/25).
Toshiba Corp. (Japan. 84-106114 [1984] from C.A. **101** [1984] No. 220352).
Tsang, W. T. (Appl. Phys. Letters **45** [1984] 1234/6).
Uwai, K.; Susa, N.; Mikami, O.; Fukui, T. (Japan. J. Appl. Phys. **23** Pt. 2 [1984] 121/3).
Wakefield, B.; Eaves, L.; Prior, K. A.; Nelson, A. W.; Davies, G. J. (J. Phys. D **17** [1984] L133/L136).
Bedair, S. M.; Tischler, M. A.; Katsuyama, T.; El-Masry, N. A. (Appl. Phys. Letters **47** [1985] 51/3).
Carey, K. W. (Appl. Phys. Letters **46** [1985] 89).
Chan, K. T.; Zhu, L. D.; Ballantyne, J. M. (Appl. Phys. Letters **47** [1985] 44/6).
Donnelly, V. M.; Brasen, D.; Appelbaum, A.; Geva, M. (J. Appl. Phys. **58** [1985] 2022/35).
Dupuis, R. D.; Temkin, H.; Hopkins, L. C. (Electron. Letters **21** [1985] 60/2).
Fry, K. L.; Kuo, C. P.; Cohen, R. M.; Stringfellow, G. B. (Appl. Phys. Letters **46** [1985] 955/7).
Furukawa Electric Co., Ltd. (Japan. 85-81092 [1985] from C.A. **103** [1985] No. 151376).
Ishikawa, M.; Ohba, Y.; Sugawara, H.; Yamamoto, M.; Nakanishi, T. (Electron. Letters **21** [1985] 1084/5).
Kuo, C. P.; Fry, K. L.; Stringfellow, G. B. (Appl. Phys. Letters **47** [1985] 855/7).
Kuo, C. P.; Cohen, R. M.; Fry, K. L.; Stringfellow, G. B. (J. Electron. Mater. **14** [1985] 231/44).
Nelson, A. W.; Wong, S.; Ritchie, S.; Sargood, S. K. (Electron. Letters **21** [1985] 838/40).

Nojima, S.; Oishi, M.; Asahi, H.; Nagai, H. (Phys. Status Solid A **90** [1985] K 215/K 218).
Razeghi, M.; Duchemin, J. P.; Portal, J. C. (Appl. Phys. Letters **46** [1985] 46/8).
Tsang, W. T. (J. Appl. Phys. **58** [1985] 1415/8).
Yuan, J. S.; Gal, M.; Taylor, P. C.; Stringfellow, G. B. (Appl. Phys. Letters **47** [1985] 405/7).
Zhu, L. D.; Chan, K. T.; Ballantyne, J. M. (Appl. Phys. Letters **47** [1985] 47/8).
Zhu, L. D.; Chan, K. T.; Ballantyne, J. M. (J. Cryst. Growth **73** [1985] 83/95).
Zhu, L. D.; Chan, K. T.; Wagner, D. K.; Ballantyne, J. M. (J. Appl. Phys. **57** [1985] 5486/92).
Zhu, L. D.; Sulewski, P. E.; Chan, K. T.; Muro, K.; Ballantyne, J. M.; Sievers, A. J. (J. Appl. Phys. **58** [1985] 3145/9).

Zilko, J. L.; van Haren, D. L.; Lu, P. Y.; Schumaker, N. E.; Leung, S. Y. (J. Electron. Mater. **14** [1985] 563/72).
Bass, S. J.; Barnett, S. J.; Brown, G. T.; Chew, N. G.; Cullis, A. G.; Pitt, A. D.; Skolnick, M. S. (J. Cryst. Growth **79** [1986] 378/85).

Clawson, A. R.; Hanson, C. M.; Vu, T. T. (J. Cryst. Growth **77** [1986] 334/9).
Dentai, A. G.; Joyner, C. H.; Tell, B.; Zyskino, J. L.; Sulhoff, J. W.; Ferguson, J. F.; Centanni, J. C.; Chu, S. N. G.; Cheng, C. L. (Electron. Letters **22** [1986] 1186/8).
Donelly, V. M.; Brasen, D.; Appelbaum, A.; Geva, M. (J. Vac. Sci. Technol. [2] A **4** [1986] 716/21 from C.A. **105** [1986] No. 52333).
Hsu, C. C.; Yuan, J. S.; Cohen, R. M.; Stringfellow, G. B. (J. Cryst. Growth **74** [1986] 535/42).
Ishikawa, M.; Ohba, Y.; Sugawara, H.; Yamamoto, M.; Nakanisi, T. (Appl. Phys. Letters **48** [1986] 207/8).
Mellet, R. (Fr. 2569207 [1986] from C.A. **105** [1986] No. 52623).
Mircea, A.; Mellet, R.; Rose, B.; Robein, D.; Thibierge, H.; Leroux, G.; Daste, P.; Godefroy, S.; Ossart, P.; Pougnet, A. M. (J. Electron. Mater. **15** [1986] 205/13).
Morizaki, M.; Hase, N. (Japan. 86-94319 [1986] from C.A. **105** [1986] No. 201008).
Ohaba, Y.; Yamamoto, M.; Ishikawa, M.; Iwamoto, M.; Nakanisi, T. (Conf. Ser. Inst. Phys. [London] No. 79 [1986] 679/84 from C.A. **105** [1986] No. 69740).
Olson, J. M.; Kibbler, A. (J. Cryst. Growth **77** [1986] 182/7).

Rose, B.; Robein, D.; Mircea, A.; Mellet, R.; Leroux, G.; Thibierge, H.; Devoldere, P. (Vide Couches Minces **41** [1986] 245/6 from C.A. **105** [1986] No. 32665).
Smeets, E. T. J. M.; Cox, A. M. W. (J. Cryst. Growth **77** [1986] 347/53).
Webb, J. B.; Halpin, C.; Noad, J. P. (J. Appl. Phys. **60** [1986] 2949/53).
Yuan, J. S.; Tsai, M. T.; Cheng, C. H.; Cohen, R. M.; Stringfellow, G. B. (J. Appl. Phys. **60** [1986] 1346/9).
Bacher, F. R.; Leigh, W. B. (J. Cryst. Growth **80** [1987] 456/8).
Bour, D. P.; Shealy, J. R. (Appl. Phys. Letters **51** [1987] 1658/60).
Briggs, A. T. R.; Butler, B. R. (J. Cryst. Growth **85** [1987] 535/42).
Buchan, N. I.; Larsen, C. A.; Stringfellow, G. B. (Appl. Phys. Letters **51** [1987] 1024/6).
Campbell, J. C.; Tsang, W. T.; Qua, G. J.; Bowers, J. E. (Appl. Phys. Letters **51** [1987] 1454/6).
Johnson, E. S.; Legg, G. E.; Schumaker, N. E. (Proc. SPIE-Intern. Soc. Opt. Eng. No. 796 [1987] 161/9; C.A. **107** [1987] No. 226181).

Kamada, M.; Ishikawa, H.; Ikeda, M.; Mori, Y.; Kojima, C. (Conf. Ser. Inst. Phys. [London] No. 83 [1987] 575/80; C.A. **107** [1987] No. 188531).
Larsen, C. A.; Buchan, N. I.; Stringfellow, G. B. (J. Cryst. Growth **85** [1987] 148/53).
Matsui, S.; Mori, K.; Asata, S. (Japan. 87-33421 [1987] from C.A. **106** [1987] No. 205706).
Miller, B. I.; Koren, U.; Capik, R. J.; Su, Y. K. (Appl. Phys. Letters **51** [1987] 2260/2).
Minagawa, S. (Japan. 87-183109 [1987] from C.A. **108** [1988] No. 66563).
Stringfellow, G. B.; Buchan, N. I.; Larsen, C. A. (Mater. Res. Soc. Symp. Proc. **94** [1987] 245/53 from C.A. **107** [1987] No. 243779).
Oba, Y.; Yamamoto, M.; Ishikawa, M. (Japan. 87-65996 [1987] from C.A. **107** [1987] No. 68652).
Tomiyama, C. (Japan. 87-213253 [1987] from C.A. **108** [1988] No. 104356).
Tsang, W. T. (Appl. Phys. Letters **50** [1987] 63/5).
Webb, J. B. (Brit. 2181461 [1987] from C.A. **107** [1987] No. 68700).

Aina, L.; Mattingly, M.; Fathimulla, A.; Matin, E. A.; Loughran, T.; Stecker, L. (J. Cryst. Growth **93** [1988] 911/8).

Andrews, D. A.; Davey, S. T.; Tuppen, C. G.; Wakefield, B.; Davies, G. J. (Appl. Phys. Letters **52** [1988] 816/8).

Bhat, R.; Hayes, J. R.; Schumacher, H.; Koza, M. A.; Hwang, D. M.; Meynadier, M. H. (J. Cryst. Growth **93** [1988] 919/23).

Bougnot, G.; Delannoy, F.; Foucaran, A.; Pascal, F.; Roumanille, F.; Grosse, P.; Bougnot, J. (J. Electrochem. Soc. **135** [1988] 1783/8).

Bour, D. P.; Shealey, J. R. (IEEE J. Quantum Electron. **24** [1988] 1856/63).

Cole, S.; Evans, J. S.; Harlow, M. J.; Nelson, A. W.; Wong, S. (J. Cryst. Growth **93** [1988] 607/12).

Eguchi, K.; Ohba, Y.; Kushibe, M.; Funamizu, M.; Nakanisi, T. (J. Cryst. Growth **93** [1988] 88/92).

Engel, M.; Bauer, R. K.; Bimberg, D.; Grützmacher, D.; Jürgensen, H. (J. Cryst. Growth **93** [1988] 359/64).

Fujisawa, M.; Morimasa, K. (Japan. 88-11598 [1988] from C.A. **108** [1988] No. 159434).

Gerrard, N. D.; Nicholas, D. J.; Williams, J. O.; Jones, A. C. (Chemtronics **3** [1988] 17/30).

Grützmacher, D.; Meyer, R.; Zachau, M.; Helgesen, P.; Zrenner, A.; Wolter, K.; Jürgensen, H.; Koch, F.; Balk, P. (J. Cryst. Growth **93** [1988] 382/8).

Horikawa, H.; Kawai, Y.; Akiyama, M.; Sakuta, M. (J. Cryst. Growth **93** [1988] 523/6).

Hotta, H.; Hino, I.; Suzuki, T. (J. Cryst. Growth **93** [1988] 618/23).

Jou, M. J.; Cherng, Y. T.; Jen, H. R.; Stringfellow, G. B. (J. Cryst. Growth **93** [1988] 62/9).

Kadokura, H.; Yako, T. (Japan. 88-55194 [1988] from C.A. **109** [1988] No. 139732).

Kamei, H.; Hashizume, K.; Murata, M.; Kuwata, N.; Ono, K.; Yoshida, K. (J. Cryst. Growth **93** [1988] 329/35).

Kobayashi, N.; Makimoto, T.; Horikoshi, Y. (Brit. 2192198 [1988] from C.A. **108** [1988] No. 230063).

Kondo, M.; Yamazaki, S.; Sugawara, M.; Okuda, H.; Kato, K.; Nakajima, K. (J. Cryst. Growth **93** [1988] 376/81).

Mircea, A.; Ougazzaden, A.; Dasté, P.; Gao, Y.; Kazmierski, C.; Bouley, J.-C.; Carenco, A. (J. Cryst. Growth **93** [1988] 235/41).

Mori, Y.; Kamada, M. (J. Cryst. Growth **93** [1988] 892/9).

Naitoth, M.; Soga, T.; Jimbo, T.; Umeno, M. (J. Cryst. Growth **93** [1988] 52/5).

Nozaki, C.; Ohba, Y.; Sugawara, H.; Yasuami, S.; Nakanisi, T. (J. Cryst. Growth **93** [1988] 406/11).

Ohba, Y.; Nishikawa, Y.; Nozaki, C.; Sugawara, H.; Nakanisi, T. (J. Cryst. Growth **93** [1988] 613/7).

Ohori, T.; Takechi, M.; Suzuki, M.; Takikawa, M.; Komeno, J. (J. Cryst. Growth **93** [1988] 905/10).

Pak, K.; Wakahara, A.; Sato, T.; Yoshida, A.; Yonezu, H.; Itoh, N.; Takagi, Y. (J. Electrochem. Soc. **135** [1988] 2358/61).

Razeghi, M.; Maurel, M.; Defour, M.; Omnes, G.; Neu, G.; Kozacki, A. (Appl. Phys. Letters **52** [1988] 117/9).

Seki, A.; Konushi, F.; Kudo, J.; Koba, M. (J. Cryst. Growth **93** [1988] 527/31).

Ueda, T.; Onozawa, S.; Akiyama, M.; Sakuta, M. (J. Cryst. Growth **93** [1988] 517/22).

Uwai, K.; Nakagome, H.; Takahei, K. (J. Cryst. Growth **93** [1988] 583/8).

York, P. K.; Kiely, C. J.; Fernandez, G. E.; Baillargeon, J. N.; Coleman, J. J. (J. Cryst. Growth **93** [1988] 512/6).

Yoshikawa, A.; Sugino, T.; Nakamura, A.; Kano, G.; Teramoto, I. (J. Cryst. Growth **93** [1988] 532/8).

Zuhoski, S. P.; Killeen, K. P.; Biefeld, R. M. (Mater. Res. Soc. Symp. Proc. **101** [1988] 313/8 from C.A. **109** [1988] No. 101633).

Wang, T. Y.; Reihlen, E. H.; Jen, H. R.; Stringfellow, G. B. (J. Appl. Phys. **66** [1989] 5376/83).
Suhr, H. (New J. Chem. **14** [1990] 523/6).
Rousina, R.; Halpin, C.; Webb, J. B. (J. Appl. Phys. **68** [1990] 2181/6).
Nilsson, A.; Gustafsson, A.; Samuelson, L. (Appl. Phys. Letters **57** [1990] 878/80).
Takeda, Y.; Araki, S.; Takemi, M.; Noda, S.; Sasaki, A. (Japan. J. Appl. Phys. **29** [1990] L 1040/L 1042).
Eguchi, K.; Kushibe, M.; Funamizu, M.; Ohba, Y. (Japan. J. Appl. Phys. **29** [1990] 1431/4).
Jou, M. J.; Jaw, D. H.; Fang, Z. M.; Stringfellow, G. B. (J. Cryst. Growth **106** [1990] 208/16).
Ludowise, M. J.; Ranganath, T. R.; Fischer-Colbrie, A. (Appl. Phys. Letters **57** [1990] 1493/5).

Schneider, R. P., Jr.; Wessels, B. W. (Appl. Phys. Letters **57** [1990] 1998/2000).
Tanbun-Ek, T.; Logan, R. A.; Temkin, H.; Chu, S. N. G.; Olsson, N. A.; Sergent, A. M.; Wecht, K. W. (IEEE J. Quantum Electron. **24** [1990] 1323/7).
Wang, T. Y.; Kimball, A. W.; Chen, G. S.; Birkedal, D.; Stringfellow, G. B. (J. Appl. Phys. **68** [1990] 3356/63).
Koch, S. M.; Acher, O.; Omnes, F.; Defour, M.; Razeghi, M.; Drevillon, B. (J. Appl. Phys. **68** [1990] 3364/9).
Acher, O.; Koch, S. M.; Omnes, F.; Defour, M.; Razeghi, M.; Drevillon, B. (J. Appl. Phys. **68** [1990] 3564/77).

1.1.1.2 Adducts of the $(CH_3)_3In \cdot D$ Type

This section comprises adducts of the general type $(CH_3)_3In \cdot D$ in which a neutral Lewis base is bonded to the Lewis acid $In(CH_3)_3$ forming a tetrahedral arrangement at the indium atom. Bonds to group 16 elements, group 15 elements, group 14 elements (carbon), and some transition metals are described in separate subsections.

The addition compounds presented in Table 5 (Section 1.1.1.2.1) and Table 6 (Section 1.1.1.2.2) were prepared by the following general methods:

Method I: From $In(CH_3)_3$ and the appropriate donor without solvent.
Trimethylindium was co-condensed with an excess of the donor component in vacuum at low temperatures and the mixture slowly warmed to room temperature. The unused portion of the volatile Lewis base was removed by reduced pressure, and the residual adduct (mostly solid at room temperature) purified by vacuum distillation or sublimation (if the thermal stability of the compound permitted) [1].

Method II: From $In(CH_3)_3$ and the appropriate donor in an aprotic solvent.
$In(CH_3)_3$ in an inert aprotic solvent (benzene, toluene, hexane, or pentane) was combined with a solution of the donor component, and refluxed with constant stirring for 30 to 60 min. Evaporation of the volatile components and vacuum distillation of the residue was the usual procedure for isolation and purification [5, 6].

Method III: From $(CH_3)_3In \cdot O(C_2H_5)_2$ (Table 5, No. 2) and the appropriate donor.
A benzene (or other inert solvent) solution or suspension of the donor component was stirred for about 30 min with a small excess of $(CH_3)_3In \cdot n\ O(C_2H_5)_2$ in benzene at room temperature. The reaction, usually only weakly exothermic, was completed by refluxing, then the volatile components were removed in vacuum, and the ether-free $(CH_3)_3In \cdot D$ was purified by distillation, sublimation, or recrystallization [3].

Method IV: From $(CH_3)_3In \cdot OC_4H_8$ (Table 5, No. 3) and D.
An excess of the new donor was added to the THF adduct and the mixture gently warmed for about 30 min. After removal of the excess donor and the displaced THF, the product was purified by distillation [7, 8].

General Remarks. Numerous cases prove equation (1) to be the best for calculating the formation enthalpy, ΔH, of (simple) adducts:

$$-\Delta H + W = E_A \cdot E_B + C_A \cdot C_B \qquad (1)$$

E and C are parameters that quantify the electrostatic and covalent interactions between a Lewis acid (A) and a Lewis base (B), and W is a constant correction factor having a value of zero for most of the systems investigated. For $In(CH_3)_3$ the pairs of parameters $E_A/C_A = 15.3/0.654$ [4] and 13.19/0.37 [9], have been reported.

These values are not very reliable, however, since they are derived from so few experimental data. Extensive lists of Lewis base parameters, E_B and C_B, as well as those of Lewis acids are to be found in [2, 4, 9].

References:

[1] Coates, G.E.; Whitcombe, R.A. (J. Chem. Soc. **1956** 3351/4).
[2] Drago, R.S.; Wayland, B.B. (J. Am. Chem. Soc. **87** [1965] 3571/7).
[3] Schindler, F.; Schmidbaur, H. (Chem. Ber. **100** [1967] 3655/63).
[4] Drago, R.S. (Struct. Bonding [Berlin] **15** [1973] 73/139).
[5] Karschin, J.; Mann, G.; Weidlein, J. (unpublished results, 1979).
[6] Krause, H.; Sille, K.; Hausen, H.-D.; Weidlein, J. (J. Organometal. Chem. **25** [1982] 253/64).
[7] Cole-Hamilton, D.J.; Gerrard, N.D.; Jones, A.C.; Holliday, A.K.; Mullin, J.B.; U.K. Secretary of State for Defence (Brit. 2125795 [1984] from C.A. **101** [1984] No. 91212).
[8] Jones, A.C.; Gerrard, N.D.; Cole-Hamilton, D.J.; Holliday, A.K. (J. Organometal. Chem. **265** [1984] 9/15).
[9] Drago, R.S.; Wong, N.; Bilgrien, C.; Vogel, G.C. (Inorg. Chem. **26** [1987] 9/14).

1.1.1.2.1 $In(CH_3)_3$ is Bonded to a Group 16 Element

The compounds in this section are adducts in which $In(CH_3)_3$ is bonded to neutral donor molecules by O or S. The ether adducts Nos. 1 to 3 show no exact stoichiometry whereas Nos. 4 to 7 are 1:1 adducts. Contradictory statements especially concerning No. 3, are primarily based on the nonstoichiometric composition.

General methods I to IV for the preparation of the compounds in Table 5 are summarized in Section 1.1.1.2 on p. 27.

Table 5
Adducts of the $(CH_3)_3In \cdot D$ Type with O and S Donor Atoms.
An asterisk indicates further information at the end of the table.

No.	donor D method of preparation (yield)	properties and remarks
*1	$O(CH_3)_2$ I (95%)	colorless liquid; dec. 20 °C [11] ^1H NMR (C_6D_6, 20 °C): 0.01 (CH_3In), 2.84 (CH_3O) [11] IR and Raman spectra on p. 30

Table 5 (continued)

No.	donor D method of preparation (yield)	properties and remarks
*2	$O(C_2H_5)_2$	colorless liquid (2:1 adduct), m.p. −15 °C [1], b.p. 56 °C/15 Torr, 73 °C/70 Torr, 139 °C/760 Torr [1, 18], b.p. 147 °C (extrapolated) [2] ^{1}H NMR ($(CH_3)_3In \cdot 0.7\ O(C_2H_5)_2$; without solvent, 30 °C): 0.02 (CH_3In), 1.73 (t, CH_3, J=7), 4.23 (q, CH_2) [11]; J(C,H) of CH_3In=125.0 in benzene or CCl_4 [6]
*3	OC_4H_8 (THF) IV (60 to 80%)	colorless oily liquid, b.p. 60 °C/vacuum [16] ^{1}H NMR (neat, 25 °C): 0.0 (CH_3In), 2.3 (m, CH_2C), 4.2 (t, CH_2O) [16] IR (film): 2960 vs, 2910 s, 2880 vs, 2840 vs,sh, 2270 vw, 1460 m, 1340 vw, 1290 vw, 1245 vw, 1180 w,br, 1150 m, 1070 s,sh, 1050 vs,sh, 1040 vs, 915 m, 880 s, 840 w,sh, 680 vs,br, 515 vw,sh, 485 vs [16] mass spectrum (70 °C, 15 eV): $[M-THF]^+$, $[M-THF-CH_3]^+$, $[M-THF-2\ CH_3]^+$ [16]
4	$OS(CH_3)_2$ III (76%)	colorless crystals, m.p. −2 °C, b.p. 74 °C/1 Torr [6] ^{1}H NMR (CCl_4, 35 °C): −0.5 (CH_3In, J(C,H)=123.0), 2.67 (CH_3S, J(C,H)=139.5) [6] IR (film): 1298 sh $\delta_s(CH_3S)$, 1140 vs $\delta_s(CH_3In)$, 1055 sh+1025 sh+1000 vs $\nu_{as}(SOIn)$, 950 vs+930 w+897 m $\varrho(CH_3S)$, 688 vs+630 sh+510 sh $\varrho(CH_3In)+\nu_s+\nu_{as}(SC_2)$, 480 vs $\nu_{as}(InC_3)$, 407 s $\nu_s(InOS)$, 336 s $\delta_{as}(CSO)$ [6]
5	$ON(CH_3)_3$ III (96%)	colorless crystals, m.p. 66 °C, b.p. 124 °C/1 Torr [6] ^{1}H NMR (CCl_4, 35 °C): −0.64 (CH_3In, J(C,H)=123.5), 3.3 (CH_3N, J(C,H)=141.0) [6] IR (solid): 1238 s $\delta(CH_3N)$, 1138 s+1133 sh $\delta_s(CH_3In)$, 950 sh+943 vs+900 sh $\nu_{as}(NOIn)$, 765 s $\nu_{as}(NC_3)$, 682 vs+668 sh+635 w $\varrho(CH_3In)$, 470 vs $\nu_{as}(InC_3)$, 527 s+478 sh+391 m $\nu_s(InON)+\delta(CNO)$, 377 m $\delta(CNC)$ [6]
6	$OP(CH_3)_3$ III (92%)	colorless crystals, m.p. 34 °C, b.p. 88 °C/1 Torr [6] ^{1}H NMR (CCl_4, 35 °C): −0.63 (CH_3In, J(C,H)=124.0), 1.55 (CH_3P, J(P,H)=13.25, J(C,H)=127.5) [6] IR (solid): 1298 vs $\delta_s(CH_3P)$, 1175 sh+1142 vs+1033 m $\delta_s(CH_3In)+\nu_{as}(POIn)$, 940 vs+853 vs $\varrho(CH_3P)$, 745 s $\nu_{as}(PC_3)$, 680 vs+665 sh+620 sh $\varrho(CH_3In)$, 515 sh+475 vs $\nu_{as}(InC_3)$, 455 sh $\nu_s(InC_3?)$, 385 s+357 s+314 w $\nu_s(InOP)+\delta(CPO)$ [6]
*7	$S(CH_3)_2$	colorless solid, m.p. 19 to 19.5 °C (dec.), b.p. (extrapolated) 185 °C [2]

* Further information:

$(CH_3)_3In \cdot n\ O(CH_3)_2$ (Table **5**, No. **1**). Based on the peak areas of the ^1H NMR signals, an acceptor-donor ratio of 1:0.85±0.05 (n=0.9) was found [11]. Determination of the dissociation enthalpy was unsuccessful because of the pronounced thermal instability [2]. For the same reason the vibration spectra of the adduct, approximate composition $(CH_3)_3In \cdot 0.9\ O(CH_3)_2$, was recorded at about 10 °C (IR) and 0 °C (Raman). The assignments (in cm^{-1}) refer to the C_s point group [11]:

IR	Raman	assignment	IR	Raman	assignment
3000 w	3023 vw, dp	$\nu_{as}(CH_3)$ (O)	1157 s	1158 s, p	$\delta_s(CH_3)$ (In)
–	2967 sh, dp	$\nu_{as}(CH_3)$ (In)	1126 w	–	$\varrho(CH_3)$ (O)
2955 s	2953 m, p	$2\times\delta_{as}(CH_3)$ (O)	1072 s	1078 w,dp	$\nu_{as}(OC_2)$ (A″)
2910 s	2913 vs, p	$\nu_s(CH_3)$ (In)	901 s	904 m, p	$\nu_s(OC_2)$ (A′)
2870 w	2876 m, p	$\nu_s(CH_3)$ (O)	695 vs, br	701 vw, dp	$\varrho(CH_3)$ (In)
2830 m	2832 w, p	$2\times\delta_{as}(CH_3)$ (In)	–	632 w, dp	$\varrho(CH_3)$ (In)
1476 sh	1476 vw	$\delta_{as}(CH_3)$ (O)	490 s	493 m, dp	$\nu_{as}(InC_3)$ (A″)
1452 m	1458 w, dp	$\delta_s(CH_3)$ (O)	477 mw	475 vvs, p	$\nu_s(InC_3)$ (A′)
1430 sh	1435 vw	$\delta_{as}(CH_3)$ (In)	425 m, br	–	$\delta(OC_2)$, $\delta(InOC)$
1248 w	1250 vw	$\varrho(CH_3)$ (O)	–	127 m, dp	$\delta(InC_3)$ (A′+A″)

$(CH_3)_3In \cdot n\ O(C_2H_5)_2$ with $0.5 \leq n < 1$ (Table **5**, No.**2**) can be obtained electrochemically (CH_3MgI, excess CH_3I, 50 to 60 V, 45 to 70 mA), as described for No. 3, in about 60% yield using $O(C_2H_5)_2$ as the solvent [15 to 17].

A colorless distillable liquid prepared from CH_3MgCl and $InCl_3$ in $O(C_2H_5)_2$ by the Grignard synthesis (50% yield) was found to be a nearly stoichiometric 2:1 adduct with D=1.241 g/cm^3, and n_D^{20}=1.480; it was found to be monomeric in benzene [1]. No stoichiometry was given in [21]. The boiling point of the adduct has been estimated to be 139 °C at normal pressure [1, 18], but even at this temperature dissociation must be considered; formation of $In(CH_3)_3$ by cracking the etherate at normal pressure is nearly complete at 100 to 150 °C [12, 20].

The ether adduct was obtained in very good yield (90 to 100%) by reaction of InX_3 (X=Cl, Br) or $(CH_3)_2InX$ (X=Cl, Br, I) with ether solutions of $LiCH_3$ at 0 °C (1 h) [18]; it was considered to be a 2:1 adduct [18]; see also [1]. Other samples obtained by this method, used for synthesizing $In(CH_3)_3$, had ^1H NMR-determined compositions of 2:1.4 [7] or 2:1.8 [4, 6]; no specific empirical formula was reported in [5].

The adduct is also formed by the reaction of metallic In with CH_3X (X=Br, I) in ether. An In/Mg alloy, a mixture of In and Mg, or even technical grade In may be used. For example, mixing a 1:3 In/Mg alloy with CH_3Br in ether produces the adduct in 87% yield. This etherate distilled between 47 and 62 °C at 12 Torr, and had the approximate composition, $(CH_3)_3In \cdot 0.8\ O(C_2H_5)_2$, based on an In analysis. **$(CD_3)_3In \cdot O(C_2H_5)_2$** can be synthesized by this method using CD_3I [3].

The etherate obtained by Method IV [14, 17] was purified by vacuum distillation (pressure unreported) at about 60 °C. Analytical or physical data to establish the stoichiometry were not reported; undoubtedly, a 1:1 adduct was not obtained in this case, either.

It can be generally assumed that tests of the physical characteristics of the diethyl etherate were not performed on samples of exactly defined composition, even though a 1:1 adduct was usually reported.

Vapor pressure measurements in the 20 to 60 °C range (p in Torr, T in K) led to log p $=8.066-2175/T$, an enthalpy of vaporization, ΔH_v, of 10.0 kcal/mol, and an extrapolated normal boiling point of 147 °C [2]. Arithmetically derived dissociation enthalpies of 16.86 [9] and 15.38 kcal/mol [19] have been reported for the (hypothetical) 1:1 etherate.

The ^1H NMR spectrum was obtained from a sample of the composition $(CH_3)_3In \cdot 0.7\ O(C_2H_5)_2$ as shown by the CH_3In-to-C_2H_5O signal ratios [11].

The ^{115}In nuclear quadrupole resonance spectrum of the etherate (formulated as the 1:1 adduct) shows transition frequencies at 43.001 (ν_1), 81.938 (ν_2), and 127.61 (ν_3) MHz; the quadrupole coupling constant is 1024.04 MHz and the asymmetry parameter value is $\eta=3.9\%$ [8].

The thermal decomposition of the etherate (author's remark: undoubtedly, the pyrolysis of $In(CH_3)_3$ in the presence of $O(C_2H_5)_2$ has been investigated) occurs at 500 °C with formation of CH_4 (31%), C_2H_6 (46%), C_2H_4 (8%), C_3H_8 (12%), and C_4H_{10} (1%); decomposition products of the donor ligand were not mentioned [10].

The adduct was selected as the precursor of pure $In(CH_3)_3$ (see Section 1.1.1.1.1, pp. 1 to 2).

According to Method III (p. 27) the adduct can be used as starting material for further trimethylindium adducts (Table 5, p. 28 and Table 6, p. 33). Compared to pure $In(CH_3)_3$, the simpler preparation and handling of the etherate makes it a preferable starting material for the synthesis of other organoindium compounds. The reactions listed in Table 4 (Section 1.1.1.1.4 on p. 13), with inorganic or organic compounds involving in the splitting of one or more $In-CH_3$ bonds, could have been carried out, almost without exception, with pure $In(CH_3)_3$ as successfully as with the etherate. The donor molecule $O(C_2H_5)_2$ having been completely eliminated, no adduct formation with the resulting substitution products was described. Only in rare cases (e.g. [4, 6, 7]) was the composition of the etherate cited.

$(CH_3)_3In \cdot OC_4H_8$ (Table **5**, No. **3**). A mixture of $Mg(CH_3)_2$ and $[N(C_2H_5)_4]ClO_4$ in THF was electrolyzed, using a platinum cathode and an indium anode at 50 V and 4 to 70 mA. After removal of excess THF, the adduct was purified by distillation. The yield is relatively small (25 to 30% based on the weight loss of the In electrode), because the products undergo simultaneous electrolytic decomposition with deposition of In metal on the Pt cathode. This side reaction can generally be overcome by using CH_3MgI as the starting material. Yields of 60 to 80% were obtained by adding excess CH_3I, which also suppressed the deposition of Mg on the cathode [15, 16]. The electrolysis residue was freed from solvent and vacuum distilled, yielding the ether adduct [17].

When purifying the adduct by distillation, extreme caution is advised, because of the possible presence of perchlorate residues!

The compound was used for the preparation of $In(CH_3)_3$ adducts with $N(C_2H_5)_3$, $P(C_2H_5)_3$, and $P(C_6H_5)_3$; see Table 6 in Section 1.1.1.2.2 [14, 16].

$(CH_3)_3In \cdot S(CH_3)_2$ (Table **5**, No. **7**). No method of preparation (probably Method I ?) is given. In the gas phase the compound is extensively dissociated. The heat of evaporation in the range 30 to 85 °C, ΔH_v, is 8.3 kcal/mol; log $p=6.825-1808/T$ (p in Torr, T in K) [2].

References:

[1] Runge, F.; Zimmermann, W.; Pfeiffer, H.; Pfeiffer, I. (Z. Anorg. Allgem. Chem. **267** [1951] 39/48).

[2] Coates, G.E.; Whitcombe, R.A. (J. Chem. Soc. **1956** 3351/4).

[3] Todt, E.; Dötzer, R. (Z. Anorg. Allgem. Chem. **321** [1963] 120/3).
[4] Schmidbaur, H.; Schindler, F. (Chem. Ber. **99** [1966] 2178/86).
[5] Clark, H.C.; Pickard, A.L. (J. Organometal. Chem. **8** [1967] 427/34).
[6] Schindler, F.; Schmidbaur, H. (Chem. Ber. **100** [1967] 3655/3).
[7] Schmidbaur, H.; Schwirten, K.; Pickel, H.H. (Chem. Ber. **102** [1969] 564/7).
[8] Svergun, V.I.; Bednova, L.M.; Okhlovysmin, O.Yu.; Semin, G.K. (Izv. Akad. Nauk SSSR Ser. Khim. **1970** 1449; Bull. Acad. Sci. USSR Div. Chem. Sci. **1970** 1377).
[9] Drago, R.S. (Struct. Bonding [Berlin] **15** [1973] 73/139).
[10] Travkin, N.N.; Skachkov, B.K.; Tonoyan, I.G.; Kozyrkin, B.I. (Zh. Obshch. Khim. **48** [1978] 2678/81; J. Gen. Chem. [USSR] **48** [1978] 2428/31).

[11] Karschin, J.; Mann. G.; Weidlein, J. (unpublished results, 1979).
[12] Efremov, E.A.; Fedorov, V.A.; Ferapontov, A.P.; Dorogova, L.A.; Kozyrkin, B.I.; Filippov, E.P.; Sredinskaya, I.A. (U.S.S.R. 810700 [1981] from C.A. **95** [1981] No. 62396).
[13] Krause, H.; Sille, K.; Hausen, H.-D.; Weidlein, J. (J. Organometal. Chem. **235** [1982] 253/64).
[14] Cole-Hamilton, D.J.; Gerrard, N.D.; Jones, A.C.; Holliday, A.K.; Mullin, J.B.; U.K. Secretary of State for Defence (Brit. 2125795 [1984] from C.A. **101** [1984] No. 91212).
[15] Great Britain, Dept. of Industry (Japan. 84-47388 [1984] from C.A. **101** [1984] No. 139779).
[16] Jones, A.C.; Gerrard, N.D.; Cole-Hamilton, D.J.; Holliday, A.K. (J. Organometal. Chem. **265** [1984] 9/15).
[17] Jones, A.C.; Holliday, A.K.; Cole-Hamilton, D.J.; Ahmad, M.M.; Gerrard, N.D. (J. Cryst. Growth **68** [1984] 1/9).
[18] Reier, F.-W.; Wolfram, P.; Schumann, H. (J. Cryst. Growth **77** [1986] 23/6).
[19] Drago, R.S.; Wong, N.; Bilgrien, C.; Vogel, G.C. (Inorg. Chem. **26** [1987] 9/14).
[20] Laube, G.; Kohler, U.; Weidlein, J.; Schloz, F.; Streubel, K.; Dieter, R.J.; Karl, N.; Gerdon, M. (J. Cryst. Growth **93** [1988] 45/51).

[21] Foster, D.F.; Rushworth, S.A.; Cole-Hamilton, D.J.; Cafferty, R.; Harrison, J.; Parkes, P. (J. Chem. Soc. Dalton Trans. **1988** 7/11).

1.1.1.2.2 $In(CH_3)_3$ is Bonded to a Group 15 Element

1.1.1.2.2.1 Compounds with One $In(CH_3)_3$ Unit

The compounds of this section can be prepared according to the general methods described in Section 1.1.1.2 on pp. 27.

General Remarks. For the calculation of the adduct formation enthalpies, see Section 1.1.1.2 on p. 28 [7, 17].

Donor exchange in the trimethylindium-amine system (adducts No. 2 to 4 in Table 6) has been thoroughly studied by ^1H NMR spectroscopy [10].

$$In(CH_3)_3^* + (CH_3)_3In \cdot D \underset{k_2}{\overset{k_1}{\leftrightarrows}} In(CH_3)_3 + (CH_3)_3In^* \cdot D \qquad (1)$$

$$(CH_3)_3In \cdot D \underset{k_2}{\overset{k_1}{\leftrightarrows}} In(CH_3)_3 + D \qquad (2)$$

At low temperatures (−60 °C) two signals are observed (in the presence of excess trimethylindium) for the CH_3In protons; the free $In(CH_3)_3$ appears at low field. At higher temperatures these signals merge, because of exchange, to a single line; the coalescence temperature and the half-width permit the determination of the activation energy and the dissociation constant for each adduct (the first column refers to the number of the adduct in Table 6) [10]:

No.	D	solvent	T* in K	ΔH in kcal/mol	ΔS in $cal \cdot mol^{-1} \cdot K^{-1}$	k_1 (exp.)
2	NH_2CH_3	CH_2Cl_2	236	11.1	−1.3	134.41 ($s^{-1} \cdot mol^{-1}$)
3	$NH(CH_3)_2$	CH_2Cl_2	256	14.9	5.3	15.6 (s^{-1})
4	$N(CH_3)_3$	$CH_3C_6H_{11}$-c	256	19.4	21.0	6.7 (s^{-1})

*) Temperature at which ΔS is calculated.

Results of kinetic studies indicate an exchange mechanism according to (1) for the rate-determining step if $D=NH_2CH_3$, whereas the rate-determining step is the dissociation (2) in the case of the bulkier bases $N(CH_3)_3$ and $NH(CH_3)_2$ [10].

The volatilities of Nos. 7, 15 to 17, 19, 40, and 43 were measured by the modified entrainment method (MEM) between about 340 and 430 K; the apparatus for MEM is shown in a figure in [83]. Weight loss (in kg/s) was estimated for various temperatures and the apparent standard entropies of vaporization, ΔS_v, were calculated, being between 110 and 190 $J \cdot mol^{-1} \cdot K^{-1}$. In some cases the adducts are volatilized with partial dissociation into their adduct components with ΔS_v values substantially above 90 to 125 $J \cdot mol^{-1} \cdot K^{-1}$, whereas low values indicate nondissociative vaporization (ΔS_v values in parentheses): No. 7 (194); No. 15 (112); No. 16 (172); No. 17 (118); No. 19 (117); No. 40 (126); No. 43 (133) [83].

Explanation for Table 6: All $InCH_3$ groups appear as singlets in the ^1H NMR spectra. The relative intensities of fragments in the mass spectra (at 70 eV) are given in % in parentheses. Preparative Methods I to IV are described on pp. 27 to 28.

Table 6
Adducts of the $(CH_3)In \cdot D$ Type with N, P, As, and Sb Donor Atoms.
An asterisk indicates further information at the end of the table.
For abbreviations and dimensions, see p. X.

No.	donor D method of preparation (yield)	properties and remarks
$(CH_3)_3$ In bonded to an N atom		
1	NH_3 I	white solid, m.p. 28.5 to 29.0 °C [1] decomposes at room temperature to give CH_4 and $\{(CH_3)_2InNH_2\}_x$ [1]
2	NH_2CH_3	preparation and properties not reported, only mentioned in studies of exchange reactions on p. 32 [10]
*3	$NH(CH_3)_2$ I (80%)	colorless crystals, m.p. 13 °C, b.p. 33 to 35 °C/10^{-4} Torr [22] ^1H NMR (at 30 °C, neat/in CCl_4/in C_6D_6): −0.51/−0.53/−0.85 (CH_3In), 1.89/1.75/-(br, HN), 2.40/2.45/1.28 (d, CH_3N, J=4.8/6.0/6.0) [22]; −0.251 (CH_3In), 1.64 (CH_3N) in C_6D_6 [33], see also [10] IR and Raman spectra on p. 41.

Table 6 (continued)

No.	donor D method of preparation (yield)	properties and remarks
*4	$N(CH_3)_3$ I (70%) [23] III (85%) [33]	colorless crystals, m.p. 66.2 to 66.4 °C [1], 65 to 66 °C [23], b.p 171 °C (extrapolated) [1], subl. 35 °C/0.1 Torr [23] ^1H NMR (30 °C): −0.081 [33], −0.09 [62] (CH_3In), 1.803 [33] (CH_3N) in C_6D_6; −0.58 (CH_3In), 2.29 (CH_3N) in CCl_4 [23] IR and Raman spectra on p. 41 mass spectra on p. 42
*5	$N(C_2H_5)_3$ IV [35, 39, 44, 45]	white solid, m.p. 94 to 96 °C [44], 90 to 91 °C [33], b.p. 184.3 °C (extrapolated) [33], subl. 80 to 100 °C/vacuum [44] ^1H NMR (C_7D_8): −0.1 (CH_3In), 0.9 (t, CH_3C), 2.3 (q, CH_2N) [44] IR (solid): 1330 vw, 1320 vw, 1290 w, 1190 m, 1170 s, 1150 s, 1090 m, 1055 m, 1025 w, 1005 w, 900 vw,br, 800 vw, 790 vw, 730 m, 680 vs, 620 sh, 550 w, 457 vs, 455 sh, 420 vw [44] mass spectrum (71 °C, 70 eV): $[In(CH_3)_2]^+$ (47.0%), $[InCH_3]^+$ (6.7%), In^+ (23.8%), $[C_6H_{15}NH]^+$ (1.8%), $[C_6H_{15}N]^+$ (22.2%), $[C_6H_{14}N]^+$ (7.6%), $[C_5H_{12}N]^+$ (100%), $[C_4H_{10}N]^+$ (1.8%), $[C_2H_5N]^+$ (25.0%) [33, 35]
*6	$N(C_4H_9\text{-}n)_3$ IV	white solid, subl. 50 to 60 °C/vacuum [44, 45] ^1H NMR (C_7D_8): 0.1 (CH_3In), 1.0 (t, CH_3C), 1.5 (m, CH_2C), 2.6 (t, CH_2N) [36, 44] mass spectrum (200 °C, 70 eV): $[C_{12}H_{27}N]^+$, $[In(CH_3)_2]^+$, $[C_9H_{20}N]^+$, $[InCH_3]^+$, $^{115}In^+$, $[C_6H_{14}N]^+$, and $[C_5H_{12}N]^+$ [44]
7	$NH(C_6H_{11}\text{-c})_2$ III (77%)	white crystals, m.p. 43 to 45 °C, subl. 55 °C/10^{-2} Torr [62] ^1H NMR (C_6D_6): 0.13 (CH_3In), 0.90 to 1.85 (m, br, CH_2+NH), 2.55 (m, br, CH) [62] mass spectrum: $[In(CH_3)_2]^+$ (46%), $[InCH_3]^+$ (6%), ^{115}In (24%), $[C_6H_{11}N]^+$ (17%), $[C_3H_5N]^+$ (100%) [62] for volatility, see general remarks [83]
8	$(CH_3)_2NCH_2N(CH_3)_2$ I	liquid, [14] ^1H NMR (C_6H_6, 0.3 M): −0.24 (CH_3In), 1.94 (CH_3N), 2.93 (CH_2N) [14] IR (C_6H_6): 520 m $\nu_{as}(InC_3)$, 478 s $\nu_s(InC_3)$, 465 sh ν(In−N) [14] monomeric in C_6H_6 [14]
9	$N_4C_6H_{12}$ (hexamethylene-tetramine) II (70%)	white crystalline solid, m.p. 135 °C (dec.) [32] ^1H NMR (in C_6D_6/in CH_2Cl_2): −0.15/−0.53 (CH_3In), 4.15/4.58 (CH_2N) [32] IR (solid): correlation diagram, only the main vibrations of the urotropine ligand [32]
10	$C_6H_4\{N(CH_3)_2\}_2\text{-1,4}$ III (88%)	white needle-like crystals, m.p. 68 to 70 °C, subl. 70 °C/0.01 Torr [74] ^1H NMR (C_6D_6): −0.20 (CH_3In), 2.45 (s, CH_3N), 6.68 (m, C_6H_4) [74]

Table 6 (continued)

No.	donor D method of preparation (yield)	properties and remarks
10 (continued)		mass spectrum: $[C_{10}H_{16}N_2]^+$, $[C_9H_{13}N_2]^+$, $[In(CH_3)_2]^+$, $[C_8H_{10}N_2]^+$, $[InCH_3]^+$, $[In]^+$ [74] dissociates at 100 °C/0.01 Torr to give pure $In(CH_3)_3$ [74]
11	$CH_2\{C_6H_4N(CH_3)_2\text{-}4\}_2$ II	no properties reported [82] dissociates at 80 to 130 °C/10^{-2} Torr to give pure $In(CH_3)_3$ [82]
12	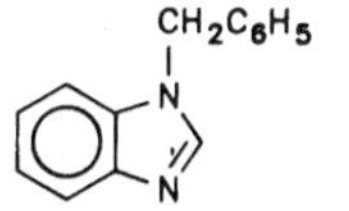 1-$CH_2C_6H_5$-benzimidazole (structure) II	m.p. 97 to 98 °C (formulated as the 1:3 adduct, $(CH_3)_3In \cdot 3$ D) [16] $\mu_D = 6.26$ D in C_6H_6 at 25 °C [16]
13	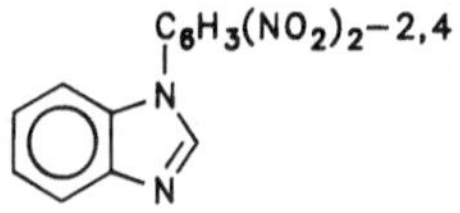 1-$C_6H_3(NO_2)_2$-2,4-benzimidazole (structure) II	m.p. 129 to 130 °C [16] $\mu_D = 3.91$ D in C_6H_6 at 25 °C [16] monomeric in C_6H_6 [16]
14	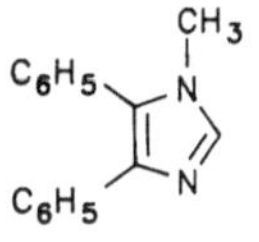 1-CH_3-4,5-(C_6H_5)$_2$-imidazole (structure) II	m.p. 147 to 148 °C (formulated as the 1:3 adduct, $(CH_3)_3In \cdot 3D$) [16] $\mu_D = 6.65$ D in C_6H_6 at 25 °C [16]
15	$NH\{CH(CH_3)CH_2\}_2CH_2$ (cis-2,6-dimethyl-piperidine) III (79%)	white solid, m.p. 37.5 to 38.5 °C, subl. 30 °C/0.1 Torr [62] ^1H NMR (C_6D_6): 0.03 (CH_3In), 0.5 to 1.5 (m, br, CH_2 and NH), 0.90 (d, CH_3C, $^3J = 6$), 2.27 (m, br, CH) [62] mass spectrum: $[In(CH_3)_3]^+$ (0.5%), $[In(CH_3)_2]^+$ (100%), $[InCH_3]^+$ (12%), $^{115}In^+$ (45%), $[NH(C_7H_{12})]^+$ (21%), $[NC_7H_{12}]^+$ (11%) [62] for volatility, see general remarks [83]
*16	$NH\{C(CH_3)_2CH_2\}_2CH_2$ (2,2,6,6-tetramethyl-piperidine) III (82%)	white solid, m.p. 60.5 to 62.5 °C, subl. 30 °C/0.1 Torr [62] ^1H NMR (C_6D_6): 0.12 (CH_3In), 0.9 to 1.4 (m, br, CH_2 and NH), 1.12 (CH_3C) [62] mass spectrum: $[M]^+$ (85%), $[149]^+$ (10%), $[In(CH_3)_2]^+$ (100%), $[NH(C_9H_{18})]^+$ (8%), $[InCH_3]^+$ (17%), $[NH(C_8H_{16})]^+$ (10%), $[NH(C_8H_{15})]^+$ (42%), $^{115}In^+$ (50%) [62] $\Delta H_{Diss} = 10 \pm 1$ kcal/mol [34] for volatility, see general remarks and further information [83]

Table 6 (continued)

No.	donor D method of preparation (yield)	properties and remarks
17	$N(CH_2CH_2)_3CH$ (quinuclidine) III (81%)	white solid, m.p. 85.5 to 91 °C, subl. 75 °C/0.1 Torr [62] ^{1}H NMR (C_6D_6): −0.07 (CH_3In), 1.03 (m, CH_2C), 1.26 (sept, CH, J = 3), 2.48 (m, CH_2N) [62] mass spectrum: $[347]^+$ (17%), $[345]^+$ (28%), $[M-CH_3]^+$ (5%), $[In(CH_3)_2]^+$ (100%), $[InCH_3]^+$ (9%), $^{115}In^+$ (28%), $[NC_7H_{12}]^+$ (55%) [62] log p(Pa) = 11.4 − 4280/T(K) [72] for volatility, see general remarks [83]
*18	$N(CH_2CH_2)_3N$ ("dabco") III (61%)	white solid, m.p. 160 to 164 °C, subl. 95 °C/0.01 Torr [62] ^{1}H NMR: see further information [62] mass spectrum: $[In(CH_3)_2]^+$ (85%), $[InCH_3]^+$ (15%), $^{115}In^+$ (64%), $[N_2C_6H_{12}]^+$ (64%), $[N_2C_5H_9]^+$ (7%), $[N_2C_4H_8]^+$ (19%) [62] log p(Pa) = 12.35 − 5280/T(K) [72]
*19	$\{N(C_2H_5)CH_2\}_3$ III (74%)	colorless liquid, b.p. 50 °C/0.1 Torr [62] ^{1}H NMR (C_7D_8): −0.13 (CH_3In), 0.89 (t, CH_3C, $^3J(H,H) = 7$), 2.33 (q, CH_2C), 3.16 (s, br, CH_2N) at 303 K; 0.09 (br, CH_3In), 0.86 (t, br, CH_3C), 2.17 (q, br, CH_2C), 2.58 and 3.56 (br, AB spectrum, CH_2N, J = 10) at 203 K; for variable temperature NMR, see further information [62] mass spectrum: $[M-CH_3]^+$ (8%), $[N_3C_9H_{21}]^+$ (25%), $[N_3C_9H_{20}]^+$ (17%), $[In(CH_3)_2]^+$ (100%), $[InCH_3]^+$ (11%), $^{115}In^+$ (34%), $[N_3C_5H_{12}]^+$ (32%) [62] for volatility, see general remarks [83]
20	4-$NC_5H_4N(CH_3)_2$ (structure drawn: pyridine ring with $N(CH_3)_2$) III (91%)	colorless crystals, m.p. 77 to 81 °C, subl. 80 °C/0.01 Torr [74] ^{1}H NMR (C_6D_6): 0.06 (CH_3In), 2.19 (CH_3N), 5.78, 7.85 (m's, C_5H_4N) [74] mass spectrum: $[M-CH_3]^+$, $[M-3CH_3]^+$, $[In(CH_3)_2]^+$, $[InCH_3]^+$, $[C_7H_{10}N_2]^+$, $[In]^+$, $[C_6H_7N_2]^+$ [74]
21	2,2′-$NC_5H_4C_5H_4N$ III (90%)	white needle-shaped crystals, no pure adduct isolated (0.4:1 to 1.7:1 mole ratio) [74]
22	3,3′-$NC_5H_4C_5H_4N$ III (90%)	white needle-shaped crystals, m.p. 101 to 103 °C, subl. 100 °C/0.01 Torr [74] ^{1}H NMR (C_6D_6): 0.14 (CH_3In), 6.61, 7.04, 8.31, 8.52 (m's, C_5H_5N) [74] mass spectrum: $[C_{10}H_8N_2]^+$, $[In(CH_3)_2]^+$, $[InCH_3]^+$, $[In]^+$ [74]
23	4,4′-$NC_5H_4C_5H_4N$ III (90%)	white needle-shaped crystals, m.p. 153 to 155 °C, subl. 70 °C/0.01 Torr [74] ^{1}H NMR (C_6D_6): 0.16 (CH_3In), 6.67, 8.37 (m's, C_5H_5N) [74] mass spectrum: see No. 22 [74]

Table 6 (continued)

No.	donor D method of preparation (yield)	properties and remarks
24	$(CH_3)_3SiN{=}C{=}NSi(CH_3)_3$ II (90%)	colorless liquid, b.p. 30 to 32 °C/0.1 Torr [80] ^1H NMR (C_6D_6): 0.06 (CH_3In), 0.20 (CH_3Si) [80] IR (film): 2189 vs ν_{as}(NCN), 737 m ν_{as}(SiN), 487 vs ν_s(SiN) + ν_{as}(InC_3), 467 vw ν_s(InC_3) [80] Raman (neat): 2190 vw ν_{as}(NCN), 1485 w ν_s(NCN), 737 vw ν_{as}(SiN), 488 m ν_s(SiN) + ν_{as}(InC_3), 468 vvs ν_s(InC_3) [80] monomeric in C_6H_6 [80]
25	$HN{=}P(C_2H_5)_3$ III (100%)	m.p. 12 to 15 °C (dec.) [11] ^1H NMR (C_6H_6, ~10%): −0.725 (CH_3In), 0.0 (NH), 1.02 (CH_3C, $^3J(P,H) = 17.1$, $^3J(H,H) = 7.5$), 1.68 (CH_2P, $^2J(P,H) = 12.0$) [11] IR (solid?): 3345 vs ν(NH), 1160 sh + 1130 m δ_s(CH_3In) + ν(P=N), 1050 m δ(NH), 675 sh + 625 m ϱ(CH_3In), 510 s ν(InC_3), 470 s ν(In–N?); other bands reported [11] should be stored at 0 °C; decomposes above 50 °C to give CH_4 and $\{(CH_3)_2InN{=}P(C_2H_5)_3\}_2$ in high yield [11]
26	$N\{Si(CH_3)_3\}{=}P(CH_3)_3$ III (90.2%)	solid, m.p. 43 to 44 °C, b.p. 67 °C/1 Torr [8, 9] ^1H NMR (CCl_4, ~5%): −0.54 (CH_3In, $J(C,H) = 124.5$), 0.11 (CH_3Si, $J(C,H) = 117.5$, $^2J(Si,H) = 6.5$), 1.59 (CH_3P, $J(C,H) = 128.5$, $^2J(P,H) = 12.6$) [8, 9] IR (solid): 1311 s + 1293 vs δ(CH_3P), 1258 s + 1247 vs δ_s(CH_3Si), 1120 vs ν(P=N), 681 s ν_{as}(SiC_3), 641 s ν_s(SiC_3), 472 s + 464 s + 458 s ν(InC_3), 520 w ν(Si–N?) [9]. monomeric in C_6H_6 [8, 9]
27	$N\{Si(CH_3)_3\}{=}P(C_2H_5)_3$ III (82%)	solid, m.p. 107 to 110 °C, b.p. 125 to 128 °C/1 Torr [9] ^1H NMR (CCl_4, 5%): −0.52 (CH_3In, $J(C,H) = 124.5$), 0.07 (CH_3Si, $J(C,H) = 118.5$, $^2J(Si,H) = 6.6$), 1.10 (CH_3C, $^3J(H,H) = 8$, $^3J(P,H) = 17.1$), 1.77 (CH_2P, $^2J(P,H) = 11.1$) [9] IR (solid): 1258 vs + 1247 vs δ(CH_3Si), 1105 vs, br ν(P=N), 852 vs + 833 vs δ(CH_3Si), 690 s + 678 m ν_{as}(SiC_3), other bands reported without assignment [9] monomeric in C_6H_6 [9]
28	$N\{Ge(CH_3)_3\}{=}P(CH_3)_3$ III (93%)	m.p. 46 to 48 °C, b.p. 93 to 95 °C/0.2 Torr [13] ^1H NMR (CCl_4): −0.65 (CH_3In), 0.36 (CH_3Ge), 1.62 (CH_3P, $J(P,H) = 12.7$) [13] IR (solid): 1310 s + 1294 vs δ(CH_3P), 1247 sh + 1238 s δ(CH_3Ge), 1090 vs ν(P=N), 748 s + 734 m ν(PC_3), 696 sh + 676 vs ϱ(CH_3In), 610 vs ν_{as}(GeC_3), 595 sh ν_s(GeC_3), 570 m + 465 vs + 460 sh ν(InC_3) + ν(Ge–N), other bands reported [13] monomeric in C_6H_6 [13]

Table 6 (continued)

No.	donor D method of preparation (yield)	properties and remarks
29	N{Ge(CH_3)$_3$}=P(C_2H_5)$_3$ III (92%)	m.p. 149 to 150 °C, subl. 140 °C/0.2 Torr [13] ^{1}H NMR (CCl_4): −0.64 (CH_3In), 0.37 (CH_3Ge), 1.09 (CH_3C, ^{3}J(P,H) = 17.15), 1.81 (CH_2P, ^{2}J(P,H) = 11.0) [13] IR (solid): 1275 m + 1262 m δ(CH_2P), 1249 m + 1238 s δ(CH_3Ge), 1138 s δ(CH_3In), 1089 vs ν(P=N), 824 vs + 784 vs ϱ(CH_3Ge), 742 s + 714 s ν_{as}(PC_3), 681 s + 670 sh ϱ(CH_3In), 602 vs + 592 sh + 581 sh + 540 sh ν(GeC_3), 464 vs + 431 sh + 420 sh ν(InC_3), other bands reported [13] monomeric in C_6H_6 [13]
30	N{Sn(CH_3)$_3$}=P(CH_3)$_3$ III	formation at −25 °C, detected by ^{1}H NMR, not isolated [13] dec. at room temperature to give {(CH_3)$_2$InN=PR$_3$}$_2$ and Sn(CH_3)$_4$ [13].
31	N{Sn(CH_3)$_3$}=P(C_2H_5)$_3$	like No. 30 [13]
32	N(=P(CH_3)$_3$)Si(CH_3)$_2$OSi(CH_3)$_3$ III (91%)	m.p. 20 to 21 °C, b.p. 74 to 76 °C/0.01 Torr [21] ^{1}H NMR (CH_2Cl_2): −0.61 (CH_3In), 0.11 (CH_3Si, (CH_3)$_2$Si), 1.60 (CH_3P, ^{2}J(P,H) = 13.1) [21] IR (film?): 1256 δ_s(CH_3Si), 1112 ν(P=N), 1024 ν(SiOSi); no other bands given [21] for further properties, see No. 33
*33	N(=P(CH_3)$_3$)Si(CH_3)$_2$N=P(CH_3)$_3$ III (98%)	white solid, m.p. 84 to 86 °C [21] ^{1}H NMR (CH_2Cl_2, 30 °C): −0.52 (CH_3In), 0.08 (CH_3Si), 1.55 (CH_3P, ^{2}J(P,H) = 12.8) [21] IR (solid): 1258 δ(CH_3Si), 1241 ν(P=N) uncoordinated, 1090 ν(P=N) coordinated; no other bands given [21]
34	N(=P(CH_3)$_3$)Si(CH_3)$_2$N(=P(CH_3)$_3$)Ga(CH_3)$_3$ III (96%)	also from No. 33 and Ga(CH_3)$_3$ (98%) [12] m.p. 82 to 84 °C [12] ^{1}H NMR (CH_2Cl_2): −0.55 (CH_3In and CH_3Ga), 0.27 (CH_3Si), 1.65 (CH_3P, ^{2}J(P,H) = 12.9) [12] could not be distilled or recrystallized; extremely air- and moisture-sensitive [12]
(CH_3)$_3$In bonded to P		
*35	PH_3 I	colorless needles at −196 °C [2, 59] dissociates slowly above −123 °C, rapid dec. at −78 °C to give (CH_3InPH)$_x$ and 2 mol CH_4 [2]
36	$PH_2C_6H_5$	probably an intermediate by the reaction of the components between 0 to 140 °C (18 h) to give 2 mol CH_4 and a yellow-orange polymer [5]
37	PH(CH_3)$_2$	like No. 36; dec. 20 to 40 °C to give {(CH_3)$_2$InPR$_2$}$_3$ and CH_4 within 3 to 5 h. No other details reported [5]
38	PH(C_2H_5)$_2$	like No. 37 [5]

Table 6 (continued)

No.	donor D method of preparation (yield)	properties and remarks
*39	$P(CH_3)_3$ I [1] III (92.5%) [60, 70]	white crystalline solid, m.p. 46.5 °C [1, 60], 44.5 °C [66], b.p. 189 °C (extrapolated) [1, 60], 60 °C/0.1 Torr [23] ^1H NMR (C_6D_6, 30 °C): −0.15 (CH_3In), 0.79 (CH_3P, $^2J(P,H)$ = 5.6) [60] ^{13}C NMR (C_6D_6, 30 °C): −10.7 (CH_3In), 11.26 (CH_3P, J(P,C) = 10.5) [60] ^{31}P NMR(C_6D_6): −58.8 [60] mass spectrum (31 °C, 70 eV): $[M+H]^+$ (1.8%), $[InPC_5H_{15}]^+$ (0.1%), $[InPC_4H_{12}]^+$ (0.1%), $[In(CH_3)_3]^+$ (0.1%), $[In(CH_3)_2]^+$ (39.6%), $[InCH_3]^+$ (4.7%), $^{115}In^+$ (21.6%), $[PC_3H_{10}]^+$ (27.4%), $[P(CH_3)_3]^+$ (4.7%), $[P(CH_3)_2]^+$ (6.5%) [33] IR and Raman spectra on p. 45
*40	$P(C_2H_5)_3$ I (90%) [76] III [42, 45] IV [35, 44]	colorless crystals, m.p. 33 to 36 °C [44], (zone refined) 39.5° [76], b.p. 218.4 °C (extrapolated) [33], 85 °C/10^{-3} Torr [76] $\Delta H = 17.71$ kcal/mol, $\Delta S = 36.03$ cal · mol^{-1} · K^{-1} [33]. ^1H NMR (C_7D_8): 0.0 (CH_3In), 0.85 (m, CH_2P), 1.2 (m, CH_3C) [44] IR (solid): 1300 vw, 1255 w, 1240 w,sh, 1145 m, 1040 s, 1010 w, 950 w, 770 s, 750 s, 730 s,sh, 720 s,sh, 685 vs, 630 w,sh, 475 vs, 460 sh [44] mass spectrum (11 °C, 70 eV): $[M-CH_3]^+$ (2.2%), $[In(CH_3)_2]^+$ (48.8%), $[InCH_3]^+$ (7.9%), $[PC_6H_{16}]^+$ (2.6%), $[PC_6H_{15}]^+$ (41.1%), $[PC_6H_{14}]^+$ (4.0%), $^{115}In^+$ (23.2%), $[PC_5H_{12}]^+$ (17.1%), $[PC_4H_{11}]^+$ (3.3%), $[PC_4H_{10}]^+$ (75.2%), $[PC_4H_9]^+$ (12.0%), $[PC_3H_8]^+$ (100%), $[PC_2H_7]^+$ (100%) [33]; the spectra at 50 °C, 25 eV [44] and at 11 °C, 30 eV [33] also show fragment peaks of 2 $In(CH_3)_3 \cdot P(C_2H_5)_3$ for volatility, see general remarks [83]
41	$P(C_6H_5)_3$ III [71] IV [39]	solid, recrystallized from C_6H_6, m.p. 130 to 133 °C [71] ^1H NMR (C_6D_6): −0.15 (CH_3In), 7.00, 7.45 (m's, C_6H_5) [71] ^{31}P NMR (C_6D_6): −8.96 [71] begins to liberate $In(CH_3)_3$ at 85 °C/0.01 Torr [71]
42	$P(C_6H_4CH_3\text{-}2)_3$ II (50%)	colorless crystals, m.p. 110 to 125 °C [71] ^1H NMR (C_6D_6): −0.15 (CH_3In), 2.45 (CH_3C), 7.01 (m, C_6H_4) [71] ^{31}P NMR (C_6D_6): −29.19 [71] begins to liberate $In(CH_3)_3$ at 40 °C/0.01 Torr [71]
43	$P\{N(CH_3)_2\}_3$ III (68%)	white solid, m.p. 139 to 142 °C, subl. 50 °C/1 Torr [62] log p(Pa) = 11.45 − 4070/T(K) [72] ^1H NMR (C_6D_6): 0.09 (CH_3In), 2.33 (CH_3N, $^3J(P,H)$ = 9.4) [62] mass spectrum: $[M-CH_3]^+$ (0.8%), $[PN_3C_6H_{18}]^+$ (36%), $[In(CH_3)_2]^+$ (67%), $[InCH_3]^+$ (8%), $[PN_2C_4H_{13}]^+$ (6%), $[PN_2C_4H_{12}]^+$ (99%), $^{115}In^+$ (35%), $[PNC_2H_7]^+$ (100%) [62] for volatility, see general remarks [83]

Table 6 (continued)

No.	donor D method of preparation (yield)	properties and remarks
	$(CH_3)_3In$ bonded to As	
44	AsH_2CH_3 I	probably an intermediate by the reaction of the components between 0 to 130 °C (14 d) to give 2 mol CH_4 and a red-orange polymer [5]
45	$AsH_2C_6H_5$ I	like No. 44 (0 to 100 °C in 12 h) [5]
46	$AsH(C_6H_5)_2$ I	not isolated; spontaneously dec. at about 30 °C to give CH_4 and $\{(CH_3)_2InAs(C_6H_5)_2\}_2$ [4]
*47	$As(CH_3)_3$ I (94%)	colorless crystals, m.p. 28.5 to 29 °C [1], 26 to 27 °C [23], b.p. 155 °C (extrapolated) [1], subl. ~20 °C/0.1 Torr [23] log p(Torr) = 8.925 − 2590/T(K), ΔH_v = 11.9 (30 to 95 °C) [1]; ΔH_v(calc.) ≧ 7.4 kcal/mol [38]. ^{1}H NMR (CCl_4, 30 °C): −0.40 (CH_3In), 1.17 (CH_3As) [23] monomeric in C_6H_6 [23]
	$(CH_3)_3In$ bonded to Sb	
48	SbH_3 I	not isolated; dec. 26 to 28 °C to give orange polymeric $(CH_3InSbH)_x$ and CH_4 [3]
*49	$Sb(CH_3)_3$ I (80%)	colorless solid at 0 °C, liquid at 26 °C, m.p. ~0 °C, b.p. 19 °C/0.1 Torr [3, 23] ^{1}H NMR (CCl_4, 20 °C): −0.23 (CH_3In), 0.87 (CH_3Sb) [23] IR (film): 1148 w $\delta_s(CH_3In)$, 692 s,br $\varrho(CH_3In)$, 525 m $\nu_{as}(SbC_3)$, 522 sh $\nu_s(SbC_3)$, 486 s $\nu_{as}(InC_3)$, 475 sh $\nu_s(InC_3)$; other bands given [23] Raman (liquid, 5 °C): 1150 vs,p $\delta_s(CH_3In)$, 718 vw, p + 638 w, dp $\varrho(CH_3In)$, 533 sh, dp $\nu_{as}(SbC_3)$, 525 vvs,p $\nu_s(SbC_3)$, 488 ms,dp $\nu_{as}(InC_3)$, 469 vvs, p $\nu_s(InC_3)$, 186 s, dp $\delta(SbC_3)$, 126 vs, p $\delta(InC_3) + \nu(InSb?)$ [23]
50	$Sb(C_2H_5)_3$ III	not isolated; solid at 5 to 10 °C [19] dissociates totally between 50 and 60 °C, dec. 300 to 500 °C to give assumed intermediates $(CH_3)_2InSb(C_2H_5)_2$ and $(CH_3InSbC_2H_5)_x$, and well-defined InSb layers [19]

* Further information:

$(CH_3)_3In \cdot NH(CH_3)_2$ (Table **6**, No. **3**) is monomeric in benzene and decomposes between 140 to 160 °C to give $\{(CH_3)_2InN(CH_3)_2\}_2$ [1, 6, 22] (see p. 255). Concentration-dependent ^{1}H NMR values between −0.04 and −0.62 ppm for the CH_3In protons in methylcyclohexane are given in [10]; see also general remarks on exchange reactions.

The IR and Raman spectra were measured on liquid samples at room temperature; the vibrations of the $In(CH_3)_3$ half of the adduct are listed in Table 7 together with the corresponding modes and assignments of the $In(CH_3)_3$ half of No. 4 [23].

Table 7
IR and Raman Frequencies of $(CH_3)_3In \cdot NH(CH_3)_2$ [23] and $(CH_3)_3In \cdot N(CH_3)_3$ [23, 33] of the $(CH_3)_3InN$ Moiety (main vibrations).

$(CH_3)_3In \cdot NH(CH_3)_2$		$(CH_3)_3In \cdot N(CH_3)_3$			
IR liquid	Raman liquid	IR solid	solution	Raman solution	assignment
3296 s	3302 m, p	–	–	–	$\nu(NH)$
2955 s	2958 mw, dp	–	2970 s	2963 w, dp	$\nu_{as}(CH_3)$
2910 s	2910 vs, p	–	2896 s, sh	2901 ms, p	$\nu_s(CH_3)$
1409 mw	1412 vw, dp	1409 vw	1410 vw	1406 vw, dp	$\delta_{as}(CH_3)$
986 ms	970 w, p	–	–	–	δ(In-N-H)
690 vs, br	703 m, br, dp	690 br	687 s	697 vw, dp	$\varrho(CH_3)$
662 sh	–	–	–	669 vw	$\varrho(CH_3)$
629 sh	630 mw, dp	–	–	625 w, dp	$\varrho(CH_3)$
523 vw	524 m, p	–	–	–	ν(InN)
484 vs	484 s, sh, dp	484 s	484 vs	483 s, dp	$\nu_{as}(InC_3)$
470 sh	471 vvs, p	(460)	465 sh	469 vvs, p	$\nu_s(InC_3)$
–	–	460 s	460 m	458 sh, p	ν(InN)
–	175 s, sh, dp	–	–	171 s, dp	δ(In-N-C)
–	125 vs, dp	–	–	121 vs, dp	$\delta(InC_3)$

$(CH_3)_3In \cdot N(CH_3)_3$ (Table **6**, No. **4**) is monomeric in benzene. The dissociation enthalpy of the adduct was determined as 19.9 ± 0.5 kcal/mol [1, 20] and 19.7 ± 1.0 kcal/mol [10], which agree very well with the value 19.84 kcal/mol calculated by [64].

Vapor pressure measurements (p in Torr) of the molten compound between 70 and 100 °C produced the equation, $\log p = 8.402 - 2460/T(K)$ and $\Delta H_v = 11.3$ kcal/mol; the values for the solid between 30 and 65 °C are $\log p = 8.802 - 2590/T(K)$, $\Delta H_s = 11.9$ [1]. Corresponding investigations by the streaming method (N_2 as carrier) at 70 to 140.5 °C led to $\log p = 8.638 - 2582.987/T(K)$, $\Delta H_v = 11.82$ kcal/mol, $\Delta S = 26.35\ cal \cdot mol^{-1} \cdot K^{-1}$ and a boiling point, extrapolated to normal pressure, of 175.5 °C [33].

The ^{115}In NQR spectrum, recorded at 77 K, shows transition frequencies of ν_1 32.64, ν_2 65.26, ν_3 97.91, and ν_4 130.49 MHz, a quadrupole coupling constant of 783.17 MHz, and an asymmetry parameter, η, reported to be 0.0% [18].

IR spectra were obtained from the solid [33], as well as from C_6H_6 and CCl_4 solutions [23]; Table 7 compares the vibrational frequencies (IR and Raman) of the $(CH_3)_3InN$ moiety of the molecule with the corresponding modes of the dimethylamine adduct No. 3.

m/e	18 eV intensity in %	70 eV	fragment ion
379	0.6	–	$[In_2(CH_3)_9N]^+$
337	6.4	2.6	$[In(CH_3)_{12}N_3]^+$
307	0.8	0.3	$[In(CH_3)_{10}N_3]^+$
277	–	0.2	$[In(CH_3)_8N_2CH_2]^+$
261	–	0.4	$[In(CH_3)_7N_2CH]^+$
219	–	0.1	$[In(CH_3)_6N]^+$
204	0.7	0.7	$[In(CH_3)_5N]^+$
175	2.2	1.1	$[In(CH_3)_3NH]^+$
160	2.2	1.1	$[In(CH_3)_3]^+$
145	71.2	59.2	$[In(CH_3)_2]^+$
130	4.8	5.7	$[InCH_3]^+$
115	11.6	32.4	In^+
113	0.6	1.4	$^{113}In^+$
59	71.7	92.3	$[N(CH_3)_3]^+$
58	100.0	100.0	$[N(CH_3)_2CH_2]^+$
57	4.5	12.4	$[N(CH_2)_2CH_3]^+$

Mass spectra were measured at ionization energies of 18, 50, and 70 eV and source temperatures of 21 °C. Along with fragments of the monomer, some fragments of the species $(CH_3)_3In \cdot 3N(CH_3)_3$, $(CH_3)_2In \cdot 2N(CH_3)_3$, and $2(CH_3)_3In \cdot N(CH_3)_3$ were also observed [33]; the relative intensities given above correspond to the isotope ^{115}In.

For exchange reactions determined by ^1H NMR measurements, see general remarks [10].

The adduct has been used to produce InP and GaInAs semiconductor layers by gas-phase epitaxy [56].

$(CH_3)_3InN(C_2H_5)_3$ and **$(CH_3)_3In \cdot N(C_4H_9\text{-}n)_3$** (Table **6**, Nos. **5** and **6**). No. **6** was also obtained directly by electrolysis of a THF solution of $Mg(CH_3)_2$ and $[N(C_4H_9\text{-}n)_4]ClO_4$ (50 mA, 120 V), similar to the method described for the THF adduct (Table 5, No. 3) [44, 45].

Vapor pressure studies on the triethylamine adduct No. 5 (conditions, see No. 4) led to the equation $\log p = 10.8585 - 3649.96/T$, and the values $\Delta H_v = 16.70$ kcal/mol, $\Delta S = 36.51$ $cal \cdot mol^{-1} \cdot K^{-1}$ [33].

$(CH_3)_3In \cdot NH\{C(CH_3)_2CH_2\}_2CH_2$ (Table **6**, No. **16**) crystallizes in the monoclinic space group $P2_1/a - C_{2h}^5$ (No. 14) with the lattice constants a = 1508.6(2), b = 1271.4(3), c = 777.1(2) pm, and $\beta = 94.15(1)°$. The unit cell contains 4 formula units, and the X-ray density was calculated as $D_c = 1.345$ g/cm^3. The molecular structure of the adduct, showing a distorted tetrahedral geometry at the In atom, is depicted in **Fig. 3** [62].

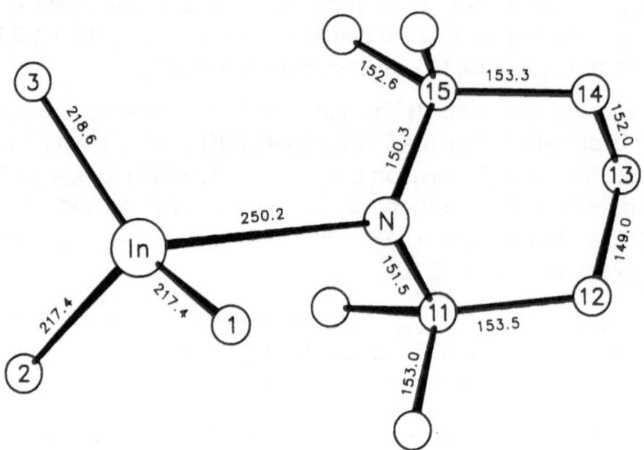

Fig. 3. Molecular structure of $(CH_3)_3In \cdot NH\{C(CH_3)_2CH_2\}_2CH_2$ [62].

Important bond angles (°)

C(3)-In-C(1)	118.9 (3)	In-N-C(15)	117.7 (3)
C(2)-In-C(1)	116.5 (3)	C(11)-N-C(15)	117.3 (3)
C(3)-In-C(2)	109.3 (3)	C(15)-C(14)-C(13)	113.6 (4)
N-In-C(3)	108.2 (2)	C(14)-C(13)-C(12)	109.8 (4)
N-In-C(2)	109.3 (2)	C(13)-C(12)-C(11)	114.1 (4)
N-In-C(1)	92.8 (2)	C(12)-C(11)-N	110.7 (4)
In-N-C(11)	114.6 (3)		

The volatility of the adduct was measured with the modified entrainment method (MEM); see also general remarks. The weight loss, estimated between 70 and 90 °C with N_2 carrier gas, can be suppressed by addition of free amine to the carrier gas, indicating dissociative evaporation of the adduct [83].

$(CH_3)_3In \cdot N(CH_2CH_2)_3N$ (Table **6**, No. **18**) very likely occurs as the monomeric 1:1 adduct on synthesis in ether solution; however, if the solvent is removed for isolation, then the residue is only slightly soluble in ether (or in pyridine, THF, or CH_3CN). The solvent-free solid is unusually air-stable; dilute solutions and the gaseous state are highly sensitive to oxygen and moisture [62].

The ^1H NMR spectrum shows only two sharp singlets at −0.15 (CH_3In) and 2.03 ppm (CH_2N), whose area ratio depends upon the nature of the solvent (C_6D_6, C_5D_5N, CD_3CN) and varies between 1:0.83 and 1:1.8 [62].

The X-ray structural analysis indicates linear chains of planar $In(CH_3)_3$ molecules connected by $N(CH_2CH_2)_3N$ units. The indium atom possesses trigonal-bipyramidal coordination; the CH_3 groups of each $In(CH_3)_3$ molecule eclipse one another, but are staggered with respect to the ethylene functions of the "dabco" ligand as shown in **Fig. 4B**. The compound crystallizes in the orthorhombic space group Pmmm$-D^1_{2h}$ (No. 47) with a = 1083.1(3), b = 783.5(1), and c = 674.0(2) pm; Z = 2 and D_c = 1.591 g/cm^3. Important distances and angles are given in **Fig. 4A**; Fig. 4B shows the conformation of the polymer chain in the crystal structure [62].

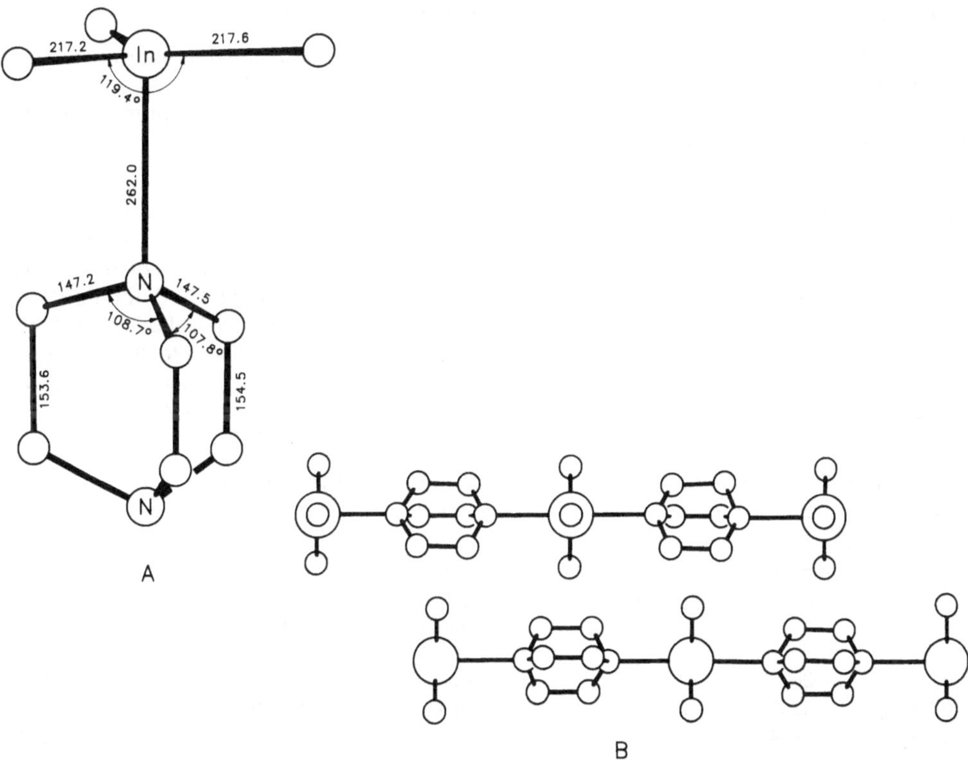

Fig. 4. A) Molecular structure of $(CH_3)_3In \cdot N(CH_2CH_2)_3N$. − B) Crystal packing of the infinite linear chains [62].

$(CH_3)_3In \cdot \{N(C_2H_5)CH_2\}_3$ (Table **6**, No. **19**). Variable temperature ^1H NMR studies show that the In-N bond is very labile and the exchange between adduct components is rapid compared with the NMR time scale even at low temperature. At low temperature the ring CH_2 group gives two signals according to axial and equatorial protons with a coalescence temperature of 263 K which is identical to that of the free amine. This indicates that the coordinated $In(CH_3)_3$ has only little or no effect on the ligand fluxionality. At room temperature equilibration occurs presumably by the chair $\leftrightarrows$ boat process of the cyclohexane like ring. However, the mean chemical shift of the methylene protons of the adduct is upfield and moves further upfield as the temperature is lowered whereas the mean shift of the pure amine shows no significant temperature dependance [62].

$(CH_3)_3In \cdot (CH_3)_3P{=}NSi(CH_3)_2N{=}P(CH_3)_3$ (Table **6**, No. **33**) is a very air- and moisture-sensitive solid, readily soluble in C_6H_6, CH_2Cl_2, and $O(C_2H_5)_2$, moderately soluble in CCl_4, and poorly soluble in cyclohexane or petroleum ether. It is monomeric in C_6H_6 solution. Between 100 and 110 °C the addition product dissociates completely into the starting materials. At −70 °C the proton resonance spectrum shows no splitting of the CH_3P signals corresponding to a coordinated and an uncoordinated $-N{=}P(CH_3)_3$ group. From this it was concluded that the $In(CH_3)_3$ group is rapidly shifting between the two nitrogen atoms in the molecule. The associated activation energy was estimated to be $\Delta H_A < 9$ kcal/mol by comparison with the data of the homologous compounds of Al (13.5 kcal/mol) and Ga (12.5 kcal/mol) [21].

The compound reacts with $Al(CH_3)_3 \cdot O(C_2H_5)_2$ to form an ionic complex with the formula $[(CH_3)_2Si(NP(CH_3)_3)_2Al(CH_3)_2][In(CH_3)_4]$ (see Section 11.2.1, p. 337) [15].

$(CH_3)_3In \cdot PH_3$ (Table **6**, No. **35**) dissociates below −123 °C, at which temperature it has a PH_3 vapor pressure of 4 mm. Rapid warming of the solid to above −78 °C can give rise to an explosive decomposition. Pyrolysis (270 to 300 °C) of the polymeric primary decomposition products, $(CH_3InPH)_x$, forms microcrystalline InP [2].

Because the $In(CH_3)_3/PH_3$ system is of great importance for producing single-crystal epitaxial InP layers, pyrolysis experiments under the conditions of gas phase epitaxy (InP substrates and H_2 or D_2 carrier gas) were conducted to elucidate the course of the reaction [59, 73]. Conspicuous is a drastic lowering of the pyrolysis temperatures of either PH_3 or $In(CH_3)_3$ in the presence of the other ($\Delta t = 225$ °C for PH_3, about 50 °C for $In(CH_3)_3$; see p. 11). For an $In(CH_3)_3/PH_3$ mixture (1:4.2 with D_2 as carrier gas) above 400 °C the only reaction product detectable by mass spectrometry was CH_4. Despite the relatively high temperature, metastable $(CH_3)_3In \cdot PH_3$ was postulated to have formed in very small amounts which were adsorbed on the substrate surface, gradually splitting off CH_4. $(CH_3)_2InPH_2$ was suggested as the first stage of this decomposition [63, 73].

$(CH_3)_3In \cdot P(CH_3)_3$ (Table **6**, No. **39**). Vapor pressure measurements (p in Torr) have been obtained by several methods, and provide the following constants for the equation, $\log p = A - B/T(K)$:

A	B	range in °C	ΔH_v in kcal/mol	ΔS_v in $cal \cdot mol^{-1} \cdot K^{-1}$	Ref.
9.015	2832	50 to 110	13.0		[1]
10.096	3312	60 to 120	15.1	33.02	[33]
5.697	1724	50 to 70			[54]

The enthalpy of dissociation into adduct components is 17.1 ± 0.8 kcal/mol [1], which agrees with the calculated value of 17.04 kcal/mol [64].

IR spectra were measured for solid [33] and dissolved (C_6H_6, CCl_4) [23] samples, and the Raman spectrum for melted samples (~50 °C); the most important frequency values are given below (obs. = obscured by solvent bands):

IR solid	solution	Raman solution	assignment
–	2970 sh	2976 m, dp	$\nu_{as}(CH_3P)$
–	2960 ms	2959 ms, dp	$\nu_{as}(CH_3In)$
–	2910 s	2911 vs, p	$\nu_s(CH_3In)$
1298 w	1310 w	1311 w, p	$\delta_s(CH_3P)$
1279 m	1290 m	–	
1136 m	1145 mw	1151 s, p	$\delta_s(CH_3In)$
950 s	959 s	958 vw	$\varrho(CH_3P)$
939 s	946 s	–	
729 m, br	736 m	740 m, dp	$\nu_{as}(PC_3)$
675 s, br	obs.	(669)	$\varrho(CH_3In)$
(675)	obs.	669 s, p	$\nu_s(PC_3)$
474 vs	474 vs	482 ms, dp	$\nu_{as}(InC_3)$
–	468 sh	466 vvs, p	$\nu_s(InC_3)$
–	322 vw	320 m, p	$\delta_s(PC_3)+\nu(InP)$?
–	–	261 m, dp	$\delta(PC_3)$
–	–	130 ms, dp	$\delta(InC_3)+\delta(CInP)$

The ^{115}In NQR spectrum, recorded at 77 K, shows transition frequencies of ν_1 28.10, ν_2 53.20, ν_3 80.21, and ν_4 107.17 MHz, a quadrupole coupling constant of 643.01 MHz, and an asymmetry parameter, η, reported to be 7.4% [18].

The UV spectrum of the adduct (0.15 Torr in H_2 at 25 °C) shows a broad absorption at $\lambda_{max}=195$ nm with $\varepsilon_{max}=15000\ L\cdot cm^{-1}\cdot mol^{-1}$. At 40 °C the absorption maximum is at 205 nm, and at 100 °C dissociation is assumed to be complete [46]. (Author's comment: therefore, values of vapor pressure, enthalpy, and entropy measured at or above 100 °C, e.g. [1, 33], should be viewed with caution).

The molecular and crystal structure of the adduct is depicted in [33], but other parameters are not available. The In-P distance is reported to be 2.683(5) Å [71] and the orientation of the methyl groups at In and at P are eclipsed [62]; all works (done in the same group) refer to the results from [34].

$(CH_3)In\cdot P(CH_3)_3$ has been used as a source material for binary and ternary III/V (13/15) semiconductor layers (InP, GaInAs, and others) in gas phase epitaxy (MOVPE); however, during the pyrolysis process additional group V (15) component has to be introduced as PH_3 (or AsH_3, see Table 6, No. 47). A list of references (not exhaustive) is [24, 26, 28 to 30, 40, 41, 47, 48, 51 to 54, 65 to 69, 77, 84]; see also the references on pp. 23 to 27.

$(CH_3)_3In\cdot P(C_2H_5)_3$ (Table **6**, No. **40**). Vapor pressure measurements by the saturation method with N_2 [33, 55] or H_2 [75] as carrier gas produced the equations:

$$N_2:\quad \log p(Pa)=12.885-3872/T(K)\ (82\ \text{to}\ 137\ °C)$$
$$H_2:\quad \log p(Pa)=7.405-2025.1/T(K)\ (50\ \text{to}\ 80\ °C)$$

Above 80 °C the two vapor pressure curves strongly diverge. Possibly, the data from [33, 55] result from a partial decomposition of the adduct at higher temperatures (T > 100 °C), leading to a high vapor pressure; this can be assumed from results obtained with similar trimethylgallium adducts [61].

The IR and Raman spectra of the melted adduct in the P-C and In-C bond vibration region (700 to 400 cm^{-1}) are discussed in detail. While the splitting of the PC_3 modes in free $P(C_2H_5)_3$ (689, 669, and 658 cm^{-1} for $\nu_{as}(PC_3)$; 618 and 609 cm^{-1} for $\nu_s(PC_3)$) argues for the existence of at least two rotational isomers, no such splitting is observed in the adduct spectrum at 690 cm^{-1} ($\nu_{as}(PC_3)+\varrho(CH_3In)$ and 621 cm^{-1} ($\nu_s(PC_3)$) [75].

The UV spectrum of the adduct (0.001 atm partial pressure in H_2 at 1 atm total pressure) shows a broad absorption band at $\lambda_{max}=210$ nm [43] (~215 nm in [46]) with $\varepsilon_0=\sim4000$ $L\cdot mol^{-1}\cdot cm^{-1}$ [43]. The change of this band with increasing temperature leads to the conclusion that the dissociation is reversible, and by about 200 °C decomposition into the components is complete [46]. From 280 °C the pyrolysis of $In(CH_3)_3$ sets in, then from ~480 °C that of the $P(C_2H_5)_3$ begins [43]. The distinctly higher thermal stability of the triethyl homologue compared to the $P(CH_3)_3$ adduct (Table 6, No. 39) explains its growing importance as an In source in gas phase epitaxy (MOVPE) [29, 31, 35, 42, 43, 49, 50, 52, 55, 58, 61, 75, 76, 78, 79, 81]. The list is not exhaustive; see also the general references on pp. 23 to 27.

$(CH_3)_3In\cdot As(CH_3)_3$ (Table **6**, No. **47**). The IR and Raman spectra, when compared with those of the pure components, show only small frequency changes caused by adduct formation. Primarily, the InC_3 bond vibration shifts to somewhat lower (5 to 10 cm^{-1} for $\nu(InC_3)$) values, and the AsC_3 vibration to higher (10 to 20 cm^{-1} for $\nu(AsC_3)$) values. The intrinsic vibrations of the various methyl groups generally remain unaltered [23]. The band contour of the symmetrical $CH_3(In)$ deformation at ~1150 cm^{-1} was used to discuss adduct formation and decomposition. IR-detectable amounts of CH_4 were observed from about 375 °C, when the adduct was pyrolyzed [38].

The NQR parameters for ^{115}In at 77 K are: ν_1 32.92, ν_2 63.87, ν_3 93.13, ν_4 128.19, e^2Qq/h 769.41 MHz, $\eta=5.5\%$; for ^{75}As: ν_1 78.16, e^2Qq/h 156.32 MHz, $\eta=0.0\%$ [18].

The adduct has been used several times as source material for epitaxy (MOVPE) of ternary $Ga_{1-x}In_xAs$ semiconductor layers [25, 27, 37, 38, 57, 75].

$(CH_3)_3In\cdot Sb(CH_3)_3$ (Table **6**, No. **49**) is completely dissociated between 65 and 100 °C [23].

The ^{115}In NQR spectrum, recorded at 77 K, shows transition frequencies of ν_1 35.13, ν_2 69.79, ν_3 104.85, and ν_4 139.66 MHz, a quadrupole coupling constant, e^2Qq/h, of 838.26 MHz, and an asymmetry parameter, η, reported to be 2.7%. The values for ^{123}Sb are ν_1 38.20, ν_2 76.00, ν_3 114.08, $e^2Qq/h=532.39$ MHz, $\eta=3.6\%$; some values are also given for ^{121}Sb [18].

References:

[1] Coates, G.E.; Whitcombe, R.A. (J. Chem. Soc. **1956** 3351/4).
[2] Didchenko, R.; Alix, J.E.; Toeniskoetter, R.H. (J. Inorg. Nucl. Chem. **14** [1960] 35/7).
[3] Harrison, B.C.; Tompkins, E.H. (Inorg. Chem. **1** [1962] 951/3).
[4] Coates, G.E.; Graham, J. (J. Chem. Soc. **1963** 233/7).
[5] Beachley, O.T.; Coates, G.E. (J. Chem. Soc. **1965** 3241/7).
[6] Beachley, O.T.; Coates, G.E.; Kohnstam, G. (J. Chem. Soc. **1965** 3248/52).
[7] Drago, R.S.; Wayland, B.B. (J. Am. Chem. Soc. **87** [1965] 3571/7).

[8] Schmidbaur, H.; Wolfsberger, W. (Angew. Chem. **78** [1966] 306/7; Angew. Chem. Intern. Ed. Engl. **5** [1966] 312).
[9] Schmidbaur, H.; Wolfsberger, W. (Chem. Ber. **100** [1967] 1000/15).
[10] Henhold, K.L.; Oliver, J.P. (Inorg. Chem. **7** [1968] 950/2).

[11] Schmidbaur, H.; Jonas, G. (Chem. Ber. **101** [1968] 1271/85).
[12] Schmidbaur, H.; Wolfsberber, W.; Schwirten, K. (Chem. Ber. **102** [1969] 556/63).
[13] Wolfsberger, W.; Schmidbaur, H. (J. Organometal. Chem. **17** [1969] 41/51).
[14] Storr, A.; Thomas, B.S. (Can. J. Chem. **48** [1970] 3667/72).
[15] Wolfsberger, W.; Schmidbaur, H. (J. Organometal. Chem. **27** [1971] 181/4).
[16] Garnovskii, A.D.; Okhlobystin, O.Yu.; Osipov, O.A.; Yunusov, K.M.; Kolodyazhnyi, Yu.V.; Golubinskaya, L.M.; Svergun, V.I. (Zh. Obshch. Khim. **42** [1972] 920/5; J. Gen. Chem. [USSR] **42** [1972] 910/4).
[17] Drago, R.S. (Struct. Bonding [Berlin] **15** [1973] 73/139).
[18] Patterson, D.B.; Carnevale, A. (J. Chem. Phys. **59** [1973] 6464/7).
[19] Nemirovskii, L.N.; Kozyrkin, B.I.; Lantsov, A.F.; Gribov, B.G.; Skvortsov, I.M.; Sredinskaya, I.A. (Dokl. Akad. Nauk SSSR **214** [1974] 590/3; Dokl. Chem. Technol. [USSR] **214** [1974] 87/9; CA **80** [1974] No. 96071).
[20] Gur'yanova, E.N.; Goldstein, I.P.; Romm, I.P. (Donor-Acceptor Bond, Wiley, New York 1975).

[21] Wolfsberger, W.; Schmidbaur, H. (J. Organometal. Chem. **122** [1976] 5/12).
[22] Mertz, K.; Schwarz, W.; Eberwein, B.; Weidlein, J.; Hess, H.; Hausen, H.-D. (Z. Anorg. Allgem. Chem. **429** [1977] 99/104).
[23] Sille, K.; Fries, W.; Weidlein, J. (unpublished results 1977).
[24] Benz, K.W.; Renz, H.; Weidlein, J.; Pilkuhn, M.H. (J. Electron. Mater. **10** [1981] 185/7).
[25] Hirtz, J.P.; Lariavain, J.P.; Duchemin, J.P.; Persall, T.P.; Bonnet, M. (Electron. Letters **16** [1980] 415/6).
[26] Renz, H.; Weidlein, J.; Benz, K.W.; Pilkuhn, M.H. (Electron. Letters **16** [1980] 228).
[27] Dietze, W.T.; Ludowise, M.J.; Cooper, C.B. (Electron. Letters **17** [1981] 698/9).
[28] Ludowise, M.J.; Cooper, C.B.; Saxena, R.R. (J. Electron. Mater. **10** [1981] 1051/68).
[29] Moss, R.H.; Evans, J.S. (J. Cryst. Growth **55** [1981] 129/34).
[30] Benz, K.W.; Haspeklo, H.; Bosch, R. (J. Phys. Colloq. [Paris] **43** [1982] C5-393/C5-399).

[31] Chatterjee, A.K.; Faktor, M.M.; Moss, R.H.; White, E.A.D. (J. Phys. Colloq. [Paris] **43** [1982] C5-491/C5-503).
[32] Krause, H.; Sille, K.; Hausen, H.-D.; Weidlein, J. (J. Organometal. Chem. **235** [1982] 253/64).
[33] Aitchison, K.A. (Diss. Univ. London 1983).
[34] Frigo, D.M. (Diss. Univ. London 1983 from [62, 71]).
[35] Mullin, J.B.; Holliday, A.K.; Cole-Hamilton, D.J.; Jones, A.C. (Eur. 80349 [1983]; C.A. **99** [1983] No. 140143).
[36] Mullin, J.B.; Holliday, A.K.; Cole-Hamilton, D.J.; Jones, A.C. (Eur. 80844 [1983]; C.A. **99** [1983] No. 212714).
[37] Speier, P.; Scholz, F.; Benz, K.W.; Renz, H.; Weidlein, J. (Electron. Letters **19** [1983] 728/9).
[38] Cheng, C.H.; Jones, K.A.; Motyl, K.M. (J. Electron Mater. **13** [1984] 703/26).
[39] Cole-Hamilton, D.J.; Gerrard, N.D.; Jones, A.C.; Holliday, A.K.; Mullin, J.B. (Brit. 2125795 [1984]; C.A. **101** [1984] No. 91212).
[40] Donnelly, V.M.; Geva, M.; Long, J.; Karlicek, R.F. (Appl. Phys. Letters **44** [1984] 951/3).

[41] Donnelly, V.M.; Karlicek, R.F. (Proc. SPIE-Intern. Soc. Opt. Eng. No. 476 [1984] 102/9; C.A. **101** [1984] No. 238003).

[42] Faktor, M.M.; Bradley, D.C.; Aitchison, K.A. (Brit. 2123422 [1984]; C.A. **101** [1984] No. 23710).
[43] Haigh, J.; O'Brien, S. (J. Cryst. Growth **68** [1984] 550/6).
[44] Jones, A.C.; Gerrard, N.D.; Cole-Hamilton, D.J.; Holliday, A.K. (J. Organometal. Chem. **265** [1984] 9/15).
[45] Jones, A.C.; Holliday, A.K.; Cole-Hamilton, D.J.; Ahmad, M.M.; Gerrard, N.D. (J. Cryst. Growth **68** [1984] 1/9).
[46] Karlicek, R.; Long, J.A.; Donnelly, V.M. (J. Cryst. Growth **68** [1984] 123/7).
[47] Long, J.A.; Riggs, V.G.; Johnston, W.D., Jr. (J. Cryst. Growth **69** [1984] 10/14).
[48] Moss, R.H. (J. Cryst. Growth **68** [1984] 78/87).
[49] Moss, R.H.; Spurdens, P.C. (Electron. Letters **20** [1984] 978/9).
[50] Moss, R.H.; Spurdens, P.C. (J. Cryst. Growth **68** [1984] 96/101).

[51] Nelson, A.W.; Westbrook, L.D. (J. Cryst. Growth **68** [1984] 102/10).
[52] Nelson, A.W.; Westbrook, L.D. (Eur. 117051 [1984]; C.A. **102** [1985] No. 53715).
[53] Nicholas, D.J.; Allsopp, D.; Hamilton, B.; Peaker, A.R.; Bass, S.J. (J. Cryst. Growth **68** [1984] 326/33).
[54] Renz, H. (Diss. Univ. Stuttgart 1984).
[55] Minagawa, S.; Nakamura, H.; Sano, H. (J. Cryst. Growth **71** [1985] 377/84).
[56] Bass, S.J.; Skolnick, M.S.; Chudzynska, H.; Smith, L. (J. Cryst. Growth **75** [1986] 221/6).
[57] Cheng, C.H.; Jones, K.A.; Motyl, K.M. (Progr. Cryst. Growth Charact. **12** [1986] 319/33).
[58] Hoercher, G.; Steiner, S.; Forchel, A.; Scholz, F.; Tränkle, G. (Advan. Mater. Telecommun. Papers 13th Symp. Eur. Mater. Res. Soc. Meeting, Strasbourg 1986, pp. 273/9).
[59] Larsen, C.A.; Stringfellow, G.B. (J. Cryst. Growth **75** [1986] 247/54).
[60] Reier, F.-W.; Wolfram, P.; Schumann, H. (J. Cryst. Growth **77** [1986] 23/6).

[61] Scholz, F.; Wiedemann, P.; Nerz, U.; Benz, K.W.; Tränkle, G.; Lach, E.; Forchel, A.; Laube, G.; Weidlein, J. (J. Cryst. Growth **77** [1986] 564/70).
[62] Bradley, D.C.; Dawes, H.; Frigo, D.M.; Hursthouse, M.B.; Hussain, B. (J. Organometal. Chem. **325** [1987] 55/67).
[63] Buchan, N.I.; Larsen, C.A.; Stringfellow, G.B. (Appl. Phys. Letters **51** [1987] 1024/5).
[64] Drago, R.S.; Wong, N.; Bilgrien, C.; Vogel, G.C. (Inorg. Chem. **26** [1987] 9/14).
[65] Kuck, M.A.; Gersten, S.W.; Baumann, J.A. (PCT Intern. Appl. 87-965 [1987]; C.A. **106** [1987] No. 225847).
[66] Lee, M.K.; Wuu, D.S.; Tung, H.H. (Appl. Phys. Letters **50** [1987] 1725/6).
[67] Lee, M.K.; Wuu, D.S.; Tung, H.H. (Appl. Phys. Letters **50** [1987] 1805/7).
[68] Lee, M.K.; Wuu, D.S.; Tung, H.H. (J. Appl. Phys. **62** [1987] 3209/11).
[69] Lee, M.K.; Wuu, D.S.; Tung, H.H. (Ts'ai Liao K'o Hsueh A **19** [1987] 35/9 from C.A. **107** [1987] No. 106603).
[70] Reier, F.W.; Wolfram, P.; Schumann, H. (Ger. Offen. 3612629 [1987]; C.A. **108** [1988] No. 186988).

[71] Bradley, D.C.; Chudzynska, H.; Faktor, M.M.; Frigo, D.M.; Hursthouse, M.B.; Hussain, B.; Smith, L.M. (Polyhedron **7** [1988] 1289/98).
[72] Bradley, D.C.; Faktor, M.M.; Frigo, D.M. (J. Cryst. Growth **89** [1988] 227/36).
[73] Buchan, N.I.; Larsen, C.A.; Stringfellow, G.B. (J. Cryst. Growth **92** [1988] 605/15).
[74] Foster, D.F.; Rushworth, S.A.; Cole-Hamilton, D.F.; Cafferty, R.; Harrison, J.; Parkes, P. (J. Chem. Soc. Dalton Trans. **1988** 7/11).
[75] Laube, G. (Diss. Univ. Stuttgart 1988)
[76] Laube, G.; Kohler, U.; Weidlein, J.; Scholz, F.; Streubel, K.; Dieter, R.J.; Karl, N.; Gerdon, M. (J. Cryst. Growth **93** [1988] 45/51).
[77] Lee, M.K.; Huang, K.C.; Wuu, D.S.; Tung, H.H.; Yu, K.Y. (J. Cryst. Growth **93** [1988] 539/42).

[78] Monserrat, K.J.; Tothill, J.N.; Haigh, J.; Moss, R.H.; Baxter, C.S.; Stobbs, W.M. (J. Cryst. Growth **93** [1988] 466/74).
[79] Streubel, K.; Scholz, F.; Laube, G.; Dieter, R.J.; Zielinski, E.; Keppler, F. (J. Cryst. Growth **93** [1988] 347/52).
[80] Lechler, R.; Hausen, H.-D.; Weidlein, J. (J. Organometal. Chem. **359** [1989] 1/12).
[81] Weber, J.; Moser, M.; Stapor, A.; Scholz, F.; Bohnert, G.; Hangleiter, A.; Hammel, A.; Wiedman, D.; Weidlein, J. (J. Cryst. Growth **104** [1990] 815/9).
[82] Cole-Hamilton, D.J.; Foster, D.F.; Rushworth, S.; Plessey Co. PLC (Brit. 2201418 [1988] from C.A. **110** [1988] No. 57844).
[83] Bradley, D.C.; Faktor, M.M.; Frigo, D.M.; Smith, L.M. (J. Cryst. Growth **92** [1988] 37/45).
[84] Wolfram, P.; Reier, F.W.; Frank, D.; Schumann, H. (J. Cryst. Growth **96** [1989] 691/92).

1.1.1.2.2.2 Compounds with Two or More $In(CH_3)_3$ Molecules

General Remarks. The products in this section are listed in Table 8. They were prepared either by co-condensation of stoichiometric amounts of $In(CH_3)_3$ with the respective Lewis base (mole ratio 2:1, 3:1, or 4:1), followed by warming to room temperature [2, 11], or by admixture of stoichiometric amounts of the dissolved starting materials, followed by mild warming [4]. The etherate of trimethylindium was sometimes used as starting material [1, 9, 13, 16]. Purification of these adducts was most commonly accomplished by recrystallization. At higher temperatures and reduced pressure the adducts usually dissociate, and this has sometimes been used to purify $In(CH_3)_3$ [8, 10].

Table 8
Compounds with Two or More $In(CH_3)_3$ Molecules.
Further information on numbers preceded by an asterisk is given at the end of the table.
Explanations, abbreviations, and units on p. X.

No.	donor (yield)	number of $In(CH_3)_3$	properties and remarks
1	$N(CH_3)_2CH_2N(CH_3)_2$	2	white solid, m.p. 53 to 56 °C [2] ^{1}H NMR (C_6H_6): −0.14 (CH_3In), 1.96 (CH_3N), 3.03 (CH_2N) [2] IR (C_6H_6): 522 s $\nu_{as}(InC_3)$, 480 vs $\nu_s(InC_3)$, 463 s $\nu(InN)$; no other bands reported (assignment doubtful) [2] about 50% dissociated in C_6H_6 [2]
*2	$NHCH_3(CH_2)_2NHCH_3$ (52%)	2	colorless crystals, m.p. 64.5 to 69.5 °C, b.p. 90 °C/0.1 Torr [9] ^{1}H NMR (C_6D_6): −0.08 (CH_3In), 0.68 (br, HN), 1.77 (CH_3N, $^3J(H,H)=7$), 2.08 (CH_2) [9] ^{13}C NMR (C_6D_6): −7.2 (CH_3In), 35.9 (CH_3N); 49.0 (CH_2) [9] mass spectrum (main fragments only): $[M-CH_3]^+$ (9%), $[M-3CH_3]^+$ (1%), $[In(CH_3)_2]^+$ (100%), $[InCH_3]^+$ (28%), $^{115}In^+$ (99%), $^{113}In^+$ (5%) [9]

Table 8 (continued)

No.	donor (yield)	number of $In(CH_3)_3$	properties and remarks
3	$N(CH_3)_2(CH_2)_2N(CH_3)_2$ (83%) [15]	2	white solid, m.p. 90 to 93 °C, monomeric in C_6H_6 [2, 15] vapor pressure: $\log p = 12.75 - 4750/T$ [14] ^1H NMR (C_6H_6): −0.24 (CH_3In), 1.76 (CH_3N), 2.26 (CH_2) [2] IR (C_6H_6): 522 s ν_{as} (InC_3), 480 vs ν_s (InC_3), 465 sh ν(InN); no other bands reported [2] mass spectrum (main fragments only): $[M-CH_3]^+$, $[M-3\,CH_3]^+$, $[In(CH_3)_2]^+$, $[InCH_3]^+$, $[C_6H_{14}N_2]^+$ [15]
4	$N(CH_3)_2(CH_2)_3N(CH_3)_2$	2	white solid, m.p. 81 to 83 °C, monomeric in benzene [2] ^1H NMR (C_6H_6): −0.21 (CH_3In), 1.76 (CH_3N and CH_2) [2] IR (C_6H_6): 520 w $\nu_{as}(InC_3)$, 478 vs $\nu_s(InC_3)$, 465 sh ν(InN); no other bands reported [2]
5	$C_6H_4\{N(CH_3)_2\}_2$-1,4	2	white, needle-shaped crystals, m.p. 88 to 92 °C [15] ^1H NMR (C_6D_6): −0.21 (s, CH_3In), 2.32 (s, CH_3N), 6.69 (m, C_6H_4) [15] mass spectrum (main fragments only): $[C_{10}H_{16}N_2]^+$, $[C_9H_{13}N_2]^+$, $[In(CH_3)_2]^+$, $[C_8H_{10}N_2]^+$, $[InCH_3]^+$, $^{115}In^+$ [15] dec. at about 100 °C under vacuum to give $In(CH_3)_3$ and the 1:1 adduct (Table 6, No. 10) [15]
*6	$N(CH_3)_2C_6H_4CH_2C_6H_4N(CH_3)_2$ 4,4′-methylene-bis(N,N′-dimethylaniline) = MBDA (89%)	2	white needle shaped crystals, m.p. 80 to 82 °C [16] ^1H NMR (C_6D_6): −0.23 (s, CH_3In), 2.33 (s, CH_3N), 3.67 (s, CH_2), 6.84 (m, C_6H_4) [16] mass spectrum: $[In \cdot MBDA]^+$, $[MBDA]^+$, $[In(CH_3)_2]^+$, $[InCH_3]^+$, $[^{115}In]^+$ [16]
7	$N_4C_6H_{12}$ (urotropine) (82%)	2	colorless crystals, m.p. 134 °C, soluble in C_6H_6, C_7H_8, CH_2Cl_2 [4] ^1H NMR: −0.18 (CH_3In), 4.01 (CH_2N) in C_6D_6; −0.48 (CH_3In), 4.52 (CH_2N) in CD_2Cl_2 [4]

Table 8 (continued)

No.	donor (yield)	number of $In(CH_3)_3$	properties and remarks
8	$N_4C_6H_{12}$ (urotropine) (92%)	3	white solid, m.p. ca. 139 °C (dec.), soluble in C_6H_6, C_7H_8, C_6H_{12} [4] ^1H NMR: −0.20 (CH_3In), 3.96 (CH_2N) in C_6D_6; −0.42 (CH_3In), 4.44 (CH_2N) in CD_2Cl_2 [4]
*9	(pyrazine)	2	not isolated [3]
10	(bipyridine-3,3′) (81%)	2	white crystals, m.p. 86 to 90 °C [15] ^1H NMR (C_6D_6): 0.08 (CH_3In), 6.52, 6.90, 8.12, 8.29 (m, C_5H_4) [15] mass spectrum (principal fragments only): $[C_{10}H_8N_2]^+$, $[In(CH_3)_2]^+$, $[InCH_3]^+$, $^{115}In^+$ [15] dec. at 150 °C/0.01 Torr to give $In(CH_3)_3$ and the 1:1 adduct (Table 6, No. 22) [15]
*11	(bipyridine-4,4′) (75%)	2	white crystals, m.p. 142 to 144 °C, subl. 80 °C/0.01 Torr, monomeric in benzene [15] ^1H NMR (C_6D_6): 0.13 (CH_3In), 6.59, 8.26 (m, C_5H_4) [15] mass spectrum: principal fragments as with No. 10 [15]
12	$(CH_3)_2Si(N{=}P(CH_3)_3)_2$ (93%)	2	solid, m.p. 84 to 86 °C, monomeric in C_6H_6 [1] ^1H NMR (CH_2Cl_2): −0.27 (CH_3In), 0.43 (CH_3Si), 1.67 (CH_3P, $^2J(P,H)=12.9$) [1]
13	$(CH_3)_2PP(CH_3)_2$ (ca. 100%)	2	white solid, m.p. 109 to 111 °C, subl. at ambient temperature/10^{-5} Torr, monomeric in C_6H_6 [11] $\log p = 14.04 - 4964/T$, $\Delta H = 23$ kcal/mol, $\Delta S = 51$ cal · mol^{-1} · K^{-1} (region 50 to 100 °C) [11] ^1H NMR (C_6H_6): −0.20 (CH_3In), 0.75 (CH_3P) [11] IR (solid): 2960 m, 2930 m, 2910 m, 2840 w, 2815 w, 1420 m, 1385 vw, 1300 s, 1145 vs, br, 1055 s, br, 945 m, 865 m, 780 w, sh, 740 m, sh, 705 m, 528 m, 475 m [11]

Table 8 (continued)

No.	donor (yield)	number of $In(CH_3)_3$	properties and remarks
13 (continued)			mass spectrum (40 °C/70 eV): $[In(CH_3)_2]^+$ (40%), $^{115}In^+$ (22%), $[P_2C_4H_{12}]^+$ (80%), $[P_2C_3H_9]^+$ (100%), $[PC_3H_8]^+$ (18%), P_2^+ or $[PC_2H_7]^+$ (40%), $[PC_2H_6]^+$ (65%), $[PC_2H_4]^+$ (42%), $[PCH_3]^+$ (48%), other fragments given; no significant difference for electron impact energies of 10 to 15 eV [11]
*14	$(C_6H_5)_2P(CH_2)_2P(C_6H_5)_2$ ("diphos")	2	large, colorless crystals, m.p. 163 to 166 °C [6, 13] ^1H NMR (C_6D_6): 0.07 (s, CH_3In), 2.51 (d, CH_2P), 6.99, 7.40 (m, C_6H_5) [13] ^{31}P NMR (C_6D_6): −15.02, no significant changes between 305 and 193 K [13] mass spectrum: $[M-In(CH_3)_3]^+$ (1.4%), $[In(CH_3)_2]^+$ (100%), $[InCH_3]^+$ (12.2%) [6]
15	$\{P(C_6H_5)_2(CH_2)_2\}_2P(C_6H_5)$ ("triphos")	3	sticky oil, attempted recrystallization to separate from pentane, benzene, or toluene was unsuccessful [6, 13] ^1H NMR (C_6H_6): 0.12 (s, CH_3In), 2.25 (m, CH_2P), 7.04, 7.47 (m, C_6H_5) [6, 13] ^{31}P NMR (C_6H_6): −17.91 (PC_6H_5), −15.29 ($P(C_6H_5)_2$, J(P,P′) = 33.6) [6, 13] mass spectrum (major peaks): $[triphos]^+$, $[In(CH_3)_2]^+$, $[InCH_3]^+$ [6] begins to liberate $In(CH_3)_3$ at 60 °C/10^{-2} Torr, up to 140 °C with 75% yield [6, 13]
16	$\{(C_6H_5)_2P(CH_2)_2PC_6H_5CH_2\}_2$ ("tetraphos") (83%)	4	colorless crystals, crystallized from C_6H_6 [13] ^1H NMR (C_6H_6): 0.02 (s, CH_3In), 2.09 (CH_2P), 6.97, 7.45 (m, C_6H_5) [13] ^{31}P NMR (C_6H_6): −18.05 ($P(C_6H_5)_2$), −15.12 PC_6H_5, J(P,P′) = 32.2, J(P′,P′) = 28.2 [13]

* Further information:

$(CH_3)_3In \cdot NHCH_3(CH_2)_2NHCH_3 \cdot In(CH_3)_3$ (Table **8**, No. **2**) crystallizes in the monoclinic space group $P2_1/n-C_{2h}^5$ (No. 14) with the lattice constants a = 1152.7(3), b = 1348.7(4), c = 1194.8(3) pm, β = 111.74(2)°; D_c = 1.570 g/cm³ and Z = 4. The structure was refined to an R value of 0.0469, in which the methyl hydrogen atoms were assumed to be in idealized positions and were refined only with isotropic temperature factors. The positions of the remaining hydrogens were experimentally determined and isotropically refined. The structure of the molecule is indicated in **Fig. 5** [9].

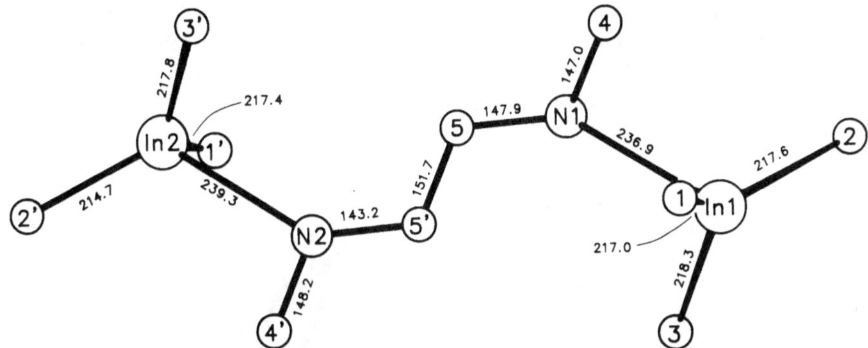

Fig. 5. Molecular structure of $(CH_3)_3In \cdot NHCH_3(CH_2)_2NHCH_3 \cdot In(CH_3)_3$ [9].

Selected bond angles (°)

C(1)-In(1)-N(1)	96.3(3)	C(1')-In(2)-N(2)	96.6(3)
C(2)-In(1)-N(1)	98.8(4)	C(2')-In(2)-N(2)	99.5(4)
C(3)-In(1)-N(1)	101.4(3)	C(3')-In(2)-N(2)	100.9(3)
C(2)-In(1)-C(1)	117.6(4)	C(2')-In(2)-C(1')	119.8(4)
C(3)-In(1)-C(2)	115.9(4)	C(3')-In(2)-C(2')	113.0(4)
C(3)-In(1)-C(1)	119.6(4)	C(3')-In(2)-C(1')	120.2(4)
C(4)-N(1)-In(1)	109.2(5)	C(4')-N(2)-In(2)	107.9(5)
C(5)-N(1)-In(1)	117.4(5)	C(5')-N(2)-In(2)	116.8(5)

$(CH_3)_3In \cdot N(CH_3)_2C_6H_4CH_2C_6H_4N(CH_3)_2 \cdot In(CH_3)_3$ (Table **8**, No. **6**) was prepared from the ether adduct in benzene solution and recrystallized from petroleum ether at −30 °C. Vapor pressures (ultrasonic method, diagram depicted) are: 4.5 Torr at 70 °C, 8 Torr at 80 °C, and 80 Torr at 150 °C [16].

It liberates $In(CH_3)_3$ on heating at 80 to 130 °C (4 h); 97% yield [16].

$(CH_3)_3In \cdot NC_4H_4N \cdot In(CH_3)_3$ (Table **8**, No. **9**) was prepared in low concentration for ESR measurements in THF solution. Reduction with Na metal produced In metal and the postulated radical complex I in very low yield [3].

$$\left[(CH_3)_2In{-}NC_4H_4N{-}In(CH_3)_2\right]^{+\cdot}\ [In(CH_3)_4]^- \qquad \left[(CH_3)_2In{-}NC_5H_4{=}C_5H_4N{-}In(CH_3)_2\right]^{+\cdot}$$

I II

$(CH_3)_3In \cdot NC_5H_4C_5H_4N \cdot In(CH_3)_3$ (Table **8**, No. **11**) was reduced with Na in THF solution with formation of the radical complex ion II [5].

$(CH_3)_3In \cdot P(C_6H_5)_2(CH_2)_2P(C_6H_5)_2 \cdot In(CH_3)_3$ (Table **8**, No. **14**) dissociates at about 100 °C. The vapor pressure of the adduct is given by log p (Torr) = 17.74 − 6895/T (K), the enthalpy of formation, ΔH°, is −31.6 ± 1 kcal/mol, and ΔS° = 68.1 ± 2 cal · mol⁻¹ · K⁻¹ [7]. The dissociation pressures have been determined by Knudsen mass loss effusion to give the thermodynamic parameters $\Delta H = 94.0 \pm 1.2$ kJ/mol and $\Delta S = 175 \pm 3$ J · mol^{-1} · K^{-1}. A plot log p versus $10^3/T$ is depicted [17].

The X-ray structure analysis shows that the "diphos" adducts of the trimethyl derivatives of aluminium, gallium, and indium are isostructural and crystallize in the triclinic space group P$\bar{1}$ – C_i^1 (No. 2). The indium derivative has the lattice constants a = 900.3(1), b = 957.6(1), c = 1133.2(1) pm, $\alpha = 69.586(9)°$, $\beta = 84.786(9)°$, $\gamma = 67.302(9)°$; Z = 1 and $D_c = 1.413$ g/cm^3. The structure of the compound is depicted in **Fig. 6** [13].

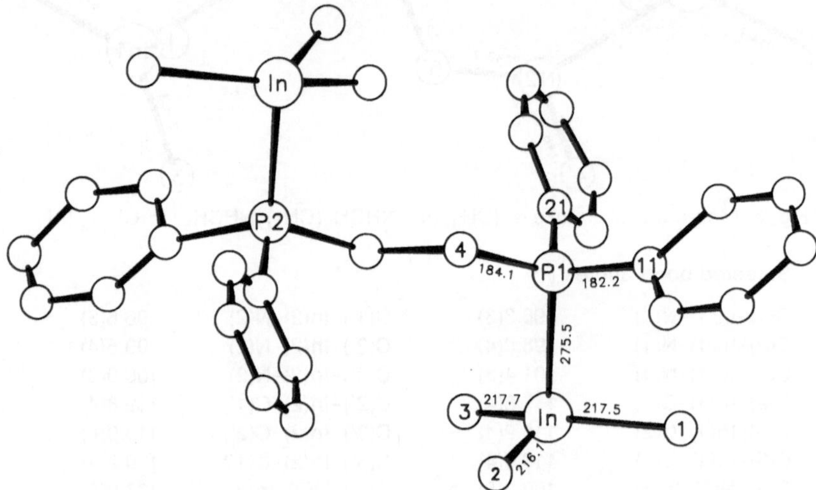

Fig. 6. Molecular structure of $(CH_3)_3In \cdot P(C_6H_5)_2(CH_2)_2P(C_6H_5)_2 \cdot In(CH_3)_3$ [13].

Selected bond angles (°)

C(1)–In–C(2)	118.8(4)	C(1)–In–P(1)	96.1(2)
C(1)–In–C(3)	116.5(4)	C(2)–In–P(1)	101.7(3)
C(2)–In–C(3)	116.7(4)	C(3)–In–P(1)	100.8(3)
C(4)–P(1)–In	116.6(2)	C(11)–P(1)–In	113.5(2)
C(11)–P(1)–C(4)	104.5(2)	C(21)–P(1)–In	112.3(2)

The low volatilities of the adduct and of the free base, as well as the relatively low temperature at which dissociation begins, appear to make the adduct a suitable choice for starting material for the production of pure base-free $In(CH_3)_3$. Splitting off of trimethylindium begins at 80 °C and 10^{-2} Torr, and gradually increasing the temperature to 120 to 130 °C gives $In(CH_3)_3$ in a total yield of about 74% [13] or 87% [6]. Because of these properties, the adduct has been used several times as source for $In(CH_3)_3$ in gas-phase epitaxy [7, 8, 10, 12].

References:

[1] Schmidbaur, H.; Wolfsberger, W.; Schwirten, K. (Chem. Ber. **102** [1969] 556/63).

[2] Storr, A.; Thomas, B. S. (Can. J. Chem. **48** [1970] 3667/2).

[3] Kaim, W. (Z. Naturforsch. **36b** [1981] 677/82).

[4] Krause, H.; Sille, K.; Hausen, H.-D.; Weidlein, J. (J. Organometal. Chem. **235** [1982] 253/64).

[5] Kaim, W. (J. Organometal. Chem. **241** [1983] 157/69).
[6] Bradley, D. C.; Chudzynska, H.; Faktor, M. M. (WO-85-4405 [1985]; C.A. **105** [1986] No. 172706).
[7] Bradley, D. C.; Faktor, M. M.; Scott, M.; White, E. A. D. (J. Cryst. Growth **75** [1986] 101/6).
[8] Moore, A. H.; Scott, M. D.; Davies, J. I.; Bradley, D. C.; Faktor, M. M.; Chudzynska, H. (J. Cryst. Growth **77** [1986] 19/22).
[9] Bradley, D. C.; Dawes, H.; Frigo, D. M.; Hursthouse, M. B.; Hussain, B. (J. Organometal. Chem. **325** [1987] 55/67).
[10] Briggs, A. T. R.; Butler, B. R. (J. Cryst. Growth **85** [1987] 535/42).
[11] Hodes, H. D.; Berry, A. D. (J. Organometal. Chem. **336** [1987] 299/305).
[12] Moss, R. H.; Bradley, D. C.; Faktor, M. M.; Frigo, D. M. (Eur. 246785 [1987] from C.A. **108** [1988] No. 47233).
[13] Bradley, D. C.; Chudzynska, H.; Faktor, M. M.; Frigo, D. M.; Hursthouse, M. B.; Hussain, B.; Smith, L. M. (Polyhedron **7** [1988] 1289/98).
[14] Bradley, D. C.; Faktor, M. M.; Frigo, D. M. (J. Cryst. Growth **89** [1988] 227/36).
[15] Foster, D. F.; Rushworth, S. A.; Cole-Hamilton, D. J.; Cafferty, R.; Harrison, J.; Parkes, P. (J. Chem. Soc. Dalton Trans. **1988** 7/11).
[16] Foster, D. F.; Rushworth, S. A.; Cole-Hamilton, D. J.; Jones, A. C.; Stagg, J. P. (Chemtronics **3** [1988] 38/43).
[17] Bradley, D. C.; Faktor, M. M.; Frigo, D. M.; Zheng, D. H. (Chemtronics **3** [1988] 53/5).

1.1.1.2.3 Adducts with Group 14 Donors

In this chapter only adducts with ylides are described; adducts with other group 14 bases, such as MR_2 compounds or the corresponding ylidic systems, $MR_2 \leftarrow ER_3$ (M = Si, Ge, Sn, Pb; E = N, P, As), have not yet been described.

$(CH_3)_3In \cdot CH_2=P(CH_3)_3$

This complex is obtained in yields of 78 to 92% from ether solutions of $In(CH_3)_3$ and $(CH_3)_3P=CH_2$ in the form of colorless, air-sensitive crystals. It melts without decomposition at 69 to 70 °C, and is a monomer in benzene [1].

^1H NMR spectrum (in C_6H_6 at 30 °C in ppm): −0.37 (s, CH_3In), −0.30 (d, CH_2, J(P,H) = 13.8 Hz), 0.38 (d, CH_3P, J(P,H) = 13.2 Hz). ^{13}C NMR spectrum (in $C_6D_5CD_3$ in ppm): −8.1 (s, CH_3In), −0.8 (d, CH_2, J(P,C) = 36.6 Hz), 12.5 (d, CH_3P, J(P,C) = 56.2 Hz) [1]. In the ^{31}P NMR spectrum (1.5 molar solution in $C_6D_5CD_3$) a decuplet was observed at −22.6 ppm. This shift was cited as evidence for fourfold coordination at the phosphorus atom [1].

The IR (C_6H_6 solution) and Raman spectra (melt) showed the following values for the significant PC_3C' and InC_3C' bond vibrations:

IR (solution)	Raman (liquid)	assignment
769 mw	769 vw, dp	$\nu_{as}(PC_3)$
755 mw	753 mw, dp	ν_{as}(P-C-In)
670 vvs	675 sh, dp	$\varrho(CH_3In)$
656 sh	657 s, p	$\nu_s(PC_3)$
455 vs	464 sh, dp	$\nu_{as}(InC_3)$
–	458 vs, p	$\nu_s(InC_3)$
437 s	446 sh	ν_s(In-C-P)

The small frequency difference between ν_{as} and $\nu_s(InC_3)$ permits an estimate of 113° to 116° for the C–In–C valence angle [3].

$(CH_3)_3In \cdot CH_2{=}P(CH_3)_2{-}N{=}P(CH_3)_3$

The adduct was prepared, analogously to the homologous gallium compound [2], by mixing $(CH_3)_3In \cdot O(C_2H_5)_2$ with the double ylide (mole ratio 1:1) at 0 °C, then slowly warming to room temperature. Pyrolysis in benzene or toluene solution occurred by splitting off methane to form the heterocyclic complex I (see p. 367); further physical or chemical properties were not reported [4].

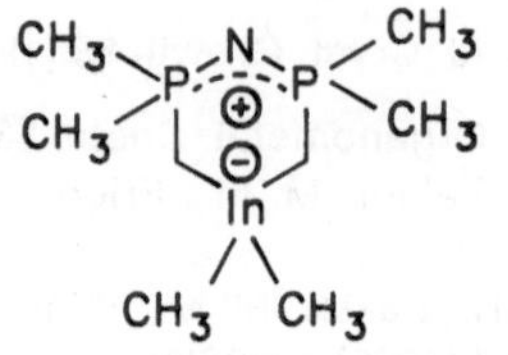

I

References:

[1] Schmidbaur, H.; Füller, H.-J.; Köhler, F. H. (J. Organometal. Chem. **99** [1975] 353/7).
[2] Schmidbaur, H.; Füller, H.-J. (Chem. Ber. **110** [1977] 3528/35).
[3] Sille, K.; Fries, W.; Weidlein, J. (unpublished results).
[4] Schmidbaur, H.; Füller, H.-J. (Ger. Offen. 2701143 [1978] from C.A. **90** [1979] No. 23243).

1.1.1.2.4 Adducts with Transition Metal Compounds

Attempts to prepare **$(CH_3)_3In \cdot (C_5H_5)Mo(CO)_3H$** and **$(CH_3)_3In \cdot (C_5H_5)W(CO)_3H$** have been undertaken. The components were condensed in stoichiometric ratios at −196 °C, then slowly warmed to room temperature. In contrast to the behavior of $Ga(CH_3)_3$ [1], no definite proof could be found for either adduct formation or intermediates of the composition $(C_5H_5M(CO)_3)_{3-n}In(CH_3)_n$ (M = Mo, W; n = 1 or 2) by splitting off methane. The final product in all cases was $In\{M(CO)_3C_5H_5\}_3$ in nearly quantitative yield [2].

For reactions with transition metal compounds, in the course of which adducts could be formed as intermediates, see Section 1.1.1.1.4.

References:

[1] Brunner, H.; Wailes, P. C.; Kaesz, H. D. (Inorg. Nucl. Chem. Letters **1** [1965] 125/9).
[2] Hsieh, A. T. T.; Mays, M. J. (J. Organometal. Chem. **37** [1972] 9/14).

1.1.2 Triethylindium and Its Adducts

1.1.2.1 $In(C_2H_5)_3$

Preparation. Generally, triethylindium can be prepared by the reactions already described for $In(CH_3)_3$ (pp. 1 to 2).

$In(C_2H_5)_3$ was obtained from $InCl_3$ and C_2H_5MgCl in diethyl ether in yields of 70%. Reduced-pressure fractional distillation of the Grignard reaction mixture gave ether-free triethylindium [1].

Direct alkylation of metallic indium by C_2H_5Br in the presence of Mg and ether as solvent formed $In(C_2H_5)_3$ in an exothermic reaction. When a mixture of In and Mg (1:5 mole ratio) was used, the yield was 72% after two distillations; when a previously prepared 1:3.5 In/Mg alloy was used, the yield was raised to 95% [15, 18, 73, 77]; see also [94].

The alkylation of $InCl_3$ with excess ethyllithium in ether produced $In(C_2H_5)_3$ in preparatively significant quantities [83], but the reaction has seldom been used (author's remark: presumably, because of the relatively low stability of LiC_2H_5 and the difficulty of preparing this alkylating agent).

More suitable, and most mentioned as an alkylating agent, is $Al(C_2H_5)_3$. Dropwise addition of $Al(C_2H_5)_3$ to a slurry of $InCl_3$ in pentane, and, after the strongly exothermic initial reaction has subsided, treatment with thoroughly dried, finely divided KCl (1:3:5 mole ratio) [12, 59], gave $In(C_2H_5)_3$ in 67% yield [12].

Studies to optimize the exchange reaction between $InBr_3$ and $Al(C_2H_5)_3$ (with addition of KCl) are described in [65].

In the electrolytic preparation $Al(C_2H_5)_3$ also served as starting material; the addition of NaF in 2:1 mole ratio gave the low-melting complex salt $Na[(Al(C_2H_5)_3)_2F]$ whose electrolytic decomposition (about 2 to 3 V and <1 A/cm^2) at an In anode and a Cu cathode formed $In(C_2H_5)_3$ in high yields [4, 9]. In [61] an electrolytic cell is described, the course of the reaction is discussed, and a collection of similar metalloalkyl preparations is presented. Individual differences in the electrochemical syntheses of gallium-, indium-, and thallium-trialkyls are compared in [52].

Electrolysis of a mixture of $Mg(C_2H_5)_2$ and $[N(C_2H_5)_4]ClO_4$ in THF (100 V, 50 mA) formed the unstable adduct $(C_2H_5)_3In \cdot OC_4H_8$, whose vacuum distillation produced the free trialkyl compound [82].

The thermal decomposition of $(C_2H_5)_2InOC_4H_9$-t at about 180 °C generated not only $In(C_2H_5)_3$, but also butane, butene, ethane, ethene, and $C_2H_5In(OC_4H_9\text{-t})_2$ which were identified by mass spectrometry [31].

Purification of $In(C_2H_5)_3$ was usually accomplished by fractional distillation of the reaction residue [1, 12, 18]. To remove any residual $(C_2H_5)_2AlCl$ from the alkylations with $Al(C_2H_5)_3$, some KCl [12] or NaF [59] was added to the last distillation. For the addition of aluminiumalkyls to remove ether, see [95]. The production of $In(C_2H_5)_3$ of high purity by decomposition of $(C_2H_5)_3In \cdot N(CH_3)_2C_6H_4CH_2C_6H_4N(CH_3)_2 \cdot In(C_2H_5)_3$ (Section 1.1.2.2) was described [100].

Treating a sample of $In(C_2H_5)_3$ contaminated with 10 ppm Si in paraffin oil at 80 °C in the presence of Na/K alloy reduced the Si content to 0.2 ppm within 2 h [89].

A combination of gas chromatography and atomic absorption spectrometry (AAS) can be used for analysis and for the determination of organometallic impurities [78]. Zinc, as an impurity in $In(C_2H_5)_3$, can be determined down to 5×10^{-3} wt% by emission spectroscopy, after conversion into the oxide [62]; $O(C_2H_5)_2$ and C_2H_5Br, as residual impurities, can be detected by gas chromatography to a limit of 5 to 8×10^{-6} wt% (silica column, He or Ar carrier, 50 to 150 °C injection temperature [57, 71]). The same research group has reported gas/liquid distribution coefficients for likely trace impurities in $In(C_2H_5)_3$ [68].

Physical Properties and Spectra. Freshly distilled $In(C_2H_5)_3$ at room temperature is a clear, colorless, and strongly refracting liquid (author's remark: after a short storage period, metallic indium appears as a slightly glistening gray-green colloid, whose quantity increases with time). The density (pycnometer) is 1.260 g/cm^3, and the refractive index is 1.538, both at 20 °C [1]. The temperature dependence of the density follows the equation $D = (1.4288 \pm 0.0003) - (1.47 \pm 0.07) \times 10^{-3}\,t$ [58]; the density of 1.3994 g/cm^3 at 20 °C was calculated by the author. For calculating the temperature dependence of the dynamic viscosity, η (in cP), the equation, $\log \eta = (2.30 \pm 0.03) + (636 \pm 22)/T$, can be used [58].

Both −32 °C [1, 85] and 237.6 K (−35.55 °C) were reported for the melting point. The enthalpy of melting, ΔH_m, is 3110±11 [49, 87] or 2021 cal/mol [68], the corresponding entropy, ΔS_m, is calculated to be 13.35 cal · mol^{-1} · K^{-1} [87]. The boiling point of 144 °C [1] or 145.2 °C [63] at normal pressure was confirmed by gas chromatographic methods [67] as 417 K (144 °C); however, other investigators report 184 °C [14, 85] and 194.9 °C [69] as extrapolated values. Other published boiling points, obtained by distillation at reduced pressure, are (in °C/Torr): 118/38 [1], 95 to 105/17 [12], 92/18 [1], 82 to 84/12 [18], 64/4 [83], and 51 to 53/3 [12].

The heat capacity of solid (60 to 230 K) and liquid (240 to 298.15 K) $In(C_2H_5)_3$ was determined [49], and an anomalous change between 160 and 200 K (C_p between 37.95 and 44.85 cal · mol^{-1} · K^{-1}) was detected. A comparison of the thermodynamic constants of a large number of organometallic compounds (including various trialkyls of Al, Ga, and In) is presented in [87].

Constants for the vapor pressure equation, log p = A − B/T, and a few values for the vaporization enthalpy, ΔH_V (in kcal/mol), and entropy, ΔS_V (in cal · mol^{-1} · K^{-1}), are as follows:

A	8.98	8.468	7.86±0.08	9.06
B	2790	2341.9	2340±30	2815
ΔH_V	12.8	10.7	10.7±0.1	–
ΔS_V	27.9	–	22.9	–
Ref.	[14]	[63]	[69]	[92]

In [92] the range from 257 to 293 K (−16 to 20 °C) was investigated, and a more extended range (−35 to 130 °C) is described in [69]. The vapor pressure diagram for the range −10 to 110 °C is presented. It also shows the vapor pressure curves of $Ga(CH_3)_3$, $Ga(C_2H_5)_3$, $In(CH_3)_3$, as well as those of their adducts with $N(CH_3)_3$, $P(CH_3)_3$, and $As(CH_3)_3$ [84].

Individual vapor pressure values have been published:

t in °C	18	20	30	34	44	53	83
p in Torr	0.24*)	0.42	0.8	1.6*)	1.2	3.0	12.0
Ref.	[92]	[84]	[86]	[88]	[85]	[85, 88]	[85, 88]

*) Given in hPa.

The ^1H NMR spectrum of $In(C_2H_5)_3$ in C_6H_6 shows signals at δ = 0.40, 0.44 (CH_2), as well as at δ = 1.24, 1.27 (CH_3) ppm [51, 86]; in C_6D_6 at δ = 0.53 and 1.42 ppm [91]; and in CH_2Cl_2 at δ = 0.68 and 1.32 ppm [19, 28]. The complex A_3B_2 system (by evaluation of the simulated spectrum) has a coupling constant, J ≈ 7.9 Hz [75].

Four frequencies are observed in the ^{115}In NQR spectrum of $In(C_2H_5)_3$ at both 77 and 296 K. The results indicate a planar InC_3 skeleton with the p_z orbital perpendicular to this plane having an occupation density near zero [32, 43, 64].

T in K	e^2Qq/h (MHz)	η (%)	transition frequencies (MHz ± 0.01)			
77	1135.2	10.7	51.44	93.06	141.40	188.50
296	1159.6	9.6	52.12	95.53	143.16	193.16

The IR spectrum of $In(C_2H_5)_3$ is pictured in [17]. Only the In–C stretching vibration at about 470 cm^{-1} was emphasized, and was used to estimate the associated force constant, $f_{In-C}=1.6$ N/cm. In [54] the Raman frequencies 494 ν_s (InC_3) and 1165 cm^{-1} (CH_2 wagging) are reported. The complete IR and Raman spectra are collected in Table 9 [75]; the assignment of C_{3h} symmetry resulted from a comparison with the spectrum of $Ga(C_2H_5)_3$ [21, 54].

Table 9
Vibrational Spectra and Assignments for $In(C_2H_5)_3$ in the Liquid State (IR) and in Benzene Solution (Raman, 1:2 mole ratio).
Wavenumbers in cm^{-1}.

IR [75]	Raman [75]	assignment [21, 54, 75]
2935 s	2932 mw, dp	ν_{as}(CH) of CH_3 and CH_2 (A'', E')
2890 s	2895 m, br, p	ν_s(CH) of CH_3 (A')
2850 s	2863 m, p	ν_s(CH) of CH_2 (A', E')
2805 m	2812 w, p	2δ(CH) of CH_2
2725 m	2728 w, p	2δ(CH) of CH_3
1460 s	1462 vw, dp	$\delta_{as}(CH_3)$ (A'+A'')
1418 m	1420 vw, dp	$\delta_s(CH_2)$ (A'+E')
1372 s	1379 vw, dp	$\delta_s(CH_3)$ (E')
1228 m	–	CH_2 wagging (A'+E')
1159 m	1166 ms, p	
998 s	1001 w, dp	ν(C–C) (E')
952 s	958 w, br, p	ν(C–C) (A')
922 m	918 vvw, dp	$\varrho(CH_3)$
645 sh	–	$\varrho(CH_2)$
618 s, br	590 vw, br	
465 s	473 m, dp	$\nu_{as}(InC_3)$ (E')
–	448 vvs, p	$\nu_s(InC_3)$ (A')
–	255 w, sh	δ(InCC) (A'+E')
–	237 mw, p	
–	202 mw	$\delta(InC_2)$ (E')

The UV spectrum of $In(C_2H_5)_3$ gas (6 Torr, 70 °C) contains a broad absorption band at $\lambda_{max}=230$ nm with $\varepsilon_{max}=3000$ $L\cdot cm^{-1}\cdot mol^{-1}$. The spectrum, together with the spectra of $Al(C_2H_5)_3$ and $Ga(C_2H_5)_3$, is illustrated in [81].

$In(C_2H_5)_3$ possesses a dipole moment of 0.6 D in heptane at 20 °C, 1.38 D in benzene, and 1.33 D in dioxane [3, 6]. In [6] the electronegativity differences between various metals (Mg, Zn, Cd, Hg, Al, Ga, In) and organic groups (CH_3, C_2H_5, C_3H_7, C_6H_5) are compared with the dipole moments of their organometallic compounds [6]. Cryoscopic molecular weight determinations in cyclohexane, benzene, and dioxane reveal the trialkyl compound to be monomeric in dilute (0.010 to 0.030 M) solutions [1, 8].

Mass Spectrum and Decomposition. The mass spectrum of $In(C_2H_5)_3$ (0.129 Torr in 152 Torr H_2, ionization energy 70 eV) is illustrated in [90], and has the following relative peak heights (in %) of the main fragments: $[In(C_2H_5)_2]^+$ (100), $[In(CH_3)C_2H_5]^+$ (95), $[^{115}In]^+$ (50), $[InC_2H_5]^+$ (45), $[In(CH_3)_2]^+$ (28), $[InCH_3]^+$ (13), and $[In(C_2H_5)_3]^+$ (4). A more detailed analysis of the mass spectrum is presented in [76]:

fragment ion	rel. intensities in %	
	20 eV	50 eV
$[M]^+$	0.3	0.2
$[In(C_2H_5)_2H_2]^+$	0.28	0.02
$[In(C_2H_5)_2H]^+$	2.7	1.65
$[M-C_2H_5]^+$	53.0	39.9
$[In(C_2H_5)H]^+$	1.40	0.9
$[InC_2H_5]^+$	6.9	8.5
$[InH_2]^+$	–	0.1
In^+	35.0	48.5

Thermal decomposition of triethylindium gas begins between 180 and 190 °C [24, 74]. A series of investigations at 200 °C shows 2.9% decomposition after 3 min and 65% after 60 min. The main products of the free radical decomposition were found to be metallic In, ethane, and ethylene with small amounts of butane and butene. C_6H_{14} and C_6H_{12} were detected only after heating for several hours [24]; higher temperatures also favor the formation of larger alkanes and alkenes [70]:

decomposition products	250 °C	350 °C in %	500 °C
CH_4	traces	1	1
C_2H_6	59	20	17
C_2H_4	41	9	14
C_3H_8	–	traces	1
C_3H_6	–	traces	2
C_4H_{10}	traces	70	65
C_4H_8	–	traces	traces

The thermogram shows an endo and an exo effect; the first at 142 °C corresponds to the boiling point; the latter at 300 °C, to the onset of decomposition [70].

Systematic study of the decomposition rate in the range 180 to 250 °C led to the equation $\log k=(12.3\pm0.5)-(7500\pm250)/T(K)$ for the rate constant, k (s^{-1}), and to $E=34.3\pm1.1$ kcal/mol for the average activation energy. The following mechanism has been suggested for the course of the radical decomposition [74]:

$$In(C_2H_5)_3 \longrightarrow C_2H_5^{\bullet} + In(C_2H_5)_2^{\bullet}$$

$$2\,C_2H_5^{\bullet} \longrightarrow C_2H_5C_2H_5$$

$$2\,C_2H_5^{\bullet} \longrightarrow C_2H_6 + C_2H_4$$

$$2\,In(C_2H_5)_2^{\bullet} \longrightarrow (In(C_2H_5)_2)_2$$

$$(In(C_2H_5)_2)_2 \longrightarrow C_2H_5^{\bullet} + (C_2H_5)_2InInC_2H_5^{\bullet}$$

Liquid $In(C_2H_5)_3$ decomposes markedly at 144 °C [74], the beginning of the decomposition is reported to be as low as 35 to 40 °C [86]. If pure $In(C_2H_5)_3$ is heated for 7 h at 195 °C under reflux in an inert gas atmosphere, metallic indium (90 to 95% yield) and small amounts of a viscous dark yellow liquid remain. The mass spectrum of the liquid shows an intense peak at m/z = 404 which was interpreted as arising from a dimeric form of triethylindium [74].

$In(C_2H_5)_3$ is sensitive to the influence of light. At room temperature in daylight the decomposition is slow [18]; at 200 °C additional photolytic treatment facilitates the formation of higher alkanes (primarily butane) and simultaneously increases the decomposition rate [24]. Decomposition of $In(C_2H_5)_3$ with separation of metallic indium occurred in a Raman excitation light (argon/krypton gas laser; 488.0 and 647.1 nm at 10 to 40 mW power) within a few minutes. Benzene solutions up to about 30 vol%, however, were sufficiently stable for Raman spectroscopy [75].

Chemical Reactions. Reactions of $In(C_2H_5)_3$ with inorganic, organic, and organometallic compounds are compiled in Table 10. Most proceed with the splitting of at least one In–C bond, and the formation of $(C_2H_5)_2InX$. These compounds are discussed in other sections, and Table 10 cites the appropriate page numbers. Adducts of triethylindium are not mentioned but are treated in Section 1.1.2.2 on p. 70.

Table 10
Reactions of $In(C_2H_5)_3$.
Further information on numbers preceded by an asterisk is given at the end of the table.
C_2H_5 groups on the In atom are abbreviated as R.

No.	reactant	products and remarks	Ref.	page
*1	O_2 (dry air)	$R_2InOOC_2H_5$, $R_2InOC_2H_5$, $(RInO)_n$	[19]	210
*2	t-$C_4H_9OOC_4H_9$-t	$R_2InOC_4H_9$-t; on irradiation high yields of $C_2H_5^{\bullet}$ radicals	[29]	186
3	H_2O	R_2InOH at 15 °C, $In(OH)_3$ at 90 °C	[1]	180
4	SO_2	$\{In(O_2SC_2H_5)_3\}_n$ O-coordinated sulfinate complex	[35]	–
5	SO_3	$(R_2InO_2S(O)C_2H_5)_n$ O-coordinated sulfonate complex	[45]	211
6	ClN_3 in N_2	$(R_2InN_3)_2$	[27]	176
7	$(SCN)_2$	$(R_2InSCN)_3$	[23]	176
8	CO_2	no reaction at room temperature; in boiling xylene insertion into one In–C bond, giving $(R_2InO_2CC_2H_5)_2$ in <5% yield	[34]	196
9	COS	no reaction at room temperature; in boiling xylene dark yellow to brown viscous products are found; no pure compound isolated	[41]	–
10	CS_2	no reaction at room temperature; at 60 to 80 °C in C_6H_6 insertion into one In–C bond, giving dark brown products; no pure compound isolated	[42]	–
11	$InCl_3$	exothermic redistribution reaction giving R_2InCl	[53]	118

Table 10 (continued)

No.	reactant	products and remarks	Ref.	page
12	MH M = Na, K	$M[InR_3H]$ in THF	[72]	347
13	MX M = Rb, Cs; X = F M = Cs; X = Cl	mole ratios 1:1 and 2:1; low-melting solid and liquid complexes, respectively, are found	[20]	–
14	$H_2C{=}CH_2$	no reaction, even after heating more than 60 h at temperatures up to 195 °C and pressures of 120 bar	[13]	–
*15	CCl_4, $CHCl_3(Br_3)$	$R_2InCl(Br)$	[26, 37, 38]	118
16	R′OH $R' = CH_3$, C_2H_5, C_4H_9-t	$(R_2InOR')_{2.5-3}$	[31, 55]	182
17	C_6H_5CHO	$(R_2InOCH(C_2H_5)C_6H_5)_2$ (in n-hexane, 69 °C, 3 h)	[28]	182
18	$(C_6H_5)_2C{=}O$	$(R_2InOCH(C_6H_5)_2)_3$ and C_2H_4 in high yield (in n-heptane, 98 °C, 28 h)	[28]	182
19	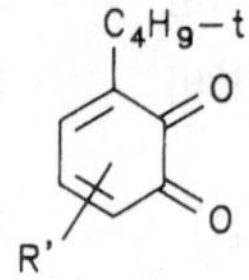 [structure: 3-C_4H_9-t-substituted o-benzoquinone with R′] R′ = 5- or 6-C_4H_9-t	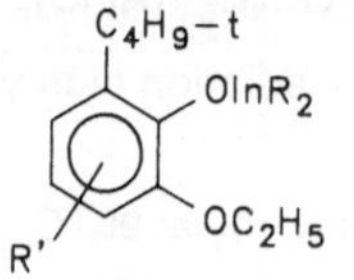 [structure: 3-C_4H_9-t-2-$OInR_2$-1-OC_2H_5 benzene with R′] formation via unstable paramagnetic intermediates	[66]	192
20	HOOCCOOH	$(R_2In)_2O_4C_2$	[46]	206
21	$CH_2(COOCH_3)_2$	$R_2InCH(COOCH_3)_2$ (in benzene at room temperature)	[36]	192
22	CH_3SO_2OH	$(R_2InOS(O)_2CH_3)_n$	[56]	216
23	R′COOH $R' = CH_3$, C_2H_5	$(R_2InO_2CR')_2$	[34, 48]	196
24	R′COSH $R' = CH_3$, C_6H_5	$R_2InOSCR'$	[39, 41, 47]	204
25	$(C_6H_5)_2C{=}S$	$(R_2InSCH(C_6H_5)_2)_2$ and C_2H_4 are formed in n-hexane within 2 to 3 min at 15 °C	[25, 28]	241
26	$CH_3C(S)SH$	$R_2InS_2CCH_3$	[40]	241
27	$(CH_3)_2NC_2H_4OH$	$(R_2InOC_2H_4N(CH_3)_2)_2$	[44]	192
28	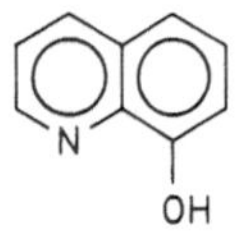 [structure: 8-hydroxyquinoline, labels N, OH]	$(R_2InOC_9H_6N)_2$	[44]	193

Table 10 (continued)

No.	reactant	products and remarks	Ref.	page
29	$CH_3CONHCH_3$	$R_2InOC(CH_3)NCH_3$	[50]	267
30	C_6H_5NCO	$(R_2InN(C_6H_5)COC_2H_5)_2$ (in petroleum ether, 48 °C, 6 h)	[28, 33]	267
31	$CH_3HNCOCONHCH_3$	$(R_2In)_2O_2C_2(NCH_3)_2$	[60]	290
32	$HCR'_2(NO_2)$ $R' = H, CH_3$	$R_2InOONCR'_2$	[43]	214
33	$HN(C_2H_5)_2$	$(R_2InN(C_2H_5)_2)_2$	[31]	255
34	$HN(CH_2)_5$ (piperidine)	$(R_2InN(CH_2)_5)_2$ formed at 120 °C as a viscous liquid	[51]	273
35	$HP(R')_2$ $R' = C_2H_5, C_4H_9\text{-t}$	$(R_2InPR'_2)_3$ ($R' = C_2H_5$ [83]) $(R_2InPR'_2)_2$ ($R' = C_4H_9\text{-t}$ [93])	[83, 93]	313
36	$HOP(O)(C_6H_5)_2$	$(R_2InOP(O)(C_6H_5)_2$	[97]	217
*37	$HSeSi(C_2H_5)_3$	1:1 mole ratio gives $(R_2InSeSi(C_2H_5)_3)_2$	[40]	252
38	$(C_2H_5)_3SnF$	$(R_2InF)_n$ and $Sn(C_2H_5)_4$	[37]	117
39	$(R'_2SnS)_3$ $R' = CH_3, C_6H_5$	$(RInS)_n$ and $Sn(CH_3)_2(C_2H_5)_2$ in the case of $R' = CH_3$	[26, 30]	245
40	GaR'_3 $R' = CH_3, C_2H_5$	alkyl exchange	[98]	–
41	$In(CH_3)_3$	1:2 mole ratio: $In(CH_3)_2R$, 2:1 mole ratio: $In(CH_3)R_2$	[86]	112
*42	$Sb(C_2H_5)_3 + C_3H_7Br$	$[(C_2H_5)_3SbC_3H_7][(C_2H_5)_3InBr]$	[22]	353
43	$IrH(P(CH_3)_3)_4$	$H(C_2H_5)(P(CH_3)_3)Ir\text{-}In(C_2H_5)_2$	[96]	332
44	$(C_5H_5Mo(CO)_3)_2$	$C_5H_5(CO)_3MoInR_2$, $(C_5H_5(CO)_3Mo)_2InR$	[99]	331

* Further information:

Reaction No. 1. $In(C_2H_5)_3$ is pyrophoric and catches fire when exposed to air, even at −20 °C [1]. Upon slow action of atmospheric oxygen the peroxide, $(C_2H_5)_2InOOC_2H_5$, is formed by insertion of O_2 into the In-C bond (maximum yield, 8%). The peroxide then either rearranges intramolecularly to $C_2H_5In(OC_2H_5)_2$, or attacks unreacted $In(C_2H_5)_3$ to form two equivalents of $(C_2H_5)_2InOC_2H_5$. A series of tests on the gas phase at 20 to 82 °C led to the postulation of a free radical reaction, influenced by the container walls. Propagation and termination reactions of this chain led to $(C_2H_5InO)_n$, $(C_2H_5)_2InOH$, C_2H_4, and CH_3CHO as end products which were identified by elemental analysis and gas chromatography [19].

Reaction No. 2. With $t\text{-}C_4H_9OOC_4H_9\text{-}t$, $In(C_2H_5)_3$ first forms a 1:1 adduct, which could not be isolated and characterized because of spontaneous decomposition. This adduct reacts in the gas phase and at temperatures between 90 and 100 °C to form $(C_2H_5)_2InOC_4H_9\text{-}t$ and $C_2H_5OC_4H_9\text{-}t$; in cyclohexane (mole ratio 1:20) below 90 °C, a free radical reaction sequence involving $(C_2H_5)_3In \cdot (t\text{-}C_4H_9OOC_4H_9\text{-}t) \rightarrow (C_2H_5)_2InOC_4H_9\text{-}t + C_2H_5^{\bullet} + t\text{-}C_4H_9CO^{\bullet}$ was proposed [29].

Reaction No. 15. Along with $(C_2H_5)_2InCl(Br)$ this reaction forms the by-products C_2H_6 (or C_2H_5Cl, Br) and 2-pentene, for whose formations the following mechanism is postulated (the *italicized* intermediates were neither isolated nor otherwise established; X = H, Cl) [37]:

$CCl_3X + In(C_2H_5)_3 \rightarrow \mathit{(C_2H_5)_2InCCl_3} + C_2H_5X$

$\mathit{(C_2H_5)_2InCCl_3} + In(C_2H_5)_3 \rightarrow \mathit{(C_2H_5)_2InCCl_2C_2H_5} + (C_2H_5)_2InCl$

$\mathit{(C_2H_5)_2InCCl_2C_2H_5} + In(C_2H_5)_3 \rightarrow \mathit{(C_2H_5)_2InCCl(C_2H_5)_2} + (C_2H_5)_2InCl$

$\mathit{(C_2H_5)_2InCCl(C_2H_5)_2} \rightarrow C_2H_5CH{=}CHCH_3 + (C_2H_5)_2InCl$

Reaction No. 37. The reaction of $In(C_2H_5)_3$ and $HSeSi(C_2H_5)_3$ (1:2 mole ratio in hexane) forms primarily $C_2H_5In(SeSi(C_2H_5)_3)_2$, which is not isolable and disproportionates at 65 °C to $Se(Si(C_2H_5)_3)_2$ and polymeric $(C_2H_5InSe)_n$. The same reactants in 1:3 mole ratio (in hexane between 0 and about 65 °C) gives polymeric $(SeInSeSi(C_2H_5)_3)_n$, along with bis(triethylsilyl)selenide and ethane [40].

Reaction No. 42. The triethyl derivatives of In and Sb do not react with each other; when a 1:1 mixture of the components at 90 to 110 °C was treated with $n\text{-}C_3H_7Br$, oxidative addition forms low-melting (ca. −67 °C) $[(C_2H_5)_3Sb(C_3H_7)]^+[(C_2H_5)_3InBr]^-$; on addition of a further equivalent of $In(C_2H_5)_3$ this salt formed $[(C_2H_5)_3Sb(C_3H_7)]^+$ $[(C_2H_5)_3InBrIn(C_2H_5)_3]^-$. Since no mention is made of any conductivity measurements, the postulated salt formation is not at all established [22].

Applications. Compared to the homologous triethyl compounds of Al and Ga, $In(C_2H_5)_3$ is less suitable as a catalyst for diene polymerization because it decomposes extensively at the necessary temperatures [13, 16]. Despite this, polymerizations of ethylene, butylene, and vinyl chloride [5, 11] have been carried out successfully using $In(C_2H_5)_3$ [2, 7, 13] with $TiCl_4$, $CuCl_2$, or $AgNO_3$ [10] as co-catalyst.

Melts of salts made from $In(C_2H_5)_3$ and alkali metal or tetramethylammonium halogenides (mole ratios of 1:1 and 2:1, halogen = F, Cl) have proven best for electrorefining indium at about 100 °C; by this method foreign element content can be kept below 10 ppm [20].

As a source material for epitactic preparation of single crystal III/V (13/15) semiconductor layers, $In(C_2H_5)_3$ has much the same importance as $In(CH_3)_3$ [79]. A collection of citations on this theme, arranged according to the year of publication is presented at the end of this section; however, the list is not complete. Further information on gas phase epitaxy with indium trialkyls can be gleaned from the corresponding $In(CH_3)_3$ section on p. 22.

References:

[1] Runge, F.; Zimmermann, W.; Pfeiffer, H.; Pfeiffer, I. (Z. Anorg. Allgem. Chem. **267** [1951] 39/48).

[2] Ziegler, K. (Brit. 713081 [1954] from C.A. **1955** 3576).

[3] Strohmeier, W.; Hümpfner, K. (Z. Electrochem. **61** [1957] 1010/3).

[4] Roetheli, B. E.; Simpson, I. B. (Brit. 797093 [1958] from C.A. **1959** 930).

[5] Sovic Soc. Anon. (Belg. 566531 [1958] from C.A. **1959** 11892).

[6] Strohmeier, W.; Nützel, K. (Z. Elektrochem. **62** [1958] 188/91).

[7] Ziegler, K. (Brit. 775384 [1957] from C.A. **1958** 12893).

[8] Strohmeier, W.; Hümpfner, K.; Miltenberger, K.; Seifert, E. (Z. Elektrochem. **63** [1959] 537/9).

[9] Ziegler, K. (Brit. 814609 [1959] from C.A. **1959** 17733).

[10] Solvic-Industria Delle Materie Plastiche S.p.A. (Ital. 696800 [1960] from C.A. **57** [1962] 6146).

[11] Solvay and Cie. (Belg. 587762 [1960] from C.A. **1961** 4052).
[12] Eisch, J. J. (J. Am. Chem. Soc. **84** [1962] 3605/10).
[13] Eisch, J. J. (J. Am. Chem. Soc. **84** [1962] 3830/6).
[14] Hartmann, H.; Lutsche, H. (Naturwissenschaften **49** [1962] 182/3).
[15] Todt, E.; Hauschildt, H. (Ger. 1136702 [1962]; C.A. **58** [1963] 4597).
[16] Huff, T.; Perry, E. (J. Polym. Sci. A **1** [1963] 1553/72).
[17] Oswald, F. (Fresenius Z. Anal. Chem. **197** [1963] 309/22).
[18] Todt, E.; Dötzer, R. (Z. Anorg. Allgem. Chem. **321** [1963] 120/3).
[19] Cullis, C. F.; Fish, A.; Pollard, R. T. (Trans. Faraday Soc. **60** [1964] 2224/33).
[20] Dötzer, R. (Chem. Ing. Tech. **36** [1964] 616/37).

[21] Chouteau, J.; Davidovics, G.; d'Amato, F.; Svaidan, K. (Compt. Rend. **260** [1965] 2759/62).
[22] Dötzer, R. (Ger. Offen. 1200817 [1965]; C.A. **63** [1965] 15896).
[23] Dehnicke, K. (Angew. Chem. **79** [1967] 942/3; Angew. Chem. Intern. Ed. Engl. **6** [1967] 947).
[24] Razuvaev, G. A.; Petukhov, G. G.; Shcherbakova, V. I.; Druzhkov, O. N.; Zhil'tsov, S. F. (Zh. Obshch. Khim. **37** [1967] 1516/20; J. Gen. Chem. [USSR] **37** [1967] 1437/41).
[25] Tada, H.; Yasuda, K.; Okawara, R. (Inorg. Nucl. Chem. Letters **3** [1967] 315/7).
[26] Yasuda, K.; Okawara, R. (Inorg. Nucl. Chem. Letters **3** [1967] 135/6).
[27] Müller, J.; Dehnicke, K. (J. Organometal. Chem. **12** [1968] 37/47).
[28] Tada, H.; Yasuda, K.; Okawara, R. (J. Organometal. Chem. **16** [1969] 215/20).
[29] Zhil'tsov, S. F.; Shcherbakov, B. I.; Druzhkov, O. N. (Zh. Obshch. Khim. **39** [1969] 1327/31; J. Gen. Chem. [USSR] **39** [1969] 1297/300).
[30] Okawara, R.; Yasuda, K. (Japan. 70-32692 [1970] from C.A. **74** [1971] No. 112621).

[31] Shcherbakov, V. I.; Zhil'tsov, S. F.; Druzhkov, O. N. (Zh. Obshch. Khim. **40** [1970] 1542/5; J. Gen. Chem. [USSR] **40** [1970] 1529/31).
[32] Svergun, V. I.; Bednova, L. M.; Okhlovysmin, O. Yu.; Semin, G. K. (Izv. Akad. Nauk SSSR Ser. Khim. **1970** 1449; Bull. Acad. Sci. USSR Div. Chem. Sci. **1970** 1377).
[33] Tada, H.; Okawara, R. (J. Org. Chem. **35** [1970] 1666/7).
[34] Weidlein, J. (Z. Anorg. Allgem. Chem. **378** [1970] 245/62).
[35] Hsieh, A. T. T. (J. Organometal. Chem. **27** [1971] 293/301).
[36] Kawakami, Y.; Tsuruta, T. (Bull. Chem. Soc. Japan. **44** [1971] 247/57).
[37] Maeda, T.; Tada, H.; Yasuda, K.; Okawara, R. (J. Organometal. Chem. **27** [1971] 13/8).
[38] Okawara, R.; Yasuda, K.; Tada, T. (Japan. 71-3567 [1971] from C.A. **74** [1971] No. 112205).
[39] Tada, H.; Okawara, R. (J. Organometal. Chem. **28** [1971] 21/4).
[40] Vyazankin, N. S.; Bochkarev, M. N.; Charov, A. I. (J. Organometal. Chem. **27** [1971] 1175/80).

[41] Weidlein, J. (J. Organometal. Chem. **32** [1971] 181/94).
[42] Weidlein, J. (Z. Anorg. Allgem. Chem. **386** [1971] 129/38).
[43] Golubinskaya, L. M.; Bregadze, E. V.; Bryuchova, E. V.; Svergun, V. I.; Semin, G. K.; Okhlobystin, O. Yu. (J. Organometal. Chem. **40** [1972] 275/9).
[44] Maeda, T.; Okawara, R. (J. Organometal. Chem. **39** [1972] 87/91).
[45] Olapinski, H.; Weidlein, J. (J. Organometal. Chem. **35** [1972] C33/C55).
[46] Schwering, H.-U.; Hausen, H.-D.; Weidlein, J. (Z. Anorg. Allgem. Chem. **391** [1972] 97/106).
[47] Hausen, H.-D.; Guder, H. J. (J. Organometal. Chem. **57** [1973] 243/53).
[48] Hausen, H.-D.; Schwering, H.-U. (Z. Anorg. Allgem. Chem. **398** [1973] 119/28).
[49] Maslova, V. A.; Novoselova, N. V.; Moseeva, E. M.; Berezhnaya, N. D.; Rabinovich, I. B. (Tr. Khim. Khim. Tekhnol. **1973** No. 2, pp. 51/2; C.A. **80** [1974] No. 95059).

[50] Schwering, H.-U.; Weidlein, J. (Chimia [Switz.] **27** [1973] 535/8).

[51] Sen, B.; White, G. L. (J. Inorg. Nucl. Chem. **35** [1973] 2207/15).

[52] Chernykh, I. N.; Tomilov, A. P. (Elektrokhimiya **10** [1974] 971/4; Soviet Electrochem. **10** [1974] 926/8).

[53] Hausen, H.-D.; Mertz, K.; Veigel, E.; Weidlein, J. (Z. Anorg. Allgem. Chem. **410** [1974] 156/64).

[54] Kurbakova, A. P.; Leimes, L. A.; Aleksanyan, V. T.; Golubinskaya, L. M.; Zorina, E. N.; Bregadze, V. I. (Zh. Strukt. Khim. **15** [1974] 1083/92; J. Struct. Chem. [USSR] **15** [1974] 961/9).

[55] Mann, G.; Olapinski, H.; Ott, R.; Weidlein, J. (Z. Anorg. Allgem. Chem. **410** [1974] 195/205).

[56] Olapinski, H.; Weidlein, J.; Hausen, H.-D. (J. Organometal. Chem. **64** [1974] 193/204).

[57] Zorin, A. D.; Umilin, V. A. (Zavodsk. Lab. **40** [1974] 502/3; Ind. Lab. [USSR] **40** [1974] 622/3 from C.A. **81** [1974] No. 99028).

[58] Efremov, E. A.; Orlov, V. Yu.; Fedorov, V. A.; Osipova, N. G.; Efremov, A. A.; Tonoyan, L. G. (Zh. Fiz. Khim. **49** [1975] 1844/5; Russ. J. Phys. Chem. **49** [1975] 1087).

[59] Ichiki, E.; Kazuo, I.; Ogura, M. (Japan. 75-53331 [1975] from C.A. **83** [1975] No. 131748).

[60] Schwering, H.-U.; Weidlein, J.; Fischer, P. (J. Organometal. Chem. **84** [1975] 17/37).

[61] Tedoradaze, G. A. (J. Organometal. Chem. **88** [1975] 1/36).

[62] Solomatin, V. S.; Zakharova, T. I.; Dzhupii, L. S.; Shatalina, L. G.; Viasova, G. I.; Lakeeva, N. I.; Krasnikova, G. V.; Nikolina, S. N. (Poluch. Anal. Veshchestv Osoboi Chist. Dokl. 5th Vses. Konf., Gorkiy 1976 [1978], pp. 240/7 from C.A. **91** [1979] No. 116916).

[63] Vanchagova, V. K.; Zorin, A. D.; Umilin, V. A. (Zh. Obshch. Khim. **46** [1976] 989/93; J. Gen. Chem. [USSR] **46** [1976] 985/8).

[64] Bancroft, G. M.; Sham, T. K. (J. Magn. Resonance **25** [1977] 83/90).

[65] Kut'in, A. M.; Frolov, I. A.; Fulkin, K. K.; Tsvetkov, V. G. (Zh. Obshch. Khim. **47** [1977] 2769/73; J. Gen. Chem. [USSR] **47** [1977] 2518/21).

[66] Razuvaev, G. A.; Abakumov, G. A.; Klimov, E. S.; Gladyshev, E. N.; Bayushkin, P. Ya. (Izv. Akad. Nauk SSSR Ser. Khim. [1977] 1128/32; Bull. Acad. Sci. USSR Div. Chem. Sci. [1977] 1034/7).

[67] Tohyama, I.; Otozai, K. (Fresenius Z. Anal. Chem **288** [1977] 286/7).

[68] Zorin, A. D.; Kut'kin, A. M.; Lokhov, N. S.; Umilin, V. A. (Poluch. Anal. Chist. Veshchestv No. 2 [1977] 33/5; C. A. **89** [1978] No. 186694).

[69] Lokhov, N. S.; Zorin, A. D.; Kuznetsova, T. V. (Poluch. Anal. Chist. Veshchestv No. 3 [1978] 83/6 from C.A. **91** [1979] No. 9655).

[70] Travkin, N. N.; Skachkov, B. K.; Tonoyan, I. G.; Kozyrkin, B. I. (Zh. Obshch. Khim. **48** [1978] 2678/81; J. Gen. Chem. [USSR] **48** [1978] 2428/31).

[71] Zorin, A. D.; Sorokina, N. M.; Lokhov, N. S.; Rachkova, O. F.; Umilin, V. A. (Zavodsk. Lab. **44** [1978] 157/9; Ind. Lab. [USSR] **44** [1978] 201/3 from C.A. **89** [1978] No. 16326).

[72] Gavrilenko, V. V.; Kolesov, V. S.; Zakharkin, L. I. (Zh. Obshch. Khim. **49** [1979] 1845/8; J. Gen. Chem. [USSR] **49** [1979] 1623/6).

[73] Lokhov, N. S.; Feshchenko, I. A.; Kuznetsova, T. V.; Zanozina, V. F. (Poluch. Anal. Chist. Veshchestv No. 4 [1979] 80/5 from C.A. **94** [1981] No. 47398).

[74] Lokhov, N. S.; Zorin, A. D.; Tomadze, A. V.; Kuznetsova, T. V.; Zanozina, V. F.; Yablokov, V. A. (Zh. Obshch. Khim **49** [1979] 1921/3; J. Gen. Chem. [USSR] **49** [1979] 1691/2).

[75] Karschin, J.; Mann, G.; Weidlein, J. (unpublished results 1979).

[76] Shushunov, N. V.; Agafonov, I. L. (Khim. Elementoorg. Soedin [Gor'kiy] No. 7 [1979] 50/2; C.A. **93** [1980] No. 70337).

[77] Lokhov, N. S.; Kuznetsova, T. V.; Zorin, A. D.; Zanozina, V. F. (Zh. Prikl. Khim. **53** [1980] 1163/6; J. Appl. Chem. [USSR] **53** [1980] 922/5 from C.A. **93** [1980] No. 186438).
[78] Shushunova, A. F.; Demarin, V. T.; Makin, G. I.; Sklemina, L. V.; Rudnevskii, N. K.; Aleksandrov, Yu. A. (Zh. Analit. Khim. **35** [1980] 349/52; J. Anal. Chem. [USSR] **35** [1980] 240/3).
[79] Manasevit, H. M. (J. Cryst. Growth **55** [1981] 1/9).
[80] Aitchison, K. A. (Diss. Univ. London 1983).

[81] Haigh, J. (J. Mater. Sci. **18** [1983] 1072/6).
[82] Mullin, J. B.; Holliday, A. K.; Cole-Hamilton, D. J.; Jones, A. C. (Eur. 80349 [1983]; C.A. **99** [1983] No. 140143).
[83] Maury, F.; Constant, G. (Polyhedron **3** [1984] 581/4).
[84] Moss, R. H. (J. Cryst. Growth **68** [1984] 78/87).
[85] Ludowise, M. J. (J. Appl. Phys. **58** [1985] R31/R55).
[86] Hui, B. C.; Lorberth, J.; Melas, A. A. (US 4720560 [1988] identical with Eur. 181706 [1986]; C.A. **105** [1986] No. 79150).
[87] Nistratov, V. P.; Rabinovich, I. B. (Primen. Metalloorg. Soedin. Poluch. Neorg. Pokrytii Mater. **1986** 34/50; C.A. **108** [1988] No. 227706).
[88] Alfa Products, Morton Thiokol Inc. (Alfa Organometallics for Vapour Phase Epitaxy, Literature and Product Review, 1986, Danvers, Mass., 1987).
[89] Kadkura, H.; Sawara, K.; Yako, T. (Brit. 2183651 [1987] from C.A. **107** [1987] No. 154514).
[90] Agnello, P. D.; Ghandhi, S. K. (J. Electrochem. Soc. **135** [1988] 1530/4).

[91] Bradley, D. C.; Chudzynska, H.; Faktor, M. M.; Frigo, D. M.; Hursthouse, M. B.; Hussain, B.; Smith, L. M. (Polyhedron **7** [1988] 1289/98).
[92] Kayser, O.; Heinecke, H.; Brauers, A.; Lüth, H.; Balk, P. (Chemtronics **3** [1988] 90/3).
[93] Alcock, N. W.; Degnan, L. A.; Wallbridge, M. G. H.; Powell, H. R.; McPartlin, M.; Sheldrick, G. M. (J. Organometal. Chem. **361** [1989] C33/C36).
[94] Imori, T.; Nakamura, K.; Ninomiya, T. (Japan. 88-130180 [1988] from C.A. **112** [1989] No. 217259).
[95] Imori, T.; Nakamura, K.; Ninomiya, T.; Hirai, T. (Japan. 88-95774 [1988] from C.A. **112** [1990] No. 158629).
[96] Thorn, D. L.; Harlow, R. L. (J. Am. Chem. Soc. **111** [1989] 2575/80).
[97] Hahn, F. H.; Schneider, B.; Reier, F.-W. (Z. Naturforsch. **45b** [1990] 134/40).
[98] Agnello, P. D.; Ghandhi, S. K. (J. Cryst. Growth **94** [1989] 311/20).
[99] Thorn, D. L. (J. Organometal. Chem. **405** [1991] 161/71).
[100] Forster, D. F.; Rushworth, S. A.; Cole-Hamilton, D. J.; Jones, A. C.; Stagg, J. P. (Chemtronics **3** [1988] 38/43).

List on individual works on MOCVD in which primarily $In(C_2H_5)_3$ is used in industrial research and application:

Manasevit, H. M.; Simpson, W. I. (J. Electrochem. Soc. **120** [1973] 135/7).
Fukui, T.; Horikoshi, Y. (Japan. J. Appl. Phys. **18** [1979] 2157/8).
Fukui, T.; Horikoshi, Y. (Japan. J. Appl. Phys. **19** [1980] L53/L56).
Fukui, T.; Horikoshi, Y. (Japan. J. Appl. Phys. **19** [1980] L551/L554).
Hirtz, J. P.; Larivain, J. P.; Duchemin, J. P.; Pearsall, T. P.; Bonnet, M. (Electron. Letters **16** [1980] 415/6).
Dietze, W. T.; Ludowise, M. J.; Cooper, C. B. (Electron. Letters **17** [1981] 698/9 from C.A. **95** [1981] No. 53520).
Duchemin, J. P. (J. Vac. Sci. Technol. **18** [1981] 753/5).

Duchemin, J. P.; Hirtz, J. P.; Razeghi, M.; Bonnet, M.; Hersee, S. D. (J. Cryst. Growth **55** [1981] 64/73).
Fukui, T.; Horikoshi, Y. (Japan. J. Appl. Phys. **20** [1981] 587/91).
Nippon Telegraph and Telephone Public Corp. (Japan. 81-137615 [1981] from C.A. **96** [1982] No. 78026).
Razeghi, M.; Hirtz, J. P.; Larivain, J. P.; Blondeau, R.; De-Cremoux, B.; Duchemin, J. P. (Electron. Letters **17** [1981] 597/8).
Sergeev, V. I.; Aitkhozhin, S. A.; Shemet, V. V.; Zorin, A. D.; Feshchenko, I. A.; Ronina, O. V. (Poluch. Anal. Chist. Veshchestv **1981** 82/5 from C.A. **97** [1982] No. 172598).
Yoshino, J.; Iwamoto, T.; Kukimoto, H. (Japan. J. Appl. Phys. **20** [1981] L290/L292).
Yoshino, J.; Iwamoto, T.; Kukimoto, H. (J. Cryst. Growth **55** [1981] 74/8).
Duchemin, J. P.; Razeghi, M.; Hirtz, J. P.; Bonnet, M. (Conf. Ser. Inst. Phys. [London] No. 63 [1982] 89/94).
Goetz, K. H.; Solomonov, A. V.; Bimberg, D.; Juergensen, H.; Razeghi, M.; Selders, J. (J. Phys. Colloq. [Paris] **43** [1982] C5-383/C5-392).
Oishi, M.; Kuoiwa, K. (Japan. J. Appl. Phys. **21** Pt. 1 [1982] 203/5).
Razeghi, M.; Blondeau, R.; Larivain, J. P.; Noel, L.; De-Cremoux, B.; Duchemin, J. P. (Electron. Letters **18** [1982] 132/3).
Semiconductor Energy Research Inst. Co., Ltd. (Japan. 82-193025 [1982] from C.A. **98** [1983] No. 82313).
Suzuki, T.; Hine, I.; Gomyo, A.; Nishida, K. (Japan. J. Appl. Phys. **21** Pt. 2 [1982] L731/L733).

Iwamoto, T.; Mori, K.; Mitzuza, M.; Kukimoto, H. (Japan. J. Appl. Phys. **22** Pt. 2 [1983] L191/L193).
Matsushita Electric Industrial Co., Ltd. (Japan. 83-125698 [1983] from C.A. **99** [1983] No. 204027).
Matsushita Electric Industrial Co., Ltd. (Japan. 83-140391 [1983] from C.A. **99** [1983] No. 222887).
Maury, F.; Combes, M.; Constant, G. (EURO CVD Four Proc. 4th Eur. Conf. Chem. Vap. Deposition, Eindhoven 1983, pp. 257/64; C.A. **99** [1983] No. 185120).
Nippon Telegraph and Telephone Public Corp. (Japan. 83-156592 [1983] from C.A. **100** [1984] No. 28351 and Japan. 83-209117 [1983] from C.A. **100** [1984] No. 113417 and Japan. 83-209118 [1983] from C.A. **100** [1984] No. 113416).
Razeghi, M.; Duchemin, J. P. (J. Vac. Sci. Technol. B **1** [1983] 262/5).
Razeghi, M.; Duchemin, J. P. (J. Cryst. Growth **64** [1983] 76/82).
Razeghi, M.; Hersee, S.; Hirtz, P.; Blondeau, R.; De-Cremoux, B.; Duchemin, J. P. (Electron Letters **19** [1983] 336/7).
Razeghi, M.; Hirtz, J. P.; Ziemelis, U. O.; Delalande, C.; Etienne, B.; Voos, M. (Appl. Phys. Letters **43** [1983] 585/7).
Razeghi, M.; Poisson, M. A.; Larivain, J. P.; Duchemin, J. P. (J. Electron. Mater. **12** [1983] 371/95).

Sogou, S.; Kameyama, A.; Katsuda, H.; Miyamoto, Y.; Furuya, K.; Suematsu, Y. (Electron. Letters **19** [1983] 1036/7).
Whiteley, J. S.; Ghandhi, S. K. (J. Electrochem. Soc. **130** [1983] 1191/5).
Whiteley, J. S.; Ghandhi, S. K. (Thin Solid Films **104** [1983] 145/52).
Ban, Y.; Ogura, M.; Morisaki, M.; Hase, N. (Japan. J. Appl. Phys. **23** Pt. 2 [1984] L606/L609).
Haigh, J.; O'Brien, S. (J. Cryst. Growth **68** [1984] 550/6).
Hess, K. L.; Kasemset, D. L.; Dapkus, P. D. (J. Electron. Mater. **13** [1984] 779/98).
Hino, I.; Suzuki, T. (J. Cryst. Growth **68** [1984] 483/9).
Ikeda, M.; Honda, M.; Mori, Y.; Kaneko, K.; Watanabe, N. (Appl. Phys. Letters **45** [1984] 964/6).

Kamijoh, T. (J. Cryst. Growth **67** [1984] 144/8).
Kasemset, D.; Hess, K. L.; Mohammed, K.; Merz, J. L. (J. Electron. Mater. **13** [1984] 655/7).

Ikeda, M.; Mori, Y.; Takiguchi, M.; Kaneko, K.; Watanabe, N. (Appl. Phys. Letters **45** [1984] 661/3).
Kawaguchi, Y.; Asahi, H.; Nagai, H. (Japan. J. Appl. Phys. **23** Pt. 2 [1984] L737/L739).
NEC Corp. (Japan. 84-128299 [1984] from C.A. **101** [1984] No. 202083).
Nippon Telegraph and Telephone Public Corp. (Japan. 84-3099 [1984] from C.A. **100** [1984] No. 201406).
Nippon Electric Co., Ltd. (Japan. 84-22320 [1984] from C.A. **101** [1984] No. 46836).
Ogura, M.; Ban, Y.; Morisaki, M.; Hase, N. (J. Cryst. Growth **68** [1984] 32/8).
Oishi, M.; Nojima, S.; Kuroiwa, K. (Japan. J. Appl. Phys. **23** Pt. 2 [1984] L625/L627).
Razeghi, M.; Bondeau, R.; Kazmierski, K.; Krakowski, M.; De-Cremoux, B.; Duchemin, J. P.; Bouley, J. C. (Appl. Phys. Letters **45** [1984] 784/6).
Razeghi, M.; De-Cremoux, B.; Duchemin, J. P. (J. Cryst. Growth **68** [1984] 389/97).
Sogou, S.; Kameyama, A.; Miyamoto, Y.; Furuya, K.; Suematsu, Y. (Japan. J. Appl. Phys. **23** Pt. 1 [1984] L1182/L1199).

Toshiba Corp. (Japan. 84-116191 [1984] from C.A. **101** [1984] No. 220354).
Uwai, K.; Mikami, O.; Susa, N. (Extend. Abstr. Conf. Solid State Devices Mater. **16** [1984] 667/70 from C.A. **101** [1984] No. 238673).
Uwai, K.; Susa, N.; Mikami, O.; Fukui, T. (Japan. J. Appl. Phys. **23** Pt. 2 [1984] L121/L123).
Chiang, P. K.; Bedair, S. M. (Appl. Phys. Letters **46** [1985] 383/5).
Chu, S. N. G.; Nakahara, S.; Long, J. A.; Riggs, V. G.; Johnston, W. D., Jr. (J. Electrochem. Soc. **132** [1985] 2795/8).
DiForte-Poisson, M. A.; Brylinski, C.; Duchemin, J. P. (Appl. Phys. Letters **46** [1985] 476/8).
Kawaguch, Y.; Asahi, H.; Nagai, H. (Japan. J. Appl. Phys. **24** Pt. 2 [1985] L221/L223).
Kobayashi, K.; Hino, S.; Suzuki, T. (Appl. Phys. Letters **46** [1985] 7/9).
Maebotoke, S.; Kobayashi, M. (Japan. 85-173829 [1985] from C.A. **104** [1986] No. 43704).
Nelson, A. W.; Wong, S.; Regnault, J. C.; Hobbs, R. E.; Murrell, D. L.; Walling, R. H. (Electron. Letters **21** [1895] 493/4).

Nippon Telegraph and Telephone Public Corp. (Japan. 85-251612 [1985] from C.A. **105** [1986] No. 15795).
Oishi, M.; Kuroiwa, K. (J. Electrochem. Soc. **132** [1985] 1209/14).
Poulain, P.; Razeghi, M.; Kazmierski, K.; Bondeau, R.; Philippe, P. (Electron. Letters **21** [1985] 441/2).
Tsang, W. T. (J. Vac. Sci. Technol. [2] B **3** [1985] 666/7).
Uwai, K.; Mikami, O.; Susa, N. (Electron. Letters **21** [1985] 131/2).
Zhu, L. D.; Chan, K. T.; Wagner, D. K.; Ballantyne, J. M. (J. Appl. Phys. **57** [1985] 5486/92).
Andre, J. P.; Menu, E. P.; Erman, M.; Meynadier, M. H.; Ngo, T. (J. Electron. Mater. **15** [1986] 71/4).
Bougnot, G. J.; Foucaran, A. F.; Marjan, M.; Etienne, D.; Bougnot, J.; Delannoy, F. M. H.; Roumanille, F. M. (J. Cryst. Growth **77** [1986] 400/7).
Chang, H. L.; Meiners, L. G.; Sa, C. J. (Appl. Phys. Letters **48** [1986] 375/7).
Ikeda, M.; Nakano, K.; Mori, Y.; Kaneko, K.; Watanabe, N. (J. Cryst. Growth **77** [1986] 380/5).

Morizaki, M.; Ban, Y.; Ogura, M.; Hase, N. (Japan. 86-21995 [1986] from C.A. **105** [1986] No. 70574).
Tischler, M. A.; Bedair, S. M. (Appl. Phys. Letters **49** [1986] 174/5).
Tsang, W. T. (PCT Intern. Appl. 86-3232 [1986] from C.A. **105** [1986] No. 124850).

Yoshida, M. (Japan. 86-279119 [1986] from C.A. **106** [1986] No. 147418).
Zeveke, T. A.; Babushkina, T. S.; Nikolaeva, L. E.; Tolomasov, V. A.; Karataev, E. N. (Poluch. Anal. Chist. Veshchestv **1986** 58/61 from C.A. **107** [1987] No. 246945).
Hasee, N.; Oshima, M.; Hirayama, N. (Japan. 87-159420 [1987] from C.A. **108** [1987] No. 196280).
Ide, Y.; Onuma, T.; Kageyama, Y. (Kobunshi Ronbunshu **44** [1987] 595/603 from C.A. **107** [1988] No. 188440).
Matsumoto, T. (Japan. 87-123099 [1987] from C.A. **107** [1987] No. 189054).
Tsang, W. T.; Campbell, J. C.; Qua, G. J. (IEEE Electron Device Letters EDL-**8** [1987] 294/6 from C.A. **107** [1987] No. 208417).
Agnello, P. D.; Ghandi, S. K. (Solar Cells **24** [1988] 117/26 from C.A. **109** [1988] No. 25239).
Ban, Y.; Kimura, S.; Morisaki, M.; Ogura, M.; Shibata, J. (J. Cryst. Growth **93** [1988] 924/8).
Daste, P.; Miyake, Y.; Cao, M.; Miyamoto, Y.; Arai, S.; Suematsu, Y. (J. Cryst. Growth **93** [1988] 365/9).
Fukui, T. (J. Cryst. Growth **93** [1988] 301/6).
Hess, K. L.; Zehr, S. W.; Cheng, W. H.; Pooladdej, J.; Buehring, K. D.; Wolf, D. L. (J. Cryst. Growth **93** [1988] 576/82).
Miyamoto, Y.; Uesaka, K.; Takadou, M.; Furuya, K.; Suematsu, Y. (J. Cryst. Growth **93** [1988] 353/8).
Ozasa, K.; Yuri, M.; Nishino, S.; Matsunami, H. (J. Cryst. Growth **93** [1988] 177/81).
Saxena, R. R.; Fouquuet, J. E.; Sardi, V. M.; Moon, R. L. (Appl. Phys. Letters **53** [1988] 304/6).
Sugiura, O.; Kameda, H.; Shiina, K.; Matsumura, M. (J. Electron. Mater. **17** [1988] 11/4).
Taskar, N. R.; Natarajan, V.; Bhat, I. B.; Ghandhi, S. K. (J. Cryst. Growth **86** [1988] 228/32).
Yokogawa, T.; Ogura, M. (Japan. 88-7620 [1988] from C.A. **108** [1988] No. 152648).

1.1.2.2 Adducts of the $(C_2H_5)_3In \cdot D$ and $((C_2H_5)_3In)_2 \cdot D\text{-}D$ Types

The tendency of the trialkyl derivatives to form stable 1:1 addition compounds decreases sharply from Al to Ga to In, which is explained easily and with general validity by their decreasing Lewis acid strengths [10]. As a result, the adducts with O and S donors are not very stable.

Adducts with Group 16 Donor Atoms. In this series no pure compounds could be isolated.

The THF adduct should be able to be prepared, by analogy to the corresponding trimethylindium compound (see Table 5, No. 3), by electrolytic methods [9]; isolation of the product, however, was hardly successful (author's comment). It was concluded from molecular weight determinations of the triethyl compound in dioxane that no stable adduct with dioxane was formed [2]. On the other hand, the frequency-lowering of the O-C bond vibration from 873 cm^{-1} in the spectrum of pure dioxane to 831 cm^{-1} in the spectrum of a solution of $In(C_2H_5)_3$ in this solvent unequivocally demonstrates an interaction [7].

The phenyl compounds RCHO, RNCO, $R_2C{=}O$, and $R_2C{=}S$ ($R = C_6H_5$) were tested as potential Lewis bases toward $In(C_2H_5)_3$. Equimolar mixtures in CH_2Cl_2 show no relevant shifts of the ethyl ^1H NMR signals at -40 to $-50\,°C$; however, on warming the bright yellow to orange-red solutions to room temperature successive reactions occur [4].

$(C_2H_5)_3In \cdot O(C_2H_5)_2$

This complex was obtained by the Grignard method (reflux, 1 h) in 93% yield [17]. The Lewis acid-base interaction between $In(C_2H_5)_3$ and ether is detectable only by a weakly exothermic change (about -290 cal/mol) when the reactants are combined. The coordination is so weak that the adduct decomposes at about 90 °C [1], and attempted distillation always resulted in separation of the ether and $In(C_2H_5)_3$ [1, 3, 13, 17].

At 77 K the ^{115}In NQR spectrum reveals changes to the spectrum of pure triethylindium, indicating adduct formation when the components are mixed. The following values were attributed to "$(C_2H_5)_3In \cdot O(C_2H_5)_2$": 43.0 ($\nu_1$), 84.94 ($\nu_2$), 127.61 ($\nu_3$), 1024.0 ($e^2Qq/h$) MHz, and $\eta = 3.9\%$ [5].

Adducts with Group 15 Donors. The few, more carefully studied examples were obtained by combining stoichiometric amounts (mole ratio 1:1 or 2:1) of $In(C_2H_5)_3$ and the appropriate base.

$(C_2H_5)_3In \cdot NH(CH_2)_5$

The adduct remained, after mixing the starting materials, as a viscous colorless liquid, which decomposes at about 120 °C by splitting off ethane to form dimeric $\{(C_2H_5)_2InN(CH_2)_5\}_2$. Adduct formation was recognized by the lowering of the N-H bond frequency by 75 wavenumbers, compared to free piperidine, to 3275 cm^{-1}. The proton resonance spectrum in C_6D_6 at about 35 °C shows signals at $\delta = 0.58$ (CH_2In), 1.03 (β- and γ-CH_2), 1.51 (CH_3), and 2.44 (α-CH_2) ppm; the signal of the HN proton could not to be detected [6].

$(C_2H_5)_3In \cdot NCC_6H_5$

The compound was formed in 89% yield from the components in boiling toluene. The adduct is stable and can be purified by distillation at 72 °C and 10^{-3} Torr. The ^1H NMR spectrum (CH_2Cl_2 solution) exhibits signals of the ethyl and phenyl groups at $\delta = 0.58$ (CH_2), 1.41 (CH_3), and 7.60 (C_6H_5) ppm [4].

$(C_2H_5)_3In \cdot AsR_3$

The compound is not isolable with either R=H or R=CH_3. The formation of the adduct with AsH_3, however, is probably indicated in the gas phase by the change in the intensity of the CH_2 wagging vibration of the ethyl group at 1150 cm^{-1} in the IR spectrum [11], and a 6 kcal/mol enthalpy of formation has been estimated [15].

With $As(CH_3)_3$ no comparable effects in the IR spectrum are observed. On the other hand, pyrolysis of $In(C_2H_5)_3$ in the presence of $As(CH_3)_3$ is identifiable by IR spectrometry already at 360 °C, while $In(C_2H_5)_3$ alone exhibits none of the typical CH_4 absorption bands even at 400 °C [11].

In metalorganic vapor phase epitaxy (MOVPE) experiments a considerably better quality surface layer results from using $In(C_2H_5)_3$ and AsH_3 as source materials than from using $In(C_2H_5)_3/As(CH_3)_3$, which was also attributed to the formation or "nonformation" of an adduct [8, 11, 12, 15].

$(C_2H_5)_3In \cdot N(CH_3)_2C_6H_4CH_2C_6H_4N(CH_3)_2 \cdot In(C_2H_5)_3$

This adduct with Arnold's base, MBDA (MBDA=4,4'-methylenebis(N,N'-dimethylaniline)), was obtained from shaking the components in petroleum ether followed by cooling to −80 °C; white crystals at low temperature, 93% yield. The crystals melted at room temperature and were heated at 70 °C to remove traces of solvents. ^1H NMR pectrum in C_6D_6 (in ppm): 0.43 (q, CH_2In), 1.34 (t, CH_3C), 2.91 (s, CH_3N), 3.61 (s, CH_2), 6.82 (m, C_6H_4). The mass spectrum exhibits the fragments $[MBDA]^+$, $[In(C_2H_5)_3]^+$, $[In(C_2H_5)_2]^+$, $[In(C_2H_5)]^+$, and $[^{115}In]^+$ [17].

The adduct liberated $In(C_2H_5)_3$ on heating in vacuum at 120 to 150 °C for 6 h; 91% yield. This is a method to prepare group 15 metal alkyls of high purity [17].

$(C_2H_5)_3In \cdot P(CH_3)_2P(CH_3)_2 \cdot In(C_2H_5)_3$

The compound is formed as a viscous colorless oil from the weakly exothermic combination of the components in 2:1 mole ratio. Neither melting point nor boiling point was determined. The vapor pressure between 50 and 100 °C is described by the equation log p (Torr) = 9.61 − 3294/T (K); enthalpy of formation, ΔH = 15 kcal/mol; entropy of formation, ΔS = 31 $cal \cdot mol^{-1} \cdot K^{-1}$ [14].

The ^1H NMR spectrum in C_6H_6 (5 to 10 mol%) shows signals at δ = 0.61 (CH_2In), 0.85 (CH_3P), and 1.45 ppm (CH_3C). In the IR spectrum of a capillary layer the following absorption bands (no assignments given) are found below 1500 cm^{-1}: 1465 m, 1425 m, 1375 m, 1300 w, 1290 w, 1230 m, 1160 w, 1000 m, 955 s, 925 m, 908 m, 895 m, 882 m, 828 vw, 745 vw, sh, 735 w, 710 vw, 680 vw, 622 s, 465 s, and 340 vw cm^{-1}. Mass spectral measurements lead to the conclusion that the compound is partially dissociated at 40 °C in vacuum. The most abundant fragments (rel. intensities in parentheses) at 40 °C and 70 eV are: $[In(C_2H_5)_2]^+$ (44%), $[InC_3H_8]^+$ (15%), $[InC_2H_5]^+$ (10%), $[InC]^+$ (12%), $^{115}In^+$ (37%), $[P_2C_3H_9]^+$ (13%), $[P_2C_2H_4]^+$ (34%), $[PC_3H_8]^+$ (34%), P_2^+, $[PC_2H_7]^+$ (100%), $[PC_2H_6]^+$ (25%), and $[PCH_3]^+$ (46%) [14].

$(C_2H_5)_3In \cdot P(C_6H_5)_2(CH_2)_2P(C_6H_5)_2 \cdot In(C_2H_5)_3$

The adduct was obtained from $In(C_2H_5)_3$ and the ligand in 74% yield. The light-sensitive compound can be recrystallized from benzene; the melting point is between 95 and 108 °C. At 90 °C/0.01 Torr $In(C_2H_5)_3$ begins splitting off. The decomposition goes rapidly between 100 and 115 °C, and produces pure $In(C_2H_5)_3$ (69% yield) [16].

The ^1H NMR spectrum in C_6D_6 exhibits signals at δ = 0.96 (q, CH_2In), 1.61 (t, CH_3), 2.51 (CH_2P), and 7.05, 7.47 ppm (m, C_6H_5) [16].

References:

[1] Runge, F.; Zimmermann, W.; Pfeiffer, H.; Pfeiffer, I. (Z. Anorg. Allgem. Chem. **267** [1951] 39/48).

[2] Strohmeier, W.; Hümpfner, K.; Miltenberger, K.; Seifert, E. (Z. Elektrochem. **63** [1959] 537/9).

[3] Todt, E.; Dötzer, R. (Z. Anorg. Allgem. Chem. **321** [1963] 120/3).

[4] Tada, H.; Yasuda, K.; Okawara, R. (J. Organometal. Chem. **16** [1969] 215/20).

[5] Golubinskaya, L. M.; Bregadze, E. V.; Bryuchova, E. V.; Svergun, V. I.; Semin, G. K.; Okhlobystin, O. Yu. (J. Organometal. Chem. **40** [1972] 275/9).

[6] Sen, B.; White, G. L. (J. Inorg. Nucl. Chem. **35** [1973] 2207/15).

[7] Nakasugi, O.; Ishimori, M.; Tsaruta, T. (Bull. Chem. Soc. Japan. **47** [1974] 871/5).

[8] Ludowise, M. J.; Cooper, C. B., III.; Saxena, R. R. (J. Electron. Mater. **10** [1981] 1051/68).

[9] Mullin, J. B.; Holliday, A. K.; Cole-Hamilton, D. J.; Jones, A. C. (Eur. 80349 [1983]; C.A. **99** [1983] No. 140143).

[10] Plakhotnaya, L. S.; Voronin, V. A.; Poil'skii, P. Ya. (Visn. L'viv. Politekh. Inst. No. 171 [1983] 22/4; C.A. **99** [1983] No. 175952).

[11] Cheng, C. H.; Jones, K. A.; Motyl, K. M. (J. Electron. Mater. **13** [1984] 703/26).

[12] Hess, K. L.; Kasemset, D. L.; Dapkus, P. D. (J. Electron. Mater. **13** [1984] 779/98).

[13] Maury, F.; Constant, G. (Polyhedron **3** [1984] 581/4).

[14] Hodes, H. D.; Berry, A. D. (J. Organometal. Chem. **336** [1987] 299/305).
[15] Agnello, P. D.; Ghandhi, S. K. (J. Electrochem. Soc. **135** [1988] 1530/4).
[16] Bradley, D. C.; Chudzynska, H.; Faktor, M. M.; Frigo, D. M.; Hursthouse, M. B.; Hussain, B.; Smith, L. M. (Polyhedron **7** [1988] 1289/98).
[17] Forster, D. F.; Rushworth, S. A.; Cole-Hamilton, D. J.; Jones, A. C.; Stagg, J. P. (Chemtronics **3** [1988] 38/43).

1.1.3 Other InR_3 Compounds with R = Alkyl, Substituted Alkyl, and Cycloalkyl

The compounds of this section are collected in Table 11. The products are mostly light-sensitive liquids or low-melting solids, and were synthesized by the general methods described below; Method I is the most convenient one.

Method I: From $InCl_3$ and the appropriate RMgCl.
$InCl_3$ and sometimes $InBr_3$ or InI_3 was reacted in ether. Reaction set in immediately, the mixture was gently warmed after awhile, and the reaction was complete in 1 to 1.5 h. After removing the ether, the trialkyl compound was distilled under reduced pressure directly from the reaction residue; usually, this distillation produced ether-free trialkylindium; see, e.g. [1].

Method II: From In/Mg alloy and the appropriate RX (X = Cl, Br).
The reaction was performed in refluxing ether. After the alkyl halogenide had been added, the ether was removed, and the product distilled from the residue; see, e.g. [5, 7].

Method III: Alkylation of $InCl_3$ by AlR_3 compounds.
The reaction can be carried out, e.g., in pentane. After completion of the initial exothermic reaction, excess KF was added, the solvent removed, and the residue heated to about 110 °C (to complex the R_2AlCl formed). After heating about 30 min, the trialkyl compound was distilled off [2].

Method IV: Alkylation of indium metal with HgR_2.
In metal was treated with a slight excess of the HgR_2 compound in sealed ampules at 100 to 140 °C; the reaction required approximately 30 h, and formed InR_3 in high yield [9].

Method V: Alkylation of $InCl_3$ with LiR compounds.
The reaction was carried out in ether at −5 to 0 °C. After the alkylating agent had been added, the reaction mixture was warmed to room temperature, the solvent evaporated in vacuum, and InR_3 distilled off [26, 30], or sublimed [25].

Finally, [29] suggested obtaining InR_3 (R = C_3H_7-n, C_4H_9-n, C_5H_{11}-n, C_5H_{11}-i, C_6H_{13}-n, and C_7H_{15}-n) in the form of the THF adduct by electrolyzing a THF solution of the appropriate tetraalkylammonium perchlorate, using an indium anode and platinum cathode. Since, however, none of the desired trialkyl indium was actually produced by this method, and no proof at all was given for the existence of the postulated addition compound, the procedure and $In(C_7H_{15}\text{-n})_3$ (cited only in this reference) are not included in the following discussion.

General Information. The molar susceptibility (χ_M), the molar refraction (R_M), and the magneto-optical rotation (ϱ_M) were determined for $In(C_4H_9)_3$, $In(C_5H_{11})_3$, and $In(C_6H_{13})_3$, and were used to deduce the increments for a single In–C bond. The results (20 °C, irradiation with λ = 578 nm) are collected below:

	$\chi_M \cdot 10^6$ in cgs	R_M	$\varrho_M \cdot 10^6$ in rad
$In(C_4H_9)_3$	−171.1	72.1	1820
$In(C_5H_{11})_3$	−202.5	85.7	2035
$In(C_6H_{13})_3$	−235.7	99.8	2272
InC (average)	-9.3 ± 0.6	5.2 ± 0.2	307.0 ± 3.0

These increments were compared to similar data for the homologous series with B, Al, and Ga as central atoms [9].

A comparative IR investigation of MR_3 compounds with M = Ga, In, and Sb, and $R = CH_3$, C_2H_5, n- and i-C_3H_7, n-, i-, s-, and t-C_4H_9 is given in [6]. All the spectra in the region from 4000 to 300 cm^{-1} are depicted and the most intense absorptions of the M−C stretching vibrations are emphasized in the illustrations [6]. With the exception of $In(CH_3)_3$ and $In(C_4H_9\text{-}i)_3$ the ν(InC) at 466 ± 1 cm^{-1} was observed in the spectra of all the indium trialkyls studied here. From this the force constant was calculated to have a value of f(InC) = 1.60 N/cm. For $In(C_4H_9\text{-}i)_3$ ν(InC) lies at 572 cm^{-1}, and the corresponding force constant is 2.4 N/cm [6].

For InR_3 ($R = C_4H_9$-n, C_5H_{11}-n, and C_6H_{13}-n; Nos. 3, 7, 10 in Table 11) the first-order thermal decomposition has been studied extensively [24]. For the range 200 to 280 °C and 65 to 200 Torr (reported as 85 to 265 hPa) the following equations were developed for k (in s^{-1}), the effective rate constants of decomposition:

$In(C_4H_9\text{-}n)_3$: $\log k = (7.7 \pm 0.3) - (26500 \pm 600)/2.3\,RT$
$In(C_5H_{11}\text{-}n)_3$: $\log k = (6.3 \pm 0.4) - (20600 \pm 800)/2.3\,RT$
$In(C_6H_{13}\text{-}n)_3$: $\log k = (6.0 \pm 0.4) - (15100 \pm 800)/2.3\,RT$

The free radical decomposition occurs according to $InR_3 \xrightarrow{-R^\cdot} InR_2^\cdot \xrightarrow{-2R^\cdot} In^\cdot$. The alkyl radical can react further following $2R^\cdot \rightarrow R_2$ or $2R^\cdot \rightarrow RH + (R-H)$, or abstract H from the starting material to give RH and $R_2In(R-H)$ [24].

Trialkyls Nos. 1 and 8 (Table 11) are used in gas phase epitaxy (MOVPE) to prepare III/V semiconductor layers [28], to dope silicon layers [36], and to prepare thin, light-sensitive In/Sn oxide films [37]. They also find use as catalysts [4, 8].

The trialkyls in this section do not form many stable, analytically pure addition compounds with Lewis bases. What little information cited in the literature concerning the assumed formation of this type of adduct is reported under each pertinent trialkyl; the more characterized adducts Nos. 20 to 29 are collected at the end of the table.

Table 11
Other InR_3 Compounds and Their Adducts.
Further information on numbers preceded by an asterisk is given at the end of the table.
Explanations, abbreviations, and units are on p. X.

No.	compound method of preparation (yield)	properties and remarks
*1	$In(C_3H_7)_3$ I (70%) [1] II (75%) [7]	colorless liquid, m.p. −51 °C [1] log p = 9.20 − 3051/T, $\Delta H_v = 13.9$ [3], log p = 8.428 − 2716.8/T, $\Delta H_v = 12.4$ [20] $D_m = 1.187$, $n_D^{20} = 1.501$ [1]
*2	$In(C_3H_7\text{-}i)_3$ I (55%) [40] I (85%) [44] II (70%) [7]	pale yellow liquid, b.p. 88 °C/12 Torr [7], 41 °C/0.1 Torr [44], 31 °C/0.1 Torr [40], b.p. (extrapol.) 205.2 °C [20] log p = 8.453 − 2665.8/T, $\Delta H_v = 12.2$ [20] ^1H NMR (C_6D_6): 1.00 (m, HC, AB_6-system), 1.42 (d, CH_3) [40, 44] ^{13}C NMR (C_6D_6): 23.4 (CH_3), 26.5 (CH) [40, 44] IR: see general remarks
*3	$In(C_4H_9)_3$ I (80%) [9] II (71%) [7]	colorless liquid, b.p. 85 to 86 °C/0.1 Torr [7], 86 to 87 °C/0.4 Torr [9], 81 °C/1 Torr [41] $D_m = 1.146$, $n_D^{20} = 1.491$ [9] log p = 8.72 − 3124/T, $\Delta H_v = 14.3$ [3], log p = 8.569 − 3055/T, $\Delta H_v = 14.0$ [20] ^1H NMR ($CDCl_3$): 0.92 (CH_2In and CH_3), 1.39 (CH_2-2), 1.72 (CH_2-3) [41] ^{13}C NMR ($CDCl_3$): 13.8 (CH_3), 22.5 (CH_2In), 28.7 (CH_2-3), 30.1 (CH_2-2) [41] IR and thermal decomposition: see general remarks
*4	$In(C_4H_9\text{-}i)_3$ II (85%) [7] III (50%) [2]	colorless liquid, b.p. 71 to 72 °C/0.05 Torr [7], 65 to 66 °C/2 Torr [2], 81 °C/7 Torr [41], 116 °C/12 Torr [3] log p = 8.81 − 3009/T, $\Delta H_v = 13.8$ [3] ^1H NMR: 0.80 (CH_2), 1.02 (CH_3), 2.22 (CH) in C_6D_6 [14, 16]; 1.14 (CH_2 and CH_3), 2.40 (CH) in $CDCl_3$ [41] ^{13}C NMR ($CDCl_3$): 24.7 (CH_2), 27.9 (CH_3), 37.6 (CH) [41]
5	$In(C_4H_9\text{-}s)_3$ II (79%)	yellow liquid, b.p. 70 to 72 °C/0.1 Torr [7] IR: see general remarks
*6	$In(C_4H_9\text{-}t)_3$ I (72%)	yellow crystals, mp. 53 to 57.5 °C, subl. 30 °C/0.01 Torr [35] ^1H NMR (C_6D_6): 1.30 (CH_3C) [35] ^{13}C NMR (C_6D_6): 32.6 (CH_3), 40.5 (CIn) [50] IR (film): 2760 m, 2705 m, 1384 m, sh, 1363 s, $\nu(C_3C)$, 1260 m, 1241 w, 1176 m, 1155 s, 1012 m, 936 m, 919 m, 804 s, 757 w, 722 w, 570 m, 501 w, 479 w, 464 w, 395 w, br, 384 m, br, 356 w, br, 251 m, sh, 247 s [35] mass spectrum (principal peaks only): $[In(C_4H_9)_3]^+$ (2%), $[In(C_4H_9)_2]^+$ (29%), $[In(C_4H_9)H]^+$ (11%), $[In(C_4H_9)]^+$ (33%), $^{115}In^+$ (100%), $[C_4H_{10}]^+$ (23%) [35]

Table 11 (continued)

No.	compound method of preparation (yield)	properties and remarks
7	$In(C_5H_{11})_3$ IV (71%)	b.p. 95 °C/0.01 Torr; dec. 140 °C [9] $D_4^{20} = 1.485$, $n_D^{20} = 1.485$ [9] for decomposition see general remarks
*8	$In(CH_2C_4H_9\text{-}t)_3$ I (92%)	colorless, crystalline solid, m.p. 54 to 55 °C, subl. 27 °C/0.01 Torr [31, 39] ^1H NMR (C_6H_6): 1.07 (s, CH_2In), 1.11 (s, CH_3C) [31, 39, 48] IR (Nujol): 1379 s, 1371 s, 1228 vs, 1212 s, 1103 s, 1091 m, sh, 1007 s, 990 m, 922 w, 905 w, 800 vw, 734 m, 685 s, 570 s, 446 m, 372 m, 275 w, sh, 260 m, sh, 250 m, 245 m, sh [31, 39]
9	$In(CH_2CH(CH_3)C_2H_5)_3$ S enantiomer I	b.p. 64 to 66 °C/0.01 Torr [10] $D_4^{25} = 1.083$ [10] $[\alpha]_D^{25} = +24.36°$ neat, +23.02° in toluene, +23.82° in decalin [10]
10	$In(C_6H_{13})_3$ IV	b.p. 149 to 150 °C/0.4 Torr [9] $D_4^{20} = 1.057$, $n_D^{20} = 1.483$ [9] for decomposition see general remarks
11	$In\{(CH_2)_2CH(CH_3)C_2H_5\}_3$ S enantiomer I	b.p. 106 to 108 °C/0.05 Torr [10] $D_4^{25} = 1.053$ [10] $[\alpha]_D^{25} = 18.24°$ neat, +18.85° in toluene, +19.14° in decalin [10]
12	$In(C_9H_{19}\text{-}n)_3$ I (60%)	pale-yellow oil, dec. 150 °C, monomeric in benzene [1] $n_D^{20} = 1.436$ [1]
*13	$In(CH_2C_5H_9\text{-}c)_3$ special	^1H NMR (C_6H_6): 0.94 (CH_2In, $^2J(H,H) = 6.8$), 1.65, 1.69, 1.75 (C_5H_9) [13]
*14	$In(CH_2C_6H_5)_3$ I (83%)	white solid, m.p. 86 to 87 °C [38], 79 to 83 °C [47] ^1H NMR (C_6D_6): 2.13 (s, CH_2In), 7.02 (m, C_6H_5) [38], 1.81 (CH_2), 6.72 to 7.07 (C_6H_5) [47] ^{13}C NMR (C_6D_6): 28.3 (CH_2), 122.7 (C-4), 127.2 (C-3,5), 128.9 (C-2,6), 145.1 (C-1) [47] IR (solid): 3060 s, 1945 w, 1860 w, 1800 w, 1730 w, 1590 s, 1485 s, 1415 w, 1300 w, 1255 m, 1200 s, 1185 w, 1150 w, 1080 sh, 1025 s, 990 m, 890 w, 795 s, 745 s, 690 s, 545 m, 430 m ν(InC) [38], see also [47] mass spectrum (main fragments only): $[M]^+$, $[M-C_7H_7]^+$, $[M-C_{14}H_{14}]^+$ [38]
*15	$In(CH_2Si(CH_3)_3)_3$ I (92%) [33] I (82%) [23]	colorless, pyrophoric liquid, b.p. 110 °C/high vacuum (?) [23, 33] ^1H NMR: 0.02 (CH_2In), 0.12 (CH_3Si) in CH_2Cl_2 [23]; −0.89 (CH_2In), 0.42 (CH_3Si) in C_6D_6 at 28 °C [33]

Table 11 (continued)

No.	compound method of preparation (yield)	properties and remarks
*15 (continued)		^{13}C NMR (C_6D_6): 12.0 (t, CH_2, $^1J(H,C)=118$), 2.98 (q, CH_3, $^1J(H,C)=118$) [33] IR (film): 2935 vs, 2885 s, sh, 1440 w, 1350 w, 1291 w, sh, 1244 vs, 920 s, 825 vs, 757 vs, 692 s, sh, 580 m, 490 m [22, 23].
16	$In(CH_2Sn(CH_3)_3)_3$ V (65%)	colorless liquid, b.p. 80 °C/14 Torr [30] ^1H NMR (C_6D_6): −0.25 (CH_2, $^2J(Sn,H)=57.9/60.9$), 0.12 (CH_3, $^2J(Sn,H)=50.5/52.8$) [30] ^{13}C NMR (C_6D_6): −14.76 (CH_2, $^1J(Sn,C)=258.9/270.9$), −7.74 (CH_3, $^1J(Sn,C)=314.6/329.2$) [30] ^{119}Sn NMR (C_6D_6): 24.4 [30] IR (film): 560 ms, 464 ms, both ν(InCSn), no other bands reported [30]
*17	$In(CH(Si(CH_3)_3)_2)_3$ V (ca. 90%)	white crystals, m.p. 106 °C [37], 107 to 109 °C [25], subl. 150 °C/10^{-3} Torr [25], 100 °C/10^{-4} Torr [37] ^1H NMR: 0.04 (CH), 0.15 (CH_3) in $CDCl_3$ [25]; 0.25 (CH_3), 0.28 (CH) in C_6D_6 [42] ^{13}C NMR: 4.62 (CH_3), 21.18 (CH) in $CDCl_3$ [25]; 6.0 (CH_3), 22.6 (CH) in C_6D_6 [42] mass spectrum (ca. 60 °C/70 eV): $[M-CH_3]^+$ (4.6%) $[M-CH(Si(CH_3)_3)_2]^+$ (100%) [42]; additional fragments $[M-2\,CH(Si(CH_3)_3)_2]^+$, $^{115}In^+$, $[CHSi(CH_3)_2]^+$ [25]
*18	$In(CF_3)_3$ special	no properties given ^{19}F NMR: −42.5 in $O(C_2H_5)_2$ at −80 °C [32], −41.63 in DMSO, −40.65 ($J(^{13}C,^{19}F)=360.1$) in THF, −44.46 in CH_3CN, −44.90 in DMF, −39.06 in pyridine; see also further information [49]
*19	$In(C_3H_5\text{-}c)_3$ IV (93%)	large colorless crystals, m.p. 33 to 34 °C, b.p. 80 °C/10^{-6} Torr [26] NMR: see further information
adducts $R_3In\cdot D$, $R_3In\cdot D\text{-}D$, and $R_3In\cdot D\text{-}D\cdot InR_3$		
20	$(C_4H_9\text{-}i)_3In\cdot HN(CH_2)_4CH_2$	colorless solid, m.p. 38.5 °C, not analytically pure [17] ^1H NMR (C_6D_6): 0.73 (CH_2In), 0.95 (m, CH_2-2,3), 1.21 (CH_3), 2.29 (CH), 2.37 (CH_2N) [17] decomposes at 120 °C with C_4H_{10} splitting off [17]
21	$(CH_2C_4H_9\text{-}t)_3In\cdot N(CH_3)_3$	colorless solid, m.p. 120 to 122 °C [39] ^1H NMR (C_6D_6): 0.82 (CH_2), 1.26 (CH_3C), 1.72 (CH_3N) [39]
22	$(CH_2C_4H_9\text{-}t)_3In\cdot TMEDA$ TMEDA = $((CH_3)_2NCH_2)_2$	colorless solid, m.p. 100 to 107 °C [39] ^1H NMR (C_6D_6): 0.95 (CH_2In), 1.21 (CH_3C), 1.88 (CH_3N), 2.33 (CH_2N) [39]

Table 11 (continued)

No.	compound method of preparation (yield)	properties and remarks
23	$(CH_2C_4H_9\text{-t})_3In \cdot HP(C_6H_5)_2$	colorless solid, impure, see No. 8 [43] ^{31}P NMR (C_6D_6): −50.6 [43]
24	$(CH_2C_5H_9\text{-c})_3In \cdot N(CH_3)_3$	^1H NMR (C_6H_6): 0.77 (m, CH_2In, $^2J(H,H)=6.8$), 1.75, 1.88 (C_5H_9), 1.93 (CH_3N) [13]
25	$(CH_2C_6H_5)_3In \cdot THF$	^1H NMR (C_6D_6): 0.96 (q, CH_2C, J(H,H)=6.5), 2.20 (s, CH_2In), 2.95 (q, CH_2O), 7.05 (m, C_6H_5) [38]
26	$2\,(CH_2C_6H_5)_3In \cdot C_4H_8O_2$	^1H NMR (C_6D_6): 2.12 (s, CH_2In), 2.88 (s, CH_2O), 7.10 (m, C_6H_5) [38]
27	$(CH_2Si(CH_3)_3)_3In \cdot N(CH_3)_3$	m.p. 78 to 81 °C [23] ^1H NMR (C_6D_6): −0.31 (CH_2), 0.46 (CH_3Si), 1.94 (CH_3N) [23]
28	$(CF_3)_3In \cdot 2\,CH_3CN$	white solid, dec.p. 80 °C [49] ^{19}F NMR: −44.03 in CH_3CN, −42.54 in ether [49]
29	$(CF_3)_3In \cdot P(CH_3)_3$	white solid [32] ^1H NMR (CD_2Cl_2): 1.38 (CH_3, J(H,P)=7.5) [32] ^{19}F NMR (CD_2Cl_2): −41.7 [32] IR (solid): 2925 m, 1395 m, 1250 m, 1120 w, 1050 vs, 1000 s, and 300 m [32]

* Further information:

$In(C_3H_7)_3$ (Table **11**, No. **1**) is monomeric in benzene. The boiling point of 178 °C at normal pressure [1] was corrected to 210 °C [3, 20] and confirmed by gas chromatography [22]. Boiling temperatures at reduced pressure are reported as 121 °C/70 Torr and 103 °C/12 Torr [1], as well as 97 °C/(vacuum?) [7].

The IR spectrum is illustrated in [6]. The most prominent vibration of the molecular skeleton is the asymmetric InC_3 mode at about 465 cm^{-1} (see general remarks on p. 74). The report of 490 or even 580 cm^{-1} for the symmetric InC_3 vibration [19] is based on a false interpretation (author's remark).

$In(C_3H_7\text{-i})_3$ (Table **11**, No. **2**) boils at 124 to 129 °C/60.5 Torr, 112 to 115 °C/36.5 Torr, and 91.5 to 92.5 °C/14 Torr [3]. The IR spectrum is illustrated in [6]; see general remarks on p. 74.

Thermal decomposition leads to metallic indium and to propane, propylene, 2-methylbutane, and 2,3-dimethylbutane. At 70 °C the compound is 50% decomposed within 10 d, at 100 °C about 60% in 8 h, and at 150 °C complete decomposition is observed in 4 h. On photolysis (probably at room temperature) decomposition was 100% in less than 20 h [11].

$In(C_4H_9)_3$ (Table **11**, No. **3**), prepared by method I in 60% yield, boils at 117 °C/3 Torr [34]. Method III was used by [15], but they make no mention of characterization. The extrapolated normal boiling point is reported as 265.6 °C [20].

Photolytic decomposition of $In(C_4H_9)_3$ in the light of a high pressure mercury lamp begins after 1 min with the formation of a metallic mirror. Complexation with $N(C_2H_5)_3$ (the adduct, formulated as $(C_4H_9)_3In \cdot N(C_2H_5)_3$, was neither isolated nor was its composition demonstrated!) slows down the decomposition in UV light [15].

Unlike $Ga(C_4H_9)_3$, the indium analog is not capable of cis/trans-isomerizing stereochemically pure octene [15]. It reacted with $(C_4H_9)_2SnO$ in refluxing xylene (3.5 h) to give $(C_4H_9)_2InOSn(C_4H_9)_3$ in 79% yield and with $\{(C_4H_9)_3Sn\}_2O$ in benzene at room temperature to give the same product along with $Sn(C_4H_9)_4$ [46].

$In(C_4H_9\text{-}i)_3$ (Table **11**, No. **4**) decomposes at 125±2 °C and at 130 °C yields isobutane (28%), isobutylene (53%), and H_2 (19%) [2, 11].

It reacted similarly to No. 3 with $(C_4H_9)_2SnO$ and $\{(C_4H_9)_3Sn\}_2O$ but resulting in the mixed compounds $(C_4H_9\text{-}i)_2InOSn(C_4H_9)_2(C_4H_9\text{-}i)$ and $(C_4H_9)_3Sn(C_4H_9\text{-}i)$ [46]. It forms adduct No. 20 with piperidine [17].

$In(C_4H_9\text{-}t)_3$ (Table **11**, No. **6**) reacted with $(C_5H_5Mo(CO)_3)_2$ under photochemical conditions to give $C_5H_5Mo(CO)_3In(C_4H_9\text{-}t)_2$ as the main product along with $CH_2{=}C(CH_3)_2$ and $HC_4H_9\text{-}t$; a bimolecular radical substitution reaction was proposed [48].

$In(CH_2C_4H_9\text{-}t)_3$ (Table **11**, No. **8**) is not pyrophoric in air and is monomeric in benzene [31, 39].

Similar to No. 6 the reaction with $(C_5H_5Mo(CO)_3)_2$ gave $C_5H_5Mo(CO)_3In(CH_2C_4H_9\text{-}t)_2$ along with a second compound assigned as $C_5H_5Mo(CO)_3CH_2C_4H_9\text{-}t$ [48]. It forms no stable adducts either with diethylether or with THF. With $N(CH_3)_3$ and $(CH_3)_2NCH_2CH_2N(CH_3)_2$ (TMEDA) it yields the adducts 21 and 22, respectively [39]. With $HP(C_6H_5)_2$ in pentane at room temperature the 1:1 adduct No. 23 was formed along with $(CH_2C_4H_9\text{-}t)_2InP(C_2H_5)_2$. After removal of the solvent a colorless solid remained partially melting at 48.2 to 48.8 °C; the solid particles melted at 150 to 155 °C. From ^{31}P NMR an adduct/product ratio of 1:1.4 was found [43].

$In(CH_2C_5H_9\text{-}c)_3$ (Table **11**, No. **13**) was formed by the intermolecular cyclization of the 1-hexenyl group if $Hg\{(CH_2)_4CH{=}CH_2\}_2$ and metallic indium were kept at 110 °C for 3 weeks in a sealed glass ampule (Method IV). The initially formed $In\{(CH_2)_4CH{=}CH_2\}_3$ could not be isolated [17].

Tris(cyclopentylmethyl)indium formed the 1:1 adduct No. 24 with $N(CH_3)_3$, which was not isolated, but was identified by means of its ^1H NMR spectrum [13].

$In(CH_2C_6H_5)_3$ (Table **11**, No. **14**) crystallizes in the monoclinic space group $P2_1/c-C^5_{2h}$ (No. 14) with the lattice constants a = 1243.6(2), b = 752.3(3), c = 1940.0(4) pm, β = 102.59(2)°; Z = 4 and D_m = 1.456 g/cm^3. The molecular structure is depicted in **Fig. 7**. The In atom is situated 24.6 pm above the plane formed by the atoms C(1), C(2), and C(3). The single molecules are arranged such that infinitely long, angular chains are formed in the [0 1 0] direction by intermolecular In···C contacts; the atoms C(25) and C(26) of one phenyl ring and the In atom of a neighboring molecule approach in such a manner, that an additional η^2 coordination can be discussed. The In···C(25) and In···C(26) contact distances are 308.7(7) and 300.2(6) pm, respectively [47]. The compound is monomeric in benzene [38].

To study the basicity, the compound was reacted with ether, THF, 1,4-dioxane for 1 h at room temperature, but only adducts with THF (1:1, No. 25) and dioxane (2:1, No. 26) could be characterized by NMR [38].

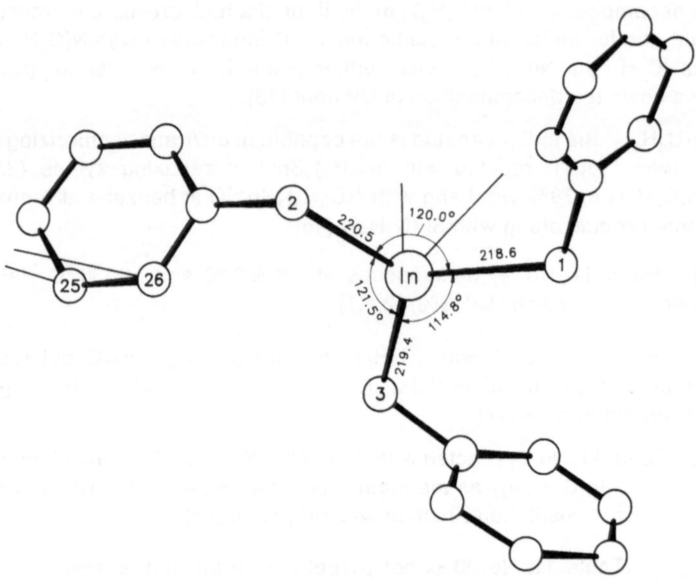

Fig. 7. Molecular structure of $In(CH_2C_6H_5)_3$ [47].

$In\{CH_2Si(CH_3)_3\}_3$ (Table **11**, No. **15**) was also prepared by Method I from $InBr_3$ in THF; however, the product was contaminated with $In\{CH_2Si(CH_3)_3\}_3 \cdot THF$. The Lewis base could not be completely removed, and it was impossible to isolate an analytically pure adduct [22].

$In\{CH_2Si(CH_3)_3\}_3$ dissolves readily in pentane, acetonitrile, benzene, and CH_2Cl_2 [22, 23].

With $N(CH_3)_3$ the stable 1:1 adduct No. 27 was obtained [23]. The compound has been used to prepare $In\{CH_2Si(CH_3)_3\}_2Cl$, $In\{CH_2Si(CH_3)_3\}Cl_2$ [23], $K[In\{CH_2Si(CH_3)_3\}_4]$, and $Na[In\{CH_2Si(CH_3)_3\}_3H]$ [27]. Reaction with $(t\text{-}C_4H_9)PH_2$ in the presence of $AgNO_3$ generated $\{(CH_3)_3SiCH_2\}_2InPHC_4H_9\text{-}t$ in 38% yield [45].

$In\{CH(Si(CH_3)_3)_2\}_3$ (Table **11**, No. **17**) also occurred as a by-product in 29% yield of the reaction between $In_2Br_4 \cdot 2$ TMEDA (N,N,N',N'-tetramethylethylenediamine) with $Li[CH(Si(CH_3)_3)_2]$ at −60 °C in ether [42].

Hexagonal prisms of the composition $In\{CH(Si(CH_3)_3)_2\}_3 \cdot O(C_2H_5)_2$ (not an adduct but a solvate) were used for crystal structure analysis. The compound crystallizes in the trigonal space group $P31c-C_{3v}^4$ (No. 159) with the lattice constants a = b = 1630.2(6) and c = 889.0(3) pm; Z = 2 and D_c = 1.083 g/cm³. **Fig. 8** shows the monomer unit. The indium atom is threefold coordinated and is located 19.1 pm above the plane formed by the three carbon atoms of the InC_3 core. This deviation from the expected planarity can not be blamed on intra- or intermolecular In ··· In, In ··· Si, or In ··· C contacts with atoms of the bulky $CH(Si(CH_3)_3)_2$ groups, since all of these distances are between 500 and 900 pm (author's comment: according to our own calculations the distance from the central In atom to the O of an ether molecule statistically embedded in the lattice is more than 800 pm.) [25].

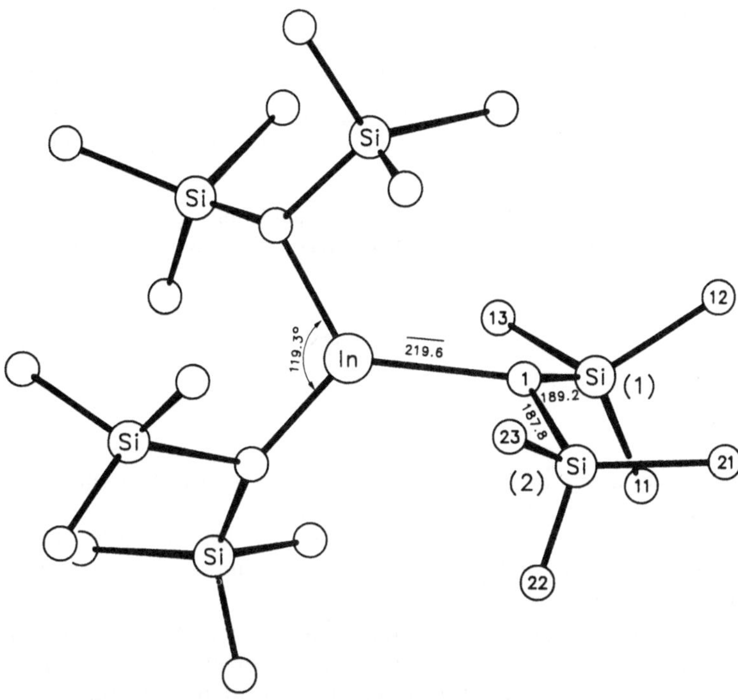

Fig. 8. Molecular structure of In{CH(Si(CH$_3$)$_2$}$_3$ [25].

Additional bond angles (°)

C(1)-Si(1)-C(11)	115.2(4)	C(1)-Si(2)-C(21)	111.2(4)
C(1)-Si(1)-C(12)	108.2(4)	C(1)-Si(2)-C(22)	112.8(3)
C(1)-Si(1)-C(13)	112.8(3)	C(1)-Si(2)-C(23)	107.8(4)

For the embedded ether molecule O-C distances of 142 and 143 pm, a C-C distance of 154 pm, and a C-O-C bond angle of 110.2° were reported [25]. "Unusual" metal alkyls, which owe their existence to the stabilizing effect of the bulky bis(trimethylsilyl)methyl ligands, CH(Si(CH$_3$)$_3$)$_2$, are discussed in [21].

In(CF$_3$)$_3$ (Table **11**, No. **18**). Prepared analogously to Tl(CF$_3$)$_3$, it was obtained by co-condensation of indium vapor and CF$_3$ radicals (prepared from C$_2$F$_6$) at −196 °C. The compound could not be examined in detail because decomposition had already begun between −40 and −20 °C [32]. The compound could also be obtained as a (CF$_3$)$_{3-x}$InCl$_x$ (x=0 to 2) mixture by trifluormethyl-chlorine exchange in the reaction of InCl$_3$ with Cd(CF$_3$)$_2$ in various donor solvents D such as DMSO (dimethylsulfoxide), THF, CH$_3$CN, DMF (dimethylformamide), or pyridine. It was supposed on the basis of ^{19}F NMR spectroscopy that the corresponding solvent adducts (1:1 or 1:2 adducts) were present in solution. The reaction in DMSO produced additionally some consecutive products; the ^{19}F NMR spectra after reaction times of 1, 7, and 21 d were depicted. In the case of D=CH$_3$CN the 1:2 adduct No. 28 (contaminated with traces of Cl and Cd containing products) could be isolated (room temperature, 1 d, 36% yield) [49].

Considerably more stable is the adduct $(CF_3)_3In \cdot P(CH_3)_3$ (No. 29), which formed from its components at −78 °C. It could be recrystallized from hexane (9% yield, based on the amount of vaporized metal) and was stable for a short time at room temperature [32].

$In(C_3H_5\text{-}c)_3$ (Table **11**, No. **19**) is weakly associated in benzene or toluene. Using nuclear resonance spectrometry, the enthalpy of dissociation of the dimer to the monomer was estimated to be ca. 4 kcal/mol [26].

The ^1H NMR spectrum has been studied in detail and correlated with the spectra of numerous monosubstituted cyclopropanes. The assignment is based upon Formula I:

H4 H2 H1 3 1 H3 2 In H5

I

Chemical shifts (13 mol% in C_6H_6) are δ (ppm) = −0.562 (H-1), 0.546 (H-3,5), and 0.791 (H-2,4). Coupling constants (Hz) are J(H-2,3) = J(H-4,5) = −3.31; J(H-2,5) = J(H-3,4) = 4.71; J(H-1,3) = J(H-1,5) = 7.24; J(H-3,5) = 7.69, and J(H-1,2) = J(H-1,4) = 9.79. The chemical shift differences, Δ[(H-1) − (H-3)] and Δ[(H-2) − (H-3)], correlate with the electronegativity of the cyclopropyl ligands [12].

The ^{13}C resonances of the α- and β-carbon atoms (C-1, C-2) exhibit strong dependences on the concentration, the temperature, and the nature of the solvent. Two signals are observed, and their chemical shifts δ (in ppm), taken from a diagram in [26], are listed below:

solvent (conc.)		−80 °C	−30 °C	+30 °C (J(C,H))	+105 °C (J(C,H))
C_7D_8	C-1	−5.5	−2.2	−0.4 (148)	2.4 (150)
(0.2 M)	C-2	3.6	3.1	2.7 (160)	2.4 (160)
C_5H_{10}-c/C_7D_8	C-1	−7.2	−5.9	−3.8	−
(0.36 M)	C-2	2.8	2.4	2.0	−

An equilibrium between monomers and dimers with rapid exchange of the cyclopropyl groups was proposed on the basis of these changes. The equilibrium constant, K = [dimer]/[monomer], for a 0.2 M toluene solution is 2.1 at 30 °C [26].

References:

[1] Runge, F.; Zimmermann, W.; Pfeiffer, H.; Pfeiffer, I. (Z. Anorg. Allgem. Chem. **267** [1951] 39/48).
[2] Eisch, J. J. (J. Am. Chem. Soc. **84** [1962] 3605/10).
[3] Hartmann, H.; Lutsche, H. (Naturwissenschaften **49** [1962] 182/3).
[4] Morton, A. (U.S. 3058912 [1962] from C.A. **60** [1964] 13084).
[5] Todt, E.; Hauschildt, H. (Ger. 1136702 [1962]; C.A. **58** [1963] 4597).
[6] Oswald, F. (Z. Anal. Chem. **197** [1963] 309/22).
[7] Todt, E.; Dötzer, R. (Z. Anorg. Allgem. Chem. **321** [1963] 120/3).
[8] Cabot Corp. (Fr. 1375984 [1984] from C.A. **62** [1965] 16403).
[9] Haran, R.; Laurent, J.-P. (Bull. Soc. Chim. France **1966** 3454/8).
[10] Lardicci, L.; Salvadori, P.; Palagi, P. (Ann. Chim. [Rome] **56** [1966] 1043/59).

[11] Razuvaev, G. A.; Petukhov, G. G.; Shcherbakova, V. I.; Druzhkov, O. N.; Zhil'tsov, S. F. (Zh. Obshch. Khim. **37** [1967] 1516/20; J. Gen. Chem. [USSR] **37** [1967] 1437/41).

[12] Scherr, P. A.; Oliver, J. P. (J. Mol. Spectrosc. **31** [1969] 109/17).

[13] Denis, J. S.; Dolzine, T.; Oliver, J. P. (J. Am. Chem. Soc. **94** [1972] 8260/1).

[14] Sen, B.; White, G. L.; Wander, J. D. (J. Chem. Soc. Dalton Trans. **1972** 447/9).

[15] Warwel, S.; Hemmerich, H.-P. (Liebigs Ann. Chem. **763** [1972] 83/6).

[16] Sen, B.; White, G. L. (J. Inorg. Nucl. Chem. **35** [1973] 497/504).

[17] Sen, B.; White, G. L. (J. Inorg. Nucl. Chem. **35** [1973] 2207/15).

[18] Dolzine, T. W.; Oliver, J. P. (J. Organometal. Chem. **78** [1974] 165/76).

[19] Kurbakova, A. P.; Leimes, L. A.; Aleksanyan, V. T.; Golubinskaya, L. M.; Zorina, E. N.; Bregadze, V. I. (Zh. Strukt. Khim. **15** [1974] 1083/92; J. Struct. Chem. [USSR] **15** [1974] 961/9).

[20] Vanchagova, V. K.; Zorin, A. D.; Umilin, V. A. (Zh. Obshch. Khim. **46** [1976] 989/93; J. Gen. Chem. [USSR] **46** [1976] 985/8).

[21] Lappert, M. F. (Pure Appl. Chem. **50** [1978] 703/8).

[22] Zorin, A. D.; Sorokina, N. M.; Lokhov, N. S.; Rachkova, O. F.; Umilin, V. A. (Zavodsk. Lab. **44** [1978] 157/9; Ind. Lab. [USSR] **44** [1978] 201/3 from C.A. **89** [1978] No. 16326).

[23] Beachley, O. T.; Rusinko, R. N. (Inorg. Chem. **18** [1979] 1966/8).

[24] Aleksandrov, Yu. A.; Druzhkov, O. N.; Baryshnikov, Yu. Yu.; Postnikova, T. K.; Makin, G. I.; Kozyrkin, B. I. (Zh. Obshch. Khim. **50** [1980] 2642/5; J. Gen. Chem. [USSR] **50** [1980] 2129/32).

[25] Carty, A. J.; Gynane, J. S.; Lappert, M. F.; Miles, S. J.; Singh, A.; Taylor, N. J. (Inorg. Chem. **19** [1980] 3637/41).

[26] Thomas, R. D.; Oliver, J. P. (Organometallics **1** [1982] 571/9).

[27] Hallock, R. B.; Beachley, O. T., Jr.; Li, Y.-J.; Sanders, W. M.; Churchill, M. R.; Hunter, W. E.; Atwood, J. L. (Inorg. Chem. **22** [1983] 3683/91).

[28] Lapidus, I. I.; Dutkina, T. A.; Malyshev, V. A.; Drozd, G. I.; Il'ina, G. P.; Bogatikov, B. F. (Izv. Akad. Nauk SSSR Neorg. Mater. **19** [1983] 711/3; Inorg. Mater. [USSR] **19** [1983] 643/5; C.A. **99** [1983] No. 14603).

[29] Mullin, J. B.; Holliday, A. K.; Cole-Hamilton, D. J.; Jones, A. C. (Eur. 80349 [1983]; C.A. **99** [1983] No. 140143).

[30] Schumann, H.; Mohtachemi, R. (Z. Naturforsch. **39b** [1984] 798/800).

[31] Beachley, O. T., Jr. (U.S. 4621147 [1986]; C.A. **106** [1986] No. 33298).

[32] Guerra, M. A.; Bierschenk, T. R.; Lagow, R. J. (Rev. Chim. Miner. **23** [1986] 701/7).

[33] Kopasz, J. P.; Hallock, R. B.; Beachley, O. T., Jr.; Andersen, R. A. (Inorg. Synth. **24** [1986] 89/91).

[34] Nomura, R.; Inazawa, S.; Matsuda, H.; Saeki, S. (Polyhedron **6** [1987] 507/12).

[35] Bradley, D. C.; Frigo, D. M.; Hursthouse, M. B.; Hussain, B. (Organometallics **7** [1988] 1112/5).

[36] Chen, J. C.; Chen, W. K.; Liu, P. L.; Maloney, J.; Beachley, O. T., Jr. (Proc. SPIE-Intern. Soc. Opt. Eng. No. 877 [1988] 21/4; C.A. **109** [1988] No. 11992).

[37] Matsuda, H.; Nomura, Y. (Japan. 87-211385 [1987] from C.A. **108** [1988] No. 86301).

[38] Barron, A. R. (J. Chem. Soc. Dalton Trans. **1989** 1625/6).

[39] Beachley, O. T., Jr.; Spiegel, E. F.; Kopasz, J. P.; Rogers, R. D. (Organometallics **8** [1989] 1915/21).

[40] Neumüller, B. (Chem. Ber. **122** [1989] 2283/7).

[41] Nomura, R.; Inazawa, S.; Kanaya, K.; Matsuda, H. (Polyhedron **8** [1989] 763/7).

[42] Uhl, W.; Layh, M.; Hiller, W. (J. Organometal. Chem. **368** [1989] 139/54).

[43] Banks, M. A.; Beachley, O. T., Jr.; Maloney, J. D. (Polyhedron **9** [1990] 335/42).

[44] Hoffmann, G. G.; Faist, R. (J. Organometal. Chem. **391** [1990] 1/5).
[45] Dembowski, U.; Noltemeyer, M.; Rockensüß, W.; Stuke, M.; Roesky, H. W. (Chem. Ber. **123** [1990] 2335/6).
[46] Nomura, R.; Matsuda, H. (Inorg. Chem. **29** [1990] 4586/8).
[47] Neumüller, B. (Z. Anorg. Allgem. Chem. **592** [1991] 42/50).
[48] Thorn, D. L. (J. Organometal. Chem. **405** [1991] 161/71).
[49] Naumann, D.; Strauß, W.; Tyrra, W. (J. Organometal. Chem. **407** [1991] 1/15).
[50] Weidlein, J. (unpublished results).

1.1.4 InR_3 Compounds and Their Adducts, where R = Alkenyl and Cycloalkenyl

The compounds of this type are listed in Table 12; their preparation was carried out by the following methods.

Method I: From metallic In and HgR_2.
HgR_2 was stirred at room temperature with a 10 to 15% excess of metallic indium for 20 to 30 h [6] or one week [5]. The reaction time can be reduced to 1 h, if finely divided In is used [2]. After separating off the volatile components (including ether or hydrocarbons used as solvent), the residue was distilled or sublimed in vacuum.

Method II: From $InCl_3$ and $M[C_5H_4R]$ or $M[C_9H_7]$.
This method was used primarily for the synthesis of cyclopentadienyl derivatives. Anhydrous $InCl_3$ (sometimes $InBr_3$) was treated with lithium or sodium cyclopentadienide in a 1:3 mole ratio. Diethyl ether or a low-boiling hydrocarbon was used as solvent; the heterogeneous reaction was carried out with constant stirring for 4 h at reflux [4]. After removing the volatile components, the most effective purification was recrystallization from benzene or toluene, because the thermal instability of the indium cyclopentadienide usually will not permit vacuum sublimation.

Method III: Reaction of InR_3 and D.
Because most of the adducts dissociated in vacuum or at higher temperatures, the reaction was carried out by vigorous stirring of a suspension of an equimolar mixture of the components in ether for several h [4].

Table 12
InR_3 Compounds and Their Adducts with R = Alkenyl and Cycloalkenyl.
Further information about compounds preceded by an asterisk is given at the end of the table.
Explanations, abbreviations, and units on p. X.

No.	compound method of preparation (yield)	properties and remarks
*1	$In(CH{=}CH_2)_3$ I (80 to 90%)	slightly yellow oil, m.p. 27 to 28 °C, b.p. 70 °C/10^{-3} Torr [6], m.p. 27.0±0.5 °C, b.p. 80 °C/10^{-6} Torr [5] NMR and IR spectra on p. 86
*2	$In(C_5H_5\text{-}c)_3$ II (65%)	bright yellow crystals, dec. ca. 150 °C [4] D = 1.6±0.1 g/cm^3 [3] ^1H NMR ($CDCl_3$, 35 °C): 5.93 (s, C_5H_5) [4]

Table 12 (continued)

No.	compound method of preparation (yield)	properties and remarks
*2 (continued)		IR (solid): 3060 vw, br, 1610 w, br, 1354 s, 1074 s, 1056 m, 1044 s, 988 s, 972 s, 905 s, 884 s, 855 s, 834 s, 819 m, 799 m, 785 s, 748 s, 739 s, 622 m, 602 s, 339 s ν(In-C), 316 m, 282 m [4] Raman (solid): 3090 w, 3074 w, 3062 w, 3038 w, 1465 m, 1414 s, 1354 s, 1336 s, 1108 s, 1068 m, 906 w, 850 w, 835 m, 812 vs, 788 m, 624 vs, 593 m, 115 w, 338 m ν(In-C), 322 s, 283 s [4]
3	$In(C_5H_4CH_3)_3$ II (56%)	yellow crystals, m.p. 124 to 126 °C (dec.) [4] ^{1}H NMR ($CDCl_3$, 35 °C): 2.23 (s, CH_3), 5.70, 6.03 (m's, C_5H_4) [4] IR (solid): 3065 w, br, 1410 m, 1072 mw, 1024 m, br, 998 s, 920 mw, 865 m, 844 m, 804 s, br, 786 vs, 739 m, 608 s, 349 m ν(In-C), 322 m [4] Raman (solid): 927 w, 795 m,br, 622 m, 610 m, 248 m, 347 sh ν(In-C), 329 m [4] mass spectrum (75 °C): $[In(C_6H_7)_3]^+$, $^{115}In^+$, $[C_6H_7]^+$, $[C_5H_4]^+$ [4]
4	$In(C_5H_4C_8H_{17}\text{-}n)_3$ I	no properties given, used for the preparation of MnR_2 [2]
5	$In(C_9H_7)_3$ C_9H_7 = indenyl II	only known as $O(C_2H_5)_2$ adduct, No. 11 [4]
adducts, $R_3In \cdot D$		
6	$(CH_2{=}CH)_3In \cdot O(CH_3)_2$ III	colorless liquid, dissociates in vacuum, monomeric in benzene [6] ^{1}H NMR (C_6D_6, ca. 30 °C): 2.91 (CH_3O), 6.78 (H-1), 6.46 (H-2), 6.11 (H-3); J(H-1,3) = 21.16, J(H-1,2) = 14.04, J(H-2,3) = 4.83 Hz (see Formula I) [6] IR and Raman (liquid): 1575 ν(C=C), 1071 ν_{as}(COC), 1027 ϱ(CH_2), 994 τ(CH_2), 940, 955 ν_s(COC), 498 $\nu_{as}(InC_3)$, 475 $\nu_s(InC_3)$, 460 γ(CH) [6]
7	$(CH_2{=}CH)_3In \cdot N(CH_3)_3$ I	not isolated and no properties given, only used in Freon 11 solution to determine the exchange behavior for the $In(C_2H_3)_3/(C_2H_3)_3In \cdot N(CH_3)_3$ system. No coalescence temperature was found even as low as −67 °C [5]
8	$(C_5H_5\text{-}c)_3In \cdot$ bipy bipy = bipyridine-2,2′ III (ca. 100%)	yellow solid, m.p. 132 to 140 °C (dec.) [4] ^{1}H NMR ($CDCl_3$): 5.98 (s, C_5H_5), 7.33 to 8.67 (m, bipy) [4] IR (solid): 290 m,br ν(In-C) [4] Raman (solid): 321 s ν(In-C), 303 w [4]

Table 12 (continued)

No.	compound method of preparation (yield)	properties and remarks
9	$(C_5H_5\text{-}c)_3In \cdot phen$ phen = phenanthroline-1,10 III	red powder, dec. 125 to 135 °C [4] ^{1}H NMR (THF): 5.55 (s, C_5H_5), 7.83 to 9.33 (m, phen) [4] IR (solid): 293 m, br ν(In-C) [4] mass spectrum (230 °C): principal peaks are $[InC_5H_5]^+$, $^{115}In^+$, $[C_{10}H_{10}]^+$, $[C_5H_5]^+$, and phen decomposition products [4]
10	$(C_5H_5\text{-}c)_3In \cdot P(C_6H_5)_3$ III	cream-colored solid, m.p. 150 to 154 °C (dec.) [4] ^{1}H NMR ($CDCl_3$): 5.97 (s, C_5H_5), 7.41 (m, C_6H_5) [4] IR (solid): 321 s ν(In-C), 303 w [4] Raman (solid): 322 m ν(In-C), 308 s, no other bands reported [4]
11	$(C_9H_7)_3In \cdot O(C_2H_5)_2$ II (74%)	thick, bright yellow oil [4] ^{1}H NMR ($CDCl_3$): 0.92 (t, CH_3C), 3.15 (q, CH_2), 5.10 (d), 5.88 (t, both C_5H_3), 7.13 (m, C_6H_4) [4] used for the preparation of $Li[In(C_9H_7)_4]$ [4]

* Further information:

$In(CH{=}CH_2)_3$ (Table **12**, No. **1**) is distinguished by a relatively high volatility (similar to that of $In(C_2H_5)_3$), by good solubility in the common organic solvents, and above all by an offensive odor; it is trimeric in benzene [6].

The typical ABC pattern of a single kind of vinyl group, found in the proton resonance spectrum in the region from 4.6 to 6.5 ppm, was interpreted as the rapid exchange between terminal and bridging $CH_2{=}CH$ groups. A complicated spectrum is provided by the three vinyl protons, which are chemically and magnetically nonequivalent, and their couplings. Their assignment is given in Formula I.

In, H3, 1, 2, H1, H2

I

The following experimental and computer-projected ^{1}H NMR data (δ in ppm, J in Hz) were reported (at ca. 30 °C in C_6D_6 [6], and at 38 °C in Freon 11 [5]):

δ(H-1)	δ(H-2)	δ(H-3)	J(H-1,3)	J(H-1,2)	J(H-2,3)	Ref.
6.63	6.13	5.88	21.47	14.40	4.13	[5]
6.66	6.47	6.02	18.86	14.28	6.77	[6]
6.742	6.444	5.989	21.46	14.37	4.08	1)

1) Calculated values.

The following ^{13}C NMR data (δ in ppm, J in Hz) were reported: 149 (C-1, ^1J(C-1,H-1) = 142.5, ^2J(C-1,H-2) = 153.8, ^2J(C-1,H-3) = 163.0), 138.2 (C-2, ^1J(C-2,H-2) = 6.5, ^1J(C-2,H-3) = 10.2, ^2J(C-2,H-1) = 15.0) [6].

The terminal and bridging CH_2=CH groups are distinguishable by vibrational spectroscopy from the characteristic C=C vibrations. The vibration frequencies of $In(CH{=}CH_2)_3$ are listed in Table 13. A comparison with the spectra of other metal vinyl compounds gives an average of $1581 \pm 7\ cm^{-1}$ for the C=C vibration of terminal vinyl ligands, and $1559 \pm 3\ cm^{-1}$ for bridging vinyls [6].

The exchange reaction between $In(CH{=}CH_2)_3$ and $In(CH_3)_3$ or $InCl_3$ (1:2 or 2:1 mole ratio, respectively) forms trimeric $(CH_3)_2In(CH{=}CH_2)$ (see Section 1.1.6 on p. 103) or the dimeric $(CH_2{=}CH)_2InCl$ (see Section 2.2.1.2 on p. 129) [6].

Table 13
IR and Raman Frequencies of $In(CH{=}CH_2)_3$ as Neat Liquid [6].
Wavenumbers in cm^{-1}, (t) and (b) indicate terminal and bridging groups, respectively.

IR	Raman	assignment	IR	Raman	assignment
3020 s	3024(20) dp	$\nu_{as}(CH_2)$	987 m	987(8) dp	$\varrho(CH_2)$ (t)
2972 sh	–	overtone	967 mw	967(5) dp	CH_2 twist (b)
2945 s	2949(80) p	ν(CH)	952 sh	959(5) dp	CH_2 twist (t)
2914 m	2920(15) p	$\nu_s(CH_2)$	940 s	–	
1572 sh	1572(20) p	ν(C=C) (t)	493 m, br	498 sh, dp	$\nu_{as}(InC_2)$ (t)
1559 m	1559(26) p	ν(C=C) (b)	468 sh	468(100) p	$\nu_s(InC_2)$ (t)
1439 w, br	–	$\delta(CH_2)$?	440 vs, br	435 sh, dp	ν(InC) (b) + γ(CH)
1390 sh	1387(51) p	$\delta_s(CH_2)$ (t)	284 sh	–	
1385 s	1383 sh, p	$\delta_s(CH_2)$ (b)	274 sh	–	δ(In-CH=C)
1266 sh	1256(58) p	δ(CH) (t)	263 m	261(74) p	$\delta_s(InC_2)$
1244 ms	–	δ(CH) (b)			+ γ(In-CH=C)
1028 ms	–	$\varrho(CH_2)$ (b)	250 w	–	
1012 m	–	$\varrho(CH_2)$ (t)	243 w	225(24) p	

$In(C_5H_5\text{-}c)_3$ (Table **12**, No. **2**) was first described in [1] and was prepared from $Na(C_5H_5)$ and $InCl_3$ in $O(C_2H_5)_2$. In an attempt to purify by vacuum sublimation at ca. 150 °C the compound decomposed to $In^I(C_5H_5\text{-}c)$ (see Section 12.2.1.1 on p. 372), so that only a trace of $In(C_5H_5\text{-}c)_3$ was recovered.

From the proton resonance spectrum a dynamic behavior of the C_5H_5 ligand was deduced, since no signal splitting could be observed in CH_2Cl_2 solution even at −90 °C. Sigma-bonded C_5H_5 residues were inferred from the large number of ν(CH) vibrations above 3000 cm^{-1} [4].

The crystal structure of $In(C_5H_5\text{-}c)_3$ was determined at −100 °C. The lattice constants reported for 25 °C deviated only slightly from the low-temperature values. The compound crystallizes in the orthorhombic space group $P2_12_12_1 - D_2^4$ (No. 19) with the unit cell parameters a = 961.6(1), b = 970.2(1), c = 1340.7(2) pm; Z = 4 and $D_c = 1.63 \pm 0.1\ g/cm^3$. $In(C_5H_5\text{-}c)_3$ is a polymer, and is built up from parallel ···$In(C_5H_5\text{-}c)_2\text{-}(C_5H_5\text{-}c)\text{-}In(C_5H_5\text{-}c)_2$··· chains. The compound contains two nonbridging C_5H_5 ligands, B and C as depicted in **Fig. 9**, which are identical and monohapto-bonded and show nonplanarity. The third C_5H_5 ring, A, is bridging and planar and connected with the neighboring In atom by C-3. Thus, the central In atom is surrounded by a distorted tetrahedron of 4 carbon atoms. A few of the most important bond distances and angles are listed below [3].

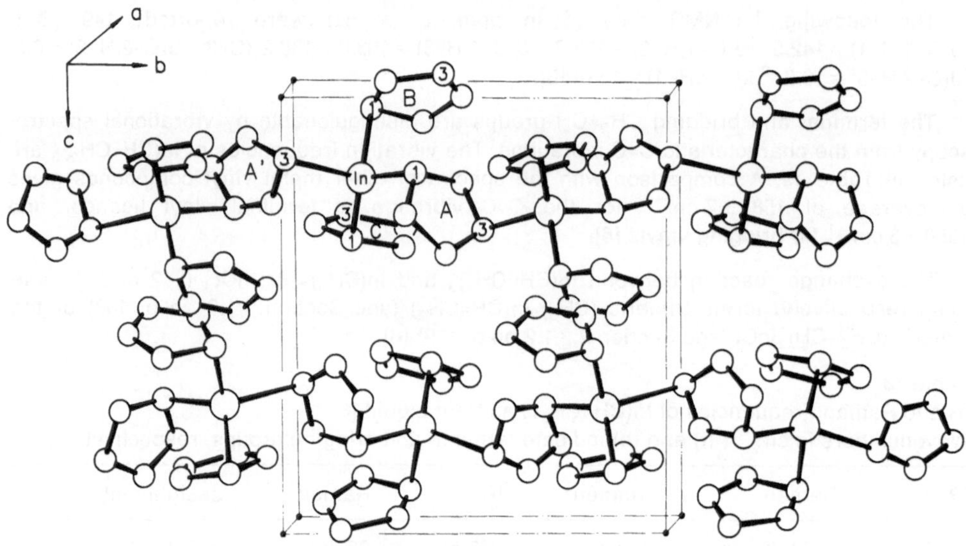

Fig. 9. The molecular structure of $In(C_5H_5\text{-}c)_3$ [3].

Distances (pm)		Angles (°)	
In–C(1)(A)	237.4(7)	C(1)(A)–In–C(1)(A′)	92.1(3)
In–C(3)(A′)	246.6(8)	C(1)(A)–In–C(1)(B)	110.2(4)
In–C(1)(B)	224.3(9)	C(1)(A)–In–C(1)(C)	117.7(4)
In–C(1)(C)	223.7(9)	C(3)(A′)–In–C(1)(B)	101.4(3)
In···In	557	C(3)(A′)–In–C(1)(C)	112.2(3)

Thermal decomposition of $In(C_5H_5\text{-}c)_3$ occurs, starting at about 150 °C, to give $In^I(C_5H_5\text{-}c)$ and $(C_5H_5\text{-}c)_2$; the cyclopentadienide anion has a catalytic function. The same mechanism was proposed for the decomposition of the (hypothetical!) $In(C_5(CH_3)_5)_3$ [7]. The mass spectrum of $In(C_5H_5\text{-}c)_3$ (inlet temperature 190 °C ?) shows $[InC_5H_5]^+$ as the base peak, along with the fragments $[In(C_5H_5)_2]^+$, $^{115}In^+$, $[(C_5H_5)_2]^+$, and $[C_5H_5]^+$ [4].

References:

[1] Fischer, E. O.; Hofmann, H. P. (Angew. Chem. **69** [1957] 639/40).
[2] Mangham, J. R. (U. S. 2969382 [1961] from C. A. **1961** 11332).
[3] Einstein, F. W. B.; Gilbert, M. M.; Tuck, D. G. (Inorg. Chem. **11** [1972] 2832/6).
[4] Poland, J. S.; Tuck, D. G. (J. Organometal. Chem. **42** [1972] 307/14).
[5] Visser, H. D.; Oliver, J. P. (J. Organometal. Chem. **40** [1972] 7/14).
[6] Fries, W.; Sille, K.; Weidlein, J. (Spectrochim. Acta A**36** [1980] 611/9).
[7] Beachley, O. T., Jr.; Blom, R.; Churchill, M. R.; Faegri, K., Jr.; Fettinger, J. C.; Pazik, J. C.; Victoriano, L. (Organometallics **8** [1989] 346/56).

1.1.5 InR_3 Compounds and Their Adducts, where R = Aryl

The compounds of this section are listed in Table 14. They were synthesized by the following methods.

Method I: From metallic In and HgR_2.
Indium metal reacted with HgR_2 in a 2:3 mole ratio at temperatures over 100 °C.

In the first preparation of $In(C_6H_5)_3$ (Table 14, No. 1) the reaction mixture was kept at 130 °C for 8 to 10 h under CO_2 as a protective gas [1]. In [2] or [35] it was heated for 48 h (under N_2) at 130 or 150 °C, nearly doubling the yield of product. In boiling toluene the reaction required longer times (~60 h), but gave similar yields [35]. By far the best results were achieved with especially finely divided ("active") indium. For this method Na (or K) was used to reduce $InCl_3$ in boiling toluene within 4 to 6 h; the resulting mixture of In and NaCl (or KCl) was treated with an equimolar amount of HgR_2, and was then refluxed another 1.5 to 3 h under argon. After filtering off the residue and working up the solution, the yield of InR_3 was nearly quantitative [27].
For many R groups (C_6F_5, C_6H_4X) the separation of InR_3 from unreacted HgR_2 presents difficulties. In such cases the filtrate of the reaction mixture (in toluene, xylene, or benzene) was treated with dioxane, precipitating slightly soluble $R_3In \cdot C_4H_8O_2$, which can be decomposed by Method III [10, 11, 21].

Method II: From $InCl_3$ and RMgCl.
A freshly prepared ether solution of RMgCl (or RMgBr) was used to alkylate $InCl_3$. When the reaction was finished, the ether phase was evaporated, and the residue dried in vacuum at 100 °C [8, 11]. More elegant for this reaction, too, is to precipitate the slightly soluble dioxane adduct, $R_3In \cdot C_4H_8O_2$, and work it up by Method III.

Method III: From decomposition of the dioxane adduct $R_3In \cdot C_4H_8O_2$.
The adduct, $R_3In \cdot C_4H_8O_2$, was precipitated by dioxane from the reaction mixtures from Methods I or II, was washed with a little cold ether, benzene, or toluene, and then was dried. Decomposition at 100 °C and 2 to 3 Torr gave base-free InR_3 in nearly quantitative yield [10, 11].

Method IV: Adducts of the type $R_3In \cdot D$ from the components or from the ether or dioxane adducts.
The reactants were mixed in equimolar amounts in an inert solvent, freed from solvent after 1 to 2 h, then purified by recrystallization [12]. Much more often the initial reagent was the ether adduct, $R_3In \cdot O(C_2H_5)_2$, or the dioxane adduct, $R_3In \cdot C_4H_8O_2$. These were mixed with an excess of the new Lewis component, D, thus eliminating the ether or dioxane and replacing it by D. The adducts formed were purified by recrystallization or sublimation [10, 11, 21]. The etherate, $R_3In \cdot O(C_2H_5)_2$, occurred when working in ether solution by Method I or II, and it was usually used in solution for further work. $R_3In \cdot C_4H_8O_2$ was obtained from such ether solutions by precipitation with dioxane [10, 11, 21].

General Remarks: $In(C_6F_5)_3$ and its various adducts have all been characterized by infrared spectroscopy. The most distinctive feature of the spectra of these compounds is the constancy of absorption frequency for the many vibrations of the C_6F_5 ligands. Assignments of the frequency values for the $In(C_6F_5)_3$ framework are given in Table 16 on p. 98 [21]; the typical C_6F_5 vibrations of the adducts described in [17, 21, 25] lie at (in cm^{-1}): 1637 ± 2 m to s, 1609 ± 1 vw, 1577 ± 4 vw to w, 1554 ± 2 sh to vw, 1534 ± 2 sh to w, 1509 ± 1 sh, 1506 ± 2 s to vs, 1461 ± 4 vs, 2 bands, 1375 ± 3 m to s, 1355 ± 5 m to s, 2 bands, 1375 ± 3 m to s, 1355 ± 5 m to s, 2 bands, 1264 ± 5 m, 1210 ± 2 vw, br, 1126 ± 5 vw to w, 1111 ± 3 vw, br, 1068 ± 5 vs, 1054 ± 1 s to vs, 1014 ± 3 sh to w, 956 ± 2 vs, 784 ± 1 w to m, 736 ± 2 vw, 717 ± 2 vw to w, 604 ± 3 w to m, 582 ± 1 vw, 486 ± 3 w, 355 ± 5 mw to m, 2 bands, 310 vw, 276 ± 1 vw to w. For the individual compounds in Table 14 only the additional absorptions will be cited.

Table 14
InR_3 Compounds and Their Adducts, $R_3In \cdot D$ where R = Aryl.
Further information on compounds whose numbers are preceded by an asterisk is given at the end of the table.
Explanations, abbreviations, and units on p. X.

No.	compound method of preparation (yield)	properties and remarks
*1	$In(C_6H_5)_3$ I (45%) [1] I (65 to 90%) [2, 35] II (ca. 50%) [11]	colorless crystals, m.p. 206 to 208 °C (from xylene) [27] ^{13}C NMR ($CDCl_3$): 128.5 (C-3,5), 129.2 (C-4), 138.4 (C-2,6), C-1 not observable [29] ^{115}In NQR (solid, 25 °C): 49.95 (ν_1), 85.17 (ν_2), 129.97 (ν_3), 173.63 (ν_4), e^2Qq/h 1043.5 MHz, η = 13.3% [31] UV (C_6H_{11}-CH_3): λ_{max} ($\varepsilon \cdot 10^3$) = 236(18), 259(5.5), 270(3) [32] IR: see further information
2	$In(C_6H_4F\text{-}3)_3$ I (71%)	white solid, m.p. 174 to 176 °C (from C_6H_6/C_6H_{12}-c) [35] mass spectrum (150 to 250 °C, 70 eV): $[M]^+$ (1%), $[M-H]^+$ (0.8%), $[M-C_6H_4F]^+$ (42.7%), $[M-2\,C_6H_4F]^+$ (2.1%), $^{115}In^+$ (53.1%) [35]
3	$In(C_6H_4F\text{-}4)_3$ I (85%)	white solid, m.p. 252 to 254 °C (from $C_6H_5CH_3$) [35] ^{115}In NQR (ca. 20 °C): ν_1 44.15, ν_2 88.25, ν_3 132.44, ν_4 179.59, e^2Qq/h 1059.3 MHz, η = 0.8% [29] mass spectrum (150 to 250 °C, 70 eV): $[M]^+$ (0.6%), $[M-H]^+$ (0.9%), $[M-C_6H_4F]^+$ (50.7%), $[M-2\,C_6H_4F]^+$ (2.0%), $^{115}In^+$ (45.8%) [35]
*4	$In(C_6F_5)_3$ special (31%)	white solid, m.p. 176 to 178 °C, subl. 140 to 150 °C/ 10^{-3} Torr, dec. 280 °C; monomeric in C_6H_6 [21, 22] ^{19}F NMR (C_6H_6): 118.4 (F-2), 150.6 (F-4), 159.5 (F-3) [21] IR: see further information
5	$In(C_6H_4Cl\text{-}4)_3$ I (31%)	white solid, m.p. 267 to 269 °C (from $C_6H_5CH_3$) [35] ^{115}In NQR (20 °C): ν_1 44.72, ν_2 87.99, ν_3 132.25, ν_4 176.36, e^2Qq/h 1058.2 MHz, η = 4.0% [29] ^{35}Cl NQR: 34.05 MHz [29] mass spectrum (150 to 250 °C, 70 eV): $[M]^+$ (0.8%), $[M-H]^+$ (0.8%), $[M-C_6H_4Cl]^+$ (39.6%), $[M-2\,C_6H_4Cl]^+$ (2.1%), $^{115}In^+$ (56.8%) [35]
6	$In(C_6H_4Br\text{-}4)_3$ III	m.p. > 200 °C, dec. 200 °C [10,11] readily soluble in dioxane, ether, $CHCl_3$ and $(CH_3)_2SO$, slightly soluble in C_6H_6, CCl_4, C_6H_{14}, and petroleum ether [10, 11]
7	$In(C_6H_4OCH_3\text{-}4)_3$ I (68%)	white solid, m.p. 140 to 142 °C (from C_6H_6/C_6H_{14}) [35] ^{115}In NQR (ca. 20 °C): ν_1 50.54, ν_2 87.21, ν_3 133.01, ν_4 177.66, e^2Qq/h 1067.3 MHz, η = 12.9% [29] mass spectrum (150 to 250 °C, 70 eV): $[M]^+$ (1.7%), $[M-H]^+$ (2.0%), $[M-C_6H_4OCH_3]^+$ (42.2%), $[M-2\,C_6H_4OCH_3]^+$ (1.2%), $^{115}In^+$ (52.9%) [35]

Table 14 (continued)

No.	compound method of preparation (yield)	properties and remarks
8	$In(C_6H_4CH_3\text{-}2)_3$ I (70%) III [10]	white solid, m.p. 151 to 153 °C (from C_6H_{12}-c) [35], 138 to 140 °C, soluble in organic solvents like No. 6 [10, 11]
9	$In(C_6H_4CH_3\text{-}3)_3$ I (ca. 65%)	white solid, m.p. 131 to 133 °C (from C_6H_6/C_6H_{14}) [35] soluble in organic solvents like No. 6, except hexane and petroleum ether [35]
10	$In(C_6H_4CH_3\text{-}4)_3$ I (90 to 100%) III	white solid, m.p. 243 to 244 °C (from $C_6H_5CH_3$) [27, 35], m.p. 225 °C [10, 11] ^{115}In NQR (ca. 20 °C): ν_1 44.40, ν_2 88.49, ν_3 132.84, ν_4 177.14, e^2Qq/h 1062.8 MHz, $\eta=2.5\%$ [29] mass spectrum (150 to 250 °C, 70 eV): $[M]^+$ (0.9%), $[M-H]^+$ (1.7%), $[M-C_6H_4CH_3]^+$ (41.9%), $[M-2\,C_6H_4CH_3]^+$ (1.7%), $^{115}In^+$ (53.8%) [35]
11	$In(C_6H_4C_2H_5\text{-}4)_3$ I (52%)	white solid, m.p. 119 to 121 °C (from C_5H_{14}) [35] ^{115}In NQR (20 °C): ν_1 46.63, ν_2 85.84, e^2Qq/h 1041.2 MHz, $\eta=9.3\%$, ν_3 and ν_4 not observable [29] mass spectrum (150 to 250 °C, 70 eV): $[M]^+$ (1.1%), $[M-H]^+$ (2.0%), $[M-C_6H_4CH_3]^+$ (50.8%), $[M-2\,C_6H_4CH_3]^+$ (1.0%), $^{115}In^+$ (44.7%) [35]
12	$In(C_6H_4C_4H_9\text{-}t)_3$ I (87%)	white solid, m.p. 230 to 233 °C (from C_6H_{12}-c) [35] ^{115}In NQR (ca. 20 °C): ν_1 45.72, ν_2 84.82, ν_3 128.30, ν_4 171.20, e^2Qq/h 1027.7 MHz, $\eta=8.7\%$ [29] mass spectrum (150 to 250 °C, 70 eV): $[M]^+$ (1.1%), $[M-H]^+$ (1.6%), $[M-C_6H_4C_4H_9]^+$ (46.5%), $[M-2\,C_6H_4C_4H_9]^+$ (0.9%), $^{115}In^+$ (50.0%) [35]
*13	$In(C_6H_2(CH_3)_3\text{-}2,4,6)_3$ II (80%)	colorless crystals, m.p. 178 to 179 °C [37] ^1H NMR (C_7D_8): 2.17 (s, CH_3-4), 2.34 (s, CH_3-2,6), 6.81 (s, C_6H_2) [37] ^{13}C NMR (C_7D_8): 21.3 (CH_3-4, J(C,H) = 126), 25.2 (CH_3-2,6, J(C,H) = 124.1), 127.8, 136.2, 140.6, 150.4 (C_6H_2); for the low temperature NMR, see further information [37] IR (solid): 1740 w, 1700 w, 1590 m, 1530 m, 1280 w, 1265 w, 1220 w, 1005 s, 940 s, 855 s, 715 m [37]
14	$In(C_{10}H_7\text{-}1)_3$ $C_{10}H_7\text{-}1=\alpha$-naphthyl III	dec. > 200 °C without melting [11] solubility like No. 6
15	$In(C_4H_3S\text{-}2)_3$ C_4H_3S = thienyl III	dec. > 200 °C without melting [11] solubility like No. 6

Table 14 (continued)

No.	compound method of preparation (yield)	properties and remarks
adducts, $R_3In \cdot D$ and $R_3In \cdot 2\,D$		
16	$(C_6H_5)_3In \cdot O(C_2H_5)_2$	not isolated, decomposition forms No. 1 [10, 11]
17	$(C_6H_5)_3In \cdot C_4H_8O_2$ II+IV (56%)	white crystalline solid, m.p. 138 to 139 °C [10, 11] soluble in all organic solvents except C_6H_{14} and petroleum ether. Used for the preparation of other adducts [10, 11]
18	$(C_6H_5)_3In \cdot NC_5H_5$ IV	white solid, m.p. 128 to 130 °C, monomeric in solution [12] ΔH_s (assumed) = 33 kcal/mol; ΔH_f = 13.8 to 13.9 [7, 12]
*19	$(C_6F_5)_3In \cdot O(C_2H_5)_2$ II+IV (34%)	white crystals, m.p. 137 to 140 °C, subl. ca. 127 °C/0.05 Torr, monomeric in C_6H_6 [8] IR: see further information
20	$(C_6F_5)_3In \cdot C_4H_8O_2$ $C_4H_8O_2$ = dioxane II+IV (41%)	white solid, m.p. 293 to 298 °C (dec.) [21] IR (solid): 1356 s, 1298 w, 1104 sh, 1086 sh, 1044 sh, 897 w, 859 s, 824 sh, 609 m, 302 w, 289 w, in addition to the C_6F_5-modes (see general remarks on p. 89) [21] low solubility in most organic solvents indicates a polymeric structure, $\cdot\cdot InR_3 \cdot OC_4H_8O \cdot InR_3 \cdot OC_4H_8O \cdot\cdot$
21	$(C_6F_5)_3In \cdot 2\,THF$ IV (ca. 100%)	white solid, m.p. 157 to 160 °C, monomeric in C_6H_6 [25] IR (solid): 1441 sh, 1176 vw, br, 1012 m, 919 vw, 852 m, br, in addition to the C_6F_5-modes (see general remarks on p. 89) [25]
22	$(C_6F_5)_3In \cdot 2\,OS(CH_3)_2$ IV (91%)	white solid, m.p. 160 to 162 °C (from ether/petroleum ether), monomeric in benzene [25] IR (solid): 1440 sh, 1418 sh, 1322 w, 1304 vw, 1160 vw, 1033 vw, 1005 vs ν (S=O), 670 vw ν(SC_2?), 413 m, in addition to the C_6F_5-modes (see general remarks on p. 89) [25]
23	$(C_6F_5)_3In \cdot OP(C_6H_5)_3$ IV (65%)	white solid, m.p. 160 to 163 °C (from C_6H_6/petroleum ether), monomeric in benzene [25] IR (solid): 1594 vw, 1498 vw, 1443 s, 1421 vw, 1340 vw, 1313 vw, 1159 s ν(P=O), 1097 m, 1029 vw, 1000 w, 747 w, 731 s, 695 s, 676 sh, 540 s, 522 sh, 470 vw, 437 vw, 414 vw ν(In–O?), 293 vw, in addition to the C_6F_5-modes (see general remarks on p. 89) [25]
24	$(C_6F_5)_3In \cdot OAs(C_6H_5)_3$ IV (81%)	white solid, m.p. 164 to 167 °C (dec.) (from C_6H_6/petroleum ether), monomeric in benzene [25] IR (solid): 1485 m, 1440 vs, 1437 sh, 1422 sh, 1341 sh, 1314 sh, 1279 vw, 1186 w, 1163 vw, 1089 sh, 1025 sh, 1000 w, 930 vw, 874 vs ν(As=O), 853 sh, 742 vs, 690 s,

Table 14 (continued)

No.	compound method of preparation (yield)	properties and remarks
24 (continued)		682 sh, 473 m, 461 m, 408 m ν(In–O), 397 sh, 290 w, 264 vw, in addition to the C_6F_5-modes (see general remarks on p. 89) [25]
25	$(C_6F_5)_3In \cdot (CH_3)_2NCH_2CH_2N(CH_3)_2$ IV (87%)	slightly brown solid, m.p. 70 to 115 °C, dimeric in benzene [25] IR (solid): 1481 sh, 1437 sh, 1420 sh, 1290 vw, 1240 sh, 1180 vw, 1170 sh, 1019 w, 1001 sh, 893 vw, 837 vw, 799 m, 773 w, 708 sh, 496 sh, 485 m, br, 446 vw, in addition to the C_6F_5-modes (see general remarks on p. 89) [25]
26	$(C_6F_5)_3In \cdot NC_5H_5$ IV (90%)	white solid, m.p. 127 to 130 °C (from petroleum ether), monomeric in benzene [25] IR (solid): 1614 s, 1494 sh, 1438 sh, 1243 vw, 1223 m, 1163 w, 1077 sh, 1019 m, 754 m, 693 s, 680 sh, 643 m, 419 w, in addition to the C_6F_5-modes (see general remarks on p. 89) [25]
27	$(C_6F_5)_3In \cdot P(C_6H_5)_3$ IV (71%)	white solid, m.p. 220 to 222 °C [25], 220 to 224 °C (dec.) [17], monomeric in benzene [25] IR (solid): 1440 s ν(CC), 1099 m ν("X-sens."q), 1005 w, br δ(CH), 749 vs γ(CH), 695 s δ(phenyl), 525 s and 495 s ν("X-sens."y), 444 m ν("X-sens."t), in addition to the C_6F_5-modes (see general remarks on p. 89) [25] oxidizes in air to give No. 23 [25]
28	$(C_6F_5)_3In \cdot As(C_6H_5)_3$ IV (60%)	white solid, m.p. 217 to 220 °C, dec. >200 °C, partly dissociates in benzene [25] IR (solid): 1486 w, 1441 sh, 1336 sh, 1309 w, 1185 w, 1159 w, 1080 sh, 1007 sh, 999 w, 920 vw, 739 vs, 693 s, 672 vw, 474 s, 468 sh, 332 m, 327 sh, in addition to the C_6F_5-modes (see general remarks on p. 89) [25]
29	$(C_6H_4Br\text{-}4)_3In \cdot C_4H_8O_2$ II + IV (34%)	white solid, m.p. ca. 200 °C [10, 11] thermal dec. gives No. 6 [10, 11]
30	$(C_6H_4CH_3\text{-}2)_3In \cdot C_4H_8O_2$ II + IV (54%)	white solid, m.p. 114 to 115 °C [10, 11] solubility like No. 6 [10, 11] thermal dec. gives No. 8 [10, 11]
31	$(C_6H_4CH_3\text{-}3)_3In \cdot 1/2\,C_4H_8O_2$ II + IV (86%)	white solid, m.p. 103 to 104 °C [11] solubility like No. 6 [11] thermal dec. gives No. 9 [11]
32	$(C_6H_4CH_3\text{-}4)_3In \cdot C_4H_8O_2$ II + IV (49%)	white solid, m.p. 132 to 133 °C [10, 11] thermal dec. gives No. 10 [10, 11]

Table 14 (continued)

No.	compound method of preparation (yield)	properties and remarks
33	$(C_{10}H_7\text{-}1)_3In \cdot C_4H_8O_2$ $C_{10}H_7\text{-}1 = \alpha$-naphthyl II + IV	solid, dec. >150 °C [11]
34	$(C_4H_3S\text{-}2)_3In \cdot C_4H_8O_2$ C_4H_3S = thienyl II + IV (60%)	yellow crystalline solid, dec. >150 °C without melting [11] thermal dec. in vacuum gives No. 15 [11]
adducts, $R_3In \cdot D\text{-}D \cdot InR_3$		
35	$2\,(C_6F_5)_3In \cdot NC_6H_4\text{-}C_6H_4N\text{-}2,2'$ IV (62%)	white solid, m.p. 188 to 189 °C, monomeric in $CHCl_3$ [25] UV (1.4×10^{-5} mol/L in $O(C_2H_5)_2$): λ_{max} ($\varepsilon \cdot 10^3$) = 245(18), 252(16), 258(14), 291(22), 299(26), 310(23) [25] IR (solid): 1613 m, 1604 m, 1586 m, 1573 w, 1481 sh, 1440 sh, 1421 sh, 1342 sh, 1319 w, 1280 vw, 1221 vw, 1181 w, 1162 w, 1021 s, 1004 sh, 892 w, 771 sh, 761 vs, 737 m, 653 m, 633 w, 418 m, 368 sh, in addition to the C_6F_5-modes (see general remarks on p. 89) [25]
36	$2\,(C_6F_5)_3In \cdot (C_6H_5)_2PCH_2CH_2P(C_6H_5)_2$ IV (ca. 100%)	white solid, m.p. 189 to 191 °C (slight dec.), monomeric in $CHCl_3$ [25] IR (solid): 1586 vw, 1486 vw, 1441 sh, 1420 vw, 1337 sh, 1310 vw, 1189 sh, 1168 vw, 1158 vw, 1099 w, 1028 vw, 1000 w, 918 vw, 891 vw, 839 vw, 751 sh, 747 m, 729 s, 706 vw, 692 m, 682 m, 674 sh, 667 sh, 513 m, 475 m, 458 sh, 454 w, 299 vw, in addition to the C_6F_5-modes (see general remarks on p. 89) [25]

* Further Information:

$In(C_6H_5)_3$ (Table **14**, No. **1**). Melting points of 208 °C [2], 206 to 208 °C [10, 11, 27], and 209 to 210 °C [35] have been reported. The value of 291 °C, given in [1], deviates so drastically that a printing error is suspected. $In(C_6H_5)_3$ can be purified by vacuum sublimation (no pressure given) at temperatures of 150 to 180 °C. With the help of gravimetric effusion techniques the heat of sublimation, ΔH_s, was determined to be 33.6 ± 0.4 kcal/mol; structural conclusions were drawn from comparison with the sublimation heats of the triphenyl compounds of the other group 13 elements [18]; the size of ΔH_s makes a planar structure plausible.

$In(C_6H_5)_3$ is, according to cryoscopic molecular weight determinations in benzene (or dioxane), strongly monomeric [5]. Ebullioscopic measurements in benzene and dioxane demonstrated the presence of only monomers [14].

The dipole moment, $\mu = 0.7$ D in benzene (0.8 D in dioxane) is very small, leading to the conclusion that the compound should be a distinctly poorer electron acceptor than the homologous triphenyl derivatives of B, Al, and Ga [4].

Solubility in g/L (mol/L) is: 4.1 (0.0118) in heptane, 49.0 (0.142) in benzene, 197 (0.569) in diethyl ether, and 51.2 (0.148) in chloroform [3].

The IR spectrum of solid $In(C_6H_5)_3$ was recorded in [23], and has been partially assigned by [28]. In [14] the spectrum between 4000 and 100 cm^{-1} was recorded, completely assigned, and compared with the spectra of $M(C_6H_5)_3$ (M = B, Al, Ga). To calculate the planar and nonplanar vibrations two molecular fragments were discussed: the InC_3 skeleton (with D_{3h} symmetry) and the $In-C_6H_5$ fragment (with C_{2v} symmetry). The force constant, k(InC) = 4.5, was obtained for the planar modes, and a value of k = 0.8 N/cm was found for the nonplanar vibrations [14]. To confirm and refine the assignments and calculations $In(C_6D_5)_3$ was measured and evaluated to establish the force constants [20]. Table 15 lists the frequencies and their assignments.

Table 15
IR Spectra of Solid $In(C_6H_5)_3$ and $In(C_6D_5)_3$ [20].
"In-sensitive" vibrations are designated by an asterisk; A_2 modes given in parentheses are calculated values.
Wavenumbers in cm^{-1}.

$In(C_6H_5)_3$	$In(C_6D_5)_3$	class	assignment	$In(C_6H_5)_3$	$In(C_6D_5)_3$	class	assignment
3090	–	A_1	ν(CH)	1020	770	A_1	δ(CH), δ(CD)
3063	–	B_1	ν(CH)	995	955	A_1	δ(CCC)
3049	2265	A_1	ν(CH), ν(CD)	985	920	B_2	γ(CH), γ(CD)
3040	2235	A_1	ν(CH), ν(CD)	(977)	(870)	A_2	γ(CH), γ(CD)
3016	–	B_1	ν(CH)	905	820	B_2	γ(CH), γ(CD)
1595	1560	B_1	ν(CC)	(817)	(643)	A_2	γ(CH), γ(CD)
–	1543	A_1	ν(CC)*	723	532	B_2	γ(CH), γ(CD)
1490	–	A_1	ν(CC)	697, 690	552	B_2	δ(CC)
1423	1310	A_1	ν(CC)*				
1330, 1290	1290	B_1	ν(CC)	673	630	A_1	ν(CCIn)*
				615	585	B_1	δ(CCC)
1250	1015	B_1	δ(CH), δ(CD)	446	400	B_2	γ(CIn)*
1185	835	B_1	δ(CH), δ(CD)	(405)	(325)	A_2	γ(CCC)
1152	825	A_1	δ(CH), δ(CD)	245	255	A_1	δ(CCIn)*
1070	1040	A_1	ν(CCIn)*	185	190	B_1	δ(CIn)*
1040	–	B_1	δ(CH)	180	–	B_2	δ(CCIn)*

$In(C_6H_5)_3$ crystallizes in the orthorhombic space group Pbcn – D_{2h}^{14} (No. 60) with the lattice constants a = 1482(3), b = 1113(2), and c = 886(2) pm; Z = 4 and D_c = 1.57 g/cm^3. The central atom is coordinated to 5 carbon atoms forming the corners of a distorted trigonal bipyramid. The three phenyl groups of the basal plane are twisted only about 3° out of the plane of the InC_3 skeleton. The vertices of the bipyramid are occupied by the ortho carbon atoms (6″ and 6‴) of two neighboring $In(C_6H_5)_3$ units at a distance of 307(2) pm as shown in **Fig. 10**. The significantly greater distances to the carbon atoms 5″ and 7″ are 349(2) and 340(2) pm, respectively [13, 19].

The mass spectrum of $In(C_6H_5)_3$ was recorded at excitation energies of 70 eV [26, 35], 20 and 10 eV [26], and at temperatures of 150 to 225 °C [35] and 290 °C [26]. At temperatures of 120 to 200 °C no fragments with two (or more) In atoms were observed. At ca. 225 °C and 70 eV, however, such ions were found (rel. abundance in %): $[In_2(C_6H_5)_5]^+$ (8), $[In_2(C_6H_5)_4]^+$ (1), $[In_2(C_6H_5)_3]^+$ (22), along with $[In(C_6H_5)_3]^+$ (2), $[In(C_6H_5)_2]^+$ (20), $[InC_6H_5]^+$ (0.4), and In^+ (7). The $[In_2(C_6H_5)_6]^+$, reported in [26], was not found in [35]. The In-C

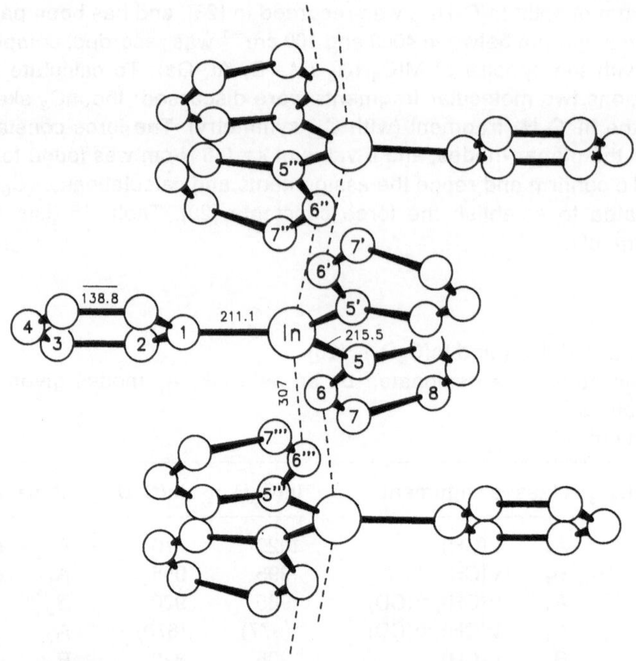

Fig. 10. Molecular structure of $In(C_6H_5)_3$ [19]; projection down [302].

Angles (°)			
C(1)-In-C(5)	121.6(3)	C(6'')-In-C(1)	97.8
C(5)-In-C(5')	116.8(6)	C(6'')-In-C(5)	87.5
In-C(1)-C(2)	123.0(8)	C(6'')-In-C(5')	84.4
In-C(5)-C(6)	120.5(11)		

dissociation energy, $D(C_6H_5\text{-}In(C_6H_5)_2) = 0.4$ eV, was estimated from the appearance potentials $AP = 9.0 \pm 0.2$ eV for $[In(C_6H_5)_3]^+$, 9.4 ± 0.2 eV for $[In(C_6H_5)_2]^+$, and 9.3 ± 0.5 eV for In^+ [26]. The relative abundances (in %) of In-containing ions are collected below, the base peak of each spectrum is that of $[C_{12}H_{10}]^+$.

fragment	70 eV		20 eV	10 eV
$[M]^+$	2.1	3.4	8.3	11.2
$[M-H]^+$	2.8			
$[M-C_6H_5]^+$	39.2	39.4	67.7	73.6
$[M-C_6H_5\text{-}H]^+$				
$[M-2\,C_6H_5]^+$	3.0	2.9	3.3	3.3
In^+	52.8	50.2	18.9	11.2
Ref.	[35]	[26]	[26]	[26]

Thermal decomposition of $In(C_6H_5)_3$ in sealed ampules in the presence of ^{14}C-labeled C_6H_6 and at temperatures of 180 to 200 °C formed mainly biphenyl. The decomposition was accompanied by phenyl-group exchange between $In(C_6H_5)_3$ and C_6H_6. Contrary to the behavior of $Ga(C_6H_5)_3$, this exchange was not observed in the photolytic decomposition of $In(C_6H_5)_3$ [9].

$In(C_6H_5)_3$ and the halogens X_2 (X=Br or I) react in benzene solution with formation of the corresponding compounds $(C_6H_5)_2InX$ or $C_6H_5InX_2$, depending on the stoichiometry, in yields up to 80% [1]. Oxidation of $In(C_6H_5)_3$ by O_2 in benzene solution followed by hydrolysis of the products produced phenol and a 20% yield of biphenyl; $\{(C_6H_5)_2In\}_2O$ was assumed to be the intermediate [2]. CO_2, although used as a protective atmosphere in the first preparation of $In(C_6H_5)_3$ [1], was slowly inserted into one of the In-C bonds; after hydrolysis, benzoic acid was isolated in 18% yield [2]. Reaction with SO_2 was at first assumed to be an insertion, too, with formation of $(C_6H_5)_2InO_2SC_6H_5$ [23], but more recent investigations of the IR spectra, and the instability of this compound, argue for a polymeric adduct $\{(C_6H_5)_3InOSO\}_n$ (see Section 4.4 on p. 211) [28].

$In(C_6H_5)_3$ and thiols (RSH with R = C_2H_5, C_3H_7-n, C_4H_9-t, C_6H_5, $C_6H_4CH_2$ [36], and $C_{12}H_{25}$ [24]) in 1:1 mole ratio react to form $(C_6H_5)_2InSR$ and C_6H_6. In a 1:2 ratio the reaction does not lead to $C_6H_5In(SR)_2$ as reported in [24], but to a mixture of $(C_6H_5)_2InSR$ and $In(SR)_3$ [36].

Double substitution was achieved, however, with dithiols $HS(CH_2)_xSH$ (x=3 to 5) and $HSCH_2SCH_2SH$, heterocycles being formed [36] (see Section 5.2.2 on p. 248).

Carboxylic acids RCOOH (R = CH_3, C_2H_5) and $In(C_6H_5)_3$ (as the 1:1 dioxane adduct) reacted in 1:1 or 1:2 mole ratio to form $(C_6H_5)_2InO_2CR$ or $C_6H_5In(O_2CR)_2$, respectively, in yields of 60 to 80%. These compounds were also obtained from the metathesis of $In(C_6H_5)_3$ and $In(O_2CR)_3$ [15].

Some reactions of $In(C_6H_5)_3$ with various benzene derivatives are described and compared with the corresponding reactions of $Ga(C_6H_5)_3$ and $Tl(C_6H_5)_3$. Thus, reaction with C_6H_5CHO led (depending on the mole ratio and the reaction duration) to $C_6H_5CH_2OH$ (ca. 80%) or $(C_6H_5)_2C{=}O$ (ca. 20%), after hydrolysis. The postulated precursors are $(C_6H_5)_2InOCH(C_6H_5)_2$ and $(C_6H_5)_2InOCH_2C_6H_5$. A 58% yield of $(C_6H_5)_3COH$ resulted from steam-distilling the product obtained from $In(C_6H_5)_3$ and $(C_6H_5)_2C{=}O$ in boiling xylene. $(C_6H_5)_2C{=}O$ was synthesized in 40% yield by hydrolysis of the reaction mixture formed from $In(C_6H_5)_3$ and C_6H_5COCl in boiling benzene. Benzalacetophenone and $In(C_6H_5)_3$ produce β,β-diphenylpropiophenone in 92% yield [2].

The failure of $In(C_6H_5)_3$ to react with metallic Hg (12 h in boiling benzene) demonstrated that preparative Method I (see p. 88) is not reversible [2].

The $In(C_6H_5)_3$ adducts 16 to 18 are collected in Table 14. Anionic adducts, $[(C_6H_5)_3In\cdot X]^-$, have been obtained for X = $[Mn(CO)_5]^-$, $[Co(CO)_4]^-$, $[C_5H_5W(CO)_3]^-$, and $[C_5H_5Fe(CO)_2]^-$ (with the counter cations $[NR_4]^+$ and $[N(P(C_6H_5)_3)_2]^+$). The formation constants of these complexes have been assessed, and the IR spectra in the region of the CO bond vibrations have been analyzed [16, 34].

$In(C_6H_5)_3$ served, together with $TiCl_4$, as a catalyst for the polymerization of ethylene; the proportions of $TiCl_4$ had a decisive influence on the degree of polymerization [30].

$Mn_2(CO)_{10}$ and $In(C_6H_5)_3$ (or $M(C_6H_5)_2$ with M = Hg, Zn) undergo an insertion reaction, ultimately with formation of $(C_6H_5)_2CO$ [33].

$In(C_6F_5)_3$ (Table **14**, No. **4**) was obtained along with InI by heating a mixture of C_6F_5I and powdered In (3:4 mole ratio) at 160 °C for 6 h in a sealed ampule. Fractional sublimation produced the analytically pure compound [21, 22]. The product obtained from $Hg(C_6F_5)_2$ by Method I was not pure even after repeated sublimations (150 °C/ca. 10^{-3} Torr), its melting range being 159 to 174 °C; however, analytically pure $(C_6F_5)_3In \cdot O(C_2H_5)_2$ was isolated by extraction with diethyl ether [21] (see Table 14, No. 19).

$In(C_6F_5)_3$ was also formed along with TlBr in the reaction of $(C_6F_5)_2TlBr$ and In metal (3:2 mole ratio, 160 °C, 24 h, sealed thick-walled ampule), but it was isolated only in the form of the $P(C_6H_5)_3$ adduct No. 27 [17].

$In(C_6F_5)_3$ is extraordinarily sensitive to moisture and instantly forms C_6F_5H, when exposed to air. The IR spectral assignments for the solid were made in conjunction with the evaluation of the spectra of C_6F_5I and $(C_6F_5)_2TlBr$ [21] and given in Table 16.

Table 16
IR Spectra of Solid $In(C_6F_5)_3$ [21].
Descriptions ν_1 to ν_{28} are taken from [6].
Wavenumbers in cm^{-1}.

$In(C_6F_5)_3$	class	assignment	$In(C_6F_5)_3$	class	assignment
1858 vw	B_2	1511+359	1110 sh		
1717 vw			1080 vs	A_1	ν_5(C-F)
1644 s	A_1+B_2	$\nu_1+\nu_{21}$ (ring)	1072 vs		
1610 vw			1050 sh		
1582 sh	B_2	1277+309	1010 m		
1558 sh	B_2	956+584	956 vs	B_2	ν_{25}(C-F)
1533 sh			834 vw		
1511 vs }	A_1+B_2	$\nu_1+\nu_{22}$ (ring)	792 sh }	A_1	"X-sens"-ν_6=
1470 vs, br }			788 m }		ν(In-C)
1446 sh	B_2	956+491	739 w		
1377 s	B_2	1080+309	718 m	A_1	2×359
1371 s	A_1	ν_3(C-F)	608 m	B_1	ν_{15}
1325 vw	B_2	956+359	584 vw	A_1	ν_7
1277 sh }	B_2	ν_4(C-F)+	491 m	A_1	ν_8
1264 s }	A_1	ν_{23} (ring)	447 vw	B_2	ν_{27}
1213 vw			359 m }	A_1+B_1	$\nu_9+\nu_{17}$
1179 w			354 m }		
1136 w }	B_2	ν_{24}(C-F)	309 vw	B_2	ν_{28}
1127 w }			267 w, br	A_1	ν_{10}

$In(C_6H_2(CH_3)_3\text{-}2,4,6)_3$ (Table **14**, No **13**) can be handled in air for several minutes without appreciable decomposition; it is soluble in ether and THF but does not form stable adducts with these bases [37].

The ortho methyl resonances in the ^1H NMR spectrum show some broadening at −90 °C. The signals of the ortho CH_3 groups in the ^{13}C NMR spectrum are equivalent at room temperature and split into two signals at −80 to −90 °C (δ=25.1 ppm, J(H,C)=124 Hz and δ=27.1 ppm, J(H,C)=102.0 Hz), indicating some agostic interaction [37].

The compound crystallizes in the monoclinic space group $P2_1/n-C^5_{2h}$ (No. 14) with the cell constants a=870.4(2), b=2188.8(6), c=1249.2(4) pm, β=95.06(2)°; Z=4 and D_c=1.323 g/cm^3; the data were collected at −85 °C. The solid consists of discrete monomeric units. As depicted in **Fig. 11** the geometry around the In atom is planar and the mesityl rings 1 to 3 are crystallographically nonequivalent. The mesityl ring planes are twisted with respect to the InC_3 core in a propeller-like fashion with the following angles: ring 1, 43.1°; ring 2, 70.2°, and ring 3, 33.6°. Remarkable are the relatively short C···C distances between adjacent ortho methyl groups, close to the van der Waals radii of 400 pm (C(19)···C(29) = 398.3(8); C(17)···C(39) = 381.0(8); C(27)···C(37) = 418.2(9) pm). The proximity of these methyl groups is probably responsible for the low temperature effects in the ^{13}C NMR spectrum [37].

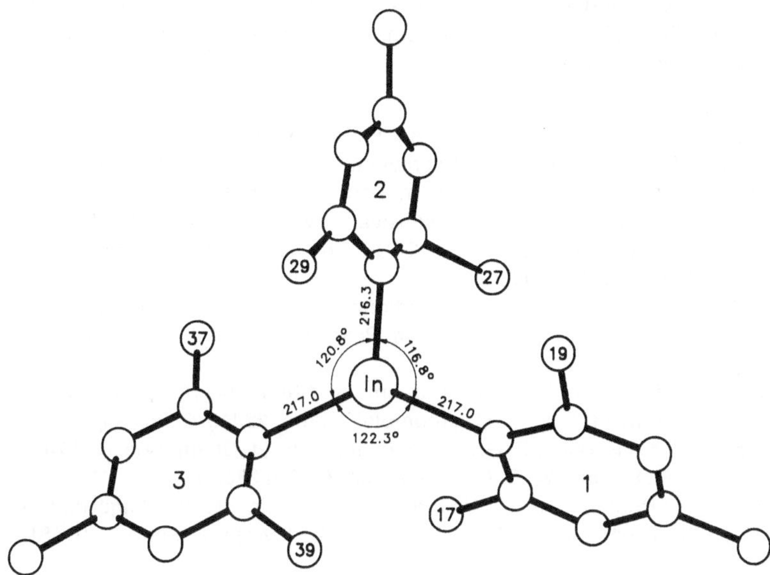

Fig. 11. Molecular structure of $In(C_6H_2(CH_3)_3\text{-}2,4,6)_3$ viewed approximately perpendicular to the plane of the InC_3 core [37].

Addition of 1 equivalent $[N(CH_3)_4]Cl$ to $In(mes)_3$ in CH_2Cl_2 solution (mes = $C_6H_2(CH_3)_3\text{-}2,4,6$) gave $[(CH_3)_4N][(mes)_3InCl]$ in 91% yield. Reaction of the compound with $InCl_3$ (2:1 or 1:2 mole ratio) in benzene solution produced $(mes)_2InCl$ (73% yield) or $mesInCl_2$ (83% yield), respectively; see Chapter 2.2.1.3 on p. 130. With 2 equivalents $[(CH_3)_3NH]Cl$ the adduct $(mes)Cl_2In \cdot N(CH_3)_3$ was obtained in 34% yield [37].

$(C_6F_5)_3In \cdot O(C_2H_5)_2$ (Table **14**, No. **19**) was obtained in 38% yield, when the residue from the reaction of In with $Hg(C_6F_5)_2$ (170 °C, 6 d) [21, 22] was extracted with ether. The adduct was eluted in 28% yield by ether from the residue of the reaction of C_6F_5I and In powder (160 °C, 6 h, see No. 4). The reported melting point was 145 to 147 °C, after sublimation at 115 °C and 5×10^{-3} Torr. The following absorption bands (in cm^{-1}) were cited for the solid (in addition to those of the C_6F_5 ligands; see general remarks on p. 89): 1442 sh, 1394 m, 1325 w, 1198 vw, 1135 w, 1030 m, 897 m, 833 w, 776 s, 524 sh, 501 vw [21].

References:

[1] Schumb, W. C.; Crane, H. I. (J. Am. Chem. Soc. **60** [1938] 306/8).

[2] Gilman, H.; Jones, R. G. (J. Am. Chem. Soc. **62** [1940] 2353/7).

[3] Strohmeier, W.; Hümpfner, K. (Chem. Ber. **90** [1957] 2339/41).

[4] Strohmeier, W.; Hümpfer, K. (Z. Elektrochem. **61** [1957] 1010/3).

[5] Strohmeier, W.; Hümpfner, K.; Miltenberger, K.; Seifert, F. (Z. Elektrochem. **63** [1959] 537/9).

[6] Long, D. A.; Steele, D. (Spectrochim. Acta **19** [1963] 1947/61).
[7] Greenwood, N. N.; Perkins, P. G. (Proc. 8th Intern. Conf. Coord. Chem., Vienna 1964, pp. 32/4; C.A. **67** [1967] 26490).
[8] Pohlmann, J. L. W.; Brinckmann, F. E. (Z. Naturforsch. **20b** [1965] 5/11).
[9] Razuvaev, G. A.; Kaplin, Yu. A.; Mitrofanova, E. V. (Izv. Akad. Nauk SSSR Ser. Khim. **1965** 1489/91; Bull. Acad. Sci. USSR Div. Chem. Sci. **1965** 1455/7).
[10] Viktorova, I. M.; Endovin, Yu. P.; Sheverdina, N. I.; Kocheshkov, K. A. (Dokl. Akad. Nauk SSSR **177** [1967] 103/4; Dokl. Chem. Proc. Acad. Sci. USSR **172/177** [1967] 967/8; C.A. **68** [1968] No. 49674).

[11] Viktorova, I. M.; Sheverdina, N. I.; Endovin, Yu. P.; Kocheshkov, K. A. (Izv. Akad. Nauk SSSR Ser. Khim. **1968** 2410/2; Bull. Acad. Sci. USSR Div. Chem. Sci. **1968** 2288/90).
[12] Greenwood, N. N.; Perkins, P. G.; Twentyman, M. E. (J. Chem. Soc. A **1969** 249/53).
[13] Malone, J. F.; McDonald, W. S. (J. Chem. Soc. D **1969** 591/2).
[14] Rodionov, A. N.; Ruch'eva, N. I.; Viktorova, I. M.; Shigorin, D. N.; Sheverdina, N. I.; Kocheshkov, K. A. (Izv. Akad. Nauk SSSR Ser. Khim. **1969** 1047/9; Bull. Acad. Sci. USSR Div. Chem. Sci. **1969** 956/8).
[15] Viktorova, I. M.; Sheverdina, N. I.; Rodionov, A. N.; Kocheshkov, K. A. (Dokl. Akad. Nauk SSSR **189** [1969] 315/7; Dokl. Chem. Proc. Acad. Sci. USSR **184/189** [1969] 889/91; C.A. **72** [1970] No. 55562).
[16] Burlitch, J. M.; Petersen, R. B. (J. Organometal. Chem. **24** [1970] C65/C67).
[17] Deacon, G. B.; Parrott, J. C. (J. Organometal. Chem. **22** [1970] 287/95).
[18] Greenwood, N. N.; Perkins, P. G.; Twentyman, M. E. (J. Chem. Soc. A **1970** 2109/11).
[19] Malone, J. F.; McDonald, W. S. (J. Chem. Soc. A **1970** 3362/7).
[20] Ruch'eva, N. I.; Rodionov, A. N.; Viktorova, I. M.; Zenina, G. V.; Shigorin, D. N.; Sheverdina, N. I. (Zh. Prikl. Spektrosk. **13** [1970] 322/6; J. Appl. Spectrosc. [USSR] **13** [1970] 1093/6).

[21] Deacon, G. B.; Parrott, J. C. (Australian J. Chem. **24** [1971] 1771/9).
[22] Deacon, G. B.; Parrott, J. C. (Inorg. Nucl. Chem. Letters **7** [1971] 329/31).
[23] Hsieh, A. T. T. (J. Organometal. Chem. **27** [1971] 293/301).
[24] Viktorova, I. M.; Sheverdina, N. I.; Kocheshkov, K. A. (Dokl. Akad. Nauk SSSR **198** [1971] 94/5; Dokl. Chem. Proc. Acad. Sci. USSR **196/201** [1971] 367/8; C.A. **75** [1971] No. 49184).
[25] Deacon, G. B.; Parrott, J. C. (Australian J. Chem. **25** [1972] 1169/77).
[26] Glockling, F.; Irwin, J. G. (J. Chem. Soc. Dalton Trans. **1973** 1424/5).
[27] Chao, L.-C.; Rieke, R. D. (Synth. React. Inorg. Metal-Org. Chem. **4** [1974] 373/8).
[28] Hsieh, A. T. T.; Deacon, G. B. (J. Organometal. Chem. **70** [1974] 39/42).
[29] Freeman, W. H.; Miller, S. B.; Brill, T. B. (J. Magn. Resonance **20** [1975] 378/87).
[30] Uzhinova, L. D.; Paleev, O. A.; Plate, N. A.; Richmond, B. G. (Vysokomol. Soedin. A **17** [1975] 1516/21 from C.A. **83** [1975] No. 164740).

[31] Bancroft, G. M.; Sham, T. K. (J. Magn. Resonance **25** [1977] 83/90).
[32] Smith, J. H.; Brill, T. B. (Inorg. Chem. **16** [1977] 20/4).
[33] Haupt, H. J.; Huber, F.; Neumann, F. (Forschungsber. Landes Nordrhein-Westfalen No. 2721 [1978] 27 pp.; C.A. **91** [1979] No. 211523).
[34] Burlitch, J. M.; Leonowicz, M. E.; Petersen, R. B.; Hughes, R. E. (Inorg. Chem. **18** [1979] 1097/105).
[35] Miller, S. B.; Jelus, B. L.; Smith, J. H.; Munson, B.; Brill, T. B. (J. Organometal. Chem. **170** [1979] 9/19).
[36] Hoffmann, G. G. (J. Organometal. Chem. **338** [1988] 305/17).
[37] Leman, J. T.; Barron, A. R. (Organometallics **8** [1989] 2214/9).

1.1.6 InR_2R' Compounds, Their Adducts and Compounds of the Type $InR_2(R'\text{-}D)$ and $InR(R'\text{-}D)_2$

The examples in this section are assembled in Table 17. InR_2R' compounds comprise Nos. 1 to 14, and the adducts $InR_2R' \cdot D$ Nos. 15 to 21. Compounds in which a nitrogen atom is located in position 4 or 5 of R′ (or R) give In containing heterocyclic systems (see general remarks) as shown in Nos. 22 to 45. The compounds were prepared by the following general methods:

Method Ia: From R_2InCl and MR′ (M = Li, Na).
A suspension of the corresponding R_2InCl in cyclohexane (No. 5 [5]), CH_2Cl_2 (No. 14 [10]), or hexane (Nos. 8 to 11 [13, 24], Nos. 42 to 45 [32]) was stirred for several hours at room temperature with the appropriate alkali organic reagent, then refluxed, usually for 1 to 2 h. The residue of LiCl or NaCl, normally very finely divided, was separated (most conveniently by centrifugation) and the InR_2R', freed from solvent, was purified by distillation, sublimation, or recrystallization. Compounds of the type $InR_2(R'\text{-}D)$ [12, 29, 30, 32] can also be made by this procedure; the Li compounds were used for Nos. 42 to 45 [32]. Starting with CH_3InCl_2, the substitution of both Cl atoms is possible to generate No. 12 [13].

Method Ib: From $RInCl_2$ and Grignard compounds.
Instead of alkali organic reagents the appropriate Grignard reagents in ether (2 h reflux temperature, 24 h at room temperature) were used. The ether was removed and the residue dissolved in pentane and worked up as described above [31].

Method II: HR elimination from InR_3 and an acid HR′.
Toluene (No. 5 [4]) or benzene (No. 7 [1]) served as reaction medium. The C-H acidity of the HR′ employed has a decisive influence on the reaction duration: e.g., 48 h for the reaction of $In(CH_3)_3$ with C_5H_6-c in boiling toluene [5], only 30 min and 50 °C for the reaction with $HC{\equiv}CC_6H_5$ in benzene [1]. By the use of the Lewis bases $(CH_3)_2NCH_2CH_2N(CH_3)_2$ [14] and $CH_3OCH_2CH_2OCH_3$ [20] as solvents, the carborane derivatives 20 and 21 could be isolated in the form of stable 1:1 adducts.

Method III: Ligand exchange between InR_3 and InR'_3.
The reaction can be carried out without a solvent (Nos. 1 and 2 [17]) or in benzene or toluene (No. 4 [10]) in mole ratios of 2:1 (forming InR_2R') or 1:2 (forming $InRR'_2$). The yield of crude product is nearly quantitative.

General Remarks. In those cases with a nitrogen atom in position 4 or 5 (compounds with other heteroatoms are not yet described), intramolecular adduct formation in $InR_2(R'\text{-}D)$ compounds occurs, resulting in heterocyclic ring systems with fourfold coordinated In atoms, as depicted in the general Formula I (Nos. 22, 23, 25 to 36) and II (No. 24). In $InR(R'\text{-}D)_2$ compounds the In atom is fivefold coordinated as shown in Formula III (Nos. 37 to 45).

Similar rings systems have also been formulated for compounds resulting from the reaction of InR_3 (R = CH_3, C_2H_5) with CH_3NO_2 or $CH_3CH_2NO_2$ (Formula IV, R′ = H, CH_3); to characterize the complexes the IR and ^{115}In NQR spectra had been measured and evaluated [3]. However, later investigations based on careful IR and Raman studies by the same group [7] showed that the chelating structure without a third In–C bond is more plausible (Formula V); for these compounds, see Section 4.4, p. 211.

IV V

Explanation for Table 17: Assignment of the vinyl protons H-2 is trans and H-3 is cis to the In atom (Nos. 4, 14). Relative intensities of mass peaks are given in % in parentheses.

Table 17
Compounds of the InR_2R' Type and Their Adducts.
Further information on numbers preceded by an asterisk is given at the end of the table.
Explanations, abbreviations, and units on p. X.

No.	compound method of preparation (yield)	properties and remarks
*1	$In(CH_3)_2C_2H_5$ Ia [19] III (ca. 100%) [17]	colorless liquid, m.p. 5 to 7 °C, b.p. 23 to 25 °C/1.5 Torr [17, 19] vapour pressure at 17 °C: 0.85 Torr [16] ^1H NMR: −0.36 (s, CH_3In), 0.37 (q, CH_2), 1.27 (t, CH_3), as neat liquid (?) [17]; 0.22, 0.52 (J = 8.4), 1.42 in C_6D_6 [19] ^{13}C NMR (C_6D_6): −0.05 (CH_3In), 13.13 (CH_3 of C_2H_5), 14.01 (CH_2) [33] IR data on p. 112
*2	$InCH_3(C_2H_5)_2$ III (ca. 100%)	colorless liquid, b.p. 33 to 35 °C/0.1 Torr [17] ^1H NMR (neat?): −0.39 (s, CH_3In), 0.39 (q, CH_2), 1.28 (t, CH_3) [17] ^{13}C NMR (C_6D_6, ca. 30 °C): −2.57 (CH_3In), 12.44 (CH_3), 12.80 (CH_2) [33]; see also further information IR and Raman data on p. 112
3	$InCH_3(CH_2C_4H_9\text{-t})_2$ Ia (62%)	colorless liquid, b.p. 55 °C/0.01 Torr; soluble in C_5H_{12}, $O(C_2H_5)_2$, and THF; degree of association in benzene 1.15 to 1.19 for 0.0598 to 0.0875 molar solutions [25] ^1H NMR (C_6D_6): −0.07 (CH_3In), 0.89 (CH_2In), 1.06 (CH_3C) [25]

Table 17 (continued)

No.	compound method of preparation (yield)	properties and remarks
3 (continued)		IR (neat): 2958 vs, 2905 s, 2860 s, 1465 m, 1457 s, 1382 w, 1360 m, 1260 w, 1234 m, 1109 m, 1090 m, 1055 w, 1012 m, 995 w, 797 w, 737 w, 688 m, 575 w, 482 m, 450 w, 372 w [25] it forms the adducts No. 15 to 18 [25]
4	$In(CH_3)_2CH{=}CH_2$ III	white solid, m.p. 67 to 68 °C, trimeric in C_6H_6 [10] ^1H NMR (C_6D_6): −0.02 (CH_3In), 6.07 (q, H-3), 6.48 (q, H-2), 6.75 (q, H-1, J(H-1,3) = 21.92, J(H-1,2) = 14.44, J(H-2,3) = 4.63) [10] ^{13}C NMR (C_6D_6): −6.33 (q, CH_3In, J(H,C) = 126.3), 153.2 (dm, C-1, ^1J(H-1,C) = 143.7, ^2J(H-3,C) = 5.5, ^2J(H-2,C) = 2.8), 142.5 (ddd, C-2, J(H-3,C) = 159.0, J(H-2,C) = 153.8, ^2J(H-1,C) = 11.6) [10] IR and Raman (solid): 1557 ν(C=C)-bridge, 1052 γ,ω(CH_2), 971, 510 $\nu_{as}(InC_2)$, 487 $\nu_s(InC_2)$, 456 γ(CH) and ν(InC) [10]
*5	$In(CH_3)_2C_5H_5$-c Ia (ca. 40%) [5] II (73%) [4]	colorless solid, m.p. 195 °C (dec.) [4], subl. 120 °C/10^{-4} Torr, dec. 130 °C [5] ^1H NMR: −0.25 (CH_3In), 6.30 (s, C_5H_5) in $CDCl_3$ [5], −1.25 and 5.63 in DMF (slow dec.) [4] IR (solid, principal absorptions): 700 s ϱ(CH_3), 530 s $\nu_{as}(InC_2)$, 470 to 410 ms, br $\nu_s(InC_2)$, 330 m ν(InC_5H_5) [4] mass spectrum (25 °C, 70 eV): M^+, $[M-CH_3]^+$, $[InC_5H_5]^+$, $[In(CH_3)_2]^+$, and $^{115}In^+$ are the principal fragments (no intensities given) [4]
*6	$In(CH_3)_2C{\equiv}CCH_3$ Ia (40%)	white crystals, m.p. 101 to 103 °C, subl. 80 °C/0.1 Torr [9] ^1H NMR (C_6D_6): 0.16 (CH_3In), 1.31 (s, CH_3C) [9] ^{13}C NMR (C_6D_6): −5.45 (q, CH_3In, J(H,C) = 128.3), 4.89 (q, CH_3, ^1J(H,C) = 131.3), 90.88 (C-1, ^3J(H,C) = 3.8), 122.43 (q, C-2, ^2J(H,C) = 8.7) [9] IR (Nujol): 2965 vs $\nu_{as}(CH_3)$, 2880 sh $\nu_s(CH_3)$, 2095 ms ν(C≡C), 1420 m, br $\delta_{as}(CH_3)$, 1162 ms δ_s(CH_3 at In), 960 s ν(C-CH_3), 718 vs, br ϱ(CH_3 at In), 685 sh, 521 s

Table 17 (continued)

No.	compound method of preparation (yield)	properties and remarks
*6 (continued)		$\nu_{as}(InC_2)$, 486 ms $\nu_s(InC_2)$, 385 s, br $\nu(InC)$ and $\delta(C{\equiv}C{-}C)$, 280 ms $\nu(InC)$ [9] Raman (solution): 2100 s, p $\nu(C{\equiv}C)$, 1375 w, p $\delta_s(CH_3)$, 963 vw $\nu(C{-}CH_3)$, 522 w, dp $\nu_{as}(InC_2)$, 490 vs, p $\nu_s(InC_2)$, 360 w, dp $\nu(InC)$ and $\delta(C{\equiv}C{-}C)$; other frequencies and Raman (solid) data are also given [9]
7	$In(CH_3)_2C{\equiv}CC_6H_5$ II	cream solid, m.p. 116 to 119 °C, soluble in hydrocarbon solvents, dimeric in C_6H_6 [1] ^1H NMR: 0.29 (CH_3In), 6.8 to 7.0 (m, H-3,4,5 of C_6H_5), 7.0 to 7.4 (m, H-2,6 of C_6H_5) in C_6D_6; −0.35 and 7.0 to 7.4 in THF [1] IR: 2050 in C_6H_6 and 2110 in THF $\nu(C{\equiv}C)$ (adduct formation?) [1]
8	$In(CH_3)_2CH_2Sn(CH_3)_3$ Ia (60%)	colorless liquid, b.p. 66 °C/14 Torr [13] ^1H NMR (C_6D_6): −0.21 (CH_2Sn, $J(^{117/119}Sn,H) = 57.4/60.1$), −0.05 ($CH_3In$), 0.19 ($CH_3Sn$, $J(^{117/119}Sn,H) = 50.6/53.0$) [13] ^{13}C NMR (C_6D_6): −14.86 (CH_2Sn, $J(^{117/119}Sn,C) = 258.8/270.8$), −10.30 ($CH_3In$), −7.1 ($CH_3Sn$, $J(^{117/119}Sn,C) = 314.5/328.9$) [13] ^{119}Sn NMR (C_6D_6): 48.4 [13]
9	$\{In(CH_3)_2CH_2\}_2Sn(CH_3)_2$ Ia (74%)	colorless, very air-sensitive liquid, b.p. 55 °C/25 Torr; dec. in short time even stored under Ar, but stable in $O(C_2H_5)_2$ [24] ^1H NMR (C_6D_6): −0.12 (CH_3In), 0.15 (CH_3Sn, $J(Sn,H) = 55.4$ and 59.5), 1.35 (CH_2Sn, $J(Sn,H) = 38.4$ and 40.9) [24] ^{119}Sn NMR (C_6D_6): −0.7 [24] IR (liquid ?): only 595 ms and 438 ms are reported and assigned as $\nu(InC)$ [24]
10	$In(C_2H_5)_2CH_2I$	not isolated, existence confirmed by methanolysis at −20 °C of the reaction mixture of $In(C_2H_5)_3$ and CH_2I_2 to give CH_3I [2]
11	$In(C_4H_9\text{-}t)_2CH_2Sn(CH_3)_3$ Ia (55%)	colorless liquid, b.p. 75 °C/11 Torr [13] ^1H NMR (C_6D_6): −0.21 (CH_2Sn, $J(^{117/119}Sn,H) = 57.3/60.0$), 0.14 ($CH_3C$), 0.18 ($CH_3Sn$, $J(^{117/119}Sn,H) = 50.7/53.1$) [13] ^{13}C NMR (C_6D_6): −14.88 (CH_2Sn, $J(^{117/119}Sn,C) = 258.7/272.0$), −10.34 ($CH_3$), 7.85 ($CH_3Sn$, $J(^{117/119}Sn,C) = 314.4/329.0$)[13] ^{119}Sn NMR (C_6D_6): 23.3 [13]

Table 17 (continued)

No.	compound method of preparation (yield)	properties and remarks
12	$InCH_3\{CH_2Sn(CH_3)_3\}_2$ Ia (60%)	colorless liquid, b.p. 84 °C/18 Torr [13] ^1H NMR (C_6D_6): −0.24 (CH_2Sn, $J(^{117/119}Sn,H)=57.7/60.3$), 0.11 ($CH_3In$), 0.14 ($CH_3Sn$, $J(^{117/119}Sn,H)=50.7/52.5$) [13] ^{13}C NMR (C_6D_6): −14.89 (CH_2Sn, $J(^{117/119}Sn,C)=258.8/270.5$), −9.44 ($CH_3In$), −7.87 ($CH_3Sn$, $J(^{117/119}Sn,C)=314.5/329.1$) [13] ^{119}Sn NMR (C_6D_6): 23.2 [13]
13	$In(C_2H_5)_2C_5H_5$-c Ia (30 to 40%)	colorless solid, dec. 130 °C [5], subl. 110 °C/10^{-4} Torr [8] ^1H NMR (C_6D_6): 0.40 (q, CH_2In), 1.41 (t, CH_3, J=8), 6.40 (s, C_5H_5) [5] ^{13}C NMR (C_6D_6): 9.74 (CH_2), 13.02 (CH_3, $^1J(H,C)=122.0$), 111.54 (C_5H_5, $^1J(H,C)=163.3$, $^2J(H,C)=7.0$) [6] IR (solid): 3110 vw $\nu_s(CH)$, 3070 m, br $\nu_{as}(CH)$, 1465 mw $\nu_{as}(CC)$, 1109 w $\nu_s(CC)$, 500 s $\nu_{as}(InC_2)$, 459 m $\nu_s(InC_2)$, 255 sh ν(InC ?) [5]
14	$In(CH{=}CH_2)_2C{\equiv}CCH_3$ Ia	white solid, m.p. 61 to 63 °C (from CH_2Cl_2), dimeric in C_6H_6 [10] ^1H NMR (C_6D_6): 1.39 (CH_3C), 6.08 (H-3), 6.41 (H-2), 6.66 (H-1), J(H-1,3)=20.95, J(H-1,2)=13.91, J(H-2,3)=3.55 [10] ^{13}C NMR (C_6D_6): 5.65 (q, CIn, $^3J(H,C)=3.8$), 89.16 (q, br, CH_3, $^1J(H,C)=132.7$), 125.50 (q, CCH_3, $^2J(H,C)=9.8$), 136.0 (ddd, C-2, J(H-2,C)=154.5, J(H-3,C)=161.3, $^2J(H-1,C)=11.1$), 146.0 (d, br, C-1, J(H-1,C)=151) [10] IR and Raman (solid): 2092 $\nu(C{\equiv}C)$, 1573 ν(C=C, terminal), 1011, 988, 945 δ, $\omega(CH_2)$, 510, 473 $\nu(InC_2)$, 450 γ(CH), 386 δ(CCC), 280 ν(InC) [10]
adducts, $InR_2R' \cdot D$		
*15	$InCH_3(CH_2C_4H_9\text{-}t)_2 \cdot OC_4H_8$	colorless liquid [25] ^1H NMR (C_6D_6): 0.01 (CH_3In), 0.89 (CH_2In), 0.99 (m, OC_4H_8), 1.21 (CH_3C), 3.29 (m, OC_4H_8) [25]
*16	$InCH_3(CH_2C_4H_9\text{-}t)_2 \cdot N(CH_3)_3$	colorless liquid [25] ^1H NMR (C_6D_6): −0.12 (CH_3In), 0.79 (CH_2In), 1.13 (CH_3C), 1.66 (CH_3N) [25]

Table 17 (continued)

No.	compound method of preparation (yield)	properties and remarks
*17	$InCH_3(CH_2C_4H_9\text{-t})_2 \cdot$ $(CH_3)_2NCH_2CH_2N(CH_3)_2$	white semisolid [25] ^1H NMR (C_6D_6): −0.04 (CH_3In), 0.78, 1.17 (CH_2In), 1.21 (CH_3C), 1.87 (CH_3N), 2.22 (CH_2N) [25]
*18	$\{InCH_3(CH_2C_4H_9\text{-t})_2\}_2 \cdot$ $(CH_3)_2NCH_2CH_2N(CH_3)_2$	white solid [25] ^1H NMR (C_6D_6): −0.04 (CH_3In), 0.78, 1.15 (CH_2In), 1.22 (CH_3C), 1.82 (CH_3N), 2.25 (CH_2N) [25] IR spectra on p. 114
19	$In(CH_3)_2(C_5H_5\text{-c}) \cdot HN(CH_3)_2$	not isolated, unstable preliminary stage of No. 5 [4]
20	$In(CH_3)_2R' \cdot (CH_3)_2NCH_2CH_2N(CH_3)_2$ R' = $C\overset{O}{-}C-C_6H_5$ with $B_{10}H_{10}$ II (50%)	m.p. 260 °C (dec.) [14]
21	$In(CH_3)_2R' \cdot CH_3OCH_2CH_2OCH_3$ R' = $C\overset{O}{-}CH$ with $B_{10}H_{10}$ II (50%)	decomposes at 290 °C without melting [20]
adducts, $InR_2(R'\text{-}D)$ and $InR(R'\text{-}D)_2$		
22	$In(CH_3)_2(CH_2)_3N(CH_3)_2$ Ia (67%) Ib [23, 26]	non-pyrophoric, air-insensitive, white solid, m.p. 12 °C, b.p. 57 °C/ca. 4 Torr, monomeric in benzene [29] vapor pressure: 0.225 Torr at 20 °C and 3.75 Torr at 75 °C [27] ^1H NMR (C_6D_6): −0.28 (CH_3In), 0.57 (t, CH_2In), 1.75 (tt, CH_2), 1.88 (CH_3N), 1.93 (t, CH_2N) [29] ^{13}C NMR (C_6D_6): −7.5 (CH_3In), 10.7 (CH_2In), 25.9 (CH_2), 46.3 (CH_3N), 64.8 (CH_2N) [29] mass spectrum (70 eV): $[M-CH_3]^+$ (84%), $[M-2\,CH_3]^+$ (6%), $[In(CH_3)_2]^+$ (11%), $[InCH_3]^+$ (2%), ^{115}In (31%), $[R'\text{-}D]^+$ (43%), $[C_3H_8N]^+$ (100%), $[C_2H_4N]^+$ (15%) [29] used for vapor phase deposition of InP [26 to 28]
23	$In(CH_3)_2(CH_2)_3N(C_2H_5)_2$ Ia (68%)	colorless liquid, b.p. 101 °C/ca. 4 Torr [29] ^1H NMR (C_6D_6): −0.27 (CH_3In), 0.50 (t, CH_2In), 0.62 (t, CH_3 of C_2H_5), 1.73 (CH_2), 2.04 (t, CH_2N), 2.29 (q, CH_2 of C_2H_5) [29]

Table 17 (continued)

No.	compound method of preparation (yield)	properties and remarks
23 (continued)		^{13}C NMR (C_6D_6): −7.3 (CH_3In), 8.4 (CH_3 of C_2H_5), 9.3 (CH_2In), 25.0 (CH_2), 44.4 (CH_2 of C_2H_5), 57.5 (CH_2N) [29] mass spectrum (70 eV): $[M-CH_3]^+$ (100%), $[In(CH_3)_2]^+$ (17%), $[InCH_3]^+$ (4%), $[R'\text{-}D]^+$ (62%), $[(R'\text{-}D)-H_2]^+$ (52%), $[C_6H_{12}N]^+$ (14%), $[C_3H_8N]^+$ (21%), $[C_3H_6N]^+$ (16%), $[C_2H_4N]^+$ (12%) [29] used for vapor phase deposition of InP [21]
24	$In(CH_3)_2CH_2C_6H_4\{CH_2N(CH_3)_2\}$-2 Ia (54%)	colorless liquid, b.p. 78 to 79 °C/0.02 Torr [30] ^1H NMR (C_6D_6): −0.37 (CH_3In), 1.67 (CH_3N), 1.90 (CH_2In), 2.96 (CH_2N), 6.71 to 7.16 (m, C_6H_4) [30] ^{13}C NMR (C_6D_6): −8.6 (CH_3In), 21.6 (CH_2In), 46.9 (CH_3N), 64.8 (CH_2N), 121.3 (C-4 of C_6H_4), 127.6 (C-6), 129.3 (C-5), 130.9 (C-3), 149.2 (C-2), 150.0 (C-1 ?) [30] mass spectrum (70 eV): $[M]^+$ (3%), $[M-CH_3]^+$ (46%), $[M-C_2H_6]^+$ (12%), $[M-NC_4H_{12}]^+$ (14%), $[LH]^+$ (24%), L^+ (30%), $[In(CH_3)_2]^+$ (11%), $[InCH_3]^+$ (4%), $[LH-C_2H_6]^+$ (11%), $^{115}In^+$ (100%); $L=CH_2C_6H_4CH_2N(CH_3)_2$ [30]
25	$In(CH_3)_2CH_2C_6H_4\{N(CH_3)_2\}$-2 Ia (71%)	colorless solid, m.p. 47 °C, b.p. 94 °C/ 0.03 Torr [30] ^1H NMR (C_6D_6): −0.42 (CH_3In), 1.76 (CH_2In), 2.28 (CH_3N), 6.76 to 7.44 (m, C_6H_4) [30] ^{13}C NMR (C_6D_6): −4.6 (CH_3In), 18.3 (CH_2In), 45.3 (CH_3N), 118.3, 123.7, 125.1, 131.2 (C_6H_4), 143.8 (C-1 of C_6H_4), 149.8 (CN of C_6H_4) [30] mass spectrum (70 eV): $[M-CH_3]^+$ (37%), $[In(CH_3)_2]^+$ (6%), $[LH]^+$ (27%), L^+ (100%), $[InCH_3]^+$ (2%), $^{115}In^+$ (30%), $[C_7H_7]^+$ (18%); $L=CH_2C_6H_4N(CH_3)_2$ [30]
26	$In(CH_3)_2C_6H_4\{CH_2N(CH_3)_2\}$-2 Ia (79%)	colorless oil, b.p. 104 °C/0.1 Torr, monomeric in benzene [12] ^1H NMR (C_7D_8): 0.05 (CH_3In), 1.95 (s, CH_3N), 3.15 (s, CH_2N), 7.75 (m, C_6H_4) [12] ^{13}C NMR (C_7D_8): −3.3 (CH_3In), 45.4 (CH_3N), 68.1 (CH_2N), 136.8 (C-6 of C_6H_4), 144.5 (C-2 of C_6H_4), 159.4 (C-1 of C_6H_4) [12]

Table 17 (continued)

No.	compound method of preparation (yield)	properties and remarks
*27	$In(CH_3)_2C_6H_4\{CH_2N(C_2H_5)_2\}$-2 Ia (68%)	colorless solid, m.p. 30 to 31 °C, b.p. 104 °C/1.5 Torr [30] ^1H NMR (C_6D_6): −0.06 (s, CH_3In), 0.53 (t, CH_3 of C_2H_5), 2.29 (q, CH_2 of C_2H_5), 3.24 (s, CH_2N), 6.86 to 7.78 (m, C_6H_4) [30] ^{13}C NMR (C_6D_6): −8.2 (CH_3In), 8.5 (CH_3 of C_2H_5), 44.2 (CH_2 of C_2H_5), 61.4 (CH_2N), 125.4, 126.6, 126.9, 138.4 (C-3,4,5,6 of C_6H_4), 144.3 (C-2), 159.9 (C-1) [30] mass spectrum (70 eV): $[M-CH_3]^+$ (12%), $[LH]^+$ (7%), $[L]^+$ (4%), $[L-CH_2]^+$ (32%), $[In(CH_3)_2]^+$ (2%), $^{115}In^+$ (9%), $[C_7H_7]^+$ (100%), $[CH_2N(C_2H_5)_2]^+$ (7%), $[C_5H_5]^+$ (9%); $L=C_6H_4CH_2N(C_2H_5)_2$ [30]
*28	$In(CH_3)_2C_6H_4\{CHCH_3N(CH_3)_2\}$-2 Ia (69%)	colorless oil, b.p. 111 °C/0.1 Torr [12] ^1H NMR (C_7D_8): 0.05 (s, CH_3In), 1.20 (s, CH_3C), 2.05 (s, CH_3N), 7.80 (m, C_6H_4) [12] ^{13}C NMR spectra on p. 115
29	$In(CH_3)_2C_6H_3\{CH_2N(CH_3)_2\}_2$-2,6	no details reported, used for UV photoelectron spectroscopic comparison with similar complexes of Ni, Pd, and Pt, structure with 5-coordinated In [18]
30	$In(C_2H_5)_2(CH_2)_3N(CH_3)_2$ Ia (62%)	colorless liquid, b.p. 80 °C/ca. 2 Torr [29] ^1H NMR (C_6D_6): 0.50 (q, CH_2 of C_2H_5), 0.57 (t, CH_2In), 1.41 (t, CH_3 of C_2H_5), 1.78 (tt, CH_2), 1.91 (CH_3N), 1.93 (t, CH_2N) [29] ^{13}C NMR (C_6D_6): 5.3 (CH_2 of C_2H_5), 7.8 (CH_2In), 13.2 (CH_3 of C_2H_5), 26.2 (CH_2), 46.3 (CH_3N), 65.3 (CH_2N) [29]
31	$In(C_2H_5)_2(CH_2)_3N(C_2H_5)_2$ Ia (68%)	colorless liquid, b.p. 85 °C/ca. 0.2 Torr [29] ^1H NMR (C_6D_6): 0.57 (q, CH_2 of C_2H_5In), 0.58 (t, CH_3 of C_2H_5N), 0.73 (t, CH_2In), 1.47 (t, CH_3 of C_2H_5In), 1.73 (tt, CH_2), 2.01 (t, CH_2N), 2.28 (q, CH_2 of C_2H_5N) [29] ^{13}C NMR (C_6D_6): 5.7 (CH_2In), 6.3 (CH_2 of C_2H_5In), 8.5 (CH_3 of C_2H_5N), 13.3 (CH_3 of C_2H_5In), 25.2 (CH_2), 44.4 (CH_2 of C_2H_5N), 57.9 (CH_2N) [29]
32	$In(C_2H_5)_2C_6H_4\{CH_2N(CH_3)_2\}$-2 Ia (57%)	colorless liquid, b.p. 115 °C/0.08 Torr [30] ^1H NMR (C_6D_6): 0.76 (q, CH_2 of C_2H_5), 1.52 (t, CH_3 of C_2H_5), 1.94 (s, CH_3N), 3.21 (s, CH_2N), 7.00 to 7.22 (m, C_6H_4) [30]

Table 17 (continued)

No.	compound method of preparation (yield)	properties and remarks
32 (continued)		^{13}C NMR (C_6D_6): 5.2 (CH_2In), 13.6 (CH_3 of C_2H_5), 45.6 (CH_3N), 68.5 (CH_2N), 125.7 (C-5), 126.5 (C-4), 126.8 (C-3), 139.0 (C-6), 144.7 (C-2), 159.4 (C-1) [30]
33	$In(C_3H_7\text{-}n)_2(CH_2)_3N(CH_3)_2$ Ia (52%)	colorless liquid, b.p. 62 °C/ca. 0.04 Torr [29] ^1H NMR (C_6D_6): 0.66 (t, CH_2In), 0.73 (t, CH_2In of C_3H_7), 1.18 (t, CH_3 of C_3H_7), 1.62 (m, CH_2 of C_3H_7), 1.84 (tt, CH_2), 1.94 (CH_3N), 1.98 (t, CH_2N) [29] ^{13}C NMR (C_6D_6): 9.2 (CH_2In), 17.9 (CH_2In of C_3H_7), 20.8 (CH_3 of C_3H_7), 22.6 (CH_2 of C_3H_7), 26.1 (CH_2), 46.3 (CH_3N), 65.1 (CH_2N) [29]
34	$In(C_3H_7\text{-}n)_2(CH_2)_3N(C_2H_5)_2$ Ia (51%)	colorless liquid, b.p. 88 °C/ca. 10^{-2} Torr [29] ^1H NMR (C_6D_6): 0.53 (t, CH_2In of C_3H_7), 0.63 (t, CH_3 of C_2H_5N), 0.65 (CH_2In), 1.10 (t, CH_3 of C_3H_7), 1.77 (m, CH_2 of C_3H_7), 1.80 (tt, CH_2), 2.06 (t, CH_2N), 2.32 (q, CH_2 of C_2H_5N) [29] ^{13}C NMR (C_6D_6): 7.7 (CH_2In), 8.5 (CH_3 of C_2H_5N), 18.0 (CH_2In of C_3H_7), 20.9 (CH_3 of C_3H_7), 22.7 (CH_2 of C_3H_7), 25.1 (CH_2), 44.4 (CH_2 of C_2H_5N), 57.9 (CH_2N) [29]
35	$In(C_3H_7\text{-}i)_2(CH_2)_3N(CH_3)_2$ Ia (55%)	colorless liquid, b.p. 54 °C/ca. 0.15 Torr [29] ^1H NMR (C_6D_6): 0.68 (CH_2In), 1.12 (sept, CHIn), 1.62 (d, CH_3C), 1.81 (tt, CH_2), 1.93 (CH_3N), 2.02 (t, CH_2N) [29] ^{13}C NMR (C_6D_6): 6.4 (CH_2In), 18.9 (CHIn), 25.9 (CH_2), 46.3 (CH_3N), 65.2 (CH_2N) [29]
36	$In(C_3H_7\text{-}i)_2(CH_2)_3N(C_2H_5)_2$ Ia (53%)	colorless liquid, b.p. 81 °C/ca. 0.2 Torr [29] ^1H NMR (C_6D_6): 0.58 (t, CH_3 of C_2H_5N), 0.59 (t, CH_2In), 1.04 (m, CHIn), 1.56 (d, CH_3 of C_3H_7), 1.75 (tt, CH_2), 2.00 (t, CH_2N), 2.36 (q, CH_2 of C_2H_5) [29] ^{13}C NMR (C_6D_6): 4.9 (CH_2In), 8.6 (CH_3 of C_2H_5N), 19.4 (CHIn), 24.8 (CH_3 of C_3H_7), 25.1 (CH_2), 44.3 (CH_2 of C_2H_5N), 57.9 (CH_2N) [29]
37	$CH_3-In \leftarrow N-CH_3$ Ib (51%)	colorless liquid, b.p. 38 °C/0.05 Torr [31] ^1H NMR (C_6D_6): −0.26 (CH_3In), 0.53 (t, CH_2In), 1.80 (tt, CCH_2C), 1.96 (CH_3N), 2.05 (t, CH_2N) [31]

Table 17 (continued)

No.	compound method of preparation (yield)	properties and remarks
37 (continued)		^{13}C NMR (C_6D_6): −8.5 (CH_3In), 11.8 (CH_2In), 26.8 (-CH_2-), 44.5 (CH_3N), 60.5 (CH_2N) [31] mass spectrum: $[M+H]^+$ (13%), $[M]^+$ (4%), $[M-CH_3]^+$ (84%), $^{115}In^+$ (100%), $[C_5H_{11}N]^+$ (84%), $[C_3H_7N]^+$ (6%), $[C_2H_4N]^+$ (13%) [31]
38	$InCH_3\{(CH_2)_3N(CH_3)_2\}_2$	from corresponding R_2InCl-compound and $LiCH_3$ in $O(C_2H_5)_2$ (60% yield) [29] colorless liquid, b.p. 57 °C/ca. 0.08 Torr [29] ^{1}H NMR (C_7D_8): −0.24 (CH_3In), 0.54 (CH_2In), 1.91 (tt, CH_2), 2.00 (CH_3N), 2.10 (t, CH_2N) [29] ^{13}C NMR (C_7D_8): −7.6 (CH_3In), 10.7 (CH_2In), 25.9 (CH_2), 46.1 (CH_3N), 64.6 (CH_2) [29]
39	$InCH_3\{(CH_2)_3N(C_2H_5)_2\}_2$	prepared in analogy to No. 38 (56%) [29] colorless liquid, b.p. 52 °C/0.02 Torr [29] ^{1}H NMR (C_6D_6): −0.23 (CH_3In), 0.52 (t, CH_2In), 0.59 (t, CH_3 of C_2H_5), 1.74 (tt, CH_2), 2.02 (t, CH_2N), 2.27 (q, CH_2 of C_2H_5) [29] ^{13}C NMR (C_6D_6): −7.3 (CH_3In), 8.3 (CH_3 of C_2H_5), 9.1 (CH_2In), 25.0 (CH_2), 44.3 (CH_2 of C_2H_5), 57.5 (CH_2N) [29] mass spectrum (70 eV): $[M-CH_3-N(C_2H_5)_2]^+$ (100%), $[In(CH_3)_2]^+$ (11%), $^{115}In^+$ (29%), $[R'-D]^+$ (41%), $[(R'-D)-H_2]^+$ (32%), $[(R'-D)-C_2H_4]^+$ (84%), $[C_3H_8N]^+$ (14), $[C_3H_6N]^+$ (9%), $[C_2H_4N]^+$ (7%) [29]
40	$InCH_3\{C_6H_4(CH_2N(CH_3)_2)\text{-}2\}_2$	prepared in analogy to No. 38 [15] white powder (77%) [15] ^{1}H NMR ($CDCl_3$, ca. 35 °C): −0.30 (CH_3In), 2.50 (CH_3N), 3.70 (CH_2N), 7.20 (m, meta and para H of C_6H_4), 7.7 (m, ortho H of C_6H_4) [15] mass spectrum: M^+ and $[M-CH_3]^+$ (100%) and several recombination fragments (relative intensities not reported) [15]
41	[structure: CH_3–In with two N-CH_3 chelate rings] Ib (47%)	colorless liquid, b.p 114 °C/1 Torr [31] ^{1}H NMR (C_6D_6): −0.27 (CH_3In), 0.54 (CH_2In), 1.78 (tt, CCH_2C), 1.87 (CH_3N), 1.99 (t, CH_2N), 2.34 (NCH_2CH_2N) [31] ^{13}C NMR (C_6D_6): −8.4 (CH_3In), 12.0 (CH_2In), 25.3 (-CH_2-), 41.9 (CH_3N), 55.7 (NCH_2CH_2N), 62.0 (CH_2N) [31]

Table 17 (continued)

No.	compound method of preparation (yield)	properties and remarks
42	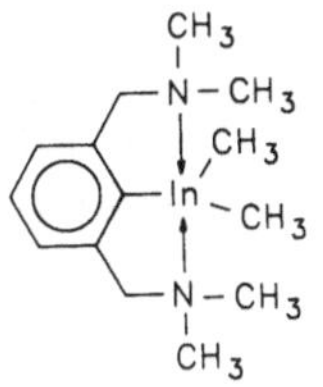 Ia (68%)	colorless crystals, m.p. 53 °C [32] ^{1}H NMR (C_6D_6): −0.09 (s, CH_3In), 2.01 (s, CH_3N), 3.20 (CH_2N), 6.82 to 7.79 (m, C_6H_3) [32] ^{13}C NMR (C_6D_6): −10.4 (CH_3In), 45.5 (CH_3N), 66.7 (CH_2N), 124.3 (C-3,5), 126.9 (C-4), 145.5 (C-2,6), 158.9 (C-1) [32] mass spectrum: $[M-N(CH_3)_2-C_2H_4]^+$ (39%), $[In(CH_3)_2]^+$ (7%), $[C_6H_4CH_2N(CH_3)_2]^+$ (100%), $[InCH_3]^+$ (2%), $^{115}In^+$ (33%), further fragments of the ligand [32]
43	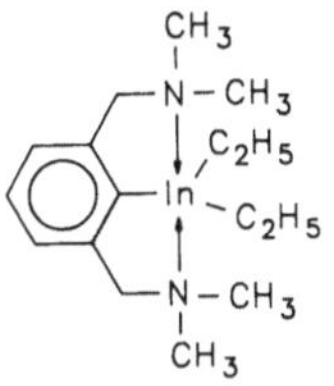 Ia (57%)	colorless crystals, m.p. 41 °C [32] ^{1}H NMR (C_6D_6): −0.79 (t, CH_3 of C_2H_5In), 1.61 (q, CH_2 of C_2H_5In), 2.02 (s, CH_3N), 3.20 (CH_2N), 6.93 to 7.16 (m, C_6H_3) [32] ^{13}C NMR (C_6D_6): 3.7 (CH_2 of C_2H_5In), 13.7 (CH_3 of C_2H_5In), 45.6 (CH_3N), 67.1 (CH_2N), 124.4 (C-3,5), 126.8 (C-4), 145.8 (C-2,6), 158.6 (C-1) [32] mass spectrum: $[M-C_2H_4]^+$ (16%), $[M-C_2H_5]^+$ (100%), $[M-2\,C_2H_4-N(CH_3)_2]^+$ (3%), $[M-C_2H_5-C_2H_4-N(CH_3)_2]^+$ (21%), $[M-In(C_2H_5)_2]^+$ (5%), $[InC_2H_5]^+$ (4%), $^{115}In^+$ (93%), further fragments of the ligand [32]
44	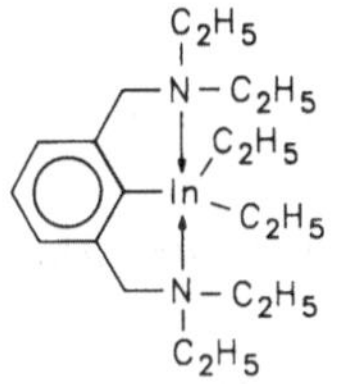 Ia (52%)	colorless liquid, b.p. 115 °C/ca. 0.04 Torr [32] ^{1}H NMR (C_6D_6): −0.47 (t, CH_3 of C_2H_5In), 0.60 (q, CH_2 of C_2H_5In), 1.32 (t, CH_3 of C_2H_5N), 2.24 (q, CH_2 of C_2H_5N), 3.13 (CH_2N), 6.59 to 7.26 (m, C_6H_3) [32] ^{13}C NMR (C_6D_6): 5.1 (CH_2 of C_2H_5In), 8.7 (CH_3 of C_2H_5N), 12.2 (CH_3 of C_2H_5In), 44.1 (CH_2 of C_2H_5N), 60.6 (CH_2N), 124.6 (C-3,5), 127.4 (C-4), 146.0 (C-2,6), 159.8 (C-1) [32]
45	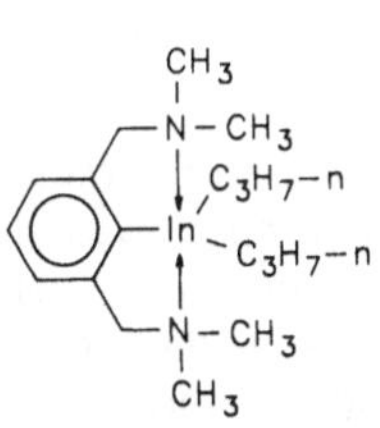 Ia (66%)	colorless liquid, b.p. 125 °C/ca. 0.03 Torr [32] ^{1}H NMR (C_6D_6): 0.75 (t, CH_2In), 1.22 (t, CH_3C), 1.86 (m, CH_2), 2.10 (s, CH_3N), 3.28 (CH_2N), 7.05 to 7.15 (m, C_6H_3) [32] ^{13}C NMR (C_6D_6): 16.1 (CH_2In), 21.4 (CH_3), 23.0 (CH_2), 45.8 (CH_3N), 67.2 (CH_2N), 124.5 (C-3,5), 126.7 (C-4), 145.7 (C-2,6), 158.9 (C-1) [32] mass spectrum: $[M-C_3H_6]^+$ (7%), $[M-C_3H_7]^+$ (26%), $[M-2\,C_3H_7]^+$ (2%), $[InC_3H_7]^+$ (3%), $[In]^+$ (85%), further ligand fragments [32]

* Further information:

$In(CH_3)_2C_2H_5$ (Table **17**, No. **1**) can not be considered as a real trialkyl compound according to [19]. Method Ib gave a product which was converted with $(C_6H_5)_2PCH_2CH_2P(C_6H_5)_2$ (dppe) into its 2:1 adduct. Pyrolysis of this adduct (130 °C, 10^{-2} Torr) followed by redistillation produced fractions with increasing C_2H_5 portions, thus indicating the instability of the compound [19].

The ^1H NMR spectrum of the product obtained from the adduct pyrolysis in C_7D_8 showed the expected pattern. However, on cooling to −63 °C broadening of the signals occurred with final splitting into a complex multiplet which was not compatible with the spectrum of a pure compound. It was postulated that mainly a 2:1 mixture of $In(CH_3)_3$ and $In(C_2H_5)_3$ was present. Similarly, the corresonding adducts with dppe or $N(CH_3)_3$ were identified as a mixture of adducts of these components [19].

Important IR frequencies (in cm^{-1}) from Nos. 1 and 2 were taken from [17] and assigned [33] below.

$In(CH_3)_2C_2H_5$ [17]	$In(C_2H_5)_2CH_3$ [17]	assignment [33]
715 m, sh	700 m, br	$\varrho(CH_3)$
670 vs	670 w, sh	
629 w, sh	630 vs	$\varrho(CH_2)$
490 vs	485 vs	$\nu_{as}(InC_3)$

$InCH_3(C_2H_5)_2$ (Table **17**, No. **2**). The compound was also synthesized in good yield from $(C_2H_5)_2InCl$ and $LiCH_3$ (Method Ia) [17].

The IR spectra of the pure substance is illustrated in [17]; for significant values and assignment, see above. The Raman spectrum of a 50% solution in C_6H_6 is not a superposition of the spectra of the components and exhibits ν(In–C) vibrations at 487 w, sh, p, 480 mw, dp, and 452 vs, p cm^{-1}. IR (neat): 480 vs, br, 455 sh cm^{-1} [33].

The ^{13}C NMR spectrum also differs from a superposition of the single InR_3 compounds, thus indicating a real compound [33].

In MOVPE experiments No. 1 and 2 serve as substitutes for $In(CH_3)_3$ or $In(C_2H_5)_3$ [16, 17, 22].

$In(CH_3)_2C_5H_5$-c (Table **17**, No. **5**) was also prepared in 90% yield from $\{(CH_3)_2InN(CH_3)_2\}_2$ and C_5H_6-c in a special procedure. The primary product of the reaction in toluene, $In(CH_3)_2C_5H_5\text{-c} \cdot HN(CH_3)_2$, is very soluble, but was not isolated and not characterized; it was thoroughly freed of $HN(CH_3)_2$ in vacuum and purified by recrystallization from toluene [4].

$In(CH_3)_2C{\equiv}CCH_3$ (Table **17**, No. **6**) formed in small amounts from Method II ($In(CH_3)_3 + HC{\equiv}CCH_3$), but at the temperature required for reaction the product was already starting to decompose [9]. The compound is only moderately soluble in C_6H_6, $C_6H_5CH_3$, CH_2Cl_2, $CHCl_3$, and CCl_4, but is very soluble in $O(CH_3)_2$ and $O(C_2H_5)_2$; adduct formation, however, was not detected. According to cryoscopic molecular weight determinations in benzene, the compound exists as a dimer in which the propyne group assumes a bridging function. The IR and Raman data were compared with the homologous compounds of Al and Ga [9]. The ring structure of C_{2h} symmetry was confirmed for the gas state by electron diffraction [11]: the four-membered In_2C_2 ring of the dimeric molecule is planar (within

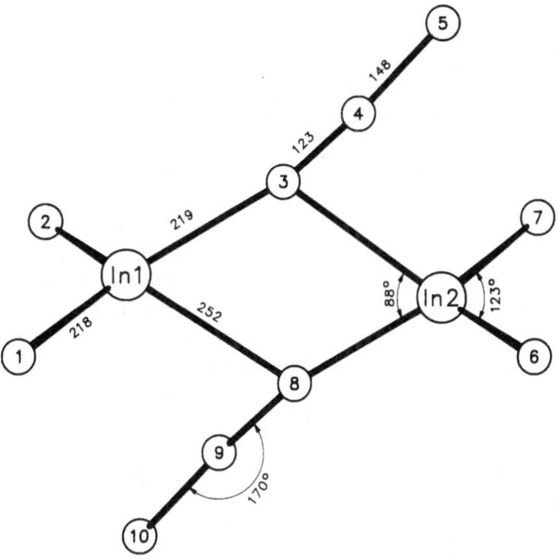

Fig. 12. Molecular structure of $In(CH_3)_2C{\equiv}CCH_3$ in the gas phase [11].

Bond angles (°)

C(1)-In(1)-C(3)	117(2)	In(1)-C(3)-C(4)	156(6)
C(1)-In(1)-C(8)	100(2)	In(1)-C(3)-In(2)	92(2)

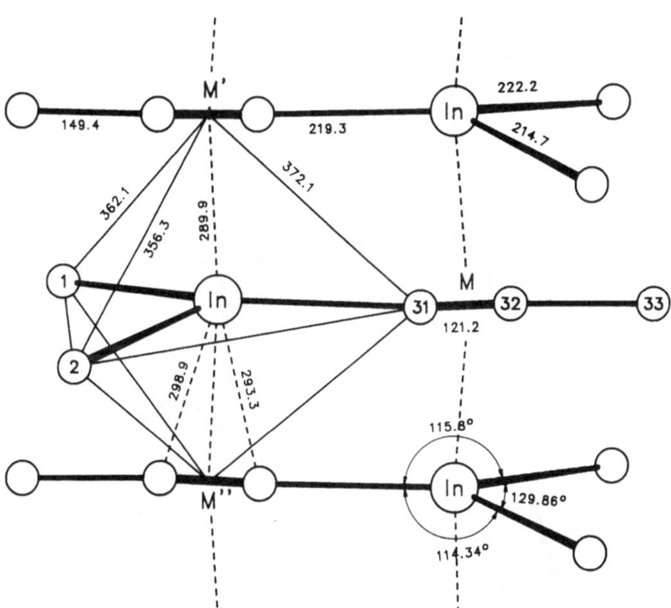

Fig. 13. Coordination polyhedron in $In(CH_3)_2C{\equiv}CCH_3$ in the solid state [9].

Bond angles (°)

In-C(31)-C(32)	117.09(107)	M′-In-M″	173.27
C(31)-C(32)-C(33)	178.63(131)		

the uncertainty of the measurements), but not rectangular. The structure in the gas phase is depicted in **Fig. 12**. The compound (as well as the homologs of Al and Ga) is described as a somewhat distorted monomeric entity, which is associated to form a loose dimer by the acceptor function of the empty p_z metal orbital and the donor action of C≡C π electrons of a neighboring unit [11].

Crystalline $In(CH_3)_2C{\equiv}CCH_3$ is a polymer. The compound crystallizes in the centrosymmetric orthorhombic space group Pnma $- D_{2h}^{16}$ (No. 62) with four monomers in the unit cell. The lattice constants, obtained at −100 °C, are a = 926.9(3), b = 578.7(4), c = 1216.0(4) pm; $D_c = 1.872$ while $D_m = 1.83$ g/cm³ [9]. The In atom is surrounded by a slightly distorted trigonal-bipyramid as depicted in **Fig. 13**. The three equatorial positions of the polyhedron are occupied by the two methyl carbon atoms C-1 and C-2 and C-31 of the propyne residue. The axial positions are occupied by the centers of the C≡C bonds of two neighboring molecules denoted as M′ and M″. This association of monomers through π bonds leads to a polymeric band structure along the b axis [9].

$InCH_3(CH_2C_4H_9\text{-}t)_2 \cdot D$ (Table **17**, No. **15**, D = THF; No. **16**, D = $N(CH_3)_3$; No. **17**, D = TMEDA) and **$\{InCH_3(CH_2C_4H_9\text{-}t)_2\}_2 \cdot TMEDA$** (Table **17**, No. **18**; TMEDA = $(CH_3)_2NCH_2CH_2N(CH_3)_2$). The adducts were prepared from No. 3 and an excess of the appropriate donor without solvent on standing for 12 h at room temperature. After removal of all volatile materials the products were not further purified and were characterized by mass spectrometry (no details) [25].

The IR spectra show the absorptions of the basic alkyl compound with highly constant positions (in cm⁻¹): 2950 ± 10 vs, 2887 ± 7 s to vs, 2860 vs, (the last two bands not given

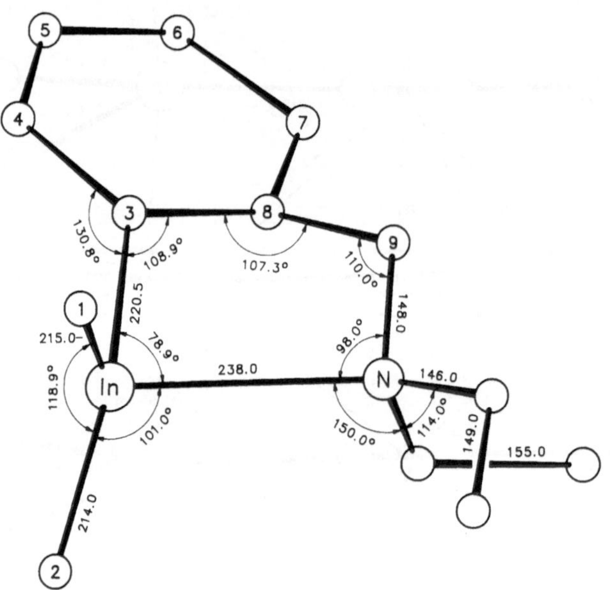

Fig. 14. Molecular structure of $In(CH_3)_2C_6H_4\{CH_2N(C_2H_5)_2\}$-2 [30].

Other bond angles (°)

C(1)-In-N	109.8(9)	C(2)-In-C(3)	130.2(9)
C(1)-In-C(3)	107.0(7)		

for the THF adduct), 1465±4 vs, 1457 m to s, 1409±1 m, 1382±1 m to s, 1358 vs, 1290±5 m, 1253±3 m to vs, 1214±2 s to vs, 1153±3 m to s, 1100±4 m to s, 1010 m to vs, 997±1 m to s, 925±3 m, 800±5 m to vs, 740±1 m to s, 680 m to vs, 586±5 m to s, 511±1 m, 475±5 w to s, 449±1 m to s, 379±1 m. The bands of the donors are given in Table 17; see also the bands of pure No. 3 [25].

$In(CH_3)_2C_6H_4\{CH_2N(C_2H_5)_2\}$-2 (Table **17**, No. **27**) crystallizes in the monoclinic space group $P2_1-C_2^2$ (No. 4) with the parameters a=941.4(2), b=1013.7(3), c=2242.6(3) pm, β= 100.89(2)°; Z=6 and D_c=1.456 g/cm^{-3}. As shown in **Fig. 14** on p. 114 the In atom of this intramolecular adduct is located 24.4(1) pm above the plane formed by the three carbon atoms at the In atom directed to the N atom. The torsion angle In-C(3)-C(8)-C(9) is 46.1° [30].

$In(CH_3)_2C_6H_4\{CHCH_3N(CH_3)_2\}$-2 (Table **17**, No. **28**). ^{13}C NMR studies between −80 and 100 °C showed that the $CH_3(In)$ resonance was split at low temperature and coalesced at ~25 °C; a structure similar to No. 27 with four-coordinated indium was deduced [12].

In pyridine the intramolecular In-N bond is replaced by an In-NC_5H_5 coordination. The results of the ^{13}C NMR measurements at various temperatures in toluene and in C_5H_5N are given below [12]:

t in °C	solvent	δ (CH_3In)	δ (CH_3N)	δ (CH_3C)	δ (CH)	δ (C-1)
30	C_7D_8	−6.0	41.4/46.9	19.7	70.9	158.6
−50	C_7D_8	−7.1/−4.8	41.5/46.7	19.3	70.7	157.8
−40	NC_5H_5	−7.7	43.0(sh)	19.0	68.6	158.2

References:

[1] Jeffery, E. A.; Mole, T. (J. Organometal. Chem. **11** [1968] 393/8).
[2] Maeda, T.; Tada, H.; Yasuda, K.; Okawara, R. (J. Organometal. Chem. **27** [1971] 13/8).
[3] Golubinskaya, L. M.; Bregadze, E. V.; Bryuchova, E. V.; Svergun, V. I.; Semin, G. K.; Okhlobystin, O. Yu. (J. Organometal. Chem. **40** [1972] 275/9).
[4] Krommes, P.; Lorberth, J. (J. Organometal. Chem. **88** [1975] 329/36).
[5] Stadelhofer, J.; Weidlein, J.; Haaland, A. (J. Organometal. Chem. **84** [1975] C1/C4).
[6] Fischer, P.; Stadelhofer, J.; Weidlein, J. (J. Organometal. Chem. **116** [1976] 65/73).
[7] Leites, L. A.; Kurbakova, A. P.; Golubinskaya, L. M.; Bregadze, V. I. (J. Organometal. Chem. **122** [1976] 1/4).
[8] Stadelhofer, J.; Weidlein, J.; Fischer, P.; Haaland, A. (J. Organometal. Chem. **116** [1976] 55/63).
[9] Fries, W.; Schwarz, W.; Hausen, H.-D.; Weidlein, J. (J. Organometal. Chem. **159** [1978] 373/84).
[10] Fries, W.; Sille, K.; Weidlein, J. (Spectrochim. Acta A **36** [1980] 611/9).
[11] Fjeldberg, T.; Haaland, A.; Seip, R.; Weidlein, J. (Acta Chem. Scand. A **35** [1981] 437/41).
[12] Jastrzebski, J. T. B. H.; van Koten, G.; Tuck, D. G.; Meinema, H. A.; Noltes, J. G. (Organometallics **1** [1982] 1492/5).
[13] Schumann, H.; Mohtachemi, R. (Z. Naturforsch. **39b** [1984] 798/800).
[14] Bregadze, V. I.; Usyatinskii, A. Ya.; Kampel', V. T.; Golubinskaya, N. N. (Izv. Akad. Nauk SSSR Ser. Khim. **1985** 1212/3; Bull. Acad. Sci. USSR Div. Chem. Sci. **1985** 1113).
[15] Steevenz, R. S.; Tuck, D. G.; Meinema, H. A.; Noltes, J. G. (Can. J. Chem. **63** [1985] 755/8).
[16] Fry, K. L.; Kuo, C. P.; Larsen, C. A.; Cohen, R. M.; Stringfellow, G. B.; Melas, A. (J. Electron. Mater. **15** [1986] 91/6).

[17] Hui, B. C.; Lorberth, J.; Melas, A. A. (Eur. 181706 [1986]; C.A. **105** [1986] No. 79150; identical with U.S. 4720560 [1988]).
[18] Oskam, A.; Stufkens, D. J.; Louwen, J. N. (J. Mol. Struct. **142** [1986] 347/9).
[19] Bradley, D. C.; Chudzynska, H.; Frigo, D. M. (Chemtronics **3** [1988] 159/61).
[20] Bregadze, V. I.; Usyatinskii, A. Ya.; Antonovich, V. A.; Godovikov, N. N. (Izv. Akad. Nauk SSSR Ser. Khim. **1988** 670/4; Bull. Acad. Sci. USSR Div. Chem. Sci. **1988** 570/3).

[21] Erdmann, D.; van Ghemen, M. E.; Pohl, L.; Schumann, H.; Hartmann, U.; Wassermann, W.; Heyen, M.; Jürgensen, H. (Ger. Offen. 3631469 [1988] from C.A. **109** [1988] No. 42300).
[22] Knauf, J.; Schmitz, D.; Strauch, G.; Jürgensen, H.; Heyen, M.; Melas, A. (J. Cryst. Growth **93** [1988] 34/40).
[23] Schumann, H.; Hartmann, U.; Dietrich, A.; Pickardt, J. (Angew. Chem. **100** [1988] 1119/20; Angew. Chem. Intern. Ed. Engl. **27** [1988] 1077/8).
[24] Schumann, H.; Motachemi, R.; Schwichtenberg, M. (Z. Naturforsch. **43b** [1988] 1510/3).
[25] Beachley, O. T.; Spiegel, E. F.; Kopasz, J. P.; Rogers, R. D. (Organometallics **8** [1989] 1915/21).
[26] Hostalek, M.; Pohl, L.; Brauers, A.; Balk, P.; Frese, V.; Hardtdegen, H.; Hövel, R.; Regel, G. K.; Molassioti, A.; Moser, M.; Scholz, F. (Thin Solid Films **174** [1989] 1/4).
[27] Molassioti, A.; Moser, M.; Stapor, A.; Scholz, F.; Hostalek, M.; Pohl, L. (Appl. Phys. Letters **54** [1989] 857/8).
[28] Frese, V.; Regel, G. K.; Hardtdegen, H.; Brauers, A.; Balk, P.; Hostalek, M.; Lokai, M.; Pohl, L.; Miklis, A.; Werner, K. (J. Electron. Mater. **19** [1990] 305/8).
[29] Schumann, H.; Hartmann, U.; Wassermann, W. (Polyhedron **9** [1990] 353/60).
[30] Schumann, H.; Hartmann, U.; Wassermann, W.; Dietrich, A.; Frank, H. G.; Pohl, L.; Hostalek, M. (Chem. Ber. **123** [1990] 2093/9).

[31] Schumann, H.; Hartmann, U.; Wassermann, W.; Just, O.; Dietrich, A.; Pohl, L.; Hostalek, M.; Lokai, M. (Chem. Ber. **124** [1991] 1113/9).
[32] Schumann, H.; Hartmann, U.; Wassermann, W. (Chem. Ber. **124** [1991] 1567/9).
[33] Imdahl, R.; Weidlein, J. (unpublished results).

2 Organoindium Halogenides

Compounds of the type X_2InCH_2X, which were not further characterized, will not be treated in this section. The compounds were formulated as intermediates in the electrochemical oxidation of In in CH_2X_2 and were finally trapped as $[X_3InCH_2X]^-$ ions (X = Br, I; see Section 11.3.4) [1].

Reference:

[1] Annan, T.H.; Tuck, D.G.; Khan, M.A.; Peppe, C. (Organometallics **10** [1991] 2159/66).

2.1 Organoindium Fluorides

Only four dialkylindium fluorides, R_2InF (R = CH_3, C_2H_5, $CH_2C_6H_5$, and $C_6H_2(CH_3)_3$-2,4,6), have been described. The compounds with R = CH_3, C_2H_5 are polymeric compounds that formed from the corresponding trialkyls by breaking one of the In-C bonds. A dimeric structure with an In_2F_2 four-membered ring was proposed for the other compounds. Whether the corresponding salt $[NR_3R'][(CH_3)_3InF]$ was actually formed by the reaction of $In(CH_3)_3$ with $[(CH_3)_3(C_6H_5CH_2)N]F$ at 28 to 32 °C was not demonstrated [3].

$(CH_3)_2InF$

The compound was formed in about 50% yield by the reaction of $In(CH_3)_3$ with excess $F_3B \cdot O(C_2H_5)_2$ in ether solution. It is a white, finely crystalline solid, purified by recrystallizing from anhydrous methanol; m.p. 295 °C (dec.) [2]. Unexpectedly it was formed in high yields by the reaction of $In(CH_3)_3$ with $CH_3N{=}PF_2OH$ in pentane, hexane, or benzene between 0 and 5 °C; a by-product is CH_4 [31].

The compound was sublimed in vacuum, but no temperature or pressure was reported. A very low solubility in aprotic solvents and the high melting point indicate a polymeric structure, and the appearance of two InC_2 stretching vibrations in the IR spectrum signals a nonlinear $(CH_3)_2In$ grouping [2].

The IR spectrum of the solid shows the following important absorptions (in cm^{-1}): 1180, 1165 m $\delta_s(CH_3)$, 720 vs, br $\varrho(CH_3)$, 548 s $\nu_{as}(InC_2)$, 493 $\nu_s(InC_2)$; no In-F vibrations were given [2].

The ^{115}In NQR spectrum (room temperature) is also in agreement with a nonlinear $In(CH_3)_2$ species in a polymer of the form $\{(CH_3)_2InF\}_n$ ($n \gg 2$) and having In-F-In bridging [16, 21, 23], which explains the correlation of the asymmetry parameters, η, and the C-In-C valence angles in a series of $(CH_3)_2In$ derivatives [23]. For $(CH_3)_2InF$ were found: 83.97 (ν_1), 129.84 (ν_2), 173.72 (ν_3), 1045.04 (e^2Qq/h) MHz, as well as $\eta = 18.5\%$ [16, 21].

$(C_2H_5)_2InF$

Indium diethyl fluoride formed (along with $Sn(C_2H_5)_4$) in 73% yield within one day when $(C_2H_5)_3SnF$ was alkylated by $In(C_2H_5)_3$ in THF at room temperature. The compound occurs as fine needles which melt between 258 and 260 °C, are insoluble in CH_2Cl_2 and C_6H_6, but are slightly soluble in THF [7].

The IR spectrum of the solid in Nujol was evaluated only in the region of the In-C vibrations (in cm^{-1}): 518 s $\nu_{as}(InC_2)$, 461 vw $\nu_s(InC_2)$. From the very low intensity of the absorption band at 461 cm^{-1} it was concluded that the C-In-C bond angle is quite large (>160 °?) [7].

$(C_6H_5CH_2)_2InF$

The compound was obtained in 73% yield from the corresponding trialkyl indium and $BF_3 \cdot O(C_2H_5)_2$ in ether (3:1 mole ratio, 6 h at room temperature); recrystallization from toluene, m.p. 149 °C [52].

The NMR spectra (shifts in ppm) were recorded in C_6D_6 solution. ^1H NMR: 2.15 (s, CH_2), 6.86 to 7.10 (m, C_6H_5). ^{13}C NMR: 27.1 (CH_2), 124.0 (C-4), 128 (C-3,5), 129.0 (C-2,6), 142.2 (C-1). ^{19}F NMR: −186.0 [52].

IR: 459 s ν_{as}(In-C), 440 m ν_s(In-C), 347 m, br ν(In-F) cm^{-1}. Raman: 437 m ν_s(In-C) cm^{-1} [52].

The mass spectrum exhibited the following fragments (relative intensities in % in parentheses): $[M-H]^+$ (5.3), $[C_6H_5CH_2CH_2C_6H_5]^+$ (69.4), $[In]^+$ (14.3), $[C_6H_5CH_2]^+$ (100) [52].

A solution of the compound in CD_3CN exhibited a ^{19}F NMR signal at −162.8 ppm which was assigned to the adduct **$(C_6H_5CH_2)_2InF \cdot n\ CD_3CN$** [52].

$(C_6H_2(CH_3)_3\text{-}2,4,6)_2InF$

The mesityl derivative was prepared similarly to the benzyl compound in 80% yield. It melts at 133 to 135 °C. By-product from the synthesis is $\{C_6H_2(CH_3)_3BO\}_3$, obtained in small amounts [52].

The NMR spectra (shifts in ppm) were recorded in C_6D_6 solution. ^1H NMR: 2.04 (s, CH_3-4), 2.30 (CH_3-2,6), 6.57 (s, C_6H_2). ^{13}C NMR: 21.2 (CH_3-4), 26.1 (CH_3-2,6), 127.5 (C-3,5), 138.3 (C-4), 144.6 (C-2,6), 148.5 (C-1). ^{19}F NMR: −173.0 [52].

IR: 536 m, 519 s ν(In-C), 385 m, br ν(In-F) cm^{-1} [52].

The mass spectrum showed the following fragmentation (relative intensities in % in parentheses, mes = $C_6H_2(CH_3)_3$-2,4,6): $[In_3mes_3F_3-3H]^+$ (0.91), $[In_3mes_4F_3]^+$ (0.21), $[In_3mes_4F_2-H]^+$ (0.54), $[In_2mes_3F_2]^+$ (58.7), $[Inmes_2]^+$ (100), $[InmesF]^+$ (12.1), $[In]^+$ (55.7) [52].

2.2 Organoindium Chlorides

2.2.1 R_2InCl Compounds and Their Adducts

2.2.1.1 R_2InCl and $RR'InCl$, where R = Alkyl and Substituted Alkyl

Table 18 lists the base compounds, InR_2Cl (Nos. 1 to 7), then the adducts, $InR_2Cl \cdot D$ (Nos. 8 to 24) and $In(R-D)_2Cl$ (Nos. 25 and 26). One compound of the type RR'InCl is placed at the end of the section. The following procedures were used to prepare the compounds:

Method I: Alkylation of $InCl_3$ with LiR.
Alkylation in the 1:2 mole ratio has been the most used procedure for producing $(CH_3)_2InCl$. $O(C_2H_5)_2$ was a suitable solvent, because no adduct forms with it; however, benzene or toluene were also used [2, 6]. The initial reaction is vigorous, but 1 to 2 days' continuous stirring or shaking is required for completion; otherwise, only $In(CH_3)_3$ and unreacted $InCl_3$ are isolated [2]. This method served also for the preparation of $(C_4H_9\text{-}t)_2InCl$ (No. 4); higher lithium alkyls were not available [33].

Method II: Ligand exchange between $InCl_3$ and InR_3.
The reaction, using a 1:2 mole ratio, has been carried out in ether, benzene, and toluene. The reaction was exothermic for dialkylindium halides No. 3, 5, and 6, and gave almost quantitative yields within 2 to 6 h [14, 27, 46].

Method III: Reaction of InR_3 compounds with HCl.
R_2InCl (along with RH) was formed in high yields when InR_3 reacted with equimolar amounts of HCl (passed in as gas or added dropwise as an ether solution). No. 1 was prepared from $(CH_3)_3In \cdot O(C_2H_5)_2$ with $O(C_2H_5)_2$ as solvent [1, 5]. $\{(CH_3)_3SiCH_2\}_2InCl$ (No. 7) was synthesized from the corresponding trialkyl in toluene [27].

Method IV: Adducts from the reaction of R_2InCl with D compounds.
Either stoichiometric amounts of the reactants in inert solvents were mixed and stirred for a short time at room temperature, or a large excess of the volatile Lewis base was condensed into dialkylindium chloride, warmed (usually) to room temperature, and the remaining volatile components removed in vacuum. The simple $R_2InCl \cdot D$ adducts can normally be assumed to have a fourfold coordination of the central atom, but adducts of the $InR_2Cl \cdot D\text{-}D$ type can also contain five- or six-coordinated In [2, 5, 6].

Table 18
R_2InCl Compounds with R = Alkyl and Their Adducts.
Further information on numbers preceded by an asterisk is given at the end of the table.
Explanations, abbreviations, and units on p. X.

No.	compound method of preparation (yield)	properties and remarks
*1	$(CH_3)_2InCl$ I (81%) II III (82%)	white crystalline solid, m.p. 218 to 219 °C, [2], 222 to 225 °C [1], subl. 100 °C in vacuum [6], 100 to 110 °C/1 Torr [1], 100 °C/10^{-4} Torr [14] D = 2.31 g/cm^3 [14] ^1H NMR: 0.4 in CCl_4, 0.28 in $CHCl_3$, −0.05 in H_2O [6], 0.27 in $(CH_3)_2SO$, 0.025 in H_2O [1] (s, CH_3), $J(^{13}C,H) = 127 \pm 1$ (saturated H_2O solution), 130 ± 1 (1-molar H_2O solution) [6] NQR, IR, and Raman spectra on p. 123
*2	$(C_2H_5)_2InCl$ special	white crystalline solid, m.p. 205 to 206 °C [7], 202 to 205 °C [4, 9], subl. 90 °C/10^{-4} Torr [14] IR and Raman spectra on p. 126
*3	$(C_3H_7\text{-i})_2InCl$ II (>95%)	colorless fine needles, m.p. 191 to 192 °C [46], 204 °C [47], soluble in pentane, THF, $O(C_2H_5)_2$, dimeric in benzene [46, 47] ^1H NMR: 1.04 (m, HCIn), 1.55 (d, CH_3C) in C_6D_6 [47], and 1.34 (CH_3C), 1.42 (HCIn) in CD_3CN [46] ^{13}C NMR: 22.0 (CH_3), 22.6 (CHIn) in C_6D_6 and 22.3 (CH_3), 28.9 (HCIn) in CD_3CN [46, 47] IR (solid): 502 m, 462 w ν(InC), 244 mw, 215 mw $\nu(In_2Cl_2)$ [46] Raman (solid): 500 m, sh, 461 vs ν(InC), 230 m, 200 m $\nu(In_2Cl_2)$, no other bands reported [46] mass spectrum (70 eV): $[M_2-C_3H_7]^+$ (7.24%), M^+ (13.6%), $[In(C_3H_7)_2]^+$ (59%), $^{115}In^+$ (100%) [46] used for DTA measurements between 60 and 410 °C [34]
*4	$(C_4H_9\text{-t})_2InCl$ I (28%) special	white solid, m.p. 212 to 212.5 °C, dec., becomes yellow at 208 °C, subl. 100 °C/0.01 Torr, dimeric in benzene [39] ^1H NMR (C_6D_6): 1.40 (CH_3C) [39] IR (solid): 2770 m, 2715 m, 1387 m, sh, 1369 s, sh, 1365 s, 1163 s, 1160 s, sh, 1016 m, 943 w, 813 m, 808 w, sh, 515 w, 389 w, sh, 381 w, 266 m, 243 m, 220 m [39] mass spectrum (^{115}In, ^{35}Cl): $[M_2]^+$ (2%), $[M_2-C_4H_9]^+$ (9%), $[M+In]^+$ (37%), $[M]^+$ (49%), $[In(C_4H_9)_2]^+$ (39%), $[InC_4H_9]^+$ (9%), other fragments reported [39]
5	$\{(C_4H_9\text{-t})CH_2\}_2InCl$ II (91%)	colorless crystals, m.p. 162 to 165 °C, subl. 110 °C/0.01 Torr, soluble in THF, pentane and $O(C_2H_5)_2$, dimeric in benzene [44] ^1H NMR (C_6D_6 ?): 1.09 (CH_3C), 1.56 (CH_2In) [44] IR (solid): 3180 w, 2950 m, 2924 m, 2860 s, 1460 s, 1376 m,

Table 18 (continued)

No. compound method of preparation (yield)	properties and remarks
5 (continued)	1360 s, 1354 s, 1259 w, 1233 s, 1120 s, 1111 s, 1088 s, 1000 s, 594 m, 574 m, 448 m, 377 m [44] reaction with $KP(C_6H_5)_2$ [57]
*6 $(C_6H_5CH_2)_2InCl$ II (95%)	white solid, m.p. 215 to 217 °C, dimeric in CH_2Cl_2 [43, 49] ^1H NMR (CD_2Cl_2): 2.57 (s, CH_2In), 7.15 (m, C_6H_5) [43]; similar values in CD_3CN [49] ^{13}C NMR (CD_3CN): 29.4 (CH_2), 123.5 (C-4), 128.3 (C-3,5), 128.8 (C-2,6), 145.0 (C-1) [49] IR (solid): 3080 m, 1950 w, 1875 w, 1805 w, 1720 w, 1600 s, 1490 s, 1325 w, 1260 w, 1215 m, 1180 w, 1155 w, 1090 m, 1050 m, 900 m, 805 m, 755 s, 715 m, 609 s, 570 w, 455 m ν(InC) [43], 225 sh, 206 m both ν(InCl) [49] Raman (solid): 447 sh ν(InC), 222 w, 206 m ν(InCl) (ring) [49] mass spectrum (main fragments only): $[M_2]^+$, $[M-C_7H_7]^+$, and $[^{115}InCl]^+$ [43]
7 $\{(CH_3)_3SiCH_2\}_2InCl$ I (ca. 100%) III	white crystalline solid, m.p. 88 to 91 °C, subl. 90 °C/in vacuum, soluble in C_5H_{12}, C_6H_6, CH_2Cl_2, CH_3CN, $O(C_2H_5)_2$, dimeric in benzene [27] ^1H NMR (CH_2Cl_2): 0.18 (CH_3Si), 0.33 (CH_2Si) [27] IR (solid): 1248 vs, 946 s, 825 vs, 757 vs, 693 m, sh, 573 m, 518 w, sh, 477 m, 360 w [27] reaction with $As\{Si(CH_3)_3\}_3$ [58]
8 $(CH_3)_2InCl \cdot OP(C_6H_5)_3$ IV	colorless solid, recrystallized from the ternary solvent system $CHCl_3/O(C_2H_5)_2$/petroleum ether, m.p. 95 °C, monomeric in nitromethane [5] IR (solid): 1145 s ν(P=O), no other bands reported [5]
9 $(CH_3)_2InCl \cdot OAs(C_6H_5)_3$ IV	colorless, needle-like crystals, softened at 63 °C, m.p. 97 to 99 °C, recrystallized from $CHCl_3/O(C_2H_5)_2$/petroleum ether-mixtures, monomeric in nitromethane [5] IR (solid): 879 ν(As=O), 860 s [5]
10 $(CH_3)_2InCl \cdot SSb(CH_3)_3$ IV (ca. 100%)	white solid, m.p. 65 to 69 °C [11] ^1H NMR (CH_2Cl_2): 0.00 (CH_3In), 1.80 (CH_3Sb) [11] IR (solid): 564, 531 ν(SbC_3), 561, 483 ν(InC_2), 404 ν(SbS) 261, 242 ν(InCl) or ν(InS), no other bands reported [11] dec. in boiling CH_3OH to give $(CH_3InS)_n$ and $(CH_3)_4SbCl$ [11]
*11 $(CH_3)_2InCl \cdot x\ NH_3$	x = 0.5; sticky gum, probably $(CH_3)_2InCl$ dissolved in the 1:1 adduct, m.p. < 20 °C, dissociation pressure 0.5 Torr at 24 °C [5] x = 1; colorless crystals, m.p. 46 to 54 °C, soluble in $O(C_2H_5)_2$, dissociation pressure 0.5 Torr at 24 °C [5]

Table 18 (continued)

No. compound method of preparation (yield)	properties and remarks
*11 (continued)	x = 2; white powder, m.p. 37 to 55 °C, dissociation pressure 27 Torr at 24 °C [5] x = 3; bulky white powder, m.p. 42 to 73 °C, soluble in large excess of NH_3, dec. under N_2, dissociation pressure at 24 °C: 30 to 50 Torr [5]
12 $(CH_3)_2InCl \cdot NH_2CH_2$-$CH_2NH_2$ IV	white solid, m.p. 135 to 145 °C (dec.), monomeric in nitromethane [5] conductivity: 16.4 $\Omega^{-1} \cdot cm^2 \cdot mol^{-1}$ [5] IR (solid): 720 vs, br $\varrho(CH_3)$, 515 s, br $\nu_{as}(InC_2)$, 481 m, br $\nu_s(InCl_2)$ [5]
13 $(CH_3)_2InCl \cdot C_{12}H_8N_2$ $C_{12}H_8N_2$ = 1,10-phenathroline IV (80%)	crystalline solid, m.p. 207 to 210 °C (dec.); monomeric in nitromethane [5] conductivity: 6.2 $\Omega^{-1} \cdot cm^2 \cdot mol^{-1}$ [5] IR (solid): 522 $\nu_{as}(InC_2)$, 483 $\nu_s(InC_2)$ [5]
14 $(CH_3)_2InCl \cdot NC_5H_5$ IV	white powder, m.p. 83 to 84 °C [6], 82 to 84.5 °C [2] IR (solid): 720 vs $\varrho(CH_3)$, 526 vs $\nu_{as}(InC_2)$, 488 m $\nu_s(InC_2)$, 254 mw $\nu(InCl)$ [2, 6] Raman (solid): 527 w $\nu_{as}(InC_2)$, 489 s $\nu_s(InC_2)$ [2, 6]
15 $(CH_3)_2InCl \cdot C_{10}H_8N_2$ $C_{10}H_8N_2$ = 2,2'-bi-pyridine IV	white solid, slowly melted over the range 109 to 115 °C; monomeric in nitromethane [5] conductivity: 12.4 $\Omega^{-1} \cdot cm^2 \cdot mol^{-1}$ [5] IR (solid): 705 vs, br $\varrho(CH_3)$, 529 s $\nu_{as}(InC_2)$, 481 m $\nu_s(InC_2)$; no other bands reported [5]
16 $(CH_3)_2InCl \cdot L$ IV (59%) L = 2-(NHCH$_3$)-pyridine (ring drawn: N, NHCH$_3$)	white solid, m.p. 83 to 86 °C, subl. 90 °C/0.01 Torr, soluble in C_6H_6 and $C_6H_5CH_3$ [37] ^{1}H NMR (C_6D_6): 0.21 (CH_3In), 2.15 (d, CH_3N, J(H,H) = 5.3), 5.7 to 7.4 (3m, C_6H_4) [37] mass spectrum: $[In(CH_3)_2]^+$ (1%), L^+ (80%), $[L-H]^+$ (100%), $[L-2H]^+$ (27%), $[L-NCH_3]^+$ (79%) [37]
17 $\{(CH_3)_2InCl\}_2 \cdot P(C_6H_5)_3$ IV	white solid, m.p. 130 to 138 °C [2] IR (solid): 730 s $\varrho(CH_3)$, 544 m, 515 s, 498 s, 492 s $\nu(InC)$, no $\nu(InCl)$ reported [2] proposed ionic structure: $[(CH_3)_2In \cdot P(C_6H_5)_3]^+$ $[(CH_3)_2InCl_2]^-$ [2]
18 $(C_2H_5)_2InCl \cdot SSb(CH_3)_3$ IV (100%)	white solid, m.p. 78 to 80 °C [11] ^{1}H NMR (CH_2Cl_2): 1.80 (CH_3Sb), C_2H_5In not reported [11] gives $(C_2H_5InS)_n$ and $(CH_3)_3C_2H_5SbCl$ in boiling CH_3OH [11]
19 $\{(CH_3)_3SiCH_2\}_2InCl$ $\cdot N(CH_3)_3$	not isolated, dissociates at room temperature [27]

Table 18 (continued)

No.	compound method of preparation (yield)	properties and remarks
*20	$(CF_3)_2InCl \cdot OS(CH_3)_2$ special	^{19}F NMR $(OS(CH_3)_2)$: −44.78 (CF_3) [51]
*21	$(CF_3)_2InCl \cdot THF$ special	^{19}F NMR $(OS(CH_3)_2)$: −42.75 $(CF_3$, J(C,F) = 360) [51]
*22	$(CF_3)_2InCl \cdot NCCH_3$ special	^{19}F NMR (CH_3CN): −45.55 (CF_3) [51]
*23	$(CF_3)_2InCl \cdot (CH_3)_2NCHO$ special	white to yellow solid, dec. ca. 80 °C [51] ^{19}F NMR $((CH_3)_2NCHO)$: −46.15 (CF_3) [51]
*24	$(CF_3)_2InCl \cdot NC_5H_5$ special	^{19}F NMR (C_5H_5N): −44.78 $(CF_3$, J(C,F) = 367.1) [51]
25	InCl N 2 CH_3 CH_3	from $InCl_3$ and $Li[(CH_2)_3N(CH_3)_2]$ (ratio 1:2) in $O(C_2H_5)_2$ at 0 °C; crystallization at −30 °C, 66% yield [47] colorless solid, m.p. 92 °C [47] ^1H NMR (C_6D_6): 0.62 (t, CH_2In), 1.71 (tt, CH_2), 1.95 (CH_3N) [47] ^{13}C NMR (C_6D_6): 9.4 (CH_2In), 23.7 (CH_2), 45.3 (CH_3N), 61.8 (CH_2N) [47] mass spectrum (70 eV): $[M-Cl]^+$ (40%), $[M-C_3H_8N]^+$ (16%), $[M-Cl-C_4H_{10}N]^+$ (34%), $[M-C_5H_{12}]^+$ (32%), $^{115}In^+$ (35%), $[C_7H_{16}N]^+$ (14%), $[(CH_2)_3N(CH_3)_2]^+$ (85%), $[C_3H_8N]^+$ (100%) [47]
26	InCl N 2 C_2H_5 C_2H_5	prepared like No. 25, 53% yield [47] colorless solid, m.p. 94 °C [47] ^1H NMR (C_6D_6): 0.59 (t, CH_2In), 0.78 (t, CH_3 of C_2H_5), 1.75 (tt, CH_2), 2.20 (t, CH_2N), 2.47 (q, CH_2 of C_2H_5) [47] ^{13}C NMR (C_6D_6): 8.4 (CH_3 of C_2H_5), 12.3 (CH_2In), 23.2 (CH_2), 43.9 (CH_2 of C_2H_5), 55.4 (CH_2N) [47]

* Further information:

$(CH_3)_2InCl$ (Table **18**, No. **1**) is soluble in $O(C_2H_5)_2$, slightly soluble in CCl_4, $CHCl_3$, CH_2Cl_2, and C_2H_5OH. A solution in cold water (0 to about 5 °C) is stable for only a few days; dissociation forms (hydrated!) $[In(CH_3)_2]^+$ ions. As soon as dilute HCl or HNO_3 is added, rapid decomposition of the aqueous solution sets in [2, 6]. The compound is dimeric in benzene solution [2].

The ^{115}In NQR spectrum, measured at room temperature and with a quadrupole coupling constant of $e^2Qq/h = 1200.16$ MHz and an asymmetry parameter of $\eta = 9.5\%$, showed three of the four expected frequencies: $\nu_1(1/2 \leftrightarrow 3/2) = 54.03$, $\nu_2(3/2 \leftrightarrow 5/2) = 98.85$, and $\nu_3(5/2 \leftrightarrow 7/2) = 143.70$ MHz [16, 23]. The resonance frequency in the ^{35}Cl NQR spectrum is at 8.80 MHz, which was interpreted as due to an $InCl_3$-like structure [16].

IR and Raman spectra for the solid were assigned for a dimeric structure with D_{2h} symmetry similar to the structure in benzene solution (in cm^{-1}). From IR and Raman spectra of aqueous solutions of the compound, the presence of linear $[CH_3InCH_3]^+$ cations in dilute solutions was deduced; assignments were made on the basis of D_{3d} symmetry for the aqueous solution species [6]:

solid IR [2]	IR [6]	Raman [6]	solution [1)] IR [6]	Raman [6]	assignment [6]
—	2999 w	3005 w	2996 s	2993 w	$\nu_{as}(CH_3)$
—	2923 m	2924 m	2927 s	2921 m	$\nu_s(CH_3)$
1178 m	1179 m	1176 m	1176 vw	1178 m	$\delta_s(CH_3)$
1160 m					
732 vs, br	737 s, br	726 w	742 m	—	$\varrho(CH_3)$
558 s	563 s	—	566 m	—	$\nu_{as}(InC_2)$
—	—	561 w	—	550 w	$\nu_{as}(InC_2)$
—	—	500 vvs	—	499 vvs	$\nu_s(InC_2)$
488 m	492 m	—	—	—	$\nu_s(InC_2)$
—	—	193 w	—	—	$\nu(In_2Cl_2)$

[1)] IR in H_2O/D_2O, Raman in H_2O

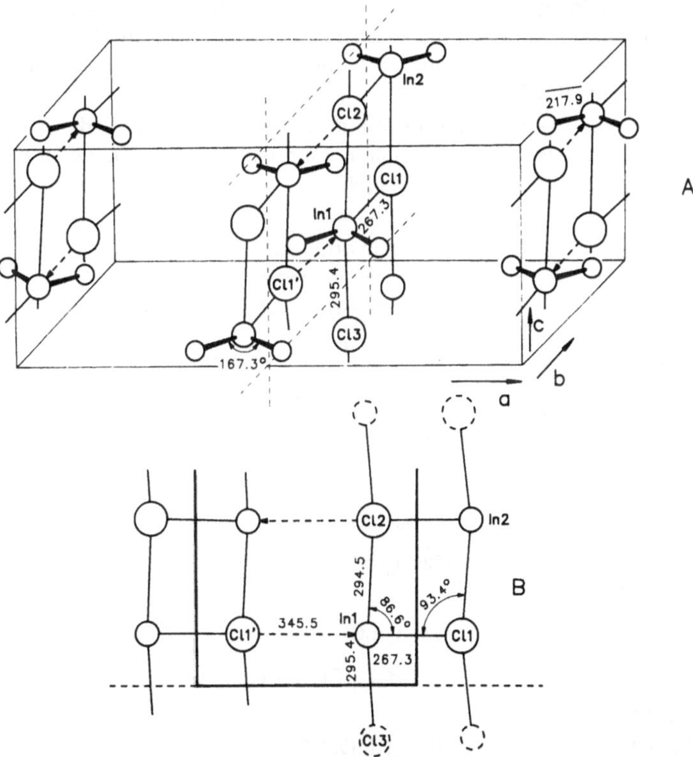

Fig. 15. Unit cell of $(CH_3)_2InCl$ (A); band structure on [200] (B) [14].

Bond distances (pm) and angles (°)

In(1)-C	217.9(7)	C-In-C	167.3
In(1)-Cl(1)	267.3(9)	In(1)-Cl(1)-In(2)	93.4
In(1)-Cl(2)	294.5(6)	Cl(2)-In(1)-Cl(1)	86.6
In(1)-Cl(3) } In(2)-Cl(1) }	295.4(6)		
In(1)-Cl(1')	345.0(9)		

Normal-coordinate analysis using the data from the aqueous solution spectra gave the force constants f(In-C) = 1.36(3) and f(C-In-C) = 0.51 mdyn/Å [6]. Later Raman spectroscopic studies of an aqueous solution of $(CH_3)_2InCl$ [14] (and similar compounds which dissociate in H_2O such as $\{In(CH_3)_2\}_2C_4O_4$ [18], $\{In(CH_3)_2\}_2SO_4$ [13], and $In(CH_3)_2SO_3CH_3$ [15]) indicated an ion of the composition $[(CH_3)_2In \cdot n\ H_2O]^+$ with $n \geqq 2$; the C-In-C angle was estimated to be 155° to 165° [14].

The compound crystallizes in the orthorhombic space group Cmcm $- D_{2h}^{17}$ (No. 63) with a = 1410.7(4), b = 612.3(2), c = 589.7(2) pm; Z = 4 and $D_c = 2.35$ g/cm^3 while $D_m = 2.31$ g/cm^3. The unit cell is depicted in **Fig. 15** (A), p. 123. It shows the presence of centrosymmetric In-Cl-In-Cl four-ring systems, but which are formed into a band structure by common In-Cl edges (B). Within one band each In atom is surrounded by three Cl atoms at different distances. A further Cl atom of the neighboring band has a distance of 345 pm to the In atom. Although this contact is appreciably larger than the sum of the covalent radii (243 pm), an interaction between In and the Cl atoms of the neighboring band can be discussed. Thus, the In atom is surrounded in a distorted octahedron by four Cl atoms and two CH_3 carbon atoms. The long In-Cl distances and the In-CH_3 distances (217.9(7) pm), which are appreciably shorter than the sum of the covalent radii, indicate some ionic character within the crystal [14].

The compound was used as starting material for other compounds. A list of reactions with inorganic, organometallic, and organic components is given below. It has been tested as source material in MOVPE experiments [56].

No.	reactant	products and remarks	Ref.	page
1	$SbCl_5$	$(CH_3)_2SbCl_3 + InCl_3$	[20]	–
2	$Tl(B_3H_8)$	unstable $(CH_3)_2In(B_3H_8)$	[25]	326
3	$AgReO_4$	$[In(CH_3)_2]ReO_4$	[22]	222
4	Ag_2SO_4	$[In(CH_3)_2]_2SO_4$; analytically not pure	[13]	222
5	$LiCH_3$	$In(CH_3)_3$, quantitatively	[6]	30
6	$NaC{\equiv}CCH_3$	$(CH_3)_2InC{\equiv}CCH_3$	[30]	112
7	MC_6H_5 (M = Li, Na)	$(CH_3)_2InC_6H_5$	[19]	103
8	$LiCH_2Sn(CH_3)_3$	$(CH_3)_2InCH_2Sn(CH_3)_3$	[33]	104
9	$(CH_3)_3MCl_2$ (M = As, Sb)	$[M(CH_3)_4][CH_3InCl_3]$	[20, 26]	352
10	$LiOM(CH_3)_3$ (M = Si, Ge)	$\{(CH_3)_2InOM(CH_3)_3\}_2$	[1]	188

Table (continued)

No.	reactant	products and remarks	Ref.	page
11	$LiM(C_4H_9\text{-}t)_2$ (M = P, As)	$\{(CH_3)_2InM(C_4H_9\text{-}t)_2\}_2$	[38]	312
12	HNR_2 (R = C_2H_5, C_3H_7-i, $Si(CH_3)_3$	in the presence of LiC_4H_9-n: $\{(CH_3)_2InNR_2\}_2$	[42]	252
13	$ClMg(CH_2)_3N(CH_3)_2$	$(CH_3)_2In(CH_2)_3N(CH_3)_2$	[45]	106
14	$LiC_6H_4\{CHXN(CH_3)_2\}$-2 (X = H, CH_3)	$(CH_3)_2InC_6H_4\{CHXN(CH_3)_2\}$-2	[32]	115
15	$LiN(CHXCH_2)_2Y$ (X = H, CH_3; Y = NCH_3, CH_2)	$\{(CH_3)_2InN(CHXCH_2)_2Y\}_2$	[36]	272
16	$LiN(C_4H_9\text{-}t)Si(CH_3)_2$-$OC_4H_9$-t	$(CH_3)_2InN(C_4H_9\text{-}t)Si(CH_3)_2OC_4H_9$-t	[41]	283
17	$(CH_3)_3P{=}CH_2$	$\{(CH_3)_2InCH_2P(CH_3)_2CH_2\text{-}\}_2$ along with $[P(CH_3)_4]Cl$	[17]	55
18	C_4H_9-t; Li; Li; t-C_4H_9-N; N; N-C_4H_9-t; Si; Si; CH_3; N; CH_3; C_4H_9-t	CH_3; In; R-N; R; N-R; N; Si; Si; CH_3; N; CH_3; R; R = C_4H_9-t + $In(CH_3)_3$	[40]	310
19	$LiC_6H_4\{CH_2N$-$(C_2H_5)_2\}$-2	$(CH_3)_2InC_6H_4\{CH_2N(C_2H_5)_2\}$-2	[48]	108
20	$K[H_2B(pz)_2]$ (pz = pyrazolyl, $C_3H_3N_2$)	$H_2B(pz_2)In(CH_3)_2$	[53]	279
21	$\{H_2B(pz)_2\}_2InCl$	$H_2B(pz)_2InCH_3Cl$	[53]	296

$(C_2H_5)_2InCl$ (Table **18**, No. **2**) has also been prepared from $In(C_2H_5)_3$ and CCl_4 (94% yield) or $CHCl_3$ (91% yield). The trialkyl indium, or its etherate, in cyclohexane was treated with $CHCl_3$ at room temperature (or with CCl_4 at 0 °C), and was stirred for several days at the same temperature; mole ratios of 1:1 to 1:2 were used. By-products were C_2H_6 (or C_2H_5Cl when CCl_4 was used) and 2-pentene; their formation is explained by the mechanism proposed on p. 64 [7, 9]. With an excess of $CHCl_3$ and at reflux temperature the reaction was complete in 10 min [4]. The starting $In(C_2H_5)_3$ can be replaced by $(C_2H_5)_2InX$ (X = OC_4H_9-t or $N(C_2H_5)_2$), and in the presence of cyclohexene the reaction of $In(C_2H_5)_3$ with $CHCl_3$ additionally produced 7,7′-dichloronorcarane [12]. The compound is soluble in $O(C_2H_5)_2$, CH_2Cl_2, $CHCl_3$, and is dimeric in benzene [7, 14].

The ^{115}In NQR spectrum at 77 K shows the following frequencies (in MHz): $\nu_1 = 57.90$, $\nu_2 = 92.45$, $\nu_3 = 142.13$; $e^2Qq/h = 1142.7 \pm 2.0$ and $\eta = 16.4 \pm 0.1\%$ [10, 24].

The vibration spectrum was assigned only in the InC_2 region in [7, 14]; a more complete assignment of the solid spectrum (in Nujol/Hostaflon) is given below (in cm^{-1}) [28]:

IR	Raman	assignment
2962 s	2955 sh	$\nu_{as}(CH_3)$
2946 s	–	$\nu_{as}(CH_2)$
–	2937 s	$\nu_s(CH_2)$
2927 s	2917 vs	$\nu_s(CH_3)$
2906 s	–	
2866 s	2880 s	$\nu_s(CH_2)$
2815 mw	2826 w	overtones
2730 vw	2741 w	
–	1470 sh	$\delta_{as}(CH_3)+$
1458 m	1462 w	$\delta_{as}(CH_2)$
1420 m	1425 w	
1379 ms	1388 vw	$\delta_s(CH_3)$
1235 mw	–	
1173 m	1180 s	$\omega(CH_2)$
1021 m	1022 w	$\nu(CC)$
970 s	969 vw	
942 mw	–	$\varrho(CH_3)$
655 vs	645 vw, br	$\varrho(CH_2)$
518 m	519 w	$\nu_{as}(InC_2)$
462 mw	468 vvs	$\nu_s(InC_2)$
–	249 wm	$\delta(InCC)$
–	237 sh	$\nu_{as}(InClIn)$
–	205 mw	$\nu_s(InClIn)$
–	157 sh	$\delta(InC_2)$
–	142 sh	

It reacted with $LiCH_3$ in ether to give $InCH_3(C_2H_5)_2$ [35] and with LiC_5H_5-c in cyclohexane at 50 to 70 °C to give $In(C_2H_5)_2C_5H_5$-c [19]. With $LiC_6H_4\{CH_2N(CH_3)_2\}$-2 arylation occurred to generate the internal base adduct $(C_2H_5)_2InC_6H_4\{CH_2N(CH_3)_2\}$-2 [48]; see Section 1.1.6. The compound was used as a catalyst for the polymerization of organic isocyanates [8].

$(C_3H_7$-i$)_2InCl$ (Table **18**, No. **3**) formed from $In(C_3H_7$-i$)_3$ and $InCl_3$ (1:1 mole ratio) as an equimolar mixture with $(C_3H_7$-i$)InCl_2$. The resulting mixture reacted with $LiNHC_4H_9$-t in ether (0 °C) to give a mixture of $\{(C_3H_7$-i$)_2InNHC_4H_9$-t$\}_2$, $\{(C_3H_7$-i$)ClInNHC_4H_9$-t$\}_2$, and $(C_3H_7$-i$)In(NHC_4H_9$-t$)_2InCl(C_3H_7$-i$)$ [50]. Pyrolysis and thermal stability is investigated in [55].

$(C_4H_9$-t$)_2InCl$ (Table **18**, No. **4**) was also obtained in 78% yield by the alkylation of $InCl_3$ with $(C_4H_9$-t$)MgCl$ in ether [39]. $(C_4H_9)_2InCl$ (C_4H_9 group not specified) also formed in an unusual way by addition of a 20% aqueous NaOH solution to a boiling mixture of $(C_4H_9)_2SnCl_2$ and $InCl_3 \cdot 4\ H_2O$ in CH_3OH/C_2H_5OH. Continous stirring under reflux followed by addition of 15% aqueous HCl to the cold reaction mixture precipitated a white product which was recrystallized from ethanol; no properties were reported. The product was used for the preparation of dibutylindium pseudohalides [29].

Treatment of the title compound with $LiOC_2H_5$, $LiNC(CH_3)_2(CH_2)_3C(CH_3)_2$ [39], or $LiCH_2Sn(CH_3)_3$ [33] gave a precipitate of LiCl with formation of the corresponding indium derivatives.

$(C_6H_5CH_2)_2InCl$ (Table **18**, No. **6**) was also prepared according to method II in toluene (5 h, 80 °C, 85% yield). It reacted with $LiNHC_4H_9$-t (conditions above) to produce $(C_6H_5CH_2)_2InNHC_4H_9$-t in 85% yield [50].

$(CH_3)_2InCl \cdot x\ NH_3$ (Table **18**, No. **11**). This series has been studied tensimetrically. Only the adduct with x = 1 is stable under inert gas at normal pressure, while the others easily dissociate with evaporation of NH_3. The 1:1 adduct dissociates in vacuum; a structure with coordination number 4 at the In atom was proposed. For the 1:2 adduct a covalent structure with fivefold coordinated In was discussed, as well as an ionic structure, such as $NH_4[(CH_3)_2InClNH_2]$ or $[(CH_3)_2In \cdot 2\ NH_3]Cl$; no decision based on IR spectroscopic data was possible [5].

$(CF_3)_2InCl \cdot D$ (Table **18**, Nos. **20** to **24**; D = DMSO, THF, CH_3CN, DMF, C_5H_5N). The reaction between $Cd(CF_3)_2$ and $InCl_3$ in various donor solvents D produced a complex mixture of the corresponding adducts $(CF_3)_{3-x}InCl_x \cdot D$; 1:1 adducts were supposed in the case of x = 1 but only the complex with D = DMF has been isolated [51].

The mixtures were studied by ^{19}F NMR spectroscopy and the signals given in Table 18 assigned to the corresponding $(CF_3)_2InCl \cdot D$ adducts [51].

On standing of the DMSO adduct mixture (No. 20) a slow (1 to 21 d) CF_3/CH_3 exchange with the solvent occurred to give a mixture of compounds of the general composition $(CF_3)_{3-x-y}(CH_3)_yInCl_x$ along with $(CF_3)_2SO$ [51].

Compounds of the Type RR'InCl

$\{((CH_3)_3Si)_2CH\}(C_3H_7\text{-}i)InCl$

A solution of $Li[CH(Si(CH_3)_3)_2]$ in ether was added dropwise to a cold solution (−30 °C) of $(C_3H_7\text{-}i)InCl_2$ (1:1 mole ratio), and the mixture stirred for 48 h at room temperature. After removal of the solvent in vacuum the product was recrystallized from pentane at −30 °C; 61.4% yield. Colorless crystals, m.p. 85 to 87 °C. The compound is very soluble in nonpolar organic solvents and is dimeric in benzene [54].

^1H NMR (C_6D_6 in ppm): 0.12 (s, CH_3Si), 0.23 (s, CHSi), 1.41 (d, CH_3C), 1.48 (m, CH). ^{13}C NMR (C_6D_6 in ppm): 3.8 (CH_3Si), 16.7 (CHSi), 22.0 (CH_3 of C_3H_7-i), 29.6 (CH of C_3H_7-i). IR/Raman (in cm^{-1}): 505/508 ν(In-C of C_3H_7), 481/479 ν(In-C of CHSi), 238/231, 210/208 $\nu(In_2Cl_2)$ [54].

The mass spectrum shows mainly fragments of the monomer unit (relative intensities); $[M-Cl-C_3H_7-Si(CH_3)_2]^+$ (2.05%), $[1/2M+H]^+$ (2.24%), $[1/2M-CH_3]^+$ (6.90%), $[1/2M-Cl]^+$ (19.23%), $[1/2M-C_3H_7]^+$ (100%) [54].

The compound crystallizes in the monoclinic space group $P2_1/c-C_{2h}^5$ (No. 14) with the cell parameters a = 1206.4(3), b = 905.7(2), c = 1591.2(5) pm, β = 101.18(2)°; Z = 2 gives $D_c = 1.374$ g/cm³. As shown in **Fig. 16** the dimeric unit forms a planar centrosymmetric

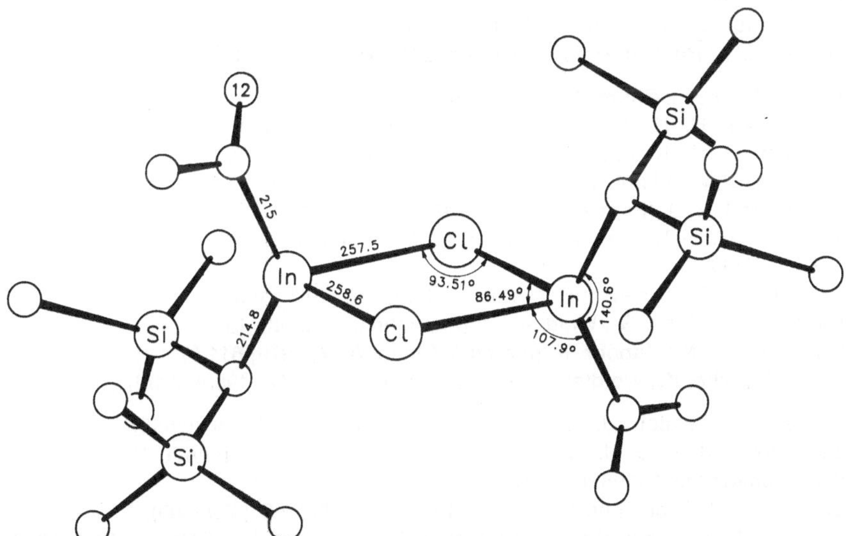

Fig. 16. Molecular structure of the dimeric $\{((CH_3)_3Si)_2CH\}(C_3H_7\text{-}i)InCl$ compound [54].

four-membered In_2Cl_2 ring with approximately equal In–Cl distances. The In · · · In distance is 376.0(1) pm. For the two C_3H_7-i groups two position were found in about equal population with C(12) directed towards the ring plane and C(12′) away from the ring plane. No further In–Cl contacts to neighboring molecules could be detected [54].

References:

[1] Armer, B.; Schmidbauer, H. (Chem. Ber. **100** [1967] 1521/35).
[2] Clark, H. C.; Pickard, A. L. (J. Organometal. Chem. **8** [1967] 427/34).
[3] Siemens-Schuckertwerke A.-G. (Fr. 1461819 [1966]; C.A. **67** [1967] No. 17386).
[4] Yasuda, K.; Okawara, R. (Inorg. Nucl. Chem. Letters **3** [1967] 135/6).
[5] Clark, H. C.; Pickard, A. L. (J. Organometal. Chem. **13** [1968] 61/71).
[6] Hobbs, C. W.; Tobias, R. S. (Inorg. Chem. **9** [1970] 1998/2004).
[7] Maeda, T.; Tada, H.; Yasuda, K.; Okawara, R. (J. Organometal. Chem. **27** [1971] 13/8).
[8] Matsui, H.; Yasuda, K.; Goto, J. (Japan. 71-9698 [1971] from C.A. **75** [1971] No. 37238).
[9] Okawara, R.; Yasuda, K.; Tada, T. (Japan. 71-3567 [1971] from C.A. **74** [1971] No. 112205).
[10] Golubinskaya, L. M.; Bregadze, E. V.; Bryuchova, E. V.; Svergun, V. I.; Semin, G. K.; Okhlobystin, O. Yu. (J. Organometal. Chem. **40** [1972] 275/9).

[11] Maeda, T.; Yoshida, G.; Okawara, R. (J. Organometal. Chem. **44** [1972] 237/41).
[12] Shcherbakov, V. I.; Zhil'tsov, S. F. (Nov. Khim. Karbenov Mater. 1st Vses. Soveshch. Khim. Karbenov Ikh Analogov, Moscow 1972 [1973], pp. 162/4 from C.A. **82** [1975] No. 43559).
[13] Olapinski, H.; Weidlein, J. (J. Organometal. Chem. **54** [1973] 87/93).
[14] Hausen, H.-D.; Mertz, K.; Veigel, E.; Weidlein, J. (Z. Anorg. Allgem. Chem. **410** [1974] 156/64).
[15] Olapinski, H.; Weidlein, J.; Hausen, H.-D. (J. Organometal. Chem. **64** [1974] 193/204).
[16] Patterson, D. B.; Carnevale, A. (Inorg. Chem. **13** [1974] 1479/83).
[17] Schmidbaur, H.; Füller, H.-J. (Chem. Ber. **107** [1974] 3674/9).
[18] Schwering, H.-U.; Olapinski, H.; Jungk, E.; Weidlein, J. (J. Organometal. Chem. **76** [1974] 315/24).
[19] Stadelhofer, J.; Weidlein, J.; Haaland, A. (J. Organometal. Chem. **84** [1975] C1/C4).
[20] Widler, H. J.; Hausen, H.-D.; Weidlein, J. (Z. Naturforsch. **30b** [1975] 645/7).

[21] Brill, T. B. (Inorg. Chem. **15** [1976] 2558/60).
[22] Schmidbaur, H.; Koth, D. (Chemiker-Ztg. **100** [1976] 290/1).
[23] Schmidbaur, H.; Koth, D. (Naturwissenschaften **63** [1976] 482/3).
[24] Bancroft, G. M.; Sham, T. K. (J. Magn. Resonance **25** [1977] 83/90).
[25] Greenwood, N. N. (Pure Appl. Chem. **49** [1977] 791/801).
[26] Widler, H.-J.; Schwarz, W.; Hausen, H.-D.; Weidlein, J. (Z. Anorg. Allgem. Chem. **435** [1977] 179/90).
[27] Beachley, O. T.; Rusinko, R. N. (Inorg. Chem. **18** [1979] 1966/8).
[28] Karschin, J.; Mann, G.; Weidlein, J. (unpublished results 1979).
[29] Srivastava, T. N.; Kapoor, K. (Indian J. Chem. A **17** [1979] 611/2).
[30] Fries, W.; Sille, K.; Weidlein, J. (Spectrochim. Acta A **36** [1980] 611/9).

[31] Schrem, H.; Weidlein, J. (Z. Anorg. Allgem. Chem. **465** [1980] 109/19).
[32] Jastrzebski, J. T. B. H.; van Koten, G.; Tuck, D. G.; Meinema, H. A.; Noltes, J. G. (Organometallics **1** [1982] 1492/5).
[33] Schumann, H.; Mohtachemi, R. (Z. Naturforsch. **39b** [1984] 798/800).
[34] Sysoeva, E. A.; Novikova, E. D.; Mittov, O. N. (Fiz. Khim. Geterog. Sist. **1984** 117/20 from C.A. **104** [1984] No. 148957).

[35] Hui, B. C.; Lorberth, J.; Melas, A. A. (Eur. 181706 [1986]; C.A. **105** [1986] No. 79150, identical with U.S. 4720560).
[36] Arif, A. M.; Bradley, D. C.; Dawes, H.; Frigo, D. M.; Hursthouse, M. B.; Hussain, B. (J. Chem. Soc. Dalton Trans. **1987** 2159/64).
[37] Bradley, D. C.; Dawes, H.; Frigo, D. M.; Hursthouse, M. B.; Hussain, B. (J. Organometal. Chem. **325** [1987] 55/67).
[38] Arif, A. M.; Benac, B. L.; Cowley, A. H.; Jones, R. A.; Kidd, K. B.; Nunn, C. M. (New J. Chem. **12** [1988] 553/7).
[39] Bradley, D. C.; Frigo, D. M.; Hursthouse, M. B.; Hussain, B. (Organometallics **7** [1988] 1112/5).
[40] Veith, M.; Goffing, F.; Huch, V. (Z. Naturforsch. **43b** [1988] 846/56).

[41] Veith, M.; Pöhlmann, J. (Z. Naturforsch. **43b** [1988] 505/12).
[42] Aitchison, K. A.; Backer-Dirks, J. D. J.; Bradley, D. C.; Faktor, M. M.; Frigo, D. M.; Hursthouse, M. B.; Hussain, B.; Short, R. L. (J. Organometal. Chem. **366** [1989] 11/23).
[43] Barron, A. R. (J. Chem. Soc. Dalton Trans. **1989** 1625/6).
[44] Beachley, O. T., Jr.; Spiegel, E. F.; Kopasz, J. P.; Rogers, R. D. (Organometallics **8** [1989] 1915/21).
[45] Hostalek, M.; Pohl, L.; Brauers, A.; Balk, P.; Frese, V.; Hardtdegen, H.; Hövel, R.; Regel, G. K.; Molassioti, A.; Moser, M.; Scholz, F. (Thin Solid Films **174** [1989] 1/4).
[46] Neumüller, B. (Chem. Ber. **122** [1989] 2283/7).
[47] Schumann, H.; Hartmann, U.; Wassermann, W. (Polyhedron **9** [1990] 353/60).
[48] Schumann, H.; Hartmann, U.; Wassermann, W.; Dietrich, A.; Görlitz, F. H.; Pohl, L.; Hostalek, M. (Chem. Ber. **123** [1990] 2093/9).
[49] Neumüller, B. (Z. Anorg. Allgem. Chem. **592** [1990] 42/50).
[50] Neumüller, B. (Z. Naturforsch. **46b** [1991] 753/61).

[51] Naumann, D.; Strauß, W.; Tyrra, W. (J. Organometal. Chem. **407** [1991] 1/15).
[52] Neumüller, B.; Gahlmann, F. (J. Organometal. Chem. to be published).
[53] Reger, D. L.; Knox, S. J.; Rheingold, A. L.; Haggerty, B. S. (Organometallics **9** [1990] 2581/7).
[54] Neumüller, B. (Z. Naturforsch. to be published).
[55] Novikova, E.D.; Gadebskaya, T.A.; Mittov, O.N.; Bezryadin, T.A. (Metallorg. Khim. **1** [1988] 650/3 from C.A. **111** [1989] No. 39423).
[56] Mori, K. (Japan. 63-085098 [1988] from C.A. **109** [1988] No. 64828).
[57] Banks, M.A.; Beachley, O.T.; Buttrey, L.A.; Churchill, M.R.; Fettinger, J.C. (Organometallics **10** [1991] 1901/6).
[58] Wells, R.L.; Jones, L.J.; McPhall, A.T.; Alvanipour, A. (Organometallics **10** [1991] 2345/8).

2.2.1.2 Compound of the Type R_2InCl with R = Alkenyl and Cycloalkenyl

$(CH_2{=}CH)_2InCl$

The compound was prepared from $In(CH{=}CH_2)_3$ and $InCl_3$ (2:1 mole ratio) in benzene solution at 40 to 50 °C and recrystallized from toluene or benzene. It is dimeric in benzene and melts at 121 °C with partial decomposition [1].

Calculated ^1H NMR shifts and coupling constants for a series of vinyl derivatives of Ga and In are collected in a table in [1]. The following values were calculated for $(CH{=}CH_2)_2InCl$ (δ values in ppm): 6.00 (H-3), 6.26 (H-2), 6.31 (H-1); J(H-1,3) = 20.60, J(H-1,2) = 12.98, and J(H-2,3) = 3.41 Hz; for the assignment, see Formula I on p. 86.

IR and Raman frequencies (in cm^{-1}): 1573±4 ν(C=C), 1008, 988, 950 δ and $\omega(CH_2)$, 515, 472 ν(InC), 465 γ(CH), 200 $\nu_{as}(In_2Cl_2$, bridge) [1].

$(C_5(CH_3)_5)_2InCl$

The compound was obtained in 77.6% yield from the reaction of $InCl_3$ with $Li[C_5(CH_3)_5]$ in pentane as a yellow crystalline solid, melting at 98.5 °C with decomposition and turning brown. At 103 °C it is converted into an opaque white mass. The compound is soluble in ether and THF and slightly soluble in pentane and benzene. It also formed as a by-product to the extent of 2 to 5% by the reaction of $Li[C_5(CH_3)_5]$ with InCl, the main product having been $InC_5(CH_3)_5$ [2].

The ^1H NMR in C_6H_6 exhibits only a singlet for the CH_3 protons at $\delta=1.88$ ppm. The IR spectrum shows the following absorptions (in cm^{-1}): 1734 w, 1728 w, 1590 vw, 1276 m, 1237 m, 1142 w, 1128 w, 1050 w, 1035 w, 1005 vw, 940 w, 798 w, 787 w, 592 m, 561 w, 420 w, 254 vs [2].

References:

[1] Fries, W.; Sille, K.; Weidlein, J.; Haaland, A. (Spectrochim. Acta A **36** [1980] 611/9).
[2] Beachley, O. T., Jr.; Blom, R.; Churchill, M. R.; Faegri, J., Jr.; Fettinger, J. C.; Pazik, J. C.; Victoriano, L. (Organometallics **8** [1989] 346/56).

2.2.1.3 R_2InCl Compounds and Their Adducts, where R = Aryl

$(C_6H_5)_2InCl$

The compound was formed on reaction of equimolar amounts of InCl and $Hg(C_6H_5)_2$ in boiling ether. After 24 h the cooled reaction mixture was separated from Hg (and unreacted InCl) and concentrated for drying; the remaining ether was removed from the raw product in vacuum, giving a white crystalline solid (m.p. 268 to 270 °C, 64 to 65% yield). A similar result was achieved by the reaction of metallic indium with $Hg(C_6H_5)_2$ and $HgCl_2$ (2:2:1) in ether solution [2].

The ^{115}In NQR spectrum was measured (in MHz) at 298 K. From the data, two crystallographically different units (with In-Cl-In bridges and almost linear $(C_6H_5)_2In$ groups) were postulated for the unit cell:

ν_1	ν_2	ν_3	ν_4	e^2Qq/h	η (%)
54.28	90.95	139.10	185.87	1117.0	14.2
58.00	91.17	140.40	187.80	1129.2	17.1

On the basis of the very large coupling constants the aryl groups were deemed to be trans, so that no dimeric four-membered ring structure with tetrahedral coordination at the In atom can exist in the crystalline form [2, 3].

Only the Raman spectrum of the solid below 700 cm^{-1} was evaluated: ν_{as}(ring, in-plane) + ν(InC) 648 vs, δ(ring, out-of-plane) + ν(InC) 445 vw, δ_s(ring, in-plane) + ν(InC) 392 vw, δ(ring, in-plane rotation) 220 mw, 209 m, 205 s, ν(In-Cl-In) 184 mw, δ(ring, out-of-plane) + δ(InC) 169 mw (see also the IR spectrum of $In(C_6H_5)_3$ on p. 95). No discrete In-C stretching vibrations, as in $(CH_3)_2InCl$ (on p. 123) for example, were observed, but one can refer to "X-sensitive" ring vibrations with some contributions of In-C(ring) valence or deformation vibrations. Simple representations of the five relevant vibrations are sketched in [2]. Likewise, a polymeric structure for the compound was inferred from the splittings of some vibration modes [2].

The adduct formation with dioxane is described below. The reaction with $TeCl_4$ (2:1 mole ratio) leads to $(C_6H_5)InCl_2$ and $(C_6H_5)_2TeCl_2$ [4]. The antimicrobial activity of the compound was evaluated in vitro for the control of *Colletrotrichium falatum*, the causal organism of red-rot of sugarcane [1].

$(C_6H_5)_2InCl \cdot O_2C_4H_8$

The 1:1 adduct formed in practically quantitative yield by dissolving $(C_6H_5)_2InCl$ in 1,4-dioxane, then precipitating the product with hexane; m.p. 258 to 260 °C. The lack of splitting in the most significant, generally constant framework vibrations in the Raman spectrum (see $(C_6H_5)_2InCl$) and the increase in frequency of the In-Cl vibration to 251 cm^{-1} indicate the monomeric character of the adduct [2].

$(C_6H_4CH_3\text{-}4)_2InCl$

No data for this compound were described; it was said to react with $TeCl_4$ in a similar manner to $(C_6H_5)_2InCl$ [4].

$\{C_6H_2(CH_3)_3\text{-}2,4,6\}_2InCl$

A mixture of $In(mes)_3$ (mes = $C_6H_2(CH_3)_3$-2,4,6) and $InCl_3$ (2:1 mole ratio) was stirred in benzene for 12 h. The resulting colorless, clear solution was evaporated to dryness and the residue extracted with pentane. The compound was obtained as colorless crystals (m.p. 178 to 179 °C, 73% yield) and is dimeric in benzene [8].

The chemical shifts were measured in C_6D_6 solution and are given in ppm. ^1H NMR: 2.11 (s, CH_3-4), 2.47 (s, CH_3-2,6), 6.68 (s, C_6H_2). ^{13}C NMR: 21.2 (CH_3-4), 26.2 (CH_3-2,6), 128.3, 139.0, 144.4, 148.6 (all C_6) [8].

The following IR bands (in cm^{-1}) were reported: 1730 w, 1725 w, 1595 m, 1545 m, 1400 m, 1290 m, 1255 s, 1100 vs, 1020 vs, 840 m, 795 m, and 695 m [8].

The compound crystallizes in the monoclinic space group $P2_1/a-C^5_{2h}$ (No. 14) with the cell constants a = 911.3(2), b = 1608.3(4), c = 1236.0(3) pm, β = 109.90(2)°. The unit cell with Z = 2 contains 1 dimeric unit; D_c = 1.515 g/cm³. The structure was resolved at −82 °C. In the centrosymmetric dimer the In atoms are in a distorted tetrahedral coordination geometry. The mesityl planes at one In atom form an angle of 86.5°. The central In_2Cl_2 plane and the C(1)-In-C(2) plane form an angle of 99.5°. The structure of the dimer is depicted in **Fig. 17**, p. 132 [8].

$\{C_6H_4(CH_2N(CH_3)_2)\text{-}2\}_2InCl$

Freshly prepared $LiC_6H_4(CH_2N(CH_3)_2)$-2 was slowly added to a suspension of $InCl_3$ (2:1 mole ratio) in ether. The strongly exothermic reaction turned brown and was completed by continuous stirring for 24 h at room temperature. The residue was separated and washed several times with C_6H_6. The compound crystallized as a white crystalline powder (m.p. 190 to 191 °C, 74% yield) after addition of petroleum ether to the combined benzene washings. The exchange reaction between $InCl_3$ and $In\{C_6H_4(CH_2N(CH_3)_2)\text{-}2\}_3$ (1:2 mole ratio) in ether led to a quantitative yield of the desired product in less than 24 h; the compound is monomeric in benzene, according to ebullioscopic molecular weight determinations [7].

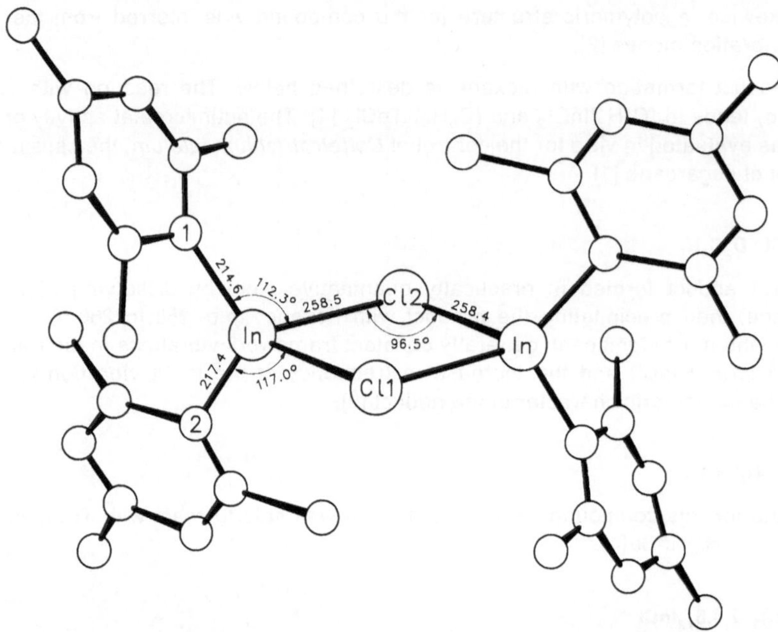

Fig. 17. Molecular structure of $\{(C_6H_2(CH_3)_3\text{-}2,4,6)_2InCl\}_2$ [8].

Other bond angles (°)

C(1)-In-Cl(1)	101.3(3)	Cl(1)-In-Cl(2)	83.5(1)
C(2)-In-Cl(2)	101.8(3)		

In order to establish the double intramolecular ring closure, the ^1H and ^{13}C NMR spectra were measured at different temperatures in $CDCl_3$ and C_7D_8 [5]. Coalescence temperatures for the CH_3 and CH_2 protons in the ^1H NMR are ca. −10 °C in $CDCl_3$ and ca. 0 °C in C_7D_8. The chemical shifts (all singlets) are:

solvent	^1H NMR t in °C	CH_3	CH_2	H-6	^{13}C NMR t in °C	CH_3	CH_2	C(1)	C(2)	C(6)
$CDCl_3$	30	2.60	3.71	7.61[2)]	30	45.5	65.9	149.5	144.9	137.2
	−40	2.49	3.28[1)]	7.67[2)]	−50	46.1	65.1	147.7	144.5	136.9
		2.78	4.29[1)]			44.5				
C_7D_8	30	2.21	3.42	7.55[2)]	30	46.1	67.1	151.2	146.4	138.5
	−35	2.06	2.81[1)]		−50	46.8	65.9	150.1	146.1	138.3
		2.63	4.33[1)]			44.3				

[1)] d's, J(AB) = 14 Hz. — [2)] m's.

The dynamic NMR spectra were explained with an equilibrium involving In–N bond dissociation with an intermediate fourfold coordinated In atom followed by recoordination [6].

The mass spectrum was measured by the field desorption method and shows the base peak at $[M-H]^+$; other important ions are $[M_2-Cl]^+$, $[M+Cl]^+$, and $[M-Cl-H]^+$ [6].

The X-ray structure analysis confirms the fivefold coordination of the In atom. The compound crystallizes in the orthorhombic space group $P2_12_12_1-D_2^4$ (No. 19) with the unit cell parameters a = 935.1(1), b = 1041.1(2), c = 1909.2(4) pm; Z = 4 and D_c = 1.497 g/cm³. The unit cell contains discrete monomeric units. As **Fig. 18** shows, the coordination around In is a trigonal bipyramid with the equatorial plane formed by Cl and the ring carbon atoms C(1) and C(2). N(1) and N(2) occupy the axial positions of the bipyramid. Departure from an ideal bipyramid is caused by the opposite deviations of C(1) and C(2) from the ideal equatorial plane. Referred to the nearly linear N(1)-In-N(2) axis, the twist amounts to ±13 to 14° [5].

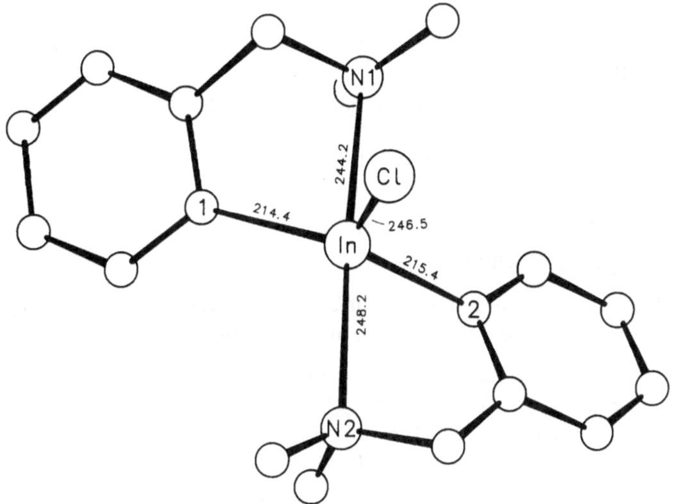

Fig. 18. Molecular structure of $\{C_6H_4(CH_2N(CH_3)_2)\text{-}2\}_2InCl$ [5].

Bond angles (°)

Cl-In-N(1)	89.4(1)	C(1)-In-C(2)	153.3(1)
Cl-In-N(2)	89.9(1)	N(1)-In-C(2)	102.7(1)
Cl-In-C(1)	102.9(1)	N(2)-In-C(2)	76.1(1)
Cl-In-C(2)	103.7(1)	N(1)-In-C(1)	77.0(1)
N(1)-In-N(2)	178.4(1)	N(2)-In-C(1)	104.6(1)

$\{C_6H_4(CHCH_3N(CH_3)_2)\text{-}2\}_2InCl$

The compound was prepared in a manner completely analogous to the above, from (S)-$LiC_6H_4(CH(CH_3)N(CH_3)_2)$-2 and $InCl_3$ (2:1 mole ratio, 79% yield, m.p. 179 °C, monomer in benzene) [6].

The ^1H and ^{13}C NMR spectra were obtained in $CDCl_3$ and C_7D_8 at both room temperature and −10 °C (all values in ppm); see above. ^1H NMR at room temperature ($CDCl_3/C_7D_8$): 1.45/1.15 (CH_3C), 2.53/2.23 (CH_3N), 7.87/7.71 (m, H-6 of C_6H_4); at −10 °C ($CDCl_3$) the CH_3N protons give signals at 2.42 and 2.67 ppm which coalesce at 12 °C. ^{13}C NMR spectrum in C_7D_8 at room temperature: 16.2 (CH_3C), 44.9 (CH_3N), 66.9 (CH), 138.8 (C-6), 150.9 (C-1), 151.3 (C-2). At −40 °C the CH_3N signal splits into two signals at 40.4 and 46.0 ppm, whereas the others remain nearly unchanged. The rate of the dynamic process described for the compound above is increased in coordinating solvents. Thus, in C_5D_5N solution at room temperature similar ^{13}C values were obtained but no splitting occurred down to −40 °C [6].

References:

[1] Srivastava, T. N.; Bajpai, K. K.; Singh, K. (Indian J. Agric. Sci. **43** [1973] 88/93 from C.A. **79** [1973] No. 122497).

[2] Miller, S. B.; Jelus, B. L.; Brill, T. B. (J. Organometal. Chem. **96** [1975] 1/14).

[3] Brill, T. B. (Inorg. Chem. **15** [1976] 2558/60).

[4] Srivastava, T. N.; Srivastava, R. C.; Kapoor, K. (J. Inorg. Nucl. Chem. **41** [1979] 413/4).

[5] Kahn, M.; Steevensz; R. C.; Tuck, D. G.; Noltes, J. G.; Corfield, P. W. R. (Inorg. Chem. **19** [1980] 3407/11).

[6] Jastrzebski, J. T. B. H.; van Koten, G.; Tuck, D. G.; Meinema, H. A.; Noltes, J. G. (Organometallics **1** [1982] 1492/5).

[7] Steevensz; R. C.; Tuck, D. G.; Meinema, H. A.; Noltes, J. G. (Can. J. Chem. **63** [1985] 755/8).

[8] Leman, J. T.; Barron, A. R. (Organometallics **8** [1989] 2214/9).

2.2.1.4 Other R_2InCl Compounds

$\{C_6H_3(CH_2N(C_2H_5)_2)_2\text{-}2,6\}CH_3InCl$

The compound was obtained from CH_3InCl_2 and $Li\{C_6H_3(CH_2N(C_2H_5)_2)_2\text{-}2,6\}$ in hexane in the form of colorless crystals (55% yield, m.p. 52 to 54 °C, b.p. 160 °C/0.1 Torr) [2].

The ^1H and ^{13}C NMR spectra were measured at room temperature in C_6D_6 (in ppm). ^1H NMR: 0.06 (s, CH_3In), 0.85 (q, CH_2 of C_2H_5), 2.59 (t, CH_3 of C_2H_5), 3.15 (s, CH_2N), 7.00 to 7.42 (m, C_6H_3). ^{13}C NMR: −9.3 (CH_3In), 8.7 (CH_2 of C_2H_5), 44.1 (CH_3 of C_2H_5), 60.2 (CH_2N), 145.0 (C-2,6), 124.4 (C-3,5), 126.7 (C-4); the signal of C-1 was not observed [2].

The following fragments were reported in the mass spectrum (70 eV?): $[M]^+$ (3.2%), $[M-CH_3]^+$ (2.8%), $[M-Cl]^+$ (92.6%), and $^{115}In^+$ (100%) [2].

The compound crystallizes in the monoclinic space group $P2_1/n-C_{2h}^5$ (No. 14) with the parameters a = 1090.0(1), b = 1354.0(2), c = 1300.0(2) pm, α = 99.99(1)°; Z = 4. The central atom is surrounded by a distorted trigonal bipyramid. The equatorial plane is formed by Cl, methyl C(1), and aryl C(3), and the central In atom; the axial positions are occupied by N(1) and N(2). The five-membered In-C(3)-C(2)-C(8)-N(1) and In-C(3)-C(4)-C(13)-N(2) rings from intramolecular adduct formation are each distinctly nonplanar. N(1) lies 75.6(3) pm below and N(2) 60.6(3) pm above the plane of the phenyl ring. The shortest intermolecular contact of 296 pm is between the Cl and the H of a methyl group on the indium atom of a neighboring molecule. The structure of the compound with relevant bond distances is depicted **Fig. 19**, p. 135 [2].

$C_4(C_6H_5)_4InCl \cdot NC_5H_5$

The compound is a yellow ether-insoluble product, formed from the reaction of dilithiated tetraphenylbutadiene-1,4 and $InCl_3 \cdot 3\ NC_5H_5$ in ether (30 h). It was purified by recrystallization from petroleum ether (50% yield, no melting or decomposition temperature was mentioned) [1].

In the ^1H NMR spectrum in CD_2Cl_2 a complex multiplet was observed between 6.6 and 8.6 ppm [1].

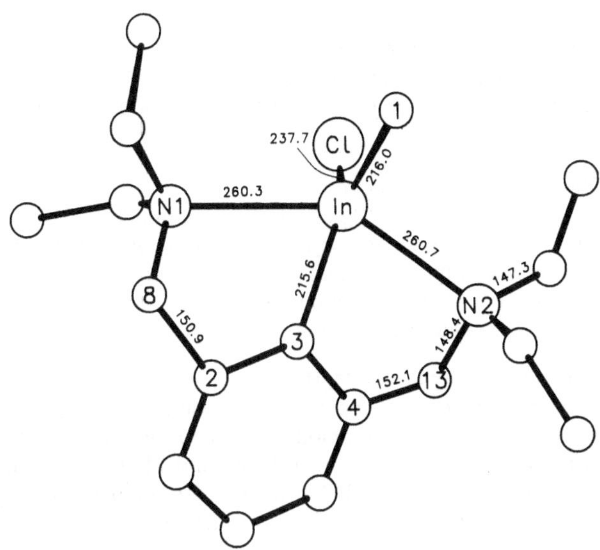

Fig. 19. Molecular structure of $\{C_6H_3(CH_2N(C_2H_5)_2)_2\text{-}2,6\}CH_3InCl$ [2].

Bond angles (°)

Cl-In-C(1)	119.2(1)	N(2)-In-N(1)	144.9(1)
Cl-In-C(3)	110.6(1)	N(1)-In-C(3)	72.7(1)
C(3)-In-C(1)	130.1(2)	N(2)-In-C(3)	72.3(1)
N(1)-In-N(2)	101.0(1)	C(13)-N(2)-In	99.9(1)
N(1)-In-C(1)	99.5(1)	C(14)-N(2)-In	103.8(2)

The Diels-Alder decomposition gave no clue as to the formation of In(I) or In(II) compounds. On the basis of the sparse analytical and physical data, an InC_4 heterocycle as shown in Formula I was postulated [1].

C_6H_5 C_6H_5 NC_5H_5 In   Cl C_6H_5 C_6H_5

I

References:

[1] Peppe, C.; Tuck, D. G. (Polyhedron **1** [1982] 549/52).
[2] Schumann, H.; Wassermann, W.; Dietrich, A. (J. Organometal. Chem. **365** [1989] 11/8).

2.2.2 $RInCl_2$ Compounds and Their Adducts, where R = Alkyl, Cycloalkenyl, Aryl

Regardless of the R group, the simple $RInCl_2$ compounds (Nos. 1 to 9) are listed in Table 19 first, and then the adducts (Nos. 10 to 32). One compound of the type $\{R\text{-}D(\text{-}D')\}InCl_2$ (No. 24) is described. The following methods have been used for preparation:

Method I: From InR_3 and HCl.
The reactions were carried out in a 1:2 mole ratio with 2 moles of RH being eliminated. Ether (No. 1 [1]) or toluene (No. 6, elimination of $Si(CH_3)_4$ [9]) were used as solvents; the yields of the $RInCl_2$ compounds were practically quantitative.

Method II: From ligand exchange reaction.
The exchange between $InCl_3$ and InR_3 compounds is carried out in a 2:1 mole ratio (No. 6 [9]) or between R_2InCl and $InCl_3$ in a 1:1 mole ratio (No. 1 [1]). The solvent was ether, and the yields of the corresponding $RInCl_2$ compounds were nearly 100%. The conditions for No. 5 and 9 were 8 h in refluxing toluene [24] and 12 h in refluxing benzene [17], respectively.

Method III: Alkylation of $InCl_3$ by LiR or MgR_2.
Ether, dioxane, or hexane were used as solvents. While $InCl_3$ with $Li(C_5(CH_3)_5)$ (1:1 mole ratio) in ether reacted without formating the etherate of No. 7 [15], the corresponding procedure with $InCl_3$ and $Mg(C_6F_5)_2$ in dioxane formed the polymeric adduct, No. 26 [2]. Intramolecular adduct No. 24 [19] arose in high yield from the reaction of $InCl_3$ with Li(R-D) compound in hexane.

Method IV: Adducts from $RInCl_2$ and the appropriate D compound.
Adducts No. 10 to 21 were prepared from 1:1 components in an inert solvent; after a short time the solution was freed of solvent, and the residue recrystallized (see [1]). The various adducts of $C_6F_5InCl_2$ (Nos. 27 to 32) used the dioxanate, $C_6F_5InCl_2 \cdot O_2C_4H_8$, as starting material; the Lewis base $O_2C_4H_8$ can be completely displaced by N-donor ligands [7].

General Remarks. All the adducts of $C_6F_5InCl_2$ (Nos. 26 to 32) except No. 31 are soluble in acetone; compounds 27 and 32 also dissolve in $CHCl_3$, but they are insoluble in $O(C_2H_5)_2$, C_6H_6, and petroleum ether. The UV spectra were recorded in Nujol emulsions, conductance measurements were made in acetone at 23 °C. IR spectra were used to infer structures with five- or six-coordinate In for Nos. 26 to 32: the position of $\nu(InCl_2)$ was used as a criterion for distinguishing between a terminal In-Cl bond and (weak) In-Cl bridging. A collection of the most important measurements is as follows [4]:

adduct No.	D	λ_{max} in nm	Λ_m in $\Omega^{-1} \cdot cm^2 \cdot mol^{-1}$ (conc. in mol/L)	$\nu(InCl_2)$ in cm^{-1}
27	$(CH_3)_2NCH_2CH_2N(CH_3)_2$	–	–	317 m, br 264 m, br
28	$C_{10}H_8N_2$ (bipyridine-2,2′)	242, 293, 311, 325	32.0 (1.35×10^{-3})	303 m, br 277 m, br
29	$C_{12}H_8N_2$ (phenanthroline-1,10)	242, 292 sh 325	36.3 (7.2×10^{-4})	314 m, br 294 s, br
31	$C_{15}H_{11}N_3$ (terpyridine-2,2′,2″)	230, 292 sh 333 sh, 350	insoluble in acetone	311 m, br 265 s, br
32	$(C_6H_5)_2P(CH_2)_2P(C_6H_5)_2$ ("diphos")	–	32.5 (1.09×10^{-3})	313 s, br 304 m, sh

In addition, adduct formation was verified by the shift of prominent characteristic vibrations of the free ligands. As in the case of $(C_6F_5)_3In \cdot D$ (Table 14, on p. 92), these spectra are also dominated by the position-constant bands of the C_6F_5 deformation and bond vibrations. These absorptions, which are omitted from the examples in Table 19, are located at: 1636±2 w, 1570 w, sh, 1509±3 s to vs, 1460±4 vs, br, 1377±1 m or m, sh, 1365±2 mw to m, 1352±3 m, 1268±4 w, 1074±4 s, 1061±3 sh, 1010 w, br, 961±4 vs, 780±3 vw to w, 718 w, sh, 606±1 w, 358±3 w to mw cm^{-1}. There are also weak to very weak bands at about 548, 489, 479, and 279 cm^{-1}, which are usually hidden by the stronger bands of the other ligands [4].

Table 19
$RInCl_2$ Compounds and Their Adducts.
Further information on numbers preceded by an asterisk is given at the end of the table.
Explanations, abbreviations, and units on p. X.

No.	compound method of preparation (yield)	properties and remarks
*1	CH_3InCl_2 I II (ca. 100%)	white crystalline solid, m.p. 165 °C [1], 164 to 166 °C [5], soluble in ether and benzene; monomeric in nitromethane [1] conductivity: 13.3 $\Omega^{-1} \cdot cm^2 \cdot mol^{-1}$ [5] IR (solid): 745 s, br, 720 sh $\varrho(CH_3)$, 523 m $\nu(InC)$ [1], 525 [5], 285 ν(InCl-terminal) [1] Raman (solid): 530 $\nu(InC)$, 280, 269 ν(InCl-terminal), 179 ν(InCl-bridge) [5]
2	$C_2H_5InCl_2$ II (97%)	white crystalline solid, m.p. 155 to 158 °C, subl. 80 °C/10^{-4} Torr [21] ^1H NMR ($C_6D_6/O(C_2D_5)_2$): 1.33 (complex m, C_2H_5) [21] IR (solid): 1459 m, 1422 w $\delta_{as}(CH_3, CH_2)$, 1376 m $\delta_s(CH_3)$, 1221 mw $\delta(CH_2)$, 1160 s $\omega(CH_2)$, 1020 mw, 997 w, 955 mw, 940 ms ν(C-C) and $\varrho(CH_3)$, 668 vs $\varrho(CH_2)$, 505 ms $\nu(InC)$, 302 vs ν(InCl-terminal), 230 vs, 212 w ν(InCl-bridge) and $\delta(InCC)$ [21] Raman (solid): 1458 mw, 1424 w $\delta_{as}(CH_3, CH_2)$, 1380 w $\delta_s(CH_3)$, 1220 w $\delta(CH_2)$, 1155 s $\omega(CH_2)$, 1020 w, 1002 w, 960 w ν(C-C) and $\varrho(CH_3)$, 504 vs $\nu(InC)$, 298 s ν(InCl-terminal), 261 mw, 230 m ν(InCl-bridge) and $\delta(InCC)$, 126 mw $\delta(InClIn')$, 106 sh, 92 w, 76 mw lattice [21]
*3	$(C_3H_7\text{-}i)InCl_2$ II (83%)	colorless needles, m.p. 165 °C (dec.) [18] ^1H NMR (C_6D_6): 1.26 (d, CH_3), 1.53 (m, CH, AB_6-system) [18] ^{13}C NMR (C_6D_6): 21.8 (CH_3), 28.1 (CH) [18] IR (solid): 486 mw $\nu(InC)$, 316 s ν(InCl-terminal), 240 m, sh, 219 s ν(InCl-bridge) [18] Raman (solid): 486 s $\nu(InC)$, 304 m ν(InCl-terminal), 249 s, br ν(InCl-bridge) [18] mass spectrum: $[M_2-Cl]^+$ (3.88%), $[M_2-2\,Cl-C_3H_7]^+$ (5.3%), $[In(C_3H_7)_2]^+$ (7%), $^{115}In^+$ (39.6%) [18]

Table 19 (continued)

No.	compound method of preparation (yield)	properties and remarks
*4	$(C_4H_9\text{-}t)CH_2InCl_2$ II (77.6%)	colorless crystals, formation of two modifications, A and B [16] ^{1}H NMR (C_4D_8O, A/B): 1.00/1.03 (CH_3C), 1.17/1.13 (CH_2) [16] IR values on p. 143
*5	$C_6H_5CH_2InCl_2$ II (98%)	white solid, m.p. 171 to 172 °C [14], 171 to 173 °C (dec.) [24] ^{1}H NMR: 2.49 (CH_2), 7.10 (m, C_6H_5) in CD_2Cl_2 [14], 2.53, 6.98 to 7.21 in CD_3CN [24] ^{13}C NMR (CD_3CN): 29.3 (CH_2), 124.7 (C-4), 128.7 (C-3,5), 129.3 (C-2,6), 142.9 (C-1) [24] IR (solid): 3020 w, 1950 w, 1870 w, 1800 w, 1765 w, 1595 m, 1410 w, 1255 s, 1090 s, 1040 sh, 1015 s, 845 w, 795 s, 755 m, 725 m, 690 m, 565 m, 445 w ν(InC, ?), 390 w [14], 279 s ν(InCl-terminal), 269 m and 240 sh ν(InCl-bridge) [24] mass spectrum (main fragments only): M_2^+, $[M_2-C_7H_7]^+$, M^+ and $[M-C_7H_7]^+$ [14]
6	$(CH_3)_3SiCH_2InCl_2$ I II (ca. 100%)	white crystals, m.p. 167 to 170 °C, subl. 150 °C/in vacuum; slightly soluble in C_6H_6 and CH_2Cl_2 [9] ^{1}H NMR (CH_2Cl_2, ca. 30 °C): 0.16 (CH_3), 0.23 (CH_2) [9] IR (solid): 1305 w, sh, 1255 vs, 1008 s, 825 vs, 775 s, 763 s, sh, 725 s, 697 m, sh, 595 m, 512 m, 308 m, sh, 265 m [9]
7	$(C_5(CH_3)_5)InCl_2$ III	orange-yellow solid, contaminated by $(C_5(CH_3)_5)_2InCl$; turned purple at 98.5 °C, by 176 °C material was black [15] ^{1}H NMR (C_6H_6): 1.84 (s, CH_3) [15] decomposes slowly in C_6H_6 to give $(C_5(CH_3)_5)_2InCl$ and $(C_5(CH_3)_5)_2$ [15]
*8	$C_6H_5InCl_2$ IV (81 to 90%)	white solid, m.p. 233 to 235 °C [7] IR and Raman (solid): 654 w, 435 s, 243 w ν(InCl-bridge?); no other bands reported [7] reacts with $(CH_3)_3SiSR'$ to give $In(SR')Cl_2$ and $(CH_3)_3SiC_6H_5$ [20]
*9	$\{C_6H_2(CH_3)_3\text{-}2,4,6\}InCl_2$ II (87%)	white solid [17] IR: 1705 w, 1600 m, 1280 w, 1250 m, 1100 vs, 1020 vs, 895 w, 845 m, 795 m, 720 m [17]
adducts, $RInCl_2 \cdot D$		
10	$CH_3InCl_2 \cdot SSb(CH_3)_3$ IV	white solid, m.p. 102 to 104 °C [3] ^{1}H NMR (CH_2Cl_2): 0.35 (CH_3In), 1.97 (CH_3Sb) [3]

Table 19 (continued)

No.	compound method of preparation (yield)	properties and remarks
10 (continued)		IR (solid): 567, 535 ν(SbC), 515 ν(InC), 395 ν(SbS), 296, 285, 275 ν(InS) and ν(InCl) [3] decomposes in boiling C_2H_5OH to give $(InSCl)_n$, In_2S_3, and $(CH_3)_4SbCl$ (33%) [3]
*11	$CH_3InCl_2 \cdot NH_2(C_4H_9\text{-t})$ II special	white needle-like crystals, m.p. 117 °C [12]
12	$CH_3InCl_2 \cdot 2\,(NC_5H_5)$ IV	white solid, recrystallized from $O(C_2H_5)_2$, softened at 75 °C, turned to a cloudy gum at 116 °C [1] IR (solid): 512 m ν(InC); no other bands reported [1]
13	$CH_3InCl_2 \cdot C_{10}H_8N_2$ $C_{10}H_8N_2$ = 2,2′-bipyridine IV	fine white powder, m.p. 197 to 224 °C, monomeric in nitromethane [1] conductivity: 26.4 $\Omega^{-1} \cdot cm^2 \cdot mol^{-1}$ [1] IR (solid): 508 vw, 496 m ν(InC), no other bands reported [1]
14	$CH_3InCl_2 \cdot C_{15}H_{11}N_3$ $C_{15}H_{11}N_3$ = 2,2′,2″-terpyridine IV	fine yellow powder, recrystallized from CH_3OH, m.p. >260 °C [1] IR (solid): 522 m, 510 w, sh ν(InC); no other bands reported [1]
15	$(C_4H_9\text{-t})CH_2InCl_2 \cdot N(CH_3)_3$ IV	tan solid, m.p. 70 to 73 °C [16] ^1H NMR (C_6D_6 ?): 1.06 (CH_3C), 1.16 (CH_2In), 1.83 (CH_3N) [16]
16	$(C_4H_9\text{-t})CH_2InCl_2 \cdot (CH_3)_2NCH_2CH_2N(CH_3)_2$ IV	colorless solid, m.p. 113 to 117 °C [16] ^1H NMR (C_6D_6 ?): 1.23 (CH_2In), 1.28 (CH_3C), 1.85 (CH_2N), 2.08 (CH_3N) [16]
17	$(CH_3)_3SiCH_2InCl_2 \cdot O(C_2H_5)_2$	Method IV; weak 1:1 adducts but no details reported [9]
18	$(CH_3)_3SiCH_2InCl_2 \cdot THF$	
19	$(CH_3)_3SiCH_2InCl_2 \cdot CH_3OCH_2CH_2OCH_3$	
20	$(CH_3)_3SiCH_2InCl_2 \cdot N(CH_3)_3$	
21	$(CH_3)_3SiCH_2InCl_2 \cdot NCCH_3$	
22	$C_6H_5CH_2InCl_2 \cdot C_{10}H_8N_2$ $C_{10}H_8N_2$ = 2,2′-bipyridine	yellow solid, prepared in good yield, by electrochemical oxidation of In metal in a solution of $C_6H_5CH_2Cl$ and $C_{10}H_8N_2$ in CH_3CN; no properties reported [10]
23	$C_6H_5InCl_2 \cdot O_2C_4H_8$ IV (ca. 100%)	white solid, m.p. 205 to 207 °C [7] IR (solid): 435 m, 329 m $\nu_{as}(InCl_2)$, 200 m [7] Raman (solid): 660 m, 402 w, 327 m $\nu_s(InCl_2)$, 163 m, no other bands reported [7]

Table 19 (continued)

No.	compound method of preparation (yield)	properties and remarks
24	[structure: 2,6-bis((dimethylamino)methyl)phenyl-$InCl_2$, N→In coordinated] III	white solid, m.p. 108 to 110 °C, subl. 100 °C/vacuum [19] ^{1}H NMR (C_6D_6): 2.20 (s, CH_3N), 3.05 (s, CH_2N), 6.77 to 7.16 (m, C_6H_3) [19] ^{13}C NMR (C_6D_6): 45.6 (CH_3N), 63.6 (CH_2N), 125.6 (C-3,5), 129.9 (C-4), 143.6 (C-2,6) [19]
25	$(C_6H_2(CH_3)_3\text{-}2,4,6)InCl_2 \cdot N(CH_3)_3$	from $In(C_6H_2(CH_3)_3\text{-}2,4,6)_3$ and $NH(CH_3)_3Cl$ (1:2 mole ratio) in C_6H_6 in 34% yield [17] colorless solid, m.p. 154 °C (dec.), monomeric in benzene [17] ^{1}H NMR (C_6D_6): 1.98 (s, CH_3N), 2.12 (s, CH_3-4), 2.47 (s, CH_3-2), 2.69 (s, CH_3-6), 6.70, 6.76 (s's, C_6H_2) [17] ^{13}C NMR (C_6D_6): 21.2 (CH_3-4), 26.4, 27.4 (CH_3-2,6), 47.1 (CH_3N), 128.2, 138.8, 144.3, 145.4 (C_6H_2) [17] IR (solid): 1720 w, 1600 w, 1360 m, 1265 s, 1095 s, 1010 s, 980 sh, 845 w, 800 s, 725 m, 600 w [17]
*26	$C_6F_5InCl_2 \cdot O_2C_4H_8$ III (60%)	polymeric solid, m.p. 256 to 259 °C, soluble in $O(C_2H_5)_2$, insoluble in petroleum ether [2] IR (solid): 1442 sh, 1401 vw, 1360 sh, 1337 vw, 1295 m, 1260 s, 1122 w, 1104 sh, 1098 s, 1085 vs, 1040 w, 899 s, 852 vs, 818 vw, 673 vw, 667 vw, 619 s, 330 m, br ν($InCl_2$), 219 m; for the C_6F_5 frequencies, see general remarks on p. 137 [2]
27	$C_6F_5InCl_2 \cdot (CH_3)_2NCH_2CH_2N(CH_3)_2$ IV	m.p. 128 °C, degree of association in $CHCl_3$: 2.5 (1.88 wt%) [4] IR (solid): 1467 sh, 1438 sh, 1295 sh, 1289 w, 1123 w, 1049 sh, 1020 m, 1000 m, 944 sh, 796 m, 763 w, 498 w, 467 w, 442 w, 366 sh; for the C_6F_5 and $InCl_2$ frequencies, see general remarks on p. 137 [4]
28	$C_6F_5InCl_2 \cdot C_{10}H_8N_2$ $C_{10}H_8N_2$ = 2,2′-bipyridine IV	m.p. 166 to 167 °C, softened without complete melting [4] IR (solid): 1612 m, 1602 s, 1580 m, 1496 sh, 1479 m, 1445 vs, 1422 sh, 1320 s, 1251 m, 1175 m, 1158 m, 1110 w, 1043 sh, 1028 s, 773 vs, 732 s, 658 sh, 653 m, 639 w, 419 m; C_6F_5 and $InCl_2$ frequencies are listed in general remarks on p. 137 [4]
29	$C_6F_5InCl_2 \cdot C_{12}H_8N_2$ $C_{12}H_8N_2$ = 1,10-phenanthroline IV	solid, softened at 145 °C, dec. to a purple solid [4] IR (solid): 1627 m, 1609 w, 1589 m, 1584 sh, 1522 s, 1495 sh, 1431 vs, 1373 w, 1224 w, 1149 m, 1109 m, 1052 sh, 992 sh, 871 m, 854 s, 734 sh, 724 vs, 684 m,

Table 19 (continued)

No. compound method of preparation (yield)	properties and remarks
29 (continued)	426 m; for the C_6F_5 and $InCl_2$ frequencies, see general remarks on p. 137 [4] forms the monohydrate No. 30 on exposure to the atmosphere for 15 min [4]
30 $C_6F_5InCl_2 \cdot N_2C_{12}H_8 \cdot H_2O$	from No. 29, dec. ≧270 °C [4] conductivity $((CH_3)_2CO)$: 29.4 $\Omega^{-1} \cdot cm^2 \cdot mol^{-1}$ [4] IR (solid): similar to No. 29, additional bands at 3450 vs, br ν(OH) [4]
31 $C_6F_5InCl_2 \cdot C_{15}H_{11}N_3$ $C_{15}H_{11}N_3$=2,2′,2″-terpyridine IV	solid, dec. >200 °C [4] IR (solid): 1596 s, 1580 s, 1564 m, 1480 s, 1442 sh, 1407 w, 1321 s, 1304 w, 1247 sh, 1194 w, 1167 m, 1119 w, 1098 w, 1048 sh, 1029 sh, 1022 s, 872 w, 833 w, 796 sh, 781 vs, 737 m, 727 w, 664 m, 652 s, 644 m, 514 w, 439 w, 432 sh, 409 w, 327 sh, for the the C_6F_5 and $InCl_2$ frequencies, see general remarks on p. 137 [4]
32 $C_6F_5InCl_2 \cdot$ $\{(C_6H_5)_2PCH_2\text{-}\}_2$ IV	m.p. 210 °C, softened at 174 °C, degree of association in $CHCl_3$:0.85 (1.28 wt%) [4] IR (solid): 1486 m, 1438 m, 1415 w, 1189 sh, 1179 w, 1110 m, 1101 sh, 1069 sh, 1029 w, 1101 w, 746 w, 732 vs, 703 sh, 693 vs, 680 sh, 514 s, 471 m, 454 m, for the C_6F_5 and $InCl_2$-frequencies, see general remarks on p. 137 [4]

* Further information:

CH_3InCl_2 (Table **19**, No. **1**) is considerably more soluble in $O(C_2H_5)_2$, benzene, and toluene than the monochloride, $(CH_3)_2InCl$ (see p. 122).

Crystals suitable for X-ray analysis were obtained by recrystallization from chlorobenzene. The compound crystallizes in the tetragonal space group $I\bar{4}-S_4^2$ (No. 82) with the unit cell constants a=b=1362.8(4), c=633.3(3) pm; Z=8 gives D_c=2.27 g/cm³. CH_3InCl_2 forms dimeric units, bridged by Cl in a trans configuration. These are loosely joined to neighboring dimers through the Cl atoms, so that a band (ribbon) structure results; the metal atoms adopt a distorted trigonal-bipyramidal coordination as depicted in **Fig. 20**, p. 142 [6].

CH_3InCl_2 has been used as a starting material for adducts No. 10 to 14. The reaction with $(CH_3)_3MCl_2$ or $(CH_3)_4MCl$ (M=As, Sb) gave ionic compounds $[M(CH_3)_4][CH_3InCl_3]$ or $[M(CH_3)_4][InCl_4]$ [8]; with $LiCH_2Sn(CH_3)_3$ the mixed alkyl $InCH_3(CH_2Sn(CH_3)_3)_2$ was accessible [11] (see p. 105); and with $Li_2[(CH_3Si(NC_4H_9\text{-}t)_2)_2]$ the complex polycycle I resulted [13] (see p. 142). With two equiv. of the salt $K[H_2B(pz)_2]$ (pz=$N_2C_3H_3$) the complex $CH_3In\{H_2B(pz)_2\}_2$ (Formula II) was prepared in 67% yield; one equiv. gave the dimeric complex III in 84% yield [22]. With the Grignard reagents $CH_3N\{(CH_2)_3MgCl\}_2$ and ClMg-

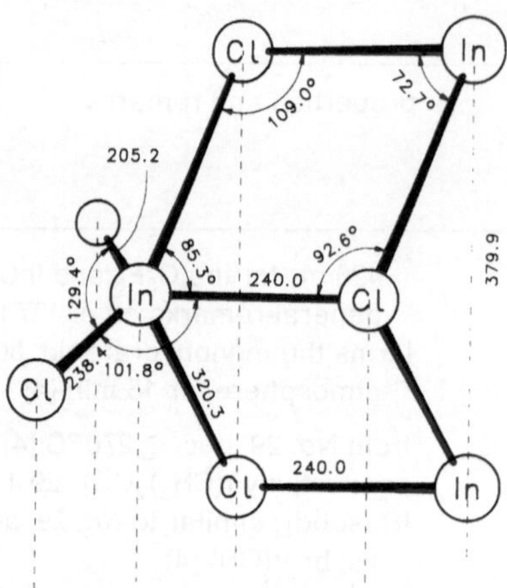

Fig. 20. Coordination at the In atom in solid CH_3InCl_2 [6].

$(CH_2)_3N(CH_3)CH_2CH_2N(CH_3)(CH_2)_3MgCl$ the In-containing heterocycles IV and V have been prepared (see on pp. 109, 110) [26]. Similarly, with $LiP(C_3H_7\text{-}i)_2$ in THF at −78 °C $\{(CH_3)_2In(\mu\text{-}P(C_3H_7\text{-}i)_2)\}_3$ was produced in 82% yield [27].

I II III

IV V

$(C_3H_7\text{-}i)InCl_2$ (Table **19**, No. **3**) formed as a 1:1 mixture with $(C_3H_7\text{-}i)_2InCl$ by treatment of $InCl_3$ with $In(C_3H_7\text{-}i)_3$; no mixed compound was obtained. Reaction of this mixture with $LiNHC_4H_9\text{-}t$ gave two possible isomers of the composition $(C_3H_7\text{-}i)_3In_2Cl(NH(C_4H_9\text{-}t))_2$ along with $\{(C_3H_7\text{-}i)ClInNHC_4H_9\text{-}t\}_2$ and $\{(C_3H_7\text{-}i)_2InNHC_4H_9\text{-}t\}_2$ [23, 25].

$(C_4H_9\text{-}t)CH_2InCl_2$ (Table **19**, No. **4**). Thorough vacuum sublimation of the crude product produced two fractions. The first fraction (form A) sublimed at 110 °C/0.01 Torr (12%) and the second fraction (form B) sublimed at 140 °C/0.01 Torr and accounted for 88% of the total product. Form A melted partially under phase change at 122 to 125 °C, form B at 208 to 210 °C. A phase transition from cis (A) into the corresponding trans (B) form was suggested. Both forms are soluble in ether and THF and only slightly soluble in C_6H_6 [16].

The IR spectra of both forms (in Nujol and Kel F mull, in cm^{-1}) differ only slightly in most of the absorption bands (remarks of the author: although form B is considered as being centrosymmetric, more bands are found than for A). Form A: 3178 w, 2960 s, 2939 s, 2865 m, 2715 m, 2310 w, 1733 w, 1464 m, 1384 m, 1386 m, 1360 s, 1302 m, 1238 s, 1168 m, 1155 m, 1120 m, 1093 m, 1015 m, 1000 m, 915 w, 890 w, 842 w, 767 w, 733 s, 590 w, 450 w, 385 w, 306 m, 280 m, 278 m. Form B: 3180 w, 2952 s, 2930 s, 2880 s, 2860 m, 2730 m, 2320 m, 1468 s, 1457 s, 1383 s, 1370 vs, 1360 vs, 1305 m, 1169 m, 1155 m, 1125 m, 1118 m, 1095 m, 1017 m, 1003 m, 970 m, 932 m, 920 w, 890 w, 848 w, 800 w, 770 w, 735 s, 599 m, 450 w, 380 w, 310 m, 300 s, 280 s, 278 vs [16].

The compound (form B) crystallizes in the orthorhombic space group $P2_12_12_1-D_2^4$ (No. 19) with the parameters a=671.7(4), b=1221.7(4), c=2265.8(7) pm; Z=8 gives D_c=1.84 g/cm³. The structure is similar to that of No. 1. Infinite strands are formed from dimeric units which are linked by chlorine bridges. No short contacts exist between the strands. The coordination geometry at the In atom is a distorted trigonal bipyramid in which the

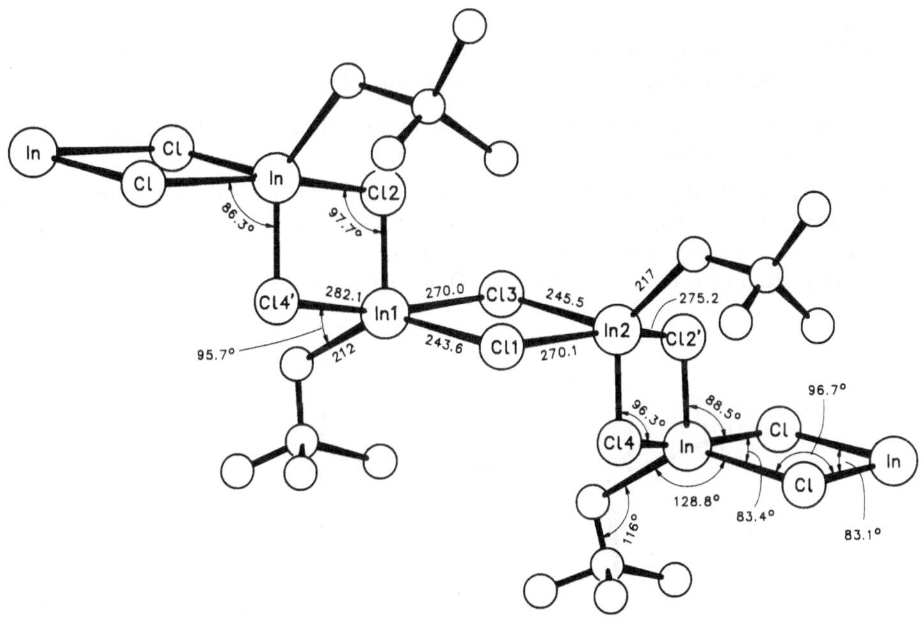

Fig. 21. Molecular structure of $(C_4H_9\text{-t})CH_2InCl_2$ [16].

Other bond angles (°)

Cl(1)-In(1)-Cl(2)	102.2(2)	Cl(3)-In(2)-Cl(4)	104.5(2)
Cl(2)-In(1)-Cl(3)	88.5(2)	Cl(2)-In(2)-Cl(3)	86.1(2)
Cl(2)-In(1)-Cl(4')	82.0(2)	Cl(4)-In(2)-Cl(2')	83.8(2)
Cl(3)-In(1)-Cl(1)	102.6(6)	Cl(1)-In(2)-C(6)	99.9(7)
Cl(3)-In(1)-Cl(4')	161.5(2)	Cl(1)-In(2)-Cl(2')	163.0(2)
Cl(2)-In(1)-Cl(1)	128.5(6)	Cl(3)-In(2)-C(6)	121.7(6)
In(1)-Cl(1)-In(2)	96.8(2)	In(1)-Cl(3)-In(2)	96.5(2)

equatorial positions are occupied by two Cl atoms and C(1); the axial atoms are bridging Cl atoms as depicted in **Fig. 21**, p. 143. Axial In-Cl distances are significantly longer (270 to 280 pm) than equatorial In-Cl distances (241 to 245 pm). The trans arrangement leads to a stair-step polymer [16].

The compound (B) forms the stable adducts No. 15 and 16, and no stable adducts with ether or THF. With $In(CH_2C_4H_9\text{-}t)_3$ comproportionation occurs to give $((C_4H_9\text{-}t)CH_2)_2InCl$. Treatment with $LiCH_3$ in ether produced a compound of the approximate composition $Li[(C_4H_9\text{-}t)(CH_3)_2InCl] \cdot n\ O(C_2H_5)_2$ [16].

$C_6H_5CH_2InCl_2$ (Table **19**, No. **5**) can also be prepared by addition of solid $[HN(CH_3)_3]Cl$ to a solution of $(C_6H_5CH_2)_2InCl$ in ether in a 1:1 mole ratio (5 h) followed by recrystallization from ether (57% yield) [14].

No stable adducts were obtained with various ethers or $N(CH_3)_3$ [14]. The compound reacts with $Na[N(Si(CH_3)_3)_2]$ at 0 °C in ether to give $\{C_6H_5CH_2(Cl)InN(Si(CH_3)_3)_2\}_2$ in 60% yield [23, 25].

$C_6H_5InCl_2$ (Table **19**, No. **8**) was prepared by arylation of In or InCl with phenyl mercury compounds. All the following variations were carried out in $O(C_2H_5)_2$ solution resulting in $C_6H_5InCl_2$ and Hg in each case: InCl + C_6H_5HgCl (1:1 mole ratio); InCl + $Hg(C_6H_5)_2$ + $HgCl_2$ (2:1:1 mole ratio); In + $Hg(C_6H_5)_2$ + $HgCl_2$ (2:1:2 mole ratio) [7].

$\{C_6H_2(CH_3)_3\text{-}2,4,6\}InCl_2$ (Table **19**, No. **9**) is insoluble in noncoordinating solvents and is probably polymeric. In ether or THF the corresponding adducts were formed [17].

The compound reacted with $Na[N(Si(CH_3)_3)_2]$ in ether at 0 °C to give $(C_6H_2(CH_3)_3\text{-}2,4,6)Cl\text{-}InN(Si(CH_3)_3)_2$ in 60% yield [23].

$CH_3InCl_2 \cdot NH_2(C_4H_9\text{-}t)$ (Table **19**, No. **11**) was obtained in high yields as very fine crystals using Method IV in toluene. Needles (along with powdery material) formed in 67% yield, when a mixture of CH_3InCl_2 and $(CH_3)_2Si(NHC_4H_9\text{-}t)_2$ in toluene (2.4:1.5 mole ratio) was hydrolyzed by adding a drop of water to form $(OSi(CH_3)_2)_n$ [12].

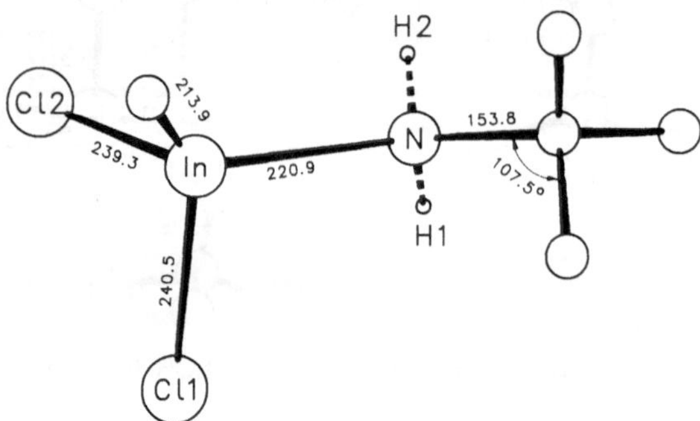

Fig. 22. Molecular structure of $CH_3InCl_2 \cdot NH_2(C_4H_9\text{-}t)$ [12].

Other bond angles (°)

Cl(1)-In-Cl(2)	100.5(1)	Cl(2)-In-N	95.9(2)
Cl(1)-In-C	115.9(3)	C-In-N	122.1(3)
Cl(1)-In-N	98.9(2)	In-N-C	120.8(4)
Cl(2)-In-C	119.8(3)		

The crystals are orthorhombic, space group $P2_12_12_1-D_2^4$ (No. 19) with the parameters a = 1544.6(3), b = 1071.4(2), and c = 628.6(1) pm; Z = 4 gives D_c = 1.75 g/cm^3. The single molecule sketched in **Fig. 22** is joined with others through N-H···Cl bridges to form a one-dimensional chain structure. The shortest intermolecular bridge bonds are Cl(2)···H(2″) = 266, H(1)···Cl(2″) = 305, and H(1)···Cl(1″) = 313 pm, while the shortest intramolecular distance, Cl(1)···H(1), is 323 pm [12].

$C_6F_5InCl_2 \cdot O_2C_4H_8$ (Table **19**, No. **26**) is poorly soluble in C_6H_6 and $CHCl_3$ and no cryoscopic molecular weight determination was possible. On exposure to the air, the moisture present caused hydrolytic decomposition. On the basis of this information, and the presence of terminal InCl bonds (typical frequency of about 300 cm^{-1}), a polymeric structure was suggested in which $C_6F_5InCl_2$ units are bridged by bidentate chair-form dioxane molecules [2].

The compound was used as starting material for adducts No. 27 to 32. With monodentate ligands, such as $(CH_3)_2SO$, NC_5H_5, and $P(C_6H_5)_3$, the adducts $Cl_3In \cdot 3$ D or $Cl_3In \cdot 2\,P(C_6H_5)_3$ were formed in high yields (>80%), along with $(C_6F_5)_3In \cdot N(C_5H_5)$, $(C_6F_5)_3In \cdot 2\,OS(CH_3)_2$, and $(C_6F_5)_3In \cdot 3\,P(C_6H_5)_3$ [4].

References:

[1] Clark, H. C.; Pickard, A. L. (J. Organometal. Chem. **13** [1968] 61/71).
[2] Deacon, G. B.; Parrott, J. C. (Australian J. Chem. **24** [1971] 1771/9).
[3] Maeda, T.; Yoshida, G.; Okawara, R. (J. Organometal. Chem. **44** [1972] 237/41).
[4] Deacon, G. B.; Parrott, J. C. (Australian J. Chem. **27** [1974] 2547/55).
[5] Hausen, H.-D.; Mertz, K.; Veigel, E.; Weidlein, J. (Z. Anorg. Allgem. Chem. **410** [1974] 156/64).
[6] Mertz, K.; Schwarz, W.; Zettler, F.; Hausen, H.-D. (Z. Naturforsch. **30b** [1975] 159/61).
[7] Miller, S. B.; Jelus, B. L.; Brill, T. B. (J. Organometal. Chem. **96** [1975] 1/14).
[8] Widler, H.-J.; Schwarz, W.; Hausen, H.-D.; Weidlein, J. (Z. Anorg. Allgem. Chem. **435** [1977] 179/90).
[9] Beachley, O. T.; Rusinko, R. N. (Inorg. Chem. **18** [1979] 1966/8).
[10] Habeeb, J. J.; Said, F. F.; Tuck, D. G. (J. Organometal. Chem. **190** [1980] 325/34).

[11] Schumann, H.; Mohtachemi, R. (Z. Naturforsch. **39b** [1984] 798/800).
[12] Veith, M.; Recktenwald, O. (J. Organometal. Chem. **264** [1984] 19/27).
[13] Veith, M.; Goffing, F.; Huch, V. (Z. Naturforsch. **43b** [1988] 846/56).
[14] Barron, A. R. (J. Chem. Soc. Dalton Trans. **1989** 1625/6).
[15] Beachley, O. T., Jr.; Blom, R.; Churchill, M. R.; Faegri, K., Jr.; Fettinger, J. C.; Pazik, J. C.; Victoriano, L. (Organometallics **8** [1989] 346/56).
[16] Beachley, O. T., Jr.; Spiegel, E. F.; Kopasz, J. P.; Rogers, R. D. (Organometallics **8** [1989] 1915/21).
[17] Leman, J. T.; Barron, A. R. (Organometallics **8** [1989] 2214/9).
[18] Neumüller, B. (Chem. Ber. **122** [1989] 2283/7).
[19] Schumann, H.; Wassermann, W.; Dietrich, A. (J. Organometal. Chem. **365** [1989] 11/8).
[20] Hoffmann, G. G.; Fischer, R. (Phosphorus, Sulfur, Silicon **41** [1990] 97/104).

[21] Imdahl, R. (Dipl.-Arbeit, Univ. Stuttgart 1990).
[22] Reger, D. L.; Knox, S. J.; Rheingold, A. L.; Haggerty, B. S. (Organometallics **9** [1990] 2581/7).
[23] Neumüller, B. (Z. Naturforsch. **45b** [1990] 1559/65).
[24] Neumüller, B. (Z. Anorg. Allgem. Chem. **592** [1990] 42/50).
[25] Neumüller, B. (Z. Naturforsch. **46b** [1991] 753/61).

[26] Schumann, H.; Hartmann, U.; Wassermann, W.; Just, O.; Dietrich, A.; Pohl, L.; Hostalek, M.; Lokai, M. (Chem. Ber. **124** [1991] 1113/9).
[27] Cowley, A. H.; Jones, R. A.; Mardones, M. A.; Nunn, C. M. (Organometallics **10** [1991] 1635/7).

2.3 Organoindium Bromides

2.3.1 R_2InBr Compounds and Their Adducts, where R = Alkyl and Aryl

Most of the examples collected in Table 20 were prepared by the reactions described for the monochlorides in Section 2.2.1.1 on p. 118.

Method I: Ligand exchange between $InBr_3$ and InR_3.
The components were allowed to react in a 1:2 mole ratio as described for the chlorides [20, 27, 29].

Method II: Reaction of InR_3 with $CHBr_3$ or CH_2Br_2.
The reaction of InR_3 or its diethyl etherate ($R = C_2H_5$ has been the most thoroughly investigated) with $CHBr_3$ [5, 7] or CH_2Br_2 [6] was usually vigorous even at room temperature and formed R_2InBr along with RBr and either 2-pentene or norcarane. $In(C_2H_5)_3$ can be replaced by $(C_2H_5)_2InX$ ($X = OC_4H_9$-t, $N(C_2H_5)_2$) in this synthesis; in [12], however, this possibility was used to form only the chloride.

Method III: From InBr and HgR_2 compounds, or In and $HgBr_2/HgR_2$.
This procedure has been used for R = aryl. InBr treated with HgR_2 in boiling ether or In metal treated with HgR_2 and $HgBr_2$ in boiling ether; the workup was carried out as described for $(C_6H_5)_2InCl$ (p. 130) [22].

Method IV: Adducts from R_2InBr and D compounds.
Combination of R_2InBr and D in equimolar amounts formed adducts No. 14 and 15 [11].

Table 20
R_2InBr Compounds and Their Adducts.
Further information on numbers preceded by an asterisk is given at the end of the table.
Explanations, abbreviations, and units on p. X.

No.	compound method of preparation (yield)	properties and remarks
*1	$(CH_3)_2InBr$ I (>90%) [20] II [11]	white crystalline solid, m.p. 220 to 223 °C [14], subl. 110 °C/in vacuum [4] ^{115}In NQR (ca. 25 °C): 50.88 (ν_1), 101.81 (ν_2), 152.87 (ν_3), 203.50 (ν_4), 1221.69 (e^2Qq/h), η = ca. 0% [16, 24] ^{79}Br NQR (ca. 25 °C): 68.67 [16, 24] IR (solid): 1170 m $\delta_s(CH_3)$, 735 vs, br $\varrho(CH_3)$, 554 s $\nu_{as}(InC_2)$, 486 w $\nu_s(InC_2)$ [4] Raman (solid): 555 vw $\nu_{as}(InC_2)$, 496 vvs $\nu_s(InC_2)$, 111 mw ν(InBr-bridge) [13, 14]

Table 20 (continued)

No.	compound method of preparation (yield)	properties and remarks
*2	$(C_2H_5)_2InBr$ I (90%) [27] II (75% with $CHBr_3$ [5], 62% with CH_2Br_2) [6]	white needle-like crystals, m.p. 168 to 170 °C [5], 169 to 171 °C [15], 184 to 186 °C (dec.) [27], subl. 85 °C/10^{-4} Torr, dimeric in benzene [6] IR (solid): 510 $\nu_{as}(InC_2)$, 460 $\nu_s(InC_2)$ [6, 14] Raman (solid): 510 w, 459 s ν(InC) [13] NQR and Raman spectra on pp. 149, 150
3	$(C_3H_7\text{-}n)_2InBr$	formed, together with $C_3H_7InBr_2$, by the reaction of In and C_3H_7Br; no properties reported [9] IR (solid): 584 m, 496 m ν(InC) [13] Raman (solid): 579 w, 565 s, 496 m, br ν(InC) [13]
4	$(C_3H_7\text{-}i)_2InBr$ I (96%)	colorless needles, m.p. 72 °C [31] ^1H NMR (C_6D_6): 1.43 (CH and CH_3) [31] ^{13}C NMR (C_6D_6): 22.5 (CH_3), 29.4 (CH) [31] IR (solid): 499, 460 $\nu(InC_2)$, other unassigned bands at 388, 235, 220 [31]
5	$(C_4H_9\text{-}n)_2InBr$	from $(C_4H_9\text{-}n)_3In_2Br_3$ and KBr at 80 to 100 °C and sublimation in vacuum [13] IR (solid): 590 w, 568 m ν(InC) [13] Raman (solid): 575 m ν(InC), 219 m ν(InBr); no other bands reported [13]
6	$\{(CH_3)_3SiCH_2\}_2InBr$ I	m.p. 93.5 to 94.5 °C [29] ^1H NMR (C_6H_6): 0.15 (CH_3Si), 0.64 (CH_2Si) [29] with $K[P(C_6H_5)_2]$ in $O(C_2H_5)_2$, dimeric $(CH_3)_2InP(C_6H_5)_2$ is formed (see p. 317) [29]
7	$(C_2H_5OC(O)CH_2)_2InBr$	from activated In metal and $C_2H_5OC(O)CH_2Br$; no properties reported [19] used for reactions with ketones [19]
*8	$(C_6H_5)_2InBr$ III (77 to 84%)	colorless solid, m.p. 255 to 257 °C [22] m.p. >315 °C [2] $n_D^{20}=1.562$ [2]. IR and Raman (solid): 650 vs, 438 vw, 392 vw, 212 s, 204 m, 169 s, 159 m ν(InBr-bridge ?); Raman spectrum depicted in this region, polymeric lattice proposed [22] mass spectrum (at 290 °C): $[M]^+$ (0.48%), $[C_6H_5InBr]^+$ (14.47%), $[(C_6H_5)_2In-H]^+$ (0.47%), $[C_6H_4InBr]^+$ (0.09%), $[(C_6H_5)_2In]^+$ (7.0%), $[(C_6H_5)_2In-H]^+$ (5.40%), $[C_6H_5In]^+$ (1.35%), $[(C_6H_5)_2]^+$ (7.22%), $[C_6H_6]^+$ (8.04%), $[C_6H_5]^+$ (11.73%), see also on p. 150 [22]
9	$(C_6H_4F\text{-}4)_2InBr$ III (73%)	colorless solid, m.p. 220 to 222 °C; no other properties reported [22]
10	$(C_6H_4CH_3\text{-}4)_2InBr$ III (95%)	colorless solid, m.p. 225 to 227 °C [22] Raman (solid): 564 w, 479 w, 260 w, 225 w, 184 m, 145 vs ν(InBr-bridge ?) [22]

Table 20 (continued)

No.	compound method of preparation (yield)	properties and remarks
11	$\{C_6H_2(CH_3)_3\text{-}2,4,6\}_2InBr$ I (78%)	forms also from $In(mes)_3$ and HBr; white solid, dimeric in solution [33]
12	$(C_6F_5)_2InBr$	from $InBr_3$ and C_6F_5MgBr in $O(C_2H_5)_2$ (43%) [3] white solid, m.p. 144 to 147 °C, subl. ca. 134 °C/0.05 Torr, moderately soluble in $C_6H_5CH_3$, dimeric in benzene [3]
13	$(C_{10}H_7)_2InBr$ $C_{10}H_7 = \alpha$-naphthyl	from $InBr_3$ and $C_{10}H_7MgBr$ in $O(C_2H_5)_2$ [2] colorless crystals, soluble in all organic solvents except C_6H_{12}-c and petroleum ether [2] $n_D^{20} = 1.61$ [2]
adducts, $R_2InBr \cdot D$		
14	$(CH_3)_2InBr \cdot SSb(CH_3)_3$ IV	no properties reported [11] dec. in boiling CH_3OH (4 h) to give $(CH_3InS)_n$ (85%) and $(CH_3)_4SbBr$ (67%) [11]
15	$(C_2H_5)_2InBr \cdot SSb(CH_3)_3$ IV	m.p. 70 to 73 °C [11] ^1H NMR (CH_2Cl_2): 1.81 (CH_3Sb), no values for (C_2H_5In) [11] IR (solid): 558, 528 ν(SbC), 496, 456 ν(InC), 407 ν(SbS), 247 ν(InS) [11] dec. in boiling CH_3OH (4 h) to give $(C_2H_5InS)_n$ (88%) and $C_2H_5(CH_3)_3SbBr$ (66%) [11]

* Further information:

$(CH_3)_2InBr$ (Table **20**, No. **1**) was formed in more than 90% yield by the reaction of $(CH_3)_2InOCH_3$ with stoichiometric amounts of HBr in ether [4]. The compound was also produced in 3 to 5 days by the reaction of In with excess CH_3Br at room temperature; however, a mixture with CH_3InBr_2 was obtained. Separation of the two products was not described [9, 10, 18].

The results of the NQR measurements were used in [24] and [25] to correlate the asymmetry parameter (η) and the C-In-C valence angle of a number of similar organoindium derivatives. The $In(CH_3)_2$ skeleton of the monobromide was found to be nearly linear and a six-coordinated central In atom was inferred from the NQR data [23].

A correlation of the frequency difference, $\Delta\nu = \nu_{as}(InC_2) - \nu_s(InC_2)$, versus the C-In-C angle was made by [20] and improved in [26] by using a larger number of examples. The bond angle is estimated by these relationships to be 172° or 174° for $(CH_3)_2InBr$. This relation between δν and the size of the InC_2 angle is nearly linear between 120° and 180° (between 109° and 120° it is parabolic [26]), but was contradicted by [30]. (Comments of the author: In [30] the determination of δν was based exclusively on IR data. Frequently, however, the IR spectrum do not permit the location of $\nu_s(InC_2)$ with any certainty, because this mode appears as a weak shoulder and is partially covered by other vibrations in the $(CH_3)_2InNR_2$ compounds measured.)

Single crystals suitable for X-ray analysis were obtained by recrystallization from 2:1 cyclohexane/bromobenzene. According to rotating crystal and KCl-calibrated Weissenberg photos, $(CH_3)_2InBr$ was found to crystallize in the tetragonal space group I4mm $-$ C_{4v}^9 (No. 107) with cell parameters a = b = 438.4(1), c = 1389.3(3) pm; Z = 2 gives D_c = 2.795 g/cm³ while D_m = 2.79 g/cm³ [20]. Later measurements with the diffractometer technique, the centrosymmetric space group I4/mmm $-$ D_{4h}^{17} (No. 139) was estimated [32]. The structure is displayed in **Fig. 23**. The indium atom is in a distorted tetragonal bipyramidal environment coordinated by two carbon and four bromine atoms. What is surprising is the difference in the In-C distances of the linear CH_3-In-CH_3 grouping, which was called on to explain the observed InC_2 valence frequencies [20]. According to later measurements, however, the In-C distances are identical at about 215 pm [32] which corresponds to the results for $(CH_3)_2TlCl$ [15].

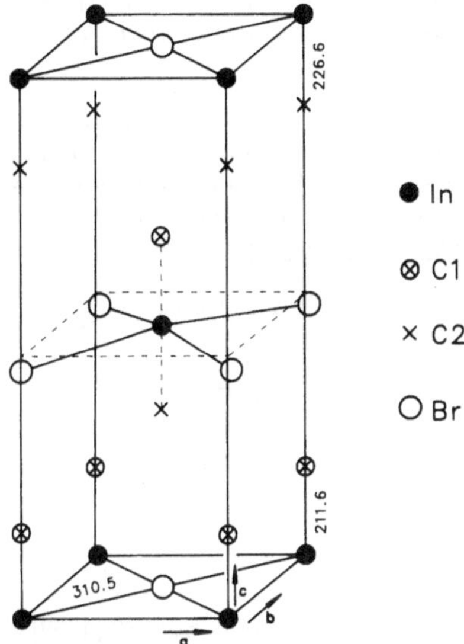

Fig. 23. Unit cell of $(CH_3)_2InBr$ [20].

The title compound reacted with $(CH_3)_3AsBr_2$ in CH_2Cl_2 to form the salt $[As(CH_3)_4]$-$[CH_3InBr_3]$ [28]. With $(CH_3)_3P{=}CH_2$ the dimeric compound $\{(CH_3)_2InCH_2P(CH_3)_2CH_2\}_2$ was formed along with $[P(CH_3)_4]Br$ [17].

$(C_2H_5)_2InBr$ (Table **20**, No. **2**) was produced after stirring a 1:1 mixture of $In(C_2H_5)_3$ and CH_2Br_2 in cyclohexane for 7 days; C_2H_5Br (15%) and norcaranes (14%) were also formed. The postulated intermediate, $(C_2H_5)_2InCH_2Br$, could not be isolated [6]. The reaction of In with excess C_2H_5Br led to a mixture of $(C_2H_5)_2InBr$ and $C_2H_5InBr_2$, which were not separated [9, 10, 18].

The ^{115}In (77 K [21]) and ^{79}Br NQR (room temperature [8]) spectra gave the following transition frequencies (in MHz) for ^{115}In: ν_1 = 54.30, ν_2 = 94.73, ν_3 = 144.25, ν_4 = 192.64, e^2Qq/h = 1157.2 [21], 1158 ± 2 [8], η = 12.2 [21], 16.4 ± 0.1% [8]; ^{79}Br: ν = 74.73 [8]. A special apparatus for low temperature measurements is described in [21]. The results of the NQR measurements with those of other organoindium derivatives have been compared [23, 25].

A discussion of the entire IR and Raman spectra is given in [27]. The spectra differ in the 1500 to 600 cm^{-1} region by an average of only -3 cm^{-1} ($+3$ to a maximum of -7 cm^{-1}) from the spectra of the homologous monochloride (see on p. 126).

The important absorption frequencies (in cm^{-1}) for the basic $(C_2InBr)_2$ skeleton are: IR (Nujol); $\nu_{as}(InC_2)$ 510 ms, $\nu_s(InC_2)$ 459 m. Raman (solid); $\nu_{as}(InC_2)$ 512 vw, $\nu_s(InC_2)$ 463 vvs, $\delta(InCC)$ 267 ms, ν_{as}(InBr bridge) and $\delta(InC_2)$ 130 sh and 110 sh, ν_s(InBr bridge) 98 m, 90 ms [27].

$(C_2H_5)_2InBr$ has been used to prepare the adduct No. 15 [11] and as a polymerization catalyst [7].

$(C_6H_5)_2InBr$ (Table **20**, No. **8**) was first obtained in 80% yield from the reaction of a benzene solution of $In(C_6H_5)_3$ and Br_2 (1:1 mole ratio). It is a brownish yellow powder which does not melt below 300 °C. It is soluble in $C_6H_5CH_3$, C_6H_6, $O(C_2H_5)_2$, NC_5H_5, slightly soluble in $CHCl_3$, and insoluble in petroleum ether [1]. Arylation of $InBr_3$ by C_6H_5MgBr in ether gave the compound in 28% yield after recrystallization from benzene/petroleum ether [2].

The mass spectrum of the compound was measured with an excitation energy of 70 eV and a probe temperature of 290 °C (see the important ions in Table 20); an except from the fragmentation pattern based on observed metastable peaks is given below. In order to determine the influence of probe temperature on the fragmentation, a series of runs were carried out between 180 and 300 °C [22].

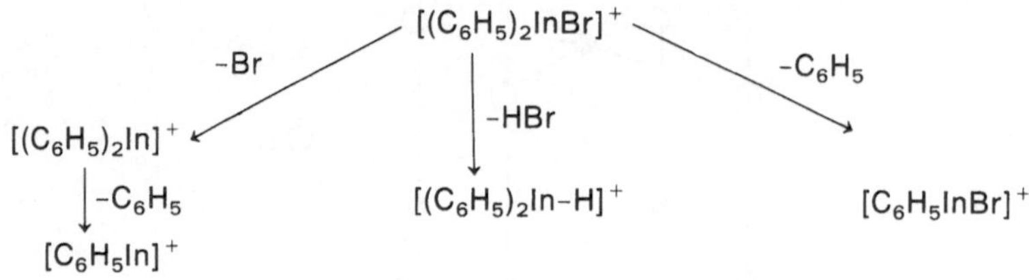

References:

[1] Schumb, W. C.; Crane, H. I. (J. Am. Chem. Soc. **60** [1938] 306/8).

[2] Runge, F.; Zimmermann, W.; Pfeiffer, H.; Pfeiffer, I. (Z. Anorg. Allgem. Chem. **267** [1951] 39/48).

[3] Pohlmann, J. L. W.; Brinckmann, F. E. (Z. Naturforsch. **20b** [1965] 5/11).

[4] Clark, H. C.; Pickard, A. L. (J. Organometal. Chem. **8** [1967] 427/34).

[5] Yasuda, K.; Okawara, R. (Inorg. Nucl. Chem. Letters **3** [1967] 135/6).

[6] Maeda, T.; Tada, H.; Yasuda, K.; Okawara, R. (J. Organometal. Chem. **27** [1971] 13/8).

[7] Okawara, R.; Yasuda, K.; Tada, T. (Japan. 71-3567 [1971] from C.A. **74** [1971] No. 112205).

[8] Golubinskaya, L. M.; Bregadze, E. V.; Bryuchova, E. V.; Svergun, V. I.; Semin, G. K.; Okhlobystin, O. Yu. (J. Organometal. Chem. **40** [1972] 275/9).

[9] Gynane, M. J. S.; Waterworth, L. G.; Worrall, I. J. (J. Organometal. Chem. **40** [1972] C9/C19).

[10] Gynane, M. J. S.; Waterworth, L. G.; Worrall, I. J. (J. Organometal. Chem. **43** [1972] 257/64).

[11] Maeda, T.; Yoshida, G.; Okawara, R. (J. Organometal. Chem. **44** [1972] 237/41).

[12] Shcherbakov, V. I.; Zhil'tsov, S. F. (Nov. Khim. Karbenov Mater. 1st Vses. Soveshch. Khim. Karbenov Ikh Analogov, Moscow 1972 [1972], pp. 162/4 from C.A. **82** [1975] No. 43559).

[13] Gynane, M. J. S.; Worrall, I. J. (J. Organometal. Chem. **81** [1974] 329/34).

[14] Hausen, H.-D.; Mertz, K.; Veigel, E.; Weidlein, J. (Z. Anorg. Allgem. Chem. **410** [1974] 156/64).

[15] Hausen, H.-D.; Veigel, E.; Guder, H.-J. (Z. Naturforsch. **29b** [1974] 269/70).

[16] Patterson, D. B.; Carnevale, A. (Inorg. Chem. **13** [1974] 1479/83).

[17] Schmidbaur, H.; Füller, H.-J. (Chem. Ber. **107** [1974] 3674/9).

[18] Waterworth, L. G.; Worrall, I. J. (J. Organometal. Chem. **81** [1974] 23/4).

[19] Chao, L.-C.; Rieke, R. D. (J. Org. Chem. **40** [1975] 2253/5).

[20] Hausen, H.-D.; Mertz, K.; Weidlein, J.; Schwarz, W. (J. Organometal. Chem. **93** [1975] 291/6).

[21] Ignatov, B. G.; Aleksandrov, A. L.; Pososhenko, L. Z.; Semin, G. K. (Izv. Akad. Nauk SSSR Ser. Fiz. **39** [1975] 2630/3; Bull. Acad. Sci. USSR Phys. Ser. **39** No. 12 [1975] 154/6; C.A. **84** [1976] No. 128662).

[22] Miller, S. B.; Jelus, B. L.; Brill, T. B. (J. Organometal. Chem. **96** [1975] 1/14).

[23] Brill, T. B. (Inorg. Chem. **15** [1976] 2558/60).

[24] Schmidbaur, H.; Koth, D. (Naturwissenschaften **63** [1976] 482/3).

[25] Bancroft, G. M.; Sham, T. K. (J. Magn. Resonance **25** [1977] 83/90).

[26] Widler, H.-J.; Schwarz, W.; Hausen, H.-D.; Weidlein, J. (Z. Anorg. Allgem. Chem. **435** [1977] 179/90).

[27] Karschin, J.; Mann, G.; Weidlein, J. (unpublished results 1979).

[28] Widler, H.-J.; Weidlein, J. (Z. Naturforsch. **34b** [1979] 18/22).

[29] Beachley, O. T., Jr.; Kopasz, J. P.; Zhang, H.; Hunter, W. E.; Atwood, J. L. (J. Organometal. Chem. **325** [1987] 69/81).

[30] Aitchison, K. A.; Backer-Dirks, J. D. J.; Bradley, D. C.; Faktor, M. M.; Frigo, D. M.; Hursthouse, M. B.; Hussain, B.; Short, R. L. (J. Organometal. Chem. **366** [1989] 11/23).

[31] Hoffmann, G. G.; Faist, R. (J. Organometal. Chem. **391** [1990] 1/5).

[32] Hausen, H.-D. (unpublished results).

[33] Leman, J. T.; Ziller, J. W.; Barron, A. R. (Organometallics **10** [1991] 1766/71).

2.3.2 $RInBr_2$ Compounds and Their Adducts with R = Alkyl and Aryl

Information about the compounds is collected in Table 21. The compounds were prepared by the procedures that are described below. The obvious method of ligand exchange between a 1:2 mole ratio of InR_3 and $InBr_3$ is doubtless also possible, but this path has not been reported.

Method I: From InBr with excess RBr.

The complex reaction of $In^{I}Br$ with excess RBr proceeded through an intensely orange colored intermediate (see equation 1 to 4) and led quantitatively to $RInBr_2$ within a week at room temperature for $R = CH_3$ and C_2H_5 [2, 5]. The reactions of n-propyl and n-butyl bromides required up to 4 weeks, and there was always an insoluble residue of In_2Br_4. With higher alkyl bromides the reaction was complete within 3 days at 50 °C [5]. The intensive study of this reaction, including the colored intermediates, which are moderately soluble in RBr, resulted in postulating the following mechanism [9]:

(1) $8\ InBr + 2\ RBr \rightarrow In_7Br_9 + R_2InBr$

(2) $In_7Br_9 + R_2InBr + 6\ RBr \rightarrow 8\ RInBr_2$

(3) $In_7Br_9 + 2\ RBr \rightarrow In_5Br_7 + 2\ RInBr_2$

(4) $In_5Br_7 + 3\ RBr \rightarrow 3\ RInBr_2 + In_2Br_4$

When R=CH_3, C_2H_5, and C_3H_7 (at 50 °C), only steps (1) and (2) are important. When R=C_3H_7 (20 °C) and C_4H_9 (20 and 50 °C), steps (3) and (4) become increasingly important, explaining the formation of the In_2Br_4 isolated.

Method II: From arylation of In or InBr with aryl mercury compounds.
All the following variations were carried out in $O(C_2H_5)_2$ solution resulting in $ArInBr_2$ and Hg in each case: InBr+ArHgBr (1:1 mole ratio); InBr+$Hg(Ar)_2$+$HgBr_2$ (2:1:1 mole ratio); In+$Hg(Ar)_2$+$HgBr_2$ (2:1:2 mole ratio); see also the preparation of the corresponding dichloroindium compounds on p. 144 [10].

Method III: Adducts from $RInBr_2$ and an appropriate base D.
The preparation of the 1:1 adducts can be achieved by adding a Lewis base D to the reaction mixture from Method I [13] or by using D as the solvent [6].

General Remarks. The ^1H NMR spectra of Nos. 1 to 4 were measured in CH_3Br (Nos. 1 and 2) or CH_2Cl_2 (Nos. 3 and 4), and the solvent was used as reference [5]. The chemical shift values given in Table 21 have been recalculated to refer to $Si(CH_3)_4$, using $\delta=2.72$ ppm for CH_3Br and $\delta=5.35$ ppm for CH_2Cl_2.

The vibrational spectra of Nos. 1 to 4 have only been evaluated in the regions of the InC and InBr bond vibrations. The number of IR and Raman bands assigned in the InC region indicates that the dimers (Nos. 1, 3, and 4) have a four-membered ring structure with a cis configuration (author's comment).

However, [5] considered a mixture of cis and trans isomers to be more likely. The spectrum of $C_2H_5InBr_2$ (No. 2) is compatible with neither the cis nor the trans form; it was postulated to have a complex associated structure with 4- and 5-coordinated In atoms [5].

The mass spectra of Nos. 1 to 4 were discussed as a group [5]. Only dimers or fragments of dimers were found for Nos. 1, 3, and 4, but the spectrum of $C_2H_5InBr_2$ (No. 2) contains additional fragments of higher association (see Table 21). The suggested fragmentation pattern for the dimeric dibromides is given below [5]:

$$[R_2In_2Br_4]^+ \xrightarrow{-R} [RIn_2Br_4]^+ \xrightarrow{-R} [In_2Br_4]^+ \xrightarrow{-Br} [In_2Br_3]^+ \xrightarrow{-In} [InBr_3]^+ \xrightarrow{-Br} [InBr_2]^+ \xrightarrow{-Br} [InBr]^+$$

$$[In_2Br_3]^+ \xrightarrow{-Br} [In_2Br_2]^+ \xrightarrow{-Br} [In_2Br]^+$$

$$[R_2In_2Br_4]^+ \xrightarrow{-Br} [R_2In_2Br_3]^+ \xrightarrow{-R} [RIn_2Br_3]^+ \xrightarrow{-Br} [RIn_2Br_2]^+ \xrightarrow{-In} [RInBr_2]^+ \xrightarrow{-Br} [RInBr]^+ \xrightarrow{-R} [InBr]^+$$

$$[R_2In_2Br_3]^+ \xrightarrow{-InBr_3} [R_2In]^+ \xrightarrow{-R} [RIn]^+$$

Table 21
$RInBr_2$ Compounds and Their Adducts.
Further information on numbers preceded by an asterisk is given at the end of the table.
Explanations, abbreviations, and units on p. X.

No.	compound method of preparation (yield)	properties and remarks
*1	CH_3InBr_2 I (ca. 100%)	white crystalline solid, m.p. 168 to 169 °C [5], 164 °C [4] ^1H NMR (CH_3Br): 1.16 (s, CH_3); see general remarks on p. 152 [5] conductivity: 21.8 $\Omega^{-1} \cdot cm^2 \cdot mol^{-1}$ [4] IR (solid): 523 s, 516 m ν(InC) [5, 7] Raman (solid): 524 s, 517 w ν(InC), 196 s, 148 m, 97 m, 85 w ν(InBr) [5, 7] mass spectrum: see general remarks on p. 152
2	$C_2H_5InBr_2$ I (ca. 100%)	white crystalline solid, m.p. 138 to 139 °C [5] ^1H NMR (CH_3Br): 1.48 (t, CH_3), 1.67 (m, CH_2); see general remarks on p. 152 [5] IR (solid): 496 m , 491 m ν(InC) [5, 7] Raman (solid): 502 m, 494 s ν(InC), 213 s, 200 s, 187 s, 135 w, 91 m ν(InBr) [5, 7] mass spectrum: $[(C_2H_5)_3In_3Br_5]^+$, $[C_2H_5In_4Br_5]^+$, $[In_4Br_4]^+$ in addition to the fragments given in the general remarks on p. 152
3	$(C_3H_7\text{-}n)InBr_2$ I (ca. 100%)	white solid, m.p. 84 °C (dec.) [5] ^1H NMR (CH_2Cl_2): 1.10 (t, CH_3), 1.87 (m, CH_2) [5] IR (solid): 578 m, br, 490 m, br ν(InC) [5, 7] Raman (solid): 584 m, br, 495 m, br ν(InC), 188 s, br ν(InBr) [5, 7] mass spectrum: see general remarks on p. 152
4	$(C_4H_9\text{-}n)InBr_2$ I	white solid, m.p. 70 °C (dec.) [5] ^1H NMR (CH_2Cl_2): 0.99 (t, CH_3); 1.65 (m, CH_2) [5] IR (solid): 583 s, br, 496 s, br ν(InC) [5, 7] Raman (solid): 584 w, br, 493 s, br ν(InC) 232 w, 189 s ν(InBr) [5, 7] mass spectrum: see general remarks on p. 152
5	$C_6H_5InBr_2$ II (81 to 90%) [10]	from $In(C_6H_5)_3$ and Br_2 in C_6H_6 (1:2 mole ratio, 80% yield) [1] white solid, m.p. 203 to 205 °C [10] soluble in C_6H_6, $O(C_2H_5)_2$, NC_5H_5, less soluble in $CHCl_3$ [10] Raman (solid): 654 m, 430 vw, 210 vw, 185 vs ν(InBr), 170 w, polymeric lattice proposed for solid state [10]
6	$(C_6H_4F\text{-}4)InBr_2$ II (82%)	white solid, m.p. 190 to 192 °C [10] Raman (solid): 572 m, 508 vw, 304 w, 243 vw, 223 vw, 207 vw, 178 vs ν(InBr); no other bands reported [10]

Table 21 (continued)

No.	compound method of preparation (yield)	properties and remarks
7	$(C_6H_4CH_3\text{-}4)InBr_2$ II (84%)	m.p. 194 to 196 °C [10] Raman (solid): 576 w, 478 w, 369 w, 240 w, 229 w, 208 w, 178 m ν(InBr); no other bands reported [10]
8	$\{C_6H_2(CH_3)_3\text{-}2,4,6\}InBr_2$ $C_6H_2(CH_3)_3\text{-}2,4,6$=mes	from $In(mes)_3$ and $InBr_3$ (1:2 mole ratio) in refluxing toluene (12 h, 55% yield) [15] white solid, m.p. 295 °C (dec.) [15] ^{1}H NMR ($CDCl_3$/THF): 2.15 (s, CH_3-4), 2.77 (s, CH_3-2,6), 6.81 (s, C_6H_2) [15] IR: 1585 w, 1550 w, 1395 w, 1285 m, 1255 w, 1015 w, 1005 w, 840 s, 800 w, 690 w, 530 m ν(InC) [15]
adducts, $RInBr_2 \cdot D$		
*9	$C_2H_5InBr_2 \cdot (CH_3)_2NCH_2CH_2N(CH_3)_2$ III (ca. 100%)	colorless crystals [13] ^{1}H NMR (CD_2Cl_2): 1.11 (q, CH_2In), 1.27 (t, CH_3, J(H,H) = 7.8), 2.57 (s, CH_3N), 2.76 (s, CH_2N) [14] ^{13}C NMR ($CDCl_3$): −0.19 (CH_3), 13.0 (CH_2In), 47.34 (CH_3N), 56.20 (CH_2N) [14]
10	$C_2H_5InBr_2 \cdot C_{10}H_8N_2$ $C_{10}H_8N_2$ = 2,2′-bipyridine	from anodic oxidation of In metal in a solution of C_2H_5Br and $C_{10}H_8N_2$ in CH_3CN [12] light yellow solid, no further properties reported [12]
11	$C_6H_5CH_2InBr_2 \cdot C_4H_8O_2$ III (ca. 100%)	white crystalline solid, no properties reported [6]
12	$C_6H_5CH_2InBr_2 \cdot C_{10}H_8N_2$ $C_{10}H_8N_2$ = 2,2′-bipyridine	prepared like No. 10 [12] solid, no properties reported [12]
13	$C_6H_5InBr_2 \cdot C_{10}H_8N_2$ $C_{10}H_8N_2$ = 2,2′-bipyridine	prepared like No. 10 [12] solid, no properties reported [12]

* Further information:

CH_3InBr_2 (Table **21**, No. **1**) also formed as a mixture with $(CH_3)_2InBr$ on reaction of metallic In with CH_3Br [3], isolation of the dibromide not being described. The compound was produced in quantitative yield by the reaction of equimolar amounts of $(CH_3)_2InBr$ and $InBr_3$ in $O(C_2H_5)_2$ [4].

The ^{115}In and ^{79}Br NQR spectra (the frequencies are given in MHz) were measured at room temperature. The splitting of the ^{79}Br signal was interpreted as resulting from the terminal and bridging Br atoms of $(CH_3InBr_2)_2$ dimers. ^{115}In: 59.82 (ν_1), 46.71 (ν_2), 68.99 (ν_3), 95.73 (ν_4), 590.15 (e^2Qq/h) and η = 56.4%. ^{79}Br: 101.59 and 102.92 (ν_1 and ν_1') [8]. The metal atoms possess distorted tetrahedral coordination [11].

The ν(InC) absorptions of the solid in the IR spectrum at 526 s cm^{-1} and Raman spectrum at 522 s cm^{-1} correspond to a dimeric trans structure [4].

$C_2H_5InBr_2 \cdot (CH_3)_2NCH_2CH_2N(CH_3)_2$ (Table **21**, No. **9**) crystallizes in the monoclinic space group $P2_1/n-C^5_{2h}$ (No. 14) with cell parameters a=805.4(3), b=1518.3(5), c=1177.6(2) pm, α=98.2(2)°; Z=4 and D_c=1.96 g/cm^3. The compound is isomorphous with the homologous iodine compound (see p. 165). The metal atom's environment is a distorted trigonal bipyramid. The C_2H_5 group along with Br(2) and N(2) of the bidentate donor ligand form the equatorial plane, while Br(1) and N(1) occupy the axial positions. The structure of the adduct is depicted in **Fig. 24** [14].

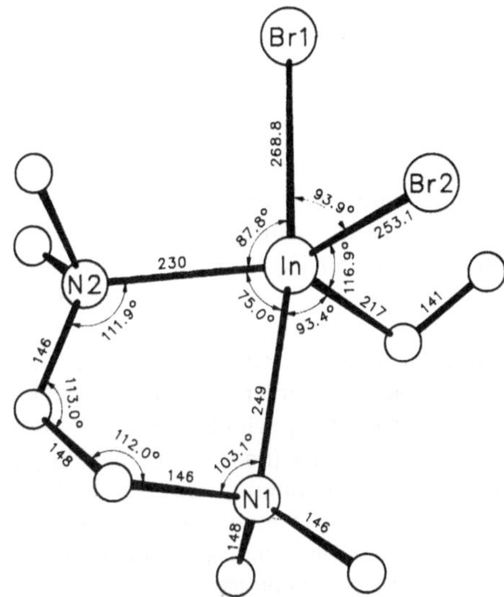

Fig. 24. Molecular structure of $C_2H_5InBr_2 \cdot (CH_3)_2NCH_2CH_2N(CH_3)_2$ [14].

Other bond angles (°)

Br(1)-In-N(1)	162.8(3)	Br(2)-In-N(2)	105.6(3)
Br(1)-In-C(3)	98.3(4)	N(2)-In-C(3)	136.4(5)
Br(2)-In-N(1)	92.0(3)	In-C(3)-C(4)	115(1)

References:

[1] Schumb, W. C.; Crane, H. I. (J. Am. Chem. Soc. **60** [1938] 306/8).

[2] Waterworth, L.; Worrall, I. J. (J. Chem. Soc. Chem. Commun. **1971** 569).

[3] Gynane, M. J. S.; Waterworth, L. G.; Worrall, I. J. (J. Organometal. Chem. **40** [1972] C9/C10).

[4] Poland, J. S.; Tuck, D. G. (J. Organometal. Chem. **42** [1972] 315/23).

[5] Gynane, M. J. S.; Waterworth, L. G.; Worrall, I. J. (J. Organometal. Chem. **43** [1972] 257/64).

[6] Gynane, M. J. S.; Waterworth, L. G.; Worrall, I. J. (Inorg. Nucl. Chem. Letters **9** [1973] 543/4).

[7] Gynane, M. J. S.; Worrall, I. J. (J. Organometal. Chem. **81** [1974] 329/34).

[8] Patterson, D. B.; Carnevale, A. (Inorg. Chem. **13** [1974] 1479/83).

[9] Waterworth, L. G.; Worrall, I. J. (J. Organometal. Chem. **81** [1974] 23/4).

[10] Miller, S. B.; Jelus, B. L.; Brill, T. B. (J. Organometal. Chem. **96** [1975] 1/14).

[11] Brill, T. B. (Inorg. Chem. **15** [1976] 2558/60).
[12] Habeeb, J. J.; Said, F. F.; Tuck, D. G. (J. Organometal. Chem. **190** [1980] 325/34).
[13] Peppe, C.; Tuck, D. G.; Victoriano, L. (J. Chem. Soc. Dalton Trans. **1982** 2165/8).
[14] Khan, A. M.; Peppe, C.; Tuck, D. G. (J. Organometal. Chem. **280** [1985] 17/25).
[15] Leman, J. T.; Ziller, J. W.; Barron, A. R. (Organometallics **10** [1991] 1766/71).

2.4 Organoindium Iodides

2.4.1 R_2InI Compounds and Their Adducts, where R = Alkyl, Cycloalkenyl, and Aryl

Table 22 contains the properties of these compounds. The procedures used for their preparation are given below. Similar methods have been used for preparing the homologous chlorides (p. 118) and bromides (p. 146):

Method I: From In and RI compounds.
Metallic indium was treated with RI, representing the oldest procedure for preparing organoindium compounds. While the first publication [1] gave no information at all about the indium compound formed, later investigations [7, 21] of the reaction using excess RI (20 to 30 °C for 2 to 3 days) usually led to a 1:1 mixture of R_2InI and $RInI_2$ (or to the sesquiiodide, $R_3In_2I_3$, with the same overall composition). Separation of the components was achieved by heating at 80 to 100 °C in vacuum after admixing an equimolar amount of KI; the volatile monohalogenide sublimed out of the mixture [12]. (Author's comment: R_2InI and $RInI_2$ form salts $K[R_2InI_2]$ and $K[RInI_3]$, with KI; see p. 349 [22]. These alkyliodoindates have a limited thermal stability (dec. <100 °C); they do not only dissociate to starting materials, but also eliminate iodine, so that separating R_2InI by heating can lead to poor yields.) The reaction of colloidal indium (obtained by co-condensing In vapor and solvents, such as diglyme, dioxane, xylene, or THF) ran much faster (20 h at reflux), but still formed R_2InI and $RInI_2$ [21].
In contrast, R_2InI and InI were formed in over 90% yield when activated In (obtained by reduction of $InCl_3$ with metallic Na or K) reacted with ca. 50% excess RI at 80 °C [9] or 150 °C [14]. (The author's comment: This method has only been demonstrated for a few examples, but is doubtless of general use.)

Method II: Ligand exchange between InR_3 and InI_3 or between InR_3 and $RInI_2$.
The exchange reaction proceeds similarly to that described for the chloro compounds on p. 118 [11]. $(CH_3)_2InI$ was obtained by modifying the procedure for making $(CH_3)_2InCl$: LiI was added to $InCl_3$ and $LiCH_3$ (1:2:1 mole ratio), giving a halogen exchange between LiI and the $(CH_3)_2InCl$, which forms first [3].

Method III: Electrochemical oxidation of metallic In in the presence of RI.
E.g., 1.2 g $(CH_3)_2InI$ formed at an In cathode during 8 h electrolysis of 30 g CH_3I and ca. 2.5 g $NaClO_4$ in 40 mL CH_3CN at 20 °C (Pt anode, 0.5 A and 0.012 A/cm^2) [10]. Other alkylindium iodides prepared by this procedure were isolated only as their 1:1 adducts with 2,2′-bipyridine [23].

Method IV: From R_2InI and the appropriate D.
The syntheses of the adducts $R_2InI \cdot D$ follow the procedures for the corresponding $R_2InCl \cdot D$ compounds on p. 118 [4].

Table 22
R_2InI Compounds and Their Adducts.
Further information on numbers preceded by an asterisk is given at the end of the table.
Explanations, abbreviations, and units on p. X.

No.	compound method of preparation (yield)	properties and remarks
*1	$(CH_3)_2InI$ I (73%) [3] II (ca. 100%) [9] III [10]	colorless crystalline solid, m.p. 212 to 218 °C, subl. ca. 110 °C/under vacuum [3] conductivity: 87 $\Omega^{-1} \cdot cm^2 \cdot mol^{-1}$ [3] ^1H NMR ($CDCl_3$): 0.75 (CH_3) [3] ^{115}In NQR, IR, and Raman spectra on p. 159
*2	$(C_2H_5)_2InI$ II (97%) special [6]	white crystalline solid, m.p. 171 to 174 °C, subl. under vacuum [6, 9] IR and Raman spectra on p. 159
3	$(C_3H_7\text{-}i)_2InI$ II (99%)	colorless needles, m.p. 136 °C [25] ^1H NMR (C_6D_6): 1.43 (m, CH and CH_3) [25] ^{13}C NMR (C_6D_6): 22.9 (CH_3), 28.6 (CH) [25] IR (solid): 499, 458 ν(InC), other bands at 388, 236, 220 (unassigned) [25]
4	$\{(CH_3)_3SiCH_2\}_2InI$ II	m.p. 64 to 65.5 °C [24] ^1H NMR (C_6D_6): 0.18 (CH_3Si), 0.64 (CH_2In) [24]
5	$(C_5H_5\text{-}c)_2InI$ II	impure, dec. at room temperature, insoluble in all common solvents [11] mass spectrum (280 °C): $[InI]^+$ (6%), $[C_{10}H_{10}]^+$ (6%), $^{127}I^+$ (7%), $^{115}In^+$ (7%), $[C_5H_5]^+$ (100%) [11]
*6	$(C_6H_5)_2InI$ I (94%) special	light yellow crystals [14], m.p. 187 to 189 °C [15], >300 °C [14] conductivity: 5.88 $\Omega^{-1} \cdot cm^2 \cdot mol^{-1}$ [14] Raman (solid): 650 s, 453 vw, 394 vw, 217 mw, 211 s, 199 mw, 177 mw, 167 ms, 141 s ν(InI-bridge ?) [14]
7	$(C_6H_4CH_3\text{-}4)_2InI$ I (93%)	light beige solid, white crystals are formed by interaction of $In(C_6H_4CH_3\text{-}4)_3$ and I_2 in xylene, no properties reported [14]
*8	$(C_6F_5)_2InI$ I (15%)	colorless at −78 °C, pale yellow solid, m.p. 136 to 140 °C, subl. 150 to 160 °C/<10^{-3} Torr [5]
*9	$\{C_6H_2(CH_3)_3\text{-}2,4,6\}_2InI$ II (67%)	colorless crystals, m.p. 140 to 141 °C (dec.) [26] ^1H NMR ($CDCl_3$): 2.29 (s, CH_3-4), 2.48 (s, CH_3-2,6), 6.80 (s, C_6H_2) [26] ^{13}C NMR ($CDCl_3$): 21.1 (CH_3-4), 26.2 (CH_3-2,6), 127.9, 138.9, 144.0, 146.3 (C_6H_2) [26] IR (solid): 1595 m, 1540 m, 1400 w, 1290 s, 1260 w, 1235 w, 1170 w, 1030 s, 920 w, 880 w, 840 s, 700 m, 580 w, 540 s ν(InC), 490 w [26]

Table 22 (continued)

No.	compound method of preparation (yield)	properties and remarks
adducts, $R_2InI \cdot D$		
10	$(CH_3)_2InI \cdot OP(C_6H_5)_3$ IV	white gummy solid [4] IR similar to $(CH_3)_2InCl \cdot OP(C_6H_5)_3$ but no bands reported (see on p. 120) [4]
11	$(CH_3)_2InI \cdot SSb(CH_3)_3$ IV	not isolated, dec. in boiling CH_3OH to give $(CH_3InS)_n$ (97%) and $(CH_3)_4SbI$ (77%) [8]
*12	$(CH_3)_2InI \cdot x\ NH_3$ IV	x=1; clear colorless liquid, m.p. <20 °C, soluble in $O(C_2H_5)_2$, dissociation pressure at 24 °C: ca. 0 Torr [4] x=2; white powder, m.p. 87 to 115 °C, soluble in $O(C_2H_5)_2$ and CH_3NO_2, dissociation pressure at 24 °C: 1 Torr [4] x=3; bulky white powder, m.p. 100 to 119 °C, soluble in NH_3, readily dec. under N_2, dissociation pressure at 24 °C: 120 to 150 Torr [4]
13	$(CH_3)_2InI \cdot NC_5H_5$ IV	white solid, m.p. 72 to 75 °C, subl. 60 °C/under vacuum [3] IR (solid): 720 vs, br $\varrho(CH_3)$, 522 s $\nu_{as}(InC_2)$, 482 m $\nu_s(InC_2)$; no other bands reported [3]
14	$(CH_3)_2InI \cdot H_2NCH_2CH_2NH_2$	white solid, m.p. 172 to 184 °C (dec.) [4] conductivity (in CH_3OH): 90.7 $\Omega^{-1} \cdot cm^2 \cdot mol^{-1}$ [4] IR (solid): 722 vs, br $\varrho(CH_3)$, 520 s, br $\nu_{as}(InC_2)$, 491 m, br $\nu_s(InC_2)$, no other bands reported [4]
15	$(CH_3)_2InI \cdot N_2C_{10}H_8$ $N_2C_{10}H_8$ = 2,2-bipyridine III [23] IV [4]	light yellow solid, m.p. 122 to 125 °C; monomeric in CH_3NO_2 [4] conductivity: 28.2 $\Omega^{-1} \cdot cm^2 \cdot mol^{-1}$ [4] IR (solid): 539 s $\nu_{as}(InC_2)$, 480 vw $\nu_s(InC_2)$; no other bands reported [4]
16	$(CH_3)_2InI \cdot P(C_6H_5)_3$ IV	white solid, insoluble in petroleum ether, slightly dissociated in C_6H_6 [3] IR (solid): 724 s $\varrho(CH_3In)$, 515 s $\nu_{as}(InC_2)$, 478 m $\nu_s(InC_2)$; no other bands reported [3]
17	$(C_2H_5)_2InI \cdot SSb(CH_3)_3$ IV	not isolated, dec. in boiling CH_3OH to give $(C_2H_5InS)_n$ (98%) and $(CH_3)_3C_2H_5SbI$ (64%) [8]
18	$(C_2H_5)_2InI \cdot N_2C_{10}H_8$ III	yellow solid, no properties reported [23]

* Further information:

$(CH_3)_2InI$ (Table **22**, No. **1**) was also obtained in 32% yield, along with CH_3InI_2, by the 20 h reflux of CH_3I in a dioxane slurry of finely divided indium (28%) [21]; it is dimeric in benzene [3].

The ^{115}In [13, 17] and ^{127}I [13] NQR spectra were measured at room temperature (frequencies in MHz, $\eta=0.0\%$ for each). ^{115}In: 49.40 (ν_1), 98.79 (ν_2), 148.17 (ν_3), 1185.47 (e^2Qq/h). ^{127}I: 71.4 (ν_1), 143.01 (ν_2), 476.41 (e^2Qq/h). The correlation of the asymmetry parameters, η, and the C-In-C bond angles [17], as well as comparison with the NQR data of other dimethylindium compounds [16, 18], led to the conclusion that $(CH_3)_2InI$ possesses linear $(CH_3)_2In$ groups similar to those of the homologous monobromide structure (see p. 149).

The IR and Raman spectra of the compound have been evaluated, mainly in the region of the In-C bond vibrations. The following IR frequencies (in cm^{-1}) have been reported for the solid: 1164 m and 1150 vw $\delta_s(CH_3In)$, 730 vs, br $\varrho(CH_3In)$, 548 s $\nu_{as}(InC_2)$, 480 vw $\nu_s(InC_2)$ [3]; similar values are given in [9, 12]. Raman: 478 s $\nu(InC_2)$ [12]. Using the linear correlation between $\delta\nu=(\nu_{as}-\nu_s)InC_2$ and the C-In-C bond angle [19] which was established for a number of samples, a C-In-C bond angle of 175° can be estimated for the title compound.

$(CH_3)_2InI$ was alkylated by $LiCH_3$ in ether to $(CH_3)_3In \cdot n\ O(C_2H_5)_2$ (see p. 30) [3] and reacted with $(CH_3)_3AsI_2$ in CH_2Cl_2 to form $[As(CH_3)_4][CH_3InI_3]$, which, however, was not isolated analytically pure [22].

$(C_2H_5)_2InI$ (Table **22**, No. **2**) formed in 72% yield on reaction of CH_2I_2 and $In(C_2H_5)_3$ (1:1 mole ratio) in 2:1 C_6H_{12}-c/C_6H_{14}-n at room temperature for one day; it is dimeric in benzene [6].

The vibrational spectrum above 550 cm^{-1} is almost identical to the spectrum of $(C_2H_5)_2InBr$ (see p. 150) [20]. The results of this and other investigations suggest for a dimeric, four-membered ring structure with bent $(C_2H_5)_2In$ groupings. The following data (in cm^{-1}) for IR/Raman bands were obtained for the solid: 1168 m/1176 ms $\delta,\omega(CH_2In)$, 655 vs/640 vvw $\varrho(CH_2In)$, 508 ms/504 vw $\nu_{as}(InC_2)$, 458 mw/462 vs $\nu_s(InC_2)$, -/130 sh and -/87 vs ν(InI-bridge); similar values were described for the ν_{as} and ν_s bands in [9, 12] for the solid and in [6] for the compound in CH_2Cl_2 solution.

$(C_6H_5)_2InI$ (Table **22**, No. **6**) was produced in 92% yield from equal moles of InI and $Hg(C_6H_5)_2$ in boiling ether [15]. The compound also formed in 80% yield (along with C_6H_5I) on direct halogenation of $In(C_6H_5)_3$ with elementary iodine in benzene [14].

The IR spectrum from 1700 to 400 cm^{-1} is illustrated in [14].

$(C_6F_5)_2InI$ (Table **22**, No. **8**) formed only as a by-product in the reaction between In and C_6F_5I (160 °C, 6 h), when the main product, $In(C_6F_5)_3$, was subjected to fractional sublimation (see p. 97) [5].

The IR spectrum is characterized chiefly by the vibrations of the C_6F_5 group and has been discussed in general with those of the corresponding triorganyl compounds (see Table 16 on p. 98) [5]; no In-I modes were given.

$\{C_6H_2(CH_3)_3\text{-}2,4,6\}_2InI$ (Table **22**, No. **9**). The starting mixture (method II) in toluene solution was refluxed for 12 h. A mole weight determination (isopiestically in CH_2Cl_2) showed a degree of association of 1.25 at 0.058 mol/L [26].

The compound crystallizes in the monoclinic space group $C2/c-C^6_{2h}$ (No. 15) with cell parameters a = 2655.6(4), b = 1712.6(3), c = 809.2(1) pm, $\beta=104.70(1)°$; Z = 4, gives $D_c=1.792\ g/cm^3$. The data were collected at 173 K. The noncentrosymmetric dimer with asymmetric iodide bridges shows a distorted geometry at the In atoms. The In_2I_2 core has a butterfly structure with a dihedral angle of 23.2° between the two InI_2 planes. The two mesityl rings at each In atom are orientated almost perpendicularly (66.4°). The rings are nearly parallel above the In_2I_2 core, while those below are almost coplanar as shown in **Fig. 25** [26].

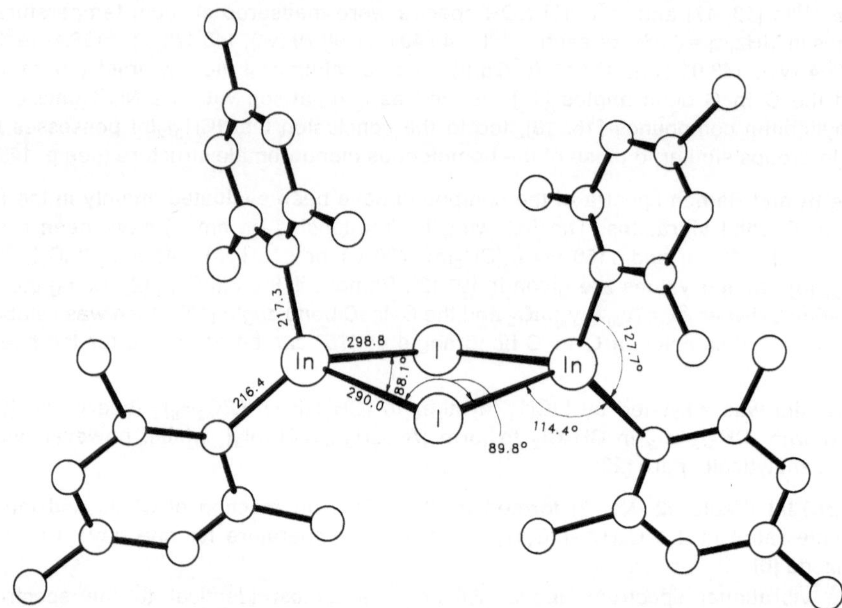

Fig. 25. Molecular structure of $\{C_6H_2(CH_3)-2,4,6\}_2InI$ [26].

$(CH_3)_2InI \cdot x\ NH_3$ (Table **22**, No. **12**). The series of compounds with x = 1 to 3 was studied by tensiometry. Under nitrogen only the 1:1 adduct was stable; the other complexes decomposed with evolution of NH_3. The 1:2 adduct adopted ionic forms, analogous to those of the homologous $(CH_3)_2InCl$ adducts (see p. 126) [4].

References:

[1] Spencer, J. F.; Wallace, M. L. (J. Chem. Soc. **93** [1908] 1827/33).
[2] Schumb, W. C.; Crane, H. I. (J. Am. Chem. Soc. **60** [1938] 306/8).
[3] Clark, H. C.; Pickard, A. L. (J. Organometal. Chem. **8** [1967] 427/34).
[4] Clark, H. C.; Pickard, A. L. (J. Organometal. Chem. **13** [1968] 61/71).
[5] Deacon, G. B.; Parrott, J. C. (Australian J. Chem. **24** [1971] 1771/9).
[6] Maeda, T.; Tada, H.; Yasuda, K.; Okawara, R. (J. Organometal. Chem. **27** [1971] 13/8).
[7] Gynane, M. J. S.; Waterworth, L. G.; Worrall, I. J. (J. Organometal. Chem. **40** [1972] C 9/C 10).
[8] Maeda, T.; Yoshida, G.; Okawara, R. (J. Organometal. Chem. **44** [1972] 237/41).
[9] Chao, L.-C.; Rieke, R. D. (J. Organometal. Chem. **67** [1974] C 64/C 66).
[10] Chernykh, I. N.; Tomilov, A. P. (Elektrokhimiya **10** [1974] 971/4; Soviet Electrochem. **10** [1974] 926/8 from C.A. **81** [1974] No. 85209).

[11] Contreras, J. G.; Tuck, D. G. (J. Organometal. Chem. **66** [1974] 405/12).
[12] Gynane, M. J. S.; Worrall, I. J. (J. Organometal. Chem. **81** [1974] 329/34).
[13] Patterson, D. B.; Carnevale, A. (Inorg. Chem. **13** [1974] 1479/83).
[14] Chao, L.-C.; Rieke, R. D. (Syn. React. Inorg. Metal.-Org. Chem. **5** [1975] 165/73).
[15] Miller, S. B.; Jelus, B. L.; Brill, T. B. (J. Organometal. Chem. **96** [1975] 1/14).
[16] Brill, T. B. (Inorg. Chem. **15** [1976] 2558/60).
[17] Schmidbaur, H.; Koth, D. (Naturwissenschaften **63** [1976] 482/3).

[18] Bancroft, G. M.; Sham, T. K. (J. Magn. Resonance **25** [1977] 83/90).
[19] Widler, H.-J.; Schwarz, W.; Hausen, H.-D.; Weidlein, J. (Z. Anorg. Allgem. Chem. **435** [1977] 179/90).
[20] Karschin, J.; Mann, G.; Weidlein, J. (unpublished results 1979).
[21] Klabunde, K. J.; Murdock, T. O. (J. Org. Chem. **44** [1979] 3901/8).
[22] Widler, H.-J.; Weidlein, J. (Z. Naturforsch. **34b** [1979] 18/22).
[23] Habeeb, J. J.; Said, F. F.; Tuck, D. G. (J. Organometal. Chem. **190** [1980] 325/34).
[24] Beachley, O. T., Jr.; Kopasz, J. P.; Zhang, H.; Hunter, W. E.; Atwood, J. L. (J. Organometal. Chem. **325** [1987] 69/81).
[25] Hoffmann, G.G.; Faist, R. (J. Organometal. Chem. **391** [1990] 1/5).
[26] Leman, J. T.; Ziller, J. W.; Barron, A. R. (Organometallics **10** [1991] 1766/71).

2.4.2 $RInI_2$ Compounds and Their Adducts with R = Alkyl, Cycloalkenyl, and Aryl

The compounds of this type are summarized in Table 23; they were prepared by the procedures for the homologous dibromides, already described in detail on p. 151.

Method I: From InI and RI.
The reaction of InI was carried out with excess RI at room temperature. Usually, $RInI_2$ formed in almost quantitative yield in less than 1 d [3, 5]. The course of the reaction was not investigated in detail; however, the mechanism should be the same as that between InBr an RBr (see p. 151). In the presence of Lewis bases the adducts $RInI_2 \cdot D$ formed directly; however, that was pursued in only a few cases (Nos. 11 and 12) [6, 18].

Method II: Electrochemical oxidation of In in the presence of RI.
The electrochemical oxidation of metallic indium in a solution of RI and an electrolyte (e.g., $NaClO_4$) in CH_3CN in the presence of a Lewis base ($C_{10}H_8N_2$) gave the adducts No. 9, 13, 14, and 16. Details are described in Section 2.4.1 on p. 156 [16].

Comments on the ^{1}H NMR, the vibrational, and the mass spectra of these compounds can be found in the general remarks in Section 2.3.2 on p. 151.

Table 23
$RInI_2$ Compounds and Their Adducts.
Further information on numbers preceded by an asterisk is given at the end of the table.
Explanations, abbreviations, and units on p. X.

No.	compound method of preparation (yield)	properties and remarks
*1	CH_3InI_2 I (98%)	white powder, m.p. 125 to 128 °C (dec.) [5] conductivity: 65 $\Omega^{-1} \cdot cm^2 \cdot mol^{-1}$ [5] ^{1}H NMR (CH_2Cl_2): 1.21 (CH_3) [3] IR (solid): 755 s, br $\varrho(CH_3In)$, 561 [5], 555 s [3] $\nu_{as}(InC_2)$ Raman (solid): 484 s $\nu_s(InC_2)$, 196 w, 185 m $\nu(InI)$, 141 s, 118 w, 73 vw, 65 m, 44 m ν and $\delta(InI)$ [3] mass spectrum (95 °C): $[M_2-CH_3]^+$, $[M_2-InI]^+$, $[InI_3]^+$, $[M]^+$, $[M-CH_3]^+$, $[M-I]^+$, $[(CH_3)_2In]^+$ [5]

Table 23 (continued)

No.	compound method of preparation (yield)	properties and remarks
2	$C_2H_5InI_2$ I (96%)	white powder, m.p. 92 to 94 °C (dec.), dimeric in C_6H_6 [5] conductivity: 27 $\Omega^{-1} \cdot cm^2 \cdot mol^{-1}$ [5] ^{1}H NMR ($CHCl_3$): 1.25 (t, CH_3), 1.88 (q, CH_2) [3] ^{115}In NQR (ca. 20 °C): ν_1 53.24, ν_2 45.26, ν_3 69.56, ν_4 95.51, e^2Qq/h 584.52 MHz, η 48.4% [10] ^{127}I NQR (ca. 20 °C): ν_1 144.77, ν_2 148.16 MHz [10] IR (solid): 490 s ν(InC) [5], 484 s [3] Raman (solid): 489 s ν(InC), 250 w ν(InI), 171 m, 144 s, 61 w; dimeric trans structure assumed [3, 5] mass spectrum (110 °C): $[M_2-C_2H_5]^+$, $[M_2-I]^+$, $[In_2I_3]^+$, $[M]^+$, $[InI_2]^+$, $[M-I]^+$, $[(C_2H_5)_2In]^+$, $[C_2H_5I]^+$ [5]
3	$(C_3H_7\text{-}n)InI_2$ I	white crystalline solid, m.p. 50 °C (dec.), dimeric in C_6H_6 [3] ^{1}H NMR (CH_2Cl_2): 1.13 (t, CH_3), 1.82 (m, CH_2), 2.10 (t, CH_2) [3] IR (solid): 572 s, br, 482 m, br ν(InC) [3] Raman (solid): 577 m, br, 485 s, br ν(InC), 196 w ν(InI), 167 m, 145 s; dimeric trans structure assumed [3]
4	$(C_3H_7\text{-}i)InI_2$	from $In(C_3H_7\text{-}i)_3$ and InI_3 in $O(C_2H_5)_2$, 98% yield [21] light yellow crystals, m.p. 94 °C [21] ^{1}H NMR (C_6D_6): 0.95 (CH_3), 1.63 (HC) [21] ^{13}C NMR (C_6D_6): 21.8 (CH_3), 24.1 (CH) [21] IR (solid): 474 ν(InC), other bands at 384, 220 [21]
5	$(C_4H_9\text{-}n)InI_2$ I (89%)	white crystalline solid, m.p. 53 °C [5], 45 °C (dec.) [3], dimeric in C_6H_6 [5] conductivity: 25 $\Omega^{-1} \cdot cm^2 \cdot mol^{-1}$ [5] ^{1}H NMR (CH_2Cl_2): 1.0 (t, CH_3), 1.65 (m, CH_2), 2.15 (t, CH_2) [5] IR (solid): 576 s, br, 482 m, br ν(InC) [3], only 489 m in [5] Raman (solid): 577 m, br ν(InC), 487 m, br [3], only 487 m in [5], 184 m, 168 m, 143 s, 87 w, 66 vw ν and δ(InI) [5] mass spectrum (150 °C): $[M_2-C_4H_9]^+$, $[M_2-I]^+$, $[In_2I_3]^+$, $[M_2-C_4H_9I]^+$, $[InI_3]^+$, $[InI_2]^+$, $[M-I]^+$, $[C_4H_9I]^+$ [5]
6	$(C_5H_5\text{-}c)InI_2$	from $In(C_5H_5\text{-}c)$ and I_2 (1:1 mole ratio) in $CHCl_3$, ca. 100% yield [7] light yellow solid, dec. after 3 to 4 days, becoming increasingly dark brown; insoluble in $CHCl_3$, $(CH_3)_2CO$, $(CH_3)_2SO$, slightly soluble in $O(C_2H_5)_2$ [7]

Table 23 (continued)

No.	compound method of preparation (yield)	properties and remarks
6	(continued)	IR and Raman (solid): 2923 w ν(C_1-H), 1388 ms, 1342 s δ(C-H), 1155 w ν(C-C), 1070, 1058 m, 1002 w, 983 s, 905 w δ(ring), 863 w, 835 w, 805 s γ(C-H), 762 s, 665 mw, 589, 315 w ν(In-C), 155 w ν(InI ?), 139 w; dimeric trans structure proposed [7] mass spectrum (100 °C): $[InI_3]^+$ (7%), $[InI_2]^+$ (15%), $[InI]^+$ (18%), $[C_{10}H_{10}]^+$ (10%), I^+ (13%), $^{115}In^+$ (36%), $[C_5H_5]^+$ (29%), $[C_5H_3]^+$ (100%) [7]
*7	$C_6H_5InI_2$	from InI, $Hg(C_6H_5)_2$, HgI_2 (2:1:1 mole ratio), 88% yield [12] m.p. 120 to 122 °C [12] Raman (solid and in C_6H_6): 654 m, 439 vw, 200 w, 167, 177 mw, 137 vs ν(InI) [12]
*8	$\{C_6H_2(CH_3)_3\text{-}2,4,6\}InI_2$ $C_6H_2(CH_3)_3\text{-}2,4,6$ = mes special	yellow needles, m.p. 213 to 214 °C [22] conductivity (CH_3NO_2): 461 $\Omega^{-1} \cdot cm^2 \cdot mol^{-1}$ [22] ^{1}H NMR ($CDCl_3$): 2.66 (s, CH_3), 6.85 (s, CH) [22] ^{13}C NMR ($CDCl_3$): 21.1 (CH_3-4), 25.8 (CH_3-2,6), 128.6, 137.7, 140.9, 143.9 (C_6H_2) [22] IR: 1710 w, 1585 w, 1550 w, 1395 w, 1295 s, 1255 m, 1170 w, 1065 w, 1025 m, 1005 m, 940 w, 925 w, 845 s, 800 w, 690 w, 575 w, 535 s ν(InC) [22]
adducts, $RInI_2 \cdot D$		
9	$CH_3InI_2 \cdot C_{10}H_8N_2$ II	yellow solid, recrystallized from CH_3CN, no other properties reported [16]
10	$C_2H_5InI_2 \cdot 2\,OS(CH_3)_2$ I (35%)	yellow solid, no other properties reported [17]
*11	$C_2H_5InI_2 \cdot (CH_3)_2NCH_2CH_2N(CH_3)_2$	pale yellow crystals [17] ^{1}H NMR (CD_2Cl_2): 1.20 (s, C_2H_5In), 2.61 (s, CH_3N), 2.80 (s, CH_2N) [19] ^{13}C NMR ($CDCl_3$): −0.05 (CH_3), 12.29 (CH_2In), 47.76 (CH_3N), 56.33 (CH_2N) [19]
12	$C_6H_5CH_2InI_2 \cdot O_2C_4H_8$ I	polymeric solid, no other properties reported [6]
13	$C_6H_5CH_2InI_2 \cdot C_{10}H_8N_2$ II	yellow solid, no other properties reported [15]
14	$(C_5H_5\text{-}c)InI_2 \cdot C_{10}H_8N_2$	from No. 6 and $C_{10}H_8N_2$ in $O(C_2H_5)_2$ [7] m.p. 198 °C (dec.) [7] conductivity (CH_3CN): 5.7 $\Omega^{-1} \cdot cm^2 \cdot mol^{-1}$ [7]

Table 23 (continued)

No. compound method of preparation (yield)	properties and remarks
14 (continued)	IR and Raman (solid): 3072 ms ν(C–H), 2965 mw, 2925 mw ν(C_1–H), 1408 s, 1385 s δ(C–H), 1105 m ν(C–C), 1072 w, 1042 mw, 1015 ms, 892 w δ(ring), 871 w, 836 m, 803 γ(C–H), 746 s, 622 m, 325 m ν(In–C), 305 m ν(In–N), 176 mw ν(In–I), 143 m, 138 m, 115 mw, 111 m, 77 w, bands due to $C_{10}H_8N_2$ not reported [7] mass spectrum (220 °C): $[C_{10}H_8N_2]^+$ (100%), I^+ (18%), $^{115}In^+$ (10%), $[C_5H_5]^+$ (22%) [7]
15 (C_5H_5-c)$InI_2 \cdot C_{12}H_8N_2$ $C_{12}H_8N_2$= 1,10-phenanthroline	from No. 6 and $C_{12}H_8N_2$ in CH_2Cl_2 [7] m.p. 200 °C (dec.) [7] conductivity (CH_3CN): 8.7 $\Omega^{-1} \cdot cm^2 \cdot mol^{-1}$ [7] IR and Raman (solid; from 3100 to 330 cm^{-1} identical with No. 14): 329 m ν(In–C), 292 m ν(In–N), 192 m ν(In–I ?), 186 m, 139 mw, 130 mw, 113 mw, 96 mw, 79 w, 62 w; bands due to $C_{12}H_8N_2$ not reported [7]
16 $C_6H_5InI_2 \cdot C_{10}H_8N_2$ II	brown solid, no other properties reported [16]

* Further information:

CH_3InI_2 (Table **23**, No. **1**) has also been prepared by ligand exchange between $(CH_3)_2InI$ and InI_3 (1:1 mole ratio) [5]. The reaction of indium metal (powdered or colloidal) with CH_3I [2, 15], however, produced the sesquiiodide, $(CH_3)_3In_2I_3$ (see on p. 169) [9].

The NQR spectrum (^{115}In and ^{127}I, in MHz, at room temperature) was measured concurrently by two research groups. The transition frequencies are given below (in MHz):

	ν_1	ν_2	ν_3	ν_4	e^2Qq/h	η(%)	Ref.
^{115}In:	54.33	105.87	159.16	212.16	1277.59	4.8	[10]
	54.36	105.55	158.68	211.65	1270.60	5.1±0.6	[11]
^{127}I:	144.77	148.16	–	–	–	–	[10]

The coupling constant, e^2Qq/h, the largest value yet found, and the very small asymmetry parameter, η, were the main factors in deciding on the ionic structure, $[In(CH_3)_2][InI_4]$ with a linear cation and a (nearly perfectly) tetrahedral anion [10, 11]. The slight distortion of the anion was accounted for by a loose interaction between the iodine atoms of the InI_4 anion and the indium atom of the cation, this indium atom displaying a coordination number of 6 (2 CH_3 + 4 I) in the crystal lattice [13, 14].

The appearance of $\nu_{as}(InC_2)$ only in the IR and $\nu_s(InC_2)$ only in the Raman spectrum confirms the linearity of the $[In(CH_3)_2]^+$ cation, while the extensive correspondence of the valence and deformation vibrations with those of $In[InI_4]$ [8] or $[N(C_2H_5)_4][InI_4]$ [1] is evidence for the structure of the anionic portion [3 to 5].

Attempts to stabilize the cation $[In(CH_3)_2]^+$ by addition of dibenzo-crown-6 (crown) produced a product of the analytical composition $In_2(CH_3)_4I_4 \cdot$ crown. From ^1H NMR measurements (two equally intense singlets at 0.60 and 0.67 ppm) and IR data (bands of $[InI_4]^-$ and bands of other InI species), however, the presence of a mixture of $[(CH_3)_2In \cdot crown][InI_4]$ and $[CH_3InI \cdot crown][InI_4]$ was deduced [17].

$C_6H_5InI_2$ (Table **23**, No. **7**) displays a few anomalies compared to the other phenylindium dihalogenides (see pp. 144 and 153). The presence of two Raman frequencies at 177 mw and 137 s cm^{-1}, in the In-I bond vibration region, is similar to CH_3InI_2 (see No. 1), and the mass spectrum at 100 °C shows the additional fragments $[InI_4]^+$ (0.13%) and $[(C_6H_5)_2InI_2]^+$ (1.17%). The conductivity in nitromethane (15.59 $\Omega^{-1} \cdot cm^2 \cdot mol^{-1}$) is significantly higher than that of the monoiodide (5.88 $\Omega^{-1} \cdot cm^2 \cdot mol^{-1}$). These facts indicate an ionic structure similar to the methyl homolog, or at least an unsymmetrical four-membered ring with an angular $(C_6H_5)_2In$ grouping as depicted in Formula I [12].

C_6H_5 I I
In In
C_6H_5 I I

I

The compound was used to prepare CH_3SInI_2 and $C_6H_5SInI_2$ by reaction with CH_3SSCH_3 or $C_6H_5SSC_6H_5$, respectively [19].

$\{C_6H_2(CH_3)_3\text{-}2,4,6\}InI_2$ (Table **23**, No. **8**) was prepared from $In(mes)_3$ and InI_3 (1:2 mole ratio) in refluxing toluene (24 h) followed by evaporating the solvent. For removing unchanged $In(mes)_3$ (about 20%) the crude material was washed with hexane and dissolved in a minimum of hot toluene; cooling to −20 °C gave the compound in 69% yield. Molecular weight determinations showed the following degrees of association: 2.03 (cryoscopically in C_6H_6 at 0.015 mol), 1.93 (isopiestically in CH_2Cl_2, at 0.020 mol). Dissolution in nitromethane results in the formation of a 1:1 electrolyte probably a result of the rearrangement reaction: 2 $(mes)InI_2 \rightarrow [In(mes)_2][InI_4]$ (see also Formula I) [22].

The ^{115}In NQR spectra (no data reported) was found to be close to those of No. 1 [22].

The crystal structure at 173 K shows the polymeric nature of the compound in the solid state. It crystallizes in the monoclinic space group $P2_1/c - C_{2h}^5$ (No. 14) with the parameters a = 823.4(2), b = 1965.7(3), c = 770.3(1) pm, β = 91.55(2)°; Z = 4 and D_c = 2.600 g/cm³. The crystal structure, of which a section is depicted in **Fig. 26** on p. 166, consists of strands of a one-dimensional polymer. The In atom has a trigonal bipyramidal environment with bridging iodine atoms in the axial positions. The mesityl groups alternate on each side of the chain and are orientated almost coplanar with respect to the equatorial I atoms [22].

$C_2H_5InI_2 \cdot (CH_3)_2NCH_2CH_2N(CH_3)_2$ (Table **23**, No. **11**) is isomorphic with the homologous dibromide (Fig. 24 on p. 155) and crystallizes in the monoclinic space group $P2_1/n - C_{2h}^5$ (No. 14) with cell constants a = 828.8(5), b = 1541.5(7), c = 1225.4(4) pm, α = 97.13(4)°; Z = 4 gives D_c = 2.20 g/cm³. The most important distances and bond angles (see Fig. 24) are given on p. 166 [20].

important distances (pm)		angles (°)	
In-I(1)	291.9(1)	I(1)-In-I(2)	93.3
In-I(2)	275.0(1)	I(1)-In-N(1)	163.9(3)
In-N(1)	244(1)	I(1)-In-N(2)	88.7(3)
In-N(2)	233(1)	I(1)-In-C(3)	98.6(5)
In-C(3)	217(1)	I(2)-In-N(1)	92.3(2)
N(1)-C(1)	150(2)	I(2)-In-N(2)	106.9(3)
N(2)-C(2)	144(2)	I(2)-In-C(3)	117.7(4)
C(1)-C(2)	146(2)	N(1)-In-N(2)	75.2(4)
C(3)-C(4)	142(2)	N(1)-In-C(3)	92.2(5)
		In-N(1)-C(1)	101.7(8)
		In-N(2)-C(2)	112.5(9)
		In-C(3)-C(4)	115(1)

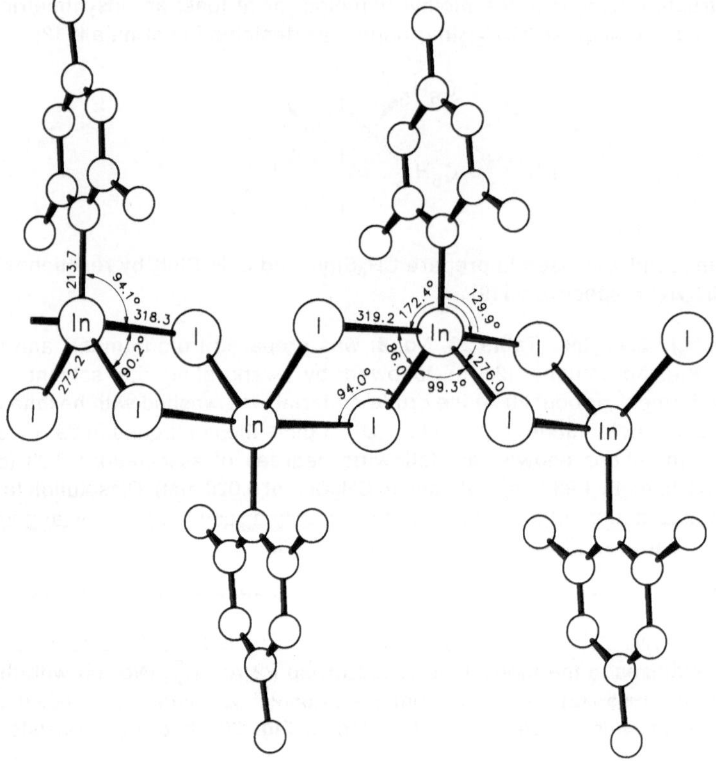

Fig. 26. Structure of a section of the polymeric (mes)InI_2 [22].

References:

[1] Gislason, J.; Lloyd, M. H.; Tuck, D. G. (Inorg. Chem. **10** [1971] 1907/10).

[2] Gynane, M. J. S.; Waterworth, L. G.; Worrall, I. J. (J. Organometal. Chem. **40** [1972] C 9/C10).

[3] Gynane, M. J. S.; Waterworth, L. G.; Worrall, I. J. (J. Organometal. Chem. **43** [1972] 257/64).

[4] Gynane, M. J. S.; Worrall, I. J. (Inorg. Nucl. Chem. Letters **8** [1972] 547/50).
[5] Poland, J. S.; Tuck, D. G. (J. Organometal. Chem. **42** [1972] 315/23).
[6] Gynane, M. J. S.; Waterworth, L. G.; Worrall, I. J. (Inorg. Nucl. Chem. Letters **9** [1973] 543/4).
[7] Contreras, J. G.; Tuck, D. G. (J. Organometal. Chem. **66** [1974] 405/12).
[8] Davies, J. E.; Waterworth, L. G.; Worrall, I. J. (J. Inorg. Nucl. Chem. **36** [1974] 805/7).
[9] Gynane, M. J. S.; Worrall, I. J. (J. Organometal. Chem. **81** [1974] 329/34).
[10] Patterson, D. B.; Carnevale, A. (Inorg. Chem. **13** [1974] 1479/83).

[11] Welsh, W. A.; Brill, T. B. (J. Organometal. Chem. **71** [1974] 23/5).
[12] Miller, S. B.; Jelus, B. L.; Brill, T. B. (J. Organometal. Chem. **96** [1975] 1/14).
[13] Brill, T. B. (Inorg. Chem. **15** [1976] 2558/60).
[14] Bancroft, G. M.; Sham, T. K. (J. Magn. Resonance **25** [1977] 83/90).
[15] Klabunde, K. J.; Murdock, T. O. (J. Org. Chem. **44** [1979] 3901/8).
[16] Habeeb, J. J.; Said, F. F.; Tuck, D. G. (J. Organometal. Chem. **190** [1980] 325/34).
[17] Taylor, M. J.; Tuck, D. G. (J. Chem. Soc. Dalton Trans. **1981** 928/32).
[18] Peppe, C.; Tuck, D. G.; Victoriano, L. (J. Chem. Soc. Dalton Trans. **1982** 2165/8).
[19] Hoffmann, G. G. (Z. Naturforsch. **39b** [1984] 352/5).
[20] Khan, A. M.; Peppe, C.; Tuck, D. G. (J. Organometal. Chem. **280** [1985] 17/25).

[21] Hoffmann, G. G.; Faist, R. (J. Organometal. Chem. **391** [1990] 1/5).
[22] Leman, J. T.; Ziller, J. W.; Barron, A. R. (Organometallics **10** [1991] 1766/71).

2.5 Other Organoindium Halogenides

2.5.1 $R_3In_2X_3$ and RIn_2X_5 Compounds with X = Br, I

In this section halogen-bridged association compounds of the general Formulas I and II are collected.

While several sesquihalogenides of Formula I are known, only one example of II has been established. Indeed, a compound with the molecular formula $CH_3In_2I_5 \cdot 4\ OS(CH_3)_2$ formed from the reaction of InI with CH_3I and $OS(CH_3)_2$ in toluene; according to the vibrational spectrum, however, it has the ionic structure $[InI_2 \cdot 4\ OS(CH_3)_2][CH_3InI_3]$ [7].

The compounds listed in Table 24 were prepared in the following ways:

Method I: From In and RX.
Finely powdered indium was allowed to react with excess RX as solvent. This procedure, already described in detail on p. 156, gave either a 1:1 mixture of R_2InX and $RInX_2$ or molecular sesquihalogenides with Formula I [1, 5]. Similar results were obtained when RX reacted with colloidal In (made by cocondensing In vapor and a solvent such as diglyme, dioxane, or toluene) [6]. Addition of allyl halides to In powder in DMF (1.5:1 mole ratio) at room temperature gave an exothermic reaction with quantitative yields of the sesquibromides No. 5 to 12 and 18 [9].

Method II: From In_2X_4 and RX.

Finely divided In_2X_4 was slowly added to excess RX and stirred at room temperature for 1 h. After removing the volatile components, RIn_2X_5 remained in quantitative yield. This method was used to prepare several gallium homologs [2, 3], but has only been used for indium in the case where $R=CH_3$ and $X=I$ to give No. 19 [4].

General Remarks. It is not always certain whether the $R_3In_2X_3$ compounds actually have structure I or are 1:1 mixtures of R_2InX an $RInX_2$. In [1] and [5] this decision was made with the help of vibrational spectroscopy, from which $(CH_3)_3In_2Br_3$ was classified as a mixture, because the IR and Raman bands (especially in the InC and InBr bond region) correspond exactly to the sums of the overlapped bands of the single components. For $(CH_3)_3In_2I_3$ (No. 13) and $(C_2H_5)_3In_2X_3$ (with X = Br or I) (Nos. 2 and 14), however, the important observed bands deviate from the summations of the component frequencies; thus, these organoindium halogenides were considered to be real sesquihalogenides of structure I. No decision was possible for the n-propyl and n-butyl homologs, since the relevant IR, and a large portion of the Raman bands, are strongly broadened and lack many of the relevant bands of the single components [5].

The ^1H NMR data in Table 24 were recalculated from the values referred to CH_2Cl_2 using $\delta = 5.35$ ppm. Two signals (or groups of signals) with an intensity ratio of 2:1 were expected in the ^1H NMR spectrum, for both the mixture and for I. This requirement was met by compounds 1 and 3 (measured in CH_2Cl_2), while ethyl derivatives 2 and 14, for example, show only one C_2H_5 multiplet. The fusing of the expected resonances to a singlet (or one group) was explained by the rapid interchange of methyl groups within molecule I. Confirming this explanation by low temperature measurements was proposed by [5], but has not been carried out.

A convincing argument came from mass spectral evidence. Mixtures with the formula $R_3In_2X_3$ typically show fragments of the $RInI_2$ vapor dimer (e.g., $(CH_3)_3In_2Br_3$ gives $[CH_3InBr_4]^+$), whereas the sesquihalogenides produce the molecule ion, $[R_3In_2X_3]^+$, and $[R_2In_2X_3]^+$. From this evidence compounds 2, 14, and 17 are believed structure I; the spectra of the other compounds lacked this characteristic peak [4, 5].

The preparation of allylic indates from various allylic sesquihalides is described in [10].

Table 24
$R_3In_2X_3$ and RIn_2X_5 Compounds with X = Br, I.
Further information on numbers preceded by an asterisk is given at the end of the table.
Explanations and abbreviations on p. X.

No.	compound method of preparation	properties and remarks
$R_3In_2X_3$ compounds		
1	$(CH_3)_3In_2Br_3$ I	1:1 mixture of $(CH_3)_2InBr$ and CH_3InBr_2, m.p. 98 to 102 °C [5] see general remarks on p. 168, and the components on pp. 148 and 154

Table 24 (continued)

No.	compound method of preparation	properties and remarks
2	$(C_2H_5)_3In_2Br_3$ I	colorless crystals, m.p. 63.5 °C [5] ^{1}H NMR (CH_2Cl_2): 1.38 (m, C_2H_5) [5] IR: 522 s, 500 m, 458 w ν(InC) [1, 5] Raman: 520 vw, 494 s, 460 s ν(InC), 243 w, 194 s ν(InBr) [5]
3	$(C_3H_7\text{-}n)_3In_2Br_3$ I	colorless liquid [5] ^{1}H NMR (CH_2Cl_2): 1.13 (m), 1.93 (m) [5] IR: 572 m, br, 558 m, br, 493 s, br ν(InC) [5] Raman: 583 s, 567 s, 494 s, br ν(InC), 189 m ν(InBr) [5]
4	$(C_4H_9\text{-}n)_3In_2Br_3$ I	colorless liquid [5] IR: 584 s, 568 m, br, 493 s, br ν(InC) [5] Raman: 589 s, 571 s, 498 s, br ν(InC), 188 s ν(InBr) [5]
*5	$\{(CH_3)HC{=}CHCH_2\}_3In_2Br_3$ I	no properties reported [9]
*6	$\{(CH_3)_2C{=}CHCH_2\}_3In_2Br_3$ I	no properties reported [9]
*7	$\{(C_3H_7\text{-}n)HC{=}CHCH_2\}_3In_2Br_3$ I	no properties reported [9]
*8	$\{(C_6H_5)HC{=}CHCH_2\}_3In_2Br_3$ I	no properties reported [9]
*9	$\{(C_6H_5)_2C{=}CHCH_2\}_3In_2Br_3$ I	no properties reported [9]
*10	$\{(C_6H_{11})CH_3C{=}CHCH_2\}_3In_2Br_3$ I	^{1}H NMR ($DCON(CD_3)_2$): 1.60, 1.62, 1.66 (s's, 3 CH_3), 1.88 to 2.12 (m, CH_2), 5.17, 5.48 (m's, 2 CH) [9]
*11	$\{(CH_3)C_6H_{11}C{=}CHCH_2\}_3In_2Br_3$ I	no properties reported [9]
*12	$(C_{10}H_{15})_3In_2Br_3$ $C_{10}H_{15}$= 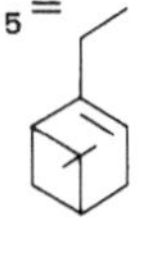 I	no properties reported [9]
13	$(CH_3)_3In_2I_3$ I	colorless crystals, m.p. 112 to 114 °C [5], 90 to 100 °C (dec.) [4] ^{1}H NMR (CH_2Cl_2): 0.98 (s, 2 CH_3), 1.16 (s, 1 CH_3) [5] IR: 1160 mw $\delta_s(CH_3)$, 1145 m, 730 vs, br $\varrho(CH_3)$, 555 s, 509 s, 485 w ν(InC) [4]

Table 24 (continued)

No.	compound method of preparation	properties and remarks
13 (continued)		Raman: 508 m, 486 m ν(InC), 161 sh, 152 s ν(InI), 114 m, br, 66 w [4] mass spectrum (130 °C): $[M-CH_3]^+$, $[M-I]^+$, $[CH_3InI_2]^+$, $[InI_2]^+$, $[CH_3InI]^+$, $[In(CH_3)_2]^+$, $[InCH_3]^+$ [4]
14	$(C_2H_5)_3In_2I_3$ I	white solid, m.p. 72 to 75 °C [5] ^1H NMR (CH_2Cl_2): 1.43 (s, C_2H_5) [5] IR: 503 s, 486 s, 450 s ν(InC) [5] Raman: 502 w, 486 s, 455 s ν(InC), 215 w, 153 s, 147 s ν(InI) [5]
15	$R_3In_2I_3$ $R=CH_2COOC_2H_5$ I	not isolated, used for reactions with ketones to give RR^1R^2COH [8]
16	$(C_3H_7\text{-}n)_3In_2I_3$ I	colorless liquid [5] ^1H NMR (CH_2Cl_2): 1.11, 1.76 (m's, complex second-order spectrum) [5] IR: 573 s, br, 557 s, br, 484 s, br ν(InC) [1, 5] Raman: 575 s, 563 s, 488 s, br ν(InC), 146 s ν(InI) [1, 5]
17	$(C_4H_9\text{-}n)_3In_2I_3$ I	liquid [5] ^1H NMR (CH_2Cl_2): 1.10, 1.71 (m's, complex second-order spectrum) [5] IR: 577 m, br, 562 m, br, 480 m, br ν(InC) [5] Raman: 584 m, br, 569 m, br, 488 m, br ν(InC), 145 s ν(InI) [5]
*18	$(H_2C{=}CHCH_2)_3In_2I_3$ I	no properties reported [9]
RIn_2X_5 compound		
19	$CH_3In_2I_5$ II	bright yellow solid, m.p. 141 °C [4] IR: 1145 s $\delta_s(CH_3)$, 730 vs, br $\varrho(CH_3)$, 509 m ν(InC) [4] Raman: 510 m ν(InC), 216 w (InI-terminal), 188 m, 140 vs, 108 w (InI-bridge) [4] mass spectrum (160 °C): $[M-CH_3]^+$, $[M-I]^+$, $[M-I_2]^+$, $[In_2I_3]^+$, $[InI_3]^+$, $[CH_3InI_2]^+$, $[InI_2]^+$, $[CH_3InI]^+$, $[CH_3I]^+$ [4]

* Further information:

$\{(CH_3)HC{=}CHCH_2\}_3In_2Br_3$ (Table **24**, No. **5**; Formula III, R-1 = CH_3, R-2 = H), **$\{(CH_3)_2C{=}CHCH_2\}_3In_2Br_3$** (Table **24**, No. **6**; Formula III, R-1 = R-2 = CH_3), **$\{(C_3H_7\text{-}n)HC{=}CHCH_2\}_3In_2Br_3$** (Table **24**, No. **7**; Formula III, R-1 = C_3H_7-n, R-2 = H), **$\{(C_6H_5)HC{=}CHCH_2\}_3In_2Br_3$** (Table **24**, No. **8**; Formula III, R-1 = C_6H_5, R-2 = H), **$\{(C_6H_5)_2C{=}CHCH_2\}_3In_2Br_3$** (Table **24**, No. **9**; Formula III, R-1 = R-2 = C_6H_5), **$\{(C_6H_{11})CH_3C{=}CHCH_2\}_3In_2Br_3$** (Table **24**, No. **10**; Formula III,

R-1 = $(CH_3)_2C{=}CHCH_2CH_2$, R-2 = CH_3), **$\{(CH_3)C_6H_{11}C{=}CHCH_2\}_3In_2Br_3$** (Table **24**, No. **11**; Formula III, R-1 = CH_3, R-2 = $(CH_3)_2C{=}CHCH_2CH_2$), **$(C_{10}H_{15})_3In_2Br_3$** (Table **24**, No. **12**), and **$(H_2C{=}CHCH_2)_3In_2I_3$** (Table **24**, No. **18**). Protolysis with dilute aqueous HCl (Nos. 8 to 11) generated the corresponding hydrocarbons IV. Similarly β-pinene was formed from No. 12. Oxygenation (Nos. 5 to 8, 10, 18) followed by reductive work-up ($NaBH_4$) produced a mixture of the alcohols V and VI in yields between 33 and 62%. The reaction with N-substituted maleimides and phthalimides with formation of various C-alkylated products was described. Treatment of Nos. 7 to 10 and 18 with $(C_4H_9)_3SnCl$ gave the stannylated products in 50 to 83% yields (Formula V, OH is replaced by $Sn(C_4H_9)_3$) [9]. For indate formation, see [10].

R^1 In_2Br_3 R^2 3 — III; R^1 R^2 — IV; R^1 R^2 OH — V; R^1 OH R^2 — VI

References:

[1] Gynane, M. J. S.; Waterworth, L. G.; Worrall, I. J. (J. Organometal. Chem. **40** [1972] C 9/C10).
[2] Lind, W.; Worrall, I. J. (J. Organometal. Chem. **36** [1972] 35/9).
[3] Lind, W.; Worrall, I. J. (J. Organometal. Chem. **40** [1972] 35/41).
[4] Poland, J. S.; Tuck, D. G. (J. Organometal. Chem. **42** [1972] 315/23).
[5] Gynane, M. J. S.; Worrall, I. J. (J. Organometal. Chem. **81** [1974] 329/34).
[6] Klabunde, K. J.; Murdock, T. O. (J. Org. Chem. **44** [1979] 3901/8).
[7] Peppe, C.; Tuck, D. G.; Victoriano, L. (J. Chem. Soc. Dalton Trans. **1982** 2165/8).
[8] Araki, S.; Ito, H.; Butsugan, Y. (Synth. Commun. **18** [1988] 453/8).
[9] Araki, S.; Shimizu, T.; Johar, P. S.; Jin, S.-J.; Butsugan, Y. (J. Org. Chem. **56** [1991] 2538/42).
[10] Araki, S.; Shimizu, T.; Jin, S.-J.; Butsugan, Y. (J. Chem. Soc. Chem. Commun. **1991** 824/5).

2.5.2 Organoindium Compounds with an X_2InCH_2 Unit

2.5.2.1 Compounds of the Type $X_2InCH_2X \cdot D$

The electrochemical oxidation of In in mixtures of CH_2X_2 (X = Cl, Br, I) and CH_3CN results in the formation of InX. Whereas InCl disproportionates, InBr and InI form the corresponding halide X_2InCH_2X with CH_2Br_2 and CH_2I_2, respectively, via oxidative insertion.

According to the general procedure a solution of $[N(C_2H_5)_4][PF_6]$ in a mixture of CH_2X_2 and CH_3CN was electrolyzed at an In anode and a Pt cathode at 20 to 30 V and 30 mA (1 h for X = Br; 6 h for X = I). The resulting solution was freed from volatile materials in vacuum, and the mostly oily residue dissolved in ether, CH_3CN, or mixtures of these solvents with other Lewis bases and again evaporated to dryness. The halides X_2InCH_2X formed in the first step could not be isolated but only the corresponding adducts with the bases $O(C_2H_5)_2$, CH_3CN, $(CH_3)_2N(CH_2)_2N(CH_3)_2$ (TMEDA), or $P(C_6H_5)_3$. Whereas ether and CH_3CN form normal adducts of the type $X_2InCH_2X \cdot D$, the adducts of TMEDA and $P(C_6H_5)_3$ rearrange to ylid like compounds of the type $X_3In^-{-}CH_2{-}D^+$ [1].

Table 25
Compounds of the Types $X_2InCH_2X \cdot D$ and X_3InCH_2-D [1].
Further information on compounds preceded by an asterisk is given at the end of the table.
Explanations, abbreviations, and units on p. X.

No.	compound (yield)	properties and remarks
	$X_2InCH_2X \cdot D$ compounds	
1	$Br_2InCH_2Br \cdot 1/2\ O(C_2H_5)_2$	colorless oil ^{1}H NMR (CD_2Cl_2): 1.39 (t, CH_3), 3.07 (s, CH_2In), 4.06 (q, CH_2O) ^{13}C NMR (CD_2Cl_2): 14.9 (CH_3), 17.6 (br, CH_2In), 68.7 (CH_2O)
2	$Br_2InCH_2Br \cdot$ TMEDA	not isolated rearranges to No. 7
3	$Br_2InCH_2Br \cdot P(C_6H_5)_3$	not isolated rearranges to No. 8
4	$I_2InCH_2I \cdot CH_3CN$ (80%)	from InI and excess CH_2I_2 in CH_3CN ^{1}H NMR (CD_3SOCD_3): 1.67, 1.77 (CH_2In), 2.05 (CH_3) ^{13}C NMR (CD_3SOCD_3): −8.3 (br, CH_2In), 1.7 (CH_3), 118.1 (CN)
5	$I_2InCH_2I \cdot$ TMEDA	not isolated rearranges to No. 9
6	$I_2InCH_2I \cdot P(C_6H_5)_3$	not isolated rearranges to No. 10
	X_3In^--CH_2-D^+ compounds	
*7	$Br_3InCH_2N(CH_3)_2(CH_2)_2N(CH_3)_2$ (70%)	white solid, recrystallized from CH_3CN conductivity: 20.6 $\Omega^{-1} \cdot cm^2 \cdot mol^{-1}$ ^{1}H NMR (D_2O): 2.98 (CH_3N), 3.07 (CH_2In), 3.31 (CH_3N^+), 3.71 (br, CH_2N), 3.90 (br, CH_2N^+) ^{13}C NMR (D_2O): 45.1 (CH_3N), 52.1 (CH_2In), 56.9 (CH_3N^+), 62.6 (CH_2N), 66.5 (CH_2N^+)
*8	$Br_3InCH_2P(C_6H_5)_3$ (67%)	white solid, recrystallized from CH_3CN conductivity (CH_3CN): 15.6 $\Omega^{-1} \cdot cm^2 \cdot mol^{-1}$ ^{1}H NMR (CD_2Cl_2): 2.3 (s, CH_2In), 7.7 (m, C_6H_5) ^{13}C NMR (CD_2Cl_2): 121.6, 122.8 (CP), 130.0, 130.2 (C-2,6), 132.9, 133.0 (C-3,5), 134.5 (C-4)
9	$I_3InCH_2N(CH_3)_2(CH_2)_2N(CH_3)_2$	colorless crystals from acetone/ethanol conductivity: 20.0 $\Omega^{-1} \cdot cm^2 \cdot mol^{-1}$
10	$I_3InCH_2P(C_6H_5)_3$	white solid, impure ^{1}H NMR (CD_2Cl_2): 2.3 (s, CH_2In), 7.7 (m, C_6H_5) ^{13}C NMR (CD_2Cl_2): 10.6, 11.4 (CH_2In), 118.0, 119.2 (CP), 131.1 (C-2,6), 133.4 (C-3,5), 136.0 (C-4)

* Further information

$Br_3InCH_2N(CH_3)_2(CH_2)_2N(CH_3)_2$ (Table **25**, No. **7**) crystallizes in the orthorhombic space group $Pna2-C_{2v}^4$ (No. 28) with the unit cell constants a = 1407.6(4), b = 761.0(2), c = 1293.1(2) pm; Z = 4 gives D_c = 2.24, while D_m = 2.30 g/cm³. The central In atom has a trigonal bipyramidal environment and the CH_2 group is in an equatorial position. The second N atom of the TMEDA ligand is axially coordinated, forming a six-membered ring as depicted in **Fig. 27** [1].

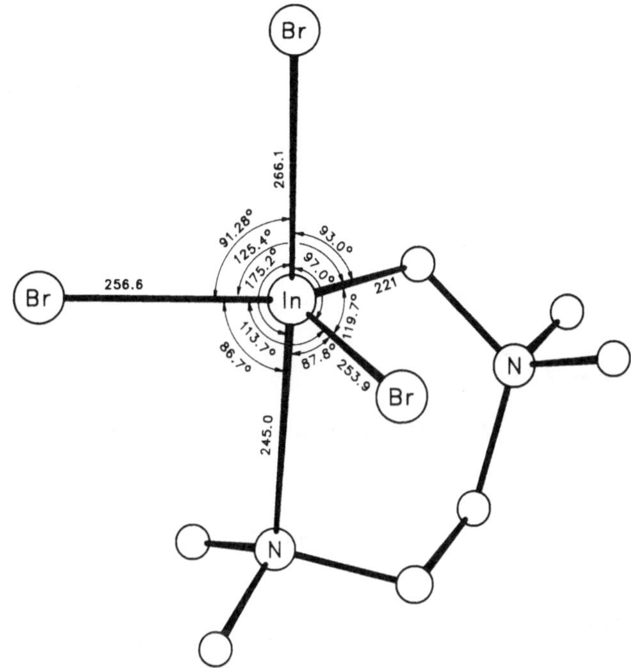

Fig. 27. Molecular structure of $Br_3InCH_2N(CH_3)_2(CH_2)_2N(CH_3)_2$ [1].

Other bond angles (°)

In-C-N	118.8(6)	N-In-C	84.7(3)
In-N-C	105.7(5)		

$Br_3InCH_2P(C_6H_5)_3$ (Table **25**, No. **8**) crystallizes in the orthorhombic space group $Pbca-D_{2h}^{15}$ (No. 61) with the unit cell constants a = 1281.4(3), b = 1572.1(4), c = 2134.3(5) pm; Z = 8 and D_c = 1.95 and D_m = 1.90 g/cm³. In contrast to No. 7 the In atom has coordination number 4 as shown in **Fig. 28** on p. 174. Further In-Br distances are 251.0 and 251.1 pm [1].

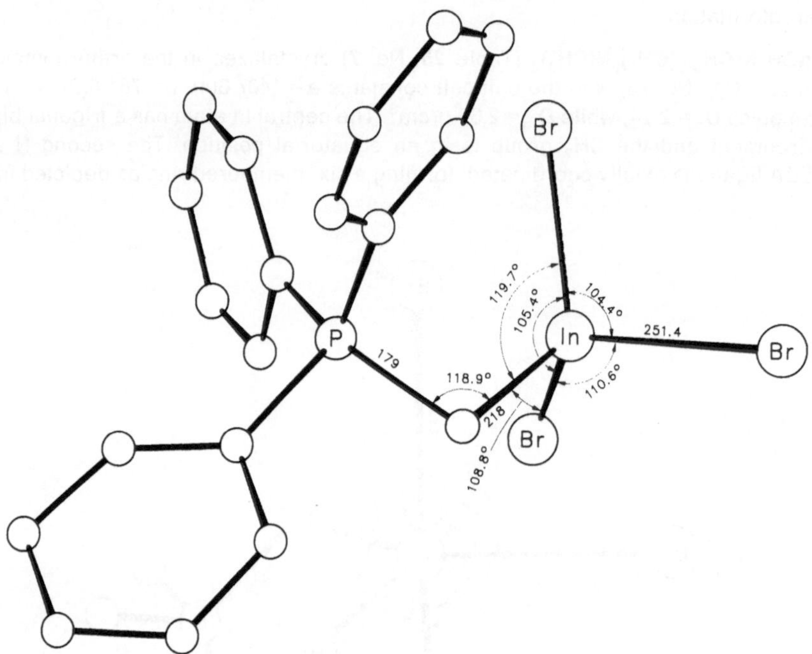

Fig. 28. Molecular structure of $Br_3InCH_2P(C_6H_5)_3$ [1].

Reference:

[1] Annan, T. A.; Tuck, D. G.; Khan, M. A.; Peppe, C. (Organometallics **10** [1991] 2159/66).

2.5.2.2 $X_2InCH_2InX_2 \cdot 2$ D Compounds

Reaction of In_2X_4 with CH_2X_2 proceeds through the intermediate X_2InCH_2X to $X_2InCH_2InX_2$. The mononuclear intermediate was postulated [1] as the first stage of the very slow reaction of In metal with CH_2I_2. Without a base to form a stabilizing adduct, however, neither the intermediate nor the end product could be isolated. Study of the first reaction mentioned above had shown that an equimolar mixture of InX and InX_3 in the presence of a Lewis base gave the best yields of the methylene-bridged dinuclear halogenide [2].

The bidentate Lewis base $(CH_3)_2NCH_2CH_2N(CH_3)_2$ used in this section is abbreviated as TMEDA.

$Cl_2InCH_2InCl_2 \cdot 2\ (CH_3)_2NCH_2CH_2N(CH_3)_2$

To prepare this compound InCl and $InCl_3$ (1:1 mole ratio) were suspended in toluene/dichloromethane (1:1) at −80 °C and treated with an excess of the base. The reaction mixture was stirred continuously and allowed to warm to room temperature over a 3 to 4 h period; the clear solution was filtered from the small quantity of metallic indium and concentrated; the solute crystallized, resulting in ca. 20% yield of air-stable, colorless crystals, no m.p. reported, $D_m = 1.74$ g/cm³ [2].

^1H NMR (CD_2Cl_2): δ = 0.40 (CH_2In), 2.8 (CH_3N), and 2.78 (CH_2N) ppm [2].

The compound crystallizes in the orthorhombic space group Pccn $-D_{2h}^{10}$ (No. 56) with the cell parameters a = 778.2(2), b = 2119.2(5), c = 1436.6(3) pm; Z = 4 gives D_c = 1.732 g/cm³. The coordination of the In atom is (distorted) trigonal bipyramidal. The central atom stands 7.7 pm out of the Cl(2)-N(1)-C(7) equatorial plane toward N(2) as shown in **Fig. 29**; the N(2)-In-Cl(1) axis deviates about 16.6° from linear. The very large, intramolecular In ··· In contact distance of 367.2 pm argues against any significant interaction of the two metal centers. The entire molecule possesses C_2 symmetry [2].

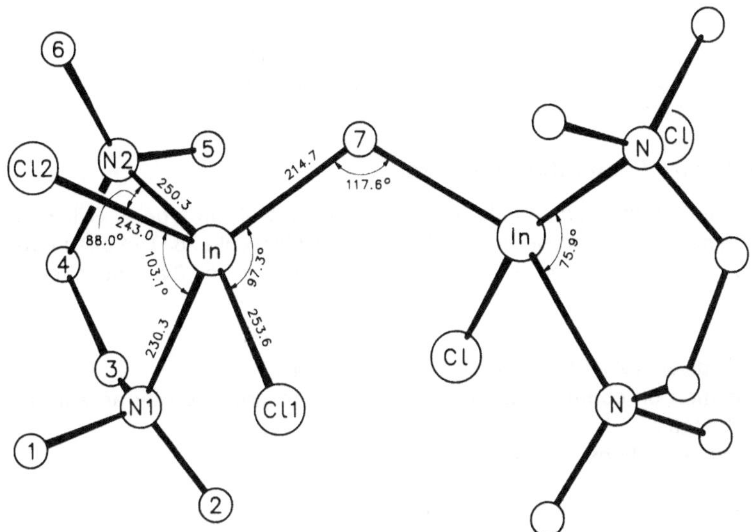

Fig. 29. Molecular structure of $Cl_2InCH_2InCl_2 \cdot 2\,(CH_3)_2NCH_2CH_2N(CH_3)_2$ [2].

Other bond angles (°)

Cl(1)-In-Cl(2)	92.0(1)	In-N(1)-C(3)	109.9(3)
Cl(1)-In-N(1)	87.9(1)	In-N(2)-C(4)	103.8(3)
Cl(1)-In-N(2)	163.4(1)	N(1)-C(3)-C(4)	112.4(5)
Cl(2)-In-C(7)	118.6(2)	N(2)-C(4)-C(3)	113.0(5)

$Cl(Br)InCH_2InCl_2 \cdot 2\,(CH_3)_2NCH_2CH_2N(CH_3)_2$

When a 1:1 mixture of InBr and $InCl_3$ was treated with CH_2Cl_2 in a procedure similar to that above, a 36% yield of colorless crystalline product of the composition $In_3Cl_6Br(CH_2) \cdot 4$ TMEDA, a mixture of $InCl_3 \cdot 2$ TMEDA and the title compound, was obtained. The latter could be picked out of the mixture as individual crystals. In CD_2Cl_2 solution the mixture shows a singlet at 0.57 ppm, typical of the In-CH_2-In bridge, and resonances at 2.64, 2.75, and 2.87 ppm, ascribed without assignment to the coordinating base [2].

The title compound is isomorphous with the preceding compound crystallizing in the orthorhombic space group Pccn $-D_{2h}^{10}$ (No. 56) with a = 784.9(2), b = 2120.5(3), c = 1437.4(4) pm; Z = 4, gives D_c = 1.839 while D_m = 1.83 g/cm³. The bond distances and angles agree very well with those of the all-chloro compound in Fig. 29. Noteworthy differences (from 2 to 5 pm beyond the error limits) are: In-Br(1) 258.7(1), In-Cl(2) 249.2(1), In-N(1) 228.8(3) pm [2].

$Cl(I)InCH_2InCl_2 \cdot 2\ (CH_3)_2NCH_2CH_2N(CH_3)_2$

The compound was formed along with $InCl_3 \cdot 2$ TMEDA in a preparation similar to that for the bromo compound described above. InI, $InCl_3$, and CH_2Cl_2 gave a crystalline yellow powder in 90% yield, but no m.p. and no analytical data were reported [2].

The ^1H NMR spectrum in CD_2Cl_2 solution displays the characteristic In-CH_2-In signal at 0.56 ppm; signals from the TMEDA ligand are observed at 2.64 and 2.78 (br) ppm. No other chemical or physical properties were mentioned [2].

References:

[1] Schumb, W. C.; Crane, H. I. (J. Am. Chem. Soc. **60** [1938] 306/8).
[2] Khan, M. A.; Peppe, C.; Tuck, D. G. (Organometallics **5** [1986] 525/30).

3 Organoindium Pseudohalogenides

Without exception the compounds in this chapter are of the type of R_2InX and are oligomers or polymers. Their properties are summarized in Table 26. The pseudohalogenides X are CN, OCN, NCS, NCSe, and N_3. The C_4H_9 groups in 4, 5, and 8 are not specified.

Table 26
R_2InX Compounds with X=CN, NCO, NCS, NCSe, and N_3.
Further information on compounds whose numbers are preceded by an asterisk is given at the end of the table.
Explanations, abbreviations, and units on p. X.

No.	compound	properties and remarks
*1	$(CH_3)_2InCN$	white solid, m.p. 147 °C, subl. 120 to 140 °C/0.05 Torr [1] IR and Raman spectra on p. 177 mass spectrum (70 eV): $[M_3-CH_3]^+$ (34%), $[M_2-CH_3]^+$ (2%), $[M_2-CN]$ (100%), $[M_2-CN-C_2H_6]^+$ (10%), $[M_2-CN-C_4H_{12}]^+$ (11%), $[^{115}InCH_3]^+$ (2%), ^{115}In (99%) [1]
*2	$(C_4H_9)_2InNCO$	white crystalline solid, m.p. >240 °C; insoluble in common organic solvents [6] IR (solid): 2170 vs ν(CN), 1300 ν(CO), 635 m and 624 m δ(NCO) [6]
*3	$(C_2H_5)_2InSCN$	yellow solid, m.p. 51 °C, dec. 180 °C in vacuum [2] IR (neat): 2128 vs ν(CN), 649 vs ν(CS), 522 s $\nu_{as}(InC_2)$, 468 s $\nu_s(InC_2)$ and δ(SCN), 450 sh δ(SCN) [2]
*4	$(C_4H_9)_2InNCS$	white crystalline solid, m.p. 240 °C; insoluble in common organic solvents [6] IR (solid): 2040 vs ν(CN), 860 s ν(CS), 475 m δ(NCS) [6]
*5	$(C_4H_9)_2InNCSe$	white crystalline solid, m.p. 116 °C; insoluble in common organic solvents [6] IR (solid): 2040 vs ν(CN), 675 m ν(CSe) [6]

Table 26 (continued)

No.	compound	properties and remarks
*6	$(CH_3)_2InN_3$	colorless solid [4] mass spectrum (70 eV): $[M_3-CH_3]^+$ (ca. 1%), $[M_2-CH_3]^+$ (32.5%), $[M_2-N_3]^+$ (11.4%), $[M_2-2\,CH_3-N_3]^+$ (3.3%), $[M_2-2\,CH_3-5\,N]^+$ (4.9%), $[M+N_3]^+$ (1.4%), $[M-CH_3]^+$ (1.8%) [4] IR and Raman spectra on p. 178
*7	$(C_2H_5)_2InN_3$	colorless crystals, m.p. 140 to 142 °C [3] IR and Raman spectra on p. 178
*8	$(C_4H_9)_2InN_3$	white crystalline solid, m.p. 165 °C [6] IR (solid): 2042 vs $\nu_{as}(N_3)$, 1320 m $\nu_s(N_3)$, 655 m $\delta(N_3)$ [6]

* Further information:

$(CH_3)_2InCN$ (Table **26**, No. **1**) was formed in practically quantitative yield by the reaction of $In(CH_3)_3$ with stoichiometric amounts of HCN in ether. After removal of the solvent, it was purified by vacuum sublimation. The compound is a tetramer in benzene, but is less soluble in the common aprotic solvents than the homologous Al or Ga compounds. It is completely decomposed by contact with strong acids [1].

From the position of the $\nu(C{\equiv}N)$ frequency at 2178 cm^{-1} (CCl_4 solution) the cyano group was considered to have a bridging function [1]. The entire vibrational spectrum (IR and Raman) of the solid compound is given in Table 27. The splitting in the region of the In-C bond vibrations was explained by intermolecular interactions (crystal splitting), and from the data a fully planar ring system was excluded [5].

Table 27
Vibrational Spectra of Solid $\{(CH_3)_2InCN\}_4$ [5].
Wavenumbers in cm^{-1}.

IR	Raman	assignment
2985 m	2995 w	$\nu_{as}(CH_3)$
2925 m	2935 m	$\nu_s(CH_3)$
2308 mw	–	$2\times\delta_s(CH_3)$
2172 s	2175 ms	$\nu(^{12}CN)$
2130 vw, sh	–	$\nu(^{13}CN)$
1660 w	–	$\delta_s(CH_3)+\nu(InC_2)$
1165 w	1162 s	$\delta(CH_3)$
730 vs	–	$\varrho(CH_3)$
690 vw, sh	685 w	
543 vw, sh	541 vw, sh	$\nu_{as}(InC_2C')$
533 s	533 m	
525 s	523 m	
494 m	494 vs	$\nu_s(InC_2)$
470 w	445 vw, sh	
–	425 w	
365 vs, br	355 w	$\nu(InCN)$
–	342 w	$\nu(InNC)$
–	281 w	
–	250 w	
235 mw		$\delta(InC_2)$
–	130 m	

$(C_4H_9)_2InNCX$ (Table **26**, Nos. **2**, **4**, and **5**; X = O, S, and Se) were obtained by the reaction of $(C_4H_9)_2InCl$ with the corresponding alkali metal salt, MNCX (M = Na, K, NH_4), in C_5H_5N/H_2O (1:1). After stirring for 1 to 3 h, the solids were filtered off, washed with ethanol/water, and dried in vacuum. In addition to the most important absorption bands of the NCX ligands given in Table 26, all the IR spectra display strong bands around 585 or weak ones around 370 cm^{-1}, which were assigned to ν(InC) or ν(InN) vibrations [6]. $(C_4H_9)_2InNCS$ (No. 4) forms (together with elementary sulfur and N_2) on refluxing $(C_4H_9)_2InN_3$ (No. 8) in CS_2 for 12 h [8].

$(C_2H_5)_2InSCN$ (Table **26**, No. **3**) was produced in quantitative yield along with C_2H_5SCN by the exothermic reaction of a benzene solution of $In(C_2H_5)_3$ and thiocyanogen, $(SCN)_2$. The resulting yellow, viscous, and very reactive oil slowly solidified to yellow crystals after standing in the cold. A planar six-membered In_3S_3 ring with D_{3h} symmetry was suggested from the IR data [2].

$(CH_3)_2InN_3$ (Table **26**, No. **6**) was formed along with $Sn(CH_3)_4$ in practically 100% yield on reaction of equimolar amounts of $In(CH_3)_3$ and $(CH_3)_3SnN_3$ [4].

A trimeric structure with a six-membered non-planar In_3N_2 ring in the solid state was deduced from IR and Raman absorptions, which are summarized in Table 28 [4].

$(C_2H_5)_2InN_3$ (Table **26**, No. **7**). ClN_3 was introduced with nitrogen as carrier gas into a benzene solution of $In(C_2H_5)_3$ for 1 h; after a short period the compound began to separate. It decomposes with gas evolution just above the melting point, but is not explosive. It is dimeric in nitrobenzene, soluble in CH_3OH, $(CH_3)_2CO$, C_5H_5N, and DMF, and insoluble in C_6H_{14}, C_6H_6, and H_2O.

The IR spectrum of a nitrobenzene solution shows no changes compared to the solid spectrum (Table 28) which could serve as weak hints of a planar four-membered In_2N_2 ring [3].

Table 28
Vibrational Spectra of Solid $\{(CH_3)_2InN_3\}_3$ [4] and Solid $\{(C_2H_5)_2InN_3\}_2$ [3].
Wavenumbers in cm^{-1}.

$\{(CH_3)_2InN_3\}_3$ IR	Raman	$\{(C_2H_5)_2InN_3\}_2$ IR	Raman	assignment
3390 w	—	3445 w	—	$\nu_{as}(N_3)+\nu_s(N_3)$
3338 vw	—	3410 m	—	
—	—	3355 w	—	
2980 w	2988 vw	2970 s	2965 vw	$\nu_{as}(CH_3)$
—	2930	2950 s	2940 s	$\nu_{as}(CH_2)$
2900 m	2910 w	2920 s	2916 vs	$\nu_s(CH_3)$
—	—	2870 s	2869 vs	$\nu_s(CH_2)$
2850 vw	—	2820 vw	2820 vw	overtones
2720 vw	—	2735 vw	2725 vw	
2650 vw	—	—	—	
2060 vs	2080 vw	2080 vs	2068 m	$\nu_{as}(N_3)$
—	2042 w	—	—	
1440 vw	—	1455 mw	1456 m	$\delta_{as}(CH_3)$ and $\delta_{as}(CH_2)$
1369 vw	—	1420 w	1411 mw	
—	—	1380 vw	—	$\delta_s(CH_2)$

Table 28 (continued)

$\{(CH_3)_2InN_3\}_3$ IR	Raman	$\{(C_2H_5)_2InN_3\}_2$ IR	Raman	assignment
1353 s	1348 s	1358 s	1352 s	$\nu_s(N_3)$
1295 m	1289 w	1302 m	1295 m	
1171 vw	1169 s	1232 m	1230 vw	$\delta_s(CH_3)$ $(InCH_3)$
1166 vw	1163 ms	1175 m	1175 vs	δ, $\omega(CH_2)$ (InC_2H_5)
–	1155 m	–	–	
–	–	1097 w	–	
–	–	1057 w	–	
–	–	1007 m	1006 m	ν(C–C)
–	–	960 m	961 mw	
–	–	940 ms	935 vw	
970 vw	–	855 vw	–	$\varrho(CH_3)$
730 vs	–	–	–	
650 m	–	655 vs	655 m	$\delta(N_3)$
–	–	602 vw	630 vw	$\varrho(CH_2)$
615 m	–	–	–	$\omega(N_3)$
560 s	545 m	518 m	520 vw	$\nu_{as}(InC_2)$
493 w	488 vs	486 s	471 vs	$\nu_s(InC_2)+\gamma(N_3)$
308 m	–	288 s	278 vs	ν(InN)
260 w	–	252 s	251 s	δ(ring)
–	–	163 s	135 m	$\delta(InC_2)$

$(C_4H_9)_2InN_3$ (Table **26**, No. **8**) was prepared from $(C_4H_9)_2InCl$ and MN_3 (M = Na, K, and NH_4); see also Nos. 2, 4, and 5. It is insoluble in common organic solvents, thermally stable, and does not explode at the melting point [6].

An ionic structure was postulated from the IR absorptions of the azide ligand [6].

Reaction of No. 8 with CS_2 gave compound No. 4. Treating the title compound with C_6H_5NCX (X = O, S) or C_6H_5CN, adducts of type I or II, respectively, were formed [8]. The reaction with $[R_4N]I$ (R = CH_3, C_2H_5) in anhydrous methanol generated the corresponding ionic products $[R_4N][(C_4H_9)_2InN_3I]$ [7]; see Section 11.3.2.

I   II

References:

[1] Coates, G. E.; Mukherjee, R. N. (J. Chem. Soc. **1963** 229/33).

[2] Dehnicke, K. (Angew. Chem. **79** [1967] 942/3; Angew. Chem. Intern. Ed. Engl. **6** [1967] 947).

[3] Müller, J.; Dehnicke, K. (J. Organometal. Chem. **12** [1968] 37/47).

[4] Röder, N.; Dehnicke, K. (Chimia [Switz] **28** [1974] 349/51).

[5] Müller, J.; Schmock, F.; Klopsch, A.; Dehnicke, K. (Chem. Ber. **108** [1975] 664/72).

[6] Srivastava, T. N.; Kapoov, K. (Indian J. Chem. A **17** [1979] 611/2).
[7] Srivastava, T. N.; Singhal, K. (J. Indian Chem. Soc. **57** [1980] 225/6).
[8] Srivastava, T. N.; Srivastava, R. C.; Singhal, K. (Indian J. Chem. A **19** [1980] 480/2).

4 Organoindium-Oxygen Compounds

4.1 R_2InOX Compounds

4.1.1 Organoindium Hydroxides of the Type R_2InOH

Compounds of this composition have not yet been isolated and characterized. Signs concerning to existence of the simplest representatives ($R = CH_3$ and C_2H_5), however, have been reported several times.

$(CH_3)_2InOH$

Unlike the homologous gallium compound [3, 6] this compound could not be isolated. According to [1], $In(CH_3)_3$ and H_2O reacted vigorously, eliminating two equivalents of CH_4 and forming $\{(CH_3)_2In\}_2O$; [5] reports the resulting product (from moist air) to be polymeric $(CH_3InO)_n$. The radical decomposition of $(CH_3)_2InOOCH_3$ gave $(CH_3)_2InOH$ along with other products, including $(CH_3InO)_n$, its polymeric condensation product [8]. $[(CH_3)_2In]_2C_4O_4$ was prepared from $(CH_3)_2InOH$, which had been made by the reaction of $In(CH_3)_3$ and H_2O (1:1 mole ratio) in benzene at 5 to 10 °C (evolution of one equivalent of methane); however, neither isolation of the hydroxo compound nor examination of the clear benzene solution was attempted [7].

$(C_2H_5)_2InOH$

This compound formed as a white solid from the reaction of $In(C_2H_5)_3$ with H_2O in ether at 15 °C (with the evolution of C_2H_6). When the reaction was carried out at 90 °C, $In(OH)_3$ was the only product. Investigation of the initiating step is not described [2]. $(C_2H_5)_2InOH$ was named as one of the possible products of the free radical decomposition of $(C_2H_5)_2InOOC_2H_5$, but it was not actually isolated [4].

References:

[1] Dennis, L. M.; Work, R. W.; Rochow, E. G. (J. Am. Chem. Soc. **56** [1934] 1047/9).
[2] Runge, F.; Zimmermann, W.; Pfeiffer, H.; Pfeiffer, I. (Z. Anorg. Allgem. Chem. **267** [1951] 39/48).
[3] Kenney, M. E.; Laubengayer, A. W. (J. Am. Chem. Soc. **76** [1954] 4839/41).
[4] Cullis, C. F.; Fish, A.; Pollard, R. T. (Trans. Faraday Soc. **60** [1964] 2224/33).
[5] Clark, H. C.; Pickard, A. L. (J. Organometal. Chem. **13** [1968] 61/71).
[6] Sprague, M. J.; Glass, G. E.; Tobias, R. S. (Inorg. Synth. **12** [1970] 67/70).
[7] Schwering, H.-U.; Olapinski, H.; Jungk, E.; Weidlein, J. (J. Organometal. Chem. **76** [1974] 315/24).
[8] Aleksandrov, Yu. A.; Chikinova, N. V.; Makin, G. I.; Kornilova, N. V.; Bregadze, V. I. (Zh. Obshch. Khim. **48** [1978] 467; J. Gen. Chem. [USSR] **48** [1978] 417).

4.1.2 Compounds of the Type R_2InOR' with $R' = $ Alkyl and Aryl

The compounds are summarized in Table 29. They were prepared by the following general procedures:

Method I: Reaction of InR_3 with R′OH.
The trialkylindium (or its etherate) was placed in ether, benzene, or hexane and, while stirring constantly, was treated with a stoichiometric amount of R′OH. The weakly exothermic reaction eliminated RH and was usually complete on heating gently for less than 2 h. The product was purified by distillation or sublimation after removal of the solvent [8, 9, 12].

Method II: From the reaction of R_2InCl with LiOR′ compounds.
According to the "salt method" equimolar portions of the starting materials were mixed in ether and refluxed for approximately 1 h. After removing the undissolved portion, the solution was worked up as described in Method I. (Comment of the author: the LiCl that precipitates is extremely finely divided, making separation difficult and coprecipitating considerable amounts of product leading to relatively low yields) [14].

General Remarks. Most of the alkoxides described in this section are dimers with a four-membered In_2O_2 core as shown in Formula I. Nos. 1 and 10 are cyclic trimers with In_3O_3 rings puckered to varying degrees.

```
        R'
        |
   R    O    R
    \  / \  /
     In   In
    /  \ /  \
   R    O    R
        |
        R'
```

I

Indium diethyloxides 6 and 7 were dimers when prepared by Method I at 5 to 6 °C. After distillation, however, cryoscopy in benzene revealed a concentration-dependent degree of association between 2.6 and 2.9.

Table 29
R_2InOR' Compounds with R′ = Alkyl and Aryl.
Further information on compounds whose numbers are preceded by an asterisk is given at the end of the table.
Explanations, abbreviations, and units on p. X.

No.	compound method of preparation (yield)	properties and remarks
*1	$(CH_3)_2InOCH_3$ I (80 to 90%)	colorless oily liquid, m.p. 3 to 5 °C, b.p. 50 °C/10^{-3} Torr [9] ^1H NMR (CCl_4): −0.12 (CH_3In), 3.53 (CH_3O) [9] IR and Raman frequencies in Table 30 on p. 184
2	$(CH_3)_2InOC_2H_5$	no properties reported, used for insertion reactions with SO_2 to give $(CH_3)_2InO_2SOC_2H_5$ [8]
*3	$(CH_3)_2InOC_4H_9$-t I (91%)	waxy white solid, m.p. 90 °C, subl. 70 °C/1.5 Torr, dimeric in benzene [2, 3]

Table 29 (continued)

No. compound method of preparation (yield)	properties and remarks
*3 (continued)	^1H NMR (CCl_4): −0.145 (CH_3In, J(C,H) = 126.7), 1.17 (CH_3C, J(C,H) = 123.7) [2, 3] IR (solid): 1160 s $\delta_s(CH_3In)$, 708 s $\varrho(CH_3In)$, 522 s $\nu_{as}(InC_2)$, 484 m $\nu_s(InC_2)$ [2, 3]
4 $(CH_3)_2InOC_6H_5$ I	no properties reported, used for insertion reactions with SO_2 [8]
5 $(CH_3)_2InOC(CF_3)_2CH_3$	prepared from $In(CH_3)_3$ and $(CF_3)_2CO$ in $O(C_2H_5)_2$ [4] colorless liquid [4] ^1H NMR: 0.09 (s, CH_3In), 1.52 (s, CH_3C) [4] ^{19}F NMR: 81 (CF_3) [4] IR (liquid): 734 vs, br $\varrho(CH_3In)$, 540 vs, 487 s ν(InC), 466 s, 444 m δ(alkoxy group) [4]
6 $(C_2H_5)_2InOCH_3$ I (80 to 90%)	oily liquid, b.p. 73 °C/10^{-4} Torr [9] ^1H NMR (neat): 0.7 (q, CH_2), 1.3 (t, CH_3, J = 8), 3.53 (s, CH_3O) [9] IR (neat): 1008 ν(OC), 455 ν(InO) [9] Raman (neat): 1047 ν(OC), 273 ν(InO) [9]
7 $(C_2H_5)_2InOC_2H_5$ I (80 to 90%)	oily liquid, b.p. 86 °C/0.5 Torr [9]
*8 $(C_2H_5)_2InOC_4H_9$-t	white solid, m.p. 175 to 177 °C [7]
*9 $(C_2H_5)_2InOCH(C_2H_5)C_6H_5$ special	colorless crystals, m.p. 44.5 °C, b.p. 171 to 172 °C/10^{-3} Torr, dimeric in benzene [5] ^1H NMR (CCl_4): 0.60 (m, CH_2In), 1.16 (t, CH_3), 1.71 (q, CH_2), 4.61 (t, CH), 7.29 (s, C_6H_5) [5]
*10 $(C_2H_5)_2InOCH(C_6H_5)_2$ special	white solid, m.p. 146 to 148 °C, trimeric in benzene [5] ^1H NMR (CCl_4): 0.38 (q, CH_2In), 1.09 (t, CH_3), 5.80 (s, CH), 7.21 (m, C_6H_5) [5]
*11 $(C_4H_9\text{-t})_2InOC_2H_5$ II (63%)	large colorless crystals, dec. ca. 177 °C, turns orange at 208 °C and melts to a red liquid, subl. 90 °C/0.01 Torr [14] log p (Torr) = 13.75 − 5690/T [13] ^1H NMR (C_6D_6): 1.15 (t, CH_3 of C_2H_5), 1.45 (s, CH_3C), 3.80 (q, CH_2 of C_2H_5, J(H,H) = 7) [14] IR (solid): 2765 m, 2705 m, 1405 s, sh, 1361 s $\nu_{as}(CC_3)$, 1260 w, br, 1160 s, 1097 s, 1061 s, 1012 m, 940 w, 892 m, 809 s, 766 w, br, 722 w, br, 534 vs, br, 521 s, sh, 490 w, sh, 379 s, 366 s, 261 s, br, 257 s, sh, 231 m [14]

Table 29 (continued)

No.	compound method of preparation (yield)	properties and remarks
*12	$(C_4H_9\text{-n})_2InOC_4H_9\text{-t}$ I (75%)	colorless liquid, b.p. 125 °C/0.01 Torr [12] ^{1}H NMR: 0.70 (t, CH_2In, J=8.4), 0.90 (t, CH_3, J=6.0), 1.17 (s, CH_2-2), 1.26 (s, C_4H_9-t), 1.62 (m, CH_2-3) [12] IR (neat ?): 480 ν(InO), no other bands reported [12]
*13	$(C_4H_9\text{-n})_2InOC_6H_5$ I (76%)	colorless liquid, b.p. 167 °C/0.01 Torr [12] ^{1}H NMR: 0.83 (t, CH_2In, J=6.6), 1.07 (t, CH_3, J=7.5), 1.30 (s, CH_2-2), 1.62 (m, $A_2B_2B'_2$-type, CH_2-3), 6.61 (d), 6.80 (d), 7.15 (t, all C_6H_5, J=7.2) [12] IR (neat ?): 510 ν(InO); no other bands reported [12]
14	$(t\text{-}C_4H_9CH_2)_2InOCH_3$	from $(t\text{-}C_4H_9CH_2)_2InP\{Si(CH_3)_3\}_2$ and CH_3OH [15]
15	$\{(CH_3)_3SiCH_2\}_2InOCH_3$	from $\{(CH_3)_3SiCH_2\}_2InP\{Si(CH_3)_3\}_2$ and CH_3OH [15] m.p. 32 to 37 °C [15] ^{1}H NMR (C_6D_6): −0.17 (s, CH_2), 0.19 (s, CH_3Si), 3.39 (s, CH_3O) [15] ^{13}C NMR (C_6D_6): 2.07, 2.55, 52.34 (CH_3O) [15] IR (solid): 2949 s, 2887 m, 2812 w, 1444 w, 1400 w, 1294 w, 1246 s, 1039 m, 1022 m, 962 m, 852 s, 827 s, 752 s, 717 m, 684 m, 609 w, 580 w, 511 w, 478 m [15]
16	$\{(CH_3)_3SiCH_2\}_2InOC_4H_9\text{-t}$	from $\{(CH_3)_3SiCH_2\}_2InP\{Si(CH_3)_3\}_2$ and $t\text{-}C_4H_9OH$ [15]

* Further information:

$(CH_3)_2InOCH_3$ (Table **29**, No. **1**) was obtained in nearly quantitative yield by a modification of Method I. $In(CH_3)_3$ and CH_3OH were cocondensed at −98 °C, then slowly warmed to room temperature [1]. The title compound is also one of the numerous decomposition products of $(CH_3)_2InOOCH_3$ (see p. 209) but was not obtained pure when prepared this way [10]. The deuterated derivatives, **$(CH_3)_2InOCD_3$**, **$(CD_3)_2InOCH_3$**, and **$(CD_3)_2InOCD_3$**, have also been synthesized by Method I [9]. The vibration spectral data of these four alkoxides are summarized in Table 30. The data indicate a slightly puckered In_3O_3 six-membered ring with C_{3v} symmetry. $(CH_3)_2InOCH_3$ is (as are the deuterated products) a colorless, oily liquid at room temperature which is stable in dry air and very soluble in organic solvents. With excess CH_3OH it does not react further to form $CH_3In(OCH_3)_2$. According to cryoscopic molecular weight determinations in benzene the compound is a trimer [9]; ebullioscopy in the same solvent reveals a degree of association of about 3 with a small concentration dependence (no details given) [1].

The compound reacts with SO_2 via insertion into the In–O bond to form $(CH_3)_2InO_2SOCH_3$ (see p. 212) [8, 9], and reacts with HBr in ether to give $(CH_3)_2InBr$ (see p. 148) in yields over 90% [4].

Table 30
Vibrational Spectra of Liquid $(CH_3)_2InOCH_3$, $(CH_3)_2InOCD_3$, $(CD_3)_2InOCH_3$, and $(CD_3)_2InOCD_3$ [9].
Wavenumbers in cm^{-1}.

$(CH_3)_2InOCH_3$		$(CH_3)_2InOCD_3$		$(CD_3)_2InOCH_3$		$(CD_3)_2InOCD_3$		assignment
IR	Raman	IR	Raman	IR	Raman	IR	Raman	
2980 sh	2984 mw	2980 m	2990 mw	—	—	—	— }	$\nu_{as}(CH_3In)$ and
2938 s	2935 vs	2921 m	2940 vs	2934 s	2933 ms	—	— }	$\nu_{as}(CH_3O)$
—	2875 mw	2870 w	2880 mw	2910 m, br	2913 w	—	— }	$\nu_s(CH_3In)$ and
2828 s	2825 m	2820 vw	2830 vvw	2820 s	2823 s, br	—	— }	$\nu_s(CH_3O)$ + overtone
—	—	2238 m	2235 m	2225 s	2230 m	—	—	$\nu_{as}(CD_3In)$
—	—	2218 mw	2225 mw	—	2220 sh	—	—	overtone
—	—	2192 mw	2202 w	—	—	2195 ms	2200 vw, br	$\nu_{as}(CD_3O)$
—	—	2130 w	2140 mw	—	—	(2120)	(2123)	overtone
—	—	—	—	2122 m	2120 vs	2120 m	2123 vs	$\nu_s(CD_3In)$
—	—	2058 s	2058 s	2055 vw	—	2055 s	2060 ms	$\nu_s(CD_3O)$
1475 sh	1470 sh	—	—	1465 w, br	1470 sh	—	—	$\delta_{as}(CH_3O)$
1450 w, br	1451 mw	—	1452 vvw	1449 mw	1452 mw	—	—	$\delta_s(CH_3O)$
1168 m	1170 vs	1163 w	1170 vs	—	—	—	—	$\delta_s(CH_3In)$
—	—	1111 sh	1115 vw	—	—	1111 sh	1112 mw	$\delta_s(CD_3O)$
—	—	1104 s	1110 vw	—	—	1106 s	1106 w	$\delta_{as}(CD_3O)$
—	—	1075 sh	1068 vw, br	—	—	1050 sh	1060 vw, br	$\delta(CD_3O)$
(1032)	1040 mw	1006 sh	1010 mw	1038 sh	1042 mw	(990)	1003 mw	$\nu(OC; A_1)$
1032 vs	1030 w, sh	996 vs	1000 sh	1027 vs	1030 sh	990 vs	990 sh	$\nu(OC; E)$
—	970 vvw	936 w	940 vw	922 vw	925 vw, sh	931 mw	924 sh	$\delta(CD_3In)$
—	—	—	—	901 m	903 vs	899 m	902 vs	$\delta_s(CD_3In)$
—	—	890 sh	898 vw	—	—	(899)	(902)	$\varrho(CD_3O)$
—	—	785 vw	—	—	—	810 vw	—	?
710 vs	—	710 s, br	—	—	—	—	—	$\varrho(CH_3In)$
680 sh	670 w, br	680 sh	674 w, br	—	—	—	—	—
—	—	—	—	548 ms	550 vvw	548 s	550 vvw	$\varrho(CD_3In)$
510 s	523 ms	518 ms	525 ms	—	—	—	—	$\nu_{as}(InC_2, CH_3)$
—	—	—	—	510 s	508 w, br	508 m	506 w, br	$\nu(InO; E)$

Table 30 (continued)

$(CH_3)_2InOCH_3$		$(CH_3)_2InOCD_3$		$(CD_3)_2InOCH_3$		$(CD_3)_2InOCD_3$		assignment
IR	Raman	IR	Raman	IR	Raman	IR	Raman	
495 s	487 vs	490 m	493 vs	–	–	–	–	$\nu_s(InC_2, CH_3)$
–	–	–	–	480 vs	480 ms	475 vs	477 ms	$\nu_{as}(InC_2, CD_3)$
470 vs	(487)	463 vs	470 sh	–	–	–	–	$\nu(InO; E)$
–	–	–	–	439 ms	442 vs	436 ms	440 vs	$\nu_s(InC_2, CD_3)$
–	350 vvw	–	340 vvw	408 sh	340 vvw	420 sh	340 vw	$\delta(COIn)$ (?)
273 mw	278 s	269 w	270 s	272 m	276 s	276 m	269 m	$\nu_s(InO; A_1)$
–	182 m	–	180 m	–	180 mw		180 mw	δ(ring)
–	125 ms	–	128 ms	–	112 m, br		112 m, sh	$\delta_s(InC_2)$
–	98 s	–	100 s	–	89 s		87 s	$\delta(InC_2)$

$(CH_3)_2InOC_4H_9$-t (Table **29**, No. **3**) was also formed by the decomposition of $(CH_3)_2InOOC_4H_9$-t at 120 °C in various solvents. The decomposition products were analyzed by gas chromatography [11].

$(C_2H_5)_2InOC_4H_9$-t (Table **29**, No. **8**) was described in [6] as a colorless solid, that melted with decomposition above ca. 100 °C and was formed from the free radical decay of the adduct resulting from $In(C_2H_5)_3$ and t-$C_4H_9OOC_4H_9$-t:

$$(C_2H_5)_3In \cdot (OC_4H_9\text{-t})_2 \xrightarrow{<90\,°C} (C_2H_5)_2InOC_4H_9\text{-t} + C_2H_5OC_4H_9\text{-t}$$

$$(C_2H_5)_3In \cdot (OC_4H_9\text{-t})_2 \xrightarrow{>90\,°C} (C_2H_5)_2InOC_4H_9\text{-t} + C_2H_5^{\bullet} + OC_4H_9\text{-t}^{\bullet}$$

It decomposes at 180 °C within 10 h to give $C_2H_5In(OC_4H_9\text{-t})_2$, $In(C_2H_5)_3$, C_2H_4, C_2H_6, C_4H_8, C_4H_{10}, and C_4H_9OH, as shown by mass spectroscopy [7].

$(C_2H_5)_2InOCHR'R''$ (Table **29**, No. **9**, $R'=C_2H_5$, $R''=C_6H_5$ and No. **10**, $R'=R''=C_6H_5$). The compounds were prepared by insertion of C_6H_5CHO or $(C_6H_5)_2CO$, respectively, into an indium-carbon bond of $In(C_2H_5)_3$. No. 9 formed in quantitative yield within 3 h in hexane at 69 °C. No. 10 was obtained (along with C_2H_4) in 79% yield after 28 h in n-heptane at 98 °C (see also No. 5). Each reactant mixture was investigated by ^1H NMR spectroscopy in CH_2Cl_2 (No. 9 is orange, No. 10 is wine red) at −40 to −50 °C. No changes in the chemical shifts of the InC_2H_5 protons which would have indicated adduct formation were observed [5].

$(C_4H_9\text{-t})_2InOC_2H_5$ (Table **29**, No. **11**) was obtained by Method II. The $LiOC_2H_5$ was formed unintentionally on reaction of LiC_4H_9-t (1.99 M in hexane) with tetramethylpiperidine-2,2,6,6 in ether, but instead of forming the N-lithiated piperidine the ether was split to form the N-ethylated piperidine and $LiOC_2H_5$ [14].

Vapor pressures at 310 to 345 K (~36 to 70 °C) were measured by the Knudsen method and the results presented graphically [13].

The mass spectrum of the gas shows dimeric molecules (molecular mass: 548.15443 measured; 548.15750 required for dimer). Additional fragments in the spectrum are (rel. intensities in % in parentheses): $[M_2+H]^+$ (1), $[M_2]^+$ (3), $[M_2-C_2H_5]^+$ (10), $[M_2-C_4H_9]^+$ (100), $[M_2-C_8H_{17}]^+$ (5), $[M+In]^+$ (5), $[M_2-3\,C_4H_9]^+$ (24), $[M+OC_3H_7]^+$ (6), $[M+OC_2H_7]^+$ (20), $[M+3\,H]^+$ (6), $[M+H]^+$ (29), $[C_8H_{18}]^+$ (57) [14].

The compound crystallizes in the triclinic space group $P\bar{1}-C_i^1$ (No. 2) with the parameters a=904.2(2), b=988.4(2), c=991.5(1) pm, α=117.83(1)°, β=121.53(2)°, γ=75.31(2)°; Z=1 (dimer) and D_c=1.363 g/cm³. The structure of the centrosymmetric dimer with C_{2h} symmetry is reproduced in **Fig. 30** on p. 187. Each of the metal atoms is coordinated by two O and two (butyl) C atoms at the vertices of a distorted tetrahedron. The four-membered In_2O_2 ring is planar [14].

$(C_4H_9\text{-n})_2InOR'$ (Table **29**, No. **12**, $R'=C_4H_9$-t; No. **13**, $R'=C_6H_5$). Both compounds were subjected to thermogravimetric analysis. No. 12 shows an exothermic peak at 280 °C (start of decomposition), while the thermogram of No. 13 displays two such peaks, at 301 and 383 °C, indicating that thermolysis occurs in two steps. The first step is explained by the loss of the two C_4H_9-n groups; the second, higher temperature step, by loss of the phenyl residue. The decrease in weight is 58.1%, setting in at 96 °C and ending at 431 °C; the thermogram is depicted [12]. Both compounds have been used to prepare thin ··· Sn-O-In ··· oxide films by pyrolysis of $(C_4H_9\text{-n})_2InOR'$ and $(C_4H_9\text{-n})_2SnO$ mixtures at 450 °C [12].

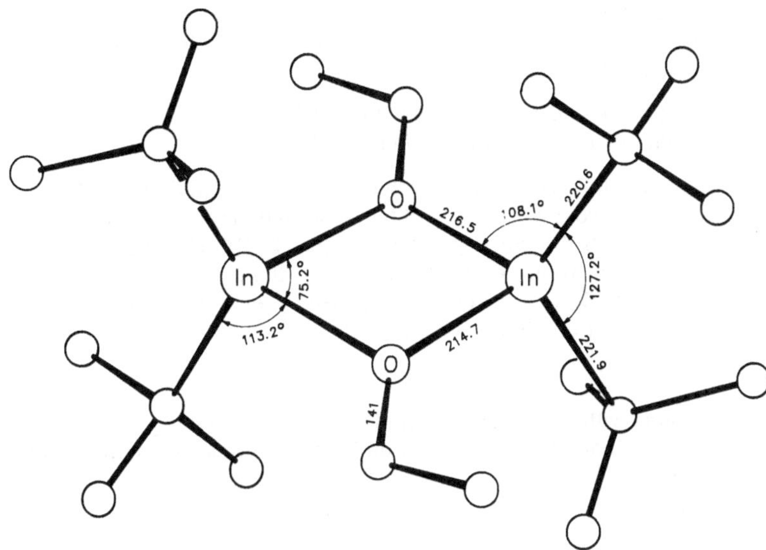

Fig. 30. Molecular structure of $\{(C_4H_9\text{-t})_2InOC_2H_5\}_2$ [14].

References:

[1] Coates, G. E.; Whitcombe, R. A. (J. Chem. Soc. **1956** 3351/4).
[2] Schmidbaur, H. (Angew. Chem. **77** [1965] 169/70; Angew. Chem. Intern. Ed. Engl. **4** [1965] 152/3).
[3] Schmidbaur, H.; Schindler, F. (Chem. Ber. **99** [1966] 2178/86).
[4] Clark, H. C.; Pickard, A. L. (J. Organometal. Chem. **8** [1967] 427/34).
[5] Tada, H.; Yasuda, K.; Okawara, R. (J. Organometal. Chem. **16** [1969] 215/20).
[6] Zhil'tsov, S. F.; Shcherbakov, B. I.; Druzhkov, O. N. (Zh. Obshch. Khim. **39** [1969] 1327/31; J. Gen. Chem. [USSR] **39** [1969] 1297/300).
[7] Shcherbakov, V. I.; Zhil'tsov, S. F.; Druzhkov, O. N. (Zh. Obshch. Khim. **40** [1970] 1542/5; J. Gen. Chem. [USSR] **40** [1970] 1529/31).
[8] Hsieh, A. T. T. (J. Organometal. Chem. **27** [1971] 293/301).
[9] Mann, G.; Olapinski, H.; Ott, R.; Weidlein, J. (Z. Anorg. Allgem. Chem. **410** [1974] 195/205).
[10] Aleksandrov, Yu. A.; Chikinova, N. V.; Makin, G. I.; Kornilova, N. V.; Bregadze, V. I. (Zh. Obshch. Khim. **48** [1978] 467; J. Gen. Chem. [USSR] **48** [1978] 417).
[11] Shushunova, A. F.; Makin, G. I.; Chikinova, N. V.; Bryukhanov, A. N.; Aleksandrov, Yu. A. (Zh. Analit. Khim. **34** [1979] 1614/7; J. Anal. Chem. [USSR] **34** [1979] 1249/52).
[12] Nomura, R.; Inazawa, S.; Matsuda, H.; Saeki, S. (Polyhedron **6** [1987] 507/12).
[13] Bradley, D. C.; Faktor, M. M.; Frigo, D. M. (J. Cryst. Growth **89** [1988] 227/36).
[14] Bradley, D. C.; Frigo, D. M.; Hursthouse, M. B.; Hussain, B. (Organometallics **7** [1988] 1112/5).
[15] Dougles, T.; Theopold, K. H. (Inorg. Chem. **30** [1991] 594/6).

4.1.3 R_2InOER_3 Compounds with E = Si, Ge, and Sn

The organoindium siloxanes, germoxanes, and stannoxanes are very similar to the alkoxides, R_2OR'. These compounds are also dimers with a four-membered In_2O_2 core and were prepared, as were the alcoholates, as described in Section 4.2.1. The few reported examples are summarized in Table 31, p. 188.

Method I: Reaction of InR_3 with $HOER_3$.
Stoichiometric quantities of the components were allowed to react as described in Method I for alcohols on p. 181 [1 to 3].

Method II: Reaction of R_2InCl compounds with $LiOER_3$.
See Method II for alcoholates on p. 181 [3].

Method III: Reaction of InR_3 compounds with $(C_4H_9)_2SnO$ or $\{(C_4H_9)_3Sn\}_2O$.
Compounds No. 5 and 6 were obtained from $(C_4H_9\text{-n})_2SnO$ in refluxing p-xylene for 3.5 h. With the corresponding $\{(C_4H_9)_3Sn\}_2O$ Nos. 5 and 7 were prepared in benzene at room temperature (3.5 h); the by-product was $Sn(C_4H_9\text{-n})_4$ or $Sn(C_4H_9\text{-n})_2(C_4H_9\text{-i})_2$, respectively. After removal of the solvent, the compounds were purified by distillation [5].

Table 31
$R_2InOER'_3$ Compounds with E = Si, Ge, Sn and R' = Alkyl and Aryl.
Further information on compounds whose numbers are preceded by an asterisk is given at the end of the table.
Explanations, abbreviations, and units on p. X.

No.	compound method of preparation (yield)	properties and remarks
*1	$(CH_3)_2InOSi(CH_3)_3$ I (92%)	colorless liquid, m.p. 16 °C, b.p. 57 to 58 °C/3 Torr; dimeric in benzene [1, 2] ^{1}H NMR (CCl_4): −0.125 (s, CH_3In, J(C,H) = 127.6), 0.013 (s, CH_3Si, J(C,H) = 117.8, J(Si,H) = 6.72) [2, 3] IR (neat): 2959 vs, 2899 s, 2833 sh $\nu(CH_3)$, 2299 vw, 1441 m $\delta_{as}(CH_3In)$, 1408 m, 1395 m, 1361; for further bands, see Table 32 on p. 189 [2]
2	$(CH_3)_2InOSi(C_6H_5)_3$ I	white solid, m.p. 154 to 157 °C; dimeric in benzene [2] ^{1}H NMR (CCl_4): −0.43 (s, CH_3In), 7.28 (m, C_6H_5) [2] IR (solid): 1188 vw $\delta_s(CH_3In)$, 1166 vw, 1156 vw, 1114 s (C_6H_5-modes), 1109 h, 1046 w, 1031 vw, 999 w, 877 vs ν(SiO), 742 s $\varrho(CH_3In$ and C_6H_5-modes), 731 m, 712 vs, 702 vs, 676 sh [2]
*3	$(CH_3)_2InOGe(CH_3)_3$ II(69%)	colorless liquid, m.p. 14 to 16 °C, b.p. 98 to 100 °C/1 Torr; dimeric in benzene [3] ^{1}H NMR (CCl_4): −0.25 (s, CH_3In, J(C,H) = 127), 0.28 (s, CH_3Ge, J(C,H) = 125.5) [3] IR see Table 32 on p. 189
4	$(CH_3)_2InOSn(C_2H_5)_3$	no properties reported [4] was said to be formed during the decomposition of the adduct $(CH_3)_3In \cdot (C_4H_9\text{-t})OOSn(C_2H_5)_3$ [4]
*5	$(C_4H_9\text{-n})_2InOSn(C_4H_9\text{-n})_3$ III (79%)	oily liquid, b.p. 25 °C/0.13 Pa [5] ^{13}C NMR ($CDCl_3$): 13.6 (CH_3 at In and Sn), 16.5 (CH_2Sn), 19.1 (CH_2In), 27.2 (CH_2-3 at Sn) 28.0 (CH_2-2 at Sn), 28.7 (CH_2-3 at In), 29.8 (CH_2-2 at In) [5] IR (liquid): ν(In-O-Sn) 635 [5]

Table 31 (continued)

No.	compound method of preparation (yield)	properties and remarks
*6	$(C_4H_9\text{-i})_2InOSn(C_4H_9\text{-n})_2(C_4H_9\text{-i})$ III (75%)	oily liquid, b.p. 140 °C/1.3 Pa [5] ^{13}C NMR ($CDCl_3$): 13.6 (CH_3 of C_4H_9-n), 17.2 (CH_2-1 of C_4H_9-n), 26.5 (CH_2-3 of C_4H_9-n), 26.8 (CH_2In and CH_2Sn of C_4H_9-i), 27.6 (CH_2-2 of C_4H_9-n), 28.2 (CH at In and Sn), 29.7 (CH_2-2 of C_4H_9-i at In and Sn) [5] IR (liquid): ν(In-O-Sn) 625 [5]
*7	$(C_4H_9\text{-i})_2InOSn(C_4H_9\text{-n})_3$ III (75%)	b.p. 150 °C/1.3 Pa [5] ^{13}C NMR ($CDCl_3$): 13.6 (CH_3 of C_4H_9-n), 16.5 (CH_2-1 of C_4H_9-n), 27.3 (CH_2-1 of C_4H_9-i), 27.4 (CH_2-3 of C_4H_9-n), 28.3 (CH_3 of C_4H_9-i and CH_2-2 of C_4H_9-n), 30.1 (CH) [5] IR (liquid): ν(In-O-Sn) 625 [5]

* Further information:

$(CH_3)_2InOE(CH_3)_3$ (Table **31**, Nos. **1** and **3**, E = Si, Ge) were characterized by IR spectroscopy on the liquids. Absorptions below 1300 cm^{-1} with their assignments are listed in Table 32. They have also been compared to the spectra of the homologous Al and Ga compounds [2, 3].

Table 32
IR Frequencies of $(CH_3)_2InOE(CH_3)_3$ (E = Si, Ge).
Wavenumbers in cm^{-1}.

$(CH_3)_2InOSi(CH_3)_3$ [2]	$(CH_3)_2InOGe(CH_3)_3$ [3]	assignment
1261 s	1244 h	$\delta_s(CH_3E)$
1250 vs	1236 m	
1163 vs	1153 m	$\delta_s(CH_3In)$
1050 vw	–	
1037 vw	–	
901 vs	–	ν(SiO)
878 vs	–	
836 vs	819 m	$\varrho(CH_3E)$
816 sh	747 m	
749 s	–	
712 vs	702 vs	$\varrho(CH_3In)$ and ν(GeO)
680 m	598 m	$\nu_{as}(EC_3)$
598 w	564 w	$\nu_s(EC_3)$
524 s	515 m	$\nu_{as}(InC_2)$
486 w	486 w	$\nu_s(InC_2)$
451 vs	–	$\nu(In_2O_2\text{-ring})$
318 s	262 m	

$(C_4H_9\text{-}n)_2InOSn(C_4H_9\text{-}n)_3$, $(C_4H_9\text{-}i)_2InOSn(C_4H_9\text{-}n)_2(C_4H_9\text{-}i)$, and **$(C_4H_9\text{-}i)_2InOSn(C_4H_9\text{-}n)_3$** (Table **31**, Nos. **5** to **7**) are dimeric in benzene solution.

Thermal analyses of Nos. 5 and 6 (600 °C, 10 °C/min in N_2 flow) gave DTA peaks at 262 and 247 °C, respectively. Acetolysis (excess) produced $In(OOCCH_3)_3$ and the corresponding $R_2R'SnOOCCH_3$ compounds. A 1:1 acetolysis of No. 6 (in benzene at room temperature, 3.5 h) produced $(C_4H_9\text{-}i)_2InOOCCH_3$ (61%) and $(C_4H_9\text{-}n)_2(C_4H_9\text{-}i)SnOOCCH_3$ (60%). FAB mass spectra showed peaks of the monomeric unit [5].

References:

[1] Schmidbaur, H. (Angew. Chem. **77** [1965] 169/70; Angew. Chem. Intern. Ed. Engl. **4** [1965] 152/3).

[2] Schmidbaur, H.; Schindler, F. (Chem. Ber. **99** [1966] 2178/86).

[3] Armer, B.; Schmidbaur, H. (Chem. Ber. **100** [1967] 1521/35).

[4] Dodonov, V. A.; Grishin, D. F.; Cherkasov, V. K. (Zh. Obshch. Khim. **52** [1982] 868/75; J. Gen. Chem. [USSR] **52** [1982] 755/61).

[5] Nomura, R.; Fujii, S; Matsuda, H. (Inorg. Chem. **29** [1990] 4586/8).

4.1.4 $R_2InOR'\text{-}D$ Compounds

This section considers compounds having an organic functional group (OR′) which contains a second donor atom (D=O or N). Unless otherwise noted all of the compounds listed in Table 33 were prepared by the following standard procedures.

Method I: The reaction of InR_3 with HOR′-D.

To a solution of InR_3 in benzene (No. 14), pentane (Nos. 8 and 18), hexane (Nos. 1, 10, and 15), or ether (Nos. 2 to 5 and 17) at room temperature, a solution of HOR′-D was added dropwise and, when alkane evolution ended, the reaction mixture was warmed gently for ~2 h to complete the reaction. Removal of all the volatile components left the desired product. After removing all the volatile components, the (usually) solid products were purified by sublimation or recrystallization. More detailed descriptions of the purification procedures were not given.

Method II: From $(CH_3)_3In \cdot x\, O(C_2H_5)_2$ and HOR′-D compounds.

The etherate, in pentane, was used as starting material for preparing No. 6, while for No. 2 this adduct was prepared (from $InCl_3$ and $LiCH_3$ in excess ether) and reacted "in situ" without isolation. The work-up procedure follows Method I.

General Remarks. In solution most of the compounds are either monomeric or dimeric. Intramolecular coordination through the donor atom (D=O or N) would close a five- or six-membered (depending on the R′ chain length) heterocyclic ring including four-coordinated In and give monomers of structure I or II. Association of these monomers through further coordination of the oxygen atom of the -OR′-D ligand would give a dimer of structure III having five-coordinated indium atoms. So far, X-ray confirmation of these structures has been limited to a series of gallium homologues [12].

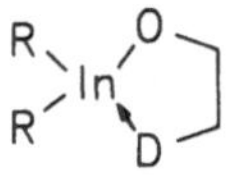

I 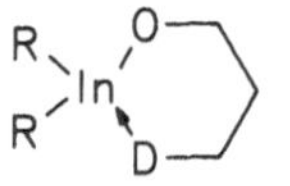 II 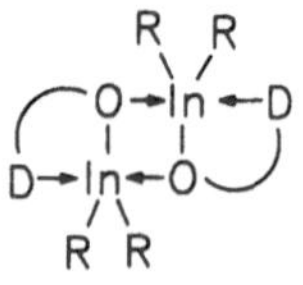 III

Table 33
$R_2InO(R'-D)$ Compounds.
Further information on compounds whose numbers are preceded by an asterisk is given at the end of the table.
Explanations, abbreviations, and units on p. X.

No.	O(R′-D) group method of preparation (yield)	properties and remarks
$(CH_3)_2InO(R'-D)$ compounds		
1	$OCH_2CH_2N(CH_3)_2$ I (95%)	colorless crystals, m.p. 110 to 111 °C; dimeric in benzene (1.82 wt%) [6] IR (solid): 694 (CH_3In), 509 and 468 (InC), no other bands reported [6] reacts slowly with CH_3I to give No. 10 [6]
*2	O=C(CH_3)–CH=C(CH_3)–O (acac, CH_3/CH_3) II	colorless crystals, m.p. 120 to 125 °C [4], 118 °C (dec.) [1], 170 to 172 °C (dec.) [3], subl. under vacuum [4] conductivity: $4.68 \times 10^{-3}\ \Omega^{-1} \cdot cm^2 \cdot mol^{-1}$ [4] IR and Raman frequencies on p. 194
*3	O=C(CF_3)–CH=C(CH_3)–O I	white solid, m.p. 93 °C [9] ^1H NMR ($CDCl_3$, ca. 35 °C): 0.4 (CH_3In), 2.43 (CH_3C), 5.88 (CH) [9] ^{19}F NMR ($CDCl_3$, relative to CF_3COOD): 2.68 (CF_3C) [9]
*4	O=C(CF_3)–CH=C(C_4H_9-t)–O I	white solid, m.p. 46 °C [9] ^1H NMR ($CDCl_3$, ca. 35 °C): 0.42 (CH_3In), 1.47 (CH_3C), 6.08 (CH) [9] ^{19}F NMR ($CDCl_3$, relative to CF_3COOD): 2.75 (CF_3C) [9]
5	O=C(CF_3)–CH=C(C_6H_5)–O I	no pure compound isolated [9] ^1H NMR ($CDCl_3$, ca. 35 °C): 0.43 (CH_3In), 6.63 (CH), 7.78 (m, C_6H_5) [9] ^{19}F NMR ($CDCl_3$, relative to CF_3COOD): 2.93 (CF_3C) [9]
6	tropolonato (O, O; ring positions 1 to 5) II	slightly yellow solid, dec. 270 °C, subl. 120 to 123 °C/3 Torr; soluble in C_6H_6, $C_6H_5CH_3$, slight solubility in C_6H_{14}, CCl_4, CH_2Cl_2, monomeric in benzene [11] ^1H NMR ($CDCl_3$): −0.07 (CH_3In), 7.01 to 7.39 (C_7H_5) [11] IR (solid): ca. 1350 and ca. 710 ν(OC), 540 $\nu_{as}(InC_2)$, 498 $\nu_s(InC_2)$, ν(InO) and δ(In-O-C) in the region of 525 to 590 [11] proposed structure, see Formula I

Table 33 (continued)

No.	O(R'-D) group method of preparation (yield)	properties and remarks
7	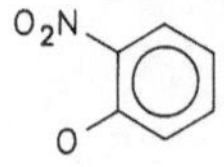 O_2N (2-nitrophenoxy structure) O I	bright orange solid, m.p. 190 °C (dec.) [2] IR (solid): 730 vs, br $\varrho(CH_3In)$, 586 s, 523 w, 497 w $\nu(InC)$ [2]
8	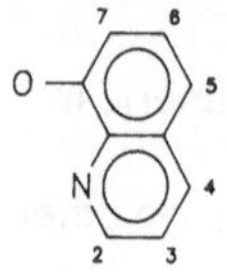 (8-quinolinolato structure, positions 2, 3, 4, 5, 6, 7) O, N I	bright yellow solid, m.p. 260 °C (dec.) [2]; air-stable, dimeric in benzene [6] IR (solid): 705 s, br $\varrho(CH_3In)$, 530 s, 498 m or 483 m $\nu(InC)$, 508 m δ(ligand) [2] UV (C_6H_6): λ_{max} (ε) = 364(3300) [2] proposed structure, see Formula III
*9	OC_6H_4CHO = salicyl I (68%)	yellow crystals, m.p. 188 to 192 °C [14] ^1H NMR ($CDCl_3$): −0.02 (s, CH_3In), 6.91 (m, 2 H of C_6H_4), 7.48 (m, 2 H of C_6H_4), 9.44 (s, CHO) [14] mass spectrum (M = monomer): $[2\,M-CH_3]^+$, $[2\,M-C_6H_3CHO]^+$, $[2\,M-C_6H_3CHO_2]^+$, $[M]^+$, $[M-CH_3]^+$, $[M-2\,CH_3]^+$, $[M-C_6H_4CHO_2]^+$ [14]
10	$OCH_2CH_2N(CH_3)_3^+I^-$ I (28%)	prepared from No. 1 and CH_3I in benzene, m.p. 218 to 221 °C (dec.) [6]
$(C_2H_5)_2InO(R'-D)$ compounds		
11	$OCH_2CH_2NH_2$ I (74%)	colorless crystals; probably dimeric [14] ^1H NMR (C_7D_8): 0.31 (t, H_2N), 0.65 (q, CH_2 of C_2H_5), 1.50 (t, CH_3), 2.22 (m, CH_2), 3.54 (t, CH_2) [14] mass spectrum (M = monomer): $[2\,M+H]^+$, $[2\,M-C_2H_5]^+$, $[2\,M-2\,C_2H_5+H_2]^+$, $[M+H]^+$, $[M-C_2H_5]^+$, $[M-C_2H_4ONH_2]^+$ [14]
12	$OCH_2CH_2N(CH_3)_2$ I (97%)	colorless crystals, m.p. 62 to 64.5 °C; dimeric in benzene [6] IR (solid): 625 $\varrho(CH_2In)$, 485 and 453 $\nu(InC)$ [6] reacts slowly with CH_3I to give No. 18 [6]
*13	$OC_6H_2(OC_2H_5\text{-}2)(C_4H_9\text{-}t)_2\text{-}3,5$ I	no properties reported [10]
*14	$OC_6H_2(OC_2H_5\text{-}2)(C_4H_9\text{-}t)_2\text{-}3,6$ I	no properties reported [10]
*15	$OC_6H_2(OC_2H_5\text{-}2)(C_4H_9\text{-}t)_2\text{-}4,6$ I	no properties reported [10]
16	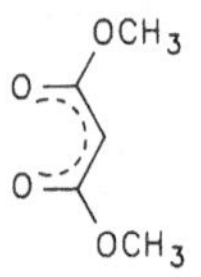 (structure) OCH_3, O, O, OCH_3 I	colorless solid [5] no properties reported; compared with acetyl-acetone compounds with $(C_2H_5)_2Al$ and $(C_4H_9)Zn$ groups [5]

Table 33 (continued)

No.	O(R′-D) group method of preparation (yield)	properties and remarks
17	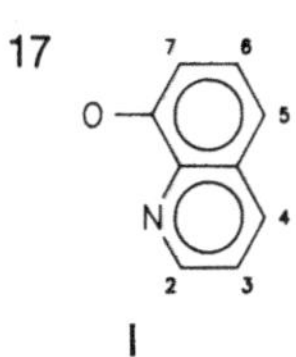 I	yellow solid, m.p. 222 to 229 °C (dec.); air-stable, dimeric in benzene [6] ^{1}H NMR (CH_2Cl_2 and C_6H_6): only $\delta(CH_2Cl_2)$-$\delta(C_6H_6)$ for the 2- and 4-protons, respectively, is discussed [6] IR (solid): 633 $\varrho(CH_2In)$, 506, 501, 460 ν(InC) [6] UV (C_6H_6, 3×10^{-4}M, 25 °C): $\lambda_{max}(\varepsilon)=370(3400)$ [6]
18	$OCH_2CH_2N(CH_3)_3^+I^-$	from No. 12 and CH_3I in C_6H_6 [6] m.p. 148 to 158 °C (dec.) [6]
$(C_4H_9\text{-n})_2InO(R'\text{-}D)$ compounds		
*19	CH_3 O O CH_3 I	white solid, m.p. 89 to 90 °C [13] ^{1}H NMR: 0.88 (t, CH_2-1, J=6.6), 0.9 to 1.8 (m, CH_2-2,3, CH_3), 1.88 (s, CH_3C), 5.24 (s, CH) [13] IR (solid): 420 ν(InO) [13]
$(C_4H_9\text{-i})_2InO(R'\text{-}D)$ compounds		
20	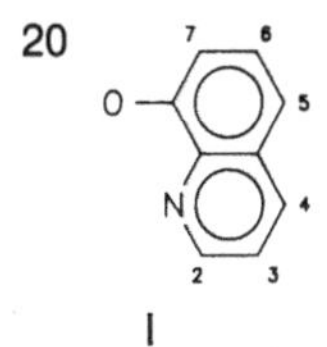 I	yellowish green solid, m.p. 160 °C [7, 8] ^{1}H NMR (C_6H_6, ca. 35 °C): 1.11 (m, CH_3), 1.22 (m, CH_2), 2.23 (m, CH), 6.72, 6.85, 7.28, 7.50, 7.56, 8.25 (m's, H-3, H-7, H-5, H-5 to H-6, H-4, H-2, J(2,3)=4.5, J(3,4)=8.0, J(2,4)=1.5, J(6,7)=8.5, and J(5,7)=4.5); the high-field portion of the spectrum is depicted in [7] IR (solid): 502 ν(InO), 308 ν(InN) [7, 8] UV: λ_{max}=380 (CT band) [7, 8] mass spectrum (70 eV): $[M]^+$ (0.4%), $[M-H]^+$ (1.2%), $[M-C_4H_8]^+$ (2.3%), $[M-C_4H_9]^+$ (13.5%), $[M-C_8H_{17}]^+$ (5.9%), $[M-C_8H_{18}]^+$ (34.0%), $[In(C_4H_9)_2]^+$ (1.8%), $[^{115}In]^+$ (100%) [7, 8]

* Further information:

$(CH_3)_2InOC(CH_3){=}CHCOCH_3$ (Table **33**, No. **2**) is monomeric in benzene solution and has a dipole moment of 2.01 D (in benzene and cyclohexane). By comparison with a variety of acetylacetonates of thallium and tin, an In-O bond moment of 3.5 D was suggested [3]. The IR and Raman spectra were measured for the solid and in CCl_4, and assignments made based on the assumption of a planar ring structure of C_{2v} symmetry (Formula II); the Raman spectrum for the solid is illustrated [4].

Table 34
IR and Raman Frequencies of $(CH_3)_2InOC(CH_3){=}CHCOCH_3$ [4].

IR solid	Raman solid	Raman CCl_4 solution	assignment
–	3073 w	3078 w	$\nu_s(CH_3C)$, A_1
2976 m	2973 m	2978 m	$\nu_{as}(CH_3In)$
2923 w	2925 s	2925 s	$\nu_s(CH_3In)$
1600 vs	1608 w	1608 w	ν(C⋯O), B_2 *)
1563 w	–	–	
1511 vs	1517 s	1515 s	ν(C⋯C⋯C), B_2
1449 w	1463 vw	1466 w	$\delta_{as}(CH_3C)$
1430 w	–	–	$\delta_{as}(CH_3In)$ *)
–	1395 s	1398 m	ν(C⋯O), A_1 *)
1381 vs	1370 s	1366 s	$\delta_s(CH_3C)$
1248 s	1252 vs	1258 vs	ν(C⋯C⋯C), A_1
1210 w	1216 vw	1208 m	δ(CH)
1200 ?	–	–	
1162 w	1170 s	1168 s	$\delta_s(CH_3In)$
1155 w	–	–	$\delta_s(CH_3In)$
1015 s	1028 m	1025 m	$\varrho(CH_3C)$
918 s	933 m	934 m	$\nu(CH_3C)$, A_1
788 m	–	–	δ(O⋯C⋯C) *)
720 vs	–	–	$\varrho(CH_3In)$
660 w	664 s	664 m	$\nu(InO_2)$ and δ(ring), A_1
552 s	564 w	561 w	δ(OCC), δ(CCC)
542 s	546 w	538 w	$\nu_{as}(InC_2)$
490 m	495 vvs	491 vvs	$\nu_s(InC_2)$
413 w	416 s	412 s	$\nu_s(InO_2)$ and γ(ring)
–	261 w	–	δ, γ(ring) *)

*) Assigned by the author.

$(CH_3)_2InOC(CF_3){=}CHCOR'$ (Table **33**, No. **3**, R' = CH_3; No. **4**, R' = C_4H_9-t). Ligand exchange with the corresponding acid, CF_3COCH_2COR', in benzene and diisobutyl ketone were investigated by ^{19}F NMR spectroscopy and given below. The exchange is first order in both components, and the values for the coalescence temperature (T_{coal}), the frequency difference $\Delta\delta$ ($\delta_{product} - \delta_{reactant}$), and the activation energy, ΔG, are given below [9].

R'	solvent	T_{coal} in °C	conc. in M	$\Delta\delta$ in Hz	ΔG in kcal/mol
CH_3	C_6H_6	32	0.34	53	13.4
	$(C_4H_9\text{-i})_2CO$	16	0.46	58	12.8
C_4H_9-t	C_6H_6	40	0.33	38	14.0
	$(C_4H_9\text{-i})_2CO$	32	0.23	45	13.3

For the formation of adducts of No. 3 with bipyridine, phenanthroline, and $(C_6H_5)_2PCH_2$-$CH_2P(C_6H_5)_2$, see Chapter 4.6 [9].

$(CH_3)_2In(OC_6H_4CHO)$ (Table **33**, No. **9**) crystallizes in the monoclinic space group $P2_1/n-C^5_{2h}$ (No. 14) with the cell parameters a = 932.2(6), b = 700.6(4), c = 1526.6(9) pm; Z = 2 and D_c = 1.80 g/cm^3. The molecular structure is shown in **Fig. 31** [14].

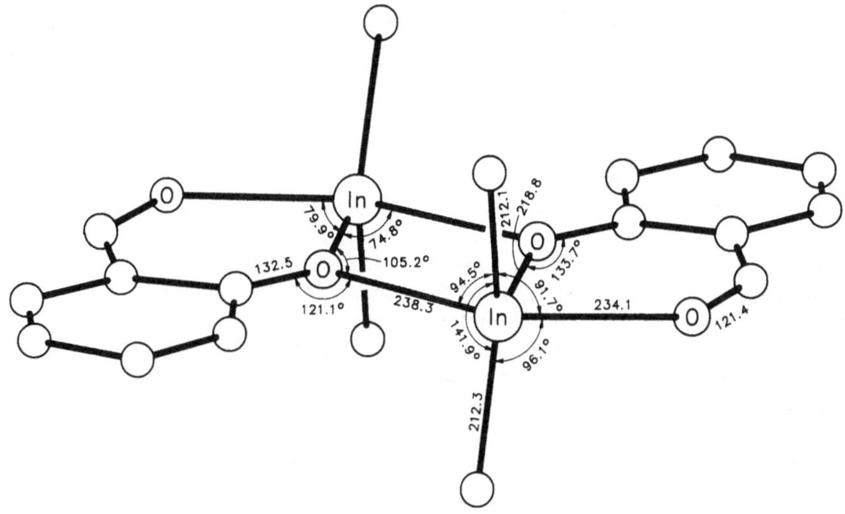

Fig. 31. Molecular structure of $(CH_3)_2In(OC_6H_4CHO)$ [14].

$(C_2H_5)_2InOC_6H_2(OC_2H_5\text{-}2)(C_4H_9\text{-}t)_2\text{-}3,5$ (Table **33**, No. **13**) usually formed with its isomer, **$(C_2H_5)_2InOC_6H_2(OC_2H_5\text{-}2)(C_4H_9\text{-}t)_2\text{-}4,6$** (Table **33**, No. **15**), from the reaction of $In(C_2H_5)_3$ with di-t-butyl-3,5-benzoquinone-1,2 in toluene. The preparative reaction is a free radical reaction and was monitored by ESR spectroscopy at −40 to −60 °C: The reaction led to a mixture of the above isomers via a green-colored, paramagnetic intermediate (the life span of the species was 15 to 20 min), according to the following scheme [10].

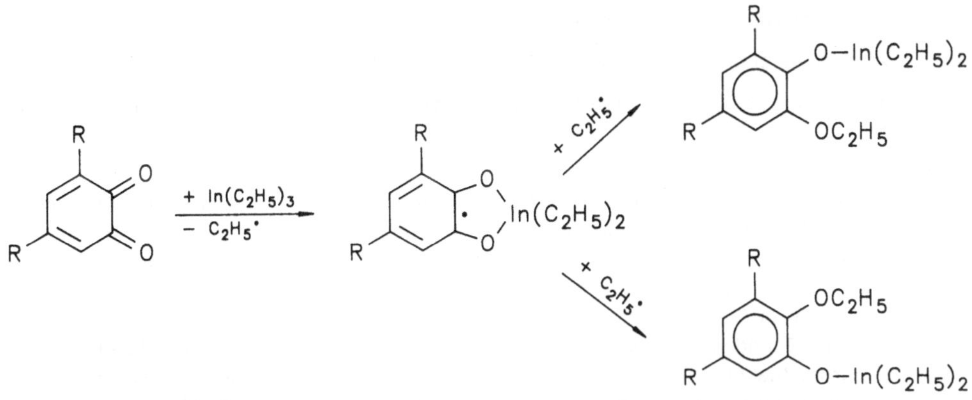

R, R, O, O
$+ In(C_2H_5)_3$ / $- C_2H_5^\bullet$
R, R, O, O, $In(C_2H_5)_2$
$+ C_2H_5^\bullet$ → R, R, $O{-}In(C_2H_5)_2$, OC_2H_5
$+ C_2H_5^\bullet$ → R, R, OC_2H_5, $O{-}In(C_2H_5)_2$

Acid hydrolysis of the compounds No. 13 and 15 produces 6-ethoxy-2,4-di-t- butylphenol and 2-ethoxy-3,5-di-t-butylphenol, respectively [10].

$(C_2H_5)_2InOC_6H_2(OC_2H_5\text{-}2)(C_4H_9\text{-}t)_2\text{-}3,6$ (Table **33**, No. **14**) formed in a similar manner from $In(C_2H_5)_3$ and di-t-butyl-3,6-benzoquinone-1,2 (15 min) to produce only one isomer in 70% yield along with 29.8% C_2H_6 [10].

$(C_4H_9\text{-}n)_2InOC(CH_3){=}CHCOCH_3$ (Table **33**, No. **19**) has three peaks in its thermogram. A sharp endothermic peak at 83 °C corresponds to the melting point. A broad exothermic peak at about 260 °C in the DTA curve is characteristic of loss of the butyl groups. Total weight loss of 67.9% begins at about 100 °C and is complete at 381 °C. The residue consisted of In_2O_3 and a small amount of indium; the thermogram is illustrated. The preparation of thin In/SnO films by pyrolyzing mixtures of No. 19 and $(C_4H_9)_2SnO$ were also reported [13]; see also [15].

References:

[1] Coates, G. E.; Whitcombe, R. A. (J. Chem. Soc. **1956** 3351/4).
[2] Clark, H. C.; Pickard, A. L. (J. Organometal. Chem. **8** [1967] 427/34 [1979] 1249/52).
[3] Moore, C. Z.; Nelson, W. H. (Inorg. Chem. **8** [1969] 143/5).
[4] Hobbs, C. W.; Tobias, R. S. (Inorg. Chem. **9** [1970] 1998/2004).
[5] Kawakami, Y.; Tsuruta, T. (Bull. Chem. Soc. Japan **44** [1971] 247/57).
[6] Maeda, T.; Okawara, R. (J. Organometal. Chem. **39** [1972] 87/91).
[7] Sen, B.; White, G. L.; Wander, J. D. (J. Chem. Soc. Dalton Trans. **1972** 447/9).
[8] Sen, B.; White, G. L. (J. Inorg. Nucl. Chem. **35** [1973] 497/504).
[9] Chung, H. L.; Tuck, D. G. (Can. J. Chem. **52** [1974] 3944/9).
[10] Razuvaev, G. A.; Abakumov, G. A.; Klimov, E. S.; Gladyshev, E. N.; Bayushkin, P. Ya. (Izv. Akad. Nauk SSSR Ser. Khim. **1977** 1128/32; Bull. Acad. Sci. USSR Div. Chem. Sci. **1977** 1034/7).

[11] Waller, I.; Halder, T.; Schwarz, W.; Weidlein, J. (J. Organometal. Chem. **232** [1982] 99/112).
[12] Gmelin Handbook "Galliumorganic Compounds" 1987, pp. 181/90.
[13] Nomura, R.; Inazawa, S.; Matsuda, H.; Saeki, S. (Polyhedron **6** [1987] 507/12).
[14] Alcock, N. W.; Degnan, I. A.; Roe, S. M.; Wallbridge, M. G. H. (J. Organometal. Chem. **414** [1991] 285/93).
[15] Nomura, R.; Fujii, S.; Matsuda, H. (Inorg. Chem. **29** [1990] 4586/8).

4.2 R_2In-Compounds of Organic Acids

4.2.1 Derivatives of Monocarboxylic Acids of the Type $R_2InOOCR'$

All the examples of this type are collected in Table 35 and were prepared by the following methods:

Method I: Reaction of InR_3 or $InR_3 \cdot D$ compounds with HOOCR′.
This has been the most used preparation. Normally, an equimolar amount of HOOCR′ was added dropwise to the indium compound dissolved in ether at room temperature. While [11] reports that an excess of acid does not result in the formation of the dicarboxylate, [5, 6] recommend exact stoichiometric ratios of the reactants in order to avoid succeeding reactions (preparation of Nos. 7 and 8 in benzene). Nos. 14 and 15 were prepared in $CHCl_3$ at −60 °C from the dioxane adduct of $In(C_6H_5)_3$ in even higher yield [4]. At the end of the vigorous evolution of RH gas, the product was filtered, washed, and purified by vacuum sublimation or recrystallization.

Method II: Ligand exchange between InR_3 and $In(OOCR')_3$ compounds.
This method was used to prepare Nos. 14 and 15 [4]. Exchange (2:1 mole ratio) was carried out in hot benzene. The yield of $R_2InOOCR'$ is generally smaller than by Method I. Workup and purification were carried out as above.

Method III: Insertion of CO_2 into one In-C bond of InR_3.
Compound No. 9 was formed in less than 5% yield by passing CO_2 into a boiling solution of $In(C_2H_5)_3$ in xylene and identified by IR spectroscopy [5]. The insertion of CO_2 into the metal-carbon bond of $In(C_6H_5)_3$ went somewhat better, but the existence of No. 16 was postulated solely on the recovery of the hydrolysis product, C_6H_5COOH [1].

General Remarks. The organoindium carboxylates are colorless solids, most of which undergo a partial unspecified decomposition below the melting temperature. They can be vacuum sublimed at temperatures between 100 and 130 °C. Their low solubility in aprotic organic solvents is not sufficient to allow cryoscopic molecular weight determinations, but according to osmometric measurements Nos. 2 to 5 are dimers or trimers in $CHCl_3$ solution. Mass spectra also established dimeric structures for Nos. 1, 2, and 8. Nos. 1 and 2 readily dissolve in water without decomposition at 0 to about 10 °C, but at higher temperatures CH_4 evolution sets in.

The crystal structures of Nos. 2 [8] and 8 [10] show that the carboxylic acid functions as a chelating ligand forming a four-membered ring with the In atom. Linear chain formation of the monomeric units in the solid state is achieved with sixfold coordinated indium atoms (Author: The structures permit the assumption that in the gas phase (MS) and in solution discrete molecules do not exist, but only fragments of the polymer chain). A dimeric structure with an eight-membered ring (Formula I on p. 202) was postulated for No. 9, based on IR and Raman studies [5].

Table 35
$R_2InOOCR'$ Compounds.
Further information on compounds whose numbers are preceded by an asterisk is given at the end of the table.
Explanations, abbreviations, and units on p. X.

No.	compound method of preparation (yield)	properties and remarks
1	$(CH_3)_2InOOCH$ I (97%)	white solid, dec. 168 °C [11], m.p. 184 to 185 °C (dec.), subl. ca. 120 °C/10^{-4} Torr [12] IR (solid): 1591 vs, 1572 vs $\nu_{as}(CO_2)$, 1368 vs, 1352 m $\nu_s(CO_2)$, 1171 w, 1160 w $\delta_s(CH_3In)$, 732 ms $\varrho(CH_3In)$, 557 m $\nu_{as}(InC_2)$, 535 m, 500 vw $\nu_s(InC_2)$ and $\delta(CO_2)$, 491 vw, 315 mw $\nu(InO_2)$ [12] Raman (solid): 1590 vw, 1568 vw $\nu_{as}(CO_2)$, 1369 m, 1358 mw $\nu_s(CO_2)$, 1174 sh, 1162 s $\delta_s(CH_3In)$, 697 vw, br $\varrho(CH_3In)$, 561 w $\nu_{as}(InC_2)$, 540 mw, 503 vs, 495 vs $\nu_s(InC_2)$, 318 mw $\nu(InO_2)$, 122 s $\delta(InC_2)$ [12] mass spectrum (70 eV, ca. 120 °C): $[M_2-CH_3]^+$ (1%), $[M_2-(O_2CH)]^+$ (1%), $[M+In]^+$ (2%), M^+ (4%), $[M-CH_3]^+$ (100%), $[M-OH]^+$ (10%), $[InOCC]^+$ (6%), $[In(CH_3)_2]^+$ (63%), $[InOC]^+$ (5%), $[InCH_3]^+$ (35%), $[InO]^+$ (51%), $^{115}In^+$ (82%) [11, 12]

Table 35 (continued)

No.	compound method of preparation (yield)	properties and remarks
*2	$(CH_3)_2InOOCCH_3$ I (97%)	colorless crystals, dec. 163 °C, subl. 130 °C/under vacuum [11]; m.p. 225 °C (dec.), subl. 120 °C/0.5 Torr [12]; dimeric in $CHCl_3$, solubility in $CHCl_3$ at 40 °C = 2.25 g/L [11] ^1H NMR: 0.05 (CH_3In), 2.08 (CH_3C) in $CHCl_3$ [11], 0.05, 2.27 in DMSO [16] and −0.07, 1.9 in D_2O [10] ^{115}In NQR (ca. 25 °C): 64.44 (ν_1), 81.56 (ν_2), 127.96 (ν_3), 171.96 (ν_4), e^2Qq/h, 1037.64 MHz, η = 26.5% [15]; see also [17] IR and Raman frequencies in Table 36 on p. 200 mass spectrum (80 eV, 90 °C): $[M_2-CH_3]^+$, $[M_2-O_2CCH_3]^+$, $[M-CH_3]^+$, $[In(CH_3)_2]^+$, $[InCH_3]^+$, $^{115}In^+$ [9]
3	$(CH_3)_2InOOCC_2H_5$ I (86 to 87%)	white solid, dec. 143 °C, dimeric in $CHCl_3$ [11] solubility ($CHCl_3$ at 35 °C): 4.8 g/L [11]
4	$(CH_3)_2InOOC(C_3H_7\text{-}i)$ I (73 to 74%)	white solid, dec. 138 °C, trimeric in $CHCl_3$ [11] solubility ($CHCl_3$ at 35 °C): 1.4 g/L [11]
5	$(CH_3)_2InOOC(C_4H_9\text{-}t)$ I (98%)	white solid, dec. 174 °C, dimeric in $CHCl_3$ [11] solubility ($CHCl_3$ at 35 °C): 3.16 g/L [11]
6	$(CH_3)_2InOOCCF_3$ I	needle like crystals, m.p. 220 to 222 °C subl. 100 °C/under vacuum, monomeric in nitromethane [3] conductivity: 2.1 $\Omega^{-1}\cdot cm^2\cdot mol^{-1}$ [3] IR (solid): 1620 vs $\nu_{as}(CO_2)$, 724 vs, br $\varrho(CH_3In)$, 564 s $\nu_{as}(InC_2)$, 518 m $\nu_s(InC_2)$, 497 w [3]
7	$(CH_3)_2InOOCC_6H_4(CHO)\text{-}2$ I (48%)	yellow, m.p. 212 to 216 °C, probably monomeric [21] ^1H NMR (CD_3SOCD_3): −0.23 (s, CH_3), 7.43, 7.74, 7.97 (m's, C_6H_4), 10.39 (s, CHO) [21] IR (solid): 1667 ν(CHO), 1541 $\nu_{as}(CO_2)$, 1458 $\nu_s(CO_2)$, 701 $\delta_s(CO_2)$ [21] mass spectrum: $[M_2-In(CH_3)_2]^+$, $[M]^+$, $[M-CH_3]^+$, $[M-C_6H_4CHOCO_2]^+$ [21]
*8	$(C_2H_5)_2InOOCCH_3$ I	colorless crystals, m.p. 172 to 174 °C (dec.) [6], 182 to 184 °C [7], subl. 105 to 110 °C/10^{-4} Torr, or 135 to 140 °C/ca. 760 Torr [6] D_m = 1.61 g/cm^3 [6] IR (solid): 1525 vs, br $\nu_{as}(CO_2)$, 1465 vs $\nu_s(CO_2)$, 1176 m $\omega(CH_2In)$, 644 vs, $\varrho(CH_2In)+\delta(CO_2)$, 515 s $\nu_{as}(InC_2)$, 466 ms $\nu_s(InC_2)$ [6] Raman (solid): 1520 vw, br $\nu_{as}(CO_2)$, 1465, 1462 s$\nu_s(CO_2)$, 1179 s $\delta_s(CH_2In)$, 640 vvw, br $\varrho(CH_2In)+\delta(CO_2)$, 518 w $\nu_{as}(InC_2)$, 469 vvs $\nu_s(InC_2)$ [6]

Table 35 (continued)

No.	compound method of preparation (yield)	properties and remarks
*9	$(C_2H_5)_2InOOCC_2H_5$ I (85%)	colorless crystalline solid, m.p. 165 °C (dec.), subl. 100 °C/10^{-4} Torr; slightly soluble in hot C_6H_{14} and CCl_4 [5] IR and Raman frequencies in Table 37 on p. 202 mass spectrum: fragmentation on p. 203
*10	$(C_4H_9\text{-}n)_2InOOCC_2H_5$ I (ca. 83%)	b.p. 140 °C/0.01 Torr [18] ^1H NMR: 0.89 (t, CH_2In, J = 7.3), 1.06 (t, CH_3, J = 7.0), 1.0 to 2.0 m, 2.22 (q, C_2H_5C) [18] IR: 1550 $\nu(CO_2)$, no other bands reported [18]
11	$(C_4H_9\text{-}n)_2InOOCCH(C_2H_5)(CH_2)_3CH_3$ I (80%)	m.p. 220 °C [18] ^1H NMR: 0.8 to 1.9 (m), 3.5 to 3.8 (m, br) [18] IR: 1555 $\nu(CO_2)$, no other bands reported [18]
12	$(C_4H_9\text{-}n)_2InOOCC_6H_5$ I (73%)	b.p. 140 °C/0.01 Torr [18] ^1H NMR: 0.88 (t, CH_2In, J = 6.3), 0.90 (t, CH_3, J = 8.1), 1.30 (m, CH_2), 1.65 (m, $A_2B_2B'_2$-type), 7.42 (dd, H-3), 7.45 (m, H-4), 7.99 (dd, H-2), J(H-1,3) = 7.5, J(H-3,4) = 2.4 [18] IR: 1550 $\nu_{as}(CO_2)$, no other bands reported [18]
13	$(C_4H_9\text{-}i)_2InOOCCH_3$	from $(C_4H_9\text{-}i)_2InOSn(C_4H_9\text{-}n)_2(C_4H_9\text{-}i)$ and CH_3COOH (1:1 mole ratio) in C_6H_6 at room temperature (3.5 h, 61%) [20] colorless needles, m.p. 95 to 96.5 °C [20] ^{13}C NMR ($CDCl_3$): 23.2 (CH_3 of CH_3COO), 27.0 (CH_2), 27.7 (CH_3), 33.2 (CH), 180.1 (COO) [20] IR (solid): 1530 s $\nu_{as}(CO_2)$, no other bands reported [20]
14	$(C_6H_5)_2InOOCCH_3$ I (85%) II (50%)	white solid, no m.p. < 300 °C [4] IR (solid): 3050 ν(CH), 1600 ν(C=C), 1550 and 1440 $\nu(CO_2)$, 530 ν(InO) 500 [4]
15	$(C_6H_5)_2InOOCC_2H_5$ I (80%) II (70%)	white solid, no m.p. < 300 °C [4] IR (solid): 3050 ν(CH), 1610 ν(C=C), 1560 and 1440 $\nu(CO_2)$, 500 ν(InO) [4]
16	$(C_6H_5)_2InOOCC_6H_5$ III	not isolated, formation proposed by hydrolysis of the reaction product; yields C_6H_5COOH [1]

* Further information:

$(CH_3)_2InOOCCH_3$ (Table **35**, No. **2**) was characterized with the aid of the IR [2, 9, 10, 12] and Raman [9, 10, 12] spectra in the 1600 to 200 cm^{-1} region. The frequencies of the two CO_2 vibrations at about 1530 (ν_{as}) and 1455 (ν_s) cm^{-1} were interpreted as resulting from equivalence of the two oxygen atoms in the bidentate carboxylate group. The frequency values in Table 36 were taken from [10] and supplemented by values from [9] and [12].

Table 36
IR and Raman Frequencies of $(CH_3)_2InOOCCH_3$ below 1600 cm^{-1} [10].
Wavenumbers in cm^{-1}.

IR solid	Raman solid	Raman H_2O solution at 5 to 10 °C	assignment
1535 vs, br	1525 vw, sh	1570 vvw	$\nu_{as}(CO_2)$
1465 sh	1464 sh	–	$\delta_{as}(CH_3C)$
1455 vs, br	1461 m, br	1425 w, p	$\nu_s(CO_2)$
1380 ms	1355 w	1354 s, p	$\delta_s(CH_3C)$
1167 mw	1163 s	1187 s, p	$\delta_s(CH_3In)$
1159 mw	1154 s		
958 m	969 m	935 m, p	$\nu(C\text{-}CH_3)$
730 ms	702 vw	–	$\varrho(CH_3In)$
681 s	680 vw	660 vw, p	$\delta_s(CO_2)$
550 s	553 m	560 vw, sh, dp	$\nu_{as}(InC_2)$
493 m	499 vs	504 vs, p	$\nu_s(InC_2)$
458 w	–	458 vw, dp	$\delta(CO_2)$
320 sh	375 w	–	
298 m	325 vw	–	$\nu(InO_2)$
–	225 m	–	
–	190 w	–	

The compound crystallized in the orthorhombic space group Pnma – D_{2h}^{16} (No. 62) with cell parameters a = 726.5(2), b = 732.5(2), and c = 1328.6(5) pm; Z = 4 gives $D_c = D_m = 1.92$ g/cm^3. In the crystal planar, four-membered InO_2C rings are joined perpendicularly (-In-O-In-O-) into parallel polymeric bands, so that each indium atom attains coordination number 6, as shown in **Fig. 32** on p. 201 [8].

The compound forms monomeric 1:1 adducts with $(CH_3)_2SO$ or NC_5H_5 having fivefold coordinated In atoms. With $NH_2CH_2CH_2NH_2$ or $(C_6H_5)_2PCH_2CH_2P(C_6H_5)_2$ monomeric addition compounds were obtained having 6-coordination at the central atom, and with 2,2′-bipyridyl or 1,10-phenanthroline it forms adducts of the type $\{(CH_3)_2InOOCCH_3\}_2 \cdot D$; see Formulas I to III on p. 224 [14]. The compound in anhydrous C_2H_5OH reacts with toluene-3,4-dithiol, evolving methane and forming dimeric $\{CH_3In(OOCCH_3)(C_7H_7S_2)\}_2$ (Section 4.8.4); in the presence of $(CH_3)_2SO$, or on reaction in $P(N(CH_3)_2)_3$, a further equivalent of CH_4 evolves and $In(OOCCH_3)(C_7H_6S_2)$ is produced [13, 16]. With $HS(CH_2)_nSH$ (n = 2, 3), salicylic acid, or phthalic acid in C_2H_5OH, the title compound also loses an equivalent of CH_4; the resulting products, however, were not investigated [16]. A solution of the compound in ice water reacted with $Na[SSCN(CH_3)_2]$ to form $(CH_3)_2InSSCN(CH_3)_2$ (68%) and in CH_3OH at room temperature to form $CH_3In\{SSCN(CH_3)_2\}_2$ (48%) (see p. 247) [7].

$(C_2H_5)_2InOOCCH_3$ (Table **35**, No. **8**) crystallized, like No. 2, in the orthorhombic space group Pnma – D_{2h}^{16} (No. 62) with unit cell parameters a = 840(2), b = 741(2), and c = 1513(3) pm; Z = 4, and $D_c = 1.64$ g/cm^3 [10]. The polymeric band structure shown in **Fig. 33** on p. 201 corresponds to that of No. 2 in Fig. 32.

A CH_3OH solution of the compound reacted with $Na[SSCN(CH_3)_2]$ to form a 23% yield of $C_2H_5In\{SSCN(CH_3)_2\}_2$ (see p. 247) [7].

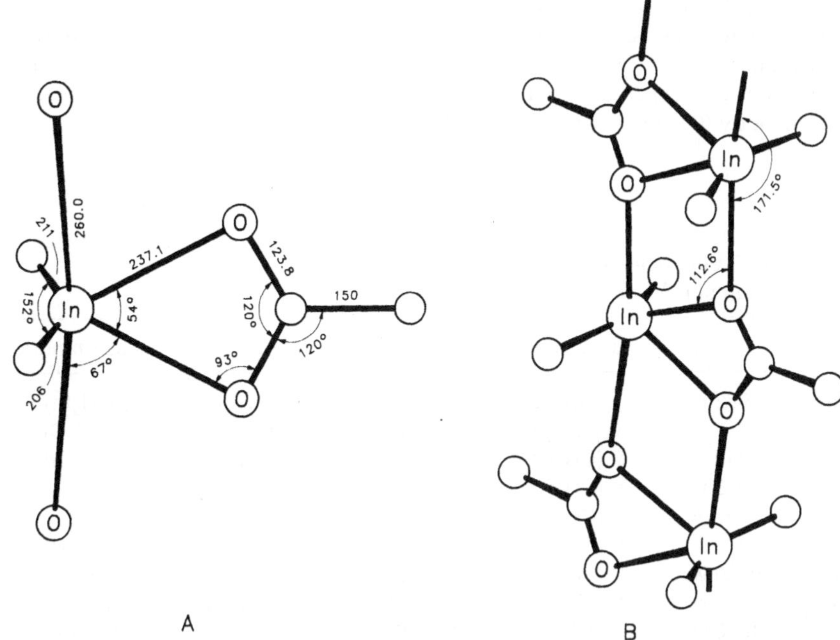

Fig. 32. Coordination at the In atom in solid $(CH_3)_2InOOCCH_3$ (A); part of the polymeric chain (B) [8].

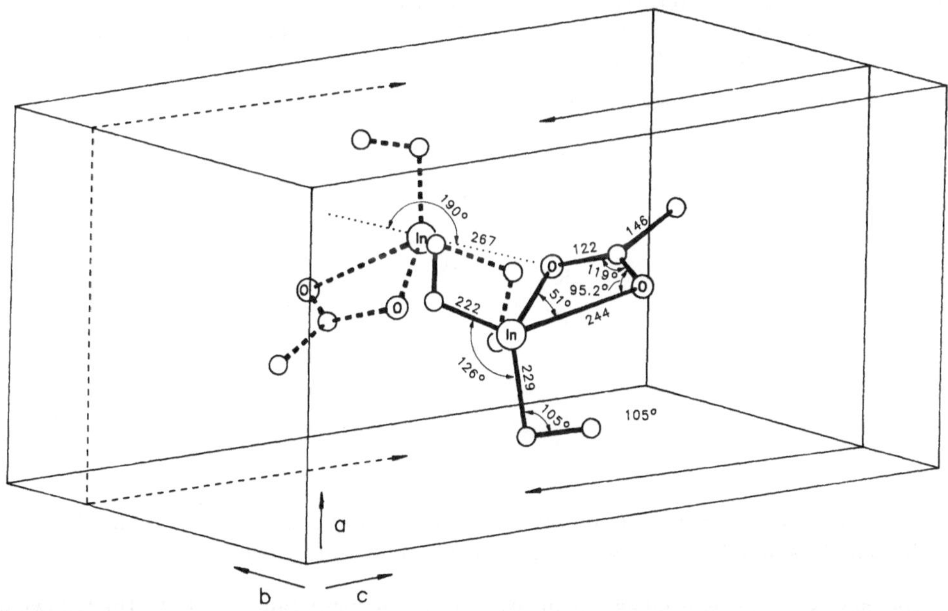

Fig. 33. Unit cell of $(C_2H_5)_2InOOCCH_3$ [10]. Only two of the four formula units are represented.

$(C_2H_5)_2InOOCC_2H_5$ (Table **35**, No. **9**) was characterized by IR and Raman spectroscopy on the solid state. The assignments in Table 37 were made on the basis of the frequencies of the most important CO_2 bond vibrations and by analogy to the homologous organogallium carboxylates. It was concluded that the dimer consists of a puckered eight-membered ring of high symmetry (C_{2h} or D_{2h}) as shown in Formula I.

Table 37
IR and Raman Frequencies of Solid $(C_2H_5)_2InOOCC_2H_5$ [5].
Wavenumbers in cm^{-1}.

IR	Raman	assignment	IR	Raman	assignment
2982 m	2982 mw	$\nu_{as}(CH_3C)$	1177 mw	1174 vs	$\delta_s(CH_2In)$
2960 s	–	$\nu_{as}(CH_2In)$	1081 ms	1078 mw	$\nu(C\text{-}C, C)$
2942 vs	2942 s	$\nu_s(CH_3C)$	1017 ms	1012 ms	$\nu(C\text{-}C\text{-}In)$
–	2928 vs	$\nu_s(CH_2C)$	962 mw	962 mw	and
2911 vs	2908 vs	$\nu_s(CH_3In)$	938 mw	–	$\varrho(CH_3)$
2868 s	2866 s	$\nu_s(CH_2In)$	910 s	907 mw	$\nu(C\text{-}C, C)$
2821 mw	2820 w	$2\times\delta(CH_2)$	815 m	813 w	$\varrho(CH_2C)$
2730 w, br	2728 w	$2\times\delta_s(CH_3)$	689 m	690 vw	$\delta_{as}(CO)_2$
1535 vs, br	1540 sh	$\nu_{as}(CO_2)$ [1]	646 vs	–	$\varrho(CH_2In)$
–	1522 w	$\nu_{as}(CO_2)$ [2]	514 s	512 w	$\nu_{as}(InC_2)$
1469 vs	1459 sh	$\nu_s(CO)_2$ [1]	468 m	468 vs	$\nu_s(InC_2)$
–	1452 s	$\nu_s(CO_2)$ [2] and	500 sh	(486)	$\delta_s(CO_2)+\delta(C\text{-}C\text{-}C)$
1442 s	1443 sh	$\delta_{as}(CH_3)$	325 m	315 w,br	$\nu(InO_2)$
1418 ms	1418 ms	$\delta(CH_2C+In)$	245 ms	241 m	$\delta(InCC)$
1380 s	1373 mw	$\delta_s(CH_3)$	225 vw	–	$\delta(InO_2)$?
1302 s	1300 mw	$\tau(CH_2C)$	162 m	168 mw	$\delta(InC_2)$
1235 mw	1250 vvw	$\delta(CH_2)$			

[1] Out-of-phase. – [2] In-phase.

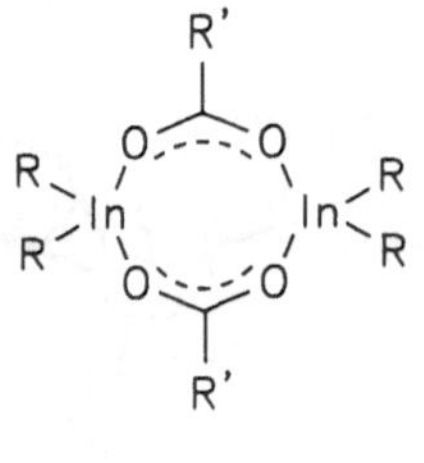

I

The compound crystallizes in the orthorhombic space group with a = 1570, b = 850, and c = 776 pm; Z = 4 and D_m = 1.58 g/cm^3. No other X-ray data were given [5].

The mass spectrum was measured at 70 eV at a probe temperature of 90 °C. The following fragmentation scheme accounts for only a part of the spectrum [5]:

$[In_2R_6O_4C_2 = M_2]^+$ (1%)
↓ −R
$[M_2 - R]^+$ (1%)
↓ −2R
$[M_2 - 3R]^+$ (1%)
−In → $[M + CO_2]^+$ (1%); −O → $[In_2R_3C_2O_3]^+$ (0%)

$[M + CO_2]^+$ (1%) —$-CO_2$→ $[M]^+$ (10%); —−R→ $[In_2R_2C_2O_4]^+$ (40%)
$[M]^+$ (10%) —−R→ $[M - R]^+$ (100%)
$[In_2R_2C_2O_4]^+$ (40%) —$-CO_2$→ $[M - R]^+$ (100%)
$[In_2R_3C_2O_3]^+$ (0%) —−R→ $[In_2R_2C_2O_3]^+$ (2%) —−In→ $[InR_2C_2O_3]^+$ (4%) —−CO→ $[M - R]^+$ (100%)

$[M - R]^+$ (100%)
↓ $-CO_2$
$[InR_2]^+$ (40%)
↓ −R
$[InR]^+$ (20%)
↓ −R
$[^{115}In]^+$ (100%)

$R = C_2H_5$

$(C_4H_9\text{-n})_2InOOCC_2H_5$ (Table **35**, No. **10**) was analyzed thermogravimetrically. The DTA diagram pictured in [18] shows exothermic peaks at 200 and 320 °C, but these could not be correlated with the simple splitting off of any of the substituents. A complicated decomposition occurs from 45 to 353 °C with a total weight loss of 50%. The compound was used to prepare thin In/SnO layers by pyrolyzing it with $(C_4H_9\text{-n})_2SnO$. By this procedure the compound $(C_4H_9\text{-n})In(OOCC_2H_5)OSn(C_4H_9\text{-n})_3$ (see on p. 233) was formed as a by-product [18, 19].

References:

[1] Gilman, H.; Jones, R. G. (J. Am. Chem. Soc. **62** [1940] 2353/7).
[2] Clark, H. C.; Pickard, A. L. (J. Organometal. Chem. **8** [1967] 427/34).
[3] Clark, H. C.; Pickard, A. L. (J. Organometal. Chem. **13** [1968] 61/71).
[4] Viktorova, I. M.; Sheverdina, N. I.; Rodionov, A. N.; Kocheshkov, K. A. (Dokl. Akad. Nauk SSSR **189** [1969] 315/7; Dokl. Chem. Proc. Acad. Sci. USSR **184/189** [1969] 889/91; C.A. **72** [1970] No. 55562).
[5] Weidlein, J. (Z. Anorg. Allgem. Chem. **378** [1970] 245/62).
[6] Hausen, H.-D. (J. Organometal. Chem. **39** [1972] C 37/C 40).
[7] Maeda, T.; Okawara, R. (J. Organometal. Chem. **39** [1972] 87/91).
[8] Einstein, F. W. B.; Gilbert, M. M.; Tuck, D. G. (J. Chem. Soc. Dalton Trans. **1973** 248/51).
[9] Habeeb, J. J.; Tuck, D. G. (J. Chem. Soc. Dalton Trans. **1973** 243/7).
[10] Hausen, H.-D.; Schwering, H.-U. (Z. Anorg. Allgem. Chem. **398** [1973] 119/28).

[11] Lindel, W.; Huber, F. (Z. Naturforsch. **28b** [1973] 517/8).

[12] Schwering, H.-U. (Diss. Univ. Stuttgart 1973).
[13] Habeeb, J. J.; Tuck, D. G. (J. Organometal. Chem. **82** [1974] C 25/C 26).
[14] Habeeb, J. J.; Tuck, D. G. (Can. J. Chem. **52** [1974] 3950/4).
[15] Patterson, D. B.; Carnevale, A. (Inorg. Chem. **13** [1974] 1479/83).
[16] Habeeb, J. J.; Tuck, D. G. (J. Organometal. Chem. **101** [1975] 1/19).
[17] Schmidbaur, H.; Koth, D. (Naturwissenschaften **63** [1976] 482/3).
[18] Nomura, R.; Inazawa, S.; Matsuda, H.; Saeki, S. (Polyhedron **6** [1987] 507/12).
[19] Nomura, R.; Fujii, S.; Kanaya, K.; Matsuda, H. (Polyhedron **9** [1990] 361/6).
[20] Nomura, R.; Fujii, S.; Matsuda, H. (Inorg. Chem. **29** [1990] 4586/8).

[21] Alcock, N. W.; Degnan, I. A.; Roe, S. M.; Wallbridge, M. G. H. (J. Organometal. Chem. **414** [1991] 285/93).

4.2.2 Derivatives of Thiocarbonic Acids, $R_2InOSCR'$

The three known compounds of this type were prepared from $In(CH_3)_3$ or $In(C_2H_5)_3$ and thioacetic acid or thiobenzoic acid.

$(CH_3)_2InOSCCH_3$

The title compound formed from $In(CH_3)_3$ and CH_3COSH (1:1 mole ratio) as a white crystalline solid, which melted with decomposition between 132 and 140 °C and sublimed at 80 °C/10^{-3} Torr. The compound is barely soluble in nonpolar organic solvents, but is very soluble in cold water; this solution keeps for only a short time [4].

IR (solid, frequencies in cm^{-1}): 1475 vs, br ν(CO), 1422 sh $\delta_{as}(CH_3In)$, 1349 w $\delta_s(CH_3C)$, 1170 sh, 1162 s, $\delta_s(CH_3In)$ and $\varrho(CH_3C)$, 985 mw ν(CC), 724 vs $\varrho(CH_3In)$, 702 m, sh ν(CS), 542 ms $\nu_{as}(InC_2)$, 531 ms δ(OCS), 485 mw, br $\nu_s(InC_2)$ and ν(InO?), 382 w, 375 w δ(CCO) and δ(CCS) [4].

Raman (solid): 1480 m ν(CO), 1425 w $\delta_{as}(CH_3C)$, 1355 mw $\delta_s(CH_3C)$, 1175 s, 1155 sh $\delta_s(CH_3In)$ and $\varrho(CH_3C)$, 990 w ν(CC), 708 s ν(CS), 540 mw $\nu_{as}(InC_2)$, 492 vs $\nu_s(InC_2)$, 387 w δ(CCO) and δ(CCS), 232 ms ν(InS) [4].

$(C_2H_5)_2InOSCCH_3$

It was obtained in 85 to 95% yields from the reaction of $In(C_2H_5)_3$ with an equimolar amount of CH_3COSH (elimination of C_2H_6) in benzene at 5 to 10 °C and was purified by sublimation at 80 °C/10^{-3} Torr. The title compound was also made by insertion of COS into one of the In-C bonds of $In(C_2H_5)_3$ at about 100 °C; because of unknown decomposition reactions, the compound could not be isolated, but was identified (in low concentrations) in the IR spectrum.

The compound melts at 125 °C and is monomeric in benzene. The IR and Raman spectra of the solid are illustrated in [2] (Author: the footnotes of the two illustrations in [2] are reversed!) and the frequencies and assignments are given in Table 38.

Table 38
IR and Raman Frequencies of Solid $(C_2H_5)_2InOSCCH_3$ [2].
Wavenumbers in cm^{-1}.

IR	Raman	assignment	IR	Raman	assignment
3022 w	3014 vw	$\nu_{as}(CH_3C)$	1009 m	1007 mw	$\nu(C\text{-}C)+$
2960 s	2960 sh	$\nu_{as}(CH_3In)$	978 ms	–	$\varrho(CH_3)$
2942 vs	–	$\nu_{as}(CH_2In)$	965 sh	966 w	
2932 sh	2928 s	$\nu_s(CH_3C)$	938 m	–	
2916 s	2908 vs	$\nu_s(CH_3In)$	699 ms	696 s	$\nu(C \cdots S)$
2861 ms	2864 ms	$\nu_s(CH_2In)$	641 vs	–	$\varrho(CH_2In)$
2818 mw	2815 w	overtone	541 s	538 w	$\delta(OCS)$
2731 w	2729 w		508 m	506 w	$\nu_{as}(InC_2)$
1483 vs	1480 ms	$\nu(C \cdots O)$	458 sh	–	$\nu(InO?)+$
1452 m, sh	1455 m, sh	$\delta_{as}(CH_3In)$	464 ms	460 vs	$\nu_s(InC_2)$
1418 mw	1414 mw	$\delta_{as}(CH_3C)+\delta(CH_2)$	382 m	381 s	$\delta(CCO)+$
1378 mw	1376 w, br	$\delta_s(CH_3In)+$	343 w, br	338 vw	$\delta(CCS)$
1349 m	1347 mw	$\delta_s(CH_3C)$	278 vw	285 sh	$\delta(InCC)$
1243 sh	–	$\delta(CH_2)$	250 mw	241 m	$\delta_s(InC_2)$
1232 mw	–		225 m	222 ms	$\nu(InS)$
1175 m, sh	1147 s	$\delta_s(CH_3In)+$	–	150 sh	$\tau(CH_3)+$
1161 vs	–	$\varrho(CH_3C)$	–	122 w	$\delta(InOS)$
–	1065 vw	?			

The compound crystallizes in the orthorhombic space group $Pnma-D_{2h}^{16}$ (No. 62) with cell parameters a = 858.8(4), b = 834.5(5), c = 1383.4(6) pm; Z = 4, gives $D_c = 1.66$ g/cm^3, whereas $D_m = 1.63$ g/cm^3 [3, 4]. Only the O atom of the thioacetate group functions as a bridge between neighboring monomers, and the indium is fivefold coordinated within a distorted trigonal bipyramid. A section of the structure is sketched in **Fig. 34** on p. 206. $In\text{-}C_\alpha = 213.5(13)$ pm and $C_\alpha\text{-}In\text{-}C_{\alpha'} = 137.7°$ [4].

The mass spectrum (70 eV) was only superficially evaluated. Along with the very weak peak of the monomeric unit, only the main fragments, $[M-C_2H_5]^+$ and $^{115}In^+$, were mentioned [2].

$(C_2H_5)_2InOSCC_6H_5$

This compound was prepared in ether solution from $In(C_2H_5)_3$ and C_6H_5COSH in a 1:1 mole ratio. After completion of the reaction by 30 min reflux, the raw product was crystallized from n-hexane/ether; white solid, m.p. 106 to 107 °C. The degree of association in C_6H_6 was concentration-dependent and lay between 1.12 and 1.28 [1].

^1H NMR (in CCl_4, in ppm): 1.2 (m, C_2H_5), 7.7 (m, C_6H_5). IR (solid): 1520 sh, 1464 s, br $\nu(CO)$; 331 m cm^{-1} $\nu(InS)$ (cyclohexane): 1547 sh, 1472 s, br $\nu(CO)$; 347 m cm^{-1} $\nu(InS)$ [1].

Acidic or alkaline hydrolysis of the compound gave benzoic acid (30 or 36%) and thiobenzoic acid (37 or 41%) along with C_2H_6. More thoroughly investigated was the thermal decomposition of samples in n-octane and propiophenone at 120 °C; along with C_2H_6, the formation of $(C_2H_5InS)_n$ and $(C_6H_5CO)_2CHCH_3$ was observed. The mechanism of the decomposition was discussed [1].

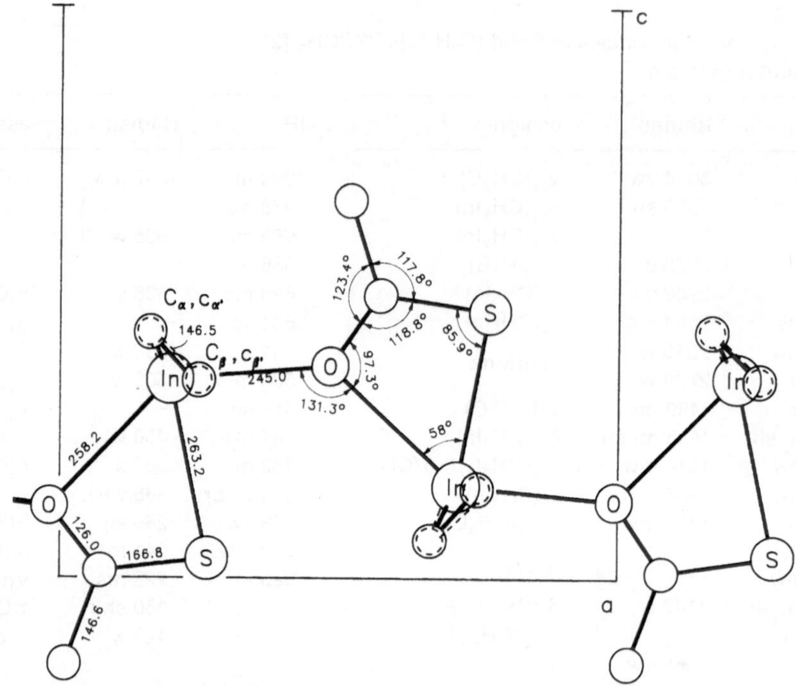

Fig. 34. Unit cell of $(C_2H_5)_2InOSCCH_3$ [4]. Section parallel to [010] in y=0.25; the ethyl groups are projected onto this plane.

References:

[1] Tada, H.; Okawara, R. (J. Organometal. Chem. **28** [1971] 21/4).
[2] Weidlein, J. (J. Organometal. Chem. **32** [1971] 181/94).
[3] Hausen, H.-D. (Z. Naturforsch. **27b** [1972] 82/3).
[4] Hausen, H.-D.; Guder, H.-J. (J. Organometal. Chem. **57** [1973] 243/53).

4.2.3 Derivatives of Oxalic and Squaric Acids

This section treats the derivatives of dicarboxylic acids; therefore they also contain two R_2In groups as shown in Formula I and II.

I   II

$(CH_3)_2InOOCCOOIn(CH_3)_2$

This compound was obtained in practically quantitative yield from the reaction of $In(CH_3)_3$ in benzene with an ether solution of oxalic acid (2:1 mole ratio). The compound began to precipitate as a white solid as soon as the acid was added. The solid was separated and purified by sublimation at 220 to 230 °C/10^{-4} Torr; no melting was observed even up to 320 °C [1].

On the basis of the high melting point, the low volatility, the low solubility in cold water and pyridine, and the similarity of the vibrational spectrum to those of alkali metal oxalates (Table 39) a largely ionic structure was proposed. Nevertheless, the spectral assignments were fitted to the assignments of the homologous gallium compound with its discrete molecular structure as shown in Formula I ($R=CH_3$) [1].

Table 39
Vibrational Spectra of Solid $(CH_3)_2InOOCCOOIn(CH_3)_2$ and $(C_2H_5)_2InOOCCOOIn(C_2H_5)_2$ [1]. Wavenumbers in cm^{-1}.

$(CH_3)_2InOOCCOOIn(CH_3)_2$		$(C_2H_5)_2InOOCCOOIn(C_2H_5)_2$	
IR	Raman	IR	assignment
2980 w, br	3002 vw	2965 mw	$\nu_{as}(CH_3)$
–	–	2952 mw	$\nu_{as}(CH_2)$
–	2943 w	2930 mw	$\nu_s(CH_3)$
–	–	2875 w	$\nu_s(CH_2)$
1698 sh	1707 vw	1680 sh	$\nu_{as}(CO_2)$ 1)
1655 vs	1656 w	1635 vs	$\nu_{as}(CO_2)$ 2)
–	1473 m	–	$\nu_s(CO_2)$ 1)
1412 vw	1430 vw	1455 vw	$\delta_{as}(CH_3In)$
–	–	1410 vw	$\delta(CH_2)$
–	–	1380 vw	$\delta_s(CH_2)$
1350 m	–	1358 vw	overtone +
1300 s	1300 vvw	1301 vs	$\nu_s(CO_2)$ 2)
1198 vw	1183 m	1180 w	$\delta_s(CH_3In)$, $\delta_s(CH_2In)$
1164 vw	1170 vs	–	
–	–	1010 w	ν(C-C, C_2H_5) and
–	–	963 vw	$\varrho(CH_3)$
–	–	940 w	
–	925 w	–	ν(C-C, C_2O_4)
800 m	–	794 s	$\delta_s(CO_2)$ 2)
738 m	–	653 s	$\varrho(CH_3In)$, $\varrho(CH_2In)$
550 mw	553 mw	529 mw	$\nu_{as}(InC_2)$
–	514 mw, sh	–	$\delta_s(CO_2)$ 1)
498 w	503 vs	475 w	$\nu_s(InC_2)$
428 w	–	411 mw	δ, $\gamma(CO_2)$
303 mw	–	–	ν(InO)?

1) In-phase. – 2) Out-of-phase.

$(C_2H_5)_2InOOCCOOIn(C_2H_5)_2$

The ethyl derivative was prepared in the same way as the methyl compound. Because of slow, progressive decomposition, it can not be sublimed in vacuum below 250 °C; at normal pressure no melting point is observed up to 320 °C [1].

The IR absorptions are listed in Table 39. The structure (Formula I, $R=C_2H_5$) was derived from these data [1].

$(CH_3)_2InOOC_4OOIn(CH_3)_2$

The derivative from the squaric acid was prepared in a two-step reaction without isolating the product of the first step. First, $In(CH_3)_3$ in benzene at 5 to 10 °C was treated with water

(1:1 mole ratio). When the methane evolution ended (formation of $(CH_3)_2InOH$, p. 180), more water as solvent was added (no further gas evolution), and the $(CH_3)_2InOH$ treated with solid squaric acid (2:1 mole ratio) with continuous stirring (30 h) at room temperature. From the aqueous phase at 5 to 6 °C the hydrate, **$(CH_3)_2InOOC_4OOIn(CH_3)_2 \cdot 4\,H_2O$**, crystallized out in the form of colorless needles and square crystals (40% yield). At room temperature this hydrate is only briefly stable; it weathers, probably under oligomerization, becoming significantly less soluble in water. The water-free compound can be obtained by simply vacuum drying the hydrate (75% overall yield). During this procedure the crystals decompose to a white powder which is stable up to 300 °C, neither melting nor subliming up to this temperature [2].

From the IR and Raman spectra a centrosymmetric structure of D_{2h} symmetry (Formula III) was postulated for the initially-formed tetrahydrate [2].

III

In aqueous solution the compound dissociates into $[C_4O_4]^{2-}$ anions (D_{4h} symmetry) and angled $[(CH_3)_2In \cdot 2\,H_2O]^+$ cations (C_{2v} symmetry). Also several vibrations of the C_4O_4 portion of the polymeric, water-free form show IR/Raman alternative behavior, resulting from a high localization symmetry in C_4O_4 [2].

The IR and Raman frequencies of the hydrated and anhydrous compounds are collected in Table 40 [2].

Table 40
Vibrational Spectra of Solid $(CH_3)_2InOOC_4OOIn(CH_3)_2 \cdot 4\,H_2O$ and $(CH_3)_2InOOC_4OOIn(CH_3)_2$ [2].
Wavenumbers in cm^{-1}.

$(CH_3)_2InOOC_4OOIn(CH_3)_2 \cdot 4\,H_2O$		$(CH_3)_2InOOC_4OOIn(CH_3)_2$		
IR	Raman	IR	Raman	assignment
3250 vs, br	–	–	–	$\nu_s, \nu_{as}(OH_2)$
–	1796 w	–	1808 w	$\nu_s(C \overline{\cdots} O)^{1)}$
1600 mw, br	–	–	–	$\delta_s(OH_2)$
–	1583 m	–	1594 m	$\nu_{as}(C \overline{\cdots} O)^{1)}$
–	–	–	1582 m	
–	–	1560 ms	–	$\nu_s(C \overline{\cdots} O)^{2)}$
1525 m, sh	–	1531 s	–	
1472 vs	–	1461 vs	–	$\nu_{as}(C \overline{\cdots} O)^{2)}$
1179 w	1179 vs	1161 w	1169 vs	$\delta_s(CH_3In)$
–	1140 s	1155 sh	1154 vs	$\nu(C \overline{\cdots} C)$
1096 ms	–	1118 w	1122 sh	
1081 ms	–	1098 m	–	
740 s	–	737 s, br	733 w, h	$\varrho(CH_3In)$
–	726 ms	–	746 s	$\nu_s(C_4)$-puls.

Table 40 (continued)

$(CH_3)_2InOOC_4OOIn(CH_3)_2 \cdot 4\,H_2O$ IR	Raman	$(CH_3)_2InOOC_4OOIn(CH_3)_2$ IR	Raman	assignment
–	644 s, br	–	658 s	$\delta(C \overset{...}{=} O)+$
–	–	640 sh	647 s	$\delta(C_4)$
560 s	560 vw	550 ms	548 mw	$\nu_{as}(InC_2)$
520 w	523 vw	–	–	?
500 vw	501 vs	500 w	495 vs	$\nu_s(InC_2)$
–	497 sh	–	–	?
395 mw	393 vw, sh	429 m	–	
362 ms	368 w, br	399 ms	–	$\delta(C \overset{...}{=} O)$
256 m	265 ms	270 m	–	

[1] In-phase. – [2] Out-of-phase.

$(C_2H_5)_2InOOC_4OOIn(C_2H_5)_2$

The ethyl derivative was prepared in 80% yield in the same way as the methyl compound. Details of the characterization, however, were not given [2].

References:

[1] Schwering, H.-U.; Hausen, H.-D.; Weidlein, J. (Z. Anorg. Allgem. Chem. **391** [1972] 97/102).

[2] Schwering, H.-U.; Olapinski, H.; Jungk, E.; Weidlein, J. (J. Organometal. Chem. **76** [1974] 315/24).

4.3 Dialkylindium Peroxides

$(CH_3)_2InOOCH_3$

This compound was formed quantitatively by O_2 oxidation of $In(CH_3)_3$ in n-heptane or n-nonane at −78 °C. The white crystalline compound decomposes above 70 °C without melting, but is stable for long periods at room temperature under O_2, N_2, or Ar [3]. The early work [1] describes the reaction of solid $In(CH_3)_3$ with O_2 at −78 °C. After 250 h, C_2H_6 was found in the gas phase along with unreacted O_2, while $(In(CH_3)_2)_2O$ was reported as the reaction product.

On warming the compound for 3 h in C_9H_{20} at 120 °C, CH_3OH, CH_4, and $(CH_3InO)_n$ formed in more than 50 mol% yield; small amounts of $(CH_3)_2InOCH_3$, $CH_3In(OCH_3)_2$, and $(CH_3)_2InOH$ were also detected. The decomposition is 1st-order with $k = 2.9 \times 10^6 \cdot e^{-15600/RT}\ s^{-1}$ [3].

The reaction with water led to the formation of CH_3OOH and $(CH_3)_2InOH$ [3].

$(CH_3)_2InOOC_4H_9$-t

The compound was prepared in 100% yield by the reaction of $In(CH_3)_3$ with $HOOC_4H_9$-t in n-heptane or n-nonane. The white crystalline solid is only moderately soluble in hydrocarbons and decomposes at temperatures over 70 °C without having melted. Decomposition in C_9H_{20} at 120 °C is 1st-order with $k = 1.1 \times 10^{12} \cdot e^{-26900/RT}\ s^{-1}$ [3]. This investigation was repeated, confirming the decomposition products by gas-liquid chromatography and iodometry [4].

$(C_2H_5)_2InOOC_2H_5$

The compound was formed in no more than 8% yield when liquid $In(C_2H_5)_3$ was oxidized by O_2. In the gas phase, formation occurred at 40 to 100 °C by a free radical mechanism with the initiation step being $In(C_2H_5)_3 + O_2 \rightarrow (C_2H_5)_2In^{\cdot} + C_2H_5OO^{\cdot}$ (wall reaction), followed by the propagation steps $(C_2H_5)_2In^{\cdot} + O_2 \rightarrow (C_2H_5)_2InOO^{\cdot}$ and $(C_2H_5)_2InOO^{\cdot} + In(C_2H_5)_3 \rightarrow (C_2H_5)_2InOOC_2H_5 + (C_2H_5)_2In^{\cdot}$. After forming by the (formal) insertion, the peroxide can undergo intramolecular and intermolecular decomposition reactions. Subsequent reactions of the above chain products, including fragmentation, led finally to $(C_2H_5InO)_n$, C_2H_4, and CH_3CHO (see also p. 63) [2].

$(C_4H_9\text{-t})_2InOOC_4H_9\text{-t}$

This compound was prepared in 60 to 70% yield by passing excess O_2 into a solution of $In(C_4H_9\text{-t})_3$ in pentane at 0 °C for about 1 min. It also formed as a by-product from the reaction of $In(C_4H_9\text{-t})_3$ with C_5H_5NO (main product is $(C_4H_9\text{-t})_2InOC_4H_9\text{-t}$). The white crystalline solid is stable at room temperature and gives no adduct with Lewis bases, e.g., pyridine [5].

^1H NMR spectrum (in C_6D_6, in ppm): 1.15 (C_4H_9O), 1.49 (C_4H_9In). The solid state IR spectrum exhibits the medium intense vibration of the peroxo group at 870 cm^{-1} [5].

The compound crystallizes in the monoclinic space group $C2/m - C^3_{2h}$ (No. 12) with the cell parameters (measured at −80 °C) a = 1674.5(10), b = 1145.3(9), c = 917.3(4) pm, β = 117.16(4)°; Z = 8 and D_c = 1.350 g/cm³. The compound is dimeric in the solid state, forming a four-membered In_2O_2 core as depicted in **Fig. 35**. No other distances and angles were reported [5].

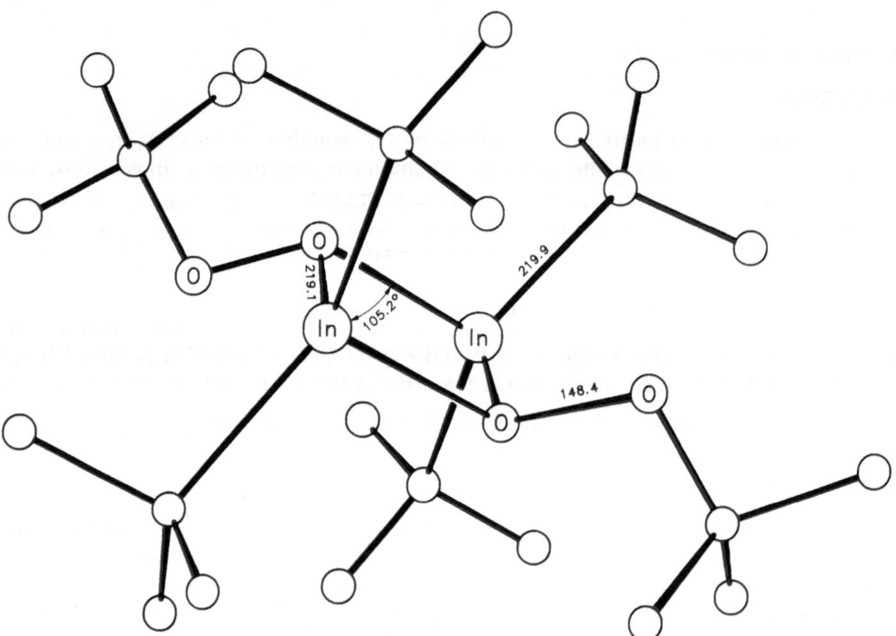

Fig. 35. Molecular structure of $(C_4H_9\text{-t})_2InOOC_4H_9\text{-t}$ [5].

References:

[1] Dennis, L. M.; Work, R. W.; Rochow, E. G. (J. Am. Chem. Soc. **56** [1934] 1047/9).

[2] Cullis, C. F.; Fish, A.; Pollard, R. T. (Trans. Faraday Soc. **60** [1964] 2224/33).

[3] Aleksandrov, Yu. A.; Chikinova, N. V.; Makin, G. I.; Kornilova, N. V.; Bregadze, V. I. (Zh. Obshch. Khim. **48** [1978] 467; J. Gen. Chem. [USSR] **48** [1978] 417).

[4] Shushunova, A. F.; Makin, G. I.; Chikinova, N. V.; Bryukhanov, A. N.; Aleksandrov, Yu. A. (Zh. Analit. Khim. **34** [1979] 1614/7; J. Anal. Chem. [USSR] **34** [1979] 1249/52).

[5] Cleaver, W. M.; Barron, A. R. (J. Am. Chem. Soc. **111** [1989] 8966/7).

4.4 $R_2InOE(X,R')_n$ Compounds with E = S, N, P, and As

This chapter is concerned with R_2In-oxygen compounds in which the O atom is bonded to an element E which is itself a donor to the indium, or (more often) is part of a ligand containing a donor (D = O, S, or N). The particular $OE(X,R')_n$ groups dealt with are OS(O)R′ (Nos. 1, 23), OS(O)OR′ (Nos. 2 to 4), $OS(O_2)R'$ (Nos. 5, 18, 19), $ON(O)=CR'_2$ (Nos. 6, 20, 21), $ON=CR'_2$ (Nos. 7, 8), ONR'_2 (No. 9), $OP(O)(X,R')_2$ (Nos. 10 to 14, 22), $OP(S)R'_2$ (No. 15), $OP(=NCH_3)R'_2$ (No. 16), and $OAs(O)R'_2$ (No. 17).

The compounds form heterocyclic ring systems through the donor atoms of the $OE(X,R')_n$ ligands; cryoscopic molecular weight determinations in benzene show that dimers (I and II) occur much more frequently than monomers (III).

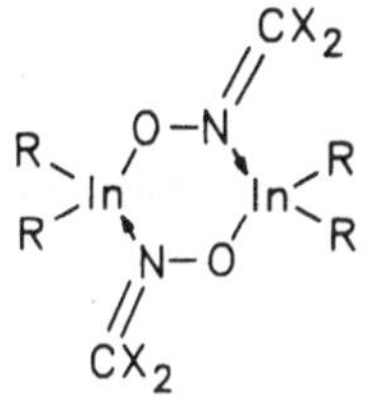

I 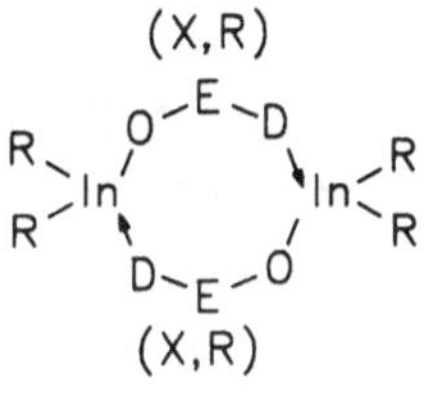 II 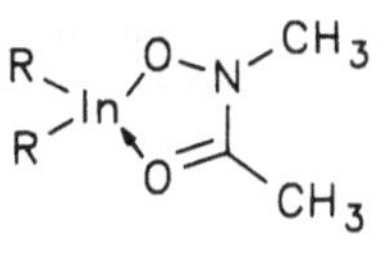 III

Preparations of the compounds in Table 41 were accomplished using the following methods.

Method I: From InR_3 and the proton acid, $HOE(X,R')_n$.

To a solution of the trialkylindium a slight excess (ca. 1 to 3%) of the dissolved acidic component was added dropwise with continuous stirring. C_6H_6 was a suitable solvent for most cases; CH_2Cl_2 for Nos. 5 and 17, pentane and ether for Nos. 6, 13, 20, and 21, hexane for No. 7, and toluene for No. 8. Depending on the acidity of the acid component, reaction temperatures were: −30 to −20 °C (Nos. 5 and 18), 0 °C (Nos. 6, 20, and 21), 5 °C (No. 11), 10 to 15 °C (Nos. 10 and 12), and 20 to 40 °C for the others. For Nos. 7 and 8 the reactants were mixed at −78 °C, but the reactions did not begin until they had been slowly warmed. After admixture, each solution was warmed at 30 to 40 °C with continuous stirring for 0.5 to 2 h to complete the reaction. The solvent was evaporated, and the product was purified by sublimation or recrystallization.

Method II: Insertion of SO_2 or SO_3 into In–C (a) or In–O (b) bonds.

a. SO_2 (diluted 3:1 with pure nitrogen) was condensed into the dilute hexane or pentane solution of $In(CH_3)_3$ at −50 °C. The end of the reaction (formation of No. 1) was recognized by the appearance of a second (denser) phase of liquid SO_2 [5]. If the reactants are co-condensed at −50 °C without solvent, the excess SO_2 must be removed in vacuum at −78 °C after a short reaction period (ca. 1 min); otherwise, polymeric $In(OOSCH_3)_3$ forms [6]. Compound No. 23 was obtained by condensing excess SO_2 at −196 °C and then warming to −45 °C. After 30 min, the residual SO_2 was removed in vacuum at −55 °C. Nos. 5 and 19 were prepared in CH_2Cl_2 at −50 to −30 °C from equimolar amounts of reactants. On warming the solution the sulfonates separate as difficultly soluble solids.

b. Compounds No. 2 to 4 result from the insertion of SO_2 (without solvent) into the In–O bond of an alcoholate, R_2InOR', not into the In–C bond. The compounds precipitate as white solids on evaporation of the large excess of SO_2 from the initially clear solution.

General Remarks. Complete and evaluated IR and Raman data were presented for most of the compounds. The absorption frequencies of Nos. 10 to 13 are detailed in Table 42, p. 219; for the remaining compounds only the most important vibrations are cited in Table 41.

The nitromethane derivatives No. 6, 20, and 21 were first reported to be of the $R_2In(^1L{-}D)$ type described in Section 1.1.6 [7]; IR investigations, however, have shown that sp^2 carbon atoms were present and the coordination to the R_2In group occurs through the oxygen atoms of the nitro group [16].

Table 41
Compounds of the $R_2InOE(X,R)_n$ Type with E=S, N, P, and As.
Further information on compounds preceded by an asterisk is given at the end of the table.
Explanations, abbreviations, and units on p. X.

No.	compound method of preparation (yield)	properties and remarks
$(CH_3)_2In$ compounds		
1	$(CH_3)_2InOOSCH_3$ IIa (90%)	colorless crystals, m.p. ca. 180 °C (dec.), subl. 145 °C/10^{-4} Torr [5] ^1H NMR: −0.03 (CH_3In), 2.07 (CH_3S) [6] IR (solid): 1030 vs $\nu_{as}(SO_2)$, 999 vs, 975 sh, 951 s, 932 m, 895 s $\nu_s(SO_2)$ and $\varrho(CH_3S)$, 717 s $\varrho(CH_3In)$, 681 s ν(C–S), 548 s, 535 sh $\nu_{as}(InC_2)$ and $\delta(SO_2)$, 491 ms $\nu_s(InC_2)$ and $\delta(SO_2)$, 411 ms ν(InO) [5] Raman (solid): 1025 mw, 993 ms $\nu_{as}(SO_2)$, 929 mw, 912 mw $\nu_s(SO_2)$ and $\varrho(CH_3S)$, 679 s ν(C–S), 545 w $\nu_{as}(InC_2)$, 490 vs $\nu_s(InC_2)$, 435 w, 404 mw ν(InO) [5] mass spectrum: $[M_2]^+$, $[M_2-CH_3]^+$, $[M_2-2\,CH_3]^+$, $[M_2-3\,CH_3]^+$, $[M_2-4\,CH_3]^+$, $[M_2-CH_3SO_2]^+$, $[M+In]^+$, $[M]^+$, $[M-CH_3]^+$, $[M-2\,CH_3]^+$, $[InSO_2]^+$, $[InSO]^+$, $[In(CH_3)_2]^+$, $[InCH_3]^+$, $^{115}In^+$ [5]; the peaks $[M+CH_3S_3O_6]^+$ and $[M+CH_3S_2O_4]^+$ were only reported in [6]
2	$(CH_3)_2InOOSOCH_3$ IIb (87%)	white solid, m.p. 140 °C (dec.), subl. 80 °C/10^{-3} Torr [6, 10]

Table 41 (continued)

No.	compound method of preparation (yield)	properties and remarks
2 (continued)		IR (solid): 1418 ms, sh, 1408 ms, 1301 m, 1070 s, 966 vvs, br, 723 s, sh, 702 vs, 570 m, sh, 528 s, br, 471 m, 422 w, 406 m [6] Raman (solid): 1290 vw $\delta_s(CH_3O)$, 1177 s $\delta_s(CH_3In)$, 1098 m, br, 1055 m $\nu_{as}(SO_2)$, 991 ms, 935 mw $\nu_s(SO_2)$ and ν(O-C) and $\varrho(CH_3O)$, 710 m, 685 ms $\varrho(CH_3In)$ and ν(S-O), 640 w, 592 vw $\delta(SO_2)$, 550 s $\nu_{as}(InC_2)$, 500 vvs $\nu_s(InC_2)$, 448 sh δ(SOC) and $\delta(SO_2)$ and ν(InO?), 432 m, 405 w, other bands for ν and $\delta(CH_3)$ reported [10] mass spectrum (100 °C): $[M_2-(CH_3)_2O]^+$, $[M_2-(CH_3)_2SO_3]^+$, $[M_2-(CH_3)_2SO_4]^+$, $[M]^+$, $[M-CH_3]^+$, $[M-(CH_3)_2O]^+$, $[In(CH_3)_2]^+$, $[InCH_3]^+$, $^{115}In^+$ [6]
3	$(CH_3)_2InOOSOC_2H_5$ IIb (85%)	IR (solid): 1475 s, 1460 m, 1408 mw, 1381 s, 1297 w, sh, 1292 ms, 1169 s, 1162 m, sh, 1115 m, sh, 1094 s, 1062 s, sh, 1045 vs, 999 vvs, 887 vs, 864 s, sh, 820 s, 724 vs, br, 560 m, sh, 532 s, br, 485 m, 466 m, 445 w, 423 m, 402 m [6]
4	$(CH_3)_2InOOSOC_6H_5$ IIb (90%)	white solid [6] IR (solid): 1247 s, sh, 1228 vs $\nu_{as}(SO_3)$, 1027 s, sh, 1001 vs $\nu_s(SO_3)$, other bands at 1594 s, 1170 ms, 1157 m, 1073 m, 932 s, 897 s, 853 s, 769 vs, 758 s, sh, 726 s, 697 vs, 624 w, 576 ms, 554 s, 518 m, 495 mw, 414 ms, 377 mw [6] mass spectrum (no data reported) indicates a monomeric structure in the gas phase [6]
5	$(CH_3)_2InOOS(O)CH_3$ I (93%) IIa (77%)	white solid, m.p. 304 °C (dec.), subl. 200 °C/10^{-4} Torr; dissociates in H_2O [9, 11] ^{1}H NMR (D_2O): −0.07 (CH_3In), 2.78 (CH_3S) [9, 11] IR (solid): 1180 vs, br $\nu_{as}(SO_3)$ and $\delta_s(CH_3In)$, 1162 ms $\nu_s(SO_3)$, 782 m ν(C-S), 571 m $\delta_s(SO_3)$, 552 s $\nu_{as}(InC_2)$, 517 m $\nu_s(InC_2)$ and $\delta_{as}(SO_3)$, 340 w δ(CSO) [11] Raman (solid): 1172 vs, 1167 mw, sh $\nu_{as}(SO_3)$ and $\delta_s(CH_3In)$, 1063 s $\nu_s(SO_3)$, 970 mw $\varrho(CH_3S)$, 783 m ν(C-S), 525 m $\nu_{as}(InC_2)$ and $\delta_s(SO_3)$, 516 sh $\delta_{as}(SO_3)$, 510 vs $\nu_s(InC_2)$, 384 w δ(CSO); frequencies for D_2O solution reported, too [11]
6	$(CH_3)_2InOON{=}CH_2$ I	^{115}In NQR (77 K): 73.07 (ν_2), 115.28 (ν_3), e^2Qq/h 945 ± 2 MHz, $\eta = 34.8 \pm 1.5\%$ [7] proposed structure, see general remarks [16]

Table 41 (continued)

No.	compound method of preparation (yield)	properties and remarks
7	$(CH_3)_2InON{=}C(CH_3)_2$ I	m.p. 57 °C, b.p. 85 °C/vacuum; dimeric in benzene [2] ^{1}H NMR: 0.03 (s, $InCH_3$), 1.90 (s, CH_3C) in CCl_4 and −0.16, 1.65, and 1.78 (s's) in C_6H_6, no significant changes between −60 and +95 °C [2] IR: 1612 m, 1366 vs, 1269 sh, 1261 m, 1156 m $\delta_s(CH_3In)$, 1076 vs, 1064 sh, 1005 sh, 985 vs, 952 s, 927 s, 810 w, 704 vs $\varrho(CH_3In)$, 646 s, 631 sh, 564 w, br, 524 s, 510 s $\nu_{as}(InC_2)$, 473 m $\nu_s(InC_2)$ as solid, 429 sh, 329 s, 323 w, 255 m, br in C_6H_6 [2] proposed structure, see Formula I on p. 211
*8	$(CH_3)_2InON{=}CH(C_5H_4N\text{-}2)$ I	pale yellow crystals, m.p. 164 °C [3] ^{1}H NMR (C_6D_6): 0.13 (CH_3In), 7.0 to 8.0 (C_6H_4N), δ(CH) not observed, but syn-configuration of the oxime postulated [3] IR (solid): 1601 s (C_5H_4N), 1535 ms ν(C=N), 1159 w $\delta_s(CH_3In)$, 1091 vs ν(N–O), 704 ms $\varrho(CH_3In)$, 516 m $\nu_{as}(InC_2)$, 482 w $\nu_s(InC_2)$; other, unassigned bands at 1581 sh, 1526 ms, 1481 m, 1340 m, 1304 vw, 1264 vw, 1230 ms, 1111 s, 1012 m, 899 s, 887 w, 788 ms, 773 ms, 752 vw, 686 s, 668 sh, 635 s, 525 s [3]
9	$(CH_3)_2InON(CH_3)C(O)CH_3$ I (66%)	crystalline solid, m.p. 163 to 166 °C, subl. 100 to 105 °C/10^{-4} Torr [15] ^{1}H NMR ($CDCl_3$): −0.13 (CH_3In), 2.13 (CH_3C), 3.43 (CH_3N) [15] IR (solid): 1609 vs ν(C=O), 1450 mw ν(C–N), 785 ms ν(N–O), no other bands reported [15]
10	$(CH_3)_2InOOPH_2$ I (85%)	white solid, m.p. 118 °C (dec.), subl. 59 °C/10^{-4} Torr, dimeric in benzene [12] ^{1}H NMR (CCl_4): −0.02 (CH_3In), 7.1 (d, HP, J(P,H) = 545) [12] ^{31}P NMR (CCl_4): −4.5 (J(H,P) = 545) [12] IR and Raman frequencies in Table 42 on p. 219
11	$(CH_3)_2InOOPF_2$ I (82%)	white solid, m.p. 46 °C, subl. 45 °C/10^{-4} Torr; dimeric in benzene [12] ^{1}H NMR (CCl_4): 0.13 (CH_3In) [12] ^{19}F NMR (CCl_4): 4.1 (d, J(F,P) = 968) [12] ^{31}P NMR (CCl_4): 22.8 (d, J(P,F) = 967) [12] IR and Raman frequencies in Table 42 on p. 219 forms a stable adduct with C_5H_5N [12]
12	$(CH_3)_2InOOPCl_2$ I (80%)	white solid, m.p. 96 to 98 °C (dec.), subl. 80 °C/10^{-4} Torr, dimeric in benzene [12] ^{1}H NMR (ether): −0.13 (CH_3In) [12]

Table 41 (continued)

No.	compound method of preparation (yield)	properties and remarks
12 (continued)		^{31}P NMR (ether): 1.1 [12] IR and Raman frequencies in Table 42 on p. 219 forms a stable adduct with C_5H_5N [12]
*13	$(CH_3)_2InOOP(CH_3)_2$ I	white solid, m.p. 75 to 76 °C, subl. 60 °C/10^{-3} Torr, 60 to 76 °C/0.01 Torr [1], dimeric in benzene [8] ^{1}H NMR (CCl_4): −0.19 (CH_3In), 1.33 (d, CH_3P, $^2J(P,H) = 14.5$) [12] ^{31}P NMR (CCl_4): −42.7 (m, $^2J(H,P) = 14.5$) [12] IR and Raman frequencies in Table 42 on p. 219
14	$(CH_3)_2InOOP(C_6H_5)_2$	from $(CH_3)_2InP(C_6H_5)_2$ and C_5H_5NO in C_6H_6(1:2 mole ratio, 91% yield) [19] colorless crystals, dimeric in C_6H_6 [19] ^{1}H NMR (C_6D_6): 0.02 (CH_3In), 7.01, 8.41 (m's, C_6H_5) [19] IR (solid): 3040 m, 2290 w, 1975 w, 1900 w, 1820 w, 1760 w, 1590 m, 1485 m, 1190 vs, 1185 sh, 1130 vs, 1075 s, 935 m, 790 s, 720 s, 690 s, 620 w, 575 m [19]
15	$(CH_3)_2InOSP(CH_3)_2$ I (60%)	white solid, m.p. 125 to 127 °C, subl. 55 °C/10^{-4} Torr; degree of association in benzene 1.66 [12] ^{1}H NMR (CCl_4): −0.05 (CH_3In), 1.87 (d, CH_3P, $^2J(H,P) = 13.4$) [12] ^{31}P NMR (CCl_4): −69.2 (m, $J(P,H) = 13.5$) [12] IR (solid, main frequencies): 1160 mw, br $\delta_s(CH_3In)$, 1089 sh, 1073 vs $\nu(P \overset{...}{=} O)$, 749 ms $\nu_{as}(PC_2)$, 723 s $\nu_s(PC_2)$ and $\varrho(CH_3In)$, 556 ms $\nu(P \overset{...}{=} S)$, 531 ms $\nu_{as}(InC_2)$, 492 mw $\nu_s(InC_2)$ and δ(OPS) [13] Raman (solid, main frequencies): 1163 vs, 1159 sh $\delta_s(CH_3In)$, 1090 sh, 1076 mw $\nu(P \overset{...}{=} O)$, 751 mw $\nu_{as}(PC_2)$, 719 m $\nu_s(PC_2)$, 558 s $\nu(P \overset{...}{=} S)$, 530 m $\nu_{as}(InC_2)$, 499 vvs $\nu_s(InC_2)$ and δ(OPS), 231, 208 w $\delta_s(PC_2)$, 192 mw ν(InS) [13]
16	$(CH_3)_2InOP(=NCH_3)(CH_3)_2$ I (60 to 70%)	white solid, m.p. 72 to 74 °C; dimeric in benzene [17] ^{1}H NMR (C_6D_6): 0.16 (s, CH_3In), 1.17 (d, CH_3P, $^2J(H,P) = 13.18$), 2.62 (d, CH_3N, $^3J(H,P) = 17.95$) [17] ^{31}P NMR (C_6D_6): −52.3 m [17] IR (solid): 1178 ms, br $\delta_s(CH_3In)$, 1080 vs, br $\nu_{as}(O \overset{...}{=} P \overset{...}{=} N)$, 1060 m, br $\varrho(CH_3N)$ and $\nu(NCH_3)$, 949 s, 919 vw $\varrho(CH_3P)$, 878 ms $\nu_s(O \overset{...}{=} P \overset{...}{=} N?)$, 866 m $\varrho(CH_3P)$, 841 mw, 748 m $\nu_{as}(PC_2)$, 688 vs, br $\varrho(CH_3In)$, 510 m $\nu_{as}(InC_2)$, 490 sh $\nu_s(InC_2)$, 481 mw ν(InN), 457 m δ(PNC), 325 vw δ(OPN) and $\delta(PC_2)$ [17]
17	$(CH_3)_2InOOAs(CH_3)_2$ I (>90%)	white solid, m.p. 168 °C, subl. 100 °C/10^{-3} Torr [8], 160 °C/1 Torr [4]; dimeric in benzene [8] ^{1}H NMR (CCl_4): −0.22 (CH_3In), 1.63 (CH_3As) [4]

Table 41 (continued)

No.	compound method of preparation (yield)	properties and remarks
17 (continued)		IR (solid): 1414 mw $\delta_{as}(CH_3)$, 1271, 1261 s $\delta_s(CH_3As)$, 1157, 1148 m $\delta_s(CH_3In)$, 905 sh $\varrho(CH_3As)$, 894, 876 vs $\nu_{as}(AsO_2)$, 823, 801 vs $\nu_s(AsO_2)$, 701 vs, br $\varrho(CH_3In)$, 640 ms $\nu_{as}(AsC_2)$, 630 s, 604 mw $\nu_s(AsC_2)$, 525 s $\nu_{as}(InC_2)$, 484 m $\nu_s(InC_2)$, 425 s $\nu_{as}(InO_2)$, 375 sh, 356 m $\delta(OAsC)$ and $\delta(AsC_2)$, 331 mw, 280 m [8] Raman (solid): 1420 vw, 1412 w $\delta_{as}(CH_3)$, 1268 vw $\delta_s(CH_3As)$, 1160, 1153 m $\delta_s(CH_3In)$, 889 vvw $\nu_{as}(AsO_2)$, 825 w $\nu_s(AsO_2)$, 645 mw $\nu_{as}(AsC_2)$, 606 m $\nu_s(AsC_2)$, 531 mw $\nu_{as}(InC_2)$, 492 vs $\nu_s(InC_2)$, 450 w $\delta_s(AsO_2)$, 420 vw $\nu_s(InO_2?)$, 357 w $\delta(OAsC)$, 320 vw, 273 vw, 220 s $\delta_s(AsC_2)$, 125 s, br $\delta(InC_2)$ [8]
$(C_2H_5)_2In$ compounds		
18	$(C_2H_5)_2InOOS(O)CH_3$ I (90%) [11]	white solid, m.p. 159 °C (dec.), subl. 140 °C/10^{-4} Torr [9] soluble in cold water, degree of association: 0.5 [9] IR (solid): 1240 sh $\nu(S{=}O)$, 1189 vs, 1160 m, sh $\nu_{as}(SO_2)$ and $\omega(CH_2In)$, 1050 s, 1040 ms $\nu_s(SO_3)$ and $\nu_s(SO_2)$, 781 ms $\nu(C{-}S)$, 563 ms $\delta(SO_3)$, 518 ms $\nu_{as}(InC_2)$, 472 w $\nu_s(InC_2)$, 337 w $\delta(CSO)$ [9] Raman (solid): 1230 vw, sh $\nu(S{=}O)$, 1180 s, 1160 w, sh $\nu_{as}(SO_2)$ and $\omega(CH_2In)$, 1055 s $\nu_s(SO_3)$, 783 m $\nu(C{-}S)$, 568 m, br $\delta(SO_3)$, 520 w $\nu_{as}(InC_2)$, 474 vs $\nu_s(InC_2)$, 338 w $\delta(CSO)$; IR and Raman for H_2O solution (partial) [9]
19	$(C_2H_5)_2InOOS(O)C_2H_5$ IIa (<50%)	white solid, dec. <120 °C; no other properties reported [9]
20	$(C_2H_5)_2InOON{=}CH_2$ I (25%)	white solid, explodes ca. 100 °C [7] IR (solid): 3174 s, 3047 m $\nu(CH_2)$, 1580 $\nu(C{=}N)$, 1255 $\nu_{as}(NO_2)$, 950 $\nu_s(NO_2)$ [16] ^{115}In NQR (77 K): 75.09 (ν_2), 118.50 (ν_3), e^2Qq/h 970±1 MHz, $\eta = 34 \pm 0.5\%$ [7] proposed structure, see general remarks [16] decomposition at 100 °C (?) in an autoclave over a period of 6 h yields (among other products) $(C_2H_5)_2InONO$ [17]
21	$(C_2H_5)_2InOON{=}C(CH_3)_2$ I (75%)	white solid, m.p. 123 to 125 °C [7] ^{115}In NQR (77 K): 72.36 (ν_1), 74.25 (ν_2), 117.0 (ν_3), 158.6 (ν_4), e^2Qq/h 962.6±1.0 MHz, $\eta = 36.7 \pm 0.1\%$ [7] proposed structure, see general remarks [16] decomposition in boiling isopropylbenzene at 152 °C, the presence of $(CH_3)_2C{=}C(CH_3)_2$ showed by gas-liquid chromatographic analyses [7]

Table 41 (continued)

No.	compound method of preparation (yield)	properties and remarks
*22	$(C_2H_5)_2InOOP(C_6H_5)_2$ I	colorless crystals [20] ^{1}H NMR (CD_2Cl_2): 0.64 (q, CH_2In), 1.16 (t, CH_3), 7.31 to 7.90 (m, C_6H_5) [20] ^{13}C NMR (CD_2Cl_2): 9.73 (CH_2In), 11.45 (CH_3), 128.48 (C-3,5, ^3J(C,P) = 12.86), 131.25 (C-4, ^4J(C,P) = 2.74), 131.66 (C-2,6, ^2J(C,P) = 10.10), 137.13 (CP, ^1J(C,P) = 140.33) [20] ^{31}P NMR (CD_2Cl_2): 24.20 [20]
$(C_6H_5)_2In$ compounds		
23	$(C_6H_5)_2InOOSC_6H_5$ IIa (98%)	white unstable solid, slowly loses SO_2 in vacuum and more rapidly when heated [6] soluble in common organic solvents with rapid formation of SO_2 [6] IR (solid): 1053 ms $\nu_{as}(SO_2)$, 854 ms $\nu_s(SO_2)$, 586 ms $\delta(SO_2)$, other bands at 1579 w, 1478 m, 1433 s, 1201 w, 1087 ms, 1071 m, 1030 s, 1003 m, 913 w, 740 vvs, 700 vs, 685 m, sh, 620 m, 473 s, sh, 469 vs [6] presumably monomeric structure [6]

* Further information:

$(CH_3)_2InON{=}CH(C_5H_4N\text{-}2)$ (Table **41**, No. **8**) crystallizes in the orthorhombic space group Pbcn $-D_{2h}^{14}$ (No. 60) with the parameters a = 3315(1), b = 954(1), c = 1430(1) pm; the unit cell contains 8 dimers and 4 intercalated benzene molecules, D_c = 1.62 g/cm^3 and D_m = 1.61 g/cm^3. The heterocyclic ring system is not planar, but is folded about 55° along the O(1)-O(2) axis. However, both pyridine rings are planar (within the limits of error) and atoms In(2), N(2), O(1), and O(2) also lie within a plane, as do In(1), N(1), O(1), and O(2). The pyridine-2-carbaldehyde oxime residues act as tridentate ligands, each indium atom possessing a distorted trigonal-bipyramidal environment. The most important bond distances and angles are cited in **Fig 36** [18].

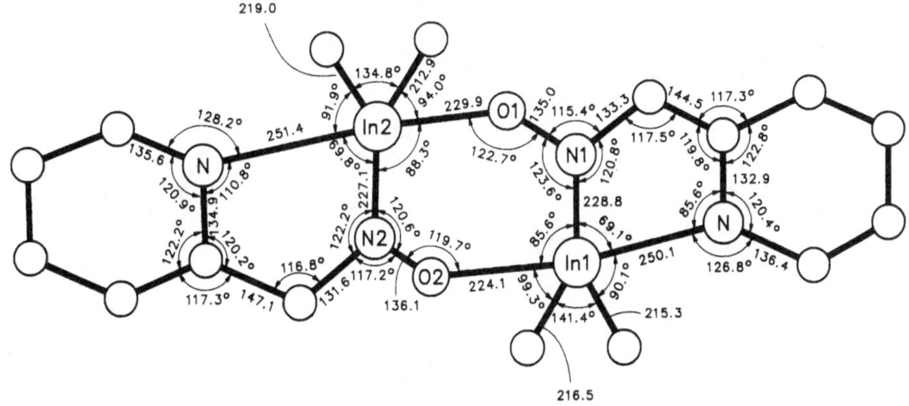

Fig. 36. Molecular structure of dimeric $(CH_3)_2InON{=}CH(C_5H_4N\text{-}2)$ [18].

$(CH_3)_2InOOP(CH_3)_2$ (Table **41**, No. **13**) is a dimer and possesses a nonplanar eight-membered ring structure as shown in Formula II on p. 211. To clarify the exact structural relationships the IR and Raman spectra of solid, molten, and dissolved (in CCl_4) samples were studied intensively in the region of the significant PO_2 bond vibrations. The distinctly different frequency values cited below were interpreted as representing a change in the eight-membered ring configuration. The solid phase and dilute solutions each contain a different centrosymmetric form (D_{2h}). In the spectrum of the melt, superposition of the solid and solution bands is obvious, but a third, acentric ring configuration (D_2) was considered probable [8].

IR solid	Raman	IR liquid	Raman	IR CCl_4 solution	Raman	assignment
1103 vs	–	1150 s	–	1156 vs	–	$\nu_{as}(PO_2)^{1)}$
–	1128 mw	1132 vs	1137 sh	–	1120 w	$\nu_{as}(PO_2)^{2)}$
–	1042 mw	–	1077 m	–	1085 m	$\nu_s(PO_2)^{2)}$
1054 vs	–	1063 vs	1060 sh	1052 vs	–	$\nu_s(PO_2)^{1)}$
522 m	–	$(490)^{3)}$	$(494)^{3)}$	478 mw	–	$\delta_s(PO_2)$
421 m	–	445 m, br	450 w	440 m	–	$\delta(CPO)$

1) Out-of-phase. – 2) In-phase. – 3) Hidden by $\nu_s(InC_2)$.

$(C_2H_5)_2InOOP(C_6H_5)_2$ (Table **41**, No. **22**) crystallizes in the monoclinic space group $C2/c-C^6_{2h}$ (No. 15) with the cell constants a = 1997.1(5), b = 859.0(2), c = 1945.7(5) pm, β = 102.91(2)°; Z = 8 (4 dimeric units) gives D_c = 1.59 g/cm³. Important angles and distances are given in **Fig. 37** [20].

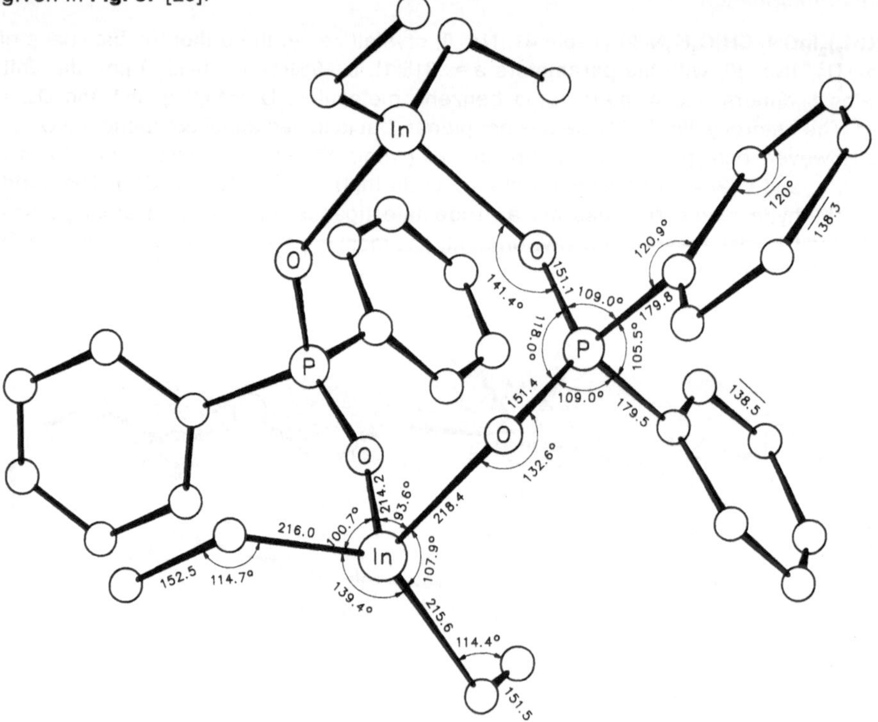

Fig. 37. Molecular structure of dimeric $(C_2H_5)_2InOOP(C_6H_5)_2$ [20].

Table 42
IR and Raman Frequencies of Solid $(CH_3)_2InOOPX_2$ Compounds, with X = H, F, Cl, CH_3.
Wavenumbers in cm^{-1}.

$(CH_3)_2InOOPH_2$ [13]		$(CH_3)_2InOOPF_2$ [14]		$(CH_3)_2InOOPCl_2$ [14]		$(CH_3)_2InOOP(CH_3)_2$ [8]		
IR	Raman	IR	Raman	IR	Raman	IR	Raman	assignment
3010 sh	3002 w, br	3000 w, br	3003 mw	3002 w, br	3013 w, br	2992 m	3000 m	$\nu_{as}(CH_3In)$,
2983 mw	2993 w, br	–	2950 sh	–	2946 sh	2942 m	–	$\nu_{as}(CH_3P)$
2925 mw, br	2937 mw	2928 w, br	2930 s	2922 mw	2937 ms	–	2933 vs	$\nu_s(CH_3In)$,
–	–	–	–	–	–	2900 sh	2895 m	$\nu_s(CH_3P)$
–	2842 vvw	–	2870 vw	2880 vvw	2880 vw, sh	–	2825 w	$2\times\delta(CH_3)$
2405 w	2413 mw	–	–	–	–	–	–	$\nu_{as}(PH_2)$
2391 ms	2391 ms	–	–	–	–	–	–	$\nu_s(PH_2)$
–	–	1435 vw	–	1440 vw	–	1427 m	1428 mw	$\delta_{as}(CH_3In)$
–	–	–	–	–	–	1416 m	1418 mw	$\delta_{as}(CH_3P)$
–	–	–	–	–	–	1305 s, 1298 s	1300 vvw	$\delta_s(CH_3P)$
1169 vs	–	1275 vs, br	–	1220 s	–	1103 vs	–	$\nu_{as}(PO_2)^{1)}$
1157 vs	(1159)	–	1266 vw	–	1192 mw	–	1128 mw	$\nu_{as}(PO_2)^{2)}$
(1169)	1167 s	1179 w	1181 vs	1180 mw	1118 m	1176 w	1176 vs	$\delta_s(CH_3In)$
–	1159 s	–	–	–	1170 m	–	–	
1145 m	1148 mw	–	–	–	–	–	–	$\delta_s(PH_2)$
–	1050 m	–	1145 mw	–	1074 w	–	1042 mw	$\nu_s(PO_2)^{2)}$
1042 m	–	1128 vs	–	1067 vs	–	1054 vs	–	$\nu_s(PO_2)^{1)}$
1086 s	1091 w, br	–	–	–	–	–	–	$\omega(PH_2)$
–	937 mw	–	–	–	–	927 w	927 w	$\tau(PH_2)$ or $\varrho(CH_3P)$
–	–	925 vs	930 mw	–	–	–	–	$\nu_{as}(PF_2)$
–	–	878 s	883 s	–	–	–	–	$\nu_s(PF_2)$
814 s	825 vw, br	–	–	–	–	–	–	$\varrho(PH_2)$
–	–	–	–	–	–	746 ms	749 mw	$\nu_{as}(PC_2)$
723 ms, br	697 vw, br	735 s, br	706 vw	740 vs, br	718 vw	725 ms	684 vw	$\varrho(CH_3In)$
–	–	–	–	–	–	695 sh	712 s	$\nu_s(PC_2)$
–	–	–	–	590 s	590 vw, br	–	–	$\nu_{as}(PCl_2)$
–	–	–	–	561 ms	562 sh	–	–	$\nu_s(PCl_2)$
549 mw	549 mw	559 ms	559 w	551 m	555 w	530 ms	532 m	$\nu_{as}(InC_2)$

Table 42 (continued)

$(CH_3)_2InOOPH_2$ [13]		$(CH_3)_2InOOPF_2$ [14]		$(CH_3)_2InOOPCl_2$ [14]		$(CH_3)_2InOOP(CH_3)_2$ [8]		
IR	Raman	IR	Raman	IR	Raman	IR	Raman	assignment
–	–	535 m	532 w	–	–	–	–	$\delta_s(PO_2)$ and $\delta(OPF)$
498 w	500 vvw	500 sh	501 vs	496 m	500 vs	490 m	494 vs	$\nu_s(InC_2)$
460 mw	458 w	–	–	422 mw	418 vw	421 m	454 vw ?	$\delta_s(PO_2)$ and $\delta(OPX)$
–	–	–	–	401 vw	404 w	–	390 w	
290 sh	289 w	373 mw	383 w	–	307 mw	–	342 mw	$\delta(OPX)$ and $\nu(InO)$
–	–	–	357 vw	–	287 w, sh	–	322 w	
–	–	–	–	–	–	–	287 sh	$\delta(InC_2$?)
–	–	–	–	–	–	–	276 mw	$\delta_{as}(PC_2)$
–	–	–	200 sh	–	230 w	–	205 mw	$\delta_s(PX_2)$
–	–	–	145 sh	–	144 sh	–	141 m	$\delta(InC_2)$ $\delta(InOC)$
–	123 s	–	124 s	–	111 s	–	115 s	
–	115 ms	–	115 s	–	101 s	–	–	

[1] Out-of-phase. – [2] In-phase.

References:

[1] Coates, G. E.; Mukherjee, R. N. (J. Chem. Soc. **1964** 1295/303).
[2] Jennings, J. R.; Wade, K. (J. Chem. Soc. A **1967** 1333/9).
[3] Pattison, I.; Wade, K. (J. Chem. Soc. A **1968** 2618/22).
[4] Schmidbaur, H.; Kammel, G. (J. Organometal. Chem. **14** [1968] P 28/P 29).
[5] Weidlein, J. (J. Organometal. Chem. **24** [1970] 63/75).
[6] Hsieh, A. T. T. (J. Organometal. Chem. **27** [1971] 293/301).
[7] Golubinskaya, L. M.; Bregadze, V. I; Bryuchova, E. V.; Svergun, V. I.; Semin, G. K.; Okhlobystin, O. Yu. (J. Organometal. Chem. **40** [1972] 275/9).
[8] Olapinski, H.; Schaible, B.; Weidlein, J. (J. Organometal. Chem. **43** [1972] 107/16).
[9] Olapinski, H.; Weidlein, J. (J. Organometal. Chem. **35** [1972] C 53/C 55).
[10] Mann, W. G. (Diss. Univ. Stuttgart 1974).

[11] Olapinski, H.; Weidlein, J.; Hausen, H.-D. (J. Organometal. Chem. **64** [1974] 193/204).
[12] Schaible, B.; Haubold, W.; Weidlein, J. (Z. Anorg. Allgem. Chem. **403** [1974] 289/300).
[13] Schaible, B.; Roessel, K.; Weidlein, J.; Hausen, H.-D. (Z. Anorg. Allgem. Chem. **409** [1974] 176/84).
[14] Schaible, B.; Weidlein, J. (Z. Anorg. Allgem. Chem. **403** [1974] 301/9).
[15] Schwering, H.-U.; Weidlein, J. (J. Organometal. Chem. **99** [1975] 223/30).
[16] Leites, L. A.; Kurbakova, A. P.; Golubinskaya, L. M.; Bregadze, V. I. (J. Organometal. Chem. **122** [1976] 1/4).
[17] Schrem, H.; Weidlein, J. (Z. Anorg. Allgem. Chem. **465** [1980] 109/19).
[18] Shearer, H. M. M.; Twiss, J.; Wade, K. (J. Organometal. Chem. **184** [1980] 309/16).
[19] Arif, A. M.; Barron, A. R. (Polyhedron **7** [1988] 2091/4).
[20] Hahn, F. E.; Schneider, B.; Reier, F.-W. (Z. Naturforsch. **45b** [1990] 134/40).

4.5 R_2In-Compounds of Inorganic Oxygen Acids

The perchlorate and the nitrate were prepared, investigated, and handled only in cold aqueous or methanolic solutions.

$(CH_3)_2InClO_4$

To a solution of $(CH_3)_2InCl$ in water at 0 °C was added an equally cold solution of $TlClO_4$, the precipitate of TlCl was filtered off, and the remaining clear solution was concentrated to saturation in vacuum. (Author: the attempt to isolate the compound by further removal of solvent and to dry it at 0 °C in vacuum led to a violent explosion! Metallic In was found on the glass shards.) The similarly conducted reaction with $AgClO_4$, on the other hand, led to oxidative decomposition of the $(CH_3)_2InCl$ [1].

The Raman spectrum of a concentrated aqueous solution of $(CH_3)_2InClO_4$ is illustrated in [1]; it shows the lines of the (hydrated) ClO_4^- and $[In(CH_3)_2]^+$ ions. Because only the symmetric but not the asymmetric InC_2 vibrations could be observed for the cation, it was postulated (Author: erroneously) to be linear. The observed frequencies (in cm^{-1}) are: 3003 w $\nu_{as}(CH_3)$, 2928 w $\nu_s(CH_3)$, 1183 m $\delta_s(CH_3)$, 502 vvs $\nu_s(InC_2)$, 936 $\nu_s(ClO_4)$, 628 and 468 w $\delta(ClO_4)$. Using the IR and Raman data, force constants were calculated for the $[In(CH_3)_2]^+$ assuming D_{3d} symmetry (see also p. 123) [1].

$(CH_3)_2InNO_3$

This compound occurred on reaction of methanolic $(CH_3)_2InI$ with an equimolar amount of aqueous $TlNO_3$. After separating the precipitated TlI, the water-methanol solution of

$(CH_3)_2InNO_3$ was used to synthesize various adducts (p. 225); however, the compound itself was not characterized [5].

$\{(CH_3)_2In\}_2SO_4$

The sulfate was obtained in yields of 95 to 100% by addition of an emulsion of 100% H_2SO_4 in CH_2Cl_2 to a solution of $In(CH_3)_3$ in pentane at −50 to −40 °C. The vigorously stirred mixture was slowly brought to room temperature and finally stirred for about an hour at 30 to 40 °C. The compound is insoluble in nonpolar organic solvents, but is slightly soluble in pyridine, acetone, acetonitrile, and water. It dissociates in water to SO_4^{2-} and $[In(CH_3)_2]^+$ ions [2].

The ^1H NMR spectrum shows only a sharp singlet of the CH_3 group at −0.08 ppm in D_2O and −0.4 ppm in CH_3CN [2].

The Raman spectrum of the aqueous solution is explained by the existence of free SO_4^{2-} and bent $[In(CH_3)_2]^+$ ions; the vibration frequencies are shown below. Because all vibration modes of the crystalline SO_4^{2-} are split in the IR and Raman spectra, a polymeric structure (Formula I) was proposed [2].

Raman H_2O solution	IR solid	Raman solid	assignment
3001 w, br	–	–	$\nu_{as}(CH_3)$
2930 s	–	–	$\nu_s(CH_3)$
1415 vvw	–	–	$\delta_{as}(CH_3)$
1185 s	–	–	$\delta_s(CH_3)$
1125 vvw	1129 vs	1122 vvw	$\nu_{as}(SO_4)$
–	1087 vs	1085 vw	
987 s	1000 sh	1016 m	$\nu_s(SO_4)$
–	–	1003 mw	
720 vw	–	–	$\varrho(CH_3)$
–	643 m	–	$\delta(SO_4)$
622 w, br	620 w, br	618 w	
–	596 s	–	
558 vvw	553 s	555 mw	$\nu_{as}(InC_2)$
502 vs	495 mw	496 vs	$\nu_s(InC_2)$
–	474 mw	446 vw	$\delta(SO_4)$ and $\nu(InO\ ?)$
–	–	264 w	

[Formula I: polymeric structure $[\{R_2In(O_2SO_2)\}_2]_n$ with bridging sulfate groups]

I

$(CH_3)_2InReO_4$

The compound formed during the reaction of $(CH_3)_2InCl$ and $AgReO_4$ in CH_3OH along with AgCl; m.p. 220 °C (dec.). The solid is only soluble in very polar solvents. From IR absorbtions at 917 $\nu(ReO_3)$, 560 $\nu_{as}(InC_2)$, and 490 $\nu_s(InC_2)$ cm^{-1}, an ionic or at most coordination polymeric structure was deduced [3, 4].

The ^{115}In NQR spectrum showed frequencies at 50.81 (ν_1), 100.79 (ν_2), and 151.16 (ν_3) MHz with the quadrupole coupling constant $e^2Qq/h = 1211.22$ MHz and $\eta = 3.5\%$. Similarities to the corresponding NQR data of $(CH_3)_2InBr$ point to an identical structure for the $In(CH_3)_2$ unit with a nearly linear (about 172°) arrangement [4].

References:

[1] Hobbs, C. W.; Tobias, R. S. (Inorg. Chem. **9** [1970] 1998/2003).
[2] Olapinski, H.; Weidlein, J. (J. Organometal. Chem. **54** [1972] 87/93).
[3] Schmidbaur, H.; Koth, D. (Chemiker Ztg. **100** [1976] 290/1).
[4] Schmidbaur, H.; Koth, D. (Naturwissenschaften **63** [1976] 482/3).
[5] Canty, A. J.; Titcombe, L. A.; Skelton, B. W.; White, A. H. (J. Chem. Soc. Dalton Trans. **1988** 35/45).

4.6 Adducts of R_2In-Oxygen Compounds

No adducts are known for R_2InOH (p. 180), R_2InOR' (Table 28), or $R_2InOER'_3$ (Table 31); for the $R_2InO(R'$-D) compounds in Table 33 an addition compound is mentioned [5] only for diketonate No. 3. According to [4] carboxylates No. 2 to 5 and 8 [1] (Table 35) also form no stable, isolable addition compounds with pyridine, 2,2′-bipyridine, and 1,10-phenanthroline in $O(C_2H_5)_2$ or $CHCl_3$, while [6] describes this series of adducts for $(CH_3)_2InOOCCH_3$ (Table 35, No. 2). IR spectra showed no adduct formation between pyridine and the oxalates (p. 206) [3], nor with pyridine and $(CH_3)_2InOOP(CH_3)_2$ (Table 41, No. 13) [2]. However, two phosphoric acid derivatives, $(CH_3)_2InOOPX_2$ (X = F and Cl, Nos. 11 and 12, Table 41), form stable and isolable pyridine adducts, although they were not characterized in any way [7].

Among the inorganic acid derivatives in Chapter 4.5, a number of mono- and dinuclear addition compounds with nitrogen donor ligands are known for $(CH_3)_2InNO_3$ [8]; they are grouped in Table 43 together with the adducts of $(CH_3)_2InOC(CF_3)$=$CHCOCH_3$ (Section 4.1.4) and $(CH_3)_2InOOCCH_3$ (Section 4.2.1). These compounds were all prepared by the same procedure. The reagents were dissolved in $CHCl_3$ [5], C_2H_5OH [6], or CH_3OH [8] and mixed in stoichiometric ratios (1:1 or 2:1 mole ratio). After a short reaction period, the product was either precipitated by addition of $O(C_2H_5)_2$ [6] or isolated simply by evaporating the solvent and drying in vacuum [5, 8]. The yield was always quantitative. Purification by recrystallization from $CHCl_3$/petroleum ether was described only for Nos. 1 to 3. Deviating from this, $(CH_3)_2InOOCCH_3$ (to prepare Nos. 4 to 6) was dissolved in excess donor ligand and at room temperature freed from the unreacted base with vacuum; No. 7 was prepared in 1:1 $C_2H_5OH/CHCl_3$ [6].

In the case of the adducts of $(CH_3)_2InNO_3$ (Nos. 17 and 18) an equimolar amount of water was incorporated along with the basic ligand from the water-alcohol solvent of the nitrate reactant [8].

General Remarks. For the adducts of $(CH_3)_2InOOCCH_3$, structures with five- or six-coordinated indium have been proposed, depending on the nature of the donor ligand. The number of CH_3In signals in the ^1H NMR spectrum in CD_3OD (1 signal for I and 2 signals for both II and III) was the determining criterion. The "dimeric" structure III with a 2:1 In-to-ligand ratio was proposed for Nos. 7 and 8 [6].

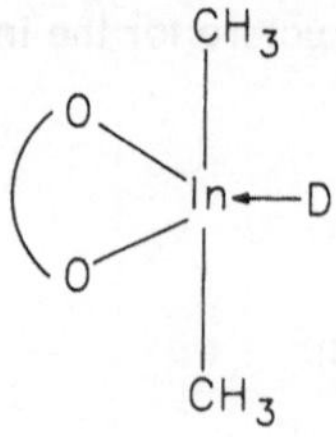

I 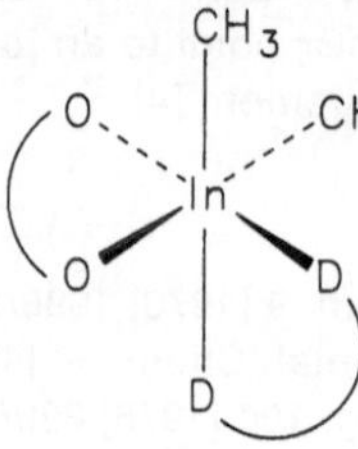 II 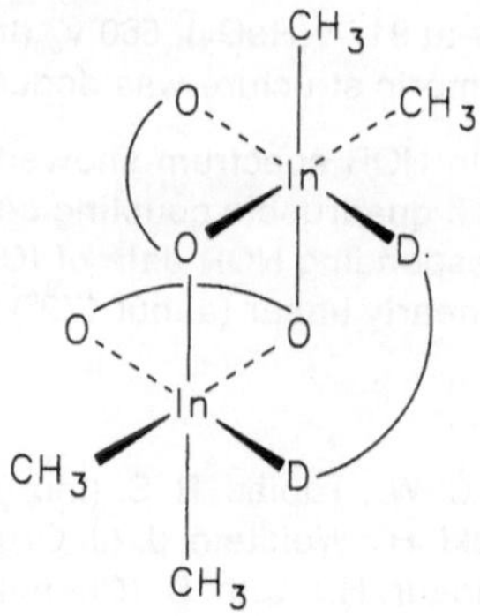 III

The crystalline compounds No. 15 and 18 show photochromic properties. In the dark they are light yellow, within about 30 min in intense illumination they turn emerald green. The return to the yellow form requires about 7 days' storage in darkness. The color change is possibly limited to the surface of the crystals and is dependent on the alteration of the In-ligand coordination InN_3 (yellow) $\leftrightarrow$ InN_2 (green) for the cations $[(CH_3)_2In \cdot D]^+$ [8]. All the adducts of $(CH_3)_2InNO_3$ are largely ionic; only rarely a very weak $In \cdots ONO_2$ coordination can be detected.

Table 43
Adducts of $(CH_3)_2InOC(CF_3)$=$CHCOCH_3$, $(CH_3)_2InOOCCH_3$, and $(CH_3)_2InNO_3$.
Further information on compounds whose numbers are preceded by an asterisk is given at the end of the table.
Explanations, abbreviations, and units on p. X.

No.	ligand D	properties and remarks
adducts of the type $(CH_3)_2InOC(CF_3)$=$CHCOCH_3 \cdot D$		
1	$C_{10}H_8N_2$-2,2′ =bipyridine	m.p. 140 °C [5] ^1H NMR ($CDCl_3$, 35 °C): 0.20 (CH_3In), 2.33 (CH_3C), 5.92 (CH), 8.1 (m, C_6H_4N) [5] ^{19}F NMR ($CDCl_3$, relative to CF_3COOD): 2.64 (CF_3) [5]
2	$C_{12}H_8N_2$-1,10 =phenanthroline	m.p. 110 °C [5] ^1H NMR ($CDCl_3$, 35 °C): 0.20 (CH_3In), 2.43 (CH_3C), 5.95 (CH), 8.0 (m, phen) [5] ^{19}F NMR ($CDCl_3$, relative to CF_3COOD): 3.79 (CF_3) [5]
3	$(C_6H_5)_2PCH_2CH_2P(C_6H_5)_2$	m.p. 66 °C [5] ^1H NMR ($CDCl_3$, 35 °C): 0.15 (CH_3In), 2.22 (CH_3C), 2.28 (d(?), CH_2P), 5.80 (CH) [5] ^{19}F NMR ($CDCl_3$, relative to CF_3COOD): 2.98 (CF_3) [5]
adducts of the type $(CH_3)_2InOOCCH_3 \cdot D$		
4	$OS(CH_3)_2$	white solid, monomeric in $OS(CH_3)_2$ [6] ^1H NMR: 0.52 (CH_3In), 2.48 (CH_3C) in $CDCl_3$ and 0.05, 2.15 in $OS(CH_3)_2$ [6]

Table 43 (continued)

No.	ligand D	properties and remarks
4 (continued)		IR (solid): 1565, 1468, 958 ν_{as}, $\nu_s(CO_2)$, and $\nu(C{-}CH_3)$, 1020 $\nu(S{=}O)$, 736 m $\varrho(CH_3In)$, 553 s $\nu_{as}(InC_2)$, 496 m $\nu_s(InC_2)$ [6] Raman (solid): 551 w $\nu_{as}(InC_2)$, 499 vs $\nu_s(InC_2)$ [6] proposed structure: Formula I
5	C_5H_5N	yellow solid, monomeric in $OS(CH_3)_2$ [6] ^1H NMR (C_5D_5N): 0.05 (CH_3In), 2.18 (CH_3C) [6] IR (solid): 555 s $\nu_{as}(InC_2)$, 498 w $\nu_s(InC_2)$, and typical bands for coordinated pyridine (no values reported) [6] Raman (solid): 548 vw $\nu_{as}(InC_2)$, 492 s $\nu_s(InC_2)$ [6] dissociates and loses C_5H_5N at room temperature [6] proposed structure: Formula I
6	$H_2NCH_2CH_2NH_2$	white solid [6] ^1H NMR ($CDCl_3$, strongly temperature-dependent): 0.47 (CH_3In, broadening with increasing temperature), 2.47 (CH_3C), 2.90 (at 0 °C), 2.50 (at 55 °C) (H_2N), 3.38 (CH_2N) [6] IR (solid): 524 m $\nu_{as}(InC_2)$, 492 w $\nu_s(InC_2)$ [6] Raman (solid): 521 w $\nu_{as}(InC_2)$, 490 s $\nu_s(InC_2)$, no other bands reported [6] rapid equilibrium between proposed structures I and II
7	$C_{10}H_8N_2$-2,2′ =bipyridine	2:1 adduct, white solid [6] ^1H NMR ($CDCl_3$, slow dec.): 0.65, 0.72 (s's, CH_3In), 2.42 (CH_3C) [6] IR (solid): 507 w [6] Raman (solid): 533 w, 503 s [6] proposed structure: Formula III
8	$C_{12}H_8N_2$-1,10 =phenanthroline	2:1 adduct, white solid [6] ^1H NMR ($CDCl_3$, slow dec.): 0.42, 0.72 (s's, CH_3In), 2.46 (CH_3C) [6] IR (solid): 530 s, 478 w [6] Raman (solid): 480 s [6] proposed structure: Formula III
9	$(C_6H_5)_2PCH_2CH_2P(C_6H_5)_2$	white solid [6] ^1H NMR (CD_2Cl_2): −0.02, 0.05 (s's CH_3In), 2.00 (CH_3C), 2.06 (t, CH_2P, J(H,P)=7.0), 7.15 (m, C_6H_5) [6] IR (solid): 554 s $\nu_{as}(InC_2)$, 511, 476, 442 [6] Raman (solid): 498 s $\nu_s(InC_2)$ [6] proposed structure: Formula II [6]
adducts of the type $(CH_3)_2InNO_3 \cdot D$		
10	$C_{10}H_8N_2$-2,2′ =bipyridine	white solid [8] conductivity (CH_3OH): 74 $\Omega^{-1} \cdot cm^2 \cdot mol^{-1}$ [8]

Table 43 (continued)

No.	ligand D	properties and remarks
10 (continued)		^{1}H NMR (CD_3OD): −0.08 (CH_3In), 7.74 (m, H-5), 8.23 (td, H-4), 8.54 (m, H-3), 8.87 (ddd, H-6), (J(H-3,4) = 7.24, J(H-5,6) = 5.2, J(H-3,5) = 1.8, J(H-4,6) ca. 1.5, J(H-3,6) ca. 0.7) [8]
11	$(C_{10}H_6N_2\text{-}2,2')(CH_3)_2\text{-}4,4'$	pale yellow solid [8] conductivity (CH_3OH): 77 $\Omega^{-1} \cdot cm^2 \cdot mol^{-1}$ [8] ^{1}H NMR (CD_3OD): −0.10 (CH_3In), 2.60 (s, CH_3), 7.59 (d, br, H-5), 8.42 (s, br, H-3), 8.67 (m, br, H-6), (J(H-5,6) ca. 5.0) [8]
12	$(C_{10}H_6N_2\text{-}2,2')(C_2H_5)_2\text{-}4,4'$	pale yellow solid [8] conductivity (CH_3OH): 77 $\Omega^{-1} \cdot cm^2 \cdot mol^{-1}$ [8] ^{1}H NMR (CD_3OD): −0.10 (CH_3In), 2.94 (t, CH_3), 4.47 (q, CH_2, J(H,H) = 7.6), 7.61 (dd, H-5), 8.45 (s, H-3), 8.71 (d, H-6), (J(H-5,6) = 5.40, J(H-3,5) = 1.55) [8]
13	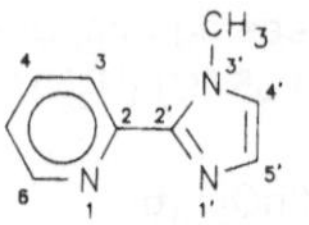	white solid [8] conductivity (CH_3OH): 76 $\Omega^{-1} \cdot cm^2 \cdot mol^{-1}$ [8] ^{1}H NMR (CD_3OD): −0.14 (CH_3In), 4.16 (s, CH_3N), 7.22, 7.45 (s's, H-4',5'), 7.62 (m, H-5), 8.14 to 8.18 (m, H-3,4), 8.70 (d, br, H-6) [8]
14	$C_{12}H_8N_2\text{-}1,10$ = phenanthroline	pale yellow solid [8] conductivity (CH_3OH): 76 $\Omega^{-1} \cdot cm^2 \cdot mol^{-1}$ [8] ^{1}H NMR (CD_3OD): −0.01 (CH_3In), 8.10 (dd, H-3,8), 8.21 (s, H-5,6), 8.84 (dd, H-4,7), 9.27 (dd, H-2,9), (J(H-3,4) = 8.2, J(H-2,3) = 4.8, J(H-2,4) = 1.5) [8]
*15	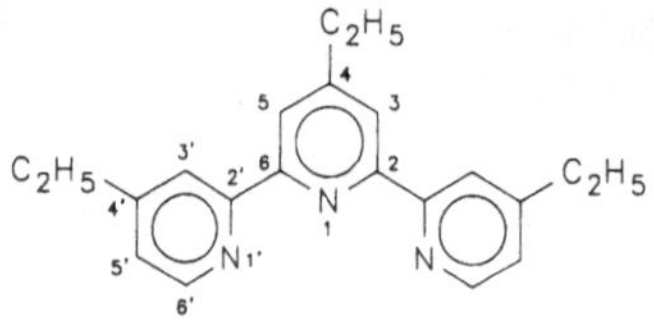	pale yellow solid when stored in darkness [8] conductivity (CH_3OH): 53 $\Omega^{-1} \cdot cm^2 \cdot mol^{-1}$ [8] ^{1}H NMR (CD_3OD): −0.16 (CH_3In), 1.41 (t, CH_3'), 1.50 (t, CH_3), 2.96 (q, CH_2', J(H,H) = 7.60), 3.09 (q, CH_2), 7.72 (d, H-5'), 8.63 (s, H-3'), 8.71 (d, H-6', J(H-5',6') ca. 5.3), 8.72 (s, H-3,5) [8]
16	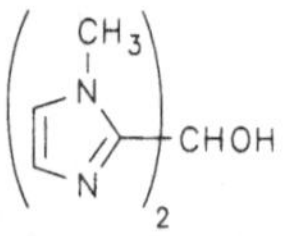	white solid [8] conductivity (CH_3OH): 88 $\Omega^{-1} \cdot cm^2 \cdot mol^{-1}$ [8] ^{1}H NMR (CD_3OD): −0.12 (CH_3In), 3.91 (s, CH_3N), 6.23 (s, HC), 7.16, 7.29 (s's, H-4,5) [8] IR (solid): 3224 m, vbr ν(OH) [8]
*17	$(C_5H_4N)_2CH_2$	1:1 hydrate, white solid [8] conductivity (CH_3OH): 83 $\Omega^{-1} \cdot cm^2 \cdot mol^{-1}$ [8] ^{1}H NMR (CD_3OD): 0.04 (CH_3In), 4.40 (s, CH_2), 7.40 (m, H-5), 7.51 (d, H-3), 7.89 (td, H-4), 8.53 (m, H-6), (J(H-3,4) = 7.9, J(H-4,5) ca. 7.7, J(H-5,6) ca. 5.2, J(H-4,6) ca. 1.8) [8] IR (solid): 3560 m, br and 3450 m, br ν(H_2O) [8]

Table 43 (continued)

No.	ligand D	properties and remarks
18	$C_{15}H_{11}N_3$ =terpyridine	1:1 hydrate, pale yellow solid when stored in darkness [8] conductivity (CH_3OH): 93 $\Omega^{-1}\cdot cm^2\cdot mol^{-1}$ [8] ^1H NMR (CD_3OD): −0.10 (CH_3In), 7.87 (dd, H-5′), 8.35 (td, H-4′), 8.58 (t, H-4), 8.72 (m, H-3′), 8.81 (d, H-3,5), 8.89 (d, br, H-6′), (J(H-3′,4′; H-4′,5′; H-3,4; H-4,5) ca. 8.0, J(H-5′,6′)=5.4, J(H-4′,6′)=1.4), for numbering, see No. 15 [8] IR (solid): 3530 m,br and 3450 m,br $\nu(H_2O)$ [8]

* Further information:

$(CH_3)_2InNO_3\cdot N_3H_8C_{15}(C_2H_5)_3$ (Table **43**, No. **15**) crystallizes in the monoclinic space group $P2_1/c-C^5_{2h}$ (No. 14) with the cell parameters a=864.1(3), b=1587.2(6), c=1724.2(7) pm, β= 90.78(3)°; Z=4, D_c=1.47 g/cm³. The InN_3 plane of this ionic compound makes an angle of 88.3° with the InC_2 plane. No bond distances or angles were given for the C_5N rings and their C_2H_5 substituents, but all the atom positions are drawn to scale. **Fig. 38** contains the most important distances and angles for the cation; N-O bond distances of 123(1), 121(1), and 119(1) pm and bond angles of 119.1(8)°, 121.5(8)°, and 119.4(8)° were reported for the planar NO_3^- ion [8].

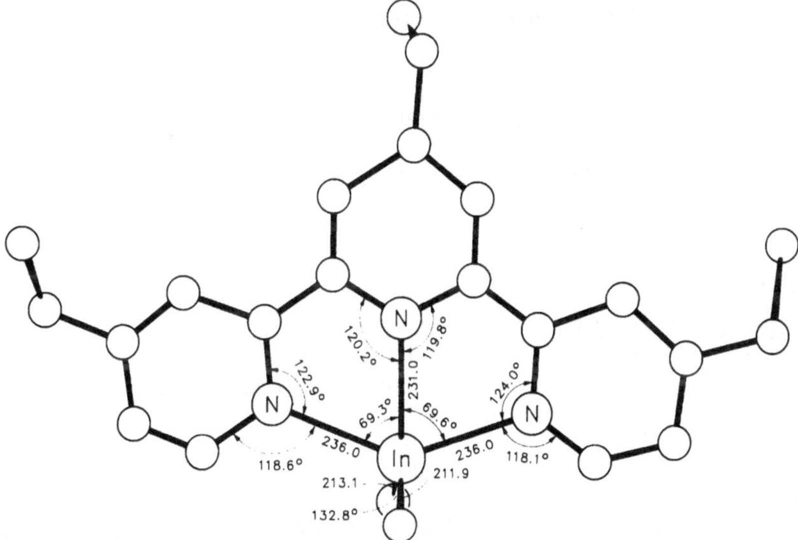

Fig. 38. Molecular structure of $(CH_3)_2InNO_3\cdot N_3H_8C_{15}(C_2H_5)_3$ [8].

$(CH_3)_2InNO_3\cdot(C_5H_4N)_2CH_2\cdot H_2O$ (Table **43**, No. **17**) crystallizes in the monoclinic space group $P2_1/c-C^5_{2h}$ (No. 14) with the cell parameters a=900.2(3), b=1253.8(7), c=1399.6(7) pm, β=91.26(3)°; Z=4, D_c=1.66 g/cm³. The nitrogen and oxygen donor atoms are arranged in an "equatorial plane" formed by InO_2N_3 with maximum deviation of 19 (N(2)) and −14 (N(1)) pm from this plane; O(2) belongs to the coordinated H_2O. The most important bond distances and angles are depicted in **Fig. 39**, p. 228 [8].

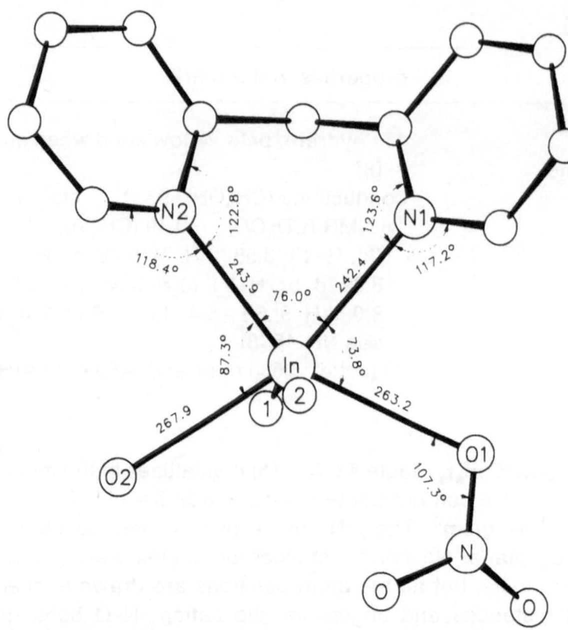

Fig. 39. Molecular structure of $(CH_3)_2InNO_3 \cdot (C_5H_4N)_2CH_2 \cdot H_2O$ [8].

Other bond angles (°) and distances (pm)

O(2)-In-O(1)	123.6(1)	C(1)-In	212.2(5)
C(1)-In-N(2)	95.2(2)	C(2)-In	212.8(6)
C(1)-In-N(1)	105.9(2)		
C(2)-In-N(2)	98.2(2)		
C(2)-In-N(1)	94.8(2)		
C(1)-In-C(2)	157.6(2)		

References:

[1] Weidlein, J. (Z. Anorg. Allgem. Chem. **378** [1970] 245/62).
[2] Olapinski, H.; Schaible, B.; Weidlein, J. (J. Organometal. Chem. **43** [1972] 107/16).
[3] Schwering, H.-U.; Hausen, H.-D.; Weidlein, J. (Z. Anorg. Allgem. Chem. **391** [1972] 97/106).
[4] Lindel, W.; Huber, F. (Z. Naturforsch. **28b** [1973] 517/8).
[5] Chung, H. L.; Tuck, D. G. (Can. J. Chem. **52** [1974] 3944/9).
[6] Habeeb, J. J.; Tuck, D. G. (Can. J. Chem. **52** [1974] 3950/4).
[7] Schaible, B.; Haubold, W.; Weidlein, J. (Z. Anorg. Allgem. Chem. **403** [1974] 289/300).
[8] Canty, A. J.; Titcombe, L. A.; Skelton, B. W.; White, A. H. (J. Chem. Soc. Dalton Trans. **1988** 35/45).

4.7 Other R_2In-Oxygen Compounds

4.7.1 $(CH_3)_2In$-Nickel Glyoximates

The starting material for the two compounds in this section was a Ni^{2+} glyoximate (Formula I, R = H, CH_3).

I

II

On reaction with 2 mol $In(CH_3)_3$ the acidic hydrogens of the OH groups were replaced by $(CH_3)_2In$ with elimination of CH_4. $In(CH_3)_3$ in C_5H_{12}, C_6H_{14}, or $C_6H_5CH_3$ was slowly added dropwise to a suspension of the glyoximate in the same solvent, and brought the reaction to completion by warming to 30 to 40 °C. The residue was filtered off, washed several times with C_5H_{12}, and recrystallized from $C_6H_5CH_3$ or 3:1 $C_6H_6/CHCl_3$. The yield of raw product was quantitative [1].

$Ni[(CH_3)_2InO_2N_2(CH)_2]_2$

This compound (Formula II, R = H) was obtained, after recrystallization, in 94% yield as small red crystals, which decomposed at 225 ± 5 °C without having melted [1].

The ^1H NMR spectrum in $CDCl_3$ contains two singlets at −0.07 (CH_3In) and 7.11 (HC) ppm. The most important IR absorptions are listed in Table 44 opposite the corresponding data for the compound described below. The structural assumptions were based on the results of the X-ray analysis of the homologous gallium derivative [1].

$Ni[(CH_3)_2InO_2N_2(CCH_3)_2]_2$

The methyl derivative (Formula II, R = CH_3) is brown and was obtained in 70% yield after purification. The compound decomposes at 205 ± 5 °C without melting.

The ^1H NMR spectrum in C_6D_6 shows singlets at 0.36 (CH_3In) and 1.68 (CH_3C) ppm. Important IR vibration frequencies are listed in Table 44.

Table 44
IR Spectra of Solid $Ni[(CH_3)_2InO_2N_2(CR)_2]_2$ (R = H, CH_3) [1].
Wavenumbers in cm^{-1}.

R = H	R = CH_3	assignment
1546 vs	1566 s	$\nu_{as}(CCN)$
1483 mw	1488 w	$\nu_s(CCN) + \nu(CC)$
(1201)	1208 vs	$\nu(C\text{-}CH_3)$
obscured	1158 mw	$\delta_s(CH_3In)$
–	1149 mw	
1201 vs, br	1084 s	$\nu(N\text{-}O) + \delta(CH)$
–	970 ms	$\varrho(CH_3C)$
852	–	$\delta(CNO)$
847 ms	845 w	
785 s	749 m	$\delta(CCN) + \gamma(CH)$
718 s, br	705 s	$\varrho(CH_3In)$
–	671 w	
578 m	547 w	$\delta(CCN)$
530 s	526 s	$\nu_{as}(InC_2)$
–	520 s	$\delta(CCCH_3)$
?	499 m	$\nu_s(InC_2)$
464 m	451 mw	$\nu_{as}(InO_2)$
423 mw	419 vw	$\nu(InN) + \delta(skeleton)$
385 m	371 m	
340 m	350 sh	
319 mw	290 sh	

Reference:

[1] Kohler, U.; Hausen, H.-D.; Weidlein, J. (J. Organometal. Chem. **272** [1984] 337/59).

4.7.2 $(C_2H_5)_2In$-Oxygen Radicals

Radicals I and II ($R=C_4H_9$-t) were produced when $In(C_2H_5)_3$ and di-t-butyl-3,5-benzoquinone-1,2 in toluene were warmed from −196 °C to room temperature [1].

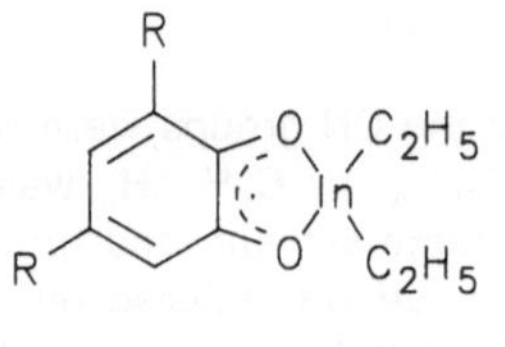

I 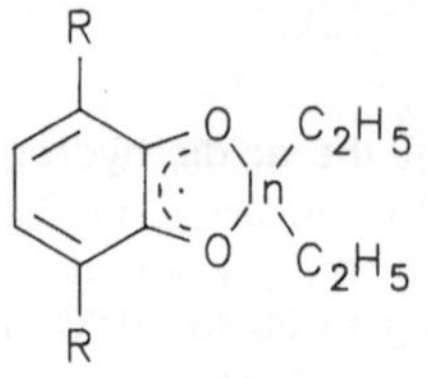 II 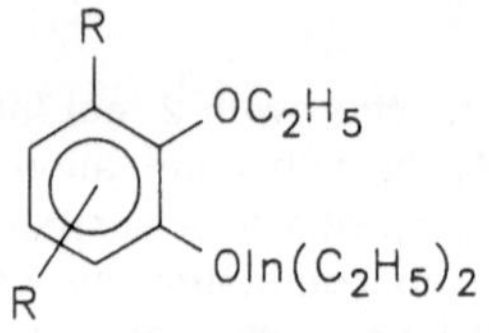 III

$In(C_2H_5)_2C_{14}H_{20}O_2^{\cdot}$

ESR parameters for the radicals were measured at room temperature in toluene: for I g=2.0021, coupling constant a=8.8 G (^{115}In); for II g=2.003, coupling constant a=8.9 G (^{115}In) and 3.8 G (H). The lifetimes of the radicals are between 15 and 20 min. The paramagnetism of the solutions disappears with their green color; III and traces of C_2H_6 are formed [1].

Reference:

[1] Razuvaev, G. A.; Abakumov, G. A.; Klimov, E. S.; Gladyshev, E. N.; Bayushkin, P. Ya. (Izv. Akad. Nauk SSSR Ser. Khim. **1977** 1128/32; Bull. Acad. Sci. USSR Div. Chem. Sci. **1977** 1034/7).

4.8 RIn-Oxygen Compounds

4.8.1 Compounds with an RInO Unit

None of the following compounds has been well characterized by physical investigations nor used for synthetic purposes. They were superficially detected as decomposition products or as intermediates in the preparation of other organoindium compounds.

A compound of the general formula $C_2H_5OC(O)CH_2In(Br)OC(R_2)CH_2C(O)OC_2H_5$ was possibly formed as an intermediate in the insertion of $R_2C{=}O$ (R is not defined) into one of the In-C bonds of $(C_2H_5OC(O)CH_2)_2InBr$ (Table 20, No. 7). The compound was not isolated; hydrolysis in acidic medium gave compounds with compositions corresponding to $HOC(R_2)CH_2C(O)OC_2H_5$ [3].

$(CH_3InO)_n$

This compound is the only compound described in this section that was isolated in preparative amounts and subjected to elemental analysis. It was produced in high yields by the reaction of $In(CH_3)_3$ with H_2O in 1:1 mole ratio (via the initial, not isolable $(CH_3)_2InOH$), 2 equivalents of CH_4 being eliminated. The white, polymeric product is insoluble in the common organic solvents [2]. It is also one of the main products (>50 mol%) in the free radical thermal decomposition of $(CH_3)_2InOOCH_3$ (see p. 209) in nonane at 120 °C [4].

$(C_2H_5InO)_n$

The ethyl derivative was obtained as one of a series of products in the radical formation of $(C_2H_5)_2InOOC_2H_5$ (see p. 210) from $In(C_2H_5)_3$ and O_2 at 40 to 100 °C. The compound could not be isolated in pure form [1].

References:

[1] Cullis, C. F.; Fish, A.; Pollard, R. T. (Trans. Faraday Soc. **60** [1964] 2224/33).
[2] Clark, H. C.; Pickard, A. L. (J. Organometal. Chem. **13** [1968] 61/71).
[3] Chao, L.-C.; Rieke, R. D. (J. Org. Chem. **40** [1975] 2253/5).
[4] Aleksandrov, Yu. A.; Chikinova, N. V.; Makin, G. I.; Kornilova, N. V.; Bregadze, V. I. (Zh. Obshch. Khim. **48** [1978] 467; J. Gen. Chem. [USSR] **48** [1978] 417).

4.8.2 Compounds with an $RInO_2$ Unit

The compounds of this section are collected in Table 45. Nos. 1 to 3 are dialcoholates that only occurred as by-products of the thermal decomposition of other organoindium compounds and were not closely scrutinized. The dicarbonic acid derivatives, Nos. 5 and 6, were derived from $In(C_6H_5)_3$; comparable derivatives of indium trialkyls are not known, and experiments in this direction led to inhomogenous mixtures [4]. Nos. 7 to 11 are ionic alcoholates and were all prepared from $(CH_3)_2InNO_3$ (see p. 221). In No. 12 two $CH_2In\{OCH_2CH_2N(CH_3)_2\}_2$ units are bonded to a $(CH_3)_2Sn$ group.

The following synthetic methods were used:

Method I: From $InR_3 \cdot D$ and HOOCR′.
To a $CHCl_3$ solution of $(C_6H_5)_3In \cdot THF$ was added, dropwise, twice the molar amount of HOOCR′ (R′ = CH_3, C_2H_5). When the gas evolution ended, the residue was filtered off, washed with petroleum ether, and dried in vacuum [2].

Method II: From $InR_3 \cdot D$ and $In(OOCR')_3$.
To a benzene solution of $(C_6H_5)_3In \cdot O_2C_4H_8$ was added twice the molar amount of the appropriate solid $In(OOCR')_3$. The mixture was refluxed for 3 h, cooled, filtered, then washed and dried in vacuum [2].

Method III: From $(CH_3)_2InNO_3$ and HOX.
Methanolic HOX was added to a freshly prepared CH_3OH/H_2O solution of the indium nitrate (from $TlNO_3$ and $(CH_3)_2InI$). At the end of the gas evolution the solution was concentrated until crystallization occurred (Nos. 7 to 11). The alcohol HOX was $HOC(R_{3-n}R'_n)$ or $HOC(R')_2C_6H_5$ (n = 0 to 3; R = I, R′ = II). Nos. 7 and 8 were recrystallized from $CH_3OH/(CH_3)_2CO$ during which the solution was subjected to an ether atmosphere in a closed system [6, 7].

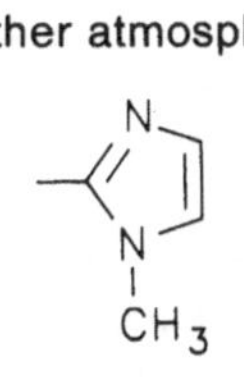

N
N
CH_3
I

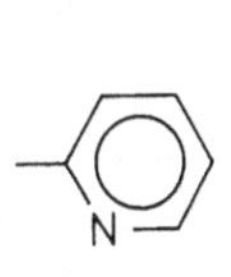

N
II

General Remarks. Nos. 7 to 11 are ionic alcoholate "pseudodimers" containing an In_2O_2 ring for the solids. The deprotonated alcohols $HOC(R_{3-n}R'_n)$ or $HOC(R')_2C_6H_5$ (see above) serve as tridentate ligands as shown in the schematic model III; one of the groups R,

R′, or C_6H_5 does not coordinate at the In atom. During intramolecular adduct formation both indium atoms of the cation coordinate with the nitrogen atoms of the ligands; additionally, for coordinative saturation one indium acquires H_2O, the other a NO_3^-; therefore, the singly-charged complex ion is unsymmetrical [7].

In aqueous or methanolic solution the nitrate group dissociates from In, resulting in a doubly-charged cation (Formula IV, S is a solvent molecule) of comparatively high conductivity [7].

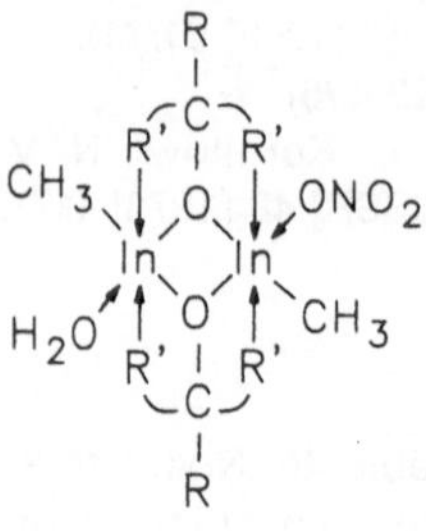

III 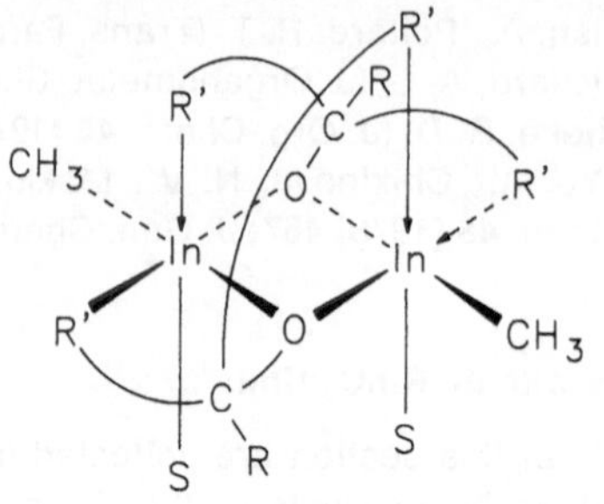 IV

In compounds 7 and 10 the chelating groups are equivalent ($R = R' = C_3H_2N_2CH_3$ or C_5H_4N), resulting in one structure; also for No. 11 with two C_5H_4N groups and one C_6H_5 group, only one structure is possible. However, several isomers must be discussed for Nos. 8 and 9 having different chelating groups ($C_3H_2N_2CH_3$ and C_5H_4N) at the OC carbon atom [7].

The proton resonance spectra of Nos. 7, 10, and 11 with the symmetrical chelating ligands show only one CH_3In signal at 20 °C with a CH_3In:ligand ratio of 1:1. At −40 °C the signals of the aromatic rings and the CH_3N groups are multiply split; at this temperature exchange between axial and equatorial positions (N(a) and N(e) of R′ in Formula IV) is no longer possible [7].

The compounds with the mixed substituents, Nos. 8 and 9, would be expected to exist as isomers (see above), which could explain the very distinct splitting of the CH_3In resonances observed at −40 °C. No. 9 gave a total of 7 CH_3In and 3 CH_3N signals of differing intensities, as well as complex multiplets for the aromatic rings. Assignment of No. 9 to a particular isomer of the six theoretically possible isomers was not possible [7].

Table 45
Compounds with an $RInO_2$ Unit.
Further information on numbers preceded by an asterisk is given at the end of the table.
Explanations, abbreviations, and units on p. X.

No.	compound method of preparation (yield)	properties and remarks
1	$CH_3In(OCH_3)_2$	by-product of the thermal decomposition of $(CH_3)_2InOOCH_3$ in nonane at 120 °C, not isolated, no data reported [5]
2	$C_2H_5In(OC_2H_5)_2$	formed as a by-product in an intramolecular rearrangement of $(C_2H_5)_2InOOC_2H_5$; not isolated, and no data reported [1]

Table 45 (continued)

No.	compound method of preparation (yield)	properties and remarks
3	$C_2H_5In(OC_4H_9\text{-}t)_2$	by-product of the thermal decomposition of $(C_2H_5)_2InO(C_4H_9\text{-}t)$ at 180 °C; not isolated, and no data reported [3]
4	$(C_4H_9\text{-}n)In(OOCC_2H_5)OSn(C_4H_9\text{-}n)_3$	from $(C_4H_9\text{-}n)_2InOOCC_2H_5$ and $(C_4H_9\text{-}n)_2SnO$ in p-xylene from 0 °C to reflux for 3.5 h [9] gummy solid, softening p. 55 °C [9] ^1H NMR ($CDCl_3$): 0.91 (t, CH_3 of C_2H_5), 0.9 to 2.0 (m, C_4H_9), 2.31 (q, CH_2 of C_2H_5) [9] ^{13}C NMR ($CDCl_3$): 10.1 (CH_3 of C_2H_5), 13.7 (δ-CH_3), 16.3 (α-CH_2), 27.1 (γ-CH_2), 27.9 (β-CH_2), 29.3 (CH_2 of C_2H_5), 184.0 (CO) [9] IR (neat?): 1560 s $\nu_{as}(CO_2)$, 650 ν(In-O-Sn) [9]
5	$C_6H_5In(OOCCH_3)_2$ I (63%) II (70%)	white solid, no m.p. below 300 °C [2] IR (solid): ca. 3050 ν(CH), ca. 1600 ν(C_6-ring), 1550, 1420 ν(CO_2), 530, 510 ν(InO); no other bands reported [2]
6	$C_6H_5In(OOCC_2H_5)_2$ I (81%)	white solid, no m.p. below 300 °C [2] IR (solid): ca. 3060 ν(CH), ca. 1600 ν(C_6-ring), 1560, 1440 ν(CO_2), 560, 500 ν(InO); no other bands reported [2]
7	$\{CH_3InNO_3OC(C_3H_2N_2CH_3)_3\}_2 \cdot 3\,H_2O$ III	white crystalline solid [7] conductivity (CH_3OH): 171 $\Omega^{-1} \cdot cm^2 \cdot mol^{-1}$ [7] ^1H NMR (CD_3OD): −0.14 (CH_3In), 3.13 (s), 3.33 (br), 3.63 (s) (CH_3N), 6.48, 6.96, 7.00, 7.32, 7.44, 7.46 (all s, H-4,5) [7] IR (solid): 3430 m, br ν(H_2O); no other bands reported [7]
*8	$\{CH_3InNO_3OC(C_3H_2N_2CH_3)_2C_5H_4N\}_2 \cdot 2\,H_2O \cdot (CH_3)_2CO$ III	colorless crystals [6, 7] conductivity (CH_3OH): 134 $\Omega^{-1} \cdot cm^2 \cdot mol^{-1}$ [6, 7] ^1H NMR (CD_3OD): −0.24 (CH_3In), 2.18 (s, $(CH_3)_2CO$), ca. 3.33 ($CH_3N + CD_3OD$), 3.62 (CH_3N), 6.0 to 9.0 (br, $C_3H_2N_2 + C_5H_4N$) at 20 °C; −0.48, −0.32, 0.04 (CH_3In), 3.20, 3.33, 3.54, 3.62 (CH_3N), 5.89, 6.63 to 8.25, 8.96 (complex m, C_6H_4N and H-4,5 of $C_3H_2N_2$) at −40 °C; similar resonances in the CH_3N-region in a spectrum in D_2O solution [6, 7] IR (solid): 3365 m,vbr ν(H_2O); no other bands reported [6, 7]

Table 45 (continued)

No.	compound method of preparation (yield)	properties and remarks
9	$\{CH_3InNO_3OC(C_3H_2N_2CH_3)(C_5H_4N)_2\}_2 \cdot 2\,H_2O$ III (36%)	white solid [7] conductivity (CH_3OH): 116 $\Omega^{-1} \cdot cm^2 \cdot mol^{-1}$ [7] ^1H NMR (CD_3OD): 0.3 (vbr, CH_3In), 3.46 (br, CH_3N), 6.5 to 8.5 (br, C_5H_4N and $C_3H_2N_2$) at 20 °C; −0.52, −0.42, −0.32, −0.27, −0.12, 0.18, 0.20 (CH_3In), 3.48, 3.56, 3.57 (all s, CH_3N), 5.72, 6.62, 6.9 to 8.9 (m's, C_5H_4N and H-4,5 of $C_3H_2N_2$) at −40 °C [7] IR (solid): 3400 m,vbr $\nu(H_2O)$; no other bands reported [7]
10	$\{CH_3InNO_3OC(C_5H_4N)_3\}_2 \cdot 5\,H_2O$ III (87%)	white solid [7] conductivity (CH_3OH): 146 $\Omega^{-1} \cdot cm^2 \cdot mol^{-1}$ [7] ^1H NMR (CD_3OD, 20 °C): −0.36 (CH_3In), 7.38 (br, H-5), 7.75 (br, H-3), 7.95 (br, H-4), 8.14 (br, H-6); variable temperature (−40, 20, 50 °C); spectrum is depicted for the region containing C_5H_4N resonances [7] IR (solid): 3370 m,vbr $\nu(H_2O)$; no other band reported [7]
*11	$\{CH_3InNO_3OC(C_5H_4N)_2C_6H_5\}_2 \cdot 1.75\,H_2O$ III (87%)	colorless crystals [7] conductivity (CH_3OH): 130 $\Omega^{-1} \cdot cm^2 \cdot mol^{-1}$ [7] ^1H NMR (CD_3OD): −0.39 (CH_3In), 7.22 to 7.24 (m, br), 7.36 to 7.46 (m, br), 8.02 (br) at 20 °C; −0.41 (s, CH_3In), 7.28 (m, br), 7.3 to 7.5 (m), 7.59 (m, br), 7.81 ('t', br), 7.89 ('t', br), 8.28, 8.82 (d, br) at −40 °C [7] IR (solid): 3380 m, vbr $\nu(H_2O)$; no other bands reported [7]
12	$\{In(OCH_2CH_2N(CH_3)_2)_2CH_2\}_2Sn(CH_3)_2$	from $(CH_3)_2Sn(CH_2Li)_2$ and $ClIn(OCH_2CH_2N(CH_3)_2)_2$ in $(C_2H_5)_2O$ at −50 °C (37% yield) [8] pale yellow, air-stable liquid, b.p. 112 to 114 °C/0.1 Torr [8] ^1H NMR (C_6D_6): −0.21 (CH_2Sn, J(Sn,H) = 54.2 and 57.0), 0.27 (CH_3Sn, J(Sn,H) = 51.4 and 54.3), 1.01 (CH_3N); no data for CH_2N and CH_2O [8]

* Further information:

$\{CH_3InNO_3OC(C_3H_2N_2CH_3)_2C_5H_4N\}_2 \cdot 2\,H_2O \cdot (CH_3)_2CO$ (Table **45**, No. **8**) crystallizes in the monoclinic space group $P2_1/c-C_{2h}^5$ (No. 14) with a = 1438.8(6), b = 1761.3(10), c = 1640.3(10) pm, β = 92.18(3)°; Z = 4 gives D_c = 1.62 g/cm³ [6, 7]. The dimeric unit is ionic with one NO_3 group and one H_2O molecule attached to the dinuclear complex, the two In atoms being therefore differently coordinated; In(1) attains coordination number 7 by the bidentate O_2NO group while In(2) is sixfold coordinated by H_2O ligation as depicted in **Fig. 40**. Each In atom is coordinated by one N atom of group R and R′ (Formula I and II on p. 231). The rings A, A′ coordinate axially (N(ax)) on the same side of the In_2O_2 core, the rings C, C′ equatorially (N(eq)) and B, B′ are not coordinated [7].

The N-O(11, 12, 13) distances of the coordinated nitrate are 117, 124, and 124 pm with ONO angles opposite to O(11, 12, 13) of 116.2°, 120.1°, and 123.7°, respectively; the NO bonds of the ionic NO_3 are 126, 126, and 111 pm and the ONO angles are 135°, 125°, and 99° [7].

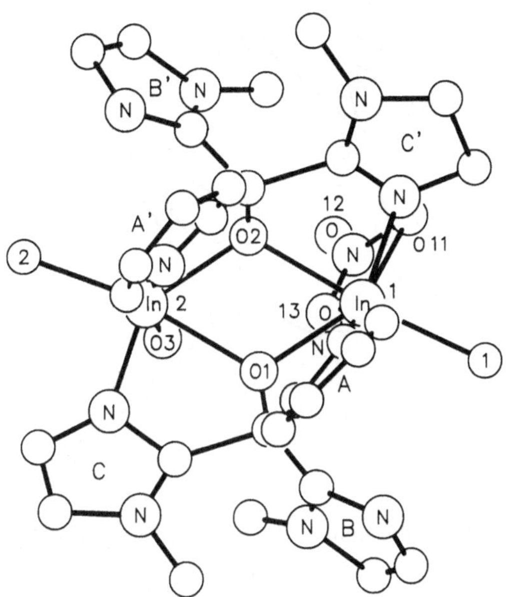

Fig. 40. Molecular structure of the cation of No. 8, $[(CH_3In)_2NO_3(OC(C_3H_2N_2CH_3)_2C_5H_4N)_2H_2O]^+$ [6, 7].

Distances (pm)		Angles (°)	
In(1)-C(1)	211.9(11)	O(1)-In(1)-O(2)	73.7(2)
In(2)-C(2)	215.3(10)	N(ax)-In(1)-N(eq)	87.1(3)
In(1)-N(eq)	225.9(8)	N(ax)-In(2)-N(eq)	88.5(3)
In(2)-N(eq)	223.4(7)	In(1)-O(1)-In(2)	106.3(2)
In(1)-N(ax)	239.1(7)	In(1)-O(2)-In(2)	107.1(2)
In(2)-N(ax)	233.4(7)		
In(1)-O(1,2)	219.3(6), 223.3(6)		
In(2)-O(1,2)	226.6(6), 220.4(6)		
In(1)···In(2)	356.9(2)		
O(1)···O(2)	265.5(8)		

$\{CH_3InNO_3OC(C_5H_4N)_2C_6H_5\}_2 \cdot 1.75\,H_2O$ (Table **45**, No. **11**) crystallizes in the triclinic space group $P\bar{1}-C_i^1$ (No. 2) with the cell parameters a = 1914.7(5), b = 1100.7(4), c = 948.6(8) pm, α = 76.14(6)°, β = 88.79(6)°, γ = 87.87(3)°; Z = 2, and D_c = 1.61 g/cm³. Its ionic structure and the coordination of the rings correspond to that of No. 8, having two differently coordinated In atoms as depicted in **Fig. 41**, p. 236. For the positions of the rings A to C, see Fig. 40 [7].

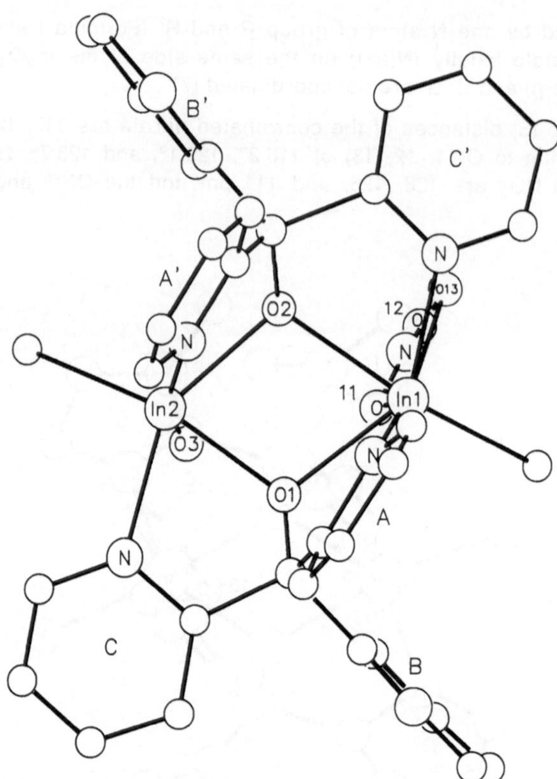

Fig. 41. Molecular structure of the cation of No. 11, $[(CH_3In)_2NO_3(OC(C_5H_4N)_2C_6H_5)_2H_2O]^+$ [7].

Distances (pm)		Angles (°)	
In(2)-O(1,2)	217.0(6), 217.7(6)	C-In(2)-N(ax, eq)	98.3(5), 100.5(4)
In(2)-O(3)	230.4(9)	O(1)-In(2)-O(2)	73.7(2)
In(1)···In(2)	349.4(1)	O(2)-In(2)-O(3)	89.6(3)
O(1)···O(2)	260.7(8)	N(ax)-In(2)-N(eq)	93.2(3)
		C-In(2)-O(1,2)	163.4(5), 118.8(4)
		N(eq)-In(2)-O(1,2)	96.3(3), 70.5(3)

The N-O(11, 12, 13) distances of the coordinated nitrate are 88, 142, and 126 pm with ONO angles opposite to O(11, 12, 13) of 102°, 136°, and 122°, respectively; the NO bonds of the ionic NO_3 are 122, 121, and 124 pm and the ONO angles are 114°, 123°, and 123° [7].

References:

[1] Cullis, C. F.; Fish, A.; Pollard, R. T. (Trans. Faraday Soc. **60** [1964] 2224/33).

[2] Viktorova, I. M.; Sheverdina, N. I.; Rodionov, A. N.; Kocheshkov, K. A. (Dokl. Akad. Nauk SSSR **189** [1969] 315/7; Dokl. Chem. Proc. Acad. Sci. USSR **184/189** [1969] 889/91).

[3] Sheherbakov, V. I.; Zhil'tsov, S. F.; Druzhkov, O. N. (Zh. Obshch. Khim. **40** [1970] 1542/5; J. Gen. Chem. [USSR] **40** [1970] 1529/31).

[4] Lindel, W.; Huber, F. (Z. Naturforsch. **28b** [1973] 517/8).
[5] Aleksandrov, Yu. A.; Chikinova, N. V.; Makin, G. I.; Kornilova, N. V.; Bregadze, V. I. (Zh. Obshch. Khim. **48** [1978] 467; J. Gen. Chem. [USSR] **48** [1978] 417).
[6] Canty, A. J.; Titcombe, L. A.; Skelton, B. W.; White, A. H. (Inorg. Chim. Acta **117** [1986] L35/L36).
[7] Canty, A. J.; Titcombe, L. A.; Skelton, B. W.; White, A. H. (J. Chem. Soc. Dalton Trans. **1988** 35/45).
[8] Schumann, H.; Motachemi, R.; Schwichtenberg, M. (Z. Naturforsch. **43b** [1988] 1510/3).
[9] Nomura, R.; Fujii, S.; Kanaya, K.; Matsuda, H. (Polyhedron **9** [1990] 361/6).

4.8.3 Compounds with an $RInO_n$ Unit

Two compounds in this series are described in which 4 InR units occupy the vertices of a cube or the bridge-heads in an adamantane-like structure with a tetrahedral arrangement of the In atoms.

$\{((CH_3)_3Si)_3CIn\}_4(O)(OH)_6$

This compound was obtained by several intermediate steps. The first step was the reaction between $InCl_3$ and $LiC(Si(CH_3)_3)_3$ in THF at −40 °C to produce $(THF)_2Li[Cl_3InC(Si(CH_3)_3)_3]$ (see p. 364). Hydrogenation with $LiAlH_4$ in THF at −78 °C gave $(THF)_2Li[In_2H_5\{C(Si(CH_3)_3)_3\}_2]$ (see p. 346), which was then converted to the title compound by CH_3OH/H_2O under H_2 evolution at room temperature. The anticipated dihydroxide, $\{(CH_3)_3Si\}_3CIn(OH)_2$, could not be trapped. After recrystallization from toluene, the title compound was obtained as colorless crystals melting at 257 °C [1].

It crystallizes in the monoclinic space group $P2_1/c-C_{2h}^5$ (No. 14) with a = 2226.3(1), b = 1495.1(7), c = 2259.7(2) pm, β = 103.73(5)°; Z = 4, and D_c = 1.36 g/cm³. The few molecular parameters cited in **Fig. 42** are averages. For the $C(Si(CH_3)_3)_3$ group the following distances (pm) and angles (°) were reported: 188(3) (C-Si), 192(5) ($Si-CH_3$), 112(1) (Si-C-Si), and 106(2) ($H_3C-Si-CH_3$) [1].

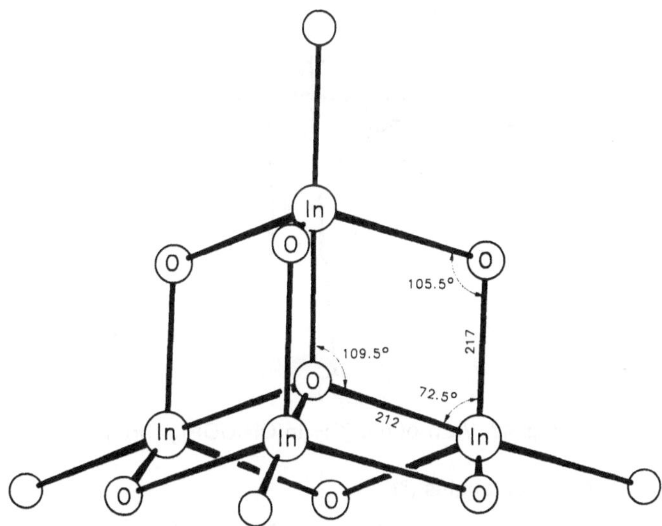

Fig. 42. Molecular structure of $\{((CH_3)_3Si)_3CIn\}_4(O)(OH)_6$; the $Si(CH_3)_3$ are omitted for clarity [1].

{$CH_3In(OH)OOP(C_6H_5)_2\}_4 \cdot 6\ C_5H_5N$

This compound was prepared as colorless crystals in 78% yield by the oxidation of dimeric $(CH_3)_2InP(C_6H_5)_2$ with C_5H_5NO (1:2 mole ratio), followed by controlled hydrolysis of the resulting pyridine-containing solution of $(CH_3)_2InOOP(C_6H_5)_2$; however, hydrolysis of pure $(CH_3)_2InOOP(C_6H_5)_2$ produced only a white powder which was slightly soluble in benzene [2].

^1H NMR (C_6D_6, in ppm): −0.17 (s, CH_3In), 3.37 (s, HO), 6.62 and 8.01 (m's, C_5H_5N), 6.91 and 8.50 (m's, C_6H_5). ^{31}P NMR (C_6D_6): 26.8 ppm. IR (Nujol mull, in cm^{-1}): 3605 m, 2710 w, 1975 w, 1900 w, 1830 w, 1785 w, 1590 m, 1305 m, 1210 w, 1150 sh, 1120 vs, 1065 m, 1030 vs, 1010 vs, 995 s, 740 m, 715 s, 685 vs, 535 m [2].

The compound crystallizes in the tetragonal $P4/n-C^3_{4h}$ (No. 85) with the cell parameters a=b=1915.3(6), c=1091.4(3) pm; Z=2. The tetrameric unit forms a cubane-like skeleton in which the In atoms are sixfold coordinated as shown in **Fig. 43**. Four of the six pyridine molecules are connected with the tetrameric unit via N···H-O bridges, the two other molecules are only included in the crystal lattice. For the bridging O-H and N-H distances 100.8(5) and 192.0(3) pm, respectively, were reported [2].

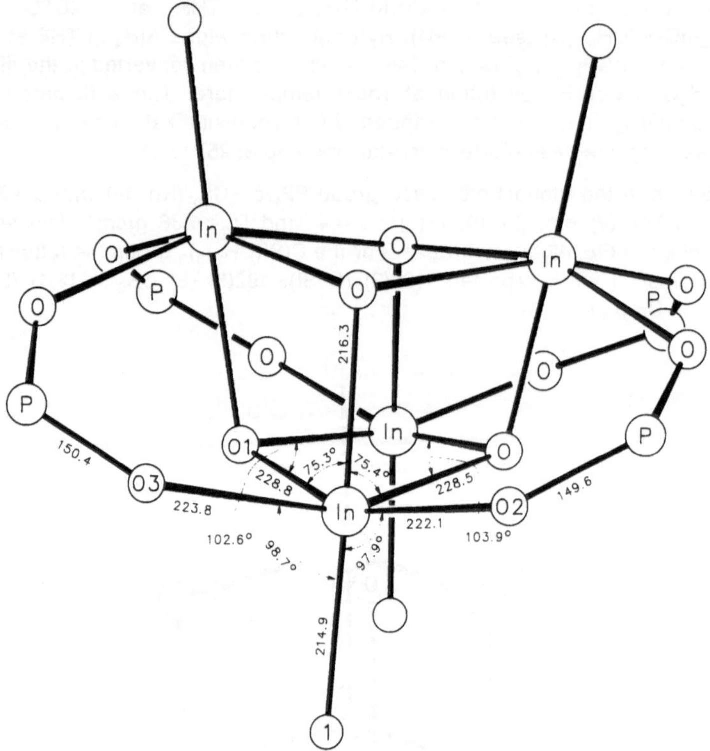

Fig. 43. View of the $[CH_3In(OH)OOP]_4$ core [2].

Other bond angles (°)

O(1)-In-O(2)	157.1(2)	O(2)-In-O(3)	94.9(1)
O(1)-In-O(3)	92.2(2)	O(2)-P-C	109.3(3)
O(1)-In-C(1)	102.4(2)		

References:

[1] Al-Juaid, S. S.; Buttrus, N. H.; Eaborn, C.; Hitchcock, P. B.; Roberts, A. T. L.; Smith, J. D.; Sullivan, A. C. (J. Chem. Soc. Chem. Commun. **1986** 908/9).

[2] Arif, A. M.; Barron, A. R. (Polyhedron **7** [1988] 2091/4).

4.8.4 Compounds with an $RIn(O_2)S$ Unit

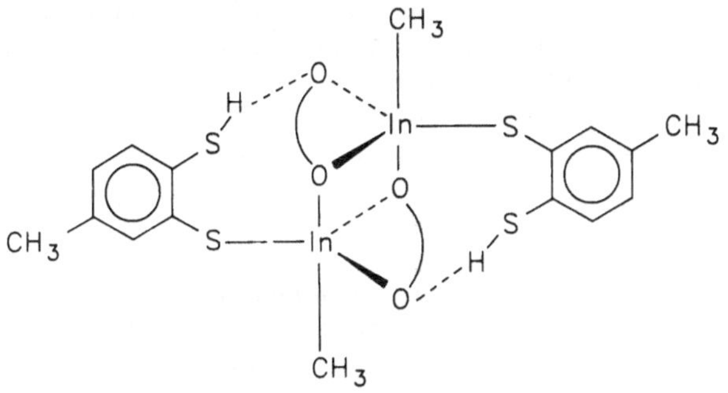

I

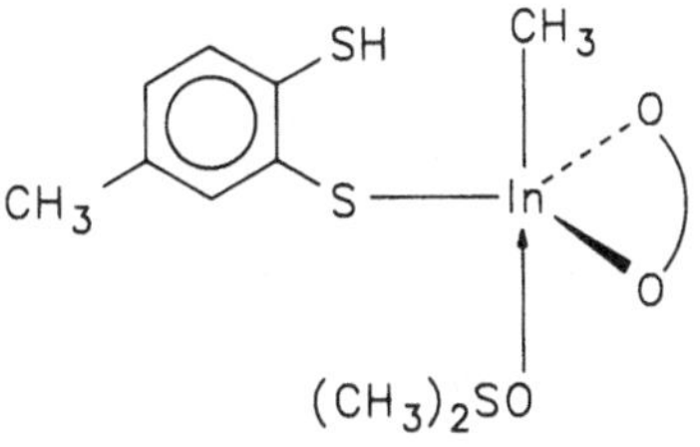

II

$CH_3In(OOCCH_3)(C_7H_7S_2)$

A compound of this composition was obtained by reaction of $(CH_3)_2InOOCCH_3$ in anhydrous C_2H_5OH with a small excess of undiluted toluene-3,4-dithiol; one equivalent of CH_4 was eliminated. The white precipitate was filtered off, washed with C_2H_5OH and $O(C_2H_5)_2$, and dried in vacuum [2]. It is dimeric in ethanol [1]. From ^1H NMR and IR spectroscopic studies (no data were quoted) [1], structure I was suggested [2].

When the compound was warmed to 40 °C with $(CH_3)_2SO$, a second equivalent of CH_4 split off to form $In(OOCCH_3)(C_7H_6S_2)$. The same product occurred when $(CH_3)_2InOOCCH_3$ reacted with $C_7H_8S_2$ in $C_2H_5OH/(CH_3)_2SO$ or in $P(N(CH_3)_2)_3$. The succeeding step was presumed to have proceeded via the monomeric DMSO adduct (II), which could not be isolated [2].

$CH_3In(OOCCH_3)\{S(CH_2)_nSH\}$

The compounds (n=2, 3) precipitated as white solids during the reaction of $(CH_3)_2$-$InOOCCH_3$ with $HS(CH_2)_nSH$ in C_2H_5OH. Warming the reaction mixture to 45 °C or addition of $(CH_3)_2SO$ did not lead to the loss of a second equivalent of CH_4; no further properties were reported [2].

References:

[1] Habeeb, J. J.; Tuck, D. G. (J. Organometal. Chem. **82** [1974] C 25/ C 26).

[2] Habeeb, J. J.; Tuck, D. G. (J. Organometal. Chem. **101** [1975] 1/9).

5 Organoindium-Sulfur Compounds

5.1 R_2InSR' and $R_2InSE(R')_n$ Compounds

The compounds in this section are mostly thiolates, R_2InSR', and are described in Table 46. In a few cases (Nos. 2 to 4, 6, and 20) chelate formation through an N or another S is possible. Derivatives of monothiocarbonic acids, $R_2InOSCR'$, have already been described as indium-oxygen compounds in Section 4.2.2 (p. 204).

The compounds were prepared by the following procedure:

Method I: From InR_3 or $R_3In \cdot D$ and the appropriate HSR′.

The appropriate HSR′ compound was added slowly with stirring to a solution of the trialkyl derivative in ether at −78 °C (Nos. 7 to 13) [14]. The mixture was warmed to room temperature and then refluxed for 2 to 6 h. No. 18 was prepared analogously from the adduct $(C_6H_5)_3In \cdot O_2C_4H_8$ at an initial temperature of −60 °C [6]. Nos. 1, 4, and 6 were prepared in benzene [7, 9, 11] at 20 to 80 °C, the phenyl derivatives No. 14 to 17 and 19 in boiling toluene [12], and No. 3 in CS_2 [2]. After evaporating or distilling off the solvent, the products were purified by sublimation (No. 1), distillation (Nos. 7 to 9 and 11 to 13), or recrystallized from CH_2Cl_2 (Nos. 14 to 17 and 19).

General Remarks. Cryoscopic or osmometric molecular weight determinations in benzene indicate that the thiolates, R_2InSR', are dimers with a structure close to Formula I. The IR and Raman spectra of solid $(CH_3)_2InSCH_3$ (No. 1) contain hints of a zigzag chain structure [9] similar to that of the homologous Al compound [4].

Vibrational spectroscopy suggested chelate structures II and III (E=S, NCH_3) for the monomeric $(CH_3)_2InSSP(CH_3)_2$ (No. 3), $(CH_3)_2InSP(NCH_3)(CH_3)_2$ (No. 4), and $(C_2H_5)_2InSSCCH_3$ (No. 6). For the monomeric Nos. 2 and 20 it was not decided whether the In was also being coordinated by another sulfur or by a nitrogen atom [8, 10].

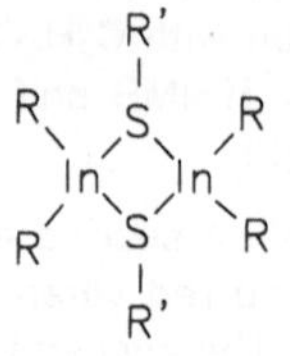

I: $R_2In(\mu\text{-}SR')_2InR_2$ II: $R_2In(S_2C\text{-}R')$ III: $R_2In(S,E)P(R')_2$

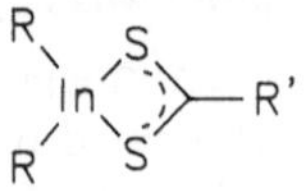

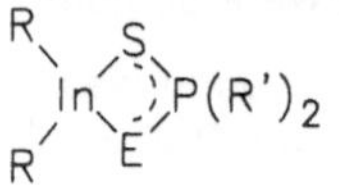

Table 46
Compounds of the R_2InSR' and $R_2InSE(R')_n$ Type.
Further information on numbers preceded by an asterisk is given at the end of the table.
Explanations, abbreviations, and units on p. X.

No.	compound method of preparation (yield)	properties and remarks
$(CH_3)_2In$ compounds		
1	$(CH_3)_2InSCH_3$ I (82%)	colorless crystals, m.p. 121 to 123 °C, 122 to 124 °C (dec.) [1], subl. 100 °C/10^{-3} Torr, dimeric in benzene [9] ^1H NMR (CCl_4): 0.07 (CH_3In), 2.22 (CH_3S) [9] IR: 1160 mw $\delta_s(CH_3In)$, 720 vs $\varrho(CH_3In)$, 695 vs $\nu(C\text{-}S)$ as solid, 510 ms $\nu_{as}(InC_2)$, 470 m $\nu_s(InC_2)$, 270 ms, 255 s $\nu(In_2S_2)$ in CCl_4 [9]

Table 46 (continued)

No.	compound method of preparation (yield)	properties and remarks
1 (continued)		Raman (CCl_4): 1165 vs $\delta_s(CH_3In)$, 725 vvw $\varrho(CH_3In)$, 700 m ν(C-S), 522 m $\nu_{as}(InC_2)$, 485 vs $\nu_s(InC_2)$, 265 s $\nu_s(In_2S_2)$, 216 w δ(InSC); other bands reported [9] a polymeric structure proposed for the solid state [9]
*2	$(CH_3)_2InSSCN(CH_3)_2$ special	white solid, m.p. 120 to 125 °C [8] IR (solid): 1515 ν(C=N), 965 ν(CS), 521 $\nu_{as}(InC_2)$, 485 $\nu_s(InC_2)$, 373 ν(InS); very similar data for the CCl_4 solution [8]
3	$(CH_3)_2InSSP(CH_3)_2$ I	white solid, m.p. 184 to 185 °C, subl. 110 to 120 °C/0.01 Torr [2] IR: 1416 ms, sh, 1406 m, sh, 1397 m, 1294 w, 1284 m, 1147 vw, 984 s, 911 s, 898 w, sh, 852 m, 738 m, 724 ms, 596 s, 508 w, 496 ms, 323 m, 313 s, 371 m, 256 m as solid; 1147 vw $\delta_s(CH_3In)$, 596 s, 496 m ν(P-S), 508 w ν(InC), 323 sh, 313 s ν(In-S) in CS_2 [2]
4	$(CH_3)_2InSP(=NCH_3)(CH_3)_2$ I (60 to 70%)	white solid, m.p. 114 °C, recrystallized from C_6H_6/C_6H_{14} (10/1), monomeric in benzene [11] ^{1}H NMR (C_6D_6): 0.23 (CH_3In), 1.45 (d, CH_3P, $^2J(H,P) = 12.82$), 2.63 (d, CH_3N, $^3J(H,P) = 21.43$) [11] ^{31}P NMR (C_6D_6): −63.8 (m) [11] IR (solid): 1158 w, 1150 vw $\delta_s(CH_3In)$, 849 s $\nu(P\overset{...}{=}N)$, 759 ms $\nu_{as}(PC_2)$, 700 vs,br $\varrho(CH_3In)+\nu_s(PC_2)$, 537 s ν(InN), 508 ms $\nu_{as}(InC_2)$, 478 s, br $\nu_s(InC_2)+\nu(P\overset{...}{=}S)$, 351 ms δ(PNC); other bands reported [11] Raman (solid): 1157 sh, 1152 s $\delta_s(CH_3In)$, 849 w $\nu(P\overset{...}{=}N)$, 763 mw $\nu_{as}(PC_2)$, 712 m $\nu_s(PC_2)$, 702 w, sh $\varrho(CH_3In)$, 536 m ν(InN), 509 m $\nu_{as}(InC_2)$, 486 vs $\nu_s(InC_2)$, 476 s $\nu(P\overset{...}{=}S)$, 368 mw δ(NPS); other bands reported [11]
$(C_2H_5)_2In$ compounds		
*5	$(C_2H_5)_2InSCH(C_6H_5)_2$ special	white crystals, m.p. 130 to 131 °C [3, 5] ^{1}H NMR (CCl_4): 0.55 (q, CH_2In), 1.05 (t, CH_3CH_2), 5.40 (s, CH), 7.15 (m, C_6H_5) [3, 5]
*6	$(C_2H_5)_2InSSCCH_3$ I (60 to 70%)	orange-yellow solid, m.p. 36 to 39 °C, b.p. 76 to 77 °C/0.1 Torr, monomeric in benzene [7] IR and Raman frequencies in Table 47 on p. 244
$(C_4H_9\text{-n})_2In$ compounds		
*7	$(C_4H_9\text{-n})_2InSC_3H_7\text{-n}$ I (87%)	colorless, viscous oil, b.p. 70 °C/10^{-3} Torr [14] ^{1}H NMR ($CDCl_3$, 10 wt%): 0.92 (CH_3 of C_4H_9), 0.97 (CH_2In), 1.1 to 1.8 (CH_2-2,3), 2.73 (CH_2S) [14] ^{13}C NMR ($CDCl_3$): 13.8 (CH_3 of C_4H_9), 16.6 (CH_2In), 28.4 (CH_2-3), 29.7 (CH_2S), 30.0 (CH_2-2) [14]

Table 46 (continued)

No.	compound method of preparation (yield)	properties and remarks
*8	$(C_4H_9\text{-n})_2InSC_3H_7\text{-i}$ I (81%)	colorless, viscous oil, b.p. 78 °C/10^{-3} Torr [14] ^{1}H NMR ($CDCl_3$, 10 wt%): 0.85 (CH_3 of C_4H_9), 0.95 (CH_2In), 1.1 to 1.8 (CH_2-2,3), 3.28 (CHS) [14] ^{13}C NMR ($CDCl_3$): 13.8 (CH_3 of C_4H_9), 17.5 (CH_2In), 28.6 (CH_2-3), 30.0 (CH_2-2), 33.8 (CHS) [14]
9	$(C_4H_9\text{-n})_2InSC_6H_{11}\text{-c}$ I (96%)	colorless, viscous oil, b.p. 78 °C/0.02 Torr [14] ^{1}H NMR ($CDCl_3$, 10 wt%): 0.93 (CH_3 of C_4H_9), 1.0 to 1.9 (CH_2In, CH_2-2,3), 3.10 (CHS) [14] ^{13}C NMR ($CDCl_3$): 13.8 (CH_3 of C_4H_9), 16.6 (CH_2In), 28.2 (CH_2-3), 29.9 (CH_2-2), 41.9 (CHS) [14]
*10	$(C_4H_9\text{-n})_2InSC_6H_5$ I (87%)	white solid, m.p. 160 °C (dec.) [14] ^{1}H NMR ($CDCl_3$, 10 wt%): 0.83 (CH_3 of C_4H_9), 0.98 (CH_2In), 1.21 (CH_2-3), 1.54 (CH_2-2) [14] ^{13}C NMR ($CDCl_3$): 13.6 (CH_3 of C_4H_9), 19.5 (CH_2In), 28.2 (CH_2-3), 29.6 (CH_2-2), no data reported for C_6H_5S [14]
$(C_4H_9\text{-i})_2In$ compounds		
*11	$(C_4H_9\text{-i})_2InSC_3H_7\text{-n}$ I (88%)	colorless, viscous oil, b.p. 132 °C/10^{-3} Torr [14] ^{1}H NMR ($CDCl_3$, 10 wt%): 0.99 (CH_3C), 1.03 (CH_2In), 2.14 (CHC), 2.75 (CH_2S) [14] ^{13}C NMR ($CDCl_3$): 26.1 (CH_2In), 27.8 (CH_3), 29.8 (CH + CH_2S) [14]
*12	$(C_4H_9\text{-i})_2InSC_3H_7\text{-i}$ I (92%)	colorless, viscous oil, b.p. 100 °C/10^{-3} Torr [14] ^{1}H NMR ($CDCl_3$, 10 wt%): 0.93 (CH_3C), 0.95 (CH_2In), 2.08 (CHC), 3.25 (CHS) [14] ^{13}C NMR ($CDCl_3$): 27.8 (CH_3), 28.0 (CH), 30.5 (CH_2In), 33.7 (CHS) [14]
13	$(C_4H_9\text{-i})_2InSC_6H_5$ I (89%)	viscous oil, b.p. > 190 °C/10^{-3} Torr [14] ^{1}H NMR ($CDCl_3$, 10 wt%): 0.86 (CH_3C), 1.01 (CH_2In), 2.00 (CHC) [14] ^{13}C NMR ($CDCl_3$): 27.0 (CH_3), 27.6 (CH), 32.7 (CH_2In), 154.2 (CS) [14]
$(C_6H_5)_2In$ compounds		
14	$(C_6H_5)_2InSC_2H_5$ I (89.5%)	white solid, m.p. 167 °C (dec.) (DTA), dimeric in benzene [12] ^{1}H NMR (CH_2Cl_2): 1.07 (t, CH_3), 2.67 (q, CH_2S), 7.18 to 7.75 (m, C_6H_5) [12] IR (CH_2Cl_2): 440 s ν(In-C_6H_5), 379 m, 250 w ν(InS); other bands at 465 s, 229 w [12]

Table 46 (continued)

No.	compound method of preparation (yield)	properties and remarks
15	$(C_6H_5)_2InSC_3H_7$-n I (83.3%)	white solid, m.p. 168 °C (DTA) [12] ^{1}H NMR (CH_2Cl_2): 0.60 (t, CH_3), 1.27 (t, CH_2), 3.23 (t, CH_2S), 7.27 to 7.83 (m, C_6H_5) [12] IR (solid): 443 s ν(In–C_6H_5), 376 m, 259 m ν(InS); other bands at 288 m, 249 m [12]
16	$(C_6H_5)_2InSC_4H_9$-t I (72.3%)	white solid, m.p. 155 °C (DTA) [12] ^{1}H NMR (CH_2Cl_2): 1.52 (s, C_4H_9), 7.11 to 7.85 (m, C_6H_5) [12] IR (CH_2Cl_2): 438 s ν(In–C_6H_5), 385 s ν(InS); other bands at 348 m, 282 w [12]
17	$(C_6H_5)_2InSCH_2C_6H_5$ I (49.3%)	white solid, m.p. 158 °C, dec. 163 °C (DTA) [12] ^{1}H NMR (CH_2Cl_2): 3.77 (s, CH_2S), 6.87 to 7.83 (m, C_6H_5) [12] IR (solid): 438 vs ν(In–C_6H_5), 336 m, 255 m ν(InS); other bands at 469 m, 444 vs, 291 m, 243 s, 220 m [12]
18	$(C_6H_5)_2InSC_{12}H_{25}$ I (83%)	solid, no properties reported [6]
19	$(C_6H_5)_2InSC_6H_5$ I (69.3%)	white solid, m.p. 174 °C (DTA) [12] ^{1}H NMR (CH_2Cl_2): 6.98 to 7.50 (m, C_6H_5) [12] IR (solid): 438 m ν(InC_6H_5), 319 w ν(InS); other band at 475 m [12]
*20	$(C_6H_5)_2InSSCN(C_2H_5)_2$ special	m.p. 142 °C [10] IR (solid): 1515 vs, 1500 vs ν(CN), 1012 vs, 1005 s ν(C–S) [10]

* Further information:

$(CH_3)_2InSSCN(CH_3)_2$ (Table **46**, No. **2**) was obtained in 68% yield from $(CH_3)_2InOOCCH_3$ (Section 4.2.1) and an equimolar amount of $NaSSCN(CH_3)_2$ in ice water. The white precipitate was filtered off and dried in vacuum. It is monomeric in benzene solution [8].

$(C_2H_5)_2InSCH(C_6H_5)_2$ (Table **46**, No. **5**) was obtained within 2 to 3 min in 78% yield from the reaction of $In(C_2H_5)_3$ with $(C_6H_5)_2C{=}S$ at 15 °C in hexane; C_2H_4 was evolved [3]. When the reaction was carried out in CH_2Cl_2 at −50 to −40 °C, a red coloration was first observed, which then disappeared on warming accompanied by increasing evolution of C_2H_4 [5]. The light yellow precipitate was washed with petroleum ether and recrystallized from CH_2Cl_2/petroleum ether [3, 5]. The product is a dimer in benzene [5] and possesses a four-ring structure as shown in Formula I on p. 240.

$(C_2H_5)_2InSSCCH_3$ (Table **46**, No. **6**). The IR and Raman spectra of the liquid were compared to those of $(CH_3)_2GaSSCCH_3$ and $(C_2H_5)_2GaSSCCH_3$. The suggested chelate structure (Formula II on p. 240) was based on the many similarities in the spectra. Frequency values appear in Table 47 [7].

Table 47
IR and Raman Frequencies of $(C_2H_5)_2InSSCCH_3$ [7].
Wavenumbers in cm^{-1}.

IR	Raman	assignment
2962 vs	2960 sh	$\nu_{as}(CH_3C)$
2943 vs	2938 m, sh	$\nu_{as}(CH_3In)$
2918 s, br	2912 s, br	$\nu_s(CH_3)$, $\nu_{as}(CH_2)$, $\nu_s(CH_2)$
2871 s	2870 ms	
2818 mw	2817 w	overtones
2728 w	–	
1461 ms	1459 mw, br	$\delta_{as}(CH_3)$
1420 m	1422 w	$\delta(CH_2)$
1378 ms	–	$\delta_s(CH_3, In)$
1357 m	1358 w	$\delta_s(CH_3, C)$
1231 m	–	$\delta(CH_2In)$
1168 m, sh	1168 s	$\omega(CH_2In)$
1140 vs	1138 m	$\nu_{as}(C \overset{...}{=} S)$
1088 sh	–	$\nu(C\text{-}CH_3)$

IR	Raman	assignment
1008 s	1001 mw	$\nu(CH_2, CH_3)$
960 m	959 w	
931 m	930 w, br	$\varrho(CH_3, In)$
872 vs	870 vw	$\varrho(CH_3, C)$
643 s	648 vw, br	$\varrho(CH_2In)$
608 m	608 vs	$\nu_s(C \overset{...}{=} S)$
498 ms	500 w	$\nu_{as}(InC_2)$
460 s	461 vs	$\nu_s(InC_2)$, $\delta(CS_2)$
386 ms	388 m, br	$\delta(CS_2)$
348 w	351 mw	
252 mw, br	243	$\delta(InCC)$
202 w	190 m	$\nu(InS_2)$
162 s, br	162 sh	$\delta(InC_2)$, $\delta(ClnS)$
145 mw	–	

$(C_4H_9\text{-n})_2InSR'$ (Table **46**, No. **7**, $R' = C_3H_7\text{-n}$; No. **8**, $R' = C_3H_7\text{-i}$; No. **10**, $R' = C_6H_5$) and **$(C_4H_9\text{-i})_2InSR'$** (Table **46**, No. **11**, $R' = C_3H_7\text{-n}$; No. **12**, $R' = C_3H_7\text{- i}$) were examined by thermogravimetry (TG) and differential thermal analysis (DTA). Nos. 7, 8, 11, and 12 decompose up to 280 °C with an exothermic DT peak giving powdery InS; for No. 10 the decomposition temperature is higher but with the same result. The breaking of the In-C and In-S bonds are the essential steps in the decomposition reactions of these organoindium thiolates. The temperature regions of the decompositions and the weight loss accompaning each are presented below. Thermal analysis was carried out with a nitrogen flow at 10 °C/min from room temperature to 400 °C. Pyrolysates were identified by comparing X-ray diffraction with known data [15].

compound	TG (°C)		weight loss (%)	DTA (°C)	
No.	initial	final		exo	endo
7	140	282	75	274	–
8	148	277	72	265	–
10	160	360	52	320	160
11	123	277	75	242	–
12	97	269	55	264	–

The pyrolysis of $(C_4H_9\text{-i})_2InSC_3H_7\text{-i}$ (No. 12) in the presence of oxygen was examined in the 200 to 500 °C range [15] and tested for producing transparent, electrically-conducting In_2O_3 films [13]; in p-xylole and in the presence of Cu-bis(dibutyldithiocarbamate) at 250 to 350 °C $CuInS_2$ films are formed [16].

$(C_6H_5)_2InSSCN(C_2H_5)_2$ (Table **46**, No. **20**) formed on adding aqueous $[NH_4][SSCN(C_2H_5)_2]$ to a solution of $(C_6H_5)_2InCl$. The white precipitate was filtered off and dried in vacuum [10].

The compound was tested for antimicrobial activity [10].

References:

[1] Coates, G. E.; Whitcombe, R. A. (J. Chem. Soc. **1956** 3351/4).
[2] Coates, G. E.; Mukherjee, R. N. (J. Chem. Soc. **1964** 1295/303).
[3] Tada, H.; Yasuda, K.; Okawara, R. (Inorg. Nucl. Chem. Letters **3** [1967] 315/7).
[4] Brauer, D. J.; Stucky, G. D. (J. Am. Chem. Soc. **91** [1969] 5462/6).
[5] Tada, H.; Yasuda, K.; Okawara, R. (J. Organometal. Chem. **16** [1969] 215/20).
[6] Viktorova, I. M.; Sheverdina, N. I.; Kocheshkov, K. A. (Dokl. Akad. Nauk SSSR **198** [1971] 94/5; Dokl. Chem. Proc. Acad. Sci. USSR **198/201** [1971] 367/8; C.A. **75** [1971] No. 49184).
[7] Weidlein, J. (Z. Anorg. Allgem. Chem. **386** [1971] 129/38).
[8] Maeda, T.; Okawara, R. (J. Organometal. Chem. **39** [1972] 87/91).
[9] Mann, W. G. (Diss. Univ. Stuttgart 1974).
[10] Srivastava, T. N.; Kumar, V.; Srivastava, O. P. (Natl. Acad. Sci. Letters [India] **1** [1978] 97/100; C.A. **89** [1978] No. 141068).

[11] Schrem, H.; Weidlein, J. (Z. Anorg. Allgem. Chem. **465** [1980] 109/19).
[12] Hoffmann, G. G. (J. Organometal. Chem. **338** [1988] 305/17).
[13] Nomura, R.; Moritake, A.; Kanaya, K.; Matsuda, H. (Thin Solid Films **167** [1988] L 27/L 29).
[14] Nomura, R.; Inazawa, S.; Kanaya, K.; Matsuda, H. (Polyhedron **8** [1989] 763/7).
[15] Nomura, R.; Inazawa, S.; Kanaya, K.; Matsuda, H. (Appl. Organometal. Chem. **3** [1989] 195/7).
[16] Nomura, R.; Kanaya, K.; Matsuda, H. (Chem. Letters **1988** 1849/50).

5.2 RIn-Sulfur Compounds

5.2.1 Compounds with an RInS Unit

The two compounds in this section have been mentioned frequently in the contexts of preparation and analysis, but have not been accurately characterized. The RInS derivatives containing an additional O_2 coordination which have been prepared from $(CH_3)_2$-$InOOCCH_3$ and some dithio acids ($HS(CH_2)_nSH$, n = 2, 3) or toluene-3,4-dithiol have already been described with the oxygen compounds in Section 4.8.4.

$(CH_3InS)_n$

This compound was prepared in 95% yield by the reaction of equimolar amounts of $In(CH_3)_3$ and $\{(CH_3)_2SiS\}_3$ in benzene [2] and in yields of 65 to 97% from $(CH_3)_2InX$ (X = Cl, Br, I) and $(CH_3)_3SbS$ (1:1 mole ratio) in less than 4 h in boiling CH_3OH via the adduct $(CH_3)_2InX \cdot SSb(CH_3)_3$; decomposition of the adduct leads to the title compound and $(CH_3)_4$-SbX [4]. The white polymeric solid is insoluble in common solvents and melts with decomposition above 270 °C [4] (> 200 °C [2]).

The compound was used to catalyze the polymerization of ethylene [2].

$(C_2H_5InS)_n$

The ethyl derivative was obtained in 96% yield from $(C_2H_5)_3In \cdot n\,O(C_2H_5)_2$ and $\{(CH_3)_2$-$SnS\}_3$ or $\{(C_6H_5)_2SnS\}_3$ in benzene solution (1:1 mole ratio) at room temperature [1]. The white polymeric solid is only slightly soluble and melts above 250 °C with decomposition. Like the preceding compound, $(C_2H_5InS)_n$ was formed (along with $(CH_3)_3C_2H_5SbX$) from $(C_2H_5)_2InX$ (X = Cl, Br, I) and $(CH_3)_3SbS$ in boiling CH_3OH. In CH_2Cl_2 at room temperature only highly viscous oily products of nonuniform composition resulted after a 70 h reaction period [4]. Finally, $(C_2H_5InS)_n$ occurred in high yield from thermal decomposition of $(C_2H_5)_2$-$InOSCC_6H_5$ in propiophenone at 120 °C (see p. 205) [3].

References:

[1] Yasuda, K.; Okawara, R. (Inorg. Nucl. Chem. Letters **3** [1967] 135/6).
[2] Okawara, R.; Yasuda, K. (Japan. 70-32692 [1970] from C.A. **74** [1971] No. 112621).
[3] Tada, H.; Okawara, R. (J. Organometal. Chem. **28** [1971] 21/4).
[4] Maeda, T.; Yoshida, G.; Okawara, R. (J. Organometal. Chem. **44** [1972] 237/41).

5.2.2 Compounds with an $RInS_2$ Unit

The compounds in Table 48 are mainly derivatives of mono- or dithioacids. With the exception of a few butylindium compounds, the compounds are white or yellow solids with offensive odors. On warming most decompose thermally before reaching the melting point and are barely soluble in nonpolar organic solvents. As a rule, a polymeric structure with 5- or 6-fold coordinated In atoms is adopted.

The following general preparative methods have been used.

Method I: From InR_3 or $R_3In \cdot D$ and HSR′ or $(HS)_2R'$.
The procedure corresponds to Method I for the monothio derivatives (p. 240). For Nos. 4 to 7, 9, and 10 InR_3 was mixed with a solution of 2 equivalents HSR′ in ether at −78 °C, stirred for 20 to 24 h at room temperature, and then refluxed for 2 to 6 h. In a similar way No. 11 was obtained from $(C_6H_5)_3In \cdot O_2C_4H_8$ [1] while [6] reports that the reaction of triphenylindium with HSR′ in 1:2 mole ratio forms only a mixture of $(C_6H_5)_2InSR'$ and $In(SR')_3$. The derivatives of the dithiols, HS(R′)SH, were obtained by reacting a 1:1 mole ratio in boiling toluene (Nos. 12 to 15) or ether (Nos. 1 and 8) for several hours with continuous stirring. Solid dithio derivatives were isolated by filtering, washing several times with ether [1, 2] or hexane [6], and then drying in vacuum; liquid products were purified by vacuum distillation [7].

Method II: From R_2InX and M^ISSCNR_2.
No. 2 formed within 24 h at room temperature from $(CH_3)_2InSSCN(CH_3)_2$ (Table 46, No. 2) in methanolic solution according to $2\,(CH_3)_2InSSCN(CH_3)_2 \rightarrow CH_3In(SSCN(CH_3)_2)_2 + In(CH_3)_3$. The expected by-product, $In(CH_3)_3$, was not detected. The homologous No. 3 was obtained directly by reaction of an ice-cold aqueous solution of $(C_2H_5)_2InOOCCH_3$ or $(C_2H_5)_2InCl$ with aqueous $Na[SSCN(CH_3)_2]$ (1:1.65 mole ratio); the precursor, $(C_2H_5)_2InSSCN(CH_3)_2$, was not isolable. Purification was by recrystallization from CH_2Cl_2/petroleum ether [3].

Method III: From $RInS_2R'$ compound and the appropriate donor D.
Adduct Nos. 17 to 19 were prepared from solutions of No. 1 in large excesses of the appropriate donors, removal of the unconverted donor by careful distillation or evaporation, and recrystallization of the residue. CH_2Cl_2 was used as solvent for preparing Nos. 20 and 21 [2]. Adduct Nos. 22 and 23 were accessible by mixing equimolar amounts of No. 16 and the bidentate base in ether solution. These adducts separate as solids and were washed with petroleum ether and dried in vacuum [4].

A number of unspecified compounds $C_6H_5In(SSCNR_2)_2$ (R = alkyl and aryl) have been prepared according to Method II and tested for antimicrobial activity, but no properties have been reported [5].

Table 48
Compounds with an $RInS_2$ Unit.
Further information on compounds whose numbers are preceded by an asterisk is given at the end of the table.
Explanations, abbreviations, and units on p. X.

No.	compound method of preparation (yield)	properties and remarks
CH_3In compounds		
*1	CH_3In (S,S-chelated 4-methylbenzene-1,2-dithiolate) I	white solid, dec. ca. 300 °C without melting, slow subl. ca. 150 °C under vacuum, monomeric in $(CH_3)_2CO$, insoluble in common solvents, soluble in donor solvents [2] ^1H NMR on p. 251 IR (solid): 724 s, br $\varrho(CH_3In)$, 510 m $\nu(InC)$, 378 w, 349 m, br $\nu(InS)$ [2] Raman (solid): 724 $\varrho(CH_3In)$, 509 s $\nu(InC)$, 378 w, 340 m, 325 m $\nu(InS)$ [2] mass spectrum (160 °C): only $[M_2]^+$ and $[M]^+$ reported [2]
2	$CH_3In(SSCN(CH_3)_2)_2$ II (48.5%)	pale yellow crystals, m.p. 206 to 208 °C, monomeric in benzene [3] IR (solid/$CHCl_3$): 1515/1490 $\nu(C{=}N)$, 979/982 $\nu(C\text{-}S)$, 505/506 $\nu(InC)$, 377/379 $\nu(InS)$; no intensities and no other bands reported [3]
C_2H_5In compounds		
3	$C_2H_5In(SSCN(CH_3)_2)_2$ II (23%)	pale yellow solid, m.p. 117 to 121 °C (dec.), monomeric in benzene [3] IR (solid/$CHCl_3$): 1510/1496 $\nu(C{=}N)$, 976/982 $\nu(C\text{-}S)$, 487/486 $\nu(InC)$, 377/378 $\nu(InS)$; no intensities and no other bands reported [3]
$(C_4H_9\text{-}n)In$ compounds		
4	$(C_4H_9\text{-}n)In(SC_3H_7\text{-}n)_2$ I	no properties reported [8] TG and DTA analyses: thermal decomposition starts at 87 °C, ends at 336 °C (49% weight loss), and gives In_2S_3 [8]
5	$(C_4H_9\text{-}n)In(SC_3H_7\text{-}i)_2$ I (44 to 58%)	colorless viscous oil, b.p. 148 °C/10^{-3} Torr; degree of association in benzene: 1.1 to 1.4 [7] ^1H NMR ($CDCl_3$, 10 wt%): 0.93 (CH_3 of C_4H_9), 1.00 (CH_2In), 1.40 (CH_2-3), 1.65 (CH_2-2), 3.46 (CH) [7] ^{13}C NMR ($CDCl_3$): 13.7 (CH_3 of C_4H_9), 17.9 (CH_2In), 28.3 (CH_2-3), 29.9 (CH_2-2), 34.1 (CH) [7]
6	$(C_4H_9\text{-}n)In(SC_6H_{11}\text{-}c)_2$ I (56%)	colorless viscous oil, b.p. 91 °C/0.04 Torr, monomeric in benzene [7]

Table 48 (continued)

No.	compound method of preparation (yield)	properties and remarks
6 (continued)		^{1}H NMR ($CDCl_3$, 10 wt%): 0.94 (CH_3 of C_4H_9), 1.3 to 2.2 (CH_2In, CH_2-2,3), 3.17 (CHS) [7] ^{13}C NMR ($CDCl_3$): 13.7 (CH_3 of C_4H_9), 19.2 (CH_2In), 25.3 (CH_2-3), 26.7 (CH_2-2), 42.6 (CHS) [7]
7	$(C_4H_9\text{-}n)In(SC_6H_5)_2$ I (67%)	white solid, m.p. 150 °C (dec.) [7] solubility too low to record NMR spectra [7]
8	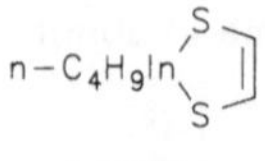 S $n\text{-}C_4H_9In$ S I (71%)	white solid, m.p. 117 to 118 °C, dimeric in benzene [7] ^{1}H NMR ($CDCl_3$, 10 wt%): 0.93 (CH_3 of C_4H_9), 1.33 (CH_2-3), 1.41 (CH_2In), 1.69 (CH_2-2), 3.16 (CHS) [7] ^{13}C NMR ($CDCl_3$): 13.8 (CH_3 of C_4H_9), 19.6 (CH_2In), 28.2 (CH_2-3), 29.6 (CH_2-2), 34.8 (CHS) [7]
$(C_4H_9\text{-}i)In$ compounds		
9	$(C_4H_9\text{-}i)In(SC_3H_7\text{-}n)_2$ I (40 to 50%)	could not be separated from $(C_4H_9\text{-}i)_2InSC_3H_7\text{-}n$ (Table 46, No. 11) [7]
10	$(C_4H_9\text{-}i)In(SC_6H_5)_2$ I (64%)	no properties reported [7]
C_6H_5In compounds		
11	$C_6H_5In(SC_{12}H_{25})_2$ I (78%)	no properties reported [1]
12	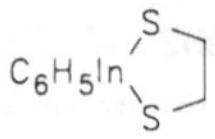 S C_6H_5In S I (89%)	white solid, m.p. 190 °C (DTA) [6] IR (solid): 439 s ν(In-C_6H_5), 257 m, 237 m ν(InS); other bands at 422 s, 317 s, 292 s, 237 m [6]
13	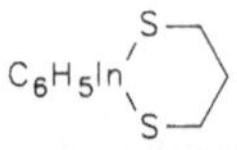 S C_6H_5In S I (90%)	white solid, m.p. 217 °C (DTA) [6] IR (solid): 441 s ν(In-C_6H_5), 277 s, 237 m ν(InS); other bands at 481 m, 464 m, 336 s [6]
14	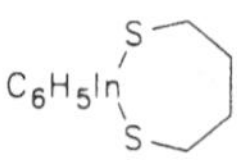 S C_6H_5In S I (91%)	white solid, m.p. 198 °C (DTA) [6] IR (solid): 443 s ν(In-C_6H_5), 264 m, 234 m ν(InS); other bands at 425 s, 270 sh [6]
15	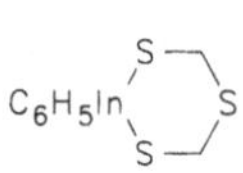 S C_6H_5In S S I (85%)	white solid, m.p. 151 °C (DTA) [6] IR (solid): 440 s ν(In-C_6H_5), 259 br, 224 m ν(InS); other bands at 464 m, 363 s, 330 m, 307 br [6]

Table 48 (continued)

No.	compound method of preparation (yield)	properties and remarks
(C_5H_5-c)In compounds		
*16	c-C_5H_5In(S$_2$C$_2$(CF$_3$)$_2$) special	yellow solid, dec. violently at 170 °C, insoluble in CCl_4, slightly soluble in C_6H_6 and $CHCl_3$, soluble in CH_3OH, $O(C_2H_5)_2$, $(CH_3)_2CO$ [4] ^{1}H NMR: 6.17 (s, C_5H_5) in $CDCl_3$; 5.75 in $(CD_3)_2SO$ [4] ^{19}F NMR ($CDCl_3$): −0.054 (s, CF_3) [4] IR (solid): 3080 w, 2940 w ν(C_5H_5), 1529 s ν(C=C), 893 s, 830 m ν(CF_3) + ν(CS), 862 s δ(C_5), 340 s ν(In-C_5) [4] Raman (solid): 337 s ν(In-C_5), no other bands reported [4]
adducts of (CH_3)In compounds		
*17	$CH_3InS_2C_6H_3CH_3 \cdot OS(CH_3)_2$ III	off-white crystals, rapid dec. 170 °C, slightly soluble in $(CH_3)_2CO$ and $CHCl_3$, soluble in donor solvents [2] conductivity (CH_3NO_2): 4.0 $\Omega^{-1} \cdot cm^2 \cdot mol^{-1}$ [2] ^{1}H NMR on p. 251 IR (solid): 722 s, br ϱ(CH_3In), 507 m ν(InC); other important bands obscured by absorptions of the donor ligand [2]
*18	$CH_3InS_2C_6H_3CH_3 \cdot N(CH_3)_3$ III	white crystals, m.p. ca. 170 °C (dec.), slightly soluble in CH_2Cl_2, $(CH_3)_2CO$, $O(C_2H_5)_2$, C_2H_5OH, soluble in donor solvents, monomeric in $(CH_3)_2CO$, dimeric in N,N′-dimethylacetamide [2] conductivity (CH_3NO_2): 1.9 $\Omega^{-1} \cdot cm^2 \cdot mol^{-1}$ [2] ^{1}H NMR on p. 251 IR (solid): 727 s, br ϱ(CH_3In), 520 s ν(InC), 378 sh, 369 m, 326 m ν(InS), 279 w ν(InN) [2]
*19	$CH_3InS_2C_6H_3CH_3 \cdot NC_5H_4$ III	white crystals, dec. 170 °C without melting, slightly soluble in $(CH_3)_2CO$ and C_2H_5OH, soluble in donor solvents, dimeric in $(CH_3)_2CO$ [2] conductivity (CH_3NO_2): 2.8 $\Omega^{-1} \cdot cm^2 \cdot mol^{-1}$ [2] ^{1}H NMR on p. 251 IR (solid): 720 sh ϱ(CH_3In), 507 s ν(InC), 383 vw, 339 mw ν(InS), 280 mw ν(InN), 270 mw [2] Raman (solid): 509 vs ν(InC), 381 m, 333 s, 324 w ν(InS), 274 s, 255 w ν(InN); no other bands reported [2]
*20	$CH_3InS_2C_6H_3CH_3 \cdot C_{10}H_8N_2$-2,2′ $C_{10}H_8N_2$-2,2′ = 2,2′-bipyridine III	yellow crystals, m.p. 220 °C (dec.), monomeric in $(CH_3)_2CO$ [2] conductivity (CH_3NO_2): 0.6 $\Omega^{-1} \cdot cm^2 \cdot mol^{-1}$ [2] ^{1}H NMR on p. 251 IR (solid): 505 m ν(InC), 372 vw ν(InS), 301 w ν(InN) [2]

Table 48 (continued)

No.	compound method of preparation (yield)	properties and remarks
*20 (continued)		Raman (solid): 505 s ν(InC), 349 s ν(InS), 242 w? ν(InN); no other bands reported [2]
*21	$CH_3InS_2C_6H_3CH_3 \cdot C_{12}H_8N_2$-1,10 $C_{12}H_8N_2$-1,10= 1,10-phenanthroline III	yellow solid, m.p. 210 °C (dec.), monomeric in $(CH_3)_2CO$ [2] conductivity (CH_3NO_2): 0.3 $\Omega^{-1} \cdot cm^2 \cdot mol^{-1}$ [2] ^1H NMR on p. 251 IR (solid): 508 s ν(InC), 381 m, 352 m ν(InS), 279 w ν(InN) [2] Raman (solid): 505 m ν(InC), 373 w, 357 m, 345 s ν(InS), 278 w ν(InN); no other bands reported [2]
adducts of $(C_5H_5\text{-c})In$ compounds		
22	$(C_5H_5\text{-c})InS_2C_2(CF_3)_2 \cdot C_{10}H_8N_2$-2,2′ III	yellow solid, dec. starts at 160 °C, turns black at 185 °C, insoluble in CH_3OH and $(C_2H_5)_2O$, slightly soluble in $CHCl_3$, soluble in C_6H_6, $(CH_3)_2SO$, $(CH_3)_2CO$, and CH_3NO_2 [4] ^1H NMR ($OS(CD_3)_2$): 6.33 (s, C_5H_5) [4] ^{19}F NMR ($OS(CD_3)_2$): −0.051 (s, CF_3) [4] IR (solid): 890 s, 830 s ν(CF_3)+ν(CS), 844 m δ(C_5), 325 m ν(In-C_5) [4] Raman (solid): 327 s ν(In-C_5), no other bands reported [4]
23	$(C_5H_5\text{-c})InS_2C_2(CF_3)_2 \cdot C_{12}H_8N_2$-1,10 III	yellow solid, dec. ca. 200 °C, properties similar to No. 22 [4] ^1H NMR ($OS(CD_3)_2$): 6.38 (s, C_6H_5) [4] ^{19}F NMR ($OS(CD_3)_2$): −0.051 (s, CF_3) [4] IR (solid): 1520 s ν(C=C), 889 s, 829 m ν(CF_3)+ν(CS), 868 m δ(C_5), 332 m ν(In-C_5) [4] Raman (solid): 335 s ν(In-C_5), no other bands reported [4] forms a monohydrate, dec. >250 °C, insoluble in most organic solvents [4]

* Further information:

$CH_3InS_2C_6H_3CH_3$-4 (Table **48**, No. **1**) and its adducts **$CH_3InS_2C_6H_3CH_3$-4 · D** (No. **17**, D=$OS(CH_3)_2$; No. **18**, D=$N(CH_3)_3$; No. **19**, D=NC_5H_4; No. **20**, D=$C_{10}H_8N_2$-2,2′; No. **21**, D=$C_{12}H_8N_2$-1,10) show splitting of the In-CH_3 signal and part of the ring-CH_3 resonances in the ^1H NMR spectra. Since the intensities of the individual lines do not change significantly between 10 and 90 °C, the splitting was explained by the existence of two stereoisomers (Formulas I and II) present in about equal amounts [2].

With the exception of the measurement on compound No. 18, the measurements had to be carried out in solvents that possess donor capability because of the low solubility in non-donor solvents. Hence, no completely certain structural clues were possible, especially for the adducts No. 20 and 21 with bidentate ligands. In addition to the following compilation of chemical shifts (δ in ppm, relative intensities in parentheses), broad multiplets at δ=5.83 to 7.5 were observed for the C_6H_3 ring protons and a singlet at 2.30 ppm for the CH_3N group of No. 18 [2].

compound	solvent	δ(CH_3In)	δ(ring CH_3)
$CH_3InS_2C_6H_3CH_3$	$(CD_3)_2SO$	−0.21(0.6)/0.03(0.4)	2.12(1.0)
	C_5D_5N	−0.33(0.35)/−0.12(0.65)	1.30(0.2)/1.43(0.8)
$CH_3InS_2C_6H_3CH_3 \cdot D$			
D=$N(CH_3)_3$	CH_2Cl_2	0.32(1.0)	2.15(1.0)
D=$C_{10}H_8N_2$-2,2′	$(CD_3)_2SO$	−0.25(0.6)/−0.05(0.4)	2.02(1.0)
D=$C_{12}H_8N_2$-1,10	$(CD_3)_2SO$	−0.26(0.4)/−0.06(0.6)	2.05(1.0)

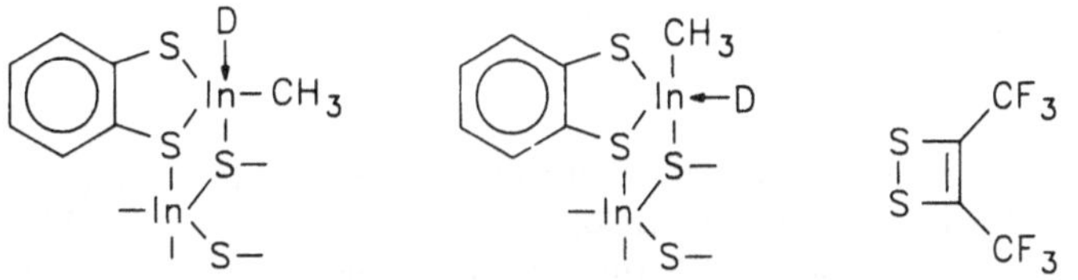

I II III

The adduct No. 18 loses $N(CH_3)_3$ in vacuum at room temperature to give $CH_3InS_2C_6H_3CH_3 \cdot 0.5\,N(CH_3)_3$; No. 21 forms a monohydrate which is stable up to ca. 300 °C [2].

$(C_5H_5$-c)$InS_2C_2(CF_3)_2$ (Table **48**, No. **16**) was formed by oxidative insertion of InC_5H_5-c into the S-S bond of the dithiet III. Thus, a $CHCl_3$ solution of indium(I) cyclopentadiene (the analytical purity of the solvent is of prime importance for the quality of the final product!) was added dropwise to an equimolar amount of a solution of III at room temperature. Precipitation of the yellow product begins immediately and was completed by adding petroleum ether to the mixture. An excess of dithiet causes a red to violet coloration. Filtration and drying in vacuum at room temperature were the final steps; the yield is practically quantitative. Traces of C_2H_5OH (added in commercial $CHCl_3$ for stabilization) led to dark brown products which could not be purified; furthermore, coloration caused by an excess of a reactant could not be removed. The properties of the compound indicated a polymeric structure for the solid; from IR spectroscopy a σ-bonded C_5H_5 ring was deduced [4].

Conductivity measurements on a freshly-prepared CH_3NO_2 solution implied a partial dissociation into $[C_5H_5]^-$ and $[InS_2C_2(CF_3)_2]^+$ ($\Lambda_m = 9.4\ \Omega^{-1} \cdot cm^2 \cdot mol^{-1}$), but the solution is stable for only a few minutes. The mass spectrum shows no molecular ion peak; the only identifiable fragments are $[InC_5H_5]^+$ and $[S_2C_2(CF_3)_2]^+$ [4].

References:

[1] Viktorova, I. M.; Sheverdina, N. I.; Kocheshkov, K. A. (Dokl. Akad. Nauk SSSR **198** [1971] 94/5; Dokl. Chem. Proc. Acad. Sci. USSR **196/201** [1971] 367/8; C.A. **75** [1971] No. 49184).

[2] Berniaz, A. F.; Tuck, D. G. (J. Organometal. Chem. **46** [1972] 243/50).

[3] Maeda, T.; Okawara, R. (J. Organometal. Chem. **39** [1972] 87/91).
[4] Berniaz, A. F.; Tuck, D. G. (J. Organometal. Chem. **51** [1973] 113/8).
[5] Srivastava, T. N.; Kumar, V.; Srivastava, O. P. (Natl. Acad. Sci. Letters [India] **1** [1978] 97/100; C.A. **89** [1978] No. 141068).
[6] Hoffmann, G. G. (J. Organometal. Chem. **338** [1988] 305/17).
[7] Nomura, R.; Inazawa, S.; Kanaya, K.; Matsuda, H. (Polyhedron **8** [1989] 763/7).
[8] Nomura, R.; Inazawa, S.; Kanaya, K.; Matsuda, H. (Appl. Organometal. Chem. **3** [1989] 195/7).

6 Organoindium-Selenium Compounds

$(C_2H_5)_2InSeSi(C_2H_5)_3$

This compound formed when $In(C_2H_5)_3$ was allowed to react with $HSeSi(C_2H_5)_3$ (1:1 mole ratio) in hexane at 0 °C, evolving C_2H_6. After removing the solvent, the residual liquid was distilled (b.p. not reported); m.p. −55 to −54 °C, 17.5% yield. The compound is dimeric in benzene (cryoscopic measurement) and was postulated to have the structure of the silanolate, $(CH_3)_2InOSi(CH_3)_3$ (p. 188), with an In_2Se_2 ring. Thermal decomposition at 170 °C led within 1 to 4 h to $(C_2H_5InSe)_n$ and $\{(C_2H_5)_2Si\}_2Se$ [1].

$(C_2H_5InSe)_n$

This species was also obtained by reaction of $In(C_2H_5)_3$ with $HSeSi(C_2H_5)_3$ in hexane (1:2 mole ratio, at 0 to 20 °C, 87.2% yield). The compound $C_2H_5In(SeSi(C_2H_5)_3)_2$, which was presumably formed by the exothermic reaction, then decomposed into $(C_2H_5InSe)_n$ (which precipitated, was filtered and washed) and $\{(C_2H_5)_3Si\}_2Se$. The compound decomposes at 280 °C [1].

It was characterized by neutron activation analysis. The sample was irradiated by a Po/Be source (10^8 neutrons/s); ^{116}In was determined by a 10-min measurement of the 0.42 MeV line and ^{77}Se determined from the 0.162 MeV line (35 s) [2].

References:

[1] Vyazankin, N. S.; Bochkarev, M. N.; Charov, A. I. (J. Organometal. Chem. **27** [1975] 175/80).
[2] Glazov, V. M. (Tr. Khim. Khim. Tekhnol. **1974** 71/3 from C.A. **83** [1975] No. 201642).

7 Organoindium-Nitrogen Compounds

7.1 R_2In Derivatives

7.1.1 $R_2InN(R'R'')$ Compounds

The dialkylindium amides collected in Table 50 were synthesized by the following two procedures:

Method I: From $R_3In \cdot NHR'R''$ compounds.
Thermal decomposition of $(CH_3)_3In \cdot NHR'R''$ (see Table 6, p. 33) at normal pressure and 140 to 160 °C [1] (155 to 160 °C [5]) or at 60 °C [7] produced No. 2 or No. 6 in nearly quantitative yield within 2 to 8 h with evolution of CH_4. Vacuum sublimation of the reaction residue gave analytically pure products. The adduct

$(CH_3)_3In \cdot NH_3$ decomposed by 70 to 80 °C, but the resulting amide (No. 1) was not obtained analytically pure [1].

Method II: From R_2InCl and Li[NR′R″] compounds.
An ether solution of Li[NR′R″], freshly prepared from LiC_4H_9-n and HNR′R″, was added to a slurry of R_2InCl at −80 °C (0 °C and $O(C_2H_5)_2/C_6H_{14}$ for No. 10). The mixture was stirred and warmed to room temperature, was freed of solvent in vacuum, and, after dissolving the residue in C_5H_{12} at room temperature, was stirred for 30 min. Finally, the LiCl was filtered off and the filtrate concentrated at −30 to −20 °C until the onset of crystallization [14, 16] or brought to dryness and vacuum sublimed [8, 10, 13].

General Remarks. Among the following dialkylindium-N-organoamides (Nos. 2 to 13), most are colorless crystalline solids, and most have been very precisely characterized for potential use as source materials for preparing binary or ternary III/IV semiconductor layers by gas phase epitaxy (MOVPE). Since the most relevant results are based on the investigations of one research group [8, 10, 11, 13], a comparative presentation of the data is possible. Additional specific information can be found in Table 50.

Vapor pressure measurements were obtained for compounds No. 2 to 5 by the Knudsen effusion method [8, 11] (method, apparatus, and theory are described in [11]); the measurement technique was not detailed in [1].

Parameters A and B of the vapor pressure equation, log p (Torr) = A − B/T (K), and the vaporization enthalpy (ΔH) are:

compound No.	A	B	range in °C	ΔH in kcal/mol	Ref.
2	5.61	1890	30 to 75	11.9	[1]
	8.86	3018	75 to 140	11.3	[1]
	12.57	4313.3	32.8 to 46.8	19.8	[8]
	10.67	3793.7	90.3 to 139.1	17.36	[8]
3	12.91	4680.9	31.8 to 60.3	21.59	[8]
	9.43	3513.9	92.8 to 129.8	16.08	[8]
4	12.05	4824.6	43.8 to 84.3	22.08	[8]
5	13.1	5850	62.3 to 100.3	–	[11]

IR absorptions under 1500 cm^{-1} were reported without assignment for Nos. 3 to 6 [13] and only a few important IR and Raman frequencies were given for No. 10 [14]. Complete IR and Raman spectra from 1500 to about 100 cm^{-1} with band assignments have been published for Nos. 2 and 3 (IR only) [15] (and for $(CH_3)_2InN(CD_3)_2$) [2, 5] (Table 51, p. 259).

For the remaining dimethylindium compounds (Nos. 3 to 7) characteristic $In(CH_3)_2$ vibrations are found at 1160 to 1170 ($\delta_s(CH_3)$), ca. 700 ($\varrho(CH_3)$), 490 to 510 ($\nu_{as}(InC_2)$), and 470 to 485 cm^{-1} ($\nu_s(InC_2)$). The In-N vibrations were observed around 450 cm^{-1}, but were frequently obscured by other vibrations [13, 15]. In contrast for No. 10, the bands at 556 and 534 cm^{-1} were assigned to the IR-active In-N vibrations [14] (Author: An increase of more than 100 wavenumbers over those of most amides is not plausible).

A systematic relationship between the C-In-C bond angle and the frequency difference, $\Delta\nu = \nu_{as}(InC_2) - \nu_s(InC_2)$, does not exist according to [8] and [13], contradictory to the general correlability of these values [6].

Cryoscopic molecular weight determinations in benzene show Nos. 2, 5, 6, and 10 to be dimers. The exact masses of the monomers, often the dimers also for Nos. 2 to 5, were determined by high-resolution mass spectrometry [10, 13]. Fragmentation is almost identical for these compounds. At a constant ionization energy of 70 eV, the changes in the relative intensities of comparable fragments reflect the influence of increasingly bulky N-substituents on the stability of the dimer, and is in such that for No. 7 at 23 eV only fragments of M_2 (but not M_2 itself) can be observed. Metastable ions were also detected for Nos. 3, 4, and 7, which helped to clarify the fragmentation scheme [8, 13, 15]. The fragmentation of amides No. 2 to 5 and 7 [8, 11, 13, 15] is outlined in Table 49.

Table 49
Mass Spectroscopic Fragmentation of the $R_2InNR'_2$ Nos. 2 to 5 and 7 ($R=CH_3$).
Relative intensities in % at 70 eV (M represents the monomeric unit).

NR'_2	$N(CH_3)_2$	$N(C_2H_5)_2$		$N(C_3H_7\text{-}i)_2$	$N(C_6H_{11}\text{-}c)_2$	$N\{Si(CH_3)_3\}_2$
No. in Table 50 [Ref.]	2 [13]	3 [13, 15]		4 [13]	5 [10]	7 [13]
source temperature	21 °C	21 °C	100 °C	56 °C	?	40 °C
$[M_2]^+$	18	4	1	1	1	–
$[M_2-R]^+$	94	46	48	16	2	–
$[M_2-NR'_2]$	40	30	39	21	4	–
$[M_2-CH_2-NR'_2]^+$	23	9	7	–	–	9 (–H)
$[M+NR'_2]^+$	12	–	–	–	–	–
$[M]^+$	7	9	9	4	2	31
$[M-H]^+$	100	100	100	20	–	–
$[M-R]^+$	14	29	40	51	9	100 (+H)
$[M-2R]^+$	22 (–H)	–	9	7	–	18 (–H)
$[M-3R]^+$	–	11	–	14	–	42
$[InR_2]^+$	92	61	91	55	13	–
$^{115}In^+$	77	71	88	51	4	63
$[HNR'_2]^+$	–	–	–	–	–	40

With the exception of Nos. 1 and 5, X-ray diffraction established that all the products have planar In_2N_2 four-ring skeletons as depicted in Formula I.

```
   R'  R''
R    \ /    R
 \    N    /
  In     In
 /    N    \
R    / \    R
   R'  R''
```

I

The average values of the most important molecular parameters are tabulated below; particular details are noted with the individual examples in Table 50. The extremely small C-In-C bond angle in No. 7 is presumed to be due to the especially bulky $N(Si(CH_3)_3)_2$ group [13].

compound No. [Ref.]	2 [5]	3 [13]	4 [13]	6 [7]	7 [13]	10 [14]
distances (pm)						
In–C	216.9	215.9	217.3	215.3	216.2	218.7
In–N	223.6	223.5	226.3	228.2	230.5	223.4
In···In	327.8	326.9	322.4	336.3	327.0	334.2
angles (°)						
C–In–C	131.3	125.9	119.0	122.4	109.1	123.5
N–In–N	85.7	86.0	87.8	85.1	89.7	83.1
In–N–In	94.3	94.0	90.4	94.9	90.3	96.9

Table 50
Compounds of the $R_2InNR'R''$ Type.
Further information on numbers preceded by an asterisk is given at the end of the table.
Explanations, abbreviations, and units on p. X.

No.	compound method of preparation (yield)	properties and remarks
$(CH_3)_2In$ compounds		
1	$(CH_3)_2InNH_2$ I	no pure compound isolated, m.p. 120 to 124 °C (dec.), forms a thick syrup which underwent further decomposition; not volatile, presumably polymeric [1] used to grow InP from a heated mixture with PH_3 [9]
*2	$(CH_3)_2InN(CH_3)_2$ I (>90%)	colorless crystals, m.p. 174 to 175 °C [1, 2], 170 to 171 °C [13], subl. 55 °C/0.01 Torr [5], 60 °C/0.01 Torr [13] ^1H NMR: −0.27 (CH_3In), 2.62 (CH_3N) in CCl_4 [4], −0.16, 2.41 in C_7D_8 [13] IR and Raman frequencies on p. 259 mass spectrum in Table 49, p. 254
*3	$(CH_3)_2InN(C_2H_5)_2$ II (83%)	colorless crystals, m.p. 80 to 90 °C, subl. 70 °C/0.01 Torr [13], 37 °C/0.001 Torr [15] ^1H NMR (C_6D_6): −0.08 (CH_3In), 0.82 (t, CH_3C), 2.87 (q, CH_2, $^2J(H,H)$ = 6.8), similar data in [15] ^{13}C NMR (C_6D_6): −10.8 (CIn), 12.6 (CH_3), 42.7 (CN) [13] IR (solid): 1162 w $\delta_s(CH_3In)$, 1146 s $\varrho(CH_3N)$, 1043 m $\nu_{as}(NC_2)$, 1002 m $\nu_s(NC_2)$, 508 m $\nu_{as}(InC_2)$, 478 m $\nu_s(InC_2)$, unassigned bands at 1511 w, 1110 w, 898 w, 850 w, 792 m, 701 m, 561 m, 543 m [15]; see also [8, 13] mass spectrum in Table 49, p. 254
*4	$(CH_3)_2InN(C_3H_7\text{-}i)_2$ II (73%)	colorless crystals, m.p. 110 to 112 °C, subl. 100 °C/0.01 Torr [8, 13] ^1H NMR (C_6D_6): 0.09 (CH_3In), 1.13 (d, CH_3C, J(H,H) = 7.0), 3.55 (sept., HCIn) [8, 13]

Table 50 (continued)

No.	compound method of preparation (yield)	properties and remarks
*4 (continued)		^{13}C NMR (C_6D_6): −3.8 (CIn), 26.3 (CH_3), 50.8 (CN) [8, 13] IR (solid): 1365 s, 1330 m, 1259 w, 1165 s, 1145 s, 1128 m, 1105 m, 1022 w, br, 1014 w, br, 969 m, 961 m, sh, 909 m, 835 w, 825 vw, 779 m, 740 s, 715 s, sh, 700 vs, br, 650 w, 622 w, 605 m, 561 m, 492 s, 471 s, 455 m, 415 m, br, 320 w, 312 w, 298 w, 288 w, 270 w, 262 w, 240 w, 219 m [8, 13] mass spectrum in Table 49, p. 254
5	$(CH_3)_2InN(C_6H_{11}\text{-c})_2$ II (56%)	fluffy white crystals, m.p. 170 to 173 °C, darkens to black at 150 to 160 °C [10] ^1H NMR (C_6D_6): 0.22 (CH_3In), 0.9 to 2.1 (m, br, CH_2), 3.15 (m, HCN) [10] IR (solid): 1451 vs, 1369 m, 1350 m, 1334 w, 1302 w, 1277 w, 1253 w, 1244 w, 1169 m, 1156 m, 1140 w, 1106 m, 1091 m, 1081 m, 1060 s, br, 1035 s, sh, 1029 s, 976 m, 943 s, sh, 936 s, 924 m, br, 900 s, 883 w, 843 m, 805 w, 789 w, 780 m, 701 vs, br, 656 m, 639 m, 582 s, 572 m, 543 m, 488 s, 471 s, 441 w, 431 m, br, 340 w, br, 232 w [10] mass spectrum in Table 49, p. 254
*6	$(CH_3)_2InN(CH_3)C_6H_5$ I (>90%)	colorless crystals, m.p. 179 to 181 °C, subl. 100 °C/under vacuum [7] ^1H NMR on p. 261 IR (solid): 1595 s, 1570 m, 1485 s, 1335 vw, 1295 w, 1230 s, 1190 m, 1165 m, 1155 w, 1090 vw, 1040 w, 1020 m, 1000 m, 890 vw, 785 m, 765 s, 700 s, 585 m, 510 m, 490 m, 410 m [7]
*7	$(CH_3)_2InN\{Si(CH_3)_3\}_2$ II	colorless crystals, m.p. 37 to 38 °C, b.p. 50 °C/0.01 Torr [8, 13] ^1H NMR (C_7D_8): 0.01 (CH_3In), 0.20 (CH_3Si) [8, 13] ^{13}C NMR (C_6D_6): 2.96 (CIn), 5.26 (CSi) [18] IR (solid): 1396 w, 1368 w, sh, 1341 vw, sh, 1285 w, sh, 1257 s, 1248 s, 1182 w, 1165 w, 1010 s, 965 s, 940 m, sh, 879 s, 864 s, 848 s, sh, 835 s, 825 sh, 789 m, 752 s, 741 m, sh, 696 m, 672 s, 615 m, 569 w, sh, 519 m, 508 m, 486 m, 463 m, br, 391 w, 357 s, 329 w, 299 w, sh, 281 w, 245 m [8, 13] mass spectrum in Table 49, p. 254
*8	$(CH_3)_2InN(SO_2CH_3)_2$ special	colorless solid, m.p. 248 °C, subl. 110 °C/ca. 4 Torr [17] ^1H NMR (CD_3CN): 0.06 (s, CH_3In), 3.02 (s, CH_3S) [17] ^{13}C NMR (CD_3CN): −0.08 (CH_3In), 43.59 (CH_3S) [17]

Table 50 (continued)

No.	compound method of preparation (yield)	properties and remarks
other R_2In compounds		
9	$(C_2H_5)_2InN(C_2H_5)_2$ I (80%)	yellow viscous liquid, b.p. 125 to 127 °C/0.5 Torr [3] dec. within 50 h at 150 °C to give polymeric $C_2H_5In(N(C_2H_5)_2)_2$ and other products; no other properties reported [3]
*10	$(C_3H_7\text{-}i)_2InNH(C_4H_9\text{-}t)$ II (67%)	colorless crystals, dec. 155 °C, dimeric in benzene [14] ^1H NMR (C_6D_6): 0.97 (C_4H_9), 1.13 (m, HCIn, AB_3C_3-system), 1.34 (HN), 1.40, 1.47 (CH_3C) [14] ^{13}C NMR (C_6D_6): 20.5 (CIn), 23.87, 23.90 (CH_3 of C_3H_7), 34.3 (CH_3 of C_4H_9), 50.8 (C of C_4H_9) [14] IR (solid): 556 ms, 534 ms ν(InN), 478 m, 459 m ν(InC) [14] Raman (solid): 524 vw ν(InN), 459 s ν(InC); no other bands reported [14] mass spectrum (70 eV): $[M_2-C_3H_7]^+$ (100%), $[M_2-3\,C_3H_7]^+$ (21.8%), $[In_2HNC_4H_9]^+$ (12.87%), $^{115}In^+$ (24.85%); no other fragments reported [14]
*11	$(C_6H_5CH_2)_2InNH(C_4H_9\text{-}t)$ II (85%)	colorless crystals; mixture of isomers A (trans) and B (cis), 1:1.25 ratio; m.p. 115 to 120 °C [16] ^1H NMR (C_6D_6 A/B): 0.59/0.68 (s, CH_3C), 1.22/1.22 (HN), 2.28/2.37 and 2.08/1.94 (CH_2, AB spin system, $^2J(H,H)=11.1$), 6.75 to 7.14 (C_6H_5); see also further information [16] ^{13}C NMR (C_6D_6): 25.4 (CH_2), 33.9 (CH_3), 51.5 (CN), 122.7 (C-4), 127.1 (C-3,5), 129.0 (C-2,6), 145.7 (C-1) of isomer A; 23.7, 26.8 (CH_2), 33.7 (CH_3), 51.8 (CN), 122.2, 123.3 (C-4), 126.7, 127.7 (C-3,5), 128.8, 129.1 (C-2,6), 144.9, 146.5 (C-1) of isomer B [16] IR (solid): 556 m, 546 m, 538 m ν(InN), 433 m, 444 m, 463 m ν(InC) of both isomers [16] Raman (solid): 551 m, 530 w ν(InN), 463 m, 468 w ν(InC) of both isomers [16] mass spectrum (isomer mixture): $[M]^+$ (2.8%), $[M-C_4H_9]^+$ (4.9%), $[M-CH_2C_6H_5]^+$ (54.3%), $[M-3\,CH_2C_6H_5]^+$ (38.3%), $[In_2(NHC_4H_9)_2]^+$ (16.5%), $[In_2(NHC_4H_9)]^+$ (50.2%), ^{115}In (74.0%), $[CH_2C_6H_5]^+$ (100%); M is the dimeric unit [16] for the structure, see Formulas II and III on p. 262
*12	$\{(CH_3)_3SiCH_2\}_2InN(SO_2CH_3)_2$ special	colorless solid, m.p. 105 °C, subl. 90 to 100 °C/ca. 7 Torr [17] ^1H NMR ($CDCl_3$): 0.09 (s, CH_3Si), 0.12 (s, CH_2In), 3.19 (s, CH_3S) [17]

Table 50 (continued)

No.	compound method of preparation (yield)	properties and remarks
*12 (continued)		^{13}C NMR ($CDCl_3$): 2.00 (CH_3Si), 8.31 (CH_2), 43.54 (CH_3S) [17]
13	$\{(CH_3)_3SiCH_2\}_2InN\{Si(CH_3)_3\}_2$ II (60%)	colorless liquid, b.p. 75 °C/10^{-3} Torr, monomeric in benzene [18] ^1H NMR (C_6D_6): 0.13 (s, CH_3SiC), 0.21 (s, CH_3SiN), 0.32 (s, CH_2In) [18] ^{13}C NMR (C_6D_6): 2.67 (q, CH_3SiC, $^1J(H,C)=118.1$), 5.36 (q, CH_3SiN, $^1J(H,C)=116.9$), 13.12 (t, CH_2In, $^1J(H,C)=119.5$) [18] ^{29}Si NMR (C_6D_6): −3.22 (m, $Si(CH_3)N$, $^2J(H,Si)=6.14$), 1.78 (m, $Si(CH_3)C$, $^2J(H,Si)=6.4$) [18]

* Further information:

$(CH_3)_2InN(CH_3)_2$ (Table **50**, No. **2**) forms large, colorless, and well developed crystals at room temperature which convert into a noncrystalline glass of dendritic habit between 70 and 80 °C. This phase change is marked by a kink in the vapor pressure curve [2] and the sublimation enthalpies of the two forms are different: ΔH_S = ca. 9 kcal/mol (T < 70 °C) and ΔH_S = 13.8 kcal/mol (T > 80 °C). The glassy form is obtained by condensation of the

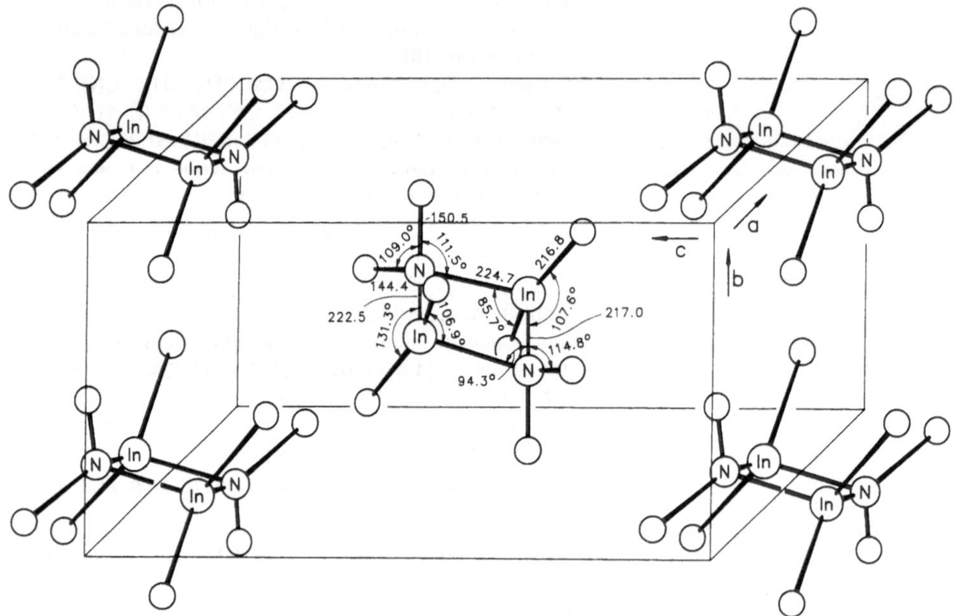

Fig. 44. Packing of the $\{(CH_3)_2InN(CH_3)_2\}_2$ molecule in the unit cell [5].

vapor at or above 80 °C; it can be undercooled to about 50 °C for a few minutes, but then converts suddenly to a polycrystalline mass. A change in the degree of association originally assumed to occur could not be confirmed [1, 2].

For IR and Raman spectra of the compound in the solid state and in CCl_4 solution and for the solid state IR spectrum of **$(CH_3)_2InN(CD_3)_2$**, see Table 51 [2, 5].

The compound crystallizes in the monoclinic space group $P2_1/c-C_{2h}^5$ (No.14) with the cell constants a=760.1(6), b=733.6(6), c=1492.7(9) pm, β=119.9(1)°; Z=4 (2 dimers), D_c=1.74 g/cm³, and D_m=1.77 g/cm³. The placement of the molecules in the unit cell and the most important bond distances and angles are shown in **Fig. 44** on p. 258 [5].

$(CH_3)_2InN(CH_3)_2$ reacts with C_5H_6-c to give $In(CH_3)_2(C_5H_5$-c) (see p. 112) via $(CH_3)_2(C_5H_5$-c)In · $NH(CH_3)_2$ as intermediate [4].

Table 51
IR and Raman Frequencies of $(CH_3)_2InN(CH_3)_2$ and IR Frequencies of $(CH_3)_2InN(CD_3)_2$ [2, 5]. Wavenumbers in cm^{-1}.

$(CH_3)_2InN(CH_3)_2$ IR [2] solid	IR [5] CCl_4 solution	Raman [5] CCl_4 solution	$(CH_3)_2InN(CD_3)_2$ IR [2] solid	assignment
1468	–	–	–	$\delta_{as}(CH_3N)$
1430	–	–	–	$\delta_s(CH_3N)$
1408	–	–	1402	$\delta_{as}(CH_3In)$
1228	–	–	–	$\varrho(CH_3N)$
1164	1160 ms	1162 vs	1162	$\delta_s(CH_3In)$
1155	1150 sh	1150 w, sh	–	$\varrho(CH_3N)$
1138	1130 s	1135 w, sh	1136	$\nu_{as}(NC_2)$
–	–	–	1071	$\delta_{as}(CD_3N)$
–	–	–	1062	$\varrho(CD_3N)$
1050	1042 s	1048 mw	1031	$\nu_s(NC_2)$
916, 891	955 s, sh	932 m	–	$\varrho(CH_3N)$
–	–	–	809, 794	$\varrho(CD_3N)$
699	700 vs	–	699	$\varrho(CH_3In)$
509	511 s	508 m	510	$\nu_{as}(InC_2)$
482	482 ms	482 vs	481	$\nu_s(InC_2)$
–	–	472 ms	–	$\nu(In_2N_2)$
443	445 s	–	405	
–	–	420 w, br	–	
–	261 m, br	265 ms	–	$\delta(NC_2)$
–	240 mw	–	–	
–	–	221 m	–	$\delta(InC_2)+\delta(In_2N_2)$
–	184 m	151 ms	–	
–	145, 120 m, br	120 s	–	

$(CH_3)_2InN(C_2H_5)_2$ (Table **50**, No. **3**). The He(I) and He(II) photoelectron spectra show in the region between 7 and 11 eV five well resolved bands at 7.78, 8.35, 8.66, 9.57, and 9.79 eV and a shoulder on the 8.35 band; the spectra are depicted. The MO calculations were based on the data of the methyl derivative No. 2 [15].

The compound crystallizes in the triclinic space group $P\bar{1}-C_i^1$ (No. 2). The cell constants have been estimated by two groups [13]/[15]: a = 754.3(2)/837.3(5), b = 834.9(1)/848.5(5), c = 846.2(2)/755.7(5) pm, α = 76.51(1)°/118.24(3)°, β = 118.37(2)°/76.00(3)°, γ = 104.02(2)°/103.60(3)°; Z = 1 (dimer), D_c = 1.60/1.58 g/cm³. Available distances and angles are given in **Fig. 45** [13].

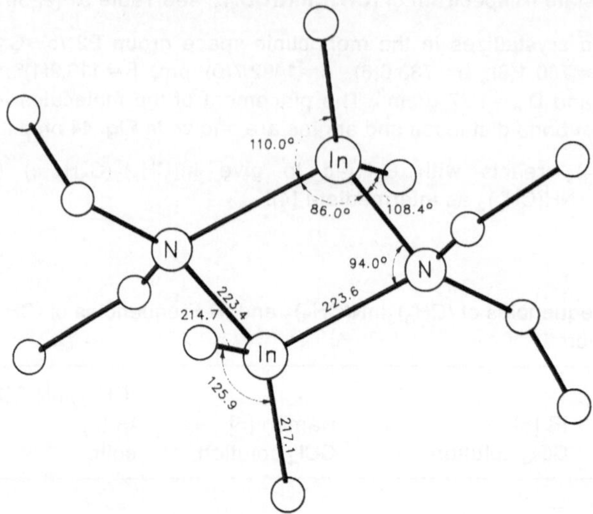

Fig. 45. Molecular structure of $\{(CH_3)_2InN(C_2H_5)_2\}_2$ [13].

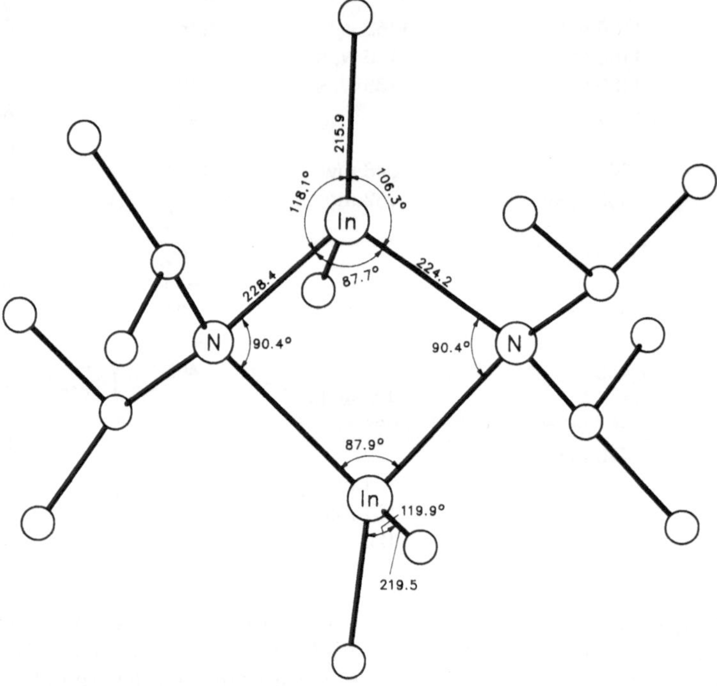

Fig. 46. Molecular structure of $\{(CH_3)_2InN(C_3H_7\text{-}i)_2\}_2$ [13].

Thermal decomposition under reduced pressure on gold foil heated at 400 °C gave pure metallic In, as shown by SIMS measurements [15]. InP layers were prepared by thermally decomposing the compound in the presence of PH_3 at 650 °C [12]. It is the starting material for the preparation of $(CH_3)_2InPR_2$ compounds ($R = C_2H_5$, C_6H_5) [15].

$(CH_3)_2InN(C_3H_7\text{-i})_2$ (Table **50**, No. **4**) crystallizes in the triclinic space group $P\bar{1}-C_i^1$ (No. 2) with the cell constants a = 763.9(2), b = 1058.7(2), c = 1381.4(2) pm, $\alpha = 85.06(1)°$, $\beta = 81.41(2)°$, $\gamma = 76.37(2)°$; Z = 4 (2 dimers), $D_c = 1.52$ g/cm³ [13]. The structure is depicted in **Fig. 46** on p. 260.

$(CH_3)_2InN(CH_3)C_6H_5$ (Table **50**, No. **6**) exists in solution as a mixture of cis and trans isomers A and B (Formulas II and III). The isomer ratio is largely independent of concentration, but strongly dependent on the nature of the solvent and the temperature. ^1H NMR measurements in C_6D_6, $C_6D_5CD_3$, and CH_2Cl_2 at temperatures from −45 to +34 °C permitted determining the cis-trans equilibrium and calculating the following thermodynamic parameters for the transformation: $\Delta H = -0.471$ kcal/mol (−1.94 kJ/mol), $\Delta S = -2.46$ cal · mol⁻¹ · K⁻¹ (−10.3 J · mol⁻¹ · K⁻¹), $T\Delta S_{298} = -0.734$ kcal/mol (−3.07 kJ/mol), and $\Delta G = 0.235$ kcal/mol (1.13 kJ/mol) [7].

The following chemical shifts (in ppm) have been obtained in various solvents at 25 °C for the methyl groups:

solvent	C_6D_6	$C_6D_5CD_3$	CH_2Cl_2
CH_3In-trans	−0.06	−0.03	−0.32
CH_3In-cis	−0.05/0.00	0.03/0.17	−0.25/−0.17
CH_3N-cis	2.97	3.03	3.10
CH_3N-trans	3.01	3.04	3.14
% cis	43.5	43.0	39.2

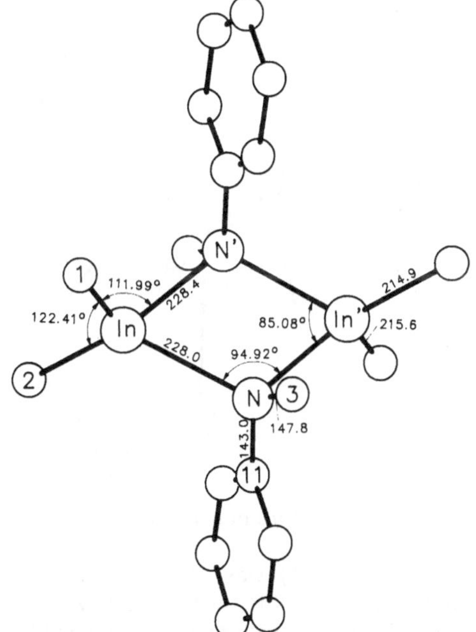

Fig. 47. Molecular structure of $\{(CH_3)_2InN(CH_3)C_6H_5\}_2$ [7].

Other bond angles (°)

N-In-C(1)	109.84(12)	In-N-C(3)	109.00(16)
N-In-C(2)	112.24(11)	In-N-C(11)	115.19(15)
C(3)-N-C(11)	113.32(21)	In′-N-C(3)	118.75(16)

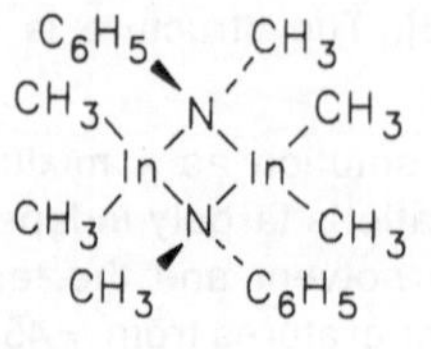

II trans A 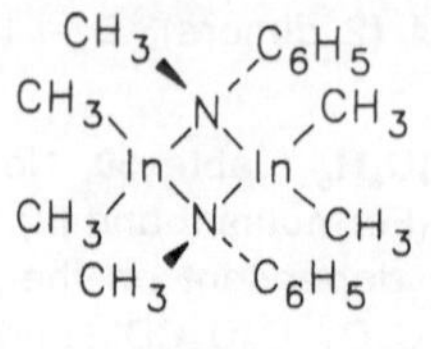   III cis B

The compound crystallizes in the triclinic space group $P\bar{1}-C_i^1$ (No. 2) with the cell constants a = 732.02(15), b = 760.95(21), c = 898.00(27) pm, α = 83.190(24)°, β = 81.800(21)°, γ = 81.986(19)°; Z = 2 (1 dimer), D_c = 1.71 g/cm³. As shown in **Fig. 47** on p. 261 only the trans form is present in the crystal [7].

$(CH_3)_2InN\{Si(CH_3)_3\}_2$ (Table **50**, No. **7**) is monomeric in C_6H_6; and it is slightly soluble even in Nujol. Thus, the IR frequencies given in Table 50 consist of bands of the monomeric and the dimeric species [18].

The compound crystallizes in the monoclinic space group $C2/c-C_{2h}^6$ (No. 15) with the cell constants a = 1552.5(4), b = 1153.8(3), c = 1693.9(8) pm, β = 111.97(3)°; Z = 8 (4 dimers), D_c = 1.24 g/cm³. The structure with the reported bond distances and angles is shown in **Fig. 48** [13].

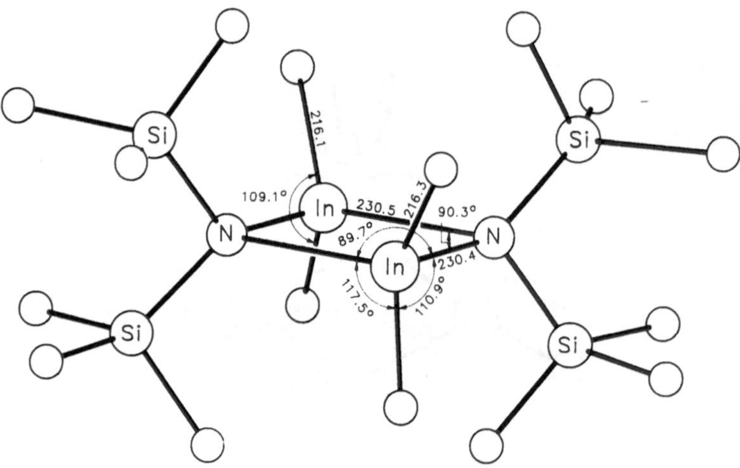

Fig. 48. Molecular structure of $(CH_3)_2InN\{Si(CH_3)_3\}_2$ [13].

$(CH_3)_2InN(SO_2CH_3)_2$ (Table **50**, No. **8**) was obtained by acidolysis of $In(CH_3)_3$ with $HN(SO_2CH_3)_2$ (1:1 mole ratio) in $O(C_2H_5)_2/CH_3CN$ (5/1) as a white precipitate in quantitative yield. It is soluble in DMSO, DMF, and THF, but only slightly soluble in CH_3CN and CH_3NO_2, and insoluble in nonpolar solvents [17].

$(C_3H_7\text{-}i)_2InNH(C_4H_9\text{-}t)$ (Table **50**, No. **10**) forms cis and trans isomers like No. 6 (Formulas II and III), but no NMR experiments were made [14]; see also [16].

The trans form crystallizes in the monoclinic space group $C2/c - C^6_{2h}$ (No. 15) with the cell constants a = 939.6(1), b = 1657.3(2), c = 1674.0(2) pm, β = 97.9(3)°; Z = 8 (4 dimers) and D_c = 1.405 g/cm^3. A disorder concerning the parameters of the propyl groups is discussed. The structure of the compound is depicted in **Fig. 49** [14].

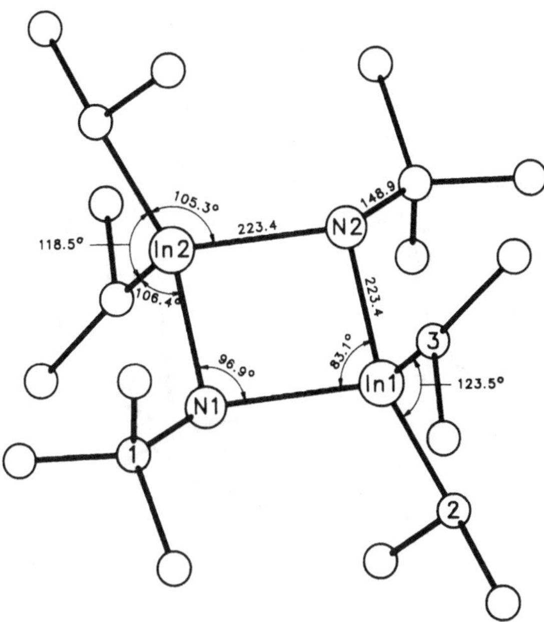

Fig. 49. Molecular structure of $(C_3H_7\text{-}i)_2InNH(C_4H_9\text{-}t)$ [14].

Other bond angles (°)

N(1)-In(1)-C(3)	116.2(3)	In(1)-N(1)-C(1)	124.4(3)
C(2)-In(1)-N(2)	115.3(2)	In(2)-N(1)-C(1)	123.0(3)

$(C_6H_5CH_2)_2InNH(C_4H_9\text{-}t)$ (Table **50**, No. **11**) shows a 1:1.25 ratio of trans (A) and cis (B) isomers (Formulas II and III) in solution. Recrystallization from pentane enriches the thermodynamically more stable trans isomer [16].

In the ^{1}H NMR spectrum the trans isomer A exhibits diastereotopic methylene protons (δ_A = 2.28, δ_B = 2.08 ppm, J = 11.1 Hz), whereas the cis isomer B shows geometrically different CH_2 protons at 2.37 and 1.94 ppm [16].

$\{(CH_3)_3SiCH_2\}_2InN(SO_2CH_3)_2$ (Table **50**, No. **12**) was prepared by dropwise addition of a slight excess of $In\{CH_2Si(CH_3)_3\}_3$ in ether to a suspension of $HN(SO_2CH_3)_2$ in ether. Evaporation of the solvent (80 °C/ca. 10 Torr) gave the compound in quantitative yield; purification by sublimation at 90 to 100 °C/ca. 7 Torr. It is soluble in polar organic solvents [17].

The compound crystallizes in the monoclinic space group $P2_1 - C^2_2$ (No. 4) with the cell constants a = 1857.9(8), b = 1207.2(5), c = 1954.8(8) pm, β = 108.60(3)°; Z = 8 monomers and

$D_c = 1.475$ g/cm^3. The asymmetric unit contains two crystallographically independent dimers. The two monomeric units are linked by N-S-O bridges to form an eight-membered $In_2N_2S_2O_2$ ring. A weaker In···O interaction (232 to 235 pm) to an exocyclic SO_2CH_3 group produces a pentacoordination at the In atom as depicted in **Fig. 50** (only one of the dimers is shown) [17].

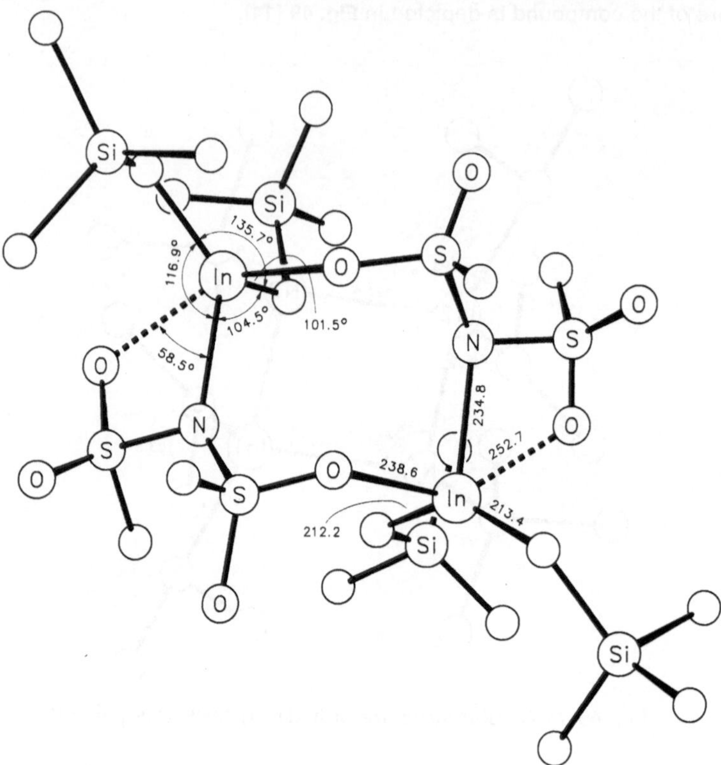

Fig. 50. Molecular structure of $\{(CH_3)_3SiCH_2\}_2InN(SO_2CH_3)_2$ [17].

References:

[1] Coates, G. E.; Whitcombe, R. A. (J. Chem. Soc. **1956** 3351/4).

[2] Beachley, O. T.; Coates, G. E.; Kohnstam, G. (J. Chem. Soc. **1965** 3248/52).

[3] Shcherbakov, V. I.; Zhil'tsov, S. F.; Druzhkov, O. N. (Zh. Obshch. Khim. **40** [1970] 1542/5; J. Gen. Chem. [USSR] **40** [1970] 1529/31).

[4] Krommes, P.; Lorberth, J. (J. Organometal. Chem. **88** [1975] 329/36).

[5] Mertz, K.; Schwarz, W.; Eberwein, B.; Weidlein, J.; Hess, H.; Hausen, H.-D. (Z. Anorg. Allgem. Chem. **429** [1977] 99/104).

[6] Widler, H.-J.; Schwarz, W.; Hausen, H.-D.; Weidlein, J. (Z. Anorg. Allgem. Chem. **435** [1977] 179/90).

[7] Beachley, O. T., Jr.; Bueno, C.; Churchill, M. R.; Hallock, R. B.; Simmons, R. G. (Inorg. Chem. **20** [1981] 2423/8).

[8] Aitchison, K. A. (Diss. Univ. London 1983).

[9] Toshiba Corp. (Japan 84-211217 [1984] from C.A. **102** [1985] No. 229835).

[10] Arif, A. M.; Bradley, D. C.; Dawes, H.; Frigo, D. M.; Hurtshouse, M. B.; Hussain, B. (J. Chem. Soc. Dalton Trans. **1987** 2159/64).

[11] Bradley, D. C.; Faktor, M. M.; Frigo, D. M. (J. Cryst. Growth **89** [1988] 227/36).
[12] Erdmann, D.; Van Ghemen, M. E.; Pohl, L.; Schumann, H.; Hartmann, U.; Wassermann, W.; Heyen, M.; Jürgensen, H. (Ger. Offen. 3631469 [1988]; C.A. **109** [1988] No. 42300).
[13] Aitchison, K. A.; Backer-Dirks, J. D. J.; Bradley, D. C.; Faktor, M. M.; Frigo, D. M.; Hursthouse, M. B.; Hussain, B.; Short, R. L. (J. Organometal. Chem. **366** [1989] 11/23).
[14] Neumüller, B. (Chem. Ber. **122** [1989] 2283/7).
[15] Rossetto, G.; Brianese, D. A. N.; Casellato, U.; Ossola, F.; Porchia, M.; Vittadini, A.; Zanella, P.; Graziani, R. (Inorg. Chim. Acta **170** [1990] 95/101).
[16] Neumüller, B. (Z. Naturforsch. **46b** [1991] to be published).
[17] Blaschette, A.; Michalides, A.; Jones, P. G. (J. Organometal. Chem. **411** [1991] 57/68).
[18] Laichinger, R. (Diploma Work Univ. Stuttgart 1991).

7.1.2 $R_2InN{=}CR'R''$ Compounds

$(CH_3)_2InN{=}C(CH_3)_2$

A solution of $(CH_3)_2C{=}NCl$ in C_6H_{12}-c was added dropwise to a suspension of $In(CH_3)_3$ (1:1 mole ratio) in the same solvent and warmed for 45 min at 40 °C. The solvent was removed in vacuum at room temperature and the residue sublimed at 40 °C/10^{-3} Torr, giving colorless leaf-shaped crystals, m.p. 83 °C, in 33% yield. Cryoscopic molecular weight determination in benzene showed it to be a dimer [2].

^{1}H NMR (C_6D_6, δ in ppm): −0.65 (CH_3In), 0.91 (CH_3C). The IR and Raman spectra of the crystalline compound are summarized in Table 52; these show a great similarity to the spectra of the homologous aluminium and gallium compounds [1] and are in accord with a molecular structure of D_{2h} symmetry [2].

The compound crystallizes in the monoclinic space group $P2_1/c-C_{2h}^5$ (No. 14) with the cell parameters a = 856(1), b = 1399(1), c = 733(1) pm, β = 106.6(1)°; Z = 4 (2 dimers). The crystallographically-determined molecular symmetry is $\bar{1}$, but the deviation from the ideal D_{2h} symmetry postulated from the vibrational spectrum is small and is based on the fact that the isopropylidene carbon atoms, C(5), C(3), and C(4), do not lie exactly in the In_2N_2 plane and the bisector of the C(1)-In-C(2) angle deviates about 5° from the In···In direction. This distortion is presumably caused by the packing of the molecules. The structure of the dimeric unit is depicted in **Fig. 51** [2].

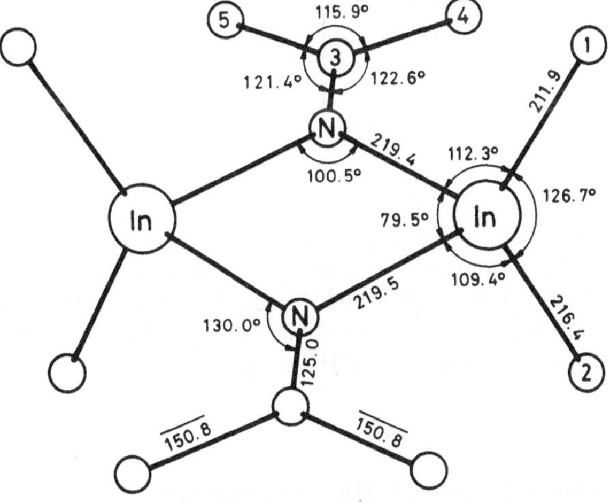

Fig. 51. Molecular structure of $(CH_3)_2InN{=}C(CH_3)_2$ [2].

The IR and Raman frequencies of the solid are assigned in Table 52 [2].

Table 52
IR and Raman Frequencies for Solid $(CH_3)_2InN=C(CH_3)_2$ [2].
Wavenumbers in cm^{-1}.

IR	Raman	assignment (symmetry class)
3315 vw	–	1675 + 1659
3295 vw	–	
–	2987 w	
2965 m	2965 vw	
–	2935 s	ν(C–H)
2913 m	–	
2850 m	2877 w	
2277 w	2290 vw	2 × 1155
1659 vs	1675 m	ν(C=N), (B_u, A_g)
1626 m	–	1155 + 482 (Fermi-res.)
–	1440 m, br	$\delta_{as}(CH_3)$
1363 s	1368 mw	$\delta_s(CH_3C)$
1214 s	1217 m	$\varrho(CH_3C)$
1152 m	1155 vs	$\delta_s(CH_3In)$
1071 m	1077 w	$\nu_{as}(CC_2)$
975 vw, 918 vw	–, 920 vw	overtones
814 w	824 mw	$\nu_s(CC_2)$, (B_u, A_g)
699 vs, br	664 w, br	$\varrho(CH_3In)$
652 vs		
511 vs	504 s	$\nu_{as}(InC_2)$, (B_{2u}, B_{1g})
493 m	–	260 + 245 (Fermi-res.)
482 s	482 vs	$\nu_s(InC_2)$, (B_{3u}, A_g)
437 m	452 m	$\nu(In_2N_2)$, (B_{3u}, A_g)
302 m		
	280 m	
260 m	245 ms	
	227 sh	skeletal
	154 s	
	115 m, 85 vs	

The field ionization mass spectrum (energy not reported) was evaluated as follows (relative intensity in %, M is the monomeric unit): $[M_2-CH_2]^+$ (10), $[M_2-CH_3]^+$ (100), $[M_2-3\,CH_3]^+$ (5), $[M_2-NC(CH_3)_2]^+$ (22), $[M+InN]^+$ (19), $[M+In]^+$ (11), $[M+In-2\,CH_3]^+$ (6), $[In_2NC(CH_3)_2]^+$ (8), $[In_2NCH_3]^+$ (4), $[M-CH_3]^+$ (11), $[InNCCH_3]^+$ (6), $[In(CH_3)_2]^+$ (76), $[InCH_3]^+$ (8), $[H_2N_2C_2(CH_3)_4]^+$ (54), $[HNC(CH_3)_2]^+$ (14), $[C(CH_3)_2]^+$ (60) [2].

References:

[1] Weller, F.; Dehnicke, K. (Chem Ber. **110** [1977] 3935/42).
[2] Weller, F.; Müller, U. (Chem Ber. **112** [1979] 2039/44).

7.1.3 $R_2InN(R')COR''$ Compounds

$(CH_3)_2InN(CH_3)COCH_3$

The compound was prepared by addition of a benzene solution of $CH_3CONHCH_3$ to a 1 to 2% excess of $In(CH_3)_3$ in benzene at 40 to 80 °C. The white crystalline product (m.p. 144 °C, subl. 130 °C/10^{-4} Torr) is very sensitive to moisture, traces of water leading to the reformation of $CH_3CONHCH_3$. The solubility in nonpolar organic solvents is so slight that no cryoscopic measurements could be made nor could an ^1H NMR spectrum be recorded. However, by analogy to the homologous gallium compound, association to a puckered eight-membered ring (Formula I) has been postulated [3].

The IR and Raman spectra agree with a centrosymmetric structure of S_2 symmetry and the most important frequencies are presented in Table 53 together with the frequencies of the ethyl derivative [3].

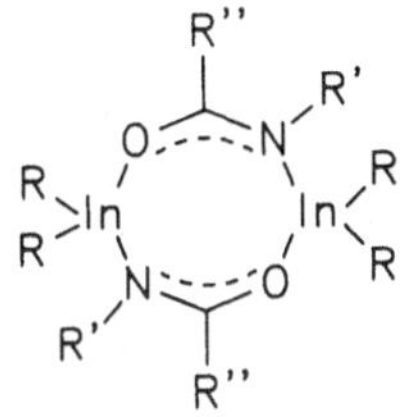

I

$(CH_3)_2In\overline{NCH_2CH_2CH_2C}{=}O$

This compound was prepared by addition of 2-pyrrolidinone to a stirred solution of $In(CH_3)_3$ (1:1 mole ratio) in toluene at room temperature, followed by heating the mixture at 50 °C for 30 min. Concentration in vacuum and addition of hexane (at −30 °C) precipitated a colorless powder in 43% yield; m.p. 130 to 132 °C. A dimeric structure analogous to I was proposed [4].

^1H NMR spectrum (in ppm, CD_3SOCD_3 solution): 0.62 (s, CH_3), 1.98 (m, CH_2, 4 H), 3.30 (t, CH_2, 2 H) [4].

The IR spectrum shows a ν(C=O) band at 1581 cm^{-1} [4].

Mass spectrum: $[M_2-CH_3]^+$, $[M_2-2\,CH_3]^+$, $[M-CH_3]^+$, $[M-2\,CH_3]^+$, $[M-C_3H_6CONH]^+$ [4].

$(C_2H_5)_2InN(CH_3)COCH_3$

The compound (m.p. 115 °C (dec.), subl. 110 °C/10^{-4} Torr) was prepared by the previous procedure from $CH_3CONHCH_3$ and $In(C_2H_5)_3$. Its structure and properties correspond largely to those of $(CH_3)_2InN(CH_3)COCH_3$. The most important IR and Raman data are included in Table 53 [3].

$(C_2H_5)_2InN(C_6H_5)COC_2H_5$

The compound was formed by insertion of C_6H_5NCO into one In-C bond of $In(C_2H_5)_3$. Thus, petroleum ether solutions of the reactants were mixed in equimolar amounts and heated at 48 °C for 6 h. After removal of the solvent, the very hygroscopic crystals (m.p. 50 to 52 °C, 89% yield) were recrystallized from n-hexane [1]. A yield of more than 80%

Table 53
IR and Raman Frequencies of Solid $(CH_3)_2InN(CH_3)COCH_3$ and Solid $(C_2H_5)_2InN(CH_3)COCH_3$ [3].
Wave numbers in cm^{-1}.

$(CH_3)_2InN(CH_3)COCH_3$ IR	Raman	$(C_2H_5)_2InN(CH_3)COCH_3$ IR	Raman	assignment
1571 vs	1581 mw	1570 vs	1547 s, br	$\nu_{as}(OCN)$ 1,2)
–	1417 ms	–	1418 m	$\nu_s(OCN)$ 1)
1392 vs	–	1395 vs	–	$\nu_s(OCN)$ 2)
1352 ms	1359 m, h	1352 s	1359 w	$\delta_s(CH_3C)$, $\delta_s(CH_3N)$
1159 m	1160 vs	1168 m	1174 vs	$\delta_s(CH_3In)$, $\delta_s(CH_2In)$
1029 mw	1030 vw	1029 m	–	$\nu(N\text{-}CH_3)$ and
1012 m	–	–	–	$\varrho(CH_3C)$
840 m	844 m	840 s	845 mw	$\nu(C\text{-}CH_3)$
518 s	519 s	493 s	494 m	$\nu_{as}(InC_2)$
483 m	485 vs	463 m	465 vs	$\nu_s(InC_2)$
355 mw, br	363 mw	365 mw	365 mw }	$\nu(In\text{-}O)$
260 mw	265 mw	267 m	– }	$\nu(In\text{-}N)$
–	243 ms	–	237 m }	$\delta_s(InCC)$ (?)

1) In-phase. – 2) Out-of-phase.

was obtained by refluxing C_6H_{14} solutions of $In(C_2H_5)_3$ and $HN(C_6H_5)COC_2H_5$ (1:1 mole ratio) for one hour [2]. Cryoscopy in benzene (?) identified dimers [1].

The ^1H NMR spectrum in CCl_4 showed the following signals (in ppm): 0.6 to 1.5 (complex m, C_2H_5In), 2.18 (q, CH_2C), 7.10 (m, C_6H_5) [1].

Only the $\nu(C{=}O)$ band at 1667 cm^{-1} was cited from the IR spectrum in Nujol mull [1]. An eight-membered ring structure like that of the above $R_2InNR'COR''$ products was assumed, but not established [1, 2].

Acid hydrolysis gives $HN(C_6H_5)COC_2H_5$; reaction with two equivalents of C_6H_5NCO leads primarily to the formation of $(C_2H_5)_2InOH$ and a substituted 1,3,5-triazine derivative [2].

References:

[1] Tada, H.; Yasuda, K.; Okawara, R. (J. Organometal. Chem. **16** [1969] 215/20).
[2] Tada, H.; Okawara, R. (J. Org. Chem. **35** [1970] 1666/7).
[3] Schwering, H.-U.; Weidlein, J. (Chimia **27** [1973] 535/8).
[4] Alcock, N. W.; Degnan, I. A.; Roe, S. M.; Wallbridge, M. G. H. (J. Organometal. Chem. **414** [1991] 285/93).

7.1.4 R_2In Derivatives of Acetamidines, Acetohydrazines, and Acetimidohydrazines

$(CH_3)_2InN(CH_3)C(CH_3){=}NCH_3$

The compound was obtained in a yield of 70 to 90% by the reaction of $In(CH_3)_3$ with $HN(CH_3)C(CH_3){=}NCH_3$ (1:1 mole ratio) in benzene at 20 to 30 °C. For purification multiple crystallization from toluene is preferable to vacuum distillation (85 °C/0.1 Torr) because partial decomposition accompanies the latter. The compound melts at 132 to 134 °C; cryoscopy in benzene reveals dimers [1].

The ^{1}H NMR spectrum in C_6D_6 shows three singlets at −0.03 (CH_3In), 1.41 (CH_3C), and 2.69 (CH_3N) ppm [1].

The vibrational spectrum was assigned by assuming a centrosymmetric (C_{2h} symmetry) puckered eight-membered ring (like Formula II) by analogy to the spectra of the homologous Al and Ga compounds. The most important frequencies are contained in Table 54 (p. 270) [1].

$(CH_3)_2InN(C_3H_7\text{-}i)C(CH_3){=}N(C_3H_7\text{-}i)$

To prepare the compound one equivalent of solid $(CH_3)_2InCl$ was added to a −20 °C slurry of $Li[N(C_3H_7\text{-}i)C(CH_3){=}N(C_3H_7\text{-}i)]$ in ether (the latter was obtained without isolation from $(C_3H_7\text{-}i)N{=}C{=}N(C_3H_7\text{-}i)$ and stoichiometric amounts of $LiCH_3$ in ether). After warming to room temperature the mixture was stirred for 30 h, the solvent removed, and the product distilled (b.p. 92 °C/14 Torr, 58% yield) from the yellow-brown residue. The pale yellow liquid is very sensitive to hydrolysis and dissolves readily in all the common organic solvents. The compound is a monomer in benzene [4].

The ^{1}H NMR spectrum in C_6D_6 exhibits the following signals (in ppm): 0.16 (s, CH_3In), 0.99 (d, CH_3 of C_3H_7), 1.53 (s, CH_3), and 3.46 (sept, HCN, J(H,H) = 6.6 Hz). Five signals were observed in the ^{13}C NMR spectrum in C_6D_6: −6.04 (CH_3In), 11.20 (CH_3), 25.98 (CH_3 of C_3H_7), 45.81 (CH of C_3H_7), and 167.79 (NCN) ppm [4].

The most important IR and Raman frequencies are listed in Table 54 (p. 270). On the basis of the vibrational spectra, this compound is considered to have a structure with a four-membered chelate ring ligand of C_{2v} symmetry (Formula I), unlike the dimeric, eight-membered ring compounds described above (see also Formula II) [4]:

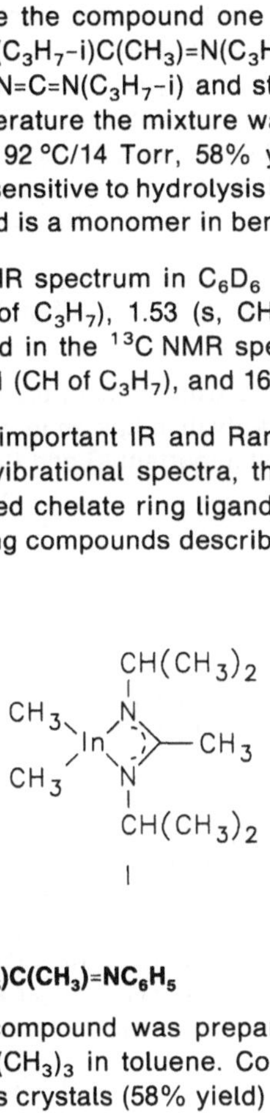

I

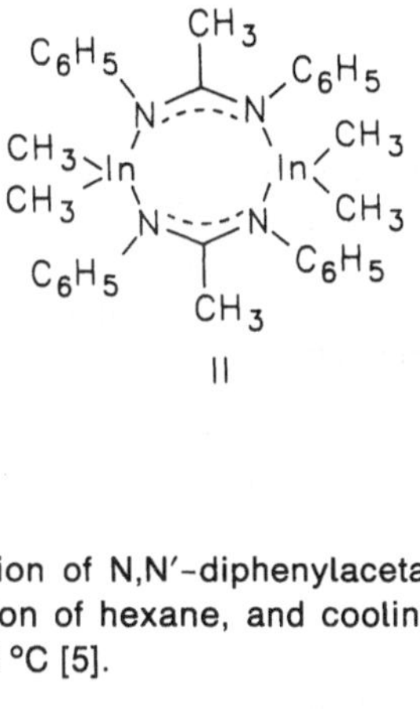

II

$(CH_3)_2InN(C_6H_5)C(CH_3){=}NC_6H_5$

The title compound was prepared by slow addition of N,N′-diphenylacetamidine to a solution of $In(CH_3)_3$ in toluene. Concentration, addition of hexane, and cooling to −20 °C gave colorless crystals (58% yield) melting at 82 to 85 °C [5].

^{1}H NMR in C_7D_8 (in ppm): 0.03 (s, CH_3In), 1.58 (s, CH_3), 6.90 to 7.13 (m, C_6H_5) [5].

The two bands in the IR spectrum at 1475 and 1528 cm^{-1} were assigned as $\nu_s(CN_2)$ and $\nu_{as}(CN_2)$ vibrations, respectively; 392 $\nu_s(InN_2)$. A dimeric structure with an eight-membered ring (Formula II) was proposed [5].

The mass spectrum showed the following fragments: $[M_2-CH_3+H]^+$, $[M_2-CH_3-In(CH_3)_2]^+$, $[M_2-In(CH_3)_4]^+$, $[M]^+$, $[M-CH_3]^+$, $[M-2\,CH_3]^+$, $[M-(C_6H_5)_2N_2HCCH_3]^+$ [5].

$(CH_3)_2InN\{Si(CH_3)_3\}C(CH_3)=NSi(CH_3)_3$

The compound was prepared in ether from $(CH_3)_3SiN=C=NSi(CH_3)_3$, $LiCH_3$, and $InCl_3$ in a three-step synthesis ($R = Si(CH_3)_3$):

(1) $RN=C=NR + LiCH_3 \rightarrow Li[N(R)C(CH_3)=NR]$
(2) $Li[N(R)C(CH_3)=NR] + InCl_3 \rightarrow Cl_2In\{N(R)C(CH_3)=NR\} + LiCl$
(3) $Cl_2In\{N(R)C(CH_3)=NR\} + 2\,LiCH_3 \rightarrow (CH_3)_2InN(R)C(CH_3)=NR + 2\,LiCl$

Silylcarbodiimide was allowed to react with $LiCH_3$ at −10 to 0 °C (eq. 1) and then, a few minutes after the addition was complete, solid $InCl_3$ was added to the lithiated intermediate at 0 to 5 °C (eq. 2). After stirring for 2 h at 5 °C, further 2 equivalents of $LiCH_3$ (in ether) were added (eq. 3), the mixture was warmed to room temperature, and stirring maintained for 30 h. After removal of the ether there remained a tough, brown mass from which the compound distilled at ca. 60 °C/0.01 Torr as a colorless viscous oil (20% yield). On cooling to about 5 °C the oil stiffened and slowly crystallized to colorless needles which melted from 23 to 28 °C. The compound is monomeric in benzene and is extremely moisture-sensitive. The low yield was caused by a side reaction: $LiCH_3 + RN=C=NR \rightarrow RCH_3 + Li[N-C\equiv N-R]$ [4].

The ^1H NMR spectrum in C_6D_6 shows three sharp singlets at 0.18 (CH_3Si), 0.22 (CH_3In), and 1.87 (CH_3C) ppm [4].

The extensive agreement of the IR and Raman spectra with those of the corresponding Al [3] and Ga compounds makes it probable that the In derivative has an identical structure of C_{2v} symmetry with a planar InN_2C framework. The most important IR and Raman frequencies are contained in Table 54 [4].

Table 54
Main IR and Raman Frequencies of $\{(CH_3)_2InN(CH_3)C(CH_3)=NCH_3\}_2$ (A), $(CH_3)_2InN(C_3H_7\text{-}i)\text{-}C(CH_3)=NC_3H_7\text{-}i$ (B), and $(CH_3)_2InN(Si(CH_3)_3)C(CH_3)=NSi(CH_3)_3$ (C).
Wavenumbers in cm^{-1}.

A (solid) [1]		B (liquid) [4]		C (solid) [4]		
IR	Raman	IR	Raman	IR	Raman	assignment
1540 vs	1520 vw, br*)	1515 vs, br	–	1481 vs, br	–	$\nu_{as}(N \cdots C \cdots N)$
1468 m, br	1469 mw, br*)	1481 sh	1478 s	1426 sh	1429 mw, br	$\nu_s(N \cdots C \cdots N)$
1353 w	1362 w	1364 w	1368 vvw	–	1362 w	$\delta(CH_3\text{-}C)$
1160 sh	1160 s	1150 w	1155 vs	1157 mw	1160 ms	$\delta_s(CH_3In)$
1152 s	1147 ms	–	–	–	–	$\varrho(CH_3\text{-}N)$ and $\nu(N\text{-}CH_3)$
1139 s	–	–	–	–	–	
1119 ms	1125 mw	–	–	–	–	
–	–	1123 w	1122 w	–	–	$\nu(C\text{-}C_2)$
–	–	1067 vw	1068 mw	–	–	
969 m	–	1025 w	–	1017 ms	–	$\varrho(CH_3\text{-}C)$
809 ms	820 m	816 w	818 w	hidden by $\varrho(CH_3Si)$		$\nu(C\text{-}CH_3)$
692 s, br	–	718 ms	–	737 m	740 vw	$\varrho(CH_3In)$
–	–	690 sh	–	720 sh	–	
–	–	–	–	685 m	688 m	$\nu_{as}(SiC_3)$
–	–	–	–	626 vw	635 s	$\nu_s(SiC_3)$

Table 54 (continued)

A (solid) [1]		B (liquid) [4]		C (solid) [4]		
IR	Raman	IR	Raman	IR	Raman	assignment
632 s	662 mw*)	625 w	–	hidden by $\nu_s(SiC_3)$		δ(N⋯C⋯N)
–	603 vw*)	–	–	–	–	and
546 m	544 m*)	560 vw	562 vw	590 w	–	γ(N⋯C⋯N)
498 s	504 ms	509 m	511 m	510 m	519 m	$\nu_{as}(InC_2)$
470 s	478 vs	482 mw	482 vs, br	480 ms	487 vs	$\nu_s(InC_2)$ and
						$\nu_{as}(InN_2)$ and
–	–	–	462 s	–	–	$\delta_s(CC_2)$
384 m	394 s*)	298 mw	310 s	340 m	342 ms	$\nu_s(InN_2)$
–	288 mw	–	278 vw	295 m	–	δ(CNR′)
255 ms, br	255 s	–	253 w	–	243 m	$\delta(SiC_3)$
219 mw	–	–	–	–	221 w	δ(skeleton)
–	141 ms	–	120 s	–	190 s, br	$\delta(InC_2)$

*) In-phase and out-of-phase vibrations.

$(CH_3)_2InN(CH_3)N(CH_3)C(CH_3)=NCH_3$ and $(CH_3)_2InN(CH_3)N(CH_3)C(CH_3)=NCH_3 \cdot In(CH_3)_3$

Both compounds were formed in the reaction of N,N′,N″-trimethylacetimidohydrazine, $HN(CH_3)N(CH_3)C(CH_3)=NCH_3$, with $In(CH_3)_3$ in toluene at 80 to 100 °C. Separation of the components was, contrary to the gallium homolog, unsuccessful, so that characterization was not possible [2].

$(CH_3)_2InN(CH_3)N(CH_3)C(CH_3)=O \cdot In(CH_3)_3$

The compound was obtained in a yield of 44% by the reaction of N,N′-dimethylacetohydrazine, $HN(CH_3)N(CH_3)COCH_3$, with $In(CH_3)_3$ (1:2 mole ratio) in toluene at 80 to 100 °C. The reaction occurs with the elimination of methane and is not quantitative when equimolar amounts of reactants are used. The remaining heavily viscous residue was purified by distillation (80 °C/0.02 Torr) and sublimation (58 to 63 °C/0.01 Torr) [2].

The colorless to pale yellow solid melts at 110 °C, and is monomeric in benzene solution. The basic, $In(CH_3)_3$-free derivative of the adduct (Formula III) is not known, but III corresponds to the X-ray structure of the Ga homolog [2].

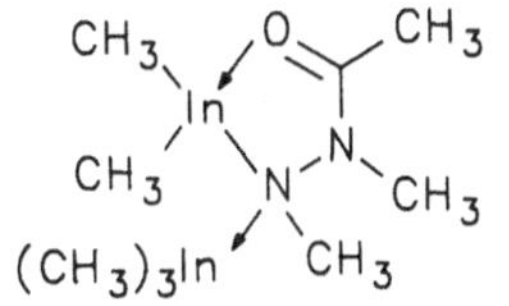

III

The ^1H NMR spectrum in C_6D_6 (in ppm) is consistent with the proposed structure: −0.04 ($In(CH_3)_3$), 0.17 ($In(CH_3)_2$), 1.37 (CH_3C), 2.47, 2.64 (CH_3N) ppm [2].

The most important IR and Raman data (in cm^{-1}) for the solid compound are: 1570 ν_{as}(OCN), 1485 ν_s(OCN), 927 ν(N-N), 526 $\nu_{as}(InC_2)$, 488 $\nu_s(InC_2)$, 472 $\nu_{as}(InC_3)$, 477 ν(OInN), 464 $\nu_s(InC_3)$ [2].

References:

[1] Gerstner, F; Weidlein, J. (Z. Naturforsch. **33b** [1978] 24/9).

[2] Gerstner, F; Hausen, H.-D.; Weidlein, J. (J. Organometal. Chem. **197** [1980] 135/46).

[3] Lechler, R.; Hausen, H.-D.; Weidlein, J. (J. Organometal. Chem. **359** [1989] 1/12).

[4] Kottmair-Maieron, D.; Lechler, R.; Weidlein, J. (Z. Anorg. Allgem. Chem. **593** [1991] 111/23).

[5] Alcock, N. W.; Degnan, I. A.; Roe, S. M.; Wallbridge, M. G. H. (J. Organometal. Chem. **414** [1991] 285/93).

7.1.5 $R_2In(NR'-c)$ Compounds with Nitrogen Heterocycles

7.1.5.1 Derivatives of Saturated Cyclic Amines

Compounds in this category are collected in Table 55 and were synthesized by the following procedures (NR′-c is a cyclic amine):

Method I: From InR_3 and H(NR′-c).
An equimolar amount of the cyclic amine was condensed into a flask containing InR_3 and (for Nos. 1 to 4 and 7) kept for several days at 120 °C. The elimination of RH was followed by measuring the gas volume [1]. The remaining white residue was purified by vacuum sublimation; No. 7 separated as a viscous liquid [2]. No. 6 was obtained in ether at room temperature [3].

Method II: From R_2InCl and Li[(NR′-c)].
The amine was first lithiated by LiC_4H_9-n in ether, the ether and excess amine removed in vacuum, the residue dissolved (or suspended) in fresh ether, and then added to a suspension of R_2InCl in ether at room temperature. After a few hours of constant stirring, the LiCl was filtered off, and the filtrate reduced to about 1/7 of its volume. The compound crystallized out within 24 h at −25 °C and was purified by vacuum sublimation [3]. For No. 8 the process was carried out in C_5H_{12} or C_6H_{14}, since in ether further reactions intrude (see p. 186). Purification was accomplished by a vacuum distillation [5].

General Remarks. Vapor pressure measurements on Nos. 3, 5, and 6 were carried out by the Knudsen effusion method in the region of 10^{-6} to 10^{-2} Torr [4]. According to cryoscopic molecular weight determinations in benzene, amines No. 2 to 7 are dimers, No. 1 is a trimer, and No. 8 a monomer. This was confirmed by the mass spectra (high-resolution mass spectra for Nos. 5, 6, and 8). A simple fragmentation scheme was reported for Nos. 1 to 4 [1], and it is essentially valid for the others also:

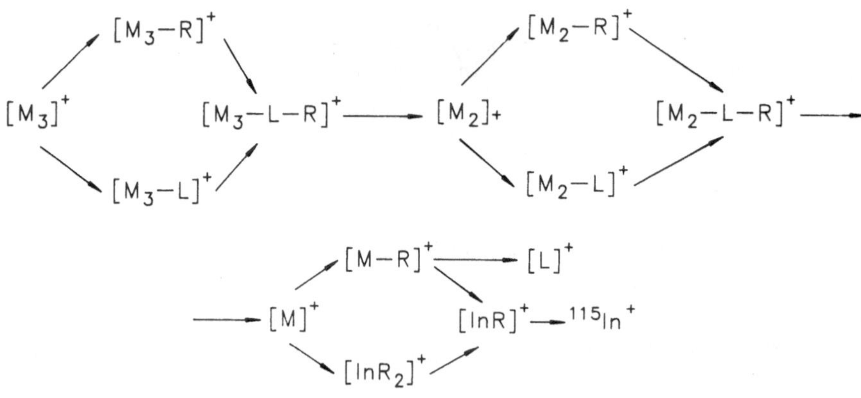

Table 55
R_2In Derivatives of Cyclic Amines.
Further information on numbers preceded by an asterisk is given at the end of the table.
Explanations, abbreviations, and units on p. X.

No.	compound method of preparation (yield)	properties and remarks
$(CH_3)_2In$ compounds		
1	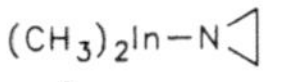 $(CH_3)_2In-N$ I	colorless crystals, m.p. 138 to 141 °C; volatile [1] ^1H NMR (C_6H_6): −0.34 (s, CH_3In), 1.46 (s, CH_2) [1]
2	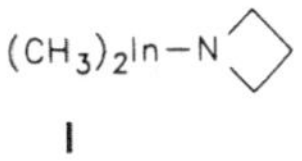 $(CH_3)_2In-N$ I	colorless crystals, m.p. 102 to 104 °C; volatile [1] ^1H NMR (C_6H_6): −0.14 (s, CH_3In), 1.93 (quint., CH_2-2), 3.51 (t, CH_2-1) [1]
3	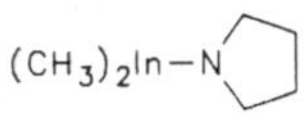 $(CH_3)_2In-N$ I	colorless crystals, m.p. 62 to 65 °C; volatile [1] log p (Torr) = 14.25 − 5330/T(K) [4] ^1H NMR (C_6H_6): −0.14 (s, CH_3In), 1.29 (m, CH_2-2), 2.93 (m, CH_2-1) [1]
4	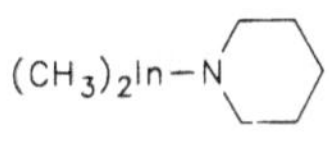 $(CH_3)_2In-N$ I	colorless crystals, m.p. 137 to 139 °C; volatile [1] ^1H NMR (C_6H_6): −0.01 (s, CH_3In), 1.33 (m, CH_2-2), 2.84 (m, CH_2-1) [1]
*5	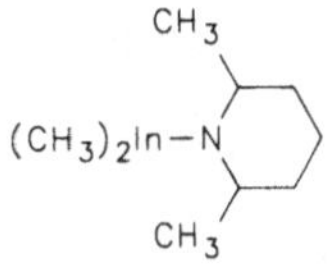 CH_3 $(CH_3)_2In-N$ CH_3 II (59%)	colorless crystals, m.p. 124.5 to 125 °C, subl. 78 °C/0.01 Torr (dec. >80 °C) [3] log p (Torr) = 14.65 − 6270/T(K) [4] ^1H NMR (C_7D_8, 80 MHz): 0.03 (3 peaks CH_3In), 1.17 (d, CH_3C, J(H,H) = 7 Hz), 0.9 to 1.7 (m, br, CH_2), 3.20 (m, br, CH); for the 400 MHz spectrum, see further information [3] IR (solid): 1444 s, 1361 m, 1358 m, 1332 w, 1316 m, 1304 w, 1212 s, 1184 w, 1163 s, 1144 s, 1130 s, 1066 s, 1022 m, 997 m, 973 m, 953 s, 930 m, 915 m, 844 s, 807 w, 743 s, 708 vs $\delta(CH_3In)$, 691 s, 664 m, br, 652 m, 621 s, 559 w, 502 s $\nu_{as}(InC_2)$, 491 s, 487 sh $\nu_s(InC_2)$, 433 w, 386 w, 353 w, 318 w, br [3] mass spectrum: $[M_2-CH_3]^+$ (14%), $[M_2-In(CH_3)_2]^+$ (20%), M^+ (5%), $[M-H]^+$ (38%), $[M-CH_3]^+$ (25%), $[In(CH_3)_2]^+$ (21%), 115 In^+ (32%) [3]
*6	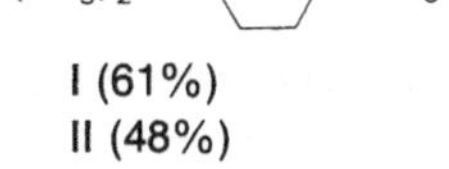 $(CH_3)_2In-N$ $N-CH_3$ I (61%) II (48%)	colorless crystals, m.p. 186 to 189 °C, subl. 90 °C/0.01 Torr [3] log p (Torr) = 14.8 − 6420/T(K) [4] ^1H NMR (C_6D_6): 0.02 (s, CH_3In), 2.08 (s, CH_3N), 2.15 (m, CH_2NIn), 3.01 (m, CH_2) [3] IR (solid): 1446 s, 1364 s, 1307 w, 1289 s, 1280 s, 1203 w, 1159 m, 1150 s, 1137 s, 1119 s, 1087 m, 1048 m, 1003 s, 969 w, br; 876 s, br, 860 m, sh, 787 m, 775 m, 699 vs, br $\delta(CH_3In)$, 668 s, 509 s $\nu_{as}(InC_2)$, 493 m, 477 s $\nu_s(InC_2)$, 456 m, 394 m, 324 w, br, 292 w [3]

Table 55 (continued)

No.	compound method of preparation (yield)	properties and remarks
*6 (continued)		mass spectrum: M_2^+ (18%), $[M_2-CH_3]^+$ (13%), $[M_2-L+H]^+$ (11%), $[M_2-L]^+$ (87%), $[M_2-L-CH_2]^+$ (16%), M^+ (9%), $[M-2\,CH_3-H]^+$ (17%), $[In(CH_3)_2]^+$ (33%), $^{115}In^+$ (30%), ($L=N_2C_5H_{11}$) [3]
other R_2In compounds		
7	$(C_2H_5)_2In-N$(piperidine ring) I	viscous liquid, air-sensitive [2] ^{1}H NMR (C_6H_6): 0.75 (CH_2In), 1.33 (CH_2-2,3), 1.47 (CH_3), 2.87 (CH_2-1) [2] mass spectrum (70 eV; principal fragments only): $[M_2-C_2H_4]^+$ (5%), $[M_2-C_2H_5]^+$ (27%), $[M_2-L-C_2H_4]^+$ (5%), $[M-H]^+$ (3.1%), $[M-C_2H_5]^+$ (1.9%), $[InL-H]^+$(3.8%), $[In(C_2H_5)_2]^+$(5%), $^{115}In^+$ (12.5%), L^+ (100%), ($L=NC_5H_{10}$) [2]
8	$(t-C_4H_9)_2In-N$(2,2,6,6-tetramethylpiperidine ring; CH_3, CH_3, CH_3, CH_3) II (81%)	orange-yellow liquid, m.p. < −25 °C; b.p. 60 °C/0.1 Torr, light-sensitive [5] ^{1}H NMR (C_6D_6): 1.13 (s, $(CH_3)_2C$), 1.34 (s, C_4H_9-t), 1.2 to 1.9 (m, br, CH_2) [5] IR (neat): 3000 m, 2985 m, 2950 s, 2925 s, 2870 s, 2830 s, 2765 m, 2710 m, 1463 s, 1453 s, 1372 s, 1366 s, 1352 s, 1340 m, 1289 m, 1239 s, 1195 w, 1175 w, 1157 s, 1136 s, 1075 m, 1051 m, 1027 s, 1013 m, 983 w, 959 m, 941 m, 923 s, 902 s, 855 m, 804 s, 733 w, 584 w, 532 w, 507 w, 490 m, 455 w, 420 w, 381 w, br, 240 s [5] mass spectrum (principal fragments only): M^+ (23%), $[M-C_4H_9]^+$ (42%), $[M-2\,C_4H_9]^+$ (8%), $[M-2\,C_4H_9-CH_2]^+$ (22%), $[LH]^+$ (11%), $[LH-CH_2]^+$ (10%), $[LH-CH_3]^+$ (100%), $^{115}In^+$ (37%), ($L=NC_9H_{18}$) [5]

* Further information:

$(CH_3)_2In\overline{NCH(CH_3)(CH_2)_3C}HCH_3$ (Table **55**, No. **5**) decomposes to a grey liquid at 125.5 °C [3].

It shows in the ^{1}H NMR spectrum (400 MHz) three singlets for the $InCH_3$ protons and two doublets of different intensities for the CCH_3 protons (the 400 MHz spectrum is illustrated in [3]). This splitting can be explained by the existence of the two conformers, I (δ CH_3In at 0.03, δ CH_3C at 1.17 ppm) and II (δ CH_3In at 0.01, 0.05, δ CH_3C at 1.18 ppm) [3]. The acentric form II produces splitting of the corresponding resonances because of the nonequivalence of the $In(CH_3)_2$ groups (designated a and b). At room temperature the I/II ratio is 5.5 and is still the same after standing four weeks. The ratio is temperature-dependent; after 180 min at 39 °C, I/II = 3.5; after 30 min at 57 °C, I/II = 1.3; and after 30 min at 71 °C, I/II = 1. On the other hand, cooling each conformer mixture does not change the ratio; the initial ratio of 5.5 is, therefore, the result of a kinetically influenced reaction. The activation energy for the configuration change was not determined [3].

I

II

$(CH_3)_2In\overline{N(CH_2)_2N(CH_3)CH_2CH_2}$ (Table **55**, No. **6**) crystallizes in the triclinic space group $P\bar{1}-C_i^1$ (No. 2) with the unit cell parameters a = 825.6(3), b = 936.3(3), c = 750.3(4) pm, α = 105.28(3)°, β = 115.68(4)°, γ = 95.226(3)°; Z = 4 (2 dimers) and D_c = 1.654 g/cm³. The most important bond distances and angles are cited in **Fig. 52** [3].

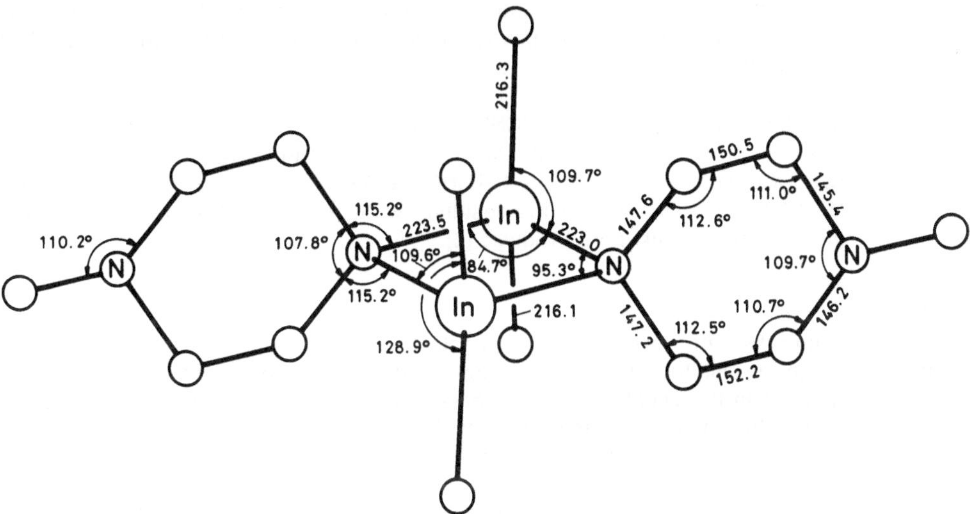

Fig. 52. Molecular structure of $\{(CH_3)_2In\overline{N(CH_2)_2N(CH_3)CH_2CH_2}\}_2$ [3].

References:

[1] Storr, A.; Thomas, B. S. (J. Chem. Soc. A **1971** 3850/4).

[2] Sen, B.; White, G. L. (J. Inorg. Nucl. Chem. **35** [1973] 2207/15).

[3] Arif, A. M.; Bradley, D. C.; Dawes, H.; Frigo, D. M.; Hursthouse, M. B.; Hussain, B. (J. Chem. Soc. Dalton Trans. **1987** 2159/64).

[4] Bradley, D. C.; Faktor, M. M.; Frigo, D. M. (J. Cryst. Growth **89** [1988] 227/36).

[5] Bradley, D. C.; Frigo, D. M.; Hursthouse, M. B.; Hussain, B. (Organometallics **7** [1988] 1112/5).

7.1.5.2 Derivatives with Unsaturated Nitrogen Heterocycles

Compounds of this type are collected in Table 56 and were prepared by the following procedures (NR′-c is an unsaturated cyclic amine).

Method I: From $In(CH_3)_3$ and H(NR′-c).
$In(CH_3)_3$ reacted in 1:1 mole ratio with the appropriate nitrogen heterocycle, using ether [3, 5, 9], benzene [1, 9], or xylene [2] as solvent. Evolution of methane was complete after 0.5 to 2 h at room temperature for Nos. 1 to 4, 9, 13 to 17 but required 24 h at reflux (ca. 110 °C) for No. 12. No. 5 (40 °C) and No. 7 (80 °C) were heated for 24 h in sealed tubes (bombs). After removing the solvent, the residue was dried in vacuum. Purification by recrystallization or vacuum sublimation is possible in many cases.

Method II: From R_2InCl and M^I[NR′-c].
For the preparation of Nos. 4 and 16 a suspension of lithiated pyrazole in ether was added to a suspension of R_2InCl ($R=CH_3$, C_2H_5). After 6 h stirring at room temperature the residue was extracted with hexane and worked up by distillation [9]. For No. 8 $(CH_3)_2InCl$ and $Ag[C_3HN_2(C_6H_5)_2]$ in 1:1 mole ratio were stirred in dry CH_2Cl_2 at room temperature as long as a voluminous white precipitate of AgCl formed. AgCl and unreacted $(CH_3)_2InCl$ were filtered off and the filtrate evaporated to dryness in vacuum. The product contains CH_2Cl_2 according to the mass spectrum [7]. No. 10 was analogously obtained from $K[(N_2C_3H_3)_2BH_2]$ in ether [8].

Method III: Insertion of $(C_4H_9)_2InN_3$ into R′CNE compounds.
Reaction of $(C_4H_9)_2InN_3$ (Table 26, p. 177, No. 8) with C_6H_5CN at 140 °C for 3 h, C_6H_5NCO at 140 °C for 2 h, or C_6H_5NCS at 120 °C for 8 h, gave the tetrazoles No. 18 to 20. Extraction of the residues with boiling hexane gave 40 to 50% yields of analytically pure products [6].

General Remarks. The derivatives of pyrazole Nos. 4 to 10, 16, and 17 are liquids or solids which can be distilled or sublimed, respectively, without decomposition. The compounds are soluble in the usual organic solvents and are dimeric. Statements about their structures [9] are based on the similarities of the spectroscopic data with those of the homologous Ga compounds, which have been studied primarily by X-ray analysis [2, 4]. The other azolides with two or more nitrogen heteroatoms are solids with high melting points; they are only slightly soluble and have a higher degree of association than the pyrazolides.

Although the vibration spectra are dominated by the bands of the heterocyclic ligands, the determination of the characteristic absorptions of the R_2In fragments, which are not very variable, is definitely possible. The following assignments are given for compounds No. 4 to 7, 11, 12, and 15 (in cm^{-1}): 1170 ± 5 $\delta_s(CH_3)$, 715 ± 3 $\varrho(CH_3)$, 523 ± 3 $\nu_{as}(InC_2)$, 490 ± 3 $\nu_s(InC_2)$; the C-In-C bond angle was estimated to be $133°\pm3°$ [9].

The fragmentation scheme in the mass spectra of compounds No. 1, 4, 15 to 17 (with exception of No. 1) show extensive agreement from which analogous basic structures have been deduced. The relative intensities (in %) of corresponding fragments at 70 eV and source temperature 25 to 35 °C (No. 15 at 100 °C) are compiled below.

NR′-c	pyrrolyl	pyrazolyl	imidazolyl	2-methylimidazolyl	1,2,4-triazolyl
No.	1	4	16	17	15
$[M_3-R]^+$	–	–	–	–	1.5
$[M_2]^+$	9.3	–	–	–	0.5
$[M_2-R]^+$	29.5	89	100	100	100
$[M_2-2R]^+$	1.7	–	14	–	12
$[M_2-3R]^+$	–	9	27	32	4
$[M_2-NR']^+$	34.1	–	–	–	2.5
$[M_2-R-NR']^+$	–	86	3	12	14
$[M_2-2R-NR']^+$	14.1	–	–	–	3
$[M_2-3R-NR']^+$	–	–	7	13	1.5
$[In_2NR']^+$	3.9	5	17	17	9.5
$[M]^+$	9.3	2	–	–	1.5
$[M-R]^+$	–	32	78	14	17
$[InNR']^+$	3.1	15	6	7	8
$[InR_2]^+$	100	68	17	14	95
$[InR]^+$	5.2	10	–	–	55
$[In]^+$	53.3	100	65	51	67
$[NR']^+$	–	18	17	27	20
$[HNR']^+$	8.7	–	–	–	–

Table 56
$R_2In(NR'\text{-}c)$ Derivatives of Unsaturated Nitrogen Heterocycles.
Further information on numbers preceded by an asterisk is given at the end of the table.
Explanations, abbreviations, and units on p. X.

No.	compound method of preparation (yield)	properties and remarks
*1	$(CH_3)_2In-N$ (pyrrolyl; positions 2, 3, 4, 5) I (91%)	colorless light-sensitive crystals, m.p. 118 to 120 °C (dec.), subl. 70 to 74°/0.1 Torr [9] ^1H NMR (C_6D_6): 0.03 (CH_3In), 6.30 (br, H-3,4), 6.82 (br, H-2,5) [9] ^{13}C NMR (C_6D_6): −5.5 (CH_3In), 110.2 (C-3,4), 129.6 (C-2,5) [9] IR/Raman (solid): 3108 vw, sh, 3085 w, 3060 vw, sh/- ν(CH-ring), 2967 mw/- $\nu_{as}(CH_3)$, 2913 w/- $\nu_s(CH_3)$, 1481 w, and 1446 mw/1450 vw, br ν(ring), 1400 w/- $\delta_{as}(CH_3)$, 1368 mw, 1295 m, and 1230 m/1364 w ν(ring), 1168 ms, 1160 m, sh, and 1152 s/1161 m and 1151 w $\delta_s(CH_3)$ and ν_s(ring), 1091 s, 1080 m, sh, 1045 m, and 1026 vs, br/1088 w δ(CH-ring, in plane), 909 ms/908 w δ(CH-ring), 878 vw, 851 s, 827 m, and 781 ms/825 w, br γ(CH-ring), 732 ms/- γ(CH-ring), 719 vs, br/- $\varrho(CH_3)$, 641 s, 616 s/- γ(ring), 531 m/530 w, br $\nu_{as}(InC_2)$, 508 m/503 vw, br ν(In-N), 482 ms/485 vvs $\nu_s(InC_2)$ and δ(ring), 408 w, 375 w/- γ(ring) (?) [9] mass spectrum, see above

Table 56 (continued)

No.	compound method of preparation (yield)	properties and remarks
*2	$(CH_3)_2In-N$ (pyrrole ring, 2-CHO) I	^1H NMR (CD_3COCD_3): −0.35 (CH_3In), 6.20 (H-4), 6.93 (H-5), 7.18 (H-3), 9.10 (HCO) [3] IR/Raman (solid): 1600/- ν(C=O), 720 s/- $\varrho(CH_3In)$, 545 s/546 m $\nu_{as}(InC_2)$, 480 m/482 m $\nu_s(InC_2)$; no other bands reported [3]
3	$(CH_3)_2In-N$ (pyrrole ring, 2-C(O)CH_3) I	solid [3] ^1H NMR (CD_3COCD_3): 1.78 (CH_3C), 2.36 (?) (CH_3In), 5.67 (H-4), 6.40 (H-5), 6.54 (H-3) [3] IR/Raman (solid): 1640 vs/- ν(C=O), 725/- $\varrho(CH_3In)$, 541 m/544 m $\nu_{as}(InC_2)$, 489 mw/490 s $\nu_s(InC_2)$; no other bands reported [3]
4	$(CH_3)_2In-N$ (pyrazole ring) I (81%) II (60%)	colorless crystals, m.p. 23 to 26 °C, b.p. 160 °C/11 Torr [9] ^1H NMR (C_6D_6): 0.19 (CH_3In), 6.24 (t, H-4), 7.43 (d, H-3,5, J(H,H) = ca. 2) [9] ^{13}C NMR: −7.0 (CH_3In), 105.1 (C-4), 140.0 (C-3,5) [9] IR/Raman (liquid): 1522 mw/vw, 1495 ms/vw, 1412 ms/vw, and 1374 s/ms ν_{as} and ν_s(ring), 1269 s/ms δ(CH-ring), 921 mw/mw δ(ring), 755 vs/vvw γ(CH), 622 s/vw γ(ring) [9] mass spectrum on p. 277 used for epitaxial growth of InP layers [9]
5	$(CH_3)_2In-N$ (pyrazole ring, 3-CH_3) I (90%)	m.p. 58 to 61 °C, subl. under high vacuum, dimeric; contrary to the Ga homologue, only one isomer [5] ^1H NMR ($CDCl_3$): −0.10 (s, CH_3In), 2.37 (s, CH_3C), 4.17 (d, H-4), 7.55 (d, H-5, J(H,H) = 2.0) [5] ^{13}C NMR (C_6D_6): −6.6 (CH_3In), 13.0 (CH_3), 105.2 (C-4), 140.8 (C-5), 149.9 (C-3) [9]
6	$(CH_3)_2In-N$ (pyrazole ring, 4-CH_3) I (87%)	colorless solid, m.p. 54 to 56 °C, b.p. 108 to 110 °C/0.2 Torr [9] ^1H NMR (C_6D_6): 0.20 (CH_3In), 1.96 (CH_3), 7.20 (d, H-3,5) [9] ^{13}C NMR (C_6D_6): −7.08 (CH_3In), 8.46 (CH_3), 115.1 (C-4), 139.5 (C-3,5) [9] IR/Raman (liquid): 1565 m/w, 1455-/vw, and 1362 ms/w ν_{as} and ν_s(ring), 1311 s/mw δ(CH), 1018 s/s δ(ring), 830 vs,br/vvw γ(CH), 622 s/w γ(ring) [9]
7	$(CH_3)_2In-N$ (pyrazole ring, 3,5-$(CH_3)_2$) I (92%)	colorless solid, m.p. 58 to 60 °C [1], 77 to 79 °C, subl. 50 to 55 °C/10^{-4} Torr [9] dipole moment (C_6H_6, 25 °C): 1.39 D [1] ^1H NMR (C_6D_6): 0.04 (CH_3In), 2.04 (CH_3), 5.70 (s, H-4) [9]

Table 56 (continued)

No.	compound method of preparation (yield)	properties and remarks
7 (continued)		^{13}C NMR (C_6D_6): −5.35 (CH_3In), 13.21 (CH_3), 106.1 (C-4), 150.8 (C-3,5) [9]
8	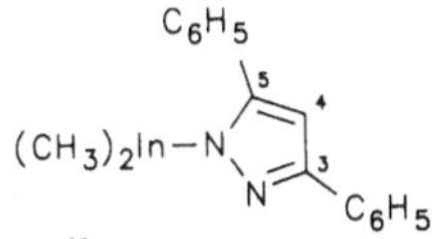 II	off-white insoluble solid; no other data reported [7]
9	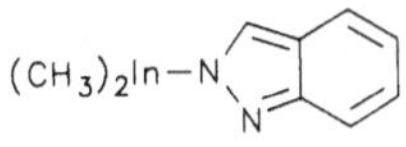 I (90 to 100%)	m.p. 108 to 110 °C, subl. under vacuum, chair conformation of the [In(N-N)$_2$In]-ring assumed [5] ^1H NMR ($CDCl_3$): −0.77 (s, CH_3In), 6.83 (m, br, $C_7H_5N_2$) [5]
*10	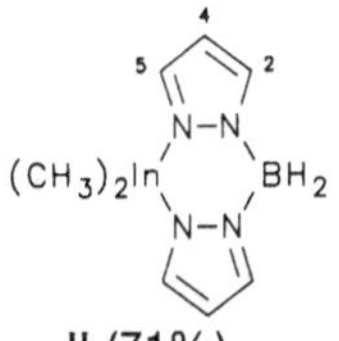 II (71%)	white solid, m.p. 53 to 56 °C, monomeric in benzene [8] ^1H NMR ($CDCl_3$): 0.09 (CH_3In), 3.6 (br, H_2B), 6.23 (t, H-4, J(H,H) = 2.1), 7.50, 7.64 (dd, H-3,5, J(H,H) = 1.9) [8] IR (C_6H_6): 2441 sh, 2432, 2408 $\nu(BH_2)$; no other bands reported [8] mass spectrum: $[M-H]^+$, $[M-CH_3]^+$, $[M-C_2H_7]^+$ [8]
11	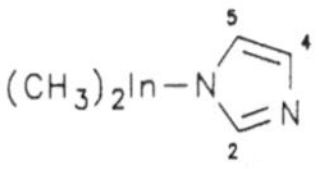 I (ca. 100%)	white solid, m.p. 150 to 152 °C [9] ^1H NMR (CCl_4): −0.10 (CH_3In), 6.89 (s, H-2,4,5) [9]
12	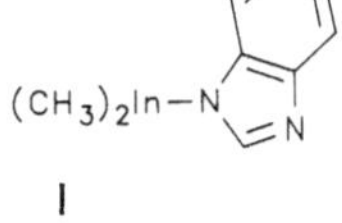 I	m.p. 221 to 222 °C (dec.), air-stable, tetrameric structure assumed [2] ^1H NMR ($CDCl_3$): −0.01 (s, CH_3In), 1.76 (s, CH_3C), 6.69 (s, $C_3H_2N_2$) [2] IR (solid): 520 m $\nu_{as}(InC_2)$, 490 m $\nu_s(InC_2)$, 420 s; no other bands reported [2]
13	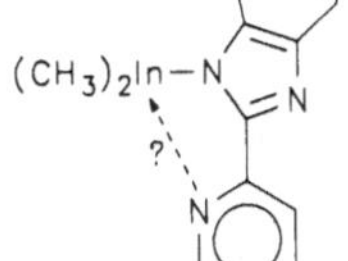 I	m.p. 108 to 109 °C, tetrameric structure assumed [1]
14	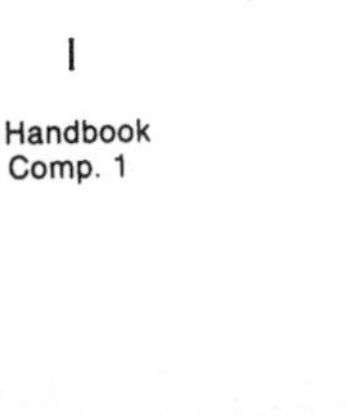 I	m.p. 251 to 252 °C [1]

Table 56 (continued)

No.	compound method of preparation (yield)	properties and remarks
15	$(CH_3)_2In-N$ (1,2,4-triazolyl, positions 3,5) I (75%)	white solid, m.p. 189 to 192 °C (dec.) [9] ^1H NMR (CD_3SOCD_3): −0.15 (CH_3In), 7.88, 7.91 (H-3,5) [9] mass spectrum on p. 277
16	$(C_2H_5)_2In-N$ (pyrazolyl, positions 3,4,5) I (90%) II (55%)	colorless oily liquid, m.p. <20 °C, b.p. 125 °C/4 Torr [9] ^1H NMR ($CDCl_3$): 0.85 (q, CH_2, J=8.7), 1.19 (t, CH_3), 6.39 (t, H-4), 7.64 (d, H-3,5, J=2) [9] ^{13}C NMR ($CDCl_3$): 6.6 (CH_2), 11.7 (CH_3), 104.7 (C-4), 139.6 (C-3,5) [9] IR/Raman (liquid): 1513 w/vw, 1491 ms/vw, 1412 ms/vw, and 1372 s/m ν_{as} and ν_s(ring), 1268 ms/ms δ(CH), 917 w/m δ(ring), 755 vvs/vvw γ(CH), 619 s/vw γ(ring) and ϱ(CH_2) [9] mass spectrum on p. 277
17	$(C_2H_5)_2In-N$ (3-CH_3-pyrazolyl, positions 3,4,5) I	colorless oily liquid, b.p. 130 °C/4 Torr [9] ^1H NMR ($CDCl_3$): 0.77 (q, CH_2, J=8.7), 1.12 (t, CH_3 of C_2H_5), 2.36 (s, CH_3), 6.15 (d, H-4), 7.52 (d, H-5, J=2) [9] ^{13}C NMR ($CDCl_3$): 6.7 (CH_2), 11.5 (CH_3 of C_2H_5), 12.9 (CH_3), 104.6 (C-4), 140.4 (C-5), 149.3 (C-3) [9] IR/Raman (liquid): 1498 ms/mw, 1489 ms/vw, 1422 mw/w, and 1373 mw/vw ν_{as} and ν_s(ring), 1321 mw/ms δ(CH), 916-/m δ(ring), 767 vs/vw γ(CH), 625 s/vw γ(ring) and ϱ(CH_2) [9] mass spectrum on p. 277
*18	$(C_4H_9)_2In-N$ (5-C_6H_5-tetrazolyl) III (40 to 50%)	white to brown solid, m.p. 121 °C, air-stable [6] ^1H NMR: 1.05 (m, C_4H_9), 7.14 (m, C_6H_5) [6] UV (CH_3OH): λ_{max} in the 245 to 268 nm region [6]
*19	$(C_4H_9)_2In-N$ (1-C_6H_5-5-oxo-tetrazolyl) III (40 to 50%)	white to brown solid, m.p. 128 °C, air-stable [6] IR (solid): 1700 ν(C=O); no other bands reported [6]
*20	$(C_4H_9)_2In-N$ (1-C_6H_5-5-thioxo-tetrazolyl) III (40 to 50%)	white to brown solid, m.p. 182 °C, air-stable [6]

* Further information:

$(CH_3)_2InNC_4H_4$ (Table **56**, No. **1**) decomposes with separation of metallic In at temperatures near the melting point as well as under the influence of light in the solid state or in solution. A polymeric structure similar to that of the isoelectronic $(CH_3)_2In(C_5H_5\text{-}c)$ (on p. 112) was deduced from spectroscopic data [9].

The compound crystallizes in the orthorhombic space group $P2_12_12_1-D_2^4$ (No. 19) with the constants a=656.6(2), b=853.7(2), c=1313.6(3) pm; Z=4 gives D_c=1.90, while D_m=ca. 1.85 g/cm³. As in $In(C_5H_5\text{-}c)_3$ (see on p. 87), parallel zigzag chains are formed in which the In atom is bonded to the N atom and additionally to a ring carbon atom. A part of the structure is presented in **Fig. 53**. The lack of good crystals, which is due to the light-sensitivity of the compound, diminishes the quality of the X-ray analysis [10].

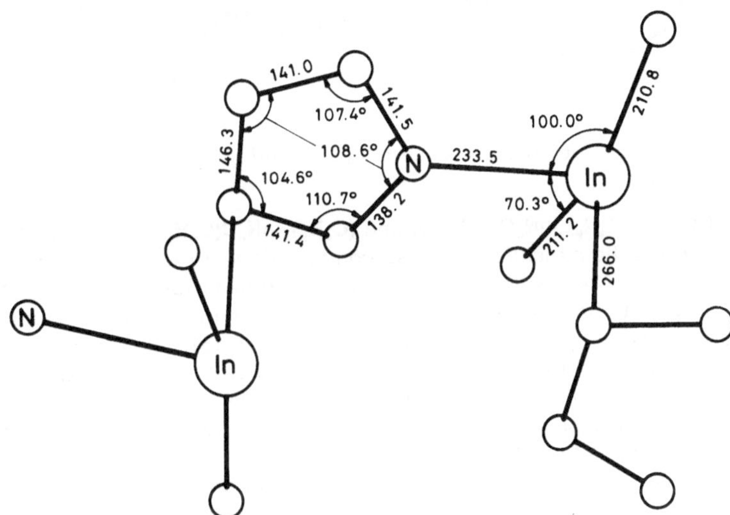

Fig. 53. Molecular structure of $\{(CH_3)_2InNC_4H_4\}_n$ [10].

$(CH_3)_2InNC_4H_3CHO$ (Table **56**, No. **2**) dissolves in $(CH_3)_2CO$, CH_3OH, C_2H_5OH, $(CH_3)_2SO$, $CHCl_3$, and hot benzene; in $O(C_2H_5)_2$, CCl_4, CS_2, $C_6H_{12}\text{-}c$, and petroleum ether it is insoluble. A mixture of it with the starting material, HNC_4H_3CHO, gives a ^1H NMR spectrum in CD_3COCD_3 at 0 °C containing two peaks (at 9.3 ppm, $\Delta\delta$=22.2 Hz) belonging to the formyl protons of the free and the bonded ligands. Coalescence occurs at 35 °C; the energy of ligand exchange was calculated to be 14.8 kcal/mol [3].

The compound forms a 1:1 adduct with 2,2′-bipyridine in acetone, but isolation in analytically pure form was not successful; the ^1H NMR spectrum of the reaction solution, however, showed the expected 6:4:8 ratio of the CH_3In, pyrrole, and bipyridyl protons [3].

$(CH_3)_2In(N_2C_3H_3)_2BH_2$ (Table **56**, No. **10**) crystallizes in the orthorhombic space group $Pn2_1a-C_{2v}^9$ (No. 33) with the parameters a=1384.2(6), b=900.8(5), c=976.1(4) pm; Z=2 and D_c=1.635 g/cm³. The InN_4B ring has the typical shallow boat configuration as shown in **Fig. 54**, p. 282. The B···In contact distance is 339.9(2) pm; an agostic B-H···In interaction could not be confirmed [8].

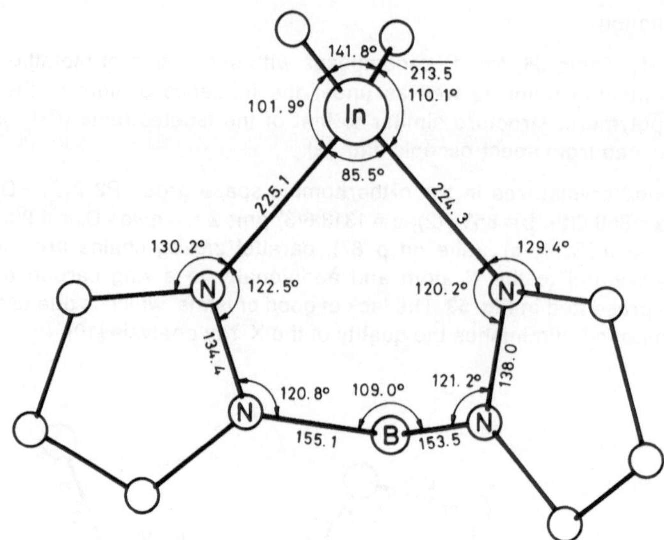

Fig. 54. Molecular structure of $(CH_3)_2In(N_2C_3H_2)_2BH_2$ [8].

$(C_4H_9)_2InN_4CC_6H_5$, $(C_4H_9)_2InN_4C(C_6H_5)O$, and **$(C_4H_9)_2InN_4C(C_6H_5)S$** (Table **56**, Nos. **18**, **19**, and **20**). The heterocyclic structures of these compounds were established mainly by the IR spectra. The strong CN-, NCO-, NCS-, or N_3-absorptions (at about 1200, 1175, 2080, or 2042 cm^{-1}) of the starting materials have disappeared; however, a new band appeared at about 1700 cm^{-1} (ν(C=O)) for No. 19 and in the 1395 to 1310 cm^{-1} region (ν(C=S)) for No. 20. The tetrazole ring has a vibration of the N=N=N fragment between 1305 and 1265 cm^{-1}, which lies in the same region as $\nu_s(N_3)$ of the reactant azide (1320 cm^{-1}); however, it is distinguishable by its considerably higher intensity. A low intensity mode of the tetrazole skeleton is found between 1100 and 1000 cm^{-1}. These interpretations were given without detailed frequency citations for the series of homologous compounds of Ga, In, and Tl [6].

References:

[1] Garnovskii, A. D.; Okhlobystin, O. Yu.; Osipov, O. A.; Yunusov, K. M.; Kolodyazhnyi, Yu, V.; Golubinskaya, L. M.; Svergun, V. I. (Zh. Obshch. Khim. **42** [1972] 920/5; J. Gen. Chem. [USSR] **42** [1972] 910/4).

[2] Breakell, K. R.; Rendle, D. F.; Storr, A.; Trotter, J. (J. Chem. Soc. Dalton Trans. **1975** 1584/9).

[3] Chung, H. L.; Tuck, D. G. (Can. J. Chem. **53** [1975] 3492/7).

[4] Rendle, D. F.; Storr, A.; Trotter, J. (Can. J. Chem. **53** [1975] 2930/43).

[5] Peterson, L. K.; Thé, K. I. (Can. J. Chem. **57** [1979] 2520/2).

[6] Srivastava, T. N.; Srivastava, R. C.; Singhal, K. (Indian J. Chem. A **19** [1980] 480/2).

[7] Byers, P. K.; Canty, A. J.; Mills, K.; Titcombe, L. (J. Organometal. Chem. **295** [1985] 401/5).

[8] Reger, D. L.; Knox, S. J.; Rheingold, A. L.; Haggerty, B. S. (Organometallics **9** [1990] 2581/7).

[9] Locke, K.; Weidlein, J.; Scholz, F.; Bouanah, N.; Brianese, N.; Zanella, P.; Gao, Y. (J. Organometal. Chem. to be published).

[10] Hausen, H.-D.; Locke, K.; Weidlein, J. (J. Organometal. Chem. to be published).

7.1.6 $R_2InNE(X,R)_n$ Compounds with E = S, P, Si

Compounds of this type are listed in Table 57. The preparative methods for each compound will be found under further information.

Table 57
$R_2InNE(X,R)_n$ Compounds with E = S, P, and Si.
Further information on numbers preceded by an asterisk is given at the end of the table.
Explanations, abbreviations, and units on p. X.

No.	compound	properties and remarks
*1	$(CH_3)_2InNS(CH_3)_2O$	white solid, m.p. 219 °C (dec.) [3] ^1H NMR (CH_2Cl_2): −0.25 (CH_3In), 3.02 (CH_3S) [3]
*2	$(CH_3)_2InN(Si(CH_3)_3)S(CH_3){=}NSi(CH_3)_3$	light yellow liquid, m.p. < −25 °C, b.p. 39 °C/ca. 0.2 Torr, monomeric in benzene [10] ^1H NMR: −0.03 (CH_3In), 0.16 (CH_3Si), 2.42 (CH_3S) in C_6D_{12}-c; −0.07, 0.01, 2.46 in $CDCl_3$ and 0.07, 0.03, 2.22 in C_6D_6 [10] ^{13}C NMR: 1.52 (CH_3Si), 53.94 (CH_3S) in C_6D_{12}-c; −0.5 (br, CH_3In), 1.76 (CH_3Si), 54.03 (CH_3S) in C_6D_6 [10] IR and Raman on p. 285
*3	$(CH_3)_2InN{=}P(CH_3)_3$	white solid, m.p. 110 to 112 °C, b.p. 118 to 120 °C/0.2 Torr [5] ^1H NMR: −0.41 (CH_3In), 1.39 ($X_9AA'X'_9$-m, CH_3P, $^2J(H,P) = 12.55$) in CH_2Cl_2; −0.56, 1.33 in CCl_4 [5] IR in Table 58 on p. 286
*4	$(CH_3)_2InN{=}P(C_2H_5)_3$	m.p. 83 to 85 °C, b.p. 192 °C/0.2 Torr [2] ^1H NMR (CCl_4): −0.55 (CH_3In), 1.08 (CH_3 of C_2H_5), 1.48 (A_3B_2X-m, CH_2P, $^2J(H,P) = 10.5$, $^3J(H,P) = 16.0$, $^3J(H,H) = 7.5$) [5] IR in Table 58 on p. 286
*5	$(CH_3)_2InN{=}P(C_6H_5)_3$	m.p. 230 to 232 °C [1, 2] ^1H NMR (CCl_4, ca. 3%): −0.83 (CH_3In), 7.41 (m, C_6H_5) [1, 2]
*6	$(CH_3)_2InN(Si(CH_3)_3)P(C_6H_5)_2{=}NSi(CH_3)_3$	m.p. 62 to 65 °C [4] ^1H NMR (CH_2Cl_2): −0.22 (CH_3Si, $^2J(H,Si) = 6.6$, $^4J(H,P) = 0.5$), 0.0 (CH_3In), 7.53 (m, C_6H_5) [4]
*7	$(CH_3)_2InN(C_4H_9\text{-t})Si(CH_3)_2OC_4H_9\text{-t}$	m.p. 5 °C, b.p. 68 to 70 °C/10^{-3} Torr [9] ^1H NMR (C_6D_6): 0.14 (CH_3In), 0.29 (CH_3Si), 1.03, 1.25 (CH_3C) [9]

Table 57 (continued)

No.	compound	properties and remarks
*8	$(CH_3)_2InN(C_4H_9\text{-}t)Si(CH_3)_2NHC_4H_9\text{-}t$	colorless liquid, m.p. 20 °C, b.p. 68 °C/10^{-3} Torr [6] ^1H NMR (C_6D_6): 0.05 (CH_3In), 0.25 (CH_3Si), 0.88, 1.12 (CH_3C) [6] mass spectrum (70 eV): $[M-CH_3]^+$ highest m/e observed [6]
*9	$(CH_3)_2InN(C_4H_9\text{-}t)Si(CH_3)_2NDC_4H_9\text{-}t$	colorless liquid, b.p. 65 °C/10^{-3} Torr [7] ^1H NMR (C_6D_6, 37 °C): 0.06 (CH_3In), 0.26 (CH_3Si), 0.91, 1.25 (CH_3C) [7] ^{13}C NMR (C_7D_8, 37 °C): −2.82 (CH_3In), 7.03 (CH_3Si), 32.13, 36.75 (CH_3 of C_4H_9), 50.82, 51.60 (C of C_4H_9) [7]
*10	$(CH_3)_2In\{(NC_4H_9\text{-}t)_2SiCH_3\}_2Li$	colorless crystals, m.p. 170 °C [8] ^1H NMR (C_6D_6): −0.35 (CH_3In), −0.24 ($Li\cdots CH_3In$), 0.63 (CH_3Si), 0.99, 1.33, 1.39 (CH_3C, ratio 1:2:1) [8]
*11	$(CH_3)_2InN(C_4H_9\text{-}t)Si(CH_3)_2N(C_4H_9\text{-}t)In(CH_3)_2$	white solid, m.p. 130 °C (dec.), subl. 95 °C/10^{-3} Torr [6] ^1H NMR (C_6D_6): 0.17 (CH_3In), 0.43 (CH_3Si), 1.00 (CH_3C) [6] mass spectrum (70 eV): $[M-CH_3]^+$ highest m/e observed [6]

* Further information:

$(CH_3)_2InNS(CH_3)_2O$ (Table **57**, No. **1**) was prepared from $(CH_3)_3In \cdot n\,O(C_2H_5)_2$ and $HNS(CH_3)_2O$ in anhydrous benzene with evolution of CH_4. The compound was produced in high yields in the form of colorless crystals that are only slightly soluble in benzene and other solvents. Because of decomposition, vacuum sublimation is not possible. Contrary to the dimeric $\{R_2MNS(CH_3)_2O\text{-}\}_2$ of the analogous Al and Ga compounds with their eight-membered rings, the In derivative was assigned a polymeric coordination structure (possibly comparable to the acetate, $(CH_3)_2InOOCCH_3$, or to the thioacetate, $(C_2H_5)_2InOSCCH_3$) [3].

$(CH_3)_2InN(Si(CH_3)_3)S(CH_3){=}NSi(CH_3)_3$ (Table **57**, No. **2**) was synthesized by the addition of $In(CH_3)_3$ to $(CH_3)_3SiN{=}S{=}NSi(CH_3)_3$ (1:1.5 mole ratio) in ether. The reactants were condensed one after the other into ether at −196 °C, slowly warmed, and stirred for 24 h at room temperature. After drawing off the solvent and unreacted diimide, the residue was purified by vacuum distillation (87% yield). It also formed in about 70% yield on reaction of solid $(CH_3)_2InCl$ with $Li[NSi(CH_3)_3S(CH_3){=}NSi(CH_3)_3]$ in ether at −78 °C. The solution was stirred for 1 h, warmed slowly to room temperature, and stirred another 24 h. After filtering off the LiCl, it was worked up as above [10].

The compound is monomeric in benzene and possesses, according to the NMR and vibrational spectral evidence, a four-ring structure (Formula I) with C_s symmetry.

$Si(CH_3)_3$ / CH_3, CH_3 / In, N, S, N / CH_3 / $Si(CH_3)_3$

I

$P(CH_3)_3$ / N / $(CH_3)_2In$, $In(CH_3)_2$ / N / $P(CH_3)_3$

II

In contrast to the analogous Al and Ga compounds, the $InCH_3$ resonances in the room temperature ^1H and ^{13}C NMR spectra are not split. An excerpt of the IR and Raman spectra of the liquid below 1500 cm^{-1} is given below [10]:

IR	Raman	assignment
1480 w	—	
1435 vw	—	$\delta_{as}(CH_3)$
1419 mw	1412 mw	(In-, Si-, SCH_3)
1406 vw	—	
1292 w	1290 vw	$\delta_s(CH_3S)$
1260 sh	1262 w	$\delta_s(CH_3Si)$
1252 s	1250 vw, sh	
1182 vw	—	?
1160 vw	1159 s	$\delta_s(CH_3In)$
1036 m	1043 vw	$\varrho(CH_3S)$
981 vs, br	—	$\nu_{as}(N \cdots S \cdots N)$
950 sh	951 mw	$\nu_s(N \cdots S \cdots N)$
932 m	—	$\varrho(CH_3S)$
—	856 vw	
838 vs	841 w	$\varrho(CH_3Si)$
781 m	783 vw	
750 m	750 vw	$\varrho(CH_3Si)$
710 mw	736 vw	$\varrho(CH_3In)$
686 m	683 mw	$\nu_{as}(SiC_3)$
667 mw	665 m	$\nu(S\text{-}CH_3)$
620 w	625 s	$\nu_s(SiC_3)$
511 ms	513 m	$\nu_{as}(InC_2)$
479 mw	482 vs	$\nu_s(InC_2)$
458 w	—	$\nu_{as}(InN_2)$
433 w	440 vw	$\nu(Si\text{-}N)$
425 vw	—	$\delta(SN_2)$
355 vw	361 ms	$\delta_s(SiC_3)+$ $\nu(Si\text{-}N)$
320 m, br	322 m	$\nu_s(InN_2)$
—	255 vvw	$\delta(SiC_3)$ and
—	236 w	$\delta(SiNS)$
—	208 w	
—	186 sh	$\delta(NSC)$
—	178 m	$\delta(CSiN)$
—	122 ms	$\delta(InC_2)$

$(CH_3)_2InN{=}P(CH_3)_3$ (Table **57**, No. **3**) was made by dropwise addition of ca. 10% excess $(CH_3)_3In \cdot 0.7\ O(C_2H_5)_2$ to $(CH_3)_3SnN{=}P(CH_3)_3$. After removing the ether and the $Sn(CH_3)_4$ formed, the compound was purified by distillation (75% yield). The initial adduct, $(CH_3)_3In \cdot N(Sn(CH_3)_3){=}P(CH_3)_3$ (Table 6, p. 38, No. 30), was detected below $-25\,^{\circ}C$ by NMR, but was not isolated. This adduct decomposes at room temperature by eliminating $Sn(CH_3)_4$, forming the dimeric (cryoscopy in benzene) title compound with suggested structure II.

The IR spectrum (together with that of No. 4) is given in Table 58 and is consistent with this structure [5].

Table 58
IR Frequencies for Solid $(CH_3)_2InN{=}P(CH_3)_3$ [5] and $(CH_3)_2InN{=}P(C_2H_5)_3$ [2].
Wavenumbers in cm^{-1}.

$(CH_3)_2InN{=}P(CH_3)_3$	assignment	$(CH_3)_2InN{=}P(C_2H_5)_3$	assignment
1303 m		1280 vw	$\delta(CH_3, CH_2)$
1290 sh	$\delta(CH_3P)$	1260 vs	
1286 s		1145 s	$\delta(CH_3In)$
1146 m	$\delta(CH_3In)$	1110 sh	
1140 m		1120 vs	$\nu(P{=}NIn_2)$
1111 vs	$\nu(P{=}N)$	1045 w	
932 s	$\varrho_1(CH_3P)$	1020 m	
857 s	$\varrho_2(CH_3P)$	1005 sh	$\nu(C{-}CP)$
847 s		990 s	
733 m	$\nu_{as}(PC_3)$	980 sh	
722 m	$\nu_s(PC_3)$	780 vs	$\varrho(CH_3, CH_2P)$
672 s	$\varrho_1(CH_3In)$	760 vs	
642 m	$\varrho_2(CH_3In)$	735 m	$\nu(PC_3)$
536 s	$\nu(InN)_2$	720 m	
481 s	$\nu(InC_2)$	675 s	$\varrho(CH_3In)$
462 m		660 sh	
		625 w	$\nu(PC_3)$
		585 sh	$\nu(InC_2)$
		540 vs	
		475 s	
		465 m	
		430 m	

$(CH_3)_2InN{=}P(C_2H_5)_3$ (Table **57**, No. **4**) formed in 98% yield from $(CH_3)_3SnN{=}P(C_2H_5)_3$ and $(CH_3)_3In \cdot 0.7\ O(C_2H_5)_2$, as described for No. 3 [5]. The compound was obtained in 83% yield from equimolar $HN{=}P(C_2H_5)_3$ and $(CH_3)_3In \cdot n\ O(C_2H_5)_2$ in ether. The initially formed adduct (Table 6, p. 38, No. 31) decomposes in a short time at the boiling temperature of the solvent, eliminating CH_4. The reaction residue was purified by distillation. The IR frequencies are listed in Table 58 and the structure of the dimer (cryoscopy in benzene) corresponds to Formula II of the methyl homolog [2].

$(CH_3)_2InN{=}P(C_6H_5)_3$ (Table **57**, No. **5**) was obtained in 52% yield from $(CH_3)_3In \cdot n\ O(C_2H_5)_2$ and $HN{=}P(C_6H_5)_3$ under the same conditions as its ethyl homolog (No. 4). Purification was accomplished by repeated crystallization from benzene. The degree of association in this solvent is about 1.5 (by cryoscopy; see Formula II) [1, 2].

$(CH_3)_2InN(Si(CH_3)_3)P(C_6H_5)_2{=}NSi(CH_3)_3$ (Table **57**, No. **6**) formed in 60.5% yield on reaction of the proton bearing part of $HN(Si(CH_3)_3)P(C_6H_5)_2{=}NSi(CH_3)_3$ with an equimolar amount of $(CH_3)_3In$ etherate and elimination of CH_4. The compound is monomeric in benzene, and has the same type of four-ring chelate structure as No. 2 (Formula I) [4].

The $\nu(N{-}H)$ IR absorption at 3370 cm^{-1} is missing, and the $\nu(P{=}N)$ vibration undergoes a strong shift to lower wavenumbers from 1305 to 1125 cm^{-1}, from which coordination of the second N atom to the In atom was deduced [4].

$(CH_3)_2InN(C_4H_9\text{-}t)Si(CH_3)_2OC_4H_9\text{-}t$ (Table **57**, No. **7**) was obtained within 1 h by adding a toluene solution of $Li[N(C_4H_9\text{-}t)Si(CH_3)_2OC_4H_9\text{-}t]$ to an equivalent of $(CH_3)_2InCl$ suspended in ether with stirring at room temperature. After filtering off the LiCl and evaporating the solvent, the oily residue was vacuum distilled (80% yield). A second procedure started with $(CH_3)_3SnN(C_4H_9\text{-t})Si(CH_3)_2OC_4H_9\text{-t}$ (Formula III, $E = OC_4H_9\text{-t}$); a benzene solution was mixed with an equimolar amount of $In(CH_3)_3$ in benzene and refluxed for 1 h. $Sn(CH_3)_4$ and solvent were removed and the residue worked up as above (90% yield). It is assumed to have an intramolecular coordination as depicted in Formula IV ($E = OC_4H_9\text{-t}$) [9].

C_4H_9−t
N
$(CH_3)_3Sn$ $Si(CH_3)_2$
E

III

C_4H_9−t
N
$(CH_3)_2In$ $Si(CH_3)_2$
E

IV

$(CH_3)_2InN(C_4H_9\text{-t})Si(CH_3)_2NHC_4H_9\text{-t}$ and **$(CH_3)_2InN(C_4H_9\text{-t})Si(CH_3)_2NDC_4H_9\text{-t}$** (Table **57**, Nos. **8** and **9**). Both compounds were made by reaction of benzene solutions of $(CH_3)_3$-$SnN(C_4H_9\text{-t})Si(CH_3)_2N(H,D)C_4H_9\text{-t}$ (Formula III, $E = N(H,D)C_4H_9\text{-t}$) and $In(CH_3)_3$. An equimolar amount of trimethylindium was used for 10 min at 40 °C for No. **8** and a large excess was used for 2 h at about 80 °C for No. **9**. Solvent, $Sn(CH_3)_4$, and any excess $In(CH_3)_3$ were allowed to evaporate, and the residue distilled under vacuum; yields were 95% [6] and 92% [7], respectively. Both products dissolve as monomers in benzene [6, 7].

The ^1H NMR spectra have been measured from −30 to +35 °C, and the spectrum of No. 8 is illustrated in [7]. Splitting of the CH_3In and CH_3Si signals at low temperatures points to an equilibrating structural rearrangement in which an open form (breaking of the In–N bond) allows interconversion between the two entiomers V and VI:

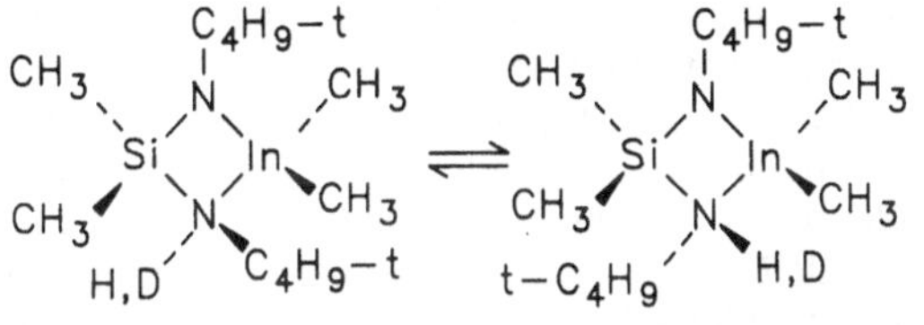

C_4H_9−t
CH_3 N CH_3
Si In
CH_3 N CH_3
H,D C_4H_9−t

V

⇌

C_4H_9−t
CH_3 N CH_3
Si In
CH_3 N CH_3
t−C_4H_9 H,D

VI

The concentration-independent coalescence temperatures (T_c) of the $In(CH_3)_2$ and $Si(CH_3)_2$ groups are somewhat different. The T_c temperatures (in K) as well as the activation energy E_a (in kJ/mol), the activation enthalpy ΔH_a (in kJ/mol), the activation entropy ΔS_a (in $J \cdot K^{-1} \cdot mol^{-1}$), and the free energy of activation ΔG_{298} (in kJ/mol), obtained from the analysis of the temperature-dependent line-shape changes of the signals, are given below. The lifetime of each structural arrangement is 0.01 (No. 8) and 0.02 (No. 9) seconds [7].

No.	T_c $(In(CH_3)_2)$	T_c $(Si(CH_3)_2)$	E_a	ΔH_a	ΔS_a	ΔG_{298}
8	255	267	32 ± 5	29 ± 3	−107 ± 10	61 ± 3
9	275	285	52 ± 7	49 ± 1	−46 ± 4	63 ± 1

The IR spectra were evaluated only in the region of the N–H and N–D stretching vibrations: ν(N–H) at 3255 and ν(N–D) at 2415 cm^{-1}. The drastic frequency lowering (compared to ν(N-H,D) of the free ligands) was explained by a cyclic structure produced by intramolecular Lewis acid-base interaction. N-H stretching frequencies for a series of homologous NH compounds of Al, Ga, and In have been correlated and graphed versus the free energy of activation ΔG_{298} [7].

$(CH_3)_2In\{(NC_4H_9\text{-}t)_2SiCH_3\}_2Li$ (Table **57**, No. **10**) was obtained by the reaction of equimolar amounts of a toluene suspension of $(CH_3)_2InCl$ at −78 °C with a toluene solution of the dilithium compound VII. Warming the mixture to room temperature, stirring for 12 h, filtering off the LiCl formed, and recrystallizing from toluene formed the product in 70% yield. The compound dissolves in $O(C_2H_5)_2$, $C_6H_5CH_3$, and C_6H_6 and is monomeric in benzene [8].

The ^1H NMR spectrum was interpreted by a cage structure (Formula VIII) in which the large line width (124 Hz) of the signals at −0.24 ppm indicates a $Li\cdots CH_3\cdots InCH_3$ two-electron three-center bridging bond [8].

VII

VIII

$(CH_3)_2InN(C_4H_9\text{-}t)Si(CH_3)_2N(C_4H_9\text{-}t)In(CH_3)_2$ (Table **57**, No. **11**) was prepared from benzene solutions of $Sn(NC_4H_9\text{-}t)_2Si(CH_3)_2$ added dropwise to a sixfold excess of $In(CH_3)_3$. The orange solution was stirred for 30 min at room temperature, the solvent and unconverted trimethylindium removed in vacuum, and the product sublimed out of the residue of polymeric $(Sn(CH_3)_2)_n$. Repeated sublimation suffices to purify the product in 94% yield. The compound was made in the same yield from No. 8 and one equivalent of $In(CH_3)_3$ in benzene (30 min reflux). When the diamine $(HNC_4H_9\text{-}t)_2Si(CH_3)_2$ reacted with 2 equivalents of $In(CH_3)_3$ in boiling toluene, the compound was formed in 72% yield with elimination of CH_4. It was also obtained in 69% yield from the dilithiated diamine, $(LiNC_4H_9\text{-}t)_2Si(CH_3)_2$, and $(CH_3)_2InCl$ in ether (ca. 2.5:1 mole ratio?); after addition, the mixture was refluxed briefly. The solid is easily deformed and is a "plastic" phase, which was characterized by the analogous Ga compound by X-ray diffraction [6].

References:

[1] Schmidbaur, H.; Kuhr, G.; Krüger, W. (Angew. Chem. **77** [1965] 866; Angew. Chem. Intern. Ed. Engl. **4** [1965] 877).
[2] Schmidbaur, H.; Jonas, G. (Chem. Ber. **101** [1968] 1271/85).
[3] Schmidbaur, H.; Kammel, G. (J. Organometal. Chem. **14** [1968] P28/P29).
[4] Schmidbaur, H.; Schwirten, K.; Pickel, H.-H. (Chem. Ber. **102** [1969] 564/7).
[5] Wolfsberger, W.; Schmidbaur, H. (J. Organometal. Chem. **17** [1969] 41/51).
[6] Veith, M.; Lange, H.; Belo, A.; Recktenwald, O. (Chem. Ber. **118** [1985] 1600/15).
[7] Veith, M.; Belo, A. (Z. Naturforsch. **42b** [1987] 525/35).
[8] Veith, M.; Goffing, F.; Huch, V. (Z. Naturforsch. **43b** [1988] 846/56).
[9] Veith, M.; Pöhlmann, J. (Z. Naturforsch. **43b** [1988] 505/12).
[10] Kottmair-Maieron, D.; Lechler, R.; Weidlein, J. (Z. Anorg. Allgem. Chem. **593** [1991] 111/23).

7.1.7 $(R_2In)_2C_2R_2O_2N_2$ and $(R_2In)_2C_2O_n(NR')_{4-n}$ Compounds with n = 2 and 4

The following compounds are derivatives of hydrazines, oxamides (n = 2), and oxamidines (n = 4). They were prepared by adding the solid NH acid (or a benzene slurry) to a benzene solution of trialkylindium (ca. 1:2 mole ratio) at 5 to 10 °C, and then warming the mixture to 30 to 60 °C; an alkane (RH) is eliminated. When gas evolution had ended, the slightly soluble product was filtered off and purified by sublimation [1, 2].

$\{(CH_3)_2In\}_2N_2(CCH_3)_2O_2$

The compound was obtained similarly from $In(CH_3)_3$ and N,N'-diacetylhydrazine in practically quantitative yield. The white microcrystalline solid decomposes above ca. 250 °C without melting and sublimes at 155 to 160 °C/10^{-4} Torr. Based on comparison with the IR and Raman spectra of the analogous Al and Ga compounds, a centrosymmetric, bicyclic structure (Formula I) was postulated [2].

I

Principal IR and Raman frequencies (in cm^{-1}) of the solid compound are presented below [2].

wavenumbers	assignment [IR and/or Raman]
1598 ms	$\nu_{as}(OCN)$ [1)] [Raman]
1548 vs	$\nu_{as}(OCN)$ [2)] [IR]
1400 s	$\nu_s(OCN)$ [1)] [Raman] + $\delta_{as}(CH_3C, In)$ [IR also]
1381 s 1330 s	$\nu_s(OCN)$ [2)] [IR] + $\delta_s(CH_3C)$ [Raman also] + combination
1160 mw	$\delta_s(CH_3In)$ [IR]
1158 s	$\delta_s(CH_3In)$ [Raman]
1075 ms	ν(N-N) [Raman]
985 mw	$\nu(C\text{-}CH_3)$ [1)] [Raman]
937 s	$\nu(C\text{-}CH_3)$ [2)] [IR]
711 vs, br	$\varrho(CH_3In)$ [IR, Raman only vw]
697 s, 601 mw	δ(OCN) [IR]
657, 644 mw	δ(OCN) [Raman]
548 ms	δ(OCN) [IR]
521 s	$\nu_{as}(InC_2)$ [IR; 529 m Raman]
486 mw	$\nu_s(InC_2)$ [IR; 492 vs Raman]
447 s	ν(OInN) [1)] [Raman]
408 m	δ(NNC) [Raman]
278 s	ν(OInN) [2)] [IR]

[1)] In-phase. – [2)] Out-of-phase.

$\{(CH_3)_2In\}_2(NCH_3)_2C_2O_2$

Following the procedure for the product above, this compound was formed in 65% yield from $In(CH_3)_3$ and N,N'-dimethyloxamide. It was purified by sublimation at 130 °C/10^{-4} Torr and melts from 160 to 163 °C [2].

The ^1H NMR spectrum (in ppm) in $CDCl_3$ at 30 °C shows a threefold splitting of the $InCH_3$ protons: −0.21, −0.10, and −0.02 (CH_3In), 2.90 (CH_3N). This splitting was explained by the existence of two different puckered (a type of chair form), five-membered bicyclic rings [2]. However, systematic spectroscopic studies of the analogous Ga compound and the X-ray analyses [3, 4] leave no doubt that this structure proposal is wrong and the In compound forms two isomers (Formulas II and III) in a 2:1 ratio. Thus, assignment of the $InCH_3$ resonances was made by analogy to the Ga compound as follows: −0.21 (CH_3In^b-cis), −0.10 (CH_3In-trans), −0.02 (CH_3In^a-cis) ppm [6].

II

III

The vibrational spectrum of the solid under 1700 cm^{-1} is given in Table 59. It was assumed in assigning bands that the fundamental structure is centrosymmetric, but puckered [2].

$\{(C_2H_5)_2In\}_2(NCH_3)_2C_2O_2$

The compound was obtained in 35% yield from $In(C_2H_5)_3$ and $(HNCH_3)_2C_2O_2$ by the same procedure as above. The white solid was purified by sublimation at 90 °C/10^{-4} Torr and melts at 133 to 136 °C. Solubility in the usual solvents was too poor for NMR examination; IR and Raman frequencies are summarized in Table 59 [2].

Table 59
IR and Raman Frequencies of Solid $\{(CH_3)_2In\}_2(NCH_3)_2C_2O_2$ and $\{(C_2H_5)_2In\}_2(NCH_3)_2C_2O_2$ (main values only) [1].
Wavenumbers in cm^{-1}.

$\{(CH_3)_2In\}_2(NCH_3)_2C_2O_2$ IR	Raman	$\{(C_2H_5)_2In\}_2(NCH_3)_2C_2O_2$ IR	Raman	assignment
–	1635 s	–	1630 s	$\nu_{as}(OCN)$ 1)
1610 vs, br	–	1596 vs	–	$\nu_{as}(OCN)$ 2)
1451 mw	1452 w	–	–	$\delta_{as}(CH_3N)$
1440 sh	–	–	–	$\delta_{as}(CH_3In)$
–	1414 s	–	1413 vs	$\nu_s(OCN)$ 1)
1402 ms	(1414)	–	–	$\delta_s(CH_3N)$ + overtone
1396 m	1398 mw	–	–	
1275 s, br	–	1280 vs	–	$\nu_s(OCN)$ 2)

Table 59 (continued)

$\{(CH_3)_2In\}_2(NCH_3)_2C_2O_2$ IR	Raman	$\{(C_2H_5)_2In\}_2(NCH_3)_2C_2O_2$ IR	Raman	assignment
1170 mw	1178 ms	–	–	$\delta_s(CH_3In)$
1156 mw	1158 ms	–	–	
1105 w	1110 vw	–	–	$\delta, \gamma(CH_3N, In(?))$
–	1085 vw	–	–	
997 m	–	994 m	–	$\nu(N\text{-}CH_3)^{2)}$
–	912 w	–	910 w	$\nu(N\text{-}CH_3)^{1)} + \nu(C\text{-}C)$
819 m	–	817 ms	–	$\delta(OCN)^{2)} + \varrho(CH_3N)$
–	810 w	–	808 w	$\delta(OCN)^{1)}$
720 ms, br	680 vw, br	–	–	$\varrho(CH_3In)$
589 m	–	582 m	–	$\delta(OCN)$
–	565 w	–	559 ms	
529 ms	527 m	510 ms	512 m	$\nu_{as}(InC_2)$
495 mw	496 vvs	465 sh	468 vs	$\nu_s(InC_2)$
490 mw	–	–	–	
461 ms	–	457 s	–	$\nu(OInN)$ and $\delta(CNC)$
421 sh	444 mw	430 sh	440 m	
–	392 vw, br	–	384 w	
275 s	280 vw	270 s	–	$\delta, \gamma(CNC)$
–	244 vw	–	260 ms	$\delta(InCC)$
–	200 w	–	134 m	$\delta(InC_2)$
–	117 ms	–	118 m	$\delta(ring)$
–	66 m	–	–	

[1] In-phase. – [2] Out-of-phase.

$\{(CH_3)_2In\}_2(NCH_3)_4C_2$

The compound formed in high yield by the previously outlined procedure; it was purified by sublimation at 60 °C/0.1 Torr. It is a white solid melting at 108 to 110 °C. Colorless crystals, suitable for X-ray analysis, were obtained by recrystallizing freshly sublimed material from toluene [5].

^1H NMR in C_6D_6 at ca. 30 °C: 0.08 (CH_3In), 2.95 (CH_3N) ppm [5].

Important IR and Raman frequencies between 1600 and 470 cm^{-1} and their assignments are given below for the solid and for the compound dissolved in CCl_4 (relative intensities in parentheses) [5].

IR solid	Raman solid	solution	assignment
–	1592 (4)	obscured	$\nu_{as}(CN_2)^{1)}$
1557 vvs, br	–		$\nu_{as}(CN_2)^{2)}$
1456 w	1474 (2)	(2, br)	$\delta_{as}(CH_3)$
–	1453 (2)	(1)	
1418 vw	1428 (3)	(4)	
1389 s	1389 sh	sh	$\delta_s(CH_3\text{-}N)$

(continued)

IR solid	Raman solid	solution	assignment
–	1383 (80)	(80, br)	$\nu_s(CN_2)$[1)]
1334 s	1340 (0)		$\nu_s(CN_2)$[2)]
1150 mw	1156 sh	(40)	$\delta_s(CH_3In)$ +
1138 mw	1148 (45)		$\varrho(CH_3N)$
–	1057 (5)	(5)	$\nu_s(N-CH_3)$[1)]
1044 s	–		$\nu_s(N-CH_3)$[2)]
921 s	–		$\nu(C-C)$
–	908 (4)	(2)	$\delta(CN_2)$[2)]
836 m	–		$\delta(CN_2)$[1)]
	794 (3)		
503 s	511 (40)	(18)	$\nu_{as}(InC_2)$
(473) br	487 (100)	(100)	$\nu_s(InC_2)$
473 s	–		$\nu(InN_2)$

[1)] In-phase. – [2)] Out-of-phase.

The compound crystallizes in the orthorhombic space group Pbca $-D_{2h}^{15}$ (No. 61) with the parameters a = 796.5(1), b = 1864.2(6), c = 2122.4(6) pm; Z = 8 and D_c = 1.81 g/cm³, while D_m = 1.75 g/cm³. The data were collected at –100 °C. Bond distances and angles are indicated in **Fig. 55**. Contrary to the analogous Al and Ga compounds, the double five-membered ring halves in this complex are twisted about 5°, so that the molecule is not centrosymmetric under the view of X-ray diffraction [5].

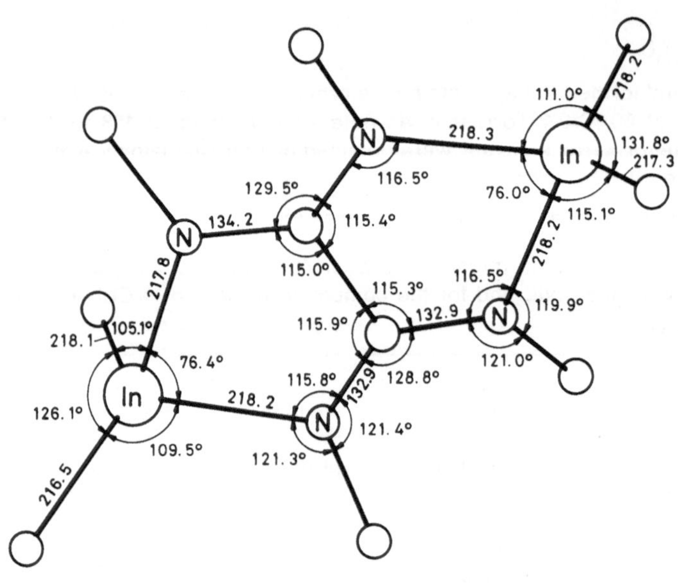

Fig. 55. Molecular structure of $\{(CH_3)_2In\}_2C_2(NCH_3)_4$ [5].

References:

[1] Schwering, H.-U.; Weidlein, J.; Fischer, P. (J. Organometal. Chem. **84** [1975] 17/37).
[2] Eberwein, B.; Lieb, W.; Weidlein, J. (Z. Naturforsch. **32b** [1977] 32/6).
[3] Fischer, P.; Gräf, R.; Stezowski, J. J.; Weidlein, J. (J. Am. Chem. Soc. **99** [1977] 6131/2).
[4] Fischer, P.; Gräf, R.; Weidlein, J. (J. Organometal. Chem. **144** [1978] 95/110).
[5] Gerstner, F.; Schwarz, W.; Hausen, H.-D.; Weidlein, J. (J. Organometal. Chem. **175** [1979] 33/47).
[6] Weidlein, J. (unpublished results).

7.1.8 $(R_2In)_2C_2S_2(NR')_2$ Compounds

The two compounds in this category are derivatives of dithiooxamides with $R=CH_3$ and $R'=H$ or CH_3. They were prepared in high yields by the reaction of $In(CH_3)_3$ with $C_2S_2(NH_2)_2$ or $C_2S_2(NHCH_3)_2$ (2:1 mole ratio) in toluene at room temperature, accompanied by the elimination of two moles of CH_4. The very moisture-sensitive products were shown to be monomeric in benzene.

As with the homologous N,N'-dimethyloxamide (p. 290), the structural isomers I and II can be formed.

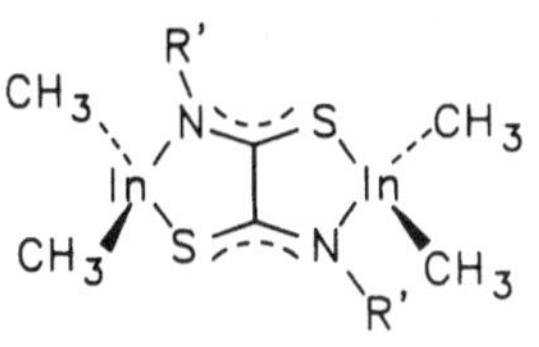

I

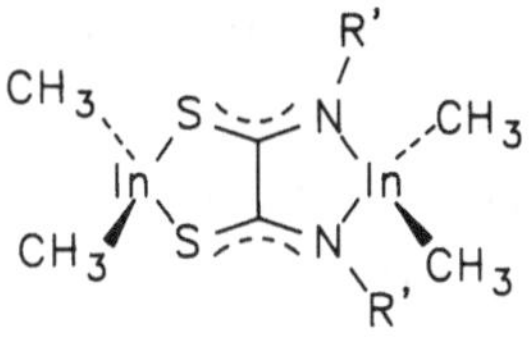

II

$\{(CH_3)_2In\}_2C_2S_2(NH)_2$

The compound was formed in 90% yield as a yellowish solid and recrystallized from $C_6H_5CH_3$, C_6H_6, or C_6H_{12}-c; m.p. ca. 70 °C (dec.) [1].

^1H NMR, ca. 0.2 m in C_6D_6 at 30 °C (δ in ppm): 0.00 (CH_3In), 8.22 (br, HN) for isomer I; −0.95, 0.95 (CH_3In), 8.22 (br, HN) ppm for isomer II. Evaluation of the line intensities suggested an estimated 5% isomer II to be in the mixture [1].

$\{(CH_3)_2In\}_2C_2S_2(NCH_3)_2$

The white solid (m.p. ca. 150 °C (dec.)) formed in 80% yield. While it can be sublimed at 90 °C/ca. 10^{-6} Torr, purification by recrystallization from one of the above solvents is preferred because of high losses by decomposition during sublimation. A second preparative possibility, reaction of a suspension of $Li_2C_2S_2(NCH_3)_2$ with two moles of $(CH_3)_2InCl$ in toluene, was mentioned but not used [1].

^1H NMR, ca. 0.2 m in C_6D_6 at 30 °C (δ in ppm): 0.04 (CH_3In), 3.18 (CH_3N) for isomer I, only [1].

The IR and Raman spectra of the solid, shown below (wavenumbers in cm^{-1}), were compared to the spectra of the analogous Al and Ga compounds, and a centrosymmetric bicyclic structure (Formula I) was assigned [1].

IR	Raman	assignment	IR	Raman	assignment
1548 vs	1544 s	$\nu(CN)$	–	684 m, br	$\delta_s(SCN)$ [1)]
1450 m	1447 mw	$\delta_{as}(CH_3N)$	725 ms	–	$\varrho(CH_3In)$
1400 m	1402 m	$\delta_s(CH_3N)$	648 w	–	$\delta_s(SCN)$ [2)]
–	1199 m }	$\nu(CS)$	522 s	528 mw	$\nu_{as}(InC_2)$
1158 mw	1163 m }	$\nu(C\text{-}C)$	497 m	–	$\nu(InN)$?
1148 sh	1154 m	$\delta_s(CH_3In)$	489 w	488 vs	$\nu_s(InC_2)$
–	1113 mw	$\nu(CS)+\nu(C\text{-}C)$	–	435 mw	$\delta(CNCH_3)$ [1)]
–	1060 mw	$\nu(N\text{-}CH_3)$ [1)]	368 w	379 m }	$\delta(NCS)$
1034 s	–	$\nu(N\text{-}CH_3)$ [2)]	–	330 w }	$\gamma(NCS)$
847 s	–	$\nu(CS)$ [2)]	–	203 mw	$\nu(InS)$

1) In-phase. – 2) Out-of-phase.

The mass spectrum (70 eV, 415 K source temperature, 305 K direct inlet) displayed the following fragments (relative intensities in %): $[M-CH_3]^+$ (100), $[M-2\,CH_3]^{++}$ (17.3; such doubly-charged fragments are typical for bicyclic chelates), $[M-3\,CH_3]^+$ (8.4), $[In_2SCH_3]^+$ (39.1), $[InCH_3]^+$ (4.0), $^{115}In^+$ (71) [1].

Reference:

[1] Halder, T.; Schwarz. W.; Weidlein, J.; Fischer, P. (J. Organometal. Chem. **246** [1983] 29/48).

7.1.9 Radicals with $R_2In\text{-}N$ Bonds

$[(CH_3)_2InN(CH)_4NIn(CH_3)_2]^{\cdot+}$

The radical cation was obtained in a sealed evacuated system by addition of freshly distilled sodium to a THF solution of $(CH_3)_3In \cdot O(C_2H_5)_2$ and pyrazine (2:1 mole ratio). The $[In(CH_3)_4]^-$ anion served as counterion for I. A gray-black precipitate of metallic indium formed rapidly. The supernatant solution gave an intense ESR signal (spectrum illustrated) that was only poorly resolved, even at high dilution [1]. Only an approximate determination of the ESR coupling constants was possible. The following values (in mT) were determined: a=0.28 (H), 0.86 (N), 0.39 (In) at 300 K; g=2.0031. The data were compared to those of the analogous Al and Ga radicals [1].

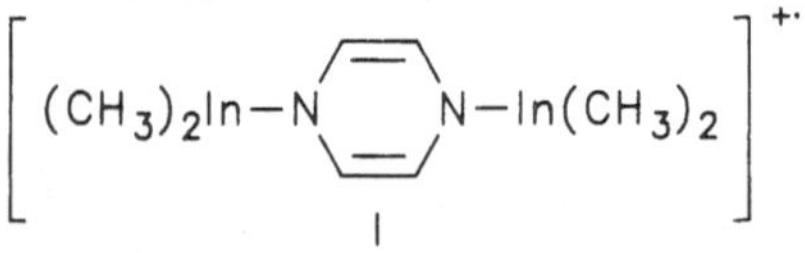

$[(CH_3)_2In-N\langle C_4H_4\rangle N-In(CH_3)_2]^{+\cdot}$ I

$[(CH_3)_2In-N\langle C_5H_4=C_5H_4\rangle N-In(CH_3)_2]^{+\cdot}$ II

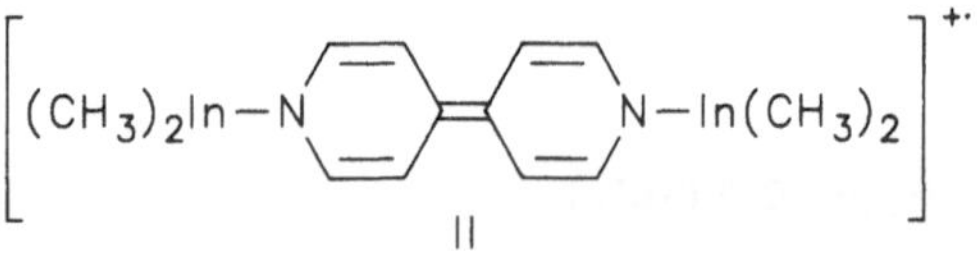

$[(CH_3)_2In(C_{10}H_8N_2\text{-}4,4')In(CH_3)_2]^{\cdot+}$

The reaction of a THF solution of $In(CH_3)_3$ and $C_{10}H_8N_2$-4,4′ (2:1 mole ratio) with metallic K in an Ar atmosphere or a sealed evacuated system succeeded in forming the radical cation II. The ESR signal, however, did not show hyperfine splitting, so that no evaluation was possible [2].

References:

[1] Kaim, W. (Z. Naturforsch. **36b** [1981] 677/82).
[2] Kaim, W. (J. Organometal. Chem. **241** [1983] 157/69).

7.2 Mixed R_2In-N/RIn-N Compounds

$\{(CH_3)_2InN(CH_3)(CH_2)_2NCH_3\}_2InCH_3$

The compound was formed by reaction of a $(CH_3)_2InCl$ suspension in ether with an ether solution of monolithiated N,N′-dimethylethylenediamine in 1:1 mole ratio. The mixture was refluxed for 1 h (formation of CH_4, LiCl, and free diamine), and after cooling to room temperature, filtered to remove LiCl. The filtrate was concentrated, the product crystallized out at −25 °C, and was sublimed at 90 °C/10^{-2} Torr; m.p. 195 to 200 °C, 63% yield [1, 2].

Data from vapor pressure measurements from 61.5 to 101.2 °C by the Knudsen effusion method are summarized by the equation $\log p\ (Torr) = 13.15 - 5690/T\ (K)$ [3].

The NMR spectra (in ppm) show multiple splitting of the CH_3In signals in C_6D_6. ^1H NMR: −0.10 (apical CH_3In), −0.03, −0.01 (s's, CH_3In proximal and distal to apical $InCH_3$), 2.50 (s, CH_3N), 2.75 (m, AA′BB′, CH_2). ^{13}C NMR: −10.7, −10.5, −10.3 (CH_3In), 43.0 (CH_3N), 54.9 (CH_2) [2].

The following bands (in cm^{-1}) are found in the IR spectrum of the solid (Nujol mull): 2790 s, 1412 w, 1358 m, 1339 m, 1299 w, br, 1276 m, 1237 m, 1189 w, 1159 m, sh, 1154 s, 1130 w, 1121 m, 1110 m, 1084 m, 1032 m, 997 s, 965 s, 851 vs, 696 vs, vbr $\varrho(CH_3In)$, 611 m, 604 m, 502 vs $\nu_{as}(InC_2)$, 481 s $\nu_s(InC_2)$, 444 s, 420 m, br, 358 s, br, 314 m, br [2].

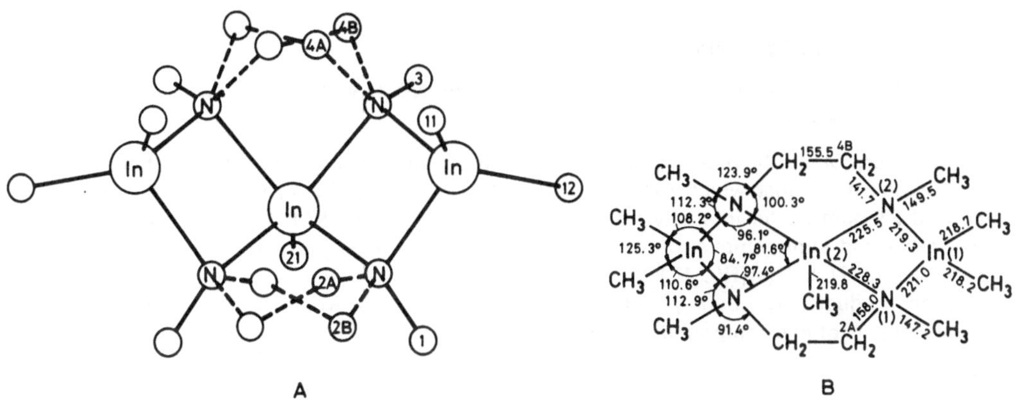

Fig. 56. Molecular structure of $\{(CH_3)_2InN(CH_3)(CH_2)_2NCH_3\}_2InCH_3$ (A); simpilfied formula (B) [1, 2].

Other bond angles (°)

C(12)-In(1)-N(2)	108.8(6)	C(3)-N(2)-In(2)	112.9(8)
C(21)-In(2)-N(2)	114.4(5)	C(4B)-N(2)-In(1)	125.2(17)
C(2A)-N(1)-In(1)	105.7(11)	N(2)-In(2)-N(2)	79.2(6)
C(2B)-N(1)-In(2)	113.3(13)	C(4A)-C(4B)-N(2)	87.2(30)
C(2B)-N(1)-C(2A)	31.8(11)	C(2A)-C(2B)-N(1)	82.5(31)
C(11)-In(1)-N(1)	111.8(6)	C(4A)-N(2)-C(3)	124.8(13)
C(21)-In(2)-N(1)	114.1(5)	C(4B)-N(2)-In(2)	117.9(14)
C(1)-N(1)-In(2)	114.8(8)	C(4B)-N(2)-C(4A)	33.5(11)
N(1)-In(2)-N(1)	78.6(5)	C(4B)-C(4A)-N(2)	59.3(28)
		C(2B)-C(2A)-N(1)	65.7(30)

The compound crystallizes in the orthorhombic space group, Pnam$-D_{2h}^{16}$ (No. 62) with the cell parameters a=1503.9(2), b=849.7(1), c=1700.6(2) pm; Z=4 and D_c=1.808 g/cm^3. The central In atom ($InCH_3$) of this cage structure is square-pyramidal coordinated and the In atom lies 93 pm above the N_4 plane. The two other In atoms ($In(CH_3)_2$) are coordinated in distorted tetrahedrons. The CH_2-CH_2 groups of the amine bridges on both sides of the cage are disordered, which results in large uncertainties in the locations of some of the atoms bound to them. **Fig. 56** (A) displays the molecular structure; in its simplified form (B) the most important parameters are designated [1, 2].

The mass spectrum shows the following fragments (relative intensities in %, L=$(CH_2NCH_3)_2$) [2]: M^+ (30), $[M-InCH_3-H]^+$ (11), $[In+L-H]^+$ (13), $[M-CH_3]^+$ (28), $[M-In(CH_3)_2-CH_4]^+$ (10), $[In+L-2\ H]^+$ (28), $[M-CH_2NCH_3]^+$ (54), $[M-2\ (InCH_3)-H]^+$ (43), $[InCH_2NCH_3]^+$ (23), $[M-CH_2N(CH_3)_2]^+$ (25), $[M-2\ In(CH_3)_2-H]^+$ (16), $[In(CH_3)_2]^+$ (26), $[M-L-CH_2]^+$ (13), $[InCH_3+LH]^+$ (11), $^{115}In^+$ (54), $[M-L-CH_3]^+$ (100), $[InCH_3+L]^+$ (75), $[LH]^+$ (25), $[M-LH_2-CH_3]^+$ (16), $[In+L]^+$ (15) [2].

References:

[1] Arif, A. M.; Bradley, D. C.; Frigo, D. M.; Hursthouse, M. B.; Hussain, B. (J. Chem. Soc. Chem. Commun. **1985** 783/4).

[2] Arif, A. M.; Bradley, D. C.; Dawes, H.; Frigo, D. M.; Hursthouse, M. B.; Hussain, B. (J. Chem. Soc. Dalton Trans. **1987** 2159/64).

[3] Bradley, D. C.; Faktor, M. M.; Frigo, D. M. (J. Cryst. Growth **89** [1988] 227/36).

7.3 RIn-Nitrogen Compounds

Simple organoindium compounds of the type $RIn(NR'_2)_2$ with R=R'=CH_3 or C_2H_5 have not been isolated pure or characterized.

However, CH_3InCl_2 reacts with two equivalents of $LiN(CH_3)_2$ in ether with formation of LiCl and a product which exhibits in the ^{1}H NMR spectrum (in C_6D_6/ether) signals at 0.0 and 2.9 ppm in the intensity ratio as expected for $\mathbf{CH_3In\{N(CH_3)_2\}_2}$. This compound is not stable and after removal of the solvent a mixture of $(CH_3)_2InN(CH_3)_2$ and $In\{N(CH_3)_2\}_3$ was present [2]. $\mathbf{C_2H_5In\{N(C_2H_5)_2\}_2}$ was formed as a polymer along with unspecified decomposition products, by heating $(C_2H_5)_2InN(C_2H_5)_2$ (Table 50, No. 9) at 150 °C for 50 h; no further details of this compound were given [1].

References:

[1] Shcherbakov, V. I.; Zhil'tsov, S. F.; Druzhkov, O. N. (Zh. Obshch. Khim. **40** [1970] 1542/5; J. Gen. Chem [USSR] **40** [1970] 1529/31).

[2] Tödtmann, J.; Weidlein, J. (unpublished results).

7.3.1 Compounds of the Type RIn(X)N(R'R'') and RIn(X)N(R'R'') · InR_2NHR'

$CH_3In(Cl)N_4C_6H_8B$

The compound was obtained by the reaction of CH_3InCl_2 (from $InCl_3$ and $Sn(CH_3)_4$) with $K[H_2B(pz)_2]$ (1:1 mole ratio; pz=pyrazolyl) in THF at room temperature (4 h). The dried mixture was extracted with benzene to give a colorless solid, m.p. 98 to 100 °C, yield 84% [3].

^{1}H NMR (in $CDCl_3$, δ in ppm): 0.59 (s, CH_3), 3.6 (br, H_2B), 6.31 (t, H-4, J=2.2), 7.73, 7.66 (d's, H-3,5, J=2.1, 2.0) [3].

The IR spectrum showed ν(BH) vibrations at 2474 and 2433 cm^{-1} and the mass spectrum had peaks for $[M-H]^+$ and $[M-Cl]^+$ ions [3].

The compound crystallizes in the triclinic space group $P\bar{1}-C_i^1$ (No. 2) with the cell constants a=760.01(11), b=861.33(14), c=1996.0(4) pm, α=79.086(15)°, β=85.966(16)°, γ=67.340(13)°; Z=4 gives D_c=1.751 g/cm^3. The dimer molecule is held together by a pair of asymmetrically bonded Cl atoms. The geometry about the In atom is a distorted trigonal bipyramid as depicted in **Fig. 57**. The unit cell contains two crystallographically independent molecules A and B which mainly differ in the In···Cl distances (320.3(1) and 306.6(1) pm); the In-C distances are 209.1(5) and 210.2(4) pm, respectively [3].

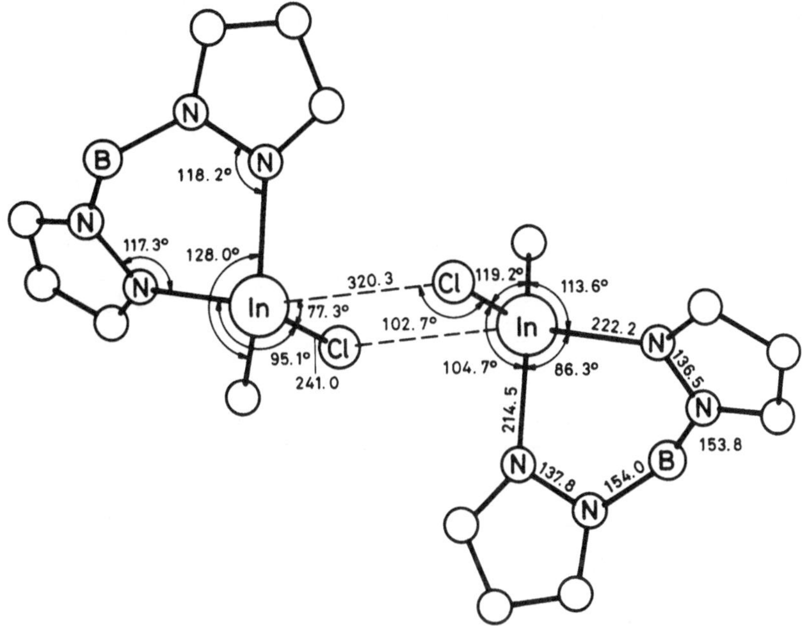

Fig. 57. Molecular structure of $CH_3In(Cl)C_6H_8N_4B$ (molecule A) [3].

$CH_3In(Cl)N(C_4H_9\text{-}t)Si(CH_3)_2OC_4H_9\text{-}t$

The compound was prepared from equimolar amounts of $(CH_3)_3SnN(C_4H_9\text{-}t)Si(CH_3)_2$-$OC_4H_9$-t and $(CH_3)_2InCl$ in benzene. After a 1 h reflux, the $Sn(CH_3)_4$ and solvent were removed in vacuum, and the product sublimed at 50 °C/10^{-3} Torr. The yield is almost quantitative [1]. The compound was obtained in 70% yield by the reaction of a toluene solution of $Li[N(C_4H_9\text{-}t)Si(CH_3)_2OC_4H_9\text{-}t]$ with a slurry of CH_3InCl_2 in ether (1:1 mole ratio). After 1 h stirring at room temperature, the LiCl was separated, the solution freed of solvent, and the residue sublimed at 80 °C/10^{-3} Torr. The product was contaminated with about 10% $Cl_2InN(C_4H_9\text{-}t)Si(CH_3)_2OC_4H_9$-t [1].

The pure compound is a monomer in the gas phase (mass spectroscopy); the ^{1}H NMR spectrum in C_6D_6 is in agreement with a chelate structure like Formula IV on p. 287: 0.30 (CH_3Si), 0.45 (CH_3In), 1.14, 1.35 (CH_3C-N,O) ppm; at 200 K there is no change in the signal sequence [1].

$(C_3H_7\text{-i})In(Cl)NHC_4H_9\text{-t}$

The title compound was prepared by slow addition of an ether suspension of $Li[NHC_4H_9\text{-t}]$ to an ether suspension of $(C_3H_7\text{-i})InCl_2$, both at 0 °C, and stirring the mixture for 12 h at room temperature. After removal of the solvent the residue was extracted with toluene and cooled to −5 °C; large colorless crystals, m.p. 110 °C, 55% yield. The degree of association in benzene (0.024 mol/L) was 1.82 [2].

The ^1H NMR spectrum in benzene shows two groups of signals which belong to two isomers of the dimeric molecule. From the possible 6 isomers either I and V, or III, IV, and VI have been proposed to be present; a decision from NMR spectroscopy was not possible. ^1H NMR (C_6D_6 in ppm), I/VI: 1.03/1.06 (C_4H_9), 1.30, 1.32/1.40 (CH_3C, diastereotopic/AB_6 system), 1.51 (HCIn), 1.85/2.04 (HN). ^{13}C NMR (C_6D_6, in ppm) I/VI: 22.47, 22.40/22.6 (CH_3 of C_3H_7-i), 25.0 (CHIn), 34.1/34.0 (CH_3 of C_4H_9), 52.7/53.0 (C of C_4H_9) [2].

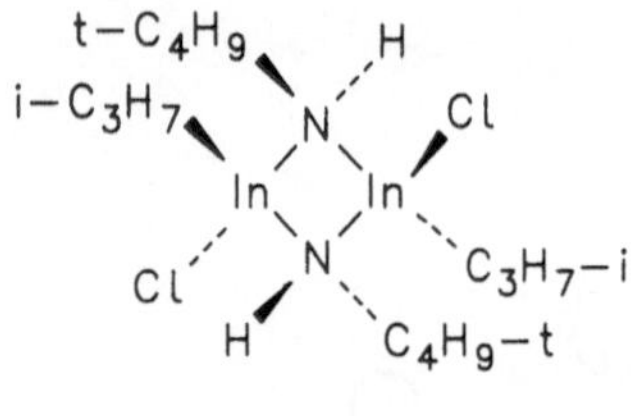

I 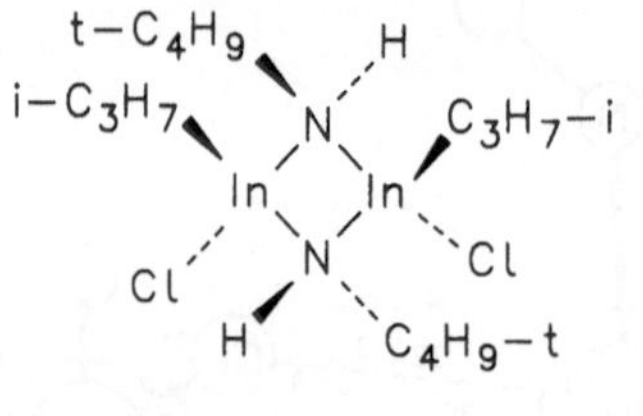 II 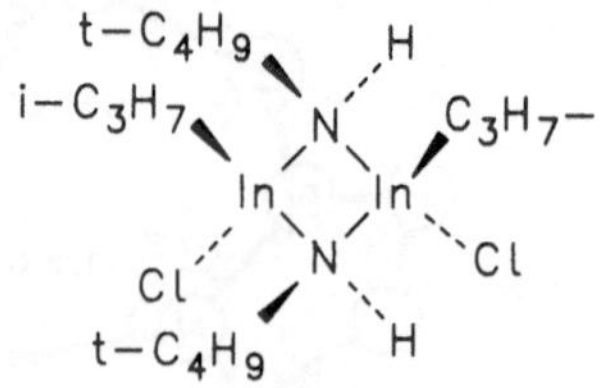 III

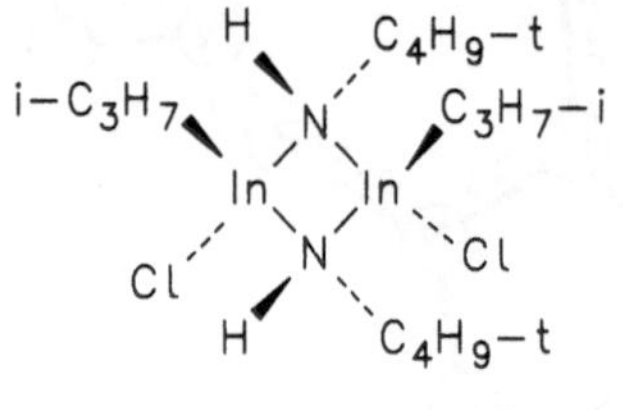

IV 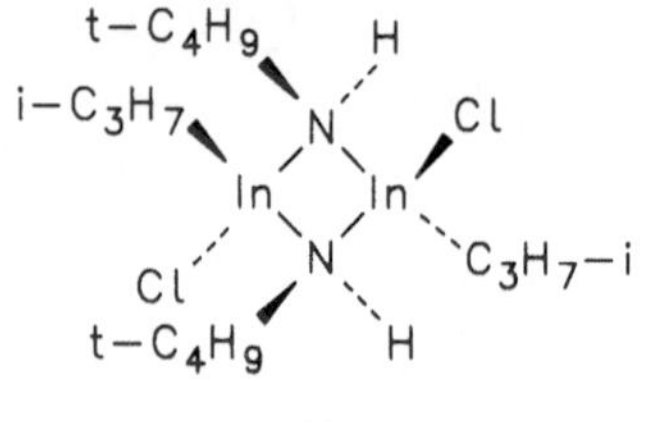 V

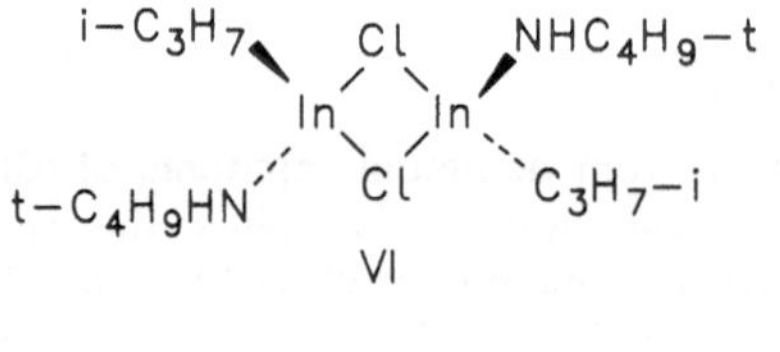

VI

Variable temperature NMR up to 75 °C in C_7D_8 showed only a temperature dependance of the chemical shift but no dependance of the isomer ratio (I,V/III,IV,VI = 1.5/1), thus excluding a monomer dimer equilibrium [2].

IR (in Nujol) and Raman (solid) vibrations have been assigned (t = terminal, b = bridge) as depicted below and interpreted in terms of the existence of different isomers although only the centrosymmetric isomer I was found by X-ray analysis [2].

IR	Raman	assignment	isomer
3195 s, br	3193	ν(NH)	I+VI
600 sh	597 vw	ν(InN-t)	VI
579 vs	577 vw	ν(InN-b)	I
550 s	550 vw	ν(InN-b)	I
503 m	495 s	ν(InC-t)	I
464 m	471 s	ν(InC-t)	VI
276 m	271 m	ν(InCl-t)	I
240 m	232 m	ν(InCl-b)	VI

The compound crystallizes in the monoclinic space group $C2/c-C^6_{2h}$ (No. 15) with the constants a = 1864.9(4), b = 1016.5(2), c = 1229.9(2) pm, β = 110.32(1)°; Z = 8 (4 dimers) and D_c = 1.613 g/cm³. The most important bond distances are shown in **Fig. 58** [2].

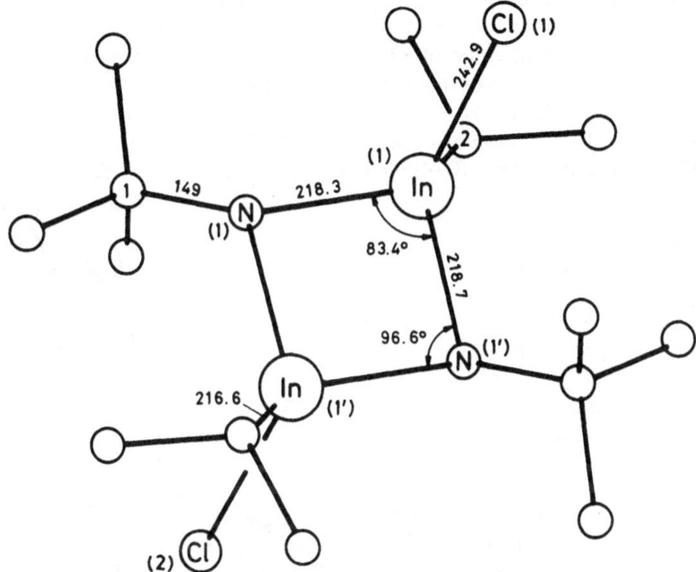

Fig. 58. Molecular structure of $\{(C_3H_7\text{-i})In(Cl)NHC_4H_9\text{-t}\}_2$ [2].

Bond angles (°)

Cl(1)-In(1)-N(1)	101.4(2)	Cl(1)-In(1)-N(1')	107.4(2)
Cl(1)-In(1)-C(2)	104.8(2)	In(1)-N(1)-C(1)	122.7(5)
N(1)-In(1)-C(2)	130.9(3)	C(1)-N(1)-In(1')	123.1(5)
C(2)-In(1)-N(1')	125.4(3)		

Mass spectrum (relative intensities in %): $[M_2-Cl-C_3H_7]^+$ (2.62), $[M_2-InNC_4H_{10}]^+$ (9.68), M^+ (9.95), $[M-NC_4H_{10}]^+$ (22.93), $[InC_3H_7]^+$ (12.05), $[InCl]^+$ (13.74), $^{115}In^+$ (84.1) [2].

$C_6H_5CH_2In(Cl)N\{Si(CH_3)_3\}_2$

This compound was prepared by addition of a solution of $Na[N\{Si(CH_3)_3\}_2]$ in ether at 0 °C to a solution of $C_6H_5CH_2InCl_2$ in the same solvent. After about 2 h the solvent was evaporated and the residue dissolved in toluene. Filtration and recrystallization from pentane gave colorless crystals in 60% yield; m.p. 72 to 75 °C. The compound is dimeric in benzene. A structure similar to VI was proposed [4].

NMR spectra were run in C_6D_6 (in ppm). ^1H NMR: 0.14 (CH_3Si), 2.50 (CH_2), 6.8 to 7.1 (C_6H_5). ^{13}C NMR: 5.4 (CH_3Si), 31.9 (CH_2), 124.9 (C-4), 127.9 (C-3,5), 129.1 (C-2,6), 139.7 (C-1) [4].

IR/Raman (in cm^{-1}): 629 mw, 618 m/633 w, br ν(InN), 445 m/439 m, br ν(InC), 244 mw/250 w, and 217 mw/212 w ν(InCl) [4].

$\{C_6H_2(CH_3)_3\text{-}2,4,6\}In(Cl)N\{Si(CH_3)_3\}_2$

This compound was prepared similarly from $mesInCl_2$ (mes = $C_6H_2(CH_3)_3$-2,4,6) and $Na[N\{Si(CH_3)_3\}_2]$ and recrystallized from pentane. Colorless solid, m.p. 105 to 107 °C, 60% yield. The degree of association in benzene (0.017 mol/L) was 1.36 [2].

The monomer:dimer (Formula I or VI) ^1H NMR signal intensity ratio was found to be 14.7:1. At 60 °C the monomer increases by a factor of 2 and the halve-width of the CH_3-2 proton decreases to a normal value of 3 Hz. The broadening at room temperature probably results from a hindered rotation of these groups. ^1H NMR (C_6D_6 at 20 °C) for the monomer/dimer: 0.25/0.20 (CH_3Si), 1.98/2.04 (CH_3-4), 2.38/2.42 (CH_3-2,6), 6.57/6.64 (C_6H_2). ^{13}C NMR (C_6D_6 at 20 °C) for the monomer/dimer: 5.7/6.0 (CH_3Si), 21.0/21.3 (CH_3-4), 26.7/26.3 (CH_3-2,6), 128.4/- (C-3,5), 139.5/- (C-4), 143.7/144.1 (C-2,6), 146.1/- (C-1) [2].

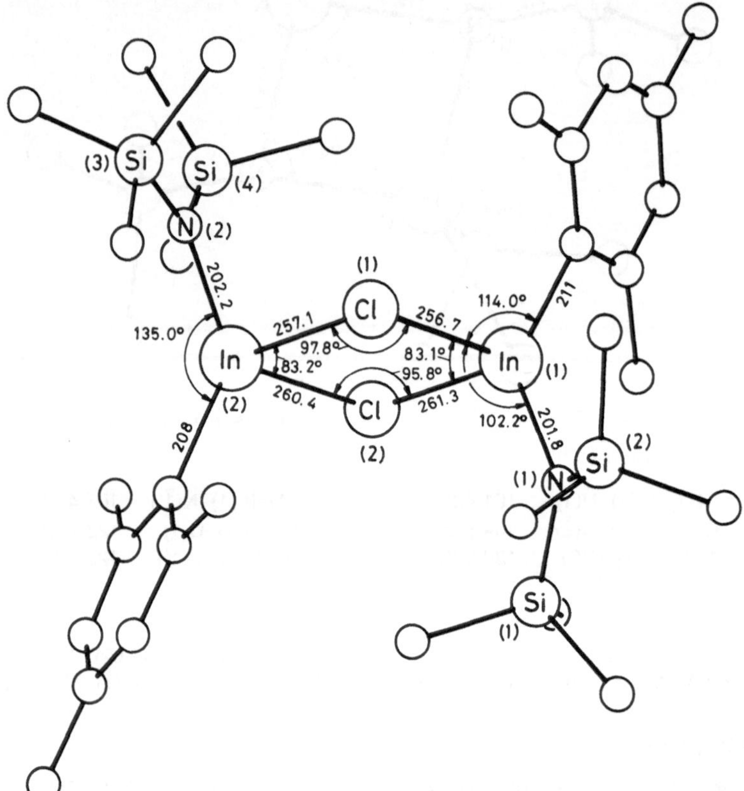

Fig. 59. Molecular structure of $(\{C_6H_2(CH_3)_3\text{-}2,4,6\}In(Cl)N\{Si(CH_3)_3\}_2)_2$ [4].

The compound crystallizes in the monoclinic space group $P2_1/c-C^5_{2h}$ (No. 14) with the constants a = 2232.3(4), b = 893.3(3), c = 2072.0(3) pm, β = 98.57(1)°; Z = 4 and D_c = 1.408 g/cm³. The molecule contains a planar In_2Cl_2 core with different InCl distances (In···In = 387.2(2) pm). The N atoms are nearly planar and the bulkiness of the ligands leads to Si(2)···In(1) and Si(3)···In(2) contacts of 317.0(4) and 314.7(4) pm, respectively, which are about 85 pm less than the van der Waals radii. No intermolecular contacts were found. The most important bond distances and angles are shown in **Fig. 59** on p. 300 [4].

The mass spectrum at 70 eV shows the following peaks (relative intensities in %): $[M-CH_3+H]$ (5.72), $[M-HNSi(CH_3)_3]$ (12.21), $[M-HN(Si(CH_3)_3)_2-H]^+$ (7.18), $[M-Cl-Si(CH_3)_3\text{-mes}]^+$ (40.48), $[InCl]^+$ (16.96), $^{115}In^+$ (22.26) [2].

$(C_3H_7\text{-i})In(Cl)NHC_4H_9\text{-t}\cdot In(C_3H_7\text{-i})_2NHC_4H_9\text{-t}$

This mixed compound was prepared by slow addition of $Li[NHC_4H_9\text{-t}]$ in ether at 0 °C to a 1:1 mixture of $(C_3H_7\text{-i})_2InCl$ and $(C_3H_7\text{-i})InCl_2$ in the same solvent. The workup procedure according to the first compound in this section gave about 32% yield of a mixture of two isomers (see Formulas II and IV, 2:1 ratio) [4].

The NMR spectra were run in C_6D_6 (δ values in ppm). ^1H NMR (isomer II): 0.97, 1.19 (C_4H_9), 1.38 to 1.58 (C_3H_7), 1.87 (NH); isomer IV: 0.98 (C_4H_9), 1.38 to 1.58 (C_3H_7), 1.87 (NH). ^{13}C NMR (isomer II): 20.0, 21.3 (CH), 22.5, 22.6, 22.9, 23.7, 23.8, 24.3 (diastereotopic CH_3 of C_3H_7), 34.1, 34.2 (CH_3 of C_4H_9), 51.3, 52.2 (C); isomer IV: 22.9, 23.8, 24.3 (geometrically different CH_3 of C_3H_7), 34.0 (CH_3 of C_4H_9), 51.2 (C) [4].

IR/Raman (in cm⁻¹, isomer mixture): 570 s/577 vw, 538 m/541 vw ν(InN), 485 m/486 s and 465 m/467 ν(InC), 250 m/259 m ν(InCl) [4].

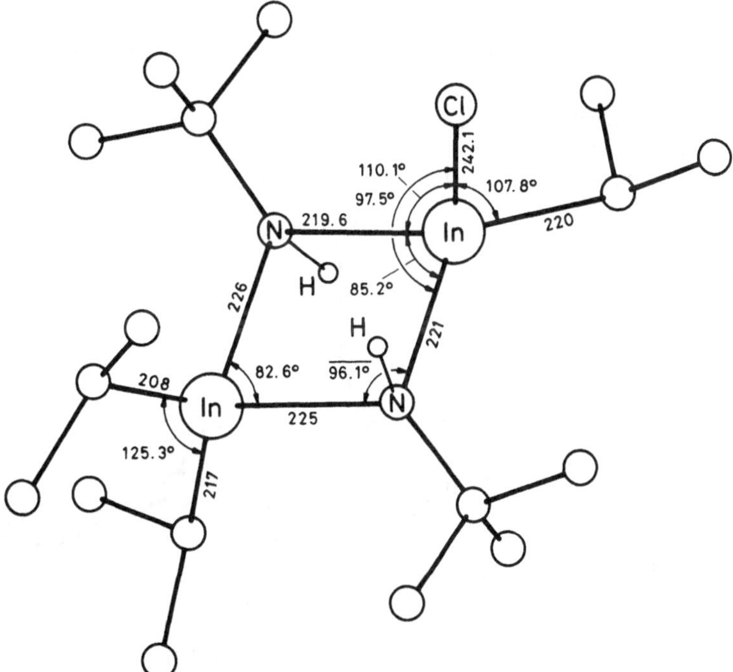

Fig. 60. Molecular structure of $(C_3H_7\text{-i})In(Cl)NHC_4H_9\text{-t}\cdot In(C_3H_7\text{-i})_2NHC_4H_9\text{-t}$ [4].

Isomer II crystallizes in the monoclinic space group $P2_1/c-C^5_{2h}$ (No. 14) with the constants a = 928.2(3), b = 2139.6(9), c = 1232(1) pm, β = 101.40(3)°; Z = 4 and D_c = 1.498 g/cm³. The molecular structure is depicted in **Fig 60**, p. 301 [4].

The mass spectrum showed fragments from the ions $[M]^+$, $[M-Cl]^+$, $[M-C_3H_7]^+$, $[M-NHC_4H_9]^+$, $[In_2NHC_4H_9]^+$, $[In]^+$ [4].

References:

[1] Veith, M.; Pöhlmann, J. (Z. Naturforsch. **43b** [1988] 505/12).

[2] Neumüller, B. (Z. Naturforsch. **45b** [1990] 1559/66).

[3] Reger, D. L.; Knox, S. J.; Rheingold, A. L.; Haggerty, B. S. (Organometallics **9** [1990] 2581/7).

[4] Neumüller, B. (Z. Naturforsch. **46b** [1991] 753/61).

7.3.2 Compounds of the Type $RIn(NR'-D)_2$ with Unsaturated N-Heterocycles

$CH_3In\{N(CH_3)C_5H_4N\}_2$

For the preparation of this compound 2-(methylamino)pyridine was condensed onto freshly sublimed $In(CH_3)_3$ (2:1 mole ratio) at −196 °C. On slow warming of the mixture the trialkyl dissolved at about 0 °C with evolution of CH_4. The resulting yellow liquid was warmed and kept at 50 °C for 30 min; the residue was recrystallized from C_5H_{12}; m.p. 140 °C (dec.), 72% yield [1, 2].

^{1}H NMR in C_6D_6 (in ppm): 0.28 (s, CH_3In), 2.71 (s, CH_3N), 6.00 to 7.68 (m's, C_5H_4N) [1, 2].

IR (Nujol, in cm^{-1}): 1600 vs, vbr, 1543 m, 1510 to 1450 (s, several), 1414 s, br, 1368 s, sh, 1336 m, 1300 s, 1290 s, 1276 m, 1165 s, 1154 s, 1127 w, br, 1082 m, 1042 w, 1029 w, 993 w, sh, 982 m, 952 w, 832 m, sh, 824 s, 819 s, 771 vs, br, 732 vs, 702 vs, vbr $\varrho(CH_3In)$, 644 s, 599 m, 574 m, br, 520 s, sh, 510 s ν(InC), 478 m, sh, 449 s, br, 419 s, 308 m, br, 283 w, sh, 243 w [1, 2].

The compound crystallizes in the monoclinic space group $P2_1/a-C^5_{2h}$ (No. 14) with the cell parameters a = 809.4(2), b = 2059.2(2), c = 880.2(3) pm, β = 94.76(4)°; Z = 4 and D_c = 1.572 g/cm³. The most important bond distances and angles are indicated in **Fig. 61** [1, 2].

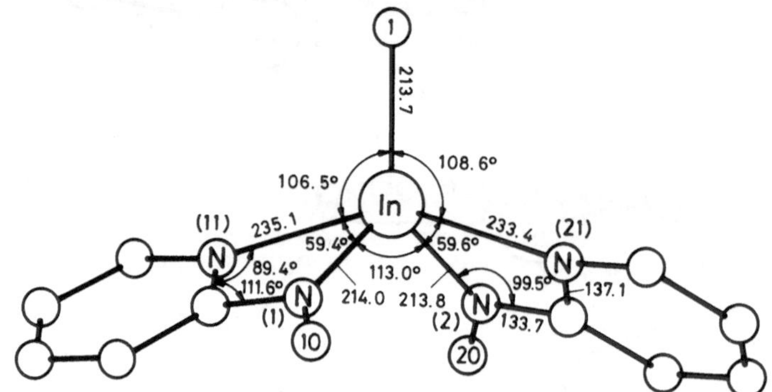

Fig. 61. Molecular structure of $CH_3In\{N(CH_3)C_5H_4N\}_2$ [1, 2].

Other angles (°)

N(21)-In-N(11)	144.8(1)	N(1)-In-N(21)	99.5(2)
C(1)-In-N(2)	123.4(3)	N(2)-In-N(11)	100.5(2)
C(1)-In-N(1)	123.6(3)	C(20)-N(2)-In	139.8(3)

The mass spectrum (70 eV) shows the following principal fragments (relative intensities in %; $L=C_5H_4N\text{-}NCH_3$): M^+ (12), $[M-L-CH_3]^+$ (47), $^{115}In^+$ (39), $[M-CH_3]^+$ (30), $[M-LH-CH_3]^+$ (25), L^+ (100), $[M-L]^+$ (30), $[In(CH_3)_2]^+$ (3), $[L-H]^+$ (58) [2].

References:

[1] Arif, A. M.; Bradley, D. C.; Frigo, D. M.; Hursthouse, M. B.; Hussain, B. (J. Chem. Soc. Chem. Commun. **1985** 783/4).

[2] Arif, A. M.; Bradley, D. C.; Dawes, H.; Frigo, D. M.; Hursthouse, M. B.; Hussain, B. (J. Chem. Soc. Dalton Trans. **1987** 2159/64).

7.3.3 RIn-Porphyrins

The compounds assembled in Table 60 were prepared in benzene or toluene by stepwise addition of a large excess of freshly prepared LiR to the appropriate ClIn-porphyrin at 5 to 6 °C. Cold (5 to 6 °C) water was added and the organic phase separated, washed with water, dried over $MgSO_4$, then freed of solvent in vacuum. The residue was taken up in $CHCl_3$ and chromatographed in the dark on a basic Al_2O_3 column. The column was eluted with CH_2Cl_2, C_6H_6, and mixtures of $CHCl_3/CH_3OH$, CH_2Cl_2/C_6H_6, CH_2Cl_2/CH_3OH, C_6H_6/CH_3OH, and C_6H_6/C_6H_{14} in various proportions [2].

RMgBr was used to arylate Cl-In-porphyrins (ca. 1:1 mole ratio) for making Nos. 9, 10, and 19 to 24. The reaction mixtures were stirred for 48 h before adding water, the chromatographic column charged with silica gel, and eluted with C_6H_6 or $C_6H_5CH_3$, only. Finally, the products were recrystallized from C_6H_6/C_7H_{16} [11].

Structure I, postulated on the basis of the ^{1}H NMR and mass spectral evidence, was confirmed by X-ray diffraction analysis of No. 11.

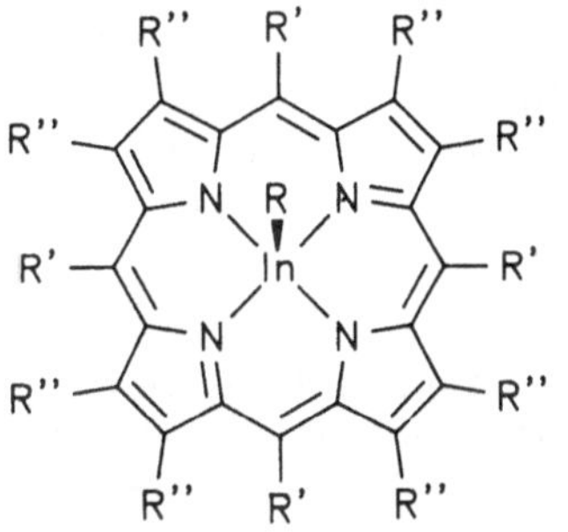

I

OEP = octaethylprophyrin : $R'=H$, $R''=C_2H_5$
TPP = tetraphenylporphyrin : $R'=C_6H_5$, $R''=H$
T(m-T)P = tetra(m-tolyl)porphyrin: $R'=C_6H_4CH_3\text{-}3$, $R''=H$
T(p-T)P = tetra(p-tolyl)porphyrin : $R'=C_6H_4CH_3\text{-}4$, $R''=H$

General Remarks. Contrary to the electronic spectrum of the starting porphyrin (ClIn-porphyrin), the spectra of the RIn-porphyrin derivatives with σ-bonded alkyl or aryl groups show the Soret band around 390 to 420 nm split into red- and blue-shifted absorptions. The blue-shifted band was assigned to the $5p_z(In)\rightarrow e_g(\pi^*)$ electron transition, and the other band to a $\pi\rightarrow\pi^*$ transition of the heterocyclic ring system. The relationship between these bands is dependent on the kind of R; for the same porphyrin residue an increase in the shift or splitting follows the sequence: $Cl < C_6F_5 \cong C_6F_4H < C_6H_5C{\equiv}C < C_6H_5CH{=}CH < CH_3 <$

$C_2H_5 < C_4H_9$-n $< C_3H_7$-i $< C_4H_9$-t. Additionally, three Q-bands, Q(2,0), Q(1,0), and Q(0,0), undergo bathochromic shifts compared to the starting compound [8, 11].

The IR data reported for Nos. 9, 10, and 19 to 24 represent exclusively vibrations of the C_6F_5 or C_6F_4H groups; bands between 1640 and 1500 cm^{-1} belong to C_6 ring vibrations, those between 1400 and 800 cm^{-1} are C–F bond motions, while C–F bendings are localized between 210 and 190 cm^{-1}. More detailed assignments are not given, nor values are for the In–C stretching [11].

The mass spectrum always contained the molecular ion peak; for Nos. 9, 10, and 17 this was also the basis peak, which was $[M-R]^+$ for Nos. 6, 11 to 16, and 18 to 24 or the recombination ion, $[M-R+H]^+$, for Nos. 1 to 5 and 8. Only for No. 7 was $[M+H]^+$ the main peak [2, 11].

The photolytic decomposition of Nos. 2, 4, 6, 7, and 12 to 17 (580 nm in OC_4H_8 at 25 °C) was followed by UV observation of the successive changes in the Q(1,0) and Q(0,0) bands, and was explained by the formation of a triplet state, followed by homolytic rupture of the In–R bond: RIn-por + hν → ^{3}[RIn-por] → [In-por] + R$^•$ (por = OEP, TPP) [10, 14].

The electrochemical behavior of all the porphyrin derivatives is very similar and was investigated by cyclic voltammetry (0.1 m $[N(C_4H_9)_4]PF_6$, 100 mV/s scan rate): No. 7 in CH_2Cl_2; Nos. 9, 10, 19 to 24 also in C_6H_5CN and C_5H_4N. Reduction occurs by the stepwise uptake of two electrons on the porphyrin macrocycle, but the In–C bond does not break. The two one-electron reductions and the accompanying half-wave potentials (in V) are given below [8, 11].

$$\text{RIn-por} \underset{}{\overset{e^-}{\rightleftharpoons}} [\text{RIn-por}]^{-\bullet} \overset{e^-}{\rightleftharpoons} [\text{RIn-por}]^{2-}$$

porphyrin	compound No.	V-1	V-2
OEP	1 to 6, 8 to 10	−1.54 to −1.45	−1.90 (estimated)
	7 (in C_6H_5CN)	−1.31	–
TPP	11 to 16, 18 to 20	−1.27 to −1.20	−1.71 to 1.60
	17 (in C_6H_5CN)	−1.04	−1.49
T(m-T)P	21, 22	−1.22, −1.22	−1.64, −1.63
T(p-T)P	23, 24	−1.24, −1.22	−1.64, −1.62

Thin-layer voltammograms of the reductions are reversible, the time-dependent electronic spectra showing the changes during the reduction especially distinctly. These spectra (λ_{max}, $\varepsilon \cdot 10^{-3}$), typical of anion radicals (step 1), are reported for all the compounds. The ESR spectra of the RInTPP complexes in their first reduction step (in C_5H_4N) give a signal at g = 2.00; the large width (150 G, 30 G for Nos. 19 and 20) of this line and its hyperfine structure (not for R = C_6F_5, C_6H_4H) point to a significant interaction with the central metal ion [8, 11].

Oxidation, also followed by cyclic voltammetry, takes place in several steps, the first of which, except for aryl derivatives 8 to 10 and 18 to 24, is irreversible, indicating rupture of the In–C bond:

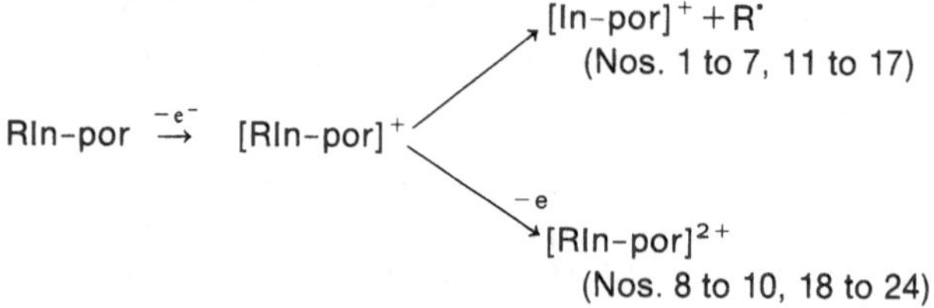

For compounds 1 to 7 and 11 to 17 (por=OEP and TPP, measured in CH_2Cl_2) the first step lies between 0.62 and 0.92 V; the other two potentials of about 1.2 and 1.5 V involve the oxidation of the $[In(por)]^+$ cation. For the remaining compounds the first stage was observed between 0.89 and 1.05 V, the second stage between 1.44 and 1.52. The oxidation was also followed by time-dependent electronic spectroscopy [8, 11].

CH_2Cl_2 solutions of 1, 5, 8, 11, 15, and 18 were converted into the corresponding sulfinates, R-S(O)OIn(por), by insertion of SO_2 and by subsequent oxidation of these sulfinates with O_2 into polymeric sulfonato complexes, R-S(O)O_2In(por) [3, 4].

On bubbling CO_2 through the solution of products from No. 1 or 11 in 2/1 benzene/pyridine and irradiating with a mercury-vapor UV lamp, insertion of CO_2 into the In-CH_3 bond produced the stable acetato complex, $CH_3C(O)OIn(por)$. These same derivatives were also obtained from the reaction of $CH_3In(por)$ with CH_3COOH (1:1 mole ratio) with elimination of CH_4 [7].

Table 60
RIn Porphyrines (see Formula I).
Further information on numbers preceded by an asterisk is given at the end of the table.
Explanations, abbreviations, and units on p. X.

No.	compound (yield)	^{1}H NMR and ^{19}F NMR in C_6D_6, UV spectra, $\Lambda_{max}(\varepsilon\cdot10^{-3})$ in nm, other remarks
1	CH_3InOEP (70%) [1] (67%) [2]	^{1}H NMR: −5.54 (s, CH_3In), 1.92 (t, CH_3), 4.14 (q, CH_2), 10.19 (s, CH) [2] UV (C_6H_6, depicted in [8]): 354(49), 428(247), 515(2.5), 554(17.2), 588(8.9) [2, 8]
2	C_2H_5InOEP (45%) [1] (40%) [2]	^{1}H NMR: −4.96 (q, CH_2In), −2.84 (t, CH_3C), 1.91 (t, CH_3 of OEP), 4.14 (q, CH_2 of OEP), 10.18 (s, CH) [2] UV (C_6H_6): 356(58.5), 435(145), 520(2.4), 558(17.7), 592(6.4) [2, 8]
3	$(C_3H_7\text{-i})InOEP$ (48%)	^{1}H NMR: −4.13 (hept., HCIn), −2.85 (d, CH_3C), 1.91 (t, CH_3 of OEP), 4.14 (q, CH_2), 10.18 (s, CH) [2] UV ($CHCl_3$): 372(88.8), 439(96.8), 525(3.4), 564(17.8), 595(5.3) [2, 8]
4	$(C_4H_9\text{-n})InOEP$ (67%) [1] (64%) [2]	^{1}H NMR: −4.68 (t, CH_2In), −2.60 (m, CH_2), −1.80 (m, CH_2), −0.61 (t, CH_3), 1.91 (t, CH_3 of OEP), 4.14 (q, CH_2), 10.17 (s, CH) [2]; −4.36 (t), −2.25 (m), −1.58 (q), −0.50 (t), no values for OEP [14] UV (C_6H_6): 365(59.3), 435(142), 520(2.4), 559(17.6), 594(6.5) [2, 8]
5	$(C_4H_9\text{-t})InOEP$ (35%) [1] (45%) [2]	^{1}H NMR: −3.05 (s, CH_3C), 1.90 (t, CH_3 of OEP), 4.15 (q, CH_2), 10.17 (s, CH) [2] UV (C_6H_6, depicted in [8]): 337(95.3), 443(61.4), 565(13.6), 594(3.6) [2, 8]

Table 60 (continued)

No.	compound (yield)	^1H NMR and ^{19}F NMR in C_6D_6, UV spectra, $\Lambda_{max}(\varepsilon \cdot 10^{-3})$ in nm, other remarks
6	$C_6H_5CH=CHInOEP$ (80%)	^1H NMR: 1.75 (d, HCIn), 1.93 (t, CH_3), 2.12 (d, HC=C), 4.15 (q, CH_2), 5.63 (m, H-2,6), 6.46 (m, H-3,4,5), 10.34 (s, CH of OEP) [2] UV (C_6H_6): 354(39.8), 426(246), 514(2.0), 554(17.6), 589(9.8) [2, 8]; 549, 586 in THF [14]
7	$C_6H_5C{\equiv}CInOEP$ (69%)	^1H NMR: 1.82 (t, CH_3), 3.97 (q, CH_2), 5.53 (m, H-2,6), 5.96 (m, H-3,4,5), 10.37 (s, CH) [2] UV (C_6H_6): 345(30.7), 415(426), 508(2.5), 547(20.0), 584(18.0) [2, 8]; 543, 581 in THF [14]
8	C_6H_5InOEP (75%) [1] (79%) [2]	^1H NMR: 1.91 (t, CH_3), 2.71 (d, H-2,6), 4.14 (q, CH_2), 5.51 (t, H-3,5), 5.84 (t, H-4), 10.21 (s, CH of OEP) [2] UV ($CHCl_3$): 354(45.1), 424(264), 512(2.8), 551(18.3), 587(9.6) [2, 8]
9	$(C_6F_4H\text{-}4)InOEP$ (ca. 85%)	^1H NMR: 1.85 (t, CH_3), 4.01 (m, CH_2), 4.70 (m, H-4, J(H,F-2) = 9.14, J(H,F-3) = 7.04), 10.50 (s, CH of OEP) [11] ^{19}F NMR: −140.69 (m, F-3,5), −125.47 (m, F-2,6), J(2,3) = 30.37, J(2,5) = 18.52, J(2,6) = 4.60, J(3,5) = 0.32 [11] UV (C_6H_6): 394(55), 414(362), 503(4), 545(20), 580(17) [11] IR (solid): 1600, 1528, 1450, 1188, 1160, 888, 210; in addition to bands found for ClInOEP [11]
10	C_6F_5InOEP (ca. 85%)	^1H NMR (depicted): 1.87 (t, CH_3), 4.03 (m, CH_2), 10.52 (s, CH) [11] ^{19}F NMR: −162.62 (m, F-3,5), −157.40 (t, F-4), −124.25 (m, F-2,6), J(2,3) = 31.73, J(2,4) = 0.32, J(2,5) = 13.29, J(3,5) = 7.54, J(3,4) = 20.30 [11] UV (C_6H_6, depicted): 393(55), 414(403), 504(4), 546(20), 580(18) [11] IR (solid): 1630, 1525, 1500, 1445, 1348, 1255, 1051, 951, 479; in addition to bands found for ClInOEP [11]
*11	CH_3InTPP (65%) [1] (91%) [2]	^1H NMR: −5.14 (s, CH_3In), 7.76 (m, H-3,4,5), 8.11, 8.31 (m, H-2,6), 8.95 (s, CH) [2] UV ($CHCl_3$): 343(34.2), 439(470), 538(4.0), 578(15.5) 623(14.6) [2, 8]
*12	C_2H_5InTPP (40%) [1] (72%) [2]	^1H NMR: −4.27 (q, CH_2In), −2.53 (t, CH_3C), 7.76 (m, H-3,4,5), 8.12, 8.29 (m, H-2,6), 8.94 (s, CH) [2] UV ($CHCl_3$): 355(36.5), 445(293), 545(3.2), 584(11.7), 628(13.8) [2, 8]; 581, 627 in THF [14]; 563, 578, 621 in $OC_4H_7CH_3$[13]
13	$(C_3H_7\text{-}i)InTPP$ (62%)	^1H NMR: −3.67 (hept., HCIn), −2.54 (d, CH_3C), 7.76 m, H-3,4,5), 8.06 (m, H-2,6), 8.96 (s, CH) [2]

Table 60 (continued)

No.	compound (yield)	^1H NMR and ^{19}F NMR in C_6D_6, UV spectra, $\Lambda_{max}(\varepsilon \cdot 10^{-3})$ in nm, other remarks
13 (continued)		UV ($CHCl_3$): 383(66.7), 451(203), 550(3.7), 590(11.7), 636(15.1) [2, 8]; 540, 580, 625 in C_6H_6 [8]; 583, 628 in THF [14]
14	$(C_4H_9\text{-}n)$InTPP (64%) [1] (68%) [2]	^1H NMR: −4.25 (t, CH_2In), −2.24 (m, CH_2), −1.53 (m, CH_2), −0.48 (t, CH_3), 7.77 (m, H-3,4,5), 8.25 (m, H-2,6), 8.95 (s, CH) [2] UV ($CHCl_3$): 356(42.4), 445(335), 544(3.6), 583(13.5), 627(15.2) [2, 8]
15	$(C_4H_9\text{-}t)$InTPP (45%) [1] (68%) [2]	^1H NMR: −2.72 (s, CH_3C), 7.76 (m, H-3,4,5), 8.17 (m, H-2,6), 8.95 (s, CH) [2] UV ($CHCl_3$): 393(100), 453(136), 551(3.8), 592(9.8), 639(136) [2, 8]; 583, 633 in THF [14]
16	$C_6H_5CH{=}CHInTPP$ (69%)	^1H NMR: 2.07 (HCIn), 2.40 (HC=C), 5.80 (m, H-2,6), 6.59 (m, H-3,4,5), 7.75 (m, H-3,4,5), 8.13, 8.31 (m, H-2,6), 8.97 (s, CH of TPP) [2] UV ($CHCl_3$): 340(32.2), 439(496), 538(3.5), 578(16.3), 621(13.8) [2, 8]; 574, 651 in THF [14]
17	$C_6H_5C{\equiv}CInTPP$ (70%)	^1H NMR: 5.83 (m, H-2,6), 6.13 (m, H-3,4,5), 7.45 (m, H-3,4,5 of TPP), 8.06 (m, H-2,6, of TPP), 9.04 (s, CH of TPP) [2] UV (C_6H_6): 322(23.1), 432(323), 528(3.7), 568(21.1), 610(12.0) [2, 8]; 565, 606 in THF [14]
18	C_6H_5InTPP (75%) [1] (94%) [2]	^1H NMR: 2.97 (d, H-2,6), 5.70 (t, H-3,5), 6.02 (t, H-4), 7.76 (m, H-3,4,5 of TPP), 8.14 (m, H-2,6 of TPP), 8.97 (s, CH of TPP) [2] UV: 343(36.2), 439(531), 535(3.4), 577(16.8), 619(14.9) [2, 8]
19	$(C_6F_4H\text{-}4)$InTPP (ca. 85%)	^1H NMR: 4.96 (m, H-4), 7.44 (m, H-3,4,5 of TPP), 8.05, 8.28 (d, H-2,6 of TPP), 9.10 (m, CH of TPP), J(H,F-2) = 9.20, J(H,F-3) = 7.11 [11] ^{19}F NMR: −139.79 (m, F-3,5), −124.72 (m, F-2,6), J(2,3) = 30.23, J(2,5) = 18.00, J(2,6) = 5.32, J(3,5) = 0.04 [11] UV (C_6H_6): 408(49), 430(634), 524(7), 566(23), 606(14) [11] IR (solid): 1605, 1503, 1458, 1190, 1162, 890, 192 in addition to the bands found for ClInTPP [11]
20	C_6F_5InTPP (ca. 85%)	^1H NMR: 7.46 (m, H-3,4,5), 8.05 (d, H-2 or H-6), 8.36 (d, H-6 or H-2), 9.12 (s, CH) [11] ^{19}F NMR: −161.69 (m, F-3,5), −157.40 (t, F-4), −123.58 (m, F-2,6), J(2,3) = 31.75, J(2,4) = 0.24, J(2,5) = 12.77, J(2,6) = 0.66, J(3,5) = 7.54, J(3,4) = 20.89 [11] UV (C_6H_6): 407(47), 429(638), 524(7), 546(23), 605(13) [11]

Table 60 (continued)

No.	compound (yield)	^1H NMR and ^{19}F NMR in C_6D_6, UV spectra, $\Lambda_{max}(\varepsilon \cdot 10^{-3})$ in nm, other remarks
20 (continued)		IR (solid, depicted): 1632, 1529, 1504, 1453, 1345, 1262, 1055, 954, 480 in addition to bands found for ClInTPP [11]
21	$(C_6F_4H\text{-}4)$InT(m-T)P	^1H NMR: 2.28, 2.34 (2 s, CH_3 of C_7H_7), 4.95 (br, H-4 of C_6F_4H), 7.38 (m, H-3,4 of C_7H_7), 7.95, 8.20 (2 m, H-2,6 of C_7H_7), 9.20 (s, CH of T(m-T)P) [11] ^{19}F NMR: −139.69 (m, F-3,5), −124.58 (m, F-2,6), J(2,3) = 29.64, J(2,5) = 18.69, J(2,6) = 0.89, J(3,5) = 1.06 [11] UV (C_6H_6): 408(45), 430(576), 525(5), 564(20), 606(12) [11] IR (solid): 1528, 1455, 1187, 1160, 888; in addition to bands found for ClInT(m-T)P [11]
22	C_6F_5InT(m-T)P (ca. 85%)	^1H NMR: 2.28, 2.34 (2 s, CH_3), 7.39 (m, H-3,4), 7.96, 8.28 (2 m, H-2,6), 9.22 (s, CH of T(m-T)P) [11] ^{19}F NMR: −161.62 (m, F-3,5), −156.08 (t, F-4), −123.41 (m, F-2,6), J(2,3) = 32.40, J(2,4) ca. 0, J(2,5) = 11.54, J(2,6) ca. 0, J(3,5) = 7.89, J(3,4) = 20.17 [11] UV (C_6H_6): 408(48), 430(609), 525(7), 566(22), 606 (13) [11] IR (solid): 1633, 1530, 1505, 1455, 1345, 1262, 1057, 957, 482; in addition to bands found for ClInT(m-T)P [11]
23	$(C_6F_4H\text{-}4)$InT(p-T)P ca. 85%	^1H NMR (depicted): 2.40 (s, CH_3), 4.95 (m, H-4 of C_6F_4H), 7.24, 7.28 (d's, H-3,5), 8.02, 8.24 (d's, H-2,6), J(H,F-2) = 9.18, J(H, F-3) = 7.08, 9.23 (s, CH of T(p-T)P) [11] ^{19}F NMR: −139.83 (m, F-3,5), −124.56 (m, F-2,6); J(2,3) = 30.16, J(2,5) = 18.48, J(2,6) = 5.27, J(3,5) = 0.28 [11] UV (C_6H_6): 409(51), 432(639), 526(8), 567(23), 608(17) [11] IR (solid): 1600, 1529, 1455, 1186, 1160, 888, 190; in addition to bands found for ClInT(p-T)P [11]
24	C_6F_5InT(p-T)P (ca. 85%)	^1H NMR: 2.40 (s, CH_3), 7.25, 7.29 (2 d, H-3,5), 8.02, 8.33 (2 d, H-2,6), 9.25 (s, CH of T(p-T)P) [11] ^{19}F NMR: −161.74 (m, F-3,5), −156.27 (t, F-4), −123.39 (m, F-2,6), J(2,3) = 31.71, J(2,4) = 0.20, J(2,5) = 12.17, J(2,6) = 0.86, J(3,5) = 7.80, J(3,4) = 20.36 [11] IR (solid): 1631, 1528, 1502, 1452, 1054, 952, 480; in addition to bands found for ClInT(p-T)P [11]

* Further information:

CH_3InTPP (Table **60**, No. **11**) crystallizes from 3:1 $CHCl_3/CH_3OH$ as intensely blue prisms in the space group $P2_1/c-C^5_{2h}$ (No. 14) with the cell parameters a = 1006.4(1), b = 1622.1(3), c = 2335.5(2) pm, β = 115.47(2)°; Z = 4 and D_c = 1.43 g/cm³. The molecular framework, $ClnN_4$, is square-pyramidal and has C_{4v} symmetry. The In atom is 78(2) pm above the N_4 basal plane and the In-C bond distance is 213(1) pm. Other relevant structural parameters are given in Fig. **62**; we have omitted the numerous data on the porphyrin ring [5].

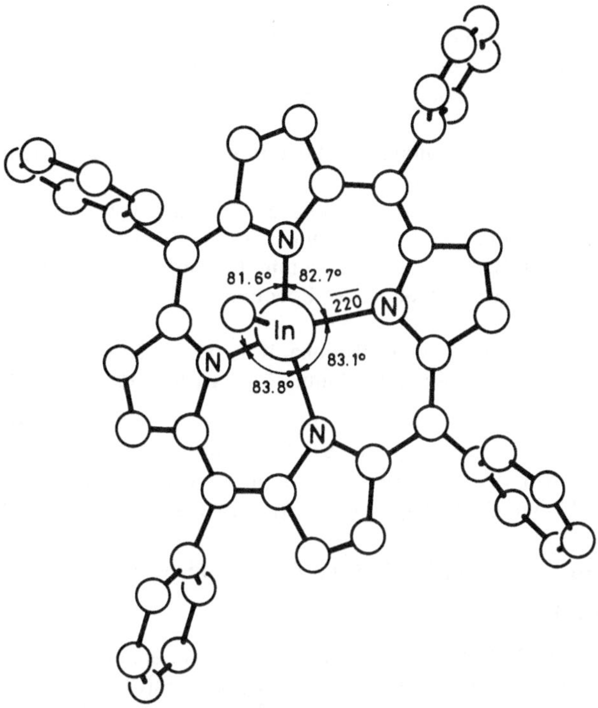

Fig. 62. Molecular structure of CH_3InTPP [5].

(C_2H_5)InTPP (Table **60**, No. **12**). Its photolytic behavior has been investigated in great depth. Momentary irradiation (580 nm) and laser flash photolysis (532 nm, 20 ns duration, ca. 100 mJ/pulse) were followed by UV/VIS, ^1H NMR, and ESR spectroscopy. The ESR spectrum in degassed benzene was weak and unresolved (g = 2.0; 7.5 G coupling constant), but was changed by traces of oxygen into a well-resolved and readable spectrum (g = 2.004(1), 27.5 G). Photolysis by sunlight for 2 d allowed the $[InTPP]^+$ cation to be isolated as ClInTPP [9].

Pyridine added to the solution to be irradiated, recognized it as an activator to obtain higher quantum yields, and postulated the formation of $^3[C_2H_5InTPP \cdot NC_5H_5]$ as the activated adduct. The changing UV (400 to 800 nm) and ESR spectrum of a sample during the irradiation are illustrated (in C_6H_6, 5 mol C_5H_5N, after 5 s irradiation with 580 nm); also given are the dependences of the quantum yield of the photolysis and the decay constant of the triplet state upon the pyridine concentration [12].

Similar investigations in $OC_4H_7(CH_3\text{-}2)$ were carried out, and are described with illustrations [10, 13]. The dependence of the quantum yield and the decay constant was measured on the amount of tetracyanoquinodimethane added. The ESR spectra of samples dissolved in OC_4H_8 and irradiated with visible light at 300 K (g = 2.004, 10.5 G) and at 115 K (g = 2.001, 113 G) are illustrated and discussed in [14] and compared with the 115 K spectrum of $[C_2H_5InTPP]^{-\bullet}$ (g = 1.968 and 2.002), the first reduction stage of the unirradiated compound.

Other laser photolysis experiments (in benzene) in the presence of ferrocene as quencher of the triplet state are described in [6].

References:

[1] Guilard, R.; Cocolios, P; Fournari, P. (J. Organometal. Chem. **129** [1977] C 11/C 13).
[2] Guilard, R.; Cocolios, P; Fournari, P. (J. Organometal. Chem. **179** [1979] 311/22).
[3] Guilard, R.; Cocolios, P; Fournari, P.; Lecomte, C.; Protas, J. (J. Organometal. Chem. **168** [1979] C 49/C 51).
[4] Cocolios, P; Fournari, P.; Guilard, R.; Lecomte, C.; Protas, J.; Boubel, J. C. (J. Chem. Soc. Dalton Trans. **1980** 2081/9).
[5] Lecomte, C.; Protas, J.; Cocolios, P; Guilard, R. (Acta Crystallogr. B **36** [1980] 2769/71).
[6] Hoshino, M. (Sci. Papers Inst. Phys. Chem. Res. **78** [1984] 113/7).
[7] Cocolios, P; Guilard, R.; Bayeul, D.; Lecomte, C. (Inorg. Chem. **24** [1985] 2058/62).
[8] Kadish, K. M.; Boisselier-Cocolios, B.; Cocolios, P; Guilard, R. (Inorg. Chem. **24** [1985] 2139/47).
[9] Hoshino, M.; Ida, H.; Yasufuku, Y.; Tanaka, K. (J. Phys. Chem. **90** [1986] 3984/7).
[10] Hoshino, M.; Yamaji, M.; Hama, Y. (Chem. Phys. Letters **125** [1986] 369/72).

[11] Tabard, A.; Guilard, R.; Kadish, K. M. (Inorg. Chem. **25** [1986] 4277/85).
[12] Hoshino, M.; Hirai, T. (J. Phys. Chem. **91** [1987] 4510/4).
[13] Yamaji, M.; Hama, Y.; Arai, S.; Hoshino, M. (Inorg. Chem. **26** [1987] 4375/8).
[14] Kadish, K. M.; Maiya, G. B.; Xu, Q. Y. (Inorg. Chem. **28** [1989] 2518/23).

7.3.4 $RIn(NR')_2E(RR'')$ and $RIn(NR'ER''NR')_2$ Compounds with E = Si

$CH_3In(NC_4H_9\text{-t})_2Si(CH_3)_2$

The reaction of $Sn(NC_4H_9\text{-t})_2Si(CH_3)_2$ with $In(CH_3)_3$ in benzene (1.5:1 mole ratio) formed a mixture of the title compound and $\{(CH_3)_2In\}_2(NC_4H_9\text{-t})_2Si(CH_3)_2$ (Table 57, p. 284, No. 11), after a 3 h reflux and after drawing off the solvent and excess Sn^{II} compound. The mixture was sublimed (110 to 120 °C/10^{-3} Torr) and dissolved in benzene for separation. The title compound crystallized out (m.p. 165 °C, 24% yield), while the dimethylindium component remained in solution. Reaction of $\{(CH_3)_2In\}_2(NC_4H_9\text{-t})_2Si(CH_3)_2$ with $Sn(NC_4H_9\text{-t})_2Si(CH_3)_2$ in benzene (1:1 mole ratio) led to a momentary red coloration of the mixture (formation of dimethylstannanes); after 10 min refluxing and slow cooling, the title compound crystallized (48%) [1].

According to cryoscopic molecular weight determinations in benzene the compound is dimeric, although M^+ was the heaviest peak observed in the mass spectrum (70 eV) [1].

^1H NMR (C_6D_6, in ppm): 0.47 (s, CH_3In), 0.48 (s, CH_3Si), 1.33 (s, CH_3C) [1].

The compound crystallizes in the monoclinic space group $P2_1/c-C_{2h}^5$ (No. 14) with the cell parameters a = 1132.3(4), b = 868.8(3), c = 1794.8(8) pm, β = 117.4(1)°; Z = 2 (one dimer) and D_c = 1.40 g/cm^3. The structure is shown in **Fig. 63** on p. 311 [2].

$CH_3In\{N(C_4H_9\text{-t})Si(CH_3)NC_4H_9\text{-t}\}_2$

This compound was formed by the reaction of the dilithium derivative, $Li_2[(N(C_4H_9\text{-t})Si(CH_3)NC_4H_9\text{-t})_2]$ (Formula VII, p. 288), with CH_3InCl_2 (1:1 mole ratio) in benzene after 12 h stirring at room temperature, separation of the LiCl formed, and purification by sublimation at 80 °C/10^{-3} Torr (colorless crystals, m.p. 235 °C, 85% yield). It also occurred when a benzene solution of the dilithium salt was added to a suspension of 2 mol $(CH_3)_2InCl$ in ether/benzene. The reaction occurs according to $Li_2L + 2\,(CH_3)_2InCl \rightarrow CH_3InL + 2$

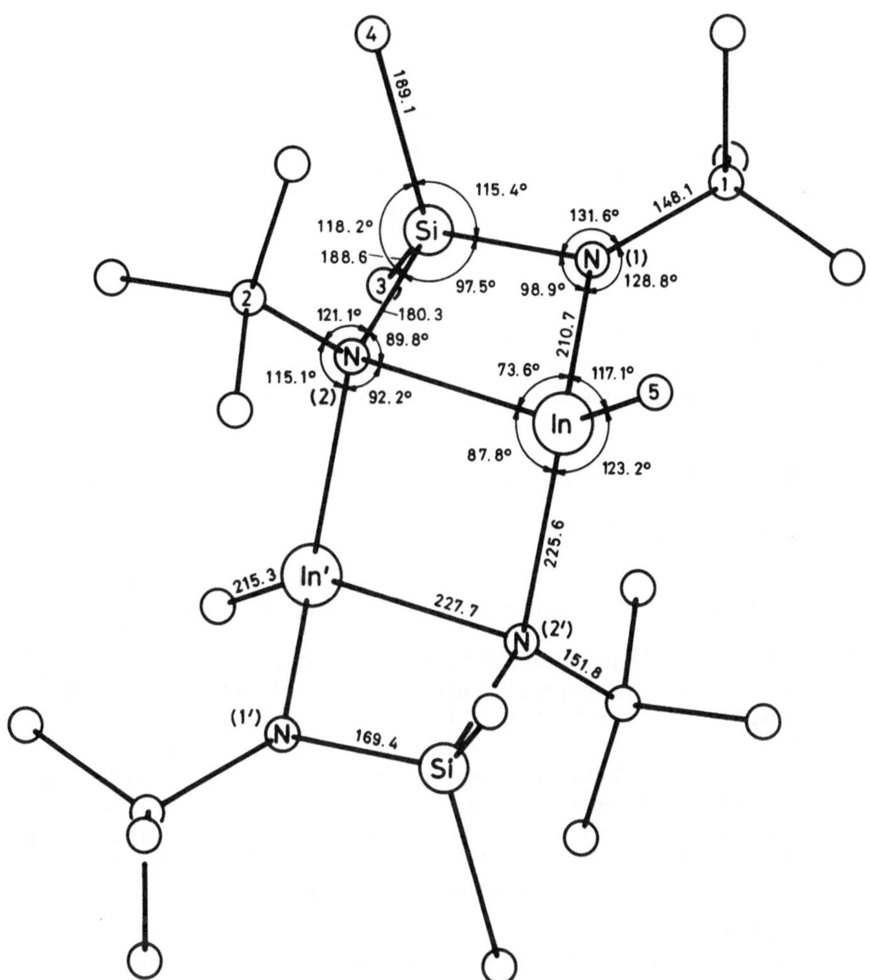

Fig. 63. Molecular structure of $\{CH_3In(NC_4H_9\text{-}t)_2Si(CH_3)_2\}_2$ [2].

Other bond angles (°)

N(1)-In-N(2')	113.6(1)	N(2)-Si-C(3)	111.0(2)
N(2)-In-C(5)	129.5(2)	C(3)-Si-C(4)	100.8(3)
N(1)-Si-C(3)	114.7(2)	In-N(2)-C(2)	119.4(2)
In'-N(2)-Si	113.1(1)		

LiCl + $In(CH_3)_3$ (L = $(N(C_4H_9\text{-}t)Si(CH_3)NC_4H_9\text{-}t)_2$); the yield of title compound is 95%. The product is a monomer in benzene and $[M-CH_3]^+$ is the heaviest fragment in the mass spectrum [3].

The proton resonance spectrum (C_7D_8, in ppm) is in agreement with the symmetrical structures I and II: 0.39 (s, CH_3In), 0.60 (s, CH_3Si), 1.29, 1.30 (s, CH_3C); the signal at 1.29 ppm is distinctly broader at −80 °C [3].

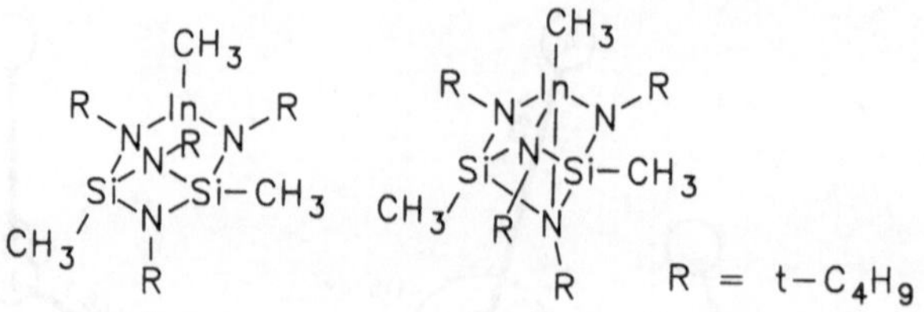

$$\text{I: } CH_3In[N(R)]_3(SiCH_3)_2N(R)\qquad \text{II}\qquad R = t\text{-}C_4H_9$$

References:

[1] Veith, M.; Lange, H.; Belo, A.; Recktenwald, O. (Chem. Ber. **118** [1985] 1600/5).
[2] Veith, M.; Lange, H.; Recktenwald, O. (J. Organometal. Chem. **294** [1985] 273/94).
[3] Veith, M.; Goffing, F.; Huch, V. (Z. Naturforsch. **43b** [1988] 846/56).

8 Organoindium-Phosphorus, -Arsenic, and -Antimony Compounds

8.1 Compounds of the Type $R_2InER'_2$

The compounds in this section are of the type $R_2InER'_2$ with E=P, As, and Sb and are collected in Table 61. The compounds were prepared by the following procedures.

Method I: Elimination of RH by thermal decomposition of $R_3In \cdot HER'_2$.
The adducts are usually not isolable pure (see Table 6, No. 37). Thus, the components were warmed for 1 to 3 h between 15 and 40 °C in a sealed, evacuated system (Nos. 2, 3, 15, 17), the alkane formed was pumped off, and the residue purified by recrystallization in benzene [2, 3]. Simpler variations of this procedure operated in open systems at 20 °C (Nos. 7 and 15), at 140 °C [8] or 170 °C [14] (No. 5, 24 or 6 h), or in aprotic solvents like pentane (No. 14, 55 °C, 3 d), hexane (No. 8, 60 °C), benzene (Nos. 9 and 14, 55 or 60 °C, 3 to 8 d), or ether (No. 6, 20 °C, 8 to 10 h). They were purified by recrystallization.

Method II: Reaction of R_2InX with KER'_2.
The synthesis was reported to be necessarily a multistep procedure [8, 14], but was spelled out only for Nos. 8 [21] and 10 [20]. R_2InBr in $O(C_2H_5)_2$ was stirred with $KP(C_6H_5)_2$ (1:1 mole ratio) for 24 h at room temperature; after removing the ether, the reaction mixture was extracted 5 times with pentane, and the collected pentane fractions concentrated and brought to crystallization at −20 °C. For No. 4 and 10 the Li salt was used (THF, −78 °C, toluene extraction) [20, 23].

Method III: From R_2InCl and $E\{Si(CH_3)_3\}_3$ (E=P, As).
The components (1:1 mole ratio) were reacted at room temperature in benzene (3 d). The solvent was removed and the residue recrystallized from pentane at −15 °C [22].

Table 61
Compounds with In-P, In-As, and In-Sb Bonds.
Further information on numbers preceded by an asterisk is given at the end of the table.
Explanations, abbreviations, and units on p. X.

No.	compound method of preparation (yield)	properties and remarks
1	$(CH_3)_2InPH_2$	not isolated, evidenced as intermediate in the decomposition reaction of $(CH_3)_3In \cdot PH_3$ (see also p. 44) [1, 10, 11]
2	$(CH_3)_2InP(CH_3)_2$ I (>95%)	colorless crystals, m.p. 115 to 120 °C (dec.), trimeric in benzene [3] IR (solid): 1425 m $\delta_{as}(CH_3)$, 1299 w and 1278 w $\delta_s(CH_3P)$, 1150 m $\delta_s(CH_3In)$, 946 s and 905 s $\varrho(CH_3P)$, 720 s $\nu(PC_2)$, 685 vs $\varrho(CH_3In)$, 493 s and 467 s $\nu(InC_2)$ [3]
3	$(CH_3)_2InP(C_2H_5)_2$ I (>95%)	colorless crystals, m.p. 100 to 104 °C (dec.), 125 to 129 °C, trimeric in benzene [3] ^1H NMR (C_6D_6): 0.12 (CH_3In), 1.09 (m, CH_3 of C_2H_5), 1.68 (q, CH_2P) [18] IR (solid): 1418 m $\delta(CH_2)$, 1261 w and 1242 w $\delta(CH_2P)$, 1149 m $\delta_s(CH_3In)$, 1040 m and 1028 m $\varrho(CH_2P)$, 995 w $\varrho(CH_3C)$, 972 w ν(C-C), 754 m and 738 m $\nu(PC_2)$, 683 vs $\varrho(CH_3In)$, 484 m and 463 m $\nu(InC_2)$ [3]
*4	$(CH_3)_2InP(C_3H_7\text{-}i)_2$ II (82%)	colorless crystals, m.p. 172 to 173 °C for the trimeric species; m.p. 159 to 160 °C [23] ^1H NMR (C_6D_6, trimer/dimer): 0.1/0.2 (s/t, CH_3In, J(P,H) = –/2.4), 1.2/1.1 (q, CH_3C, J(H,H) = 6.9), 2.2/2.1 (sept., CH) [23] ^{31}P NMR (C_6D_6, trimer/dimer): –36.3/-2.1 [23] mass spectrum: $[M]^+$, $[M-CH_3]^+$, $[M-2\,CH_3]^+$ (for M = trimer); $[M]^+$, $[M-CH_3]^+$, $[M-P(C_3H_7)_2]^+$ (for M = dimer) [23]
*5	$(CH_3)_2InP(C_4H_9\text{-}t)_2$ I (ca. 50%) [8] I (82%) [14]	colorless crystals, m.p. 280 to 300 °C (dec.), subl. 100 °C/10^{-2} Torr [8], subl. 160 °C/10^{-2} Torr, dec. 240 °C without melting [14] ^1H NMR (C_6D_6): 0.29 (t, CH_3In, ^3J(H,P) = 0.29), 1.30 (t, CH_3C, J(H,P) = 6.9) [8] ^{13}C NMR (C_6D_6): –1.77 (m, CH_3In), 33.09 (t, CH_3 of C_4H_9, J = 2), 35.66 (t, C of C_4H_9, J = 3) [8] ^{31}P NMR (C_6D_6): 39.24 [8] IR (solid): 1387 m, 1367 s, 1362 m, sh, 1344 w, sh, 1310 w, br, 1262 m, 1204 w, 1173 s $\delta_s(CH_3In)$, 1147 s, 1130 w, 1021 s, 1011 m, 975 vw, br, 953 w; 936 m, 933 m, sh, 930 to 910 m, vbr, 862 w, br, 819 m, 802 m, 717, 690 s $\delta(CH_3In)$, 657 m, br, 602 w, 578 w, sh, 534 w, 503 w $\nu_{as}(InC_2)$, 478 vs $\nu_s(InC_2)$, 444 m, 371 w [14]

Table 61 (continued)

No.	compound method of preparation (yield)	properties and remarks
*5 (continued)		mass spectrum (41 eV, 98 °C): $[M_2+H]^+$ (6%), $[M_2]^+$ (3%), $[M_2-CH_3]^+$ (94%), $[M_2-PCH_2]^+$ (10%), $[M+HP(C_4H_9)_2]^+$ (12%), $[M+In(CH_3)_2]^+$ (76%), $[M+In]^+$ (19%), $[M-CH_3]^+$ (27%), $[M-2\ CH_3]^+$ (10%), $[HP(C_4H_9)_2]^+$ (9), $[In(CH_3)_2]^+$ (18%), $^{115}In^+$ (65%) [8, 14]
*6	$(CH_3)_2InP(C_6H_5)_2$ I (81%) [14] special [7]	colorless crystals, m.p. 245 °C, darkens at 185 °C [14] ^{1}H NMR (C_6D_6): 0.18 (CH_3In), 6.8 to 7.7 (m's, C_6H_5), and 0.17 (CH_3In), 6.95 (m, C_6H_5), 7.39 (m, C_6H_5), respectively [14]; see also [18] ^{13}C NMR (C_6D_6): −1.1 (CH_3), 125 to 128 and 133.2 to 133.4 (m, C_6H_5) [14] ^{31}P NMR (C_6D_6): −53.1 [14], −56.6 [7] IR (solid): 1578 m, 1562 m, 1472 s, 1429 s, 1325 m, 1305 m, 1265 w, 1188 m, 1160 w $\delta_s(CH_3In)$, 1090 w, br, 1069 w, 1035 m, 1000 m, 970 w, 916 w, br, 845 w, 752 s, 745 s, 695 vs $\varrho(CH_3In)$, 508 s $\nu_{as}(InC_2)$, 496 m, 472 m $\nu_s(InC_2)$, 470 m, sh, 435 w, br, 380 w, 250 vw [14]; main data between 3100 and 500 cm^{-1} also given in [7] and [18] mass spectrum (165 °C): $[M_2+H]^+$ (0.2%), $[M_2-2\ C_6H_5-CH_3]^+$ (2%), $[(C_6H_5)_2PP(C_6H_5)_2]^+$ (100%), $[P(C_6H_5)_3]^+$ (7%), $[P(C_6H_5)_2]^+$ (100%), $[In(CH_3)_2]^+$ (52%), $[InCH_3]^+$ (6%), $^{115}In^+$ (30%), $[PC_6H_5]^+$ (92%) [14], see also [18]
7	$(C_2H_5)_2InP(C_2H_5)_2$ I	viscous liquid, trimeric in gas phase (mass spectrum) [4] ^{1}H NMR ($CDCl_3$): 0.63 (q, CH_2In), 1.22 (m, $CH_3(P)$, J(H,P) = 14.5), 1.29 (t, $CH_3(In)$, J(H,H) = 7.8), 1.83 (q, CH_2P, J(H,H) = 7.5) [4] ^{31}P NMR ($CDCl_3$): −74.3 [4] IR (film): 1418, 1242, 1160, 1040, 1025, 995, 970, 952, 919, 886, 754, 735, 620, 467, 450 [4] mass spectrum (180 to 200 °C): $[M_3-C_2H_5]^+$ (5%), $[M_2+P]^+$ (4%), $[M_2+P-C_2H_5]^+$ (2%), $[M_2]^+$ (7%), $[M_2-C_2H_5]^+$ (47%), $[M_2-3\ C_2H_5]^+$ (7%), $[M_2-P(C_2H_5)_2]^+$ (13%), $[M_2-P(C_2H_5)_3]^+$ (1%), $[M_2-P(C_2H_5)_3-C_2H_5]^+$ (11%), $[M+In-2\ C_2H_5]^+$ (9%), $[M-C_2H_5]^+$ (12%), $[M-2\ C_2H_5]^+$ (4%), $[In(C_2H_5)_3-H]^+$ (12%), $[In(C_2H_5)_2]^+$ (75%) [4]
*8	$(C_2H_5)_2InP(C_4H_9\text{-}t)_2$ I (ca. 100%)	colorless solid [15] ^{1}H NMR (C_6D_6): 1.04 (q, CH_2In), 1.58 (t, CH_3 of C_2H_5), 1.32 (2 t→"q", CH_3C, J(H,P) = 7) [15] ^{31}P NMR (C_6D_6): 44.8 [15] IR (solid): 728 ν(PC), 641, 620 $\varrho(CH_2In)$, 463 ν(InC); no other bands reported [15]

Table 61 (continued)

No.	compound method of preparation (yield)	properties and remarks
*9	$\{(C_4H_9\text{-t})CH_2\}_2InP(C_6H_5)_2$ I (69.3%)	white solid, m.p. 138 to 143 °C, forms a glass, dec. 143 to 150 °C; degree of association in C_6H_6: 2.01 (0.175 M) to 1.80 (0.077 M) [6] ^1H NMR (C_6D_6): 1.03 (s, CH_3), 1.10 (s, CH_3), 1.47 (t, CH_2, J = 2.4), ratio 1:3.5:1 for 0.04 M solution; 1.03, 1.07, and 1.47, ratio 1:4.9:1.2 for 0.02 M solution [6] ^{31}P NMR (C_6D_6, 0.28 M): −49.40 (dimer), −29.95 (monomer), 1:5.2 for 0.28 M; 1:5.4 for 0.02 M solution [6] IR (solid): 1580 m, 1567 w, 1478 s, sh, 1470 s, 1437 s, 1440 m, sh, 1385 m, 1365 s, 1354 s, 1297 m, 1270 w, br, 1230 s, 1205 w, sh, 1170 w, 1160 w, sh, 1150 w, 1125 vw, 1110 w, 1096 w, 1090 w, 1061 w, 1040 vw, 1020 w, 1009 w, 998 w, 965 w, br, 907 w, 890 w, 840 w, 765 w, br, 736 sh, 729 vs, 717 m, 687 vs, 616 s, 609 m, 568 m, 556 m, sh, 540 w, 497 m, 468 m, 445 w, 430 w, 375 vw, 345 w, 225 w [6]
10	$\{(C_4H_9\text{-t})CH_2\}_2InP\{Si(CH_3)_3\}_2$	
	II	no properties reported [20]
	III	methanolysis gives R_2InOCH_3 and $HP\{Si(CH_3)_3\}_2$ [20]
*11	$\{(CH_3)_3SiCH_2\}_2InP(H)C_4H_9\text{-t}$	
	I (38%)	colorless crystals, m.p. 146 °C [17] ^1H NMR (C_6D_6): −0.15 to 0.1 (m, CH_2In), 0.26 (s, CH_3Si), 1.20 (m, CH_3C) [17] ^{31}P NMR (C_6D_6): −55.6 (weak), −53.4 [17] ^{29}Si NMR (C_6D_6): 2.66 (t, ^3J(Si,P) = 3.2) [17] IR (solid): 1297 s, 1246 vs, 957 s, 854 vs, 828 s, 752 s, 721 s [17] mass spectrum: $[M-H]^+$ (1%), $[R_2In]^+$ (100%) [17]
*12	$\{(CH_3)_3SiCH_2\}_2InP(C_6H_5)_2$ I (83 to 94%) II (63%)	colorless crystals, m.p. 152 °C (dec.) [6] ^1H NMR (C_6D_6): 0.04 (s, CH_3Si), 0.13 (t, CH_2In, J = 1.8) [6] ^{13}C NMR (C_6D_6): −0.18 (t of t, CH_2In, J_1 = 117, J_2 = 8), 2.98 (q, CH_3Si, J(H,Si) = 119) [6] ^{31}P NMR (C_6D_6): −29.1 (monomer) in 0.01 M solution; −50.30 (dimer), −29.15 (monomer), 1:12.23 ratio in 0.44 M solution [6] IR (solid): 1581 m, 1480 s, 1432 s, 1353 m, 1349 m, 1322 vw, 1300 w, 1252 m, 1240 vs, 1180 w, 1155 w, 1092 m, 1065 w, 1024 m, 967 s, 960 s, sh, 955 s, 940 s, 927 m, 910 w, 905 vw, sh, 845 vs, 837 vs, 821 vs 740 s, sh, 734 vs, 712 s, 689 s, 681 m, sh, 607 vw, 570 m, 552 m, 500 m, 489 m, 471 m, 440 w, 345 vw, 270 w [6]

Table 61 (continued)

No.	compound method of preparation (yield)	properties and remarks
13	$\{(CH_3)_3SiCH_2\}_2InP\{Si(CH_3)_3\}_2$ II III	m.p. 244 to 248 °C [20] ^1H NMR (C_6D_6): 0.19 (s, CH_2), 0.31 (s, CH_3SiC), 0.45 (t, CH_3SiP, J(P,H) = 2.5) [20] ^{13}C NMR (C_6D_6): 3.7, 4.1, 5.3 (t, J(P,C) = 4) [20] methanolysis gives R_2InOCH_3 and $HP\{Si(CH_3)_3\}_2$ [20]
*14	$(CH_3)_2InAs(CH_3)_2$ I (>95%)	glassy white solid, m.p. 84 °C (dec.); trimeric in benzene [3] IR (solid): 1420 m $\delta_{as}(CH_3)$, 1252 w and 1245 w $\delta_s(CH_3As)$, 1146 m $\delta_s(CH_3In)$, 897 s and 862 s $\varrho(CH_3As)$, 685 vs $\varrho(CH_3In)$, 584 m and 569 m $\nu(AsC_2)$, 492 s and 462 s $\nu(InC_2)$ [3]
*15	$(CH_3)_2InAs(C_4H_9\text{-}t)_2$ I	colorless crystals, m.p. 250 to 252 °C (dec.) [8] ^1H NMR (C_6D_6): 0.33 (s, CH_3In), 1.36 (s, CH_3C) [8] ^{13}C NMR (C_6D_6): −3.40 (s, CH_3In), 33.46 (s, CH_3), 39.32 (s, C of C_4H_9) [8] IR (solid): 1160 w, br, 860 s, 520 w, no other bands reported [8] mass spectrum: $[M_2-CH_3]^+$, M^+, $[M-CH_3]^+$, $[AsC_8H_{19}]^+$, $[In(CH_3)_2]^+$, $[AsC_4H_{10}]^+$, $[InCH_3]^+$, $^{115}In^+$ [8]
16	$(CH_3)_2InAs(C_6H_5)_2$ I	white solid, m.p. 192 to 195 °C (dec.); dimeric in benzene, not hygroscopic, air-stable [2]
*17	$\{(CH_3)_3SiCH_2\}_2InAs\{Si(CH_3)_3\}_2$ III (75%)	colorless crystals, decomposes above 210 °C to a black solid [22] ^1H NMR (C_6D_6): 0.24 (s, CH_2), 0.32, 0.49 (s's, CH_3Si) [22] ^{13}C NMR (C_6D_6): 3.60, 5.32 (CH_3Si), 3.91 (CH_2) [22]
*18	$(CH_3)_2InSb(C_4H_9\text{-}t)_2$ special	pale orange crystals, m.p. 139 to 141 °C [16] ^1H NMR (C_6D_6): 0.28 (CH_3In), 1.52 (CH_3C) [16] ^{13}C NMR (C_6D_6): 34.77 (C of C_4H_9), 35.12 (CH_3 of C_4H_9), CH_3In not observed [16]

* Further information:

$(CH_3)_2InP(C_2H_5)_2$ (Table **61**, No. **3**) was also prepared by slow addition of $HP(C_2H_5)_2$ to a solution of $(CH_3)_2InN(C_2H_5)_2$ in toluene with removal of $HN(C_2H_5)_2$. After the mixture was stirred for 1 d at room temperature followed by 6 h at reflux temperature all volatile material was evaporated and the glassy residue dried in vacuum for 2 h [18].

The mass spectrum exhibits the following fragments (100 °C source temperature, relative intensities in %): $[M_3-CH_3]^+$ (5), $[M_3-6\,CH_3]^+$ (2), $[M_3-4\,C_2H_5-4\,CH_3]^+$ (7), $[M_2]^+$ (2), $[M_2-CH_3]^+$ (30), $[M_2-3\,CH_3]^+$ (3), $[M_2-C_2H_5-3\,CH_3]^+$ (4), $[M_2-P(C_2H_5)_2]^+$ (26), $[M_2-P(C_2H_5)_2-2\,CH_3]^+$ (8), $[M_2-P(C_2H_5)_3-3\,CH_3]^+$ (2), $[InP_2(C_2H_5)_4]^+$ (3), $[M]^+$ (2),

$[M-CH_3]^+$ (33), $[M-2\,CH_3]^+$ (7), $[P_2(C_2H_5)_4]^+$ (3), $[P_2(C_2H_5)_3]^+$ (6), $[In(CH_3)_2]^+$ (38), $[^{115}In]^+$ (100) [18].

$(CH_3)_2InP(C_3H_7\text{-}i)_2$ (Table **61**, No. **4**) is trimeric when prepared according to Method II. On vacuum sublimation at 110 °C/10^{-2} Torr it can be quantitatively converted into the dimeric species [23].

$(CH_3)_2InP(C_4H_9\text{-}t)_2$ (Table **61**, No. **5**) was also obtained in high yields from $In(CH_3)_3$ and $In(P(C_4H_9\text{-}t)_2)_3$ (2:1 mole ratio) in toluene; however, the procedure is described in detail only for the corresponding Ga compound [8].

Vapor pressure determinations by the Knudsen effusion method (apparatus illustrated) led to the equation: log p (Torr) = 14.2 − 6790/T (K), giving $\Delta H_v = 130 \pm 0.9$ kJ/mol and $\Delta S_v = 216 \pm 1.6\ J \cdot K^{-1} \cdot mol^{-1}$ [9].

The compound crystallizes in the monoclinic space group $C2/c-C^6_{2h}$ (No. 15) with the cell parameters a = 1384.7(4), b = 1286.0(4), c = 1540.8(1) pm, β = 103.59(1)°; Z = 8 (4 dimers) and $D_c = 1.16$ g/cm^3. Only those distances and angles shown in **Fig. 64** are reported in [14].

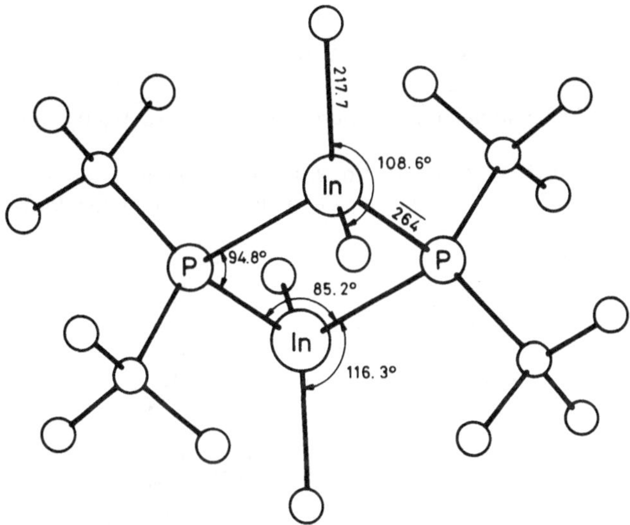

Fig. 64. Molecular structure of $(CH_3)_2InP(C_4H_9\text{-}t)_2$ [14].

The usefulness of the compound as a two-component source for the preparation of InP semiconductor layers has been investigated [12, 19].

$(CH_3)_2InP(C_6H_5)_2$ (Table **61**, No. **6**) was obtained from $(CH_3)_2InCl$ and an equimolar amount of $(CH_3)_3SiP(C_6H_5)_2$ in toluene at room temperature with the elimination of $(CH_3)_3SiCl$. After concentrating the solution and cooling to −20 °C, the crystallized product was filtered off (72% yield) [7]. A good yield of the title compound was also obtained from the reaction between $(CH_3)_2InN(C_2H_5)_2$ with $HP(C_6H_5)_2$ in boiling toluene ($HN(C_2H_5)_2$ elimination) [18]. The compound, dimeric in benzene, is not split into monomers by Lewis bases; stoichiometric amounts of water lead to the formation of CH_4, $HP(C_6H_5)_2$, and insoluble $(CH_3InO)_n$. Reaction with C_6H_5NO gave $\{(CH_3)_2InOOP(C_6H_5)_2\}_2$ [7].

$(C_2H_5)_2InP(C_4H_9\text{-t})_2$ (Table **61**, No. **8**) is monoclinic, space group $C2/m-C^3_{2h}$ (No. 12), with the parameters a = 1486.9(3), b = 1160.5(2), c = 900.3(1) pm, β = 91.38(2)°; Z = 4 (2 dimers) and D_c = 1.361 g/cm³. The distances and angles reported are presented in **Fig. 65** [15].

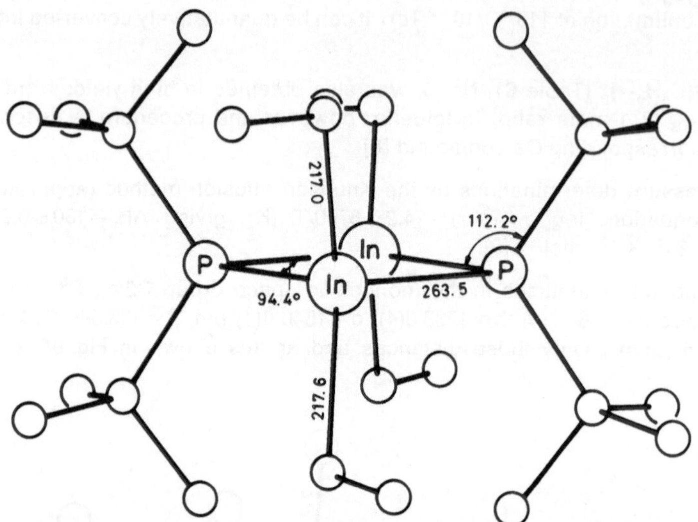

Fig. 65. Molecular structure of $(C_2H_5)_2InP(C_4H_9\text{-t})_2$ [15].

$\{(C_4H_9\text{-t})CH_2\}_2InP(C_6H_5)_2$ (Table **61**, No. **9**) forms a monomer-dimer equilibrium mixture in solution, as indicated by a concentration dependence of the molecular weight; it is trimeric in the solid state [21].

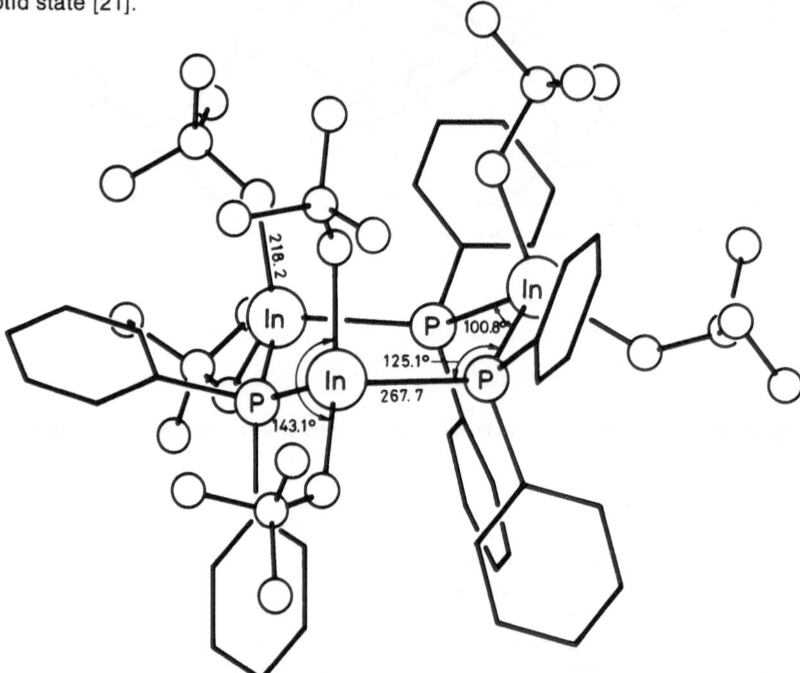

Fig. 66. Molecular structure of $\{((C_4H_9\text{-t})CH_2)_2InP(C_6H_5)_2\}_3$ [21].

It crystallizes in the rhombohedral space group $R\bar{3}-C_{3i}^2$ (No. 148) with the lattice constants a = b = 2087.3(5), c = 2903.7(4) pm, $\alpha = \beta = 90°$, $\gamma = 120°$; Z = 6 (trimers) and $D_c = 1.21$ g/cm^3. The In_3P_3 core of the trimeric compound has a chair conformation. Both In and P atoms are in a distorted tetrahedral environment as shown in **Fig. 66** on p. 318 [21].

$\{(CH_3)_3SiCH_2\}_2InPHC_4H_9$-t (Table **61**, No. **11**) crystallizes in the monoclinic space group $C2/c-C_{2h}^6$ (No. 15) with the lattice constants a = 1579.6(3), b = 1289.0(2), c = 2137.5(4) pm, $\beta = 109.91(2)°$; Z = 8 (4 dimers) gives $D_c = 1.23$ g/cm^3. The dimeric compound is centrosymmetric and has a planar In_2P_2 ring with the C_4H_9 groups in trans positions. Important bond distances and angles are shown in **Fig. 67** [17].

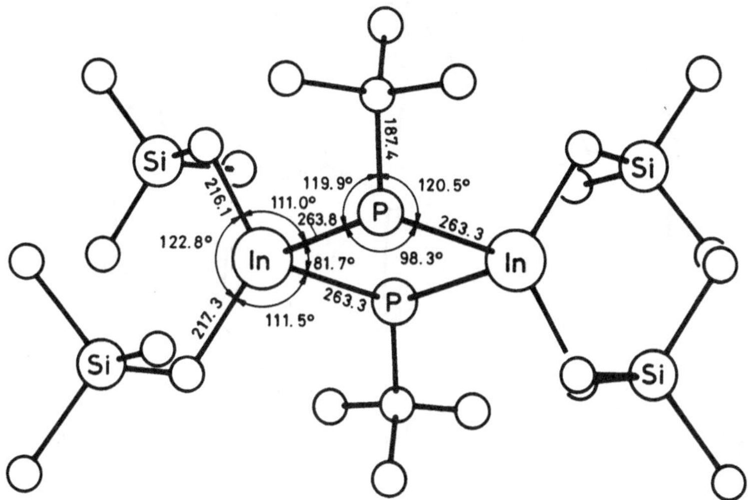

Fig. 67. Molecular structure of $\{((CH_3)_3SiCH_2)_2InPHC_4H_9\text{-t}\}_2$ [17].

$\{(CH_3)_3SiCH_2\}_2InP(C_6H_5)_2$ (Table **61**, No. **12**) exhibits, according to cryoscopic molecular weight determinations, concentration-dependent association; the degree of association is between 1.58 and 1.29 (for 0.1414 to 0.0565 molal solutions) and shows partial dissociation into monomers. In the crystalline state, however, only dimers are evident [6].

The compound crystallizes in the triclinic space group $P\bar{1}-C_i^1$ (No. 2) with the lattice constants a = 1032.3(4), b = 1111.3(5), c = 2150.9(8) pm, $\alpha = 83.85(5)°$, $\beta = 86.66(6)°$, $\gamma = 83.27(5)°$; Z = 4 (2 dimers) and $D_c = 1.29$ g/cm^3. The unit cell contains two somewhat different dimers. In molecule A (**Fig. 68**) the In···In′ distance is 396.70 and the P···P′ distance is 352.4 pm, while the corresponding distances in B are 399.92 and 346.4 pm. The most notable differences lie in the bond angles as shown on p. 320 [6].

$(CH_3)_2InAs(CH_3)_2$ (Table **61**, No. **14**) crystallizes in the triclinic space group $P\bar{1}-C_i^1$ (No. 2) with the lattice constants a = 1060.1(3), b = 1392.1(3), c = 1796.1(4) pm, $\alpha = 76.07(2)°$, $\beta = 85.38(2)°$, $\gamma = 76.48(2)°$; Z = 12 (4 trimers) and $D_c = 1.991$ g/cm^3. The most remarkable feature of the structure is the presence of two independent molecules in the unit cell. One trimer (**Fig. 69** (A) on p. 321) is planar to a large extent (the deviations of the In_3As_3 ring atoms

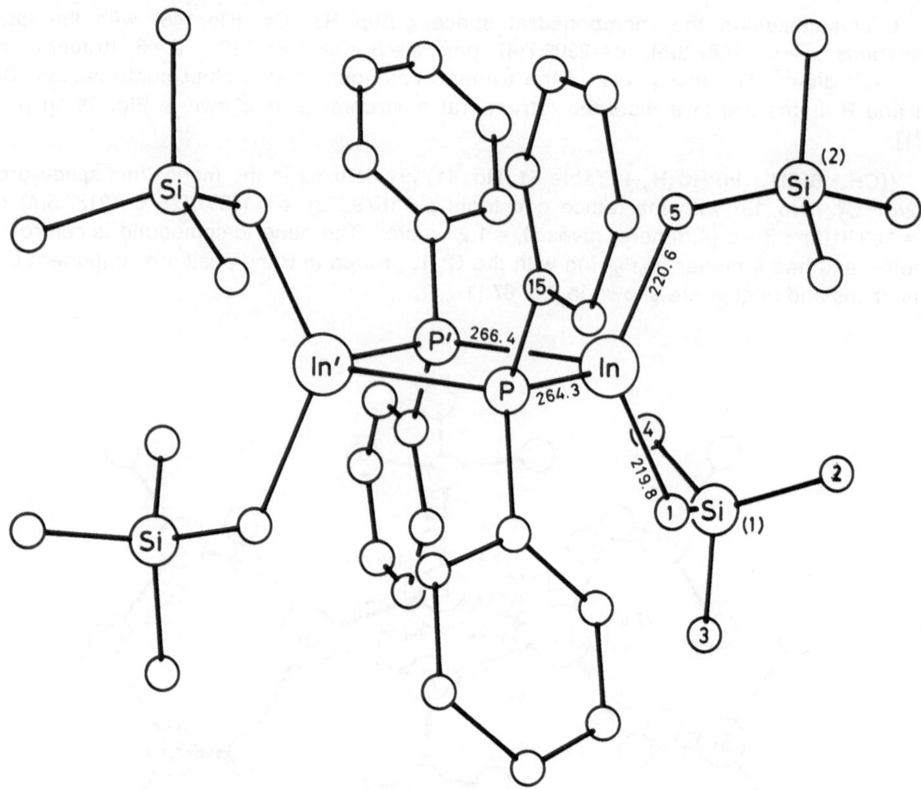

Fig. 68. Molecular structure of $\{((CH_3)_3SiCH_2)_2InP(C_6H_5)_2\}_2$, molecule A [6].

Bond angles (°)	Molecule A	Molecule B
P-In-P′	83.23(7)	81.80(7)
In-P-In′	96.77(7)	98.20(8)
P′-In-C(1)	110.1(2)	112.3(2)
P-In-C(5)	104.9(2)	106.1(2)
P′-In-C(5)	109.6(2)	106.0(2)
C(9)-P-C(15)	104.3(4)	105.0(4)
C(1)-In-C(5)	127.8(3)	125.9(3)
C(1)-Si(1)-C(2)	109.8(5)	112.1(5)
In-C(1)-Si(1)	117.3(4)	118.9(4)
In-C(5)-Si(2)	117.0(4)	115.8(4)

from a plane is 9.7(4) to 13.9(3) pm). The other entity (**Fig. 69** (B)) has a puckered In_3As_3 ring; the three In atoms and the As(3) atom are coplanar, but As(2) is 103.7 pm above, and As(1) is 125.2 pm below this plane. Thus, it has neither the classic chair nor the boat conformation [13].

Since no splitting of the ^1H NMR signals is observed down to −80 °C, either the planar form exists alone in solution or the conformation change is too rapid, even at low temperatures, in the NMR time scale [13].

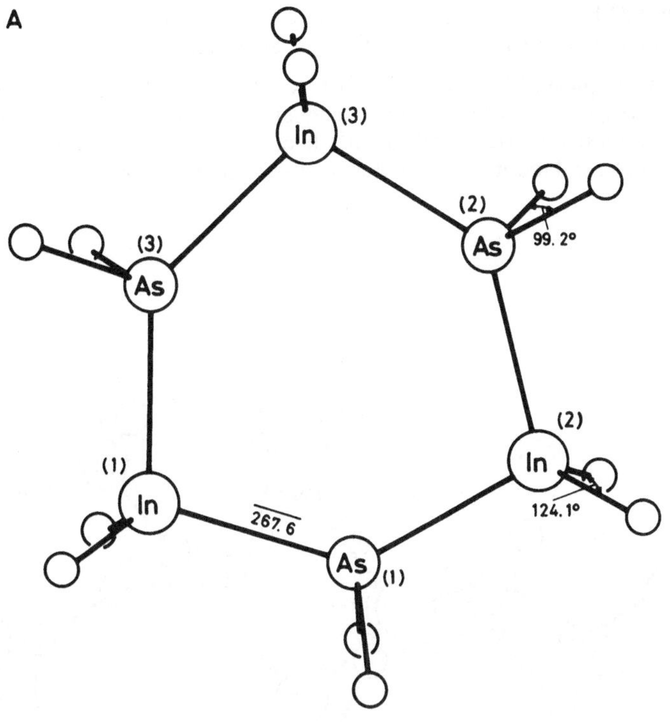

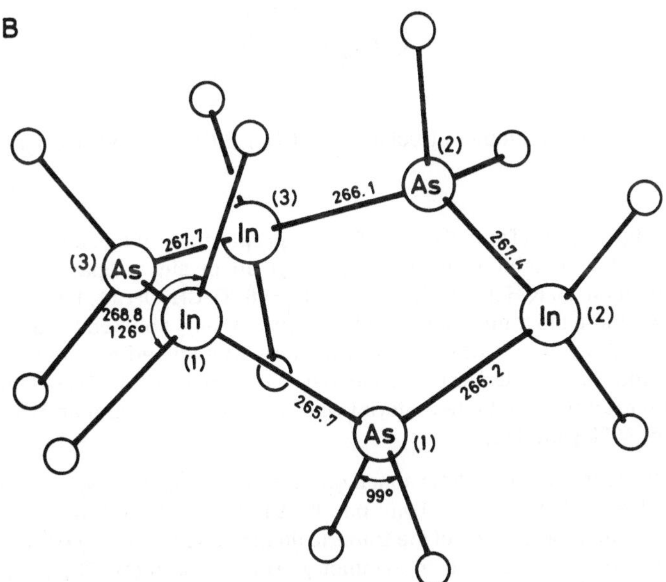

Fig. 69. Molecular structures of the two independent molecules, A and B, of $\{(CH_3)_2InAs(CH_3)_2\}_3$ [13].

$(CH_3)_2InAs(C_4H_9\text{-t})_2$ (Table **61**, No. **15**) crystallizes in the monoclinic space group C2/c with 4 dimeric molecules in the unit cell and is isostructural with the Ga-homolog [5].

$\{(CH_3)_3SiCH_2\}_2InAs\{Si(CH_3)_3\}_2$ (Table **61**, No. **17**) crystallizes in the orthorhombic space group Pbcn$-D_{2h}^{14}$ (No. 60) with the unit cell constants a = 1292.0(1), b = 2161.8(2), c = 1879.7(2) pm; Z = 4 (dimers) and D_c = 1.292 g/cm^3. It contains a planar four-membered ring with slightly unequal bond lengths and angles as depicted in **Fig. 70** [22].

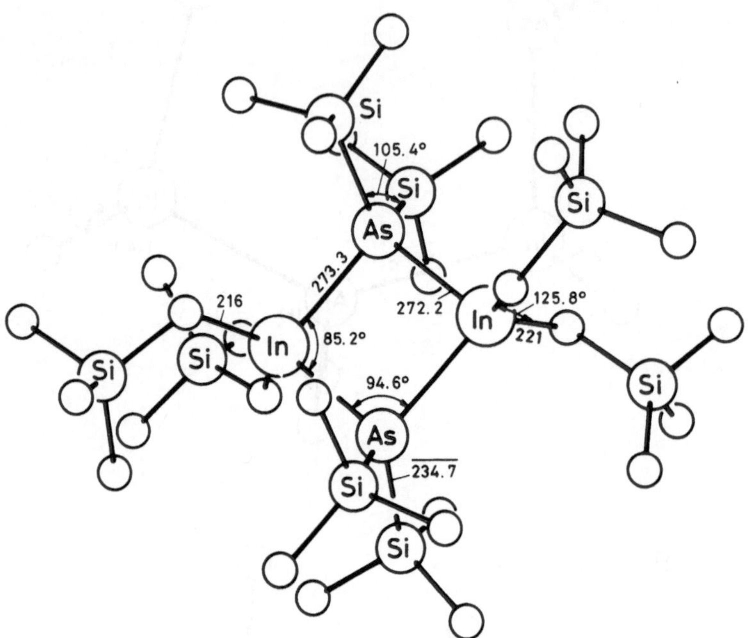

Fig. 70. Molecular structure of $(\{(CH_3)_3SiCH_2\}_2InAs\{Si(CH_3)_3\}_2)_2$ [22].

$(CH_3)_2InSb(C_4H_9\text{-t})_2$ (Table **61**, No. **18**). The preparation of this compound follows Method II, but due to the instability of $Li[Sb(C_4H_9\text{-t})_2]$ an in situ preparation is required. Thus, to a THF solution of $(CH_3)_3SiSb(C_4H_9\text{-t})_2$ at −78 °C CH_3OH (1:1 mole ratio) was added. After brief warming (15 min) of the resulting $HSb(C_4H_9\text{-t})_2$ solution it was recooled to −78 °C and treated with $LiCH_3$ (in hexane). The mixture was stirred at room temperature, cooled to −78 °C, and then treated with a suspension of $(CH_3)_2InCl$ in THF. After stirring for 1.5 h at room temperature all volatile material was removed in vacuum and the residue extracted with hexane; 46% yield [16].

It crystallizes in the monoclinic space group $P2_1/c-C_{2h}^5$ (No. 14) with the lattice constants a = 963.3(3), b = 2228.4(6), c = 2069.0(6) pm, β = 91.75°; Z = 12 (4 trimeric units) and D_c = 1.71 g/cm^3. The six-membered ring of the trimeric (in the solid state) compound adopts a distorted twist-boat conformation devoid of symmetry. The atoms In(1), Sb(2), Sb(3), and In(3) are approximately in one plane with atoms In(2) and Sb(1) about 49.0 and 124.3 pm, respectively, out of this plane. The molecular structure with important bond distances and angles is depicted in **Fig. 71** [16].

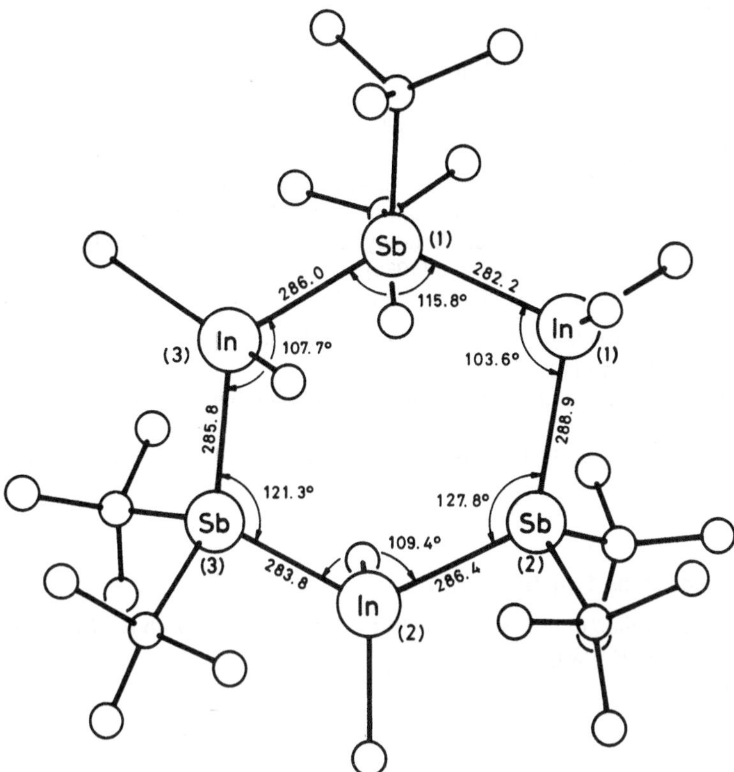

Fig. 71. Molecular structure of $\{(CH_3)_2InSb(C_4H_9\text{-}t)_2\}_3$ [16].

References:

[1] Didchenko, R.; Alix, J. E.; Toeniskoetter, R. H. (J. Inorg. Nucl. Chem. **14** [1960] 35/7).

[2] Coates, G. E.; Graham, J. (J. Chem. Soc. **1963** 233/7).

[3] Beachley, O. T.; Coates, G. E. (J. Chem. Soc. **1965** 3241/7).

[4] Maury, F.; Constant, G. (Polyhedron **3** [1984] 581/4).

[5] Arif, A. M.; Benac, B. L.; Cowley, A. H.; Geerts, R.; Jones, R. A.; Kidd, K. B.; Power, J. M.; Schwab, S. T. (J. Chem. Soc. Chem. Commun. **1986** 1543/5).

[6] Beachley, O. T., Jr.; Kopasz, J. P.; Zhang, H.; Hunter, W. E.; Atwood, J. L. (J. Organometal. Chem. **325** [1987] 69/81).

[7] Arif, A. M.; Barron, A. R. (Polyhedron **7** [1988] 2091/4).

[8] Arif, A. M.; Benac, B. L.; Cowley, A. H.; Jones, R. A.; Kidd, K. B.; Nunn, C. M. (New J. Chem. **12** [1988] 553/7).

[9] Bradley, D. C.; Faktor, M. M.; Frigo, D. M. (Chemtronics **3** [1988] 50/3).

[10] Buchan, N. I.; Larsen, C. A.; Stringfellow, G. B. (J. Cryst. Growth **92** [1988] 591/604).

[11] Buchan, N. I.; Larsen, C. A.; Stringfellow, G. B. (J. Cryst. Growth **92** [1988] 605/15).

[12] Cowley, A. H.; Benac, B. L.; Ekerdt, J. G.; Jones, R. A.; Kidd, K. B.; Lee, J. Y.; Miller, J. E. (J. Am. Chem. Soc. **110** [1988] 6248/9).

[13] Cowley, A. H.; Jones, R. A.; Kidd, K. B.; Nunn, C. M.; Westmoreland, D. L. (J. Organometal. Chem. **341** [1988] C 1/C 5).

[14] Aitchison, K. A.; Backer-Dirks, J. D. J.; Bradley, D. C.; Faktor, M. M.; Frigo, D. M.; Hursthouse, M. B.; Hussain, B.; Short, R. L. (J. Organometal. Chem. **366** [1989] 11/23).

[15] Alcock, N. W.; Degnan, L. A.; Wallbridge, M. G. H.; Powell, H. R.; McPartlin, M.; Sheldrick, G. M. (J. Organometal. Chem. **361** [1989] C 33/C 36).
[16] Cowley, A. H.; Jones, R. A.; Nunn, C. M.; Westmoreland, D. L. (Chem. Mater. **2** [1990] 221/2).
[17] Dembowski, U.; Noltemeyer, M.; Rockensüß, W.; Stuke, M.; Roesky, H. W. (Chem. Ber. **123** [1990] 2335/6).
[18] Rossetto, G.; Ajo, D.; Brianese, N.; Casellato, U.; Ossola, F.; Porchia, M.; Vittadini, A.; Zanella, P.; Graziani, R. (Inorg. Chim. Acta **170** [1990] 95/101).
[19] Andrews, D. A.; Davies, G. J.; Bradley, D. C.; Faktor, M. M.; Frigo, D. M.; White, E. A. D. (Semicond. Sci. Technol. **3** [1988] 1053/6).
[20] Douglas, T.; Theopold, K. H. (Inorg. Chem. **30** [1991] 594/6).

[21] Banks, M. A.; Beachley, O. T., Jr.; Buttrey, L. A.; Churchill, M. R.; Fettinger, J. C. (Organometallics **10** [1991] 1901/6).
[22] Wells, R. L.; Leonidas, J. J.; McPhail, A. T.; Alvanipour, A. (Organometallics **10** [1991] 2345/8).
[23] Cowley, A. H.; Jones, R. A.; Mardones, M. A.; Nunn, C. M. (Organometallics **10** [1991] 1635/7).

8.2 Compounds of the Type RIn(Cl)ER′$_2$

Only two compounds of this type have been described, one with $R = C_5(CH_3)_5$ (σ-bonded ligand) and one with $R = CH_2Si(CH_3)_3$.

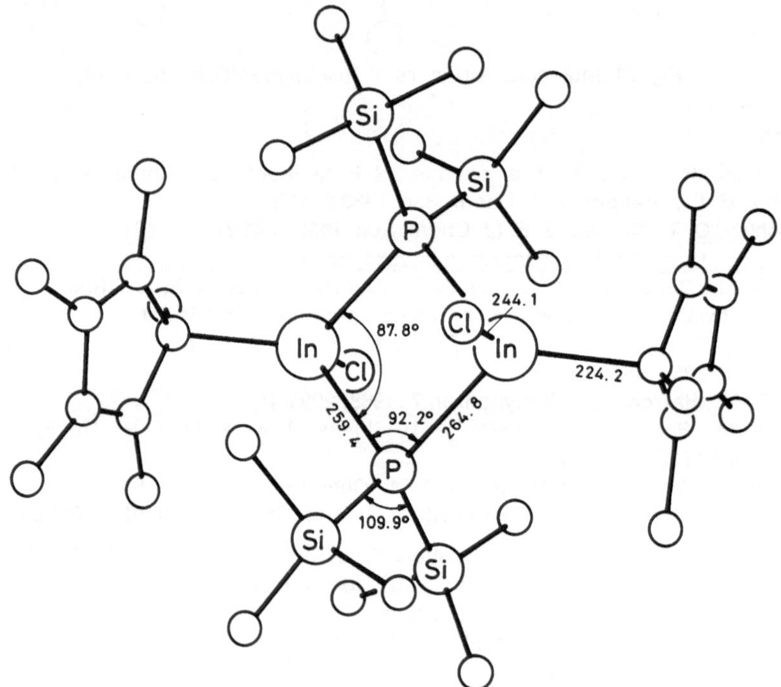

Fig. 72. Molecular structure of $\{C_5(CH_3)_5In(Cl)P\{Si(CH_3)_3\}_2\}_2$ [1].

$C_5(CH_3)_5In(Cl)P\{Si(CH_3)_3\}_2$

This compound was prepared by combining $C_5(CH_3)_5InCl_2$ and $Li[P\{Si(CH_3)_3\}_2]$ or $P\{Si(CH_3)_3\}_3$ in a 1:1 mole ratio; m.p. 126 to 135 °C (dec.) [1].

The NMR spectra were measured in C_6D_6 and the signals recorded in ppm. ^1H NMR: 0.51 (t, CH_3Si, J(P,H) = 3.0), 1.97 (s, CH_3C). ^{13}C NMR: 4.7 (s, CH_3Si), 13.4 (s, CH_3C), 121.7 (s, C_5). ^{31}P NMR: −148.6. Rapid rotation of the $C_5(CH_3)_5$ ring on the NMR time scale occurs in solution [1].

The following IR bands were reported (in cm^{-1}): 2955 m, 2901 m, 2852 m, 1442 w, 1406 w, 1375 w, 1282 m, 1261 m, 1246 s, 1145 w, 1039 w, 943 w, 848 s, 825 s, 752 w, 688 w, 626 m, 453 m [1].

It crystallizes (from pentane) in the monoclinic space group $P2_1/n-C_{2h}^2$ (No. 11) with the cell parameters a = 972.2(3), b = 1807.3(5), c = 1298.3(11) pm, β = 97.9(3)°; Z = 4. The dimeric molecule exhibits crystallographic inversion symmetry with a nearly perfect In_2P_2 square, as shown in **Fig. 72** on p. 324 [1].

Methanolysis at −70 °C gives hydrocarbons and the ether $CH_3OSi(CH_3)_3$ [1].

$(CH_3)_3SiCH_2In(Cl)P\{Si(CH_3)_3\}_2$

This compound was prepared analogously to the preceding compound, using $(CH_3)_3$-$SiCH_2InCl_2$ and $Li[P\{Si(CH_3)_3\}_2]$ or $P\{Si(CH_3)_3\}_3$; no physical data were reported [1].

Reference:

[1] Douglas, T.; Theopold, K. H. (Inorg. Chem. **30** [1991] 594/6).

8.3 Compounds of the Type $R_2InER'_2 \cdot InR_2Cl$

Only one compound of this type has been described for E = As; see also a similar compound with E = N in Section 7.3.1, p. 301.

$\{(CH_3)_3SiCH_2\}_2InAs\{Si(CH_3)_3\}_2In(Cl)\{CH_2Si(CH_3)_3\}_2$

This compound was prepared analogously to Method III in Chapter 8.1, p. 312, but the starting R_2InCl and the AsR'_3 compounds ($R = CH_2Si(CH_3)_3$, $R' = Si(CH_3)_3$) were used in a 2:1 molar ratio. Removal of all volatile material and dissolution of the solid in pentane followed by cooling to −15 °C gave colorless crystals in 16% yield; m.p. 68 to 71 °C. It is unstable at room temperature and decomposes to an orange-brown solid within a few hours [1].

The NMR spectra were run in C_6D_6 solution (δ values in ppm). ^1H NMR: 0.27 (s, 44 H, $CH_2Si(CH_3)_3$), 0.39 (s, CH_3SiAs). ^{13}C NMR: 2.83, 4.69 (CH_3), 7.75 (CH_2) [1].

The compound crystallizes in the monoclinic space group $C2/c-C_{2h}^6$ (No. 15) with the lattice constants a = 1923.3(3), b = 1079.8(2), c = 2081.9(3) pm, β = 105.91(1)°; Z = 4 and D_c = 1.334 g/cm^3. The Cl and As atoms lie on a crystallographic C_2 symmetry axis and the In_2AsCl ring is strictly planar. Important bond angles and distances are given in **Fig. 73** [1].

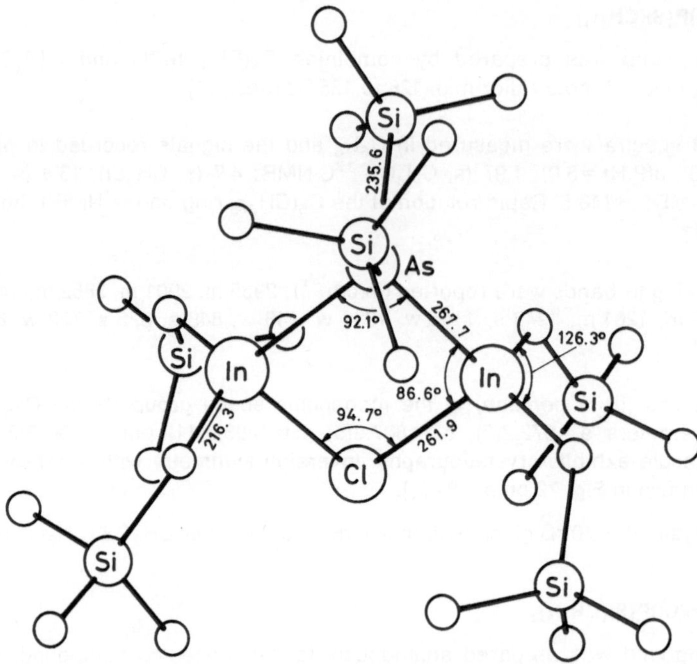

Fig. 73. Molecular structure of $\{(CH_3)_3SiCH_2\}_2InAs\{Si(CH_3)_3\}_2In(Cl)\{CH_2Si(CH_3)_3\}_2$ [1].

Reference:

[1] Wells, R. L.; Leonidas, J. J.; McPhail, A. T.; Alvanipour, A. (Organometallics **10** [1991] 2345/8).

9 Organoindium-Boron Compounds

9.1 Borane Derivatives

$(CH_3)_2InB_3H_8$

The reaction of $(CH_3)_2InCl$ with $Tl[B_3H_8]$ in monoglyme forms the title compound as shiny white crystals that decompose rapidly at room temperature [3, 4].

The ^{11}B NMR spectra in CD_2Cl_2 at different temperatures indicate fluxional behavior for the B_3H_8 group: at −43 °C a broad signal (w1/2 = 157 Hz) is δ = 29.6 ppm upfield from $F_3B \cdot O(C_2H_5)_2$; at −30 °C the resonance sharpened (w1/2 = 124 Hz) and individual signals began to emerge; at −14 °C the spectrum had the appearance of that for free $B_3H_8^-$ (seven peaks of the nonet at δ = 30.8 ppm, J = 31 Hz) [3, 4].

$[In(CH_3)_2][In(CH_3)_2B_{10}H_{12}]$

Benzene solutions of $In(CH_3)_3$ and $B_{10}H_{14}$ were mixed in 2:1 mole ratio. Over a few hours CH_4 evolved slowly and a light yellow air-sensitive solid separated. Filtration, washing with small portions of C_6H_6, and drying in vacuum gave the product in 96% yield [2].

The molar conductivity in acetone (0.1 mmol/L) is 136 $\Omega^{-1} \cdot cm^2 \cdot mol^{-1}$, which establishes the ionic character of the complex [2].

The ^1H NMR spectrum in CD_3COCD_3 is illustrated. It shows a singlet of the cation at 0.07 ppm; the splitting of the $(CH_3)_2In$ resonance of the anion into two separate signals (−0.4 and −0.73 ppm) is explained by the different environments of these methyl groups (see Formula I). The signal at −5.3 ppm is broad and represents all of the B-H components together [2].

^{11}B NMR spectrum in CD_3COCD_3: −7.6, −3.8, 3.5, 7.2, 32.0 (J = 140 Hz) ppm. The relative intensities of these peaks are 2:1:3:2:2. Because of broad line widths, an ^{11}B-^1H coupled spectrum delivered no additional information. Both spectra are illustrated [2].

IR spectrum (solid?): 2540 sh, 2511 vs, 2360 vs, br ν(B-H, terminal), 1880 w ν(B-H-B, bridge), 730 vs, br ϱ(CH_3), 570 s $\nu_{as}(InC_2$, cation), 519 vs $\nu_{as}(InC_2$, anion), 490 s $\nu_s(InC_2$, anion); other bands at 1150 s, 1084 ms, 1055 s, 1008 s, 997 s, 931 w, 915 w, 870 w, 821 w, 789 sh, 760 sh, 635 w cm^{-1} [2]; see also [4].

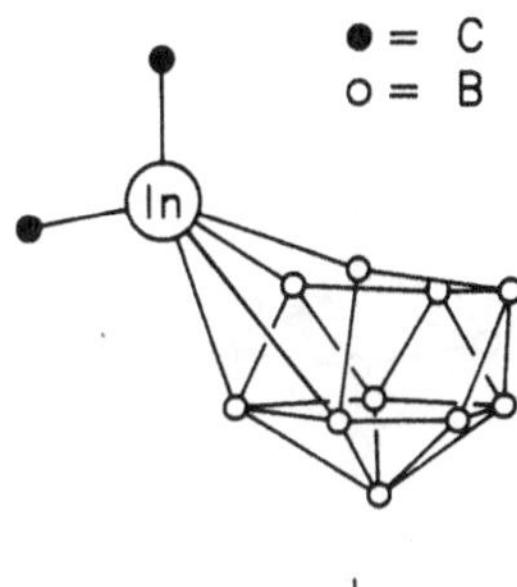

I

$CH_3InB_{10}H_{12}$

When $In(CH_3)_3$ and $B_{10}H_{14}$ are mixed in 1:1 mole ratio, the title compound can be isolated as a pale yellow solid in about 9% yield after separation of the above ionic compound (along with excess $B_{10}H_{14}$) from the filtrate. Its molar conductivity in acetone is 27 $\Omega^{-1} \cdot cm^2 \cdot mol^{-1}$, which testifies to a covalent character [2].

^1H NMR spectrum (δ values in ppm): 0.09 in C_6D_6, 0.10 in $OC(CD_3)_2$ for the CH_3In protons; a very broad unresolved peak was recorded for $B_{10}H_{12}$ [2].

^{11}B NMR spectrum (in $OC(CD_3)_2$, spectrum illustrated, δ values in ppm): −5.5, −1.3, 4.2, 11.2, and 31.9 (J = 140 Hz) with relative intensities of 2:2:3:1:2 [2].

IR spectrum (solid): 2520 s, br ν(B-H, terminal), 1890 w ν(B-H-B), 721 s, br ϱ(CH_3In), 528 w ν(InC), other unassigned bands at 1009 m, 885 w cm^{-1} [2].

9.2 Carborane Derivatives

$CH_3InC_2B_4H_6$

The compound was obtained within 8 h by the reaction of $In(CH_3)_3$ with $C_2B_4H_8$-2,3 (1.05:1 mole ratio) in a Carius tube at 100 °C. After removing the volatile $B(CH_3)_3$, the gray residue was vacuum sublimed at 100 °C; 60% yield. The compound quite probably has structure I of the analogous Ga compound [1].

The NMR spectra were run in $CDCl_3$ (δ values in ppm): ^{1}H NMR: 0.25 (CH_3In), 6.40 (HC); no signal was reported for the (usually broad and poorly-resolved) HB protons. ^{11}B NMR (see structure I for nomenclature): = −46.3 (d, B-7; J = 173 Hz), 0.31 (d, B-5; J = 143 Hz), 4.5 (d, B-4; J = 131 Hz) [1].

IR spectrum in $CDCl_3$: 3025 m, 2985 s, 2950 vs, 2923 s, 1930 m, 1510 s, 1346 m, 1290 s, 1155 m, 1060 s, 1022 m cm^{-1} [1].

The mass spectrum shows the following fragments (relative intensity in %): $[M]^+$ (5.3), $[InCH_3]^+$ (5.6), $^{115}In^+$ (100), $[C_2B_4H_8]^+$ (4.6); other peaks of lower intensity appear at m/e = 203, 202, 201 and 70, 69, belonging to several fragments in common [1].

At 215 °C the compound decomposes to the extent of 20% in 72 h, forming nonvolatile black products of unknown composition and volatile $C_2B_4H_8$ [1].

Reactions with Br_2 in CS_2 and with HCl liberated H_2 in both cases [1].

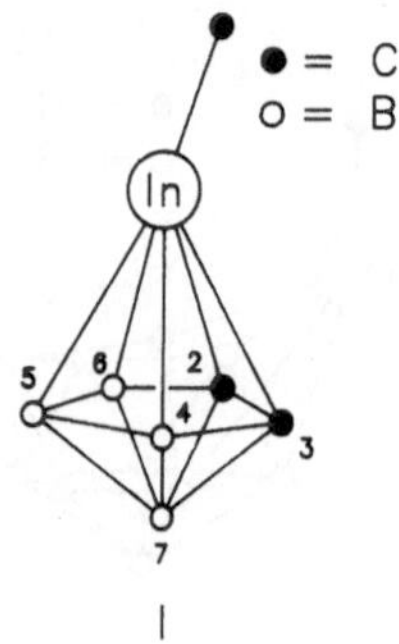

I

$(C_3H_7\text{-}i)InC_2\{Si(CH_3)_3\}_2B_4H_4$

The compound was obtained by treatment of the salt Na(THF)Li[2,3-$\{(CH_3)_3Si\}_2C_2B_4H_4$] with $(C_3H_7\text{-}i)InI_2$ in THF (1:1 mole ratio, 2 h at 0 °C followed by 1 h at room temperature). It was purified by removal of the solvent and distillation or sublimation of the residue at 160 °C into a trap. 2,3-$\{(CH_3)_3Si\}_2C_2B_4H_6$ was removed by pumping off at −10° (12 h); the title compound (m.p. 25 °C) remained in 39% yield. It is soluble in both polar and nonpolar solvents [6].

The NMR data were obtained in C_6D_6 solution (^{13}C NMR in $CDCl_3$, δ values in ppm). ^{1}H NMR: 0.35 (s, CH_3Si), 0.91 (q, 1 HB apical, ^{1}J(B,H) = 126), 1.28 (d, CH_3, J(H,H) = 5.75), 1.41 (m, CH), 3.2 to 4.9 (m, 3 HB basal). ^{13}C NMR: 2.57 (CH_3Si, J(H,C) = 118.6), 22.68 (CH_3 of C_3H_7-i, J(H,C) = 126.0), 31.23 (CHIn, J(H,C) = 127.7), 124.43 (C_2B_4). ^{11}B NMR: −44.77 (d, 1 B apical, ^{1}J(H,B) = 125), 9.63 (m, 1 B basal), 16.33 (m, 2 B basal) [6].

Unassigned IR data (in $CDCl_3$) were given between 2952 and 491 cm^{-1}; assigned bands are 2576 vs, 2433 s, sh ν(BH), 1456 ms and 1406 ms $\delta_{as}(CH_3)$, 1252 vs $\delta_s(CH_3Si)$, 833 vvs, br ϱ(CH_3Si) [6].

The compound crystallizes in the triclinic space group $P\bar{1}-C_i^1$ (No. 2) with the lattice constants a = 1101.8(3), b = 1239.4(3), c = 1448.5(4) pm, α = 73.01(2)°, β = 87.67(2)°, γ = 89.90(2)°; Z = 4 gives D_c = 1.32 g/cm³. The dimer consists of two crystallographically independent monomer units, and pairs of dimers are related by a center of inversion. The In atom of each monomeric unit seems to be η^5-bonded to the C_2B_3 carborane face; however, significant slipping toward an η^3-bonding (borallyl) is found. The structure of the dimer with important distances is shown in **Fig. 74** [6].

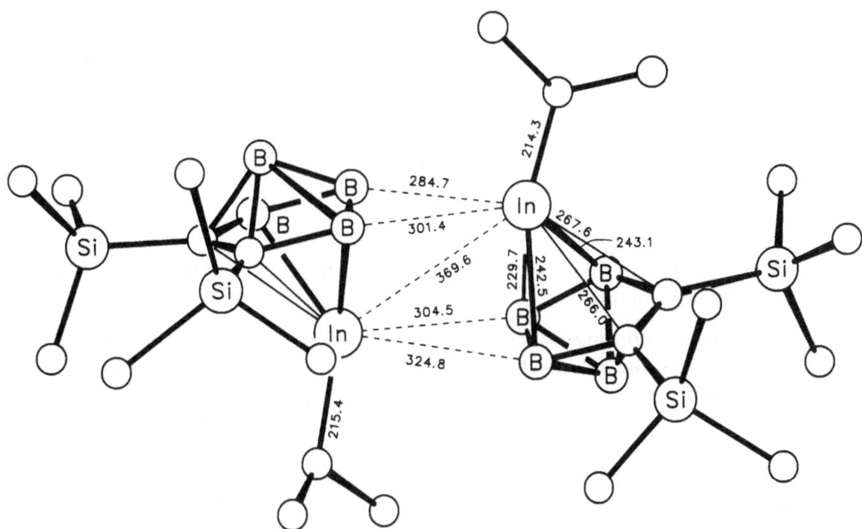

Fig. 74. Molecular structure of the dimer of $(C_3H_7\text{-}i)InC_2\{Si(CH_3)_3\}_2B_4H_4$ [6].

$(CH_3)_2InC_2(C_6H_5)B_{10}H_{10} \cdot N(CH_3)_2(CH_2)_2N(CH_3)_2$

The compound is a carborane derivative, but there are no In-B bonds present, only In-C bonds [5]. For this reason the product is listed in Section 1.1.6 (Table 17, p. 106, No. 20).

References:

[1] Grimes, R. N.; Rademaker, W. J.; Denniston, M. L.; Bryan, R. F.; Greene, P. T. (J. Am. Chem. Soc. **94** [1972] 1865/9).

[2] Greenwood, N. N.; Thomas, B. S.; Waite, D. W. (J. Chem. Soc. Dalton Trans. **1975** 299/304).

[3] Word, I. M. (Diss. Univ. Leeds 1975 from [4]).

[4] Greenwood, N. N. (Pure Appl. Chem. **49** [1977] 791/801).

[5] Bregadze, V. I.; Usyatinskii, A. Ya.; Kampel', V. Ts.; Golubinskaya, L. M.; Godovikov, N. N. (Izv. Akad. Nauk SSSR Ser. Khim. **1985** 1212/3; Bull Acad. Sci. USSR Div. Chem. Sci. **34** [1985] 1113).

[6] Hosmane, N. S.; Lu, K.-J.; Zhang, H.; Cowley, A. H.; Mardones, M. A. (Organometallics **10** [1991] 392/3).

10 Organoindium-Transition Metal Compounds

The compounds described in Table 62 contain In-M bonds (M = transition metal, L_n represent ligands which achieve the rare gas configuration upon bonding to M) and have the following structures:

$R_3In \leftarrow ML_n$ (I) $\qquad R_2In{-}ML_n$ (II)

$R(X)In{-}ML_n$ (III) $\qquad (R(X)In)_2{-}ML_n$ (IV) $\qquad R{-}In(ML_n)_2$ (V)

Formula I represents adducts of $In(C_6H_5)_3$ with anionic carbonyl complexes; neutral adducts of this type are unknown.

Compounds with 3-coordinated In (Formula II and III) have been reported for M = Ir, Pt, and Mo. Type IV is known only for $ML_n = Fe(CO)_4$, in which the Fe and both halogen ligands (X) function as bridging elements in a dinuclear complex. In V two In–M bonds are formed.

The compounds were synthesized by the following procedures:

Method I: From $In(C_6H_5)_3$ and a transition metal anion, $[ML_n]^-$.
A CH_2Cl_2 solution of the appropriate carbonyl salt, $[NR'_m]^+[ML_n]^-$, was added to one equivalent of $In(C_6H_5)_3$ in the same solvent. After 1 to 2 h stirring, the resulting bright yellow solution was treated with toluene (No. 4), hexane (Nos. 1 and 2), or THF (No. 3) and slowly cooled to −35 to −65 °C. The product crystallized out, was collected, and was dried in vacuum [1, 3].

Method II: From InR_3 or R_2InX and an ML_n complex.
InR_3 (or R_2InX) reacted with $R'ML_n$ ($R'Ir(P(CH_3)_3)_4$, $R' = H, CH_3$) to produce the compounds via the nonisolable adduct $R_3In \cdot R'ML_{n-1}$ (or $R_2InX \cdot R'ML_{n-1}$). The electron deficient intermediate adduct stabilizes itself by isomerization, followed by alkyl migration from In to the transition metal M. The rearrangement of the postulated adduct leads to the formation of mer and fac isomers, e.g., for Nos. 5 and 6. A 1:1.2 mole ratio of $CH_3Ir(P(CH_3)_3)_4$ and $In(CH_3)_3$ etherate in benzene formed a mixture of products that crystallized out at −30 °C, after removal of solvent and uptake of the residue in warm C_5H_{12}. The separation of the isomers (Nos. 5 and 6) was not described. For No. 14 the same transition metal component was mixed with $(CH_3)_2InBr$ (about 1:1 mole ratio) in ether and stirred for 30 min. The white residue, remaining after removal of the solvent, was the pure fac isomer. The mer isomer (No. 15), was obtained by purifying the product in C_6H_6 and allowing it to slowly crystallize from the filtered solution. Compound No. 8 arose analogously from $In(C_2H_5)_3$ and $HIr(P(CH_3)_3)_4$ in C_6H_6 and was isolated from the initial oily reaction residue by dissolving it in C_5H_{12} and cooling to −30 °C. The yields for these reactions were evaluated from the area ratios of the ^1H NMR signals [6].

Method III: Photolysis of InR_3 and $\{C_5H_5Mo(CO)_3\}_2$.
In a typical run, the appropriate InR_3 and $\{C_5H_5Mo(CO)_3\}_2$ were mixed in C_6D_6 in a 1:1 mole ratio and photolyzed for about 1 min using a 1000 watt Hg/Xe lamp filtered through 20 cm water and then through a 577 nm or 435 nm interference filter. This procedure produced (in the case of $R = C_2H_5$) a mixture of

$R_2InC_5H_5Mo(CO)_3$ (85%), $RIn\{C_5H_5Mo(CO)_3\}_2$ (5%), $In\{C_5H_5Mo(CO)_3\}_3$, $C_5H_5Mo(CO)_3C_2H_5$ (10%), C_2H_4, and C_2H_6. The mixture was analyzed by ^1H NMR; the product ratio depends on the Mo:In ratio [7].

General Remarks. With the exception of No. 4 the ionic compounds of type I, are air- and moisture-sensitive products, whose formation reactions and approximate equilibrium constants, K_f, have been inferred from IR spectra. $K_f = 5 \times 10^3$ for No. 1, $>10^5$ for Nos. 2 and 3, and 6×10^2 for No. 4 [3].

(1) $InR_3 + [ML_n]^- \xrightleftharpoons{K_f} [R_3In \leftarrow ML_n]^-$

Comparable equilibria were also found for Nos. 5 to 8, 14, and 15 using ^1H NMR spectra. Nos. 5 and 6 each react with excess $P(CH_3)_3$ to reform the starting material. Only 2/3 of the complex $Pt(CH_3)_2(N_2C_{10}H_8)$ could be converted into No. 7 by $In(CH_3)_3$ [6].

Compounds No. 14 and 15 isomerize in pyridine and within 2 d each forms a mixture containing 75% of 14 and 25% of 15 [6].

Table 62
Compounds with Indium-Transition Metal Bonds.
Further information on numbers preceded by an asterisk is given at the end of the table.
Explanations, abbreviations, and units on p. X.

No.	compound method of preparation (yield)	properties and remarks
	$[R_3In\text{-}ML_n]^-$ type (Formula I)	
1	$[(C_6H_5)_3InWC_5H_5(CO)_3]^-$ I (80%)	$[N(C_4H_9\text{-}n)_4]^+$ salt, yellow solid, m.p. 165 to 166 °C [1, 3] IR: 1931 vs, 1840 s, 1820 s as solid; 1933 s, 1839 s, 1819 vs in CH_2Cl_2 [1, 3]
2	$[(C_6H_5)_3InMn(CO)_5]^-$ I (71%)	$[N(C_3H_7\text{-}n)_4]^+$ salt, colorless crystals, m.p. 150 to 153 °C (dec.) [3] IR: 2049 m, 1963 m, 1949 vs, 1927 vs, 1907 w, 1895 w, 1858 vw, sh as solid; 2051 mw, 1960 m, sh, 1942 vs, 1935 vs in CH_2Cl_2 [3]
3	$[(C_6H_5)_3InFeC_5H_5(CO)_2]^-$ I (63%)	$[N(C_2H_5)_4]^+$ salt, orange-yellow crystals, m.p. 159.5 to 161 °C [3] IR (solid): 1937 vs, 1917 w, sh, 1871 vs, 1843 w [3]
4	$[(C_6H_5)_3InCo(CO)_4]^-$ I (80%)	$[N(P(C_6H_5)_3)_2]^+$ salt, colorless needle-shaped crystals, m.p. 182 to 184 °C [3] IR: 2045 m, 1968 ms, 1942 s, 1925 vs as solid; 2042 mw, 1960 m, sh, 1939 vs in CH_2Cl_2 [3]
	$R_2In\text{-}ML_n$ type (Formula II)	
5	$(CH_3)_2InIr(CH_3)_2(P(CH_3)_3)_3$-fac II (88%)	mixture with mer isomer No. 6 [6] ^1H NMR (C_6D_6): 0.12 (d with J = 11.9 of q with J = 3.5, CH_3Ir), 0.38 (s, CH_3In), 0.90 (d, CH_3P, J = 7.0), 1.14 (d, CH_3P, J = 6.8) [6]

Table 62 (continued)

No.	compound method of preparation (yield)	properties and remarks
6	$(CH_3)_2InIr(CH_3)_2(P(CH_3)_3)_3$-mer II (12%)	mixture with fac isomer No. 5 [6] ^{1}H NMR (C_6H_6): −0.11 (d (J=6.9) of t (J=6.0), CH_3Ir), 0.25 (t (J=8.2) of d (J=4.6), CH_3Ir), 0.38 (s, CH_3In), 1.10 (t, CH_3P, J=2.9), 1.13 (d, CH_3P, J=7.4) [6]
*7	$(CH_3)_2InPt(CH_3)_3(N_2C_{10}H_8)$ $N_2C_{10}H_8$=bipyridine special	no pure compound isolated [6] ^{1}H NMR (C_6D_6): −0.20 (CH_3In), 0.18, 1.67 (CH_3Pt) [6]
*8	$(C_2H_5)_2InIr(H)C_2H_5(P(CH_3)_3)_3$ II	pale yellow crystals, m.p. ca. 20 °C [6] ^{1}H NMR (C_6D_6): −11.6 (d (J=123) of t (J=17), HIr), 1.16, 1.17, 1.25 (3 d, CH_3P, J_1=7.5, J_2=7.3, J_3=7.4), 1.1 (m, br), 1.3 (m, br), 1.86 (q, br, C_2H_5, J=8) [6] ^{13}C NMR (C_6D_6): −22.11 (d (J=64.4) of t (J=6.3), CH_2Ir), 12.80 (s, CH_3 of C_2H_5In), 18.92 (d, CH_2In, J=21.9), 21.75, 22.75, 28.46 (d's, CH_3P, J_1=23.2, J_2=25.9, J_3=20.8), 30.19 (d, CH_3 of C_2H_5Ir, J=13.2) [6]
*9	$(CH_3)_2InMoC_5H_5(CO)_3$ III	white solid [7]
*10	$(C_2H_5)_2InMoC_5H_5(CO)_3$ III	white solid [7] ^{1}H NMR (C_6D_6): 0.95 (m, br, CH_2), 1.40 (t, CH_3, J=8), 4.95 (C_5H_5) [7]
*11	$(C_4H_9\text{-}t)_2InMoC_5H_5(CO)_3$ III	white solid [7] ^{1}H NMR (C_6D_6): 1.39 (CH_3), 4.75 (C_5H_5) [7]
*12	$\{(C_4H_9\text{-}t)CH_2\}_2InMoC_5H_5(CO)_3$ II	white solid [7] ^{1}H NMR (C_6D_6): 1.14, 1.45 (CH_2 and CH_3), 4.64 (C_5H_5) [7]
*13	$(CH_3)_3Si$–Pt(R'_2P–PR'_2, chelate)–In$\{CH_2Si(CH_3)_3\}_2$ R' R' / P / $\{(CH_3)_3SiCH_2\}_2In-Pt$ / P / R' R' R'=C_6H_{11}-c special	colorless crystals, m.p. 97 °C [8] ^{1}H NMR (C_6D_6): 0.28 (s, CH_2In), 0.33 (s, CH_3Si at Pt), 0.39 (s, CH_3Si at In), 1.05 to 2.20 (m's, br, C_6H_{11} + CH_2Pt) [8] ^{13}C NMR (C_6D_6): 3.75, 4.50, 19.31 (d, CH_2Pt, $^2J(P,C)$=18.5, trans), 22.2 to 36.8 (complex m's) [8] ^{31}P NMR (C_6D_6 at ca. 5 °C): 72.1 ($^1J(^{195}Pt,P)$=1838), 77.5 ($^1J(^{195}Pt,P)$=2112) [8]

RInX–ML$_n$ type (Formula III)

No.	compound method of preparation (yield)	properties and remarks
14	$CH_3InBrIr(CH_3)_2(P(CH_3)_3)_3$-fac II (57%)	white powder [6] ^{1}H NMR (C_5D_5N): 0.14 (d (J=11.5) of q (J=3.2), CH_3Ir), 0.42 (d, CH_3In, J=1.5), 1.10, 1.53 (d's, CH_3P, J=7.5) [6]

Table 62 (continued)

No.	compound method of preparation (yield)	properties and remarks
15	$CH_3InBrIr(CH_3)_2(P(CH_3)_3)_3$-mer II	white solid [6] ^1H NMR (C_5D_5N): −0.11 (q, CH_3Ir, J=6.7), 0.06 (t (J=8.1) of d (J=4.6), CH_3Ir), 0.30 (s, CH_3), 1.42 (t, CH_3P, J=3.2), 1.48 (d, CH_3P, J=7.7) [6]
$(RInX)_2$-ML_n type (Formula IV)		
*16	$\{((CH_3)_3Si)_3CInCl\}_2Fe(CO)_4$ special	yellow crystals, m.p. 235 °C [4] ^1H NMR (C_6D_6): 0.32 (s, CH_3Si) [4] IR(solid): 2030, 2000, 1990, 1980 ν(CO)[4]
$RIn(ML_n)_2$ Type (Formula V)		
17	$C_2H_5In\{MoC_5H_5(CO)_3\}_2$ III	white solid [7] ^1H NMR (C_6D_6): 1.21 (q, CH_2), 1.68 (t, CH_3, J=8), 4.74 (C_5H_5) [7]

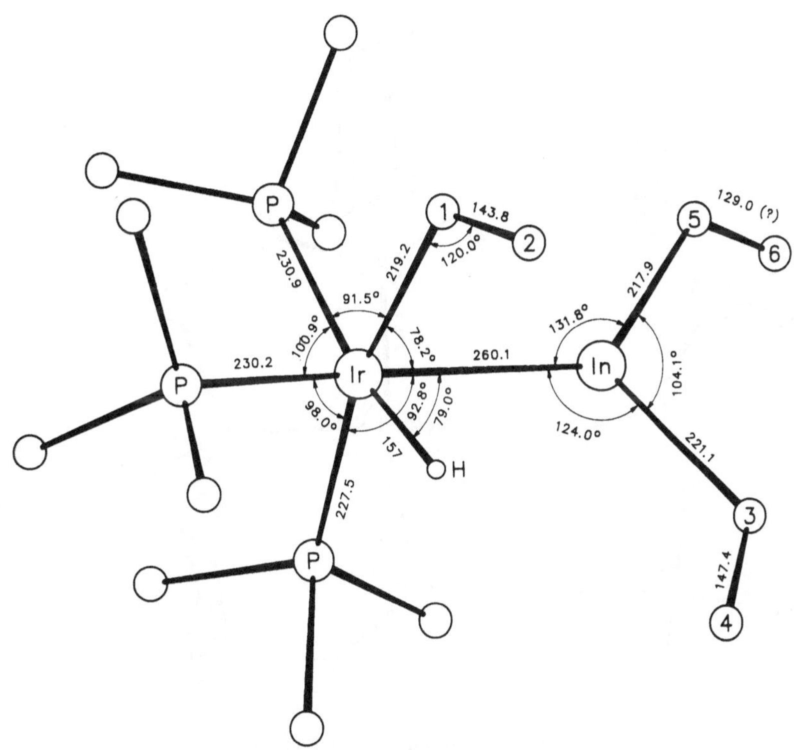

Fig. 75. Molecular structure of $(C_2H_5)_2InIr(H)C_2H_5(P(CH_3)_3)_3$ [6].

* Further information:

$(CH_3)_2InPt(CH_3)_3(N_2C_{10}H_8)$ (Table **62**, No. **7**) was formed as a moderately-soluble solid by the reaction of $(CH_3)_2Pt(N_2C_{10}H_8)$ ($N_2C_{10}H_8$=bipyridine) with excess $In(CH_3)_3$ etherate in C_6D_6. After recrystallization from hot benzene, the ^1H NMR spectrum showed about 33% unconverted starting material along with the title compound [6].

$(C_2H_5)_2InIr(H)C_2H_5(P(CH_3)_3)_3$ (Table **62**, No. **8**) crystallizes in the monoclinic space group $P2_1/n-C^5_{2h}$ (No. 14) with the lattice constants a=1483.6(3), b=1465.8(3), c=1102.2(2) pm, β=90.22(1)°; Z=4 gives D_c=1.727 g/cm³. The data were collected at −100 °C. For a few atoms thermal motion leads to unrealistic distances, e.g., 129(2) pm for the C(5)-C(6) bond distance as shown in **Fig. 75**, p. 333 [6].

$(R)_2InMoC_5H_5(CO)_3$ (Table **62**, No. **9**, R=CH_3; No. **10**, R=C_2H_5; No. **11**, R=C_4H_9-t; No. **12**, R=$CH_2C_4H_9$-t). The methyl derivative was reported to be formed from $\{C_5H_5Mo(CO)_3\}_2$ and excess $In(CH_3)_3$ in cold hexane. However, even in the solid state rearrangement occurs with formation of $In\{MoC_5H_5(CO)_3\}_3$ and $In(CH_3)_3$ at room temperature after several hours. Relative quantum yields were estimated for R=C_2H_5 and $CH_2C_4H_9$-t. A bimolecular radical substitution mechanism was proposed [7].

$\{(CH_3)_3SiCH_2\}_2InPtCH_2Si(CH_3)_3\{(C_6H_{11}\text{-c})_2PCH_2CH_2P(C_6H_{11}\text{-c})_2\}$ (Table **62**, No. **13**). A suspension of $(CH_3)_3SiCH_2PtH\{(C_6H_{11}\text{-c})_2PCH_2CH_2P(C_6H_{11}\text{-c})_2\}$ in about sevenfold excess of $In\{CH_2Si(CH_3)_3\}_3$ was heated at 80 °C in an evacuated glass vessel for 1 h. All volatile materials were removed in vacuum and the residue washed with cold pentane; recrystallization from hot benzene gave the compound in 72% yield [8].

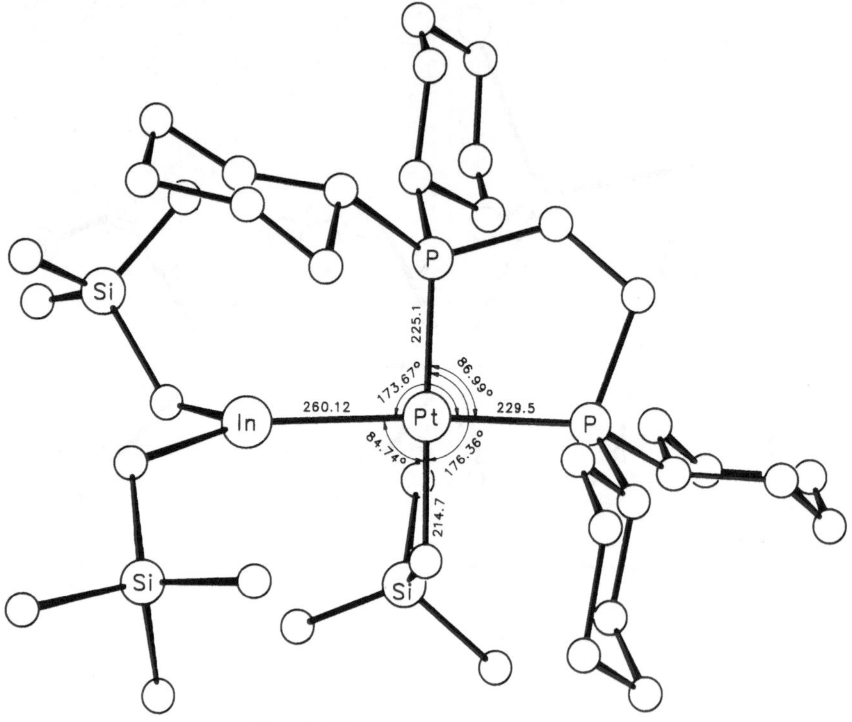

Fig. 76. Molecular structure of $\{(CH_3)_3SiCH_2\}_2InPtCH_2Si(CH_3)_3\{(C_6H_{11}\text{-c})_2PCH_2CH_2P(C_6H_{11}\text{-c})_2\}$ [8].

The structure of the compound with important bond distances and angles (no cell parameters were given) is depicted in **Fig. 76** [8].

$\{((CH_3)_3Si)_3CInCl\}_2Fe(CO)_4$ (Table **62**, No. **16**) was obtained by adding a solution of $[Li(OC_4H_8)_3][((CH_3)_3Si)_3CInCl_3]$ in THF to a suspension of $Na_2[Fe(CO)_4] \cdot 1.5\ C_4H_8O_2$ (dioxane) at −78 °C (96% yield). On warming the mixture turned orange; the solvent was removed and the residue extracted by toluene. The title compound crystallized from the extract at −20 °C [4].

The compound crystallizes in the triclinic space group $P\bar{1}-C_i^1$ (No. 2) with the unit cell constants a = 906.3(7), b = 1375.1(6), c = 1730.7(5) pm, α = 78.89(3)°, β = 80.49(5)°, γ = 79.45(6)°; Z = 2 gives D_c = 1.50 g/cm³. The Si atoms of one $C(Si(CH_3)_3)_3$ group are disordered; occupation densities of 0.35 and 0.65 were reported for the two orientations found. The most important distances and angles are included in **Fig. 77** [4].

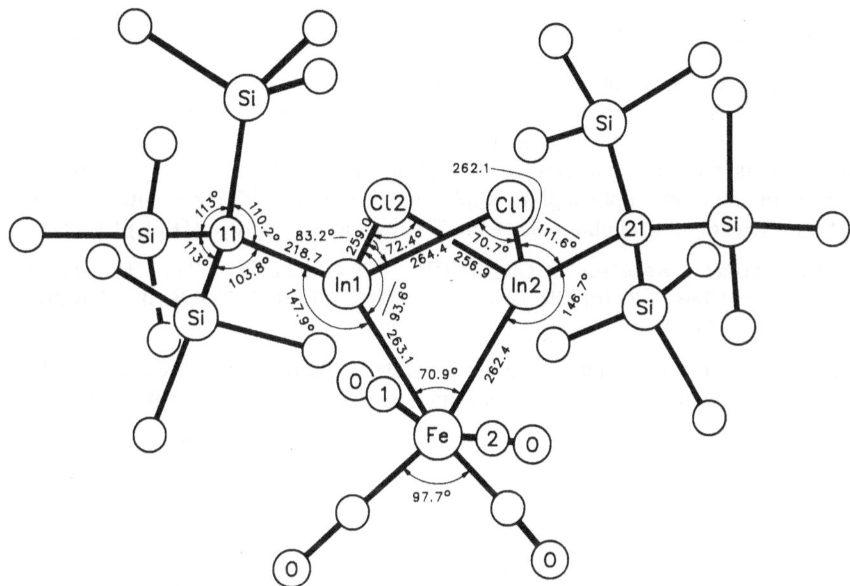

Fig. 77. Molecular structure of $\{((CH_3)_3Si)_3CInCl\}_2Fe(CO)_4$ [4].

Other bond angles (°)

Fe-In(1)-Cl(2)	91.2(1)	Cl(2)-In(2)-C(21)	112.3(2)
Cl(1)-In(1)-C(11)	111.5(3)	C(1)-Fe-C(2)	165.7(6)
Fe-In(2)-Cl(2)	90.9(1)	Cl(1)-In(2)-Cl(2)	83.2(1)

References:

[1] Burlitch, J. M.; Petersen, R. B. (J. Organometal. Chem. **24** [1970] C 65/C 67).

[2] Hsieh, A. T. T.; Mays, M. J. (J. Organometal. Chem. **37** [1972] 9/14).

[3] Burlitch, J. M.; Leonowicz, M. E.; Petersen, R. B.; Hughes, R. E. (Inorg. Chem. **18** [1979] 1097/105).

[4] Atwood, J. L.; Bott, S. G.; Hitchcock, P. B.; Eaborn, C.; Shariffudin, R. S.; Smith, J. D.; Sullivan, A. C. (J. Chem. Soc. Dalton Trans. **1987** 747/55).

[5] Clarkson, L. M.; Clegg, W.; Norman, N. C.; Tucker, A. J.; Webster, P. M. (Inorg. Chem. **27** [1988] 2653/60).
[6] Thorn, D. L.; Harlow, R. L. (J. Am. Chem. Soc. **111** [1989] 2575/80).
[7] Thorn, D. L. (J. Organometal. Chem. **405** [1991] 161/71).
[8] Fischer, R. A.; Behm, J. (J. Organometal. Chem. **413** [1991] C 10/C 14).

11 Compounds with Organoindium Ions

11.1 Compounds with Organoindium Cations

Compounds with organoindium cations are rare but have been postulated in some cases. Thus, for CH_3InI_2 (see on p. 164) it was reported that the ionic formulation $[In(CH_3)_2][InI_4]$ could also be possible. For other compounds in which an organoindium fragment is part of a cation, see Section 4.5 on p. 221.

$[(C_3H_7\text{-}i)_2In][BF_4]$

The compound was prepared by dropwise addition of a solution of $BF_3 \cdot O(C_2H_5)_2$ in ether (slight excess) to a solution of $In(C_3H_7\text{-}i)_3$ in the same solvent followed by stirring the mixture for 20 h. The ether was removed and the residue washed with pentane to give a colorless solid in 79% yield; m.p. 105 to 107 °C (dec.). It is very sensitive to the influence of moisture and only slightly soluble in toluene or THF. The compound was recrystallized from THF/ether and obtained as the THF adduct $\mathbf{[(C_3H_7\text{-}i)_2In(THF)_2][BF_4]}$ [1].

The NMR spectra were run in CD_3CN (^1H, ^{13}C) and in THF (^{19}F, ^{11}B), and the shifts given in ppm. ^1H NMR: 1.21 (m, CH, CH_3). ^{13}C NMR: 21.9 (CH_3), 28.0 (CH). ^{19}F NMR: −151.4 (q, BF_4, J(B,F) = 1.3). ^{11}B NMR: −0.4 [1].

The IR spectrum of the title compound shows $\nu_{as}(InC)$ and $\nu_s(InC)$ bands at 519 and 474 cm^{-1}, respectively. The Raman spectrum shows one band at 472 cm^{-1} [1].

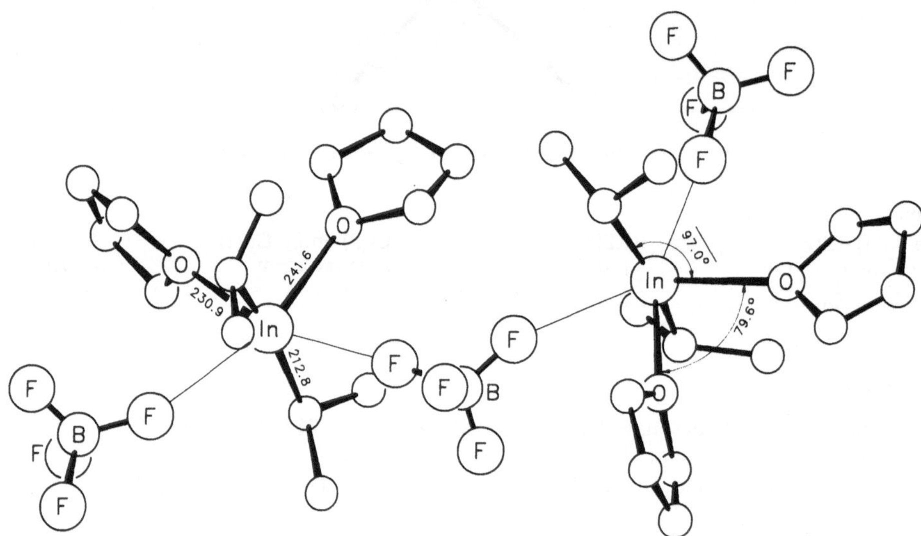

Fig. 78. Part of the chainlike structure of $[(C_3H_7\text{-}i)_2In(THF)_2][BF_4]$ [1].

The THF adduct $[(C_3H_7\text{-}i)_2In(THF)_2][BF_4]$ crystallizes in the orthorhombic space group Pnma$-D_{2h}^{16}$ (No. 62) with the unit cell parameters a=1264.1(4), b=1047.0(2), c=1441.5(4) pm; Z=4 and D_c=1.515 g/cm^3. The indium atom is in a distorted octahedral environment with the isopropyl groups in the axial positions with a C-In-C angle of 160.3(4)°. Two $[(C_3H_7\text{-}i)_2In(THF)_2]$ units are linked by two F atoms of different $[BF_4]^-$ molecules. The In···F interactions of 264.6(4) and 259.1(6) pm were considered as being mainly electrostatic in nature with In-F-B angles of 173.8(7)° and 155.0(7)°. A part of the chainlike structure is depicted in **Fig. 78**.

Reference:

[1] Neumüller, B.; Gahlmann, F. (J. Organometal. Chem. **414** [1991] 271/83).

11.2 Compounds with Organoindium Anions

11.2.1 M[InR$_4$] Compounds

This section describes various anions in which the In atom is surrounded by four equal ligands R. Mixed indates of the type Li[InR$_3$(allyl)] and Li[InR$_2$(allyl)$_2$] were obtained by alkylation of (allyl)$_3$In$_2$X$_3$ compounds with LiR compounds, but no data of these compounds were reported [13].

Compounds in this category are presented in Table 64; they were prepared by the following procedures:

Method I: From InR$_3$ and alkali metal.
Finely-divided alkali metal was suspended in ether and treated with a 20 to 40% excess of InR$_3$ or R$_3$In · O(C$_2$H$_5$)$_2$ at −60 °C, slowly warmed to room temperature, and stirred for 8 to 15 h. Finely divided In and unconverted alkali metal were filtered off, the clear filtrate freed of solvent, and the microcrystalline residue dried at 60 °C in vacuum. For Nos. 4 and 5, THF was added before the filtration and then worked up as described [2]. Nos. 14 to 16 were stirred for 48 h, treated with THF, filtered and the filtrate concentrated, treated with some C_6H_{14}, and dried. Vacuum redrying was done at 20 °C [5]. The yield in each case was practically quantitative (Author: the difficulty of separating the extremely fine, almost colloidal indium precipitate often leads to slightly contaminated products).

Method II: Reaction of InR$_3$ with LiR compounds.
To a solution of LiR in O(C$_2$H$_5$)$_2$ was added a solution of InR$_3$ (30% excess, ether solvent) at 20 °C; the mixture was stirred for 2 h and worked up as described above [2]. The yield is quantitative.

Method III: From InR$_3$ and E(CH$_3$)$_5$ (E=As, Sb).
Equimolar amounts of In(CH$_3$)$_3$ and As(CH$_3$)$_5$ or Sb(CH$_3$)$_5$ in benzene were mixed. Indates No. 6 and 7, which precipitated in high purity and almost 100% yield, were filtered, and dried in vacuum [8].

General Remarks. Indates of formula M^I[InR$_4$] are colorless microcrystalline solids that are not pyrophoric, but in the presence of moisture they rapidly decompose by hydrolysis (sometimes even igniting spontaneously). With the exception of Nos. 8 and 9 slow crystallization from ether or THF produces cubes or needles. The compounds are very soluble in

THF, soluble in ether, and moderately soluble in benzene; chlorinated hydrocarbons lead (except for Nos. 8 and 9) to decomposition.

X-ray structures have been determined for Nos. 1 to 5 and 13 to 15; key results are presented in Table 63. More structural parameters are to be found in Figures **79** to **81**. The M···R contact distance argues for isolated ions in the crystal lattice. The anions of Nos. 1 and 2 are definitely tetrahedral (T_d symmetry), while those of Nos. 3 to 5, 14, and 15 show a marked deformation in the D_{2d} direction. The following compounds are isostructural with the same space group: Nos. 1 and 2 (cubic), Nos. 3 and 4 (tetragonal), and Nos. 14 and 15 (tetragonal). The monoclinic structure of No. 13 is described on p. 334.

Table 63
X-Ray Structural Data for Various $M^I[InR_4]$ Salts with $R=CH_3$ and C_6H_5.

R	CH_3					C_6H_5	
M^I	Li	Na	K	Rb	Cs	Li	Na
No. [Ref]	1 [2]	2 [2]	3 [4]	4 [4]	5 [4]	14 [5]	15 [5]
Fig.	–	79	80	–	81	–	83
space group	$P\bar{4}3m$ (T_d)	or P23 (T^1)	$I4_1/amd$	(D_{4h}^{19})	$P\bar{4}2m$ (D_{2d}^1)	$P\bar{4}2_1/c$ (D_{2d}^1)	
system	cubic		tetragonal		tetragonal	tetragonal	
a (pm)	539.4(2)	568.2(2)	990.4(2)	1020.8(2)	745.9(1)	1219.6(6)	1196.4(6)
c (pm)	–	–	813.2(2)	814.9(2)	416.3(1)	649.1(5)	688.6(3)
Z	1	1	4	4	1	2	2
D_c (g/cm³)	1.924	1.791	1.782	2.037	2.207	1.480	1.503
In-C (pm)	222.3(4)	219.5(4)	223.9(3)	–	226(2)	–	223.0(3)
M···C (pm)	244.8	272.6	331.7	–	347(2)	–	277.0(av)
C-In-C (°)	109.47	109.47	111.45(av)	–	111.6(av)	–	109.15(av)

av = average

1,4- and 1,2-addition of $Li[InR_4]$ ($R=CH_3$, C_4H_9) to α,β-unsaturated carbonyl compounds and allylation reactions with various allyl bromides are described. Mixed indates of the type $Li[InR_3(allyl)]$ and $Li[InR_2(allyl)_2]$ were also used in reactions with various allyl bromides to give compounds of the type $R^1R^2C{=}CHCH_2CR^3R^4CH{=}CH_2$ [13].

Table 64
$M[InR_4]$ Compounds.
Further information on numbers preceded by an asterisk is given at the end of the table.
Explanations, abbreviations, and units on p. X.

No.	cation method of preparation	properties and remarks
$[In(CH_3)_4]^-$ anion		
*1	Li^+ salt II	colorless crystals [2], m.p. ca. 255 °C (dec.) [14] ^{1}H NMR (DME): −1.17 (s, CH_3In) [6] IR (solid): 1045 $\delta_s(CH_3)$, 750 $\varrho(CH_3)$, 435 $\nu(InC)$; no other bands reported [2]

Table 64 (continued)

No.	cation method of preparation	properties and remarks
*2	Na^+ salt I	dec. >100 °C [8] D_m (flotation method) = 1.8 g/cm^3 [2] ^1H NMR ($O(C_2H_5)_2$): −1.30 (s, CH_3In) [14] IR and Raman frequencies on p. 342
*3	K^+ salt I	tetragonal prismatic colorless crystals [2] IR (solid): 1103 $\delta_s(CH_3)$, 702 $\varrho(CH_3)$, 500 w $\nu(InC)$; no other bands reported [2]
*4	Rb^+ salt I	IR (solid): 1106 $\delta_s(CH_3)$, 699 $\varrho(CH_3)$, 485 w, 443 $\nu(InC)$; no other bands reported [2]
*5	Cs^+ salt I	needlelike, colorless crystals [2] IR (solid): 1105 $\delta_s(CH_3)$, 694 $\varrho(CH_3)$, 483 w, 440 $\nu(InC)$; no other bands reported [2]
*6	$[As(CH_3)_4]^+$ salt III	white powder, m.p. 126 to 128 °C (dec.) [8] IR and Raman frequencies in Table 65 on p. 343
*7	$[Sb(CH_3)_4]^+$ salt III	white powder, m.p. ca. 68 °C (dec.) [8] IR and Raman frequencies in Table 65 on p. 343
*8	$[(CH_3)_2Si(N{=}P(CH_3)_3)_2Al(CH_3)_2]^+$ salt special	white solid, m.p. 105 to 107 °C (dec.) [1] ^1H NMR (CH_2Cl_2): −0.75 (s, CH_3In), −0.64 (CH_3Al), 0.51 (CH_3Si), 1.79 ($A_9XX'A'_9$-m, CH_3P, $^2J(H,P)$ = 13.05) [1] IR (solid): 1091 $\delta_s(CH_3In)$, 670 $\varrho(CH_3)$, 465 $\nu_{as}(InC_4)$ [1]
*9	$[(CH_3)_2Si(N{=}P(CH_3)_3)Al(CH_3)_2(N{=}P(CH_3)_2C_2H_5)]^+$ salt special	white solid, m.p. 119 to 121 °C (dec.) [1] ^1H NMR (CH_2Cl_2): −0.82 (CH_3In), −0.65 (CH_3Al), 0.51 (CH_3Si), 1.75 (d, $(CH_3)_2P$, $^2J(H,P)$ = 12.8), 1.79 (d, $(CH_3)_3P$, $^2J(H,P)$ = 13.35); C_2H_5-signals are obscured [1]
$[In(C_2H_5)_4]^-$ anion		
10	Na^+ salt	from $Na[In(C_2H_5)_3H]$ by thermal decomposition [9] colorless viscous liquid [9], colorless crystals (Method I), m.p. 62 to 65 °C [14] ^1H NMR (C_6H_6): −0.47 (q, CH_2), 1.08 (t, CH_3, J = 6 to 8) [9] Raman (solid): 1118 s $\delta_s(CH_2)$, 458 vs $\nu_s(InC_4)$, 449 sh $\nu_{as}(InC_4)$, 185 w $\delta(InC_4)$ [14]
11	K^+ salt	from $K[In(C_2H_5)_3H]$ by thermal decomposition [9] colorless viscous liquid [9]
$[In\{CH_2Si(CH_3)_3\}_4]^-$ anion		
12	Na^+ salt	from $In(CH_2Si(CH_3)_3)_3$ and NaH; no pure compound isolated (see No. 13) [11]

Table 64 (continued)

No.	cation method of preparation	properties and remarks
12 (continued)		^{1}H NMR (C_6D_6): −1.11 (CH_2In), −1.00 (impurity), 0.25 (CH_3Si) [11]
*13	K^+ salt special	colorless crystals, m.p. 85 °C (glassy), 107 °C (clear liquid) [11] ^{1}H NMR (C_6D_6): −1.05 (CH_2In), 0.30 (CH_3Si) [11] IR (solid): 1250 m, 1240 m; 1220 m, sh, 1012 vw, 930 m, 917 m, 900 m, sh, 875 m, sh, 850 vs, 818 vs, 744 m, 720 m, 673 w, 662 w, 600 vw, 552 vw, 530 vw, 453 m, 436 w, sh, 387 w [11]
$[In(C_6H_5)_4]^-$ anion		
*14	Li^+ salt I	thin, needlelike colorless crystals [5] IR (solid): 448, 436 ν(InC); no other bands reported [5]
*15	Na^+ salt I	colorless, nearly cubic crystals [5] IR (solid): 450, 439 ν(InC); no other bands reported [5]
*16	K^+ salt I	white, microcrystalline powder [5] IR (solid): 451, 439, 434 sh ν(InC); no other bands reported [5]
*17	Rb^+ salt I	like No. 16 [5]
*18	Cs^+ salt I	like No. 16 [5]
$[In(C_9H_7)_4]^-$ anion (C_9H_7 = indenyl)		
*19	Li^+ salt special	pale yellow solid, m.p. 120 to 125 °C (dec.) [3] ^{1}H NMR ($CDCl_3$): 5.45 (d), 6.17 (t, C_5-ring), 7.38 (m, C_6-ring) [3] IR (solid): 371 s, 352 m, 340 m, 270 w ν(InC) [3] Raman (solid): 352 vs ν(InC); no other bands reported [3]
$[In(C_4(C_6H_5)_4)_2]^-$ anion		
*20	$[As(C_6H_5)_4]^+$ salt special	yellow solid [10] ^{1}H NMR (CD_2Cl_2): 6.75 to 6.90 (m, C_6H_5 anion), 7.45 to 7.80 (m, C_6H_5 cation) [10]

* Further information:

$M[In(CH_3)_4]$ (Table **64**, Nos. **1** to **5**; M = Li, Na, K, Rb, Cs). The Li salt formed in nearly quantitative yield from the reaction of $InCl_3$ and $LiCH_3$ (1:4 mole ratio) in ether. The initially formed $[Li(O(C_2H_5)_2)][In(CH_3)_4]$ was freed of ether in vacuum at room temperature [12]. Decomposition of $Li[In(CH_3)_3H]$ also gave the title compound, along with In, LiH, and H_2 [9].

The unit cell parameters are listed in Table 63, p. 338, and the structures are illustrated in **Fig. 79** (M = Li, Na) [2], **Fig. 80** (M = K, Rb), and **Fig. 81** (M = Cs) [4].

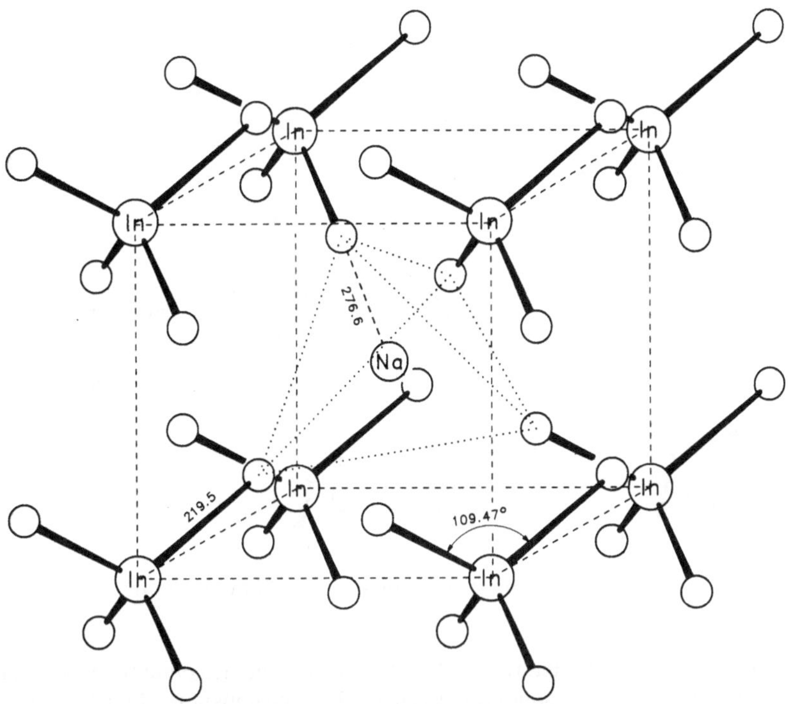

Fig. 79. Model of the unit cell of $M[In(CH_3)_4]$ (M = Li, Na; values are given for M = Na) [2].

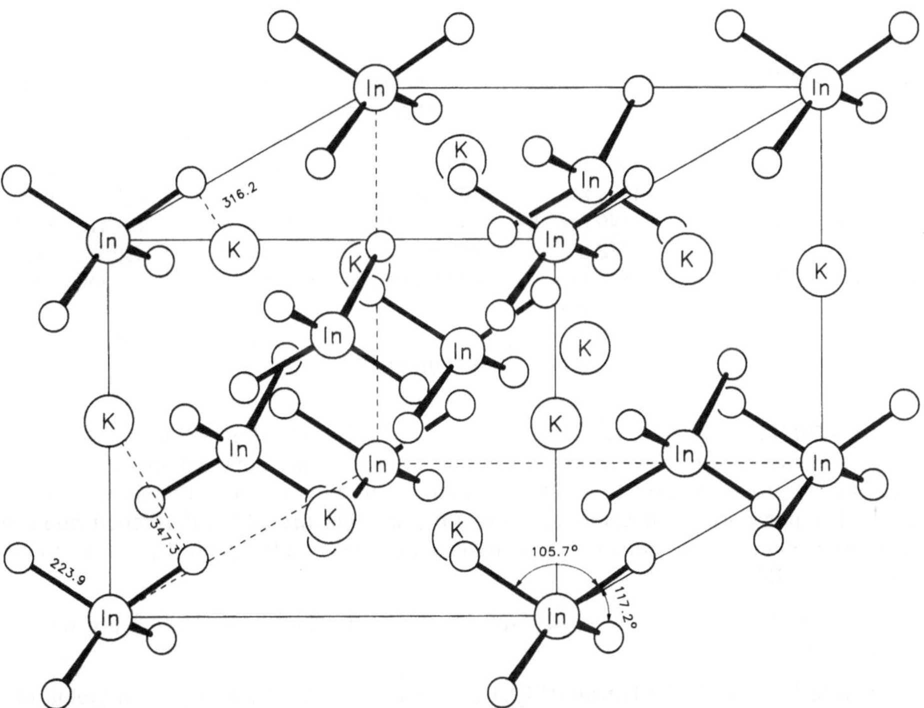

Fig. 80. Model of the unit cell of $M[In(CH_3)_4]$ (M = K, Rb; values are given for M = K) [4].

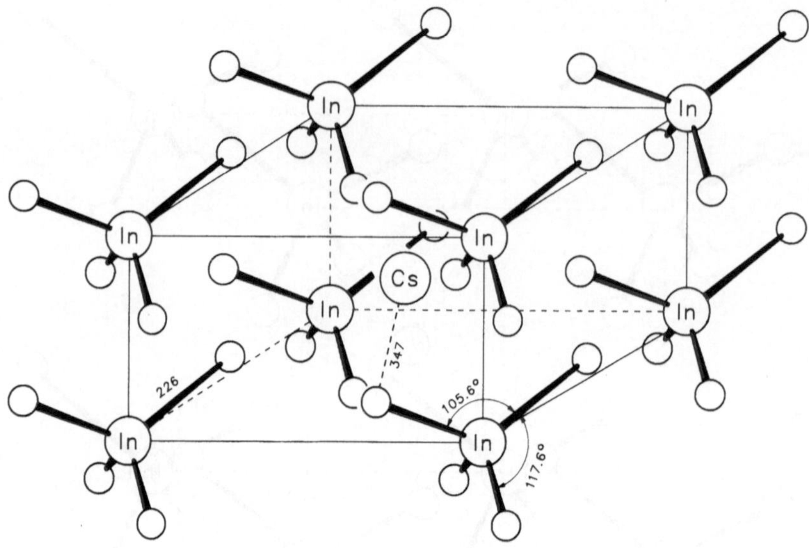

Fig. 81. Model of the unit cell of $Cs[In(CH_3)_4]$ [4].

The IR and Raman data for the Na salt No. 2 are given and assigned below (wavenumbers in cm^{-1}) [8]; similar IR values (solid) are given in [2]. Force constants: f(InC) = 1.565, f'(InC) = 0.087, d(CInC) = 0.64, d(InCH) = 0.250, d(HCH) = 0.45 N/cm [8].

IR solid	THF solution	Raman solid	THF solution	calculated	assignment
1438 w	–	–	–	1440.0, 1439.9	$\delta_{as}(CH_3)$, E+F_2
–	–	1094 ms	1113 ms	1113.1	$\delta_s(CH_3)$, A_1
1080 vs	1110 w	1074 s	1105 sh	1108.4	$\delta_s(CH_3)$, F_2
743 vs, br	668 sh	740 mw, br	643 vw, br	671.3, 668.0	$\varrho(CH_3)$, E+F_2
–	–	472 s	445 to 449 s, br	440.0	$\nu_s(InC_4)$, A_1
455 s, br	437 s	454 ms	439 w, sh	435.0	$\nu_{as}(InC_4)$, F_2
–	–	224 ms	–		
–	–	178 m	–	138.9, 128.5	$\delta(InC_4)$, E+F_2
–	–	160 m	139 m, br		

$[E(CH_3)_4][In(CH_3)_4]$ (Table **64**, No. **6**, E = As, No. **7**, E = Sb). The IR and Raman spectra of No. 6, together with the spectra of indates No. 2 and 7, were used to compare the $M(CH_3)_4$ compounds of groups 14 and 15 and to calculate the force constants of these isoelectronic species. For this purpose a computer program (that also provides information about the "electrotransmittance" interaction between methyl groups and the central atom) was developed and used [8].

The IR and Raman frequencies for solid Nos. 6 and 7 below 1500 cm^{-1} are collected in Table 65 [8].

$[(CH_3)_2Si(N{=}P(CH_3)_3)_2Al(CH_3)_2][In(CH_3)_4]$ (Table **64**, No. **8**) was obtained in yields of 91 to 92% either by reaction of $In(CH_3)_3 \cdot (N{=}P(CH_3)_3)_2Si(CH_3)_2$ (Table 6, p. 38, No. 33) with

Table 65
IR and Raman Frequencies for Solid $[As(CH_3)_4]^+[In(CH_3)_4]^-$ and $[Sb(CH_3)_4]^+[In(CH_3)_4]^-$ below 1500 cm^{-1} [8].
Wavenumbers in cm^{-1}.

$[As(CH_3)_4][In(CH_3)_4]$ IR	Raman	$[Sb(CH_3)_4][In(CH_3)_4]$ IR	Raman	assignment (T_d symmetry)
1412 w, br	—	1420 w	—	$\delta_{as}(CH_3As, Sb)$
1304 w	1297 vw, br	—	1245 m	$\delta_s(CH_3As, Sb)$, A_1
1278 m	—	1223 m	1223 vw	$\delta_s(CH_3As, Sb)$, F_2
—	1128 m	—	1100 vs	$\delta_s(CH_3In)$, A_1
1100 m	1111 ms	1090 s, br	1095 sh	$\delta_s(CH_3In)$, F_2
925 vs	930 vw, br	845 vs, br	860 vw, br	$\varrho(CH_3As, Sb)$
660 vs, br	—	670 vs, br	668 w, br	$\varrho(CH_3In)+$
—	656 ms	—	—	$\nu_{as}(AsC_4)$, F_2
—	—	572 s	575 ms	$\nu_{as}(SbC_4)$, F_2
—	590 s	—	534 vs	$\nu_s(As, SbC_4)$, A_1
—	—	462 sh	470 sh	$\nu_s(InC_4)$, A_1
—	451 vs	—	459 vs	
441 vs	445 s	448 vs	449 s	$\nu_{as}(InC_4)$, F_2
—	225 m, br	—	176 mw, br	$\delta(As, SbC_4)$, $E+F_2$
—	143 ms	—	139 ms	$\delta(InC_4)$, $E+F_2$

an equimolar amount of $(CH_3)_3Al \cdot O(C_2H_5)_2$ or from the same adduct of $Al(CH_3)_3$ and $(CH_3)_3In \cdot 0.8\ O(C_2H_5)_2$ in C_6H_6. With warming, two liquid phases appeared; the denser one was No. 8 with C_6H_6 and $O(C_2H_5)_2$. Removal of the solvent in vacuum produced a solid raw product, which gave the title compound after washing with benzene/petroleum ether and redrying. It is practically insoluble in slightly polar solvents (C_6H_{12}, C_6H_6, CCl_4, $O(C_2H_5)_2$, petroleum ether), but clear solutions, presumably containing ion pairs, are obtained in CH_2Cl_2. The suggested structure (Formula I) was derived from NMR results [1]:

$$\left[\begin{array}{c} P(CH_3)_3 \\ \vdots \\ N \\ (CH_3)_2Si(+)Al(CH_3)_2 \\ N \\ \vdots \\ P(CH_3)_3 \end{array}\right]^+ [In(CH_3)_4]^-$$

I

$[(CH_3)_2Si(N{=}P(CH_3)_3)Al(CH_3)_2(N{=}P(CH_3)_2C_2H_5)][In(CH_3)_4]$ (Table **64**, No. **9**) was formed by the same method as the above compound from $(CH_3)_3In \cdot 0.8\ O(C_2H_5)_2$ and $(CH_3)_3Al \cdot N{=}P(CH_3)_3Si(CH_3)_2N{=}P(CH_3)_2C_2H_5$ in 95% yield. Properties and structure are consistent with Formula I [1].

$K[In(CH_2Si(CH_3)_3)_4]$ (Table **64**, No. **13**) formed in small amounts as a by-product of the reaction of $In(CH_2Si(CH_3)_3)_3$ and KH. The exothermic reaction of the solvent-free reactants formed a black-brown mixture, which was treated with C_5H_{12} and allowed to stand for 2 d. From the mass grew colorless crystals which could be isolated only mechanically and were identified as the title compound [11].

The salt crystallizes in the monoclinic space group $P2_1/n-C_{2h}^5$ (No. 14) with the lattice constants a = 1110.6(4), b = 1971.2(6), c = 1288.2(3) pm, β = 91.65(2)°; Z = 4 gives D_c = 1.18 g/cm³. The structure of the anion is reproduced in **Fig. 82** [11]:

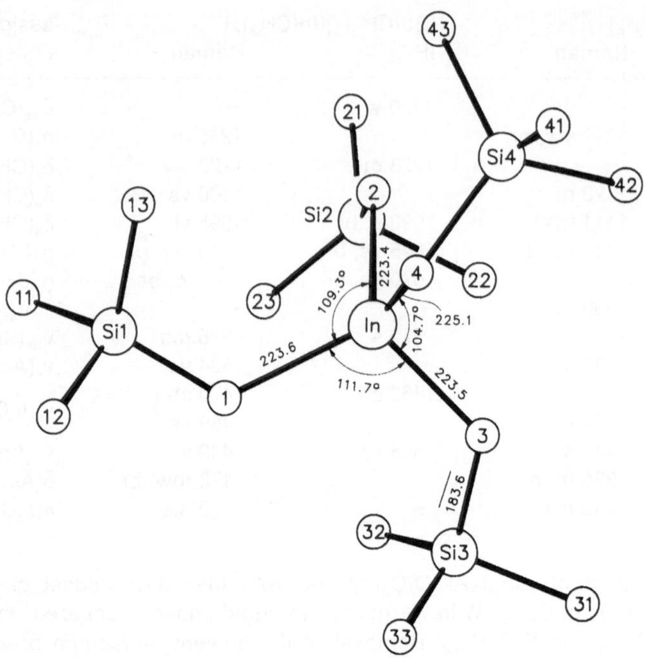

Fig. 82. Molecular structure of the anion $[In\{CH_2Si(CH_3)_3\}_4]^-$ [11].

Other bond distances (pm) and angles (°)

C(2)-Si(2)	184.0(4)	C(1)-In-C(4)	109.26(14)
K···C(2)	329.4(4)	C(2)-In-C(3)	112.00(14)
K···C(31)	328.4(5)	C(2)-In-C(4)	109.78(13)
K···C(4)	312.6(4)	In-C(2)-Si(2)	118.60(18)
K···In	353.7(1)	In-C(4)-Si(4)	118.29(18)

$M[In(C_6H_5)_4]$ (Table **64**, Nos. **14** to **18**, M = Li, Na, K, Rb, Cs). A model of the unit cell of $M[In(C_6H_5)_4]$ compounds is shown in **Fig. 83**. Lattice constants and important parameters are given in Table 63, p. 338 [5].

$Li[In(C_9H_7)_4]$ (Table **64**, No. **19**) was formed in 42% yield in the reaction of $LiCH_3$ with indene (C_9H_8) in C_6H_6, followed by addition of $InCl_3$ (4:4:1 mole ratio). After 12 h of constant stirring at 50 °C and separation of the LiCl, the filtrate was evaporated to dryness and the residue was recrystallized from C_6H_6. When this same procedure was carried out in ether, the bright yellow etherate **$Li[In(C_9H_7)_4] \cdot O(C_2H_5)_2$** resulted; 58% yield, m.p. 120 to 125 °C [3].

^1H NMR in $CDCl_3$ at about 35 °C (δ values in ppm): 0.83 (t, CH_3), 3.12 (q, CH_2), 5.07 (d), and 5.90 (t, C_5-ring), 7.15 (m, C_6-ring) [3].

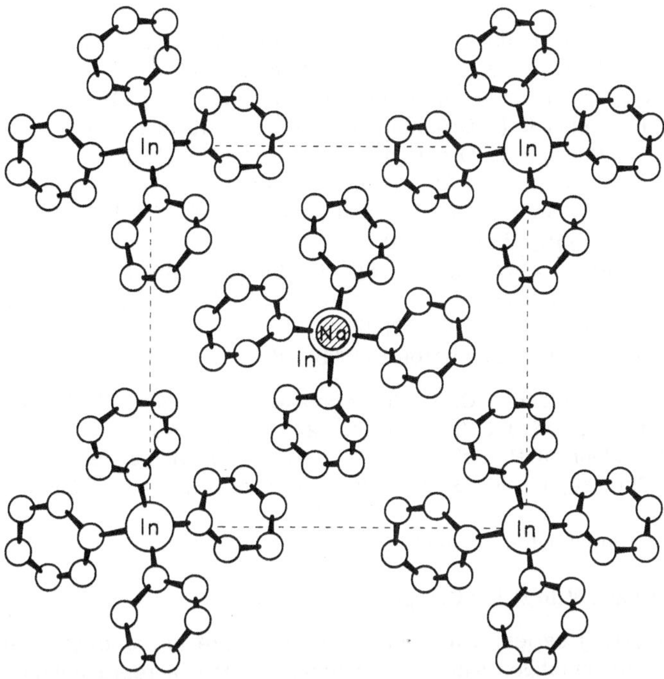

Fig. 83. Model of the unit cell of $M[In(C_6H_5)_4]$ for M = Li, Na (projection on [001] values are given for M = Na) [5].

The IR and Raman frequencies of the In-C bond vibrations are identical (within the error limits) to those of the ether-free compound. From this agreement the ether appears to be coordinated to the Li rather than In [3].

The main fragments in the mass spectrum (80 °C source) were reported to be $[InC_9H_7]^+$, $[C_9H_7]^+$, $^{115}In^+$, and $[O(C_2H_5)_2]^+$ [3].

$[As(C_6H_5)_4][In(C_4(C_6H_5)_4)_2]$ (Table **64**, No. **20**) was produced by the reaction of 1,4-dilithiated 1,2,3,4-tetraphenylbutadiene with $[As(C_6H_5)_4][InCl_4]$ (3:1 mole ratio) in ether. After stirring for 48 h, the yellow product was recrystallized in CH_2Cl_2 [10].

The ^{13}C NMR spectrum (15 lines from 121 to 151.6 ppm, but no δ values reported) and the formation of a Diels-Alder product with 2 equivalents of maleic anhydride indicate a spiro structure (Formula II, $R = C_6H_5$) for the anion; this agrees with the structure of the isoelectronic $Sn(C_4(C_6H_5)_4)_2$ [10].

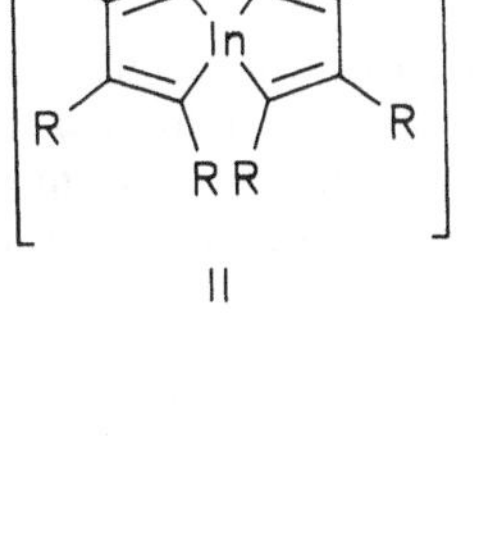

II

References:

[1] Wolfsberger, W.; Schmidbaur, H. (J. Organometal. Chem. **27** [1971] 181/4).
[2] Hoffmann, K.; Weiss, E. (J. Organometal. Chem. **37** [1972] 1/8).
[3] Poland, J. S.; Tuck, D. G. (J. Organometal. Chem. **42** [1972] 307/14).
[4] Hoffmann, K.; Weiss, E. (J. Organometal. Chem. **50** [1973] 17/24).
[5] Hoffmann, K.; Weiss, E. (J. Organometal. Chem. **50** [1973] 25/31).
[6] Weibel, A. T.; Oliver, J. P. (J. Organometal. Chem. **74** [1974] 155/66).
[7] Gavrilenko, V. V.; Kolesov, V. S.; Zakharkin, L. I. (Zh. Obshch. Khim. **47** [1977] 964; J. Gen. Chem. [USSR] **47** [1977] 881).
[8] Tatzel, G.; Schrem, H.; Weidlein, J. (Spectrochim. Acta **A34** [1978] 549/59).
[9] Gavrilenko, V. V.; Kolesov, V. S.; Zakharkin, L. I. (Zh. Obshch. Khim. **49** [1979] 1845/8; J. Gen. Chem. [USSR] **49** [1979] 1623/6).
[10] Peppe, C.; Tuck, D. G. (Polyhedron **1** [1982] 549/62).

[11] Hallock, R. B.; Beachley, O. T., Jr.; Yong-Ji, L.; Sanders, W. M.; Churchill, M. R.; Hunter, W. E.; Atwood, J. L. (Inorg. Chem. **22** [1983] 3683/91).
[12] Reier, F. W.; Wolfram, P.; Schumann, H. (J. Cryst. Growth **93** [1988] 41/4).
[13] Araki, S.; Shimizu, T.; Jin, S.-J.; Butsugan, Y. (J. Chem. Soc. Chem. Commun. **1991** 824/5).
[14] Weidlein, J. (unpublished results).

11.2.2 $M[InR_3H]$ and $M[In_2R_2H_5]$ Compounds

In the $M[InR_nH_{4-n}]$ series of In compounds there are (in contrast to Al and Ga) only a few known representatives with n=3. $Na[In(C_2H_5)_2H_2]$, a representative for n=2, has only been postulated and only as an unstable, unisolable intermediate [2]. One compound of the type $M[In_2R_2H_5]$ is described at the end of this section.

Preparation. The preparation of the compounds listed in Table 66 resulted from the reaction of excess MH with InR_3. For Nos. 1 to 5 THF or $O(C_2H_5)_2$ served as solvents. The reactants were stirred for 3 to 10 h at room temperature, the dark reaction mixture was filtered from insoluble (and undetermined) decomposition products, and the filtrate evaporated to dryness in vacuum at 20 to 45 °C; No. 1 completely and No. 2 slightly decomposed under these conditions. Yields were not reported [1, 2]. Attempts to prepare No. 2 in $CH_3O(CH_2)_2OCH_3$ at 130 °C and No. 3 in C_6H_6 at room temperature failed. In these cases only metallic In, CH_4, and insoluble gray-black products were obtained [3].

Compounds 6 to 8 were prepared in yields of 80%, although No. 6 was isolated only in impure form (presumably contaminated with $Na[In(CH_2Si(CH_3)_3)_4]$ and unidentified decomposition products) [4].

General Remarks. According to [1] and [2], the alkylhydridoindates (No. 1 at 20 °C, No. 2 in pentane at 50 °C, No. 3 at 110 °C) decomposed within a few minutes: $4\,M[InR_3H] \rightarrow 3\,M[InR_4] + MH + In + 1.5\,H_2$. An initial disproportionation $4\,M[InR_3H] \rightarrow 3\,M[InR_4] + M[InH_4]$, followed by a rapid decay of the (hypothetical) InH_4^- anion, was proposed.

The more plausible mechanism proposed by [4] consists of an initial dissociation to MCH_3 and $In(CH_3)_2H$ and rapid disproportionation of the latter to $In(CH_3)_3$ and InH_3. Then $In(CH_3)_3$ reacts with MCH_3 to give $M[In(CH_3)_4]$ and InH_3 decomposes to In and 1.5 H_2.

Nos. 6 and 8 decompose slowly, even at room temperature, and in solution. $Si(CH_3)_4$ was formed in addition to $M[In(CH_2Si(CH_3)_3)_4]$, In, and H_2. The decomposition was followed at constant temperatures (85 to 110 °C, 16 h to 27 d) in C_6H_6 (biphenyl was also produced) and in C_6H_{12}-c. The studies included determinations of the H_2, $Si(CH_3)_4$, and In metal formed [4].

Table 66
M[InR_3H] Compounds.
Further information on numbers preceded by an asterisk is given at the end of the table.
Explanations, abbreviations, and units on p. X.

No.	cation	properties and remarks
[$In(CH_3)_3H]^-$ anion		
1	Li^+	not isolated, dec. in THF solution at ca. 20 °C [2] ^1H NMR (OC_4H_8): −0.75 (CH_3In) [2]
2	Na^+	white solid, dec. ca. 100 °C [2] ^1H NMR (OC_4H_8): −0.82 (CH_3In) [2]
3	K^+	white solid, dec. ca. 110 °C [1, 2] ^1H NMR (OC_4H_8): −0.97 [1], −0.93 (CH_3In) [2] IR (solid): 1600, 1450, 1375 vbr ν(InH); no other bands reported [1]
[$In(C_2H_5)_3H]^-$ anion		
4	Na^+	viscous, colorless liquid; dec. ca. 100 °C within 5 min (90%) [2] gaseous hydrolysis products are C_2H_6 and H_2 (3:1 mole ratio) [2]
5	K^+	viscous, colorless liquid, dec. as No. 4 [2]
[$In(CH_2Si(CH_3)_3)_3H]^-$ anion		
6	Na^+	off-white solid, 80 to 85% purity, dec. 107 to 115 °C [4] ^1H NMR (C_6D_6): −1.01 (unassigned), −0.80 (CH_2In), 0.25 (CH_3Si) [4] IR (impure solid): 1700 to 1320 m, vbr ν(InH), 1285 w, 1239 vs, 950 m, sh, 930 m, sh, 915 s, 845 vs, 822 vs, 742 s, 708 s, 677 m, 600 w, 552 m, 475 m δ(InH), 434 m [4]
7	K^+	colorless crystals, dec. 109 °C (turns black), 125 °C (liquid) [4] ^1H NMR (C_6D_6): −0.83 (CH_2In), 0.33 (CH_3Si) [4] IR (solid): 1915 vw, 1835 vs, 1700 to 1300 s, vbr ν(InH), 1280 w, 1240 vs, 905 vs, br, 850 vs, br, 820 vs, br, 742 vs, 719 s, 706 s, sh, 672 s, 612 m, 595 m, sh, 578 m, 511 w, 453 s, 435 m, sh δ(InH), 381 w, 252 m [4]
[$In(CH_2Si(CH_3)_3)_3D]^-$ anion		
8	K^+	colorless crystals [4] ^1H NMR (C_6D_6): −0.82 (CH_2In), 0.33 (CH_3In) [4] IR (solid): 1280 w, 1251 m, sh, 1240 s, 1025 m, br ν(InD), 913 s, 892 m, sh, 850 s, 820 s, 743 s, 720 m, 710 m, sh, 674 m, 576 m, 562 w, sh, 454 m, 395 w, sh δ(InD), 381 m, 252 w [4]

$[Li(OC_4H_8)_2][In_2(C\{Si(CH_3)_3\}_3)_2H_5]$

This compound was prepared by adding a suspension of $LiAlH_4$ in THF to a solution of $[Li(OC_4H_8)_3][In(C\{Si(CH_3)_3\}_3)Cl_3]$ in THF at −50 °C. After 1 h the mixture was brought to room temperature, stirred for several hours, and freed of solvent. Extraction of the gray residue with $C_6H_5CH_3$ produced a rubbery mass that became powdery after washing with C_5H_{12}. A final recrystallization from $C_6H_5CH_3$ at −20 °C gave colorless crystals in 85% yield; m.p. 115 °C (dec.) [5, 7].

^1H NMR ($C_6D_5CD_3$?): 0.55 (CH_3Si), 1.38 m and 3.55 (m, OC_4H_8), 4.75 (vbr, w1/2=900 Hz from N.O.E) ppm. The very broad ^1H NMR signal at 4.7 to 4.75 ppm was assigned to In-H-Li bridge bonding. The large half-width of the signal, however, did not permit an unambiguous assignment, nor was it possible to determine the number of bridging H atoms by the nuclear Oberhauser effect (N.O.E.) [6]. ^7Li NMR (C_6D_6?): −0.8 (relative to $LiNO_3$-C_6D_6) ppm [6, 7].

The number and intensities of the IR absorption bands at 1725, 1695, 1660, 1635 (all s) cm^{-1} were interpreted as terminal and bridging H atoms and, together with the NMR results, suggested structure I (D=THF, R=$C\{Si(CH_3)_3\}_3$) [5, 7]:

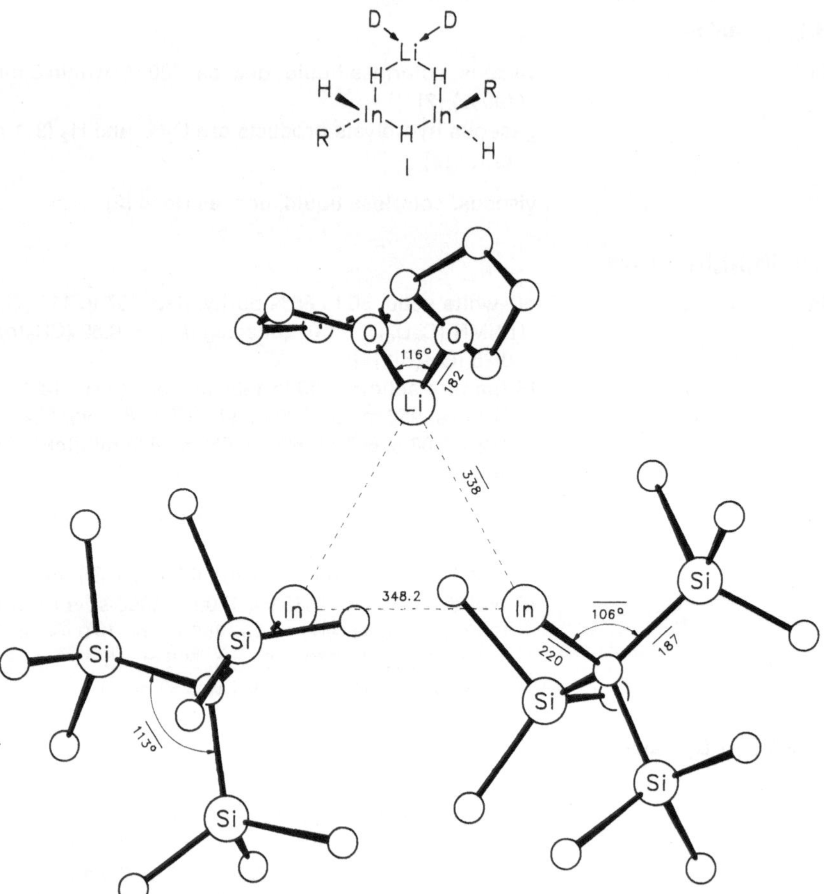

Fig. 84. Molecular structure of $[Li(OC_4H_8)_2][In_2(C\{Si(CH_3)_3\}_3)_2H_5]$ [5].

The salt crystallizes in the monoclinic space group $P2_1/n-C_{2h}^5$ (No. 14) with the lattice constants a=1292.1(3), b=1446.6(2), c=2540.0(5) pm, β=95.73(2)°; Z=4 gives D_c=1.19 g/cm^3. Each of the In and Li atoms is in a distorted tetrahedral environment; the In-C bonds are inclined by about 34° from and on opposite sides of the six-membered ring (Formula I). The LiO_2 plane is twisted about 66° with respect to the six-membered ring. The terminal and bridging H atoms were not located. The $C(Si(CH_3)_3)_3$ groups were disordered (2 possible sites with an occupancy ratio of 7:3), and **Fig. 84** reproduces only the orientation with the higher occupancy [5].

References:

[1] Gavrilenko, V. V.; Kolesov, V. S.; Zakharkin, L. I. (Zh. Obsch. Khim. **47** [1977] 964; J. Gen. Chem. [USSR] **47** [1977] 881).

[2] Gavrilenko, V. V.; Kolesov, V. S.; Zakharkin, L. I. (Zh. Obsch. Khim. **49** [1979] 1845/8; J. Gen. Chem. [USSR] **49** [1979] 1623/6).

[3] Beachley, O. T., Jr.; Tressier-Youngs, C.; Simmons, R. G.; Hallock, R. B. (Inorg. Chem. **21** [1982] 1970/3).

[4] Hallock, R. B.; Beachley, O. T., Jr.; Yong Ji, L.; Sanders, W. M.; Churchill, M. R.; Hunter, W. E.; Atwood, J. L. (Inorg. Chem. **22** [1983] 3683/91).

[5] Avent, A. G.; Eaborn, C.; Hitchcock, P. B.; Smith, J. D.; Sullivan, A. C. (J. Chem. Soc. Chem. Commun. **1986** 988/9).

[6] Avent, A. G.; Eaborn, C.; El-Kheli, M. N. A.; Molla, M. E.; Smith, J. D.; Sullivan, A. C. (J. Am. Chem. Soc. **108** [1986] 3854/5).

[7] Atwood, J. L.; Bott, S. G.; Hitchcock, P. B.; Eaborn, C.; Shariffudin, R. S.; Smith, J. D.; Sullivan, A. C. (J. Chem. Soc. Dalton Trans. **1987** 747/55).

11.2.3 Anions with Indium-Halogen Bonds

11.2.3.1 $M[InR_nX_{4-n}]$ Compounds

The compounds collected in Table 67 were prepared by the following procedures:

Method I: Reaction of R_nInX_{3-n} (R=alkyl or aryl, X=halogen, n=1 to 3) with $[ER'_4]X$ or ER'_3X_2 (E=N, P, As, Sb). Depending on n, different reactions were carried out:

a. $InR_3+[ER'_4]X \rightarrow [ER'_4][InR_3X]$
b. $R_2InX+[ER'_4]X \rightarrow [ER'_4][InR_2X_2]$
c. $InR_3+ER'_3X_2 \rightarrow [ER'_3R][InR_2X_2]$
d. $RInX_2+[ER'_4]X \rightarrow [ER'_4][InRX_3]$
e. $R_2InX+ER'_3X_2 \rightarrow [ER'_3R][InRX_3]$

In most cases these reactions were conducted by mixing equimolar amounts of reactants in CH_2Cl_2 at room temperature. After stirring for 1 to 3 h, any residue present was filtered off, the solution was cooled to 0 to 10 °C to precipitate the salt, and the solid purified by recrystallization from CH_2Cl_2 [9, 10]. Longer reaction times were required for the iodine compounds, 28 and 30 [10]. Compound 1 was obtained free of solvent by combining equimolar amounts of starting materials, while for No. 6 anhydrous CH_3OH and for No. 10 a mixture of $CHCl_3$ and $O(C_2H_5)_2$ were used as solvents. Addition of ether (No. 6) or simple concentration (No. 10) precipitated the complex salts [3].

Method II: Electrochemical oxidation of metallic In in the presence of RX.
The oxidation was carried out in a solution of RX and $[NR'_4]X$ in CH_3CN (see

also p. 156). The preparation of No. 26 can serve as an example. A solution of 2.0 g $[N(C_2H_5)_4]Br$ and 10 mL C_2H_5Br in 50 mL CH_3CN was electrolyzed for 20 h at 5 V and 80 mA on an In anode and a Pt cathode. From the clear colorless solution a white solid was precipitated by adding $O(C_2H_5)_2$, was washed carefully with $CHCl_3$, and then dried in vacuum. The yield, based on the weight loss of the In electrode, was (and for the other products, as well) between 80 and 90%. During electrolysis of the iodine-containing solutions (Nos. 29, 31, 33, and 38) the solution turned light yellow and after several minutes the anode became covered with a red-violet solid (presumably In^I iodides) [11].

General Remarks. The ionic complex compounds, $M[InR_nX_{4-n}]$, are usually hygroscopic and increasingly air-sensitive as n becomes larger. They decompose before reaching the melting point (Author: the iodides evolve I_2). The IR and Raman spectra of the anions were assigned assuming C_{3v} ($[InR_3X]^-$ and $[InRX_3]^-$) or C_{2v} ($[InR_2X_2]^-$) symmetry. The weighted averages of the In-C bond frequencies have been correlated with the chemical shifts (δ ^1H in Hz) of the indium-methyl protons and the relationships are graphed in [9, 10]. Simple molecular force fields have been determined by least-squares calculations based on the IR and Raman spectra of the chloro compounds No. 2, 3, 7, 8, 13, and 14 and their deutero analogs No. 4, 9, and 15. Some of the derived force constants (in mdyn/Å) are given below [14].

	$[In(CH_3)_3Cl]^-$	$[In(CH_3)_2Cl_2]^-$	$[In(CH_3)Cl_3]^-$
f(In-C)	1.94	2.19	2.41
f (In-Cl)	0.66	0.98	1.38
d(C-In-C)	0.12	0.09	–
d(C-In-Cl)	0.58	0.45	0.36
d(Cl-In-Cl)	–	0.50	0.54

Table 67
$M[InR_nX_{4-n}]$ Compounds with X = Halogen.
Further information on numbers preceded by an asterisk is given at the end of the table. Explanations, abbreviations, and units on p. X.

No.	anion method of preparation (yield)	cation properties and remarks
with X = F		
1	$[In(CH_3)_3F]^-$ Ia	$[N(CH_3)_3CH_2C_6H_5]^+$ salt reaction of the components at 28 to 32 °C [2] conductivity: $1.96 \times 10^{-2}\ \Omega^{-1} \cdot cm^2 \cdot mol^{-1}$ [2]
with X = Cl		
*2	$[In(CH_3)_3Cl]^-$ Ia (ca. 100%)	$[As(CH_3)_4]^+$ salt colorless crystals, m.p. 120 ± 5 °C (dec.) [9] $D_m = 1.60\ g/cm^3$ [9] ^1H NMR (CD_2Cl_2): −0.67 (CH_3In) [9]

Table 67 (continued)

No. anion method of preparation (yield)	cation properties and remarks
*2 (continued)	IR (solid): 1144 s, 1139 m, br $\delta_s(CH_3)$, 688 vs $\varrho(CH_3)$, 487 s, 477 ms $\nu_{as}(InC_3)$ 468 m $\nu_s(InC_3)$, 211 s, br $\nu(InCl)$, 144 s $\delta(ClInC)$, 130 sh $\delta(InC_3)$ [9] Raman (solid/CH_2Cl_2): 1149 s, 1139 m $\delta_s(CH_3)$, 690 vw, br $\varrho(CH_3)$, 485 m, 475/480 mw $\nu_{as}(InC_3)$, 467 vs/469 vs $\nu_s(InC_3)$, 218 m, sh $\nu(InCl)$, 145 sh/139 sh $\delta(ClInC)$, 131 ms $\delta(InC_3)$; see also [14]; force constants, see general remarks on p. 350 [9]
3 $[In(CH_3)_3Cl]^-$ Ia (ca. 100%)	$[Sb(CH_3)_4]^+$ salt colorless crystals, m.p. 115±5 °C (dec.) [9] spectra similar to No. 2
4 $[In(CD_3)_3Cl]^-$ Ia (74%)	$[As(CH_3)_4]^+$ salt IR and Raman (solid): 2218 $\nu_{as}(CD_3)$, 2120 $\nu_s(CD_3)$, 885 $\delta_s(CD_3)$, 517 $\varrho(CD_3)$, 436 $\nu_{as}(InC_3)$, 412 $\nu_s(InC_3)$, 197 $\nu(InCl)$, 135 $\delta(ClInC)$, 119 $\delta(ClInC)$; for force constants, see No. 2 [14]
*5 $[In(C_6H_2(CH_3)_3\text{-}2,4,6)_3Cl]^-$ Ia (91%)	$[N(CH_3)_4]^+$ salt white powder, m.p. 280 °C (dec.) [20] ^{1}H NMR (CD_3CN): 2.67 (s, CH_3-4), 2.72 (s, CH_3-2,6), 2.82 (s, CH_3N), 7.7 (s, C_6H_2) [20] ^{13}C NMR (CD_3CN): 21.9 (CH_3-4), 22.1 (CH_3-2,6), 40.1 (CH_3N), 126.9, 134.9, 140.7, 151.3 (C_6H_2) [20] IR (solid): 1740 w, 1700 w, 1590 m, 1530 m, 1280 w, 1265 w, 1220 w, 1005 s, 940 s, 855 s, 715 m [20]
6 $[In(CH_3)_2Cl_2]^-$ Ib	$[N(C_2H_5)_4]^+$ salt white solid, softened and melted 41 to 62 °C [3] conductivity (CH_3NO_2): 83.9 $\Omega^{-1}\cdot cm^2\cdot mol^{-1}$ [3] IR (solid): 715 vs, br $\varrho(CH_3)$, 512 s $\nu_{as}(InC_2)$, 483 m, 465 vw $\nu_s(InC_2)$; no other bands reported [3]
7 $[In(CH_3)_2Cl_2]^-$ Ic (ca. 100%)	$[As(CH_3)_4]^+$ salt colorless crystals, m.p. 115±5 °C (dec.) [9] ^{1}H NMR (CD_2Cl_2): −0.13 (CH_3In) [9] IR (solid): 1170 vw, 1157 vw $\delta_s(CH_3)$, 723 vs, br, 676 sh $\varrho(CH_3)$, 522 s $\nu_{as}(InC_2)$, 493 ms $\nu_s(InC_2)$, 260 mw $\nu_s(InCl_2)$, 241 s $\nu_{as}(InCl_2)$, 130 sh $\delta(InC_2)$, 118 s, br $\delta(InCl_2)$ and $\delta(ClInC)$ [9] Raman (solid/CH_2Cl_2): 1172 w, 1159 s $\delta_s(CH_3)$, 677 w, br $\varrho(CH_3)$, 522 m/523 w $\nu_{as}(InC_2)$, 499 vs/487 vs $\nu_s(InC_2)$, 262 m/265 m $\nu_s(InCl_2)$, 245 sh/250 sh $\nu_{as}(InCl_2)$, 145 sh, 130 m $\delta(ClInC)$ and $\delta(InC_2)$, 116 ms, 108 sh $\delta(InCl_2)$ and $\delta(ClInC)$ [9]; see also [14]; force constants, see general remarks on p. 350

Table 67 (continued)

No.	anion method of preparation (yield)	cation properties and remarks
8	$[In(CH_3)_2Cl_2]^-$ Ic (ca. 100%)	$[Sb(CH_3)_4]^+$ salt white solid, m.p. 118 ± 5 °C (dec.) [9] spectra similar to No. 7
9	$[In(CD_3)_2Cl_2]^-$ Ib (84%)	$[As(CH_3)_4]^+$ salt IR and Raman (solid): 2210 $\nu_{as}(CD_3)$, 2098 $\nu_s(CD_3)$, 904, 898 $\delta_s(CD_3)$, 549, 512 $\varrho(CD_3)$, 486 $\nu_{as}(InC_2)$, 442 $\nu_s(InC_2)$, 252 $\nu_s(InCl_2)$, 235 $\nu_{as}(InCl_2)$, 130 $\delta(InC_2)$, $\delta(ClInC)$, 114, 96, 86; force constants, see No. 7 [14]
10	$[In(CH_3)_2Cl_2]^-$ Ib	$[As(C_6H_5)_4]^+$ salt white solid, m.p. 148 to 152 °C [3] conductivity (CH_3NO_2): 71.0 $\Omega^{-1} \cdot cm^2 \cdot mol^{-1}$ [3] IR (solid/$CHCl_3$): 512 m/511 m $\nu_{as}(InC_2)$, 489 w/470 (s?) $\nu_s(InC_2)$; no other bands reported [3]
*11	$[(C_5H_6)InCl_2 \cdot InCl_4]^{2-}$ (cyclopentadiene ring drawn) special	2 $[N(C_2H_5)_4]^+$ salt no properties reported [19]
12	$[In(CH_3)Cl_3]^-$ II [11]	$[N(C_2H_5)_4]^+$ salt ^{1}H NMR ($OC(CD_3)_2$): 0.07 (CH_3In) [13] IR (solid): 522 s, 290 vs, br, 206 m, 120 s; no other bands reported [13]
*13	$[In(CH_3)Cl_3]^-$ Ie	$[As(CH_3)_4]^+$ salt colorless crystals, m.p. 134 ± 5 °C (dec.) [6] $D_m = 1.92\ g/cm^3$ [7] ^{1}H NMR (CH_2Cl_2): 0.28 (CH_3In), 2.12 (CH_3As) [6] IR (solid): 1160 w $\delta_s(CH_3)$, 748 vs $\varrho(CH_3)$, 530 s $\nu(InC)$, 307 m, sh $\nu_s(InCl_3)$, 298, 291 vs $\nu_{as}(InCl_3)$, 130 sh $\delta_s(InCl_3)$, 122, 119 s $\delta_{as}(InCl_3)$, 108 sh $\delta(ClInC)$ [6] Raman (solid): 1159 s $\delta_s(CH_3)$, 743 vw,br $\varrho(CH_3)$, 529 vs $\nu(InC)$, 309 ms $\nu_s(InCl_3)$, 293 mw $\nu_{as}(InCl_3)$, 135 s $\delta_s(InCl_3)$, 125 m $\delta_{as}(InCl_3)$, 107 m $\delta(ClInC)$; see also in [9] and [14] with slightly different assignment; force constants, see general remarks on p. 350 [6]
*14	$[In(CH_3)Cl_3]^-$ Ie (ca. 100%)	$[Sb(CH_3)_4]^+$ salt colorless crystals, m.p. 119 ± 5 °C (dec.) [6] $D_m = 2.06\ g/cm^3$ [7] spectra similar to No. 13
15	$[In(CD_3)Cl_3]^-$ Ie (88%)	$[As(CH_3)_3CD_3]^+$ salt IR and Raman (solid): 2208 $\nu_{as}(CD_3)$, 2121 $\nu_s(CD_3)$, 898 $\delta_s(CD_3)$, 557 $\varrho(CD_3)$, 481 $\nu(InC)$, 307 $\nu_s(InCl_3)$, 290 $\nu_{as}(InCl_3)$, 123, 106 $\delta(ClInCl)$ and $\delta(ClInC)$; force constants, see No. 13 [14]

Table 67 (continued)

No.	anion method of preparation (yield)	cation properties and remarks
16	$[In(CH_2C_6H_5)Cl_3]^-$ II	$[N(C_2H_5)_4]^+$ salt white solid; no other properties reported [11]
*17	$[In\{C(Si(CH_3)_3)_3\}Cl_3]^-$ special	$[Li(OC_4H_8)_3]^+$ salt colorless crystals, m.p. 152 to 155 °C [17] ^1H NMR (C_6D_6): 0.54 (CH_3Si), 1.39, 3.54 (m, OC_4H_8) [17] ^7Li NMR (C_6D_6): −0.94 (s, relative to $LiNO_3/C_6D_6$) [17] mass spectrum (100 °C): $[InRCl_2-CH_3]^+$ and $[InRCl_2-Cl]^+$ (R = $C(Si(CH_3)_3)_3$) [17]
18	$[In\{C(Si(CH_3)_3)_2C_6H_5\}Cl_3]^-$	$[Li(OC_4H_8)_3]^+$ salt prepared like No. 17 from LiR and $InCl_3$ in 83% yield [17] colorless needles, m.p. 143 °C [17] ^1H NMR (C_6D_6): 1.03 (CH_3Si), 1.4, 3.6 (m, OC_4H_8) [17] ^7Li NMR ($C_6D_5CD_3$): −0.78 s [17] mass spectrum (150 °C): $[anion-Cl+Li]^+$ (40%), $[anion-2\,Cl]^+$ (100%) [17]
19	$[In(C_6H_5)Cl_3]^-$ II	$[N(C_2H_5)_4]^+$ salt white solid, no other properties reported [11]
with X = Br		
20	$[In(CH_3)_3Br]^-$ Ia (ca. 100%)	$[As(CH_3)_4]^+$ salt colorless crystals, m.p. 109 °C [10] ^1H NMR (CD_2Cl_2): −0.5 (CH_3In) [10] IR (solid): 1138 ms $\delta_s(CH_3)$, 690 vs, br $\varrho(CH_3)$, 485, 476 vs $\nu_{as}(InC_3)$, 465 s $\nu_s(InC_3)$, 160 s, sh $\nu(InBr)$, 144 vs, br $\delta(BrInC)$, 120 m, sh $\delta(InC_3)$ [10] Raman (solid and CH_2Cl_2): 1151, 1144 s $\delta_s(CH_3)$, 690 vw, br $\varrho(CH_3)$, 486, 476 s $\nu_{as}(InC_3)$, 466 vs $\nu_s(InC_3)$, 160 sh $\nu(InBr)$, 142 s $\delta(BrInC)$, 120 s $\delta(InC_3)$ [10]
*21	$[In(C_2H_5)_3Br]^-$ special	$[Sb(C_2H_5)_3C_3H_7\text{-}n]^+$ salt colorless oily liquid, m.p. < −67 °C [1]
*22	$[In(CH_3)_2Br_2]^-$ Ic (ca. 100%)	$[As(CH_3)_4]^+$ salt colorless crystals, m.p. 112 °C (dec.) [10] D_m = 2.09 g/cm³ [8] ^1H NMR (CD_2Cl_2): 0.18 (CH_3In) [10] IR (solid): 1174, 1165, 1153 vw $\delta_s(CH_3)$, 720 vs, br $\varrho(CH_3)$, 520 s $\nu_{as}(InC_2)$, 484 s $\nu_s(InC_2)$, 169 vs $\nu_{as}(InBr_2)$, 129 s $\delta(BrInC)$, 105 sh $\delta(InBr_2)$ [10] Raman (solid and CH_2Cl_2): 1169 ms $\delta_s(CH_3)$, 725 vw, br $\varrho(CH_3)$, 522 s $\nu_{as}(InC_2)$, 486 vs $\nu_s(InC_2)$, 178 vs $\nu_s(InBr_2)$, 169 w, br $\nu_{as}(InBr_2)$, 133 s, sh $\delta(BrInC)$, 124 vs $\delta(InC_2)$, 106 s $\delta(InBr_2)$ [10]

Table 67 (continued)

No.	anion method of preparation (yield)	cation properties and remarks
*23	$[In\{CH(Si(CH_3)_3)_2\}_2Br_2]^-$ special	$[Li(N(CH_3)_2CH_2)_2]^+$ salt colorless crystals, m.p. 168 to 169 °C (dec.) [21] ^{1}H NMR (C_6D_6): −0.05 (HCIn), 0.48 (CH_3Si), 1.65 (CH_2N), 1.98 (CH_3N) [21] ^{13}C NMR (C_6D_6): 5.5 (CH_3Si), 14.0 (CHIn), 46.5 (CH_3N), 57.5 (CH_2N) [21] UV (C_5H_{12}, $c=2.6\times10^{-2}$ mol/L): $\lambda_{max}(\varepsilon)=250(60)$, 270(50) [21] IR (solid): 1018 s δ(CH), 948 m $\varrho(CH_3Si)$, 926 m, 846 vs, 789 m, 775 m, 753 m, 730 m, 730 m, 688 m, 666 m, 608 w $\nu(SiC_3)$, 439 w, 405 w $\nu(InC_2)$ [21]
24	$[In(CH_3)Br_3]^-$ Id	$[N(C_4H_9)_4]^+$ salt white needles (from CH_3OH) [5] IR/Raman (solid): 507 m/507 s ν(InC), −/202 s ν(InBr); no other bands reported [5]
25	$[In(CH_3)Br_3]^-$ Ie (ca. 100%)	$[As(CH_3)_4]^+$ salt colorless crystals, m.p. 161 °C (dec.) [10] ^{1}H NMR (CD_2Cl_2): 0.43 (CH_3In) [10] IR (solid): 1151 mw $\delta_s(CH_3)$, 731 s, br $\varrho(CH_3)$, 520 s ν(InC), 207 vs $\nu_{as}(InBr_3)$, 203 w $\nu_s(InBr_3)$ [10] Raman (solid and CH_2Cl_2): 1154 s $\delta_s(CH_3)$, 737 vw, br $\varrho(CH_3)$, 520 vs ν(InC), 210 w $\nu_{as}(InBr_3)$ 204 vs $\nu_s(InBr_3)$, 148 m, sh δ(BrInC), 127 w, 86 m $\delta(InBr_3)$ [10]
26	$[In(C_2H_5)Br_3]^-$ II	$[N(C_2H_5)_4]^+$ salt white solid, insoluble in $CHCl_3$, C_6H_6, slightly soluble in CH_3CN [11] ^{1}H NMR (CD_3OD): 0.8 (br), 2.2 (s, C_2H_5In), 1.35, 3.35 (C_2H_5N) [11]
27	$[In(C_6H_5)Br_3]^-$ II	$[N(C_2H_5)_4]^+$ salt yellow solid [11] ^{1}H NMR (CD_3OD): 1.0, 3.0 (C_2H_5N), 7.3 (br, C_6H_5) [11]
	with X = I	
28	$[In(CH_3)_3I]^-$ Ia	$[As(CH_3)_4]^+$ salt white solid, m.p. 90 °C (dec.) [10] ^{1}H NMR (CD_2Cl_2): −0.3 (CH_3In) [10] IR (solid): 680 vs, br $\varrho(CH_3)$, 484 vs $\nu_{as}(InC_3)$, 466 s $\nu_s(InC_3)$ [10] Raman (solid and CH_2Cl_2): 682 vw $\varrho(CH_3)$, 483 mw $\nu_{as}(InC_3)$, 456 vs $\nu_s(InC_3)$, 133 s, 118 s δ(IInC) and $\delta(InC_3)$, 106 s ν(InI) [10]

Table 67 (continued)

No.	anion method of preparation (yield)	cation properties and remarks
29	$[In(CH_3)_2I_2]^-$ II	$[N(C_4H_9)_4]^+$ salt white solid [11] ^1H NMR (CD_3CN): 0.43 (CH_3In), 0.75, 1.25, 1.75, 1.75, 2.85 (C_4H_9N) [11]
30	$[In(CH_3)_2I_2]^-$ Ic	$[As(CH_3)_4]^+$ salt white solid, m.p. 85 °C (dec.) [10] ^1H NMR (CD_2Cl_2): 0.33 (CH_3In) [10] IR (solid): 1165 w $\delta_s(CH_3)$, 715 vs $\varrho(CH_3)$, 517 s $\nu_{as}(InC_2)$, 480 s $\nu_s(InC_2)$, 130 vs $\nu_{as}(InI_2)$, 110 sh, 99 sh $\delta(InC_2)$ and $\delta(IInC)$ [10] Raman (solid and CH_2Cl_2): 1162 m, 1157 m $\delta_s(CH_3)$, 520 m $\nu_{as}(InC_2)$, 483 vs $\nu_s(InC_2)$, 141 vs $\nu_s(InI_2)$, 130 sh $\nu_{as}(InI_2)$, 114 m, 96 mw $\delta(InC_2)$ and $\delta(IInC)$ [10]
31	$[In(CH_3)I_3]^-$ II	$[N(C_2H_5)_4]^+$ salt pale yellow solid [11] ^1H NMR ($CDCl_3$): 0.72 (CH_3In), 1.45, 3.35 (C_2H_5N) [11]
32	$[In(CH_3)I_3]^-$	$[Sb(CH_3)_4]^+$ salt from $Sb(CH_3)_5$ and InI_3 (1:1 ratio) in CH_2Cl_2 (3 d, 20 °C), no pure compound isolated [10] ^1H NMR (CD_2Cl_2): 0.68 (CH_3In) [10] IR (solid): 1150 w $\delta_s(CH_3)$, 710 s, br $\varrho(CH_3)$, 508 s $\nu(InC)$, 160 vs, br $\nu_{as}(InI_3)$, 150 s $\nu_s(InI_3)$ [10] Raman (solid): 1150 mw $\delta_s(CH_3)$, 718 vw, br $\varrho(CH_3)$, 508 vs $\nu(InC)$, 165 w, sh $\nu_{as}(InI_3)$, 153 vs $\nu_s(InI_3)$, 141 mw, 118 m, sh, 108 mw $\delta(IInC)$ and $\delta(InI_3)$ [10]
33	$[In(C_2H_5)I_3]^-$ II	$[N(C_2H_5)_4]^+$ salt white solid, no other properties reported [11]
*34	$[In(C_2H_5)I_3]^-$ special	$[P(C_6H_5)_4]^+$ salt colorless crystals [15], pale yellow prisms [16] ^1H NMR (CD_2Cl_2): 1.03 (t, CH_3), 1.16 (q, CH_2In; J(H,H)=7.8), 7.63 to 7.94 (m, C_6H_5) [15, 16]
35	$[In(C_4H_9)I_3]^-$	$[P(C_6H_5)_4]^+$ salt prepared like No. 34 in 52% yield; no properties reported [15]
36	$[In(C_5H_5\text{-}c)I_3]^-$ Id (ca. 100%)	$[N(C_3H_7\text{-}n)_4]^+$ salt solid, insoluble in common solvents except CH_3NO_2 [4] IR and Raman (solid, anion-bands only): 3045 $\nu(CH)$, 2965 s, 2940 s $\nu(HC\text{-}1)$, 1095 ms, 982 s $\nu(C\text{-}C)$, 800 s $\gamma(CH)$, 751 s, 660 mw, 321 w $\nu(InC)$, 190 mw $\nu_{as}(InI_3?)$, 140 vs $\nu_s(InI_3)$, other bands at 114 m, 43 m [4]

Table 67 (continued)

No.	anion method of preparation (yield)	cation properties and remarks
37	$[In(C_6H_5)I_3]^-$ II	$[N(C_2H_5)_4]^+$ salt brown crystals [11] ^1H NMR ($CDCl_3$): 1.35, 3.20 (C_2H_5N), 6.76 (m), 7.1 (br), 7.2 (br, all C_6H_5) [11]
38	$[In(C_6H_5)I_3]^-$ II	$[N(C_4H_9)_4]^+$ salt yellow solid [11] ^1H NMR ($CDCl_3$): 0.9, 1.1, 1.7, 3.1 (C_4H_9N), 7.1 (br), 7.3 (br), 7.6 (br, all C_6H_5In) [11]
with X = Cl, X′ = I		
*39	$[In(CH_3)_2ClI][In(CH_3)ClI_2]^{2-}$ special	2 [K(18-crown-6]$^+$ salt colorless crystals [18]
40	$[In(C_4H_9)_2ClI]^-$ I b	$[N(CH_3)_4]^+$ salt crystalline solid, m.p. 112 °C [12] conductivity ($C_6H_5NO_2$): 24 to 29 $\Omega^{-1} \cdot cm^2 \cdot mol^{-1}$ [12]

* Further information:

$[As(CH_3)_4][In(CH_3)_3Cl]$ (Table **67**, No. **2**) crystallizes in the monoclinic space group $P2_1/c-C^5_{2h}$ (No. 14) with the lattice constants a = 726.3(4), b = 1585.4(8), c = 1500.6(6) pm, β = 128.18(3)°; Z = 4 gives D_c = 1.616 g/cm³. The molecular parameters are given in **Fig. 85** [9].

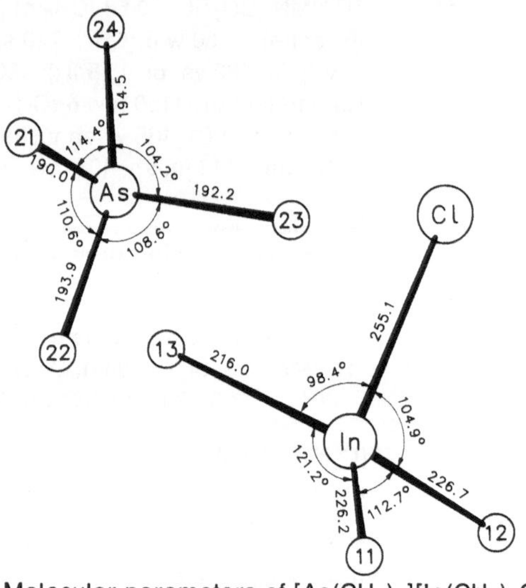

Fig. 85. Molecular parameters of $[As(CH_3)_4][In(CH_3)_3Cl]$ [9].

Other bond angles (°)

C(12)-In-C(13)	115.2	C(21)-As-C(23)	110.7
C(11)-In-Cl	100.4	C(22)-As-C(24)	111.2

$[N(CH_3)_4][In(C_6H_2(CH_3)_3\text{-}2,4,6)_3Cl]$ (Table **67**, No. **5**) was recrystallized from CH_3CN for the X-ray structure determination and contained one solvent molecule per salt unit. This solvate is monoclinic, space group $P2_1/n-C^5_{2h}$ (No. 14) with the unit cell constants a = 1693.8(5), b = 915.4(5), c = 2085.9(5) pm, β = 94.49(2)°; Z = 4 gives D_c = 1.283 g/cm³. The three mesityl rings are twisted like a propeller; in relation to the In-Cl vector the following vector angles were reported: C(11)-C(16) = 52.9°, C(21)-C(26) = 42.6°, and C(31)-C(36) = 51.0°. The structure is depicted in **Fig. 86** [20].

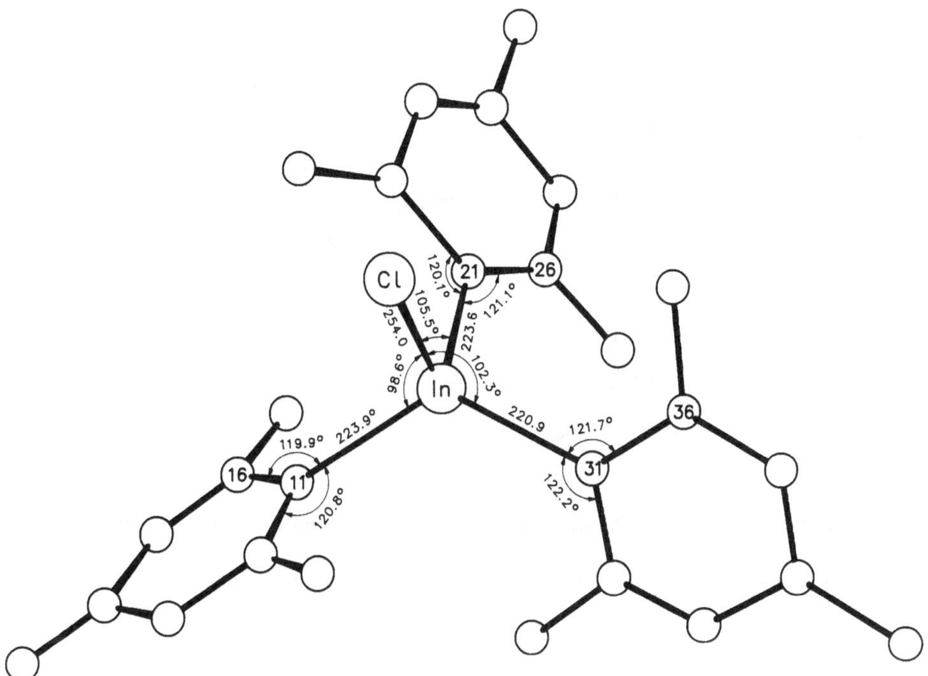

Fig. 86. Molecular structure of the anion of $[N(CH_3)_4][In(C_6H_2(CH_3)_3\text{-}2,4,6)_3Cl]$ [20].

$[N(C_2H_5)_4]_2[C_4H_4InCl_2 \cdot InCl_4]$ (Table **67**, No. **11**). This compound with an indium metallacycle was prepared by the reaction of the zirconium metallacycle $(C_5H_5\text{-}c)_2ZrC_4H_4$ with two equivalents of $[N(C_2H_5)_4][InCl_4]$ in THF (89% yield). The (white?) precipitate was purified by recrystallization from CH_2Cl_2. Presumably, the In atoms of the anion are connected by Cl bridges, so that each metal attains a coordination number of 4. The compound was not characterized by elemental analysis or spectroscopy [19].

$[E(CH_3)_4][In(CH_3)Cl_3]$ (Table **67**, No. **13**, E = As; No. **14**, E = Sb). The two compounds are isostructural and crystallize in the monoclinic space group $Pc-C^2_s$ (No. 7); Z = 2. The individual cell parameters are (As/Sb): a = 726.9(1)/743(2), b = 670.5(2)/676(1), c = 1349.2(2)/1357(3) pm, β = 94.26(1)°/94.3(1)°; D_c = 1.88/2.04 g/cm³. The distances and angles designated in **Fig. 87** are those of the As compound; the corresponding values for the Sb analog are given below [7].

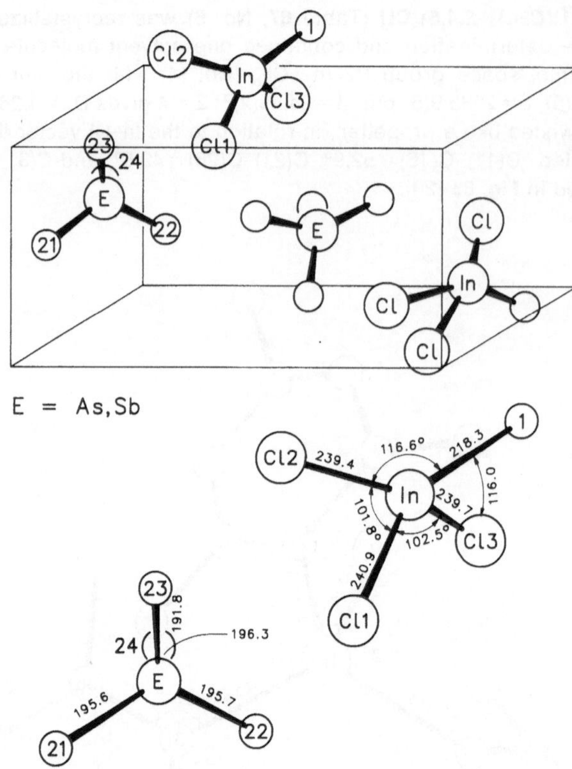

Fig. 87. Unit cell of $[E(CH_3)_4][In(CH_3)Cl_3]$ and molecular parameters for No. 13 (E = As) [7].

Distances (pm) and angles (°) of the Sb compound No. 14 (E = Sb)

In-Cl(1)	238(3)	Cl(1)-In-Cl(2,3)	106
In-Cl(2)	241(2)	C(1)-In-Cl(1,3)	114
In-Cl(3)	240(2)	Cl(2)-In-Cl(3)	101
In-C(1)	215(5)	Cl(2)-In-C(1)	115
Sb-C(21)	211(5)	C(21)-Sb-C(22)	112
Sb-C(22)	212(4)	C(21)-Sb-C(23)	113
Sb-C(23)	212(4)	C(21)-Sb-C(24)	114
Sb-C(24)	210(4)	C(22)-Sb-C(23)	116
		C(24)-Sb-C(22,23)	100

$[Li(OC_4H_8)_3][In\{C(Si(CH_3)_3)_3\}Cl_3]$ (Table **67**, No. **17**) was made from $InCl_3$ and $Li[C\{Si(CH_3)_3\}_3]$ in THF at −40 °C. The mixture was stirred for several h and warmed to room temperature. THF was removed and the residue extracted with hexane; concentration and cooling gave the compound in 87% yield [17].

The compound crystallizes in the monoclinic space group $P2_1/c - C_{2h}^5$ (No. 14) with the unit cell parameters a = 919.1(4), b = 1468.2(4), c = 2605.1(9) pm, β = 94.52(3)°; Z = 4 gives D_c = 1.29 g/cm³. The anion and cation are joined by an In-Cl-Li bridge; **Fig. 88** contains the most important distances and angles [17].

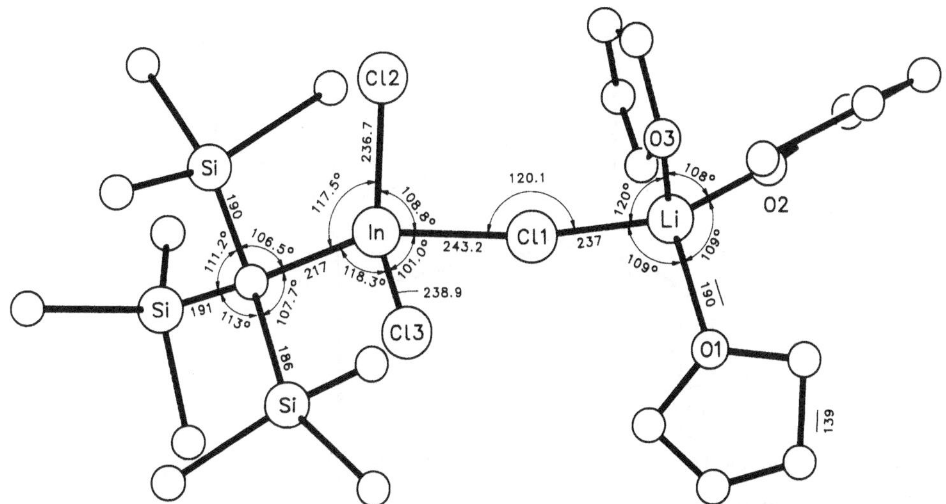

Fig. 88. The structure of $[Li(OC_4H_8)_3][In\{C(Si(CH_3)_3)_3\}Cl_3]$ [17].

Other bond angles (°)

Cl(1)-In-C = 113.8(5)	Cl(1)-Li-O(2) = 102(1)
Cl(2)-In-Cl(3) = 102.7(3)	O(1)-Li-O(3) = 109(2)

The compound was used as starting material to prepare the hydrido derivative, $[Li(OC_4H_8)_2][In_2\{C(Si(CH_3)_3)_3\}_2H_5]$ (see p. 348) [17].

$[Sb(C_2H_5)_3C_3H_7\text{-}n][In(C_2H_5)_3Br]$ (Table **67**, No. **21**) was obtained by adding C_3H_7Br to a heated (90 to 110 °C) mixture of $In(C_2H_5)_3$ and $Sb(C_2H_5)_3$ (1:1:1 mole ratio). The ionic nature was demonstrated by the distinct increase in the electrical conductivity from 4×10^{-6} for the trialkyl mixture at 50 °C to $1.85 \times 10^{-2}\ \Omega^{-1} \cdot cm^{-1}$ at 100 °C for the liquid salt. The product is thermally stable to about 125 °C and, contrary to $In(C_2H_5)_3$, is not light-sensitive. Elemental analysis and spectroscopic results were not reported [1].

$[As(CH_3)_4][In(CH_3)_2Br_2]$ (Table **67**, No. **22**) crystallizes in the orthorhombic space group $P2_12_12_1-D_2^4$ (No. 19) with the lattice constants a = 765.0(1), b = 1139.2(1), c = 1561.1(2) pm; Z = 4 and D_c = 2.14 g/cm³. The data were collected at −150 °C. The tetrahedron of the anion is definitely distorted (see **Fig. 89** on p. 360), while the polyhedron of the cation corresponds closely to those of Nos. 2 and 13 [8].

$[Li(N(CH_3)_2CH_2)_2][In\{CH(Si(CH_3)_3)_2\}_2Br_2]$ (Table **67**, No. **23**) was produced by reacting an ether solution of $InBr_3$ with a solution of $Li[CH(Si(CH_3)_3)_2]$ in $O(C_2H_5)_2$ (1:2 mole ratio) in the presence of a slight excess of $(CH_3)_2NCH_2CH_2N(CH_3)_2$ at 0 °C. After warming to room temperature and stirring for 1 h, the solvent was evaporated, and the residue extracted with C_5H_{12}; the product crystallized at −50 °C in 68% yield. The title compound is only slightly soluble in C_5H_{12}, and is monomeric in C_6H_6. In nonpolar solvents slow decomposition (precipitation of LiBr) sets in at room temperature; dec. p. 168 to 169 °C [21].

The salt crystallizes in the orthorhombic, space group $P2_12_12_1-D_2^4$ (No. 19) with the cell constants a = 1315.0(8), b = 1368.4(5), c = 2077.5(7) pm; Z = 4. The In and Li atoms are connected by two common Br atoms; the resulting $LiBr_2In$ ring is nearly planar with a

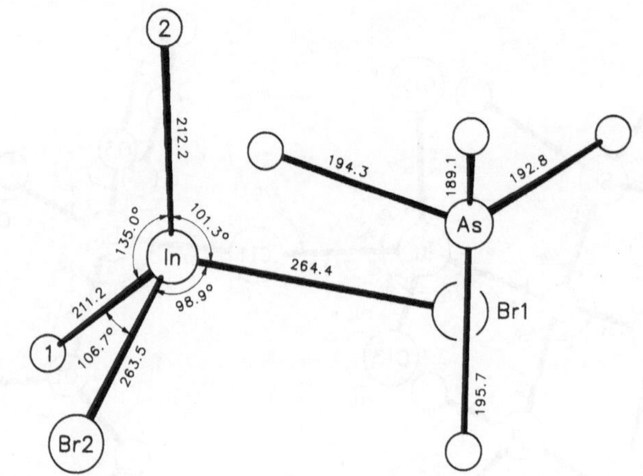

Fig. 89. Molecular parameters of $[As(CH_3)_4][In(CH_3)_2Br_2]$ [8].

Other bond angles (°)

Br(1)-In-C(1)	103.7(5)	Br(2)-In-C(2)	105.7(4)

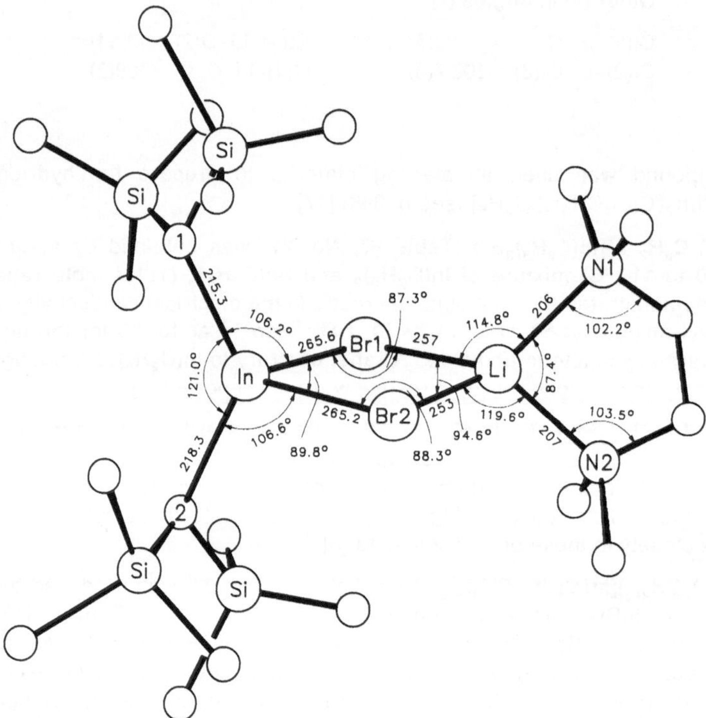

Fig. 90. Molecular structure of $[Li(N(CH_3)_2CH_2)_2][In\{CH(Si(CH_3)_3)_2\}_2Br_2]$ [21].

Other bond angles (°)

Br(1)-In-C(2)	114.2(2)	Br(1)-Li-N(2)	118.1(8)
Br(2)-In-C(1)	114.8(2)	Br(1)-Li-N(1)	124.9(8)

1.2° dihedral angle along the Br···Br axis. The C(1)-In-C(2) plane and the N(1)-Li-N(2) plane form angles of 83.8° and 86.4° with the $LiBr_2In$ ring, respectively. The $HC(Si(CH_3)_3)_2$ ligands are twisted with respect to one another, but do not lie ideally in the gaps. The structure of the molecule is depicted in **Fig. 90** on p. 360 [21].

$[P(C_6H_5)_4][In(C_2H_5)I_3]$ (Table **67**, No. **34**) was made by the reaction of $C_2H_5InI_2 \cdot 2\,OS(CH_3)_2$ (Table 23, No. 10) with $[P(C_6H_5)_4]I$ in CH_2Cl_2, then precipitation by the addition of $O(C_2H_5)_2$; 67% yield [15].

It crystallizes in the tetragonal space group $I4_1/a-C_{4h}^6$ (No. 88) with the cell parameters a = 1476.1(3), b = 1332.9(3) pm; Z = 4 and D_c = 1.98 g/cm³. The most important anion parameters are shown in **Fig. 91**; the cation parameters are also reported in [16].

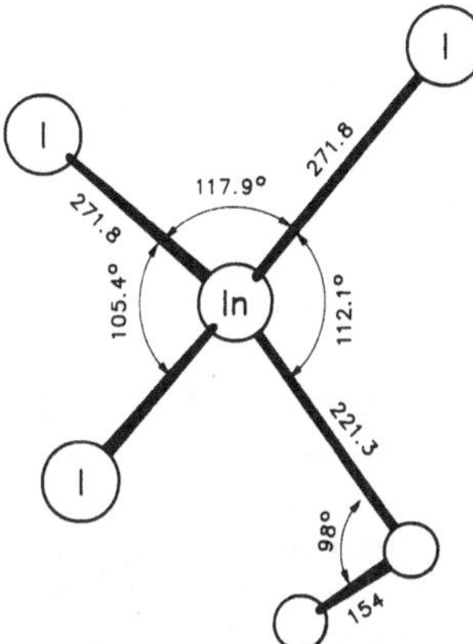

Fig. 91. The molecular parameters of the anion of $[P(C_6H_5)_4][In(C_2H_5)I_3]$ [16].

$[K(18\text{-crown-}6)]_2[In(CH_3)_2ClI][In(CH_3)ClI_2]$ (Table **67**, No. **39**). First, $InCl_3$ and $LiCH_3$ (1:3 mole ratio) in $O(C_2H_5)_2$ were stirred for 24 h at room temperature. The ether was removed in vacuum, the residue was extracted with C_6H_6, and the filtered extract treated with KI and 18-crown-6 ether (0.5 mole per mole of $InCl_3$). After another 18 h of stirring, the KI had dissolved and two liquid phases had formed. The denser phase was liquid clathrate, and after a time air-sensitive crystals separated in 60% yield. These crystals reformed the liquid clathrate when C_6H_6 or $C_6H_5CH_3$ was added, and the portion of aromatic solvent in the clathrate was determined by ^1H NMR [18].

The compound is monoclinic, space group $P2_1/c-C_{2h}^5$ (No. 14) with the lattice constants a = 1950.8(9), b = 850.3(5), c = 2943.7(9) pm, β = 96.55(3)°; Z = 4 gives D_c = 1.83 g/cm³. The crystal contains two different, independent anions, whose halogen atoms have the following contact distances from the K atoms: K(1)···I(2) = 362.3 pm, K(1)···I(3) = 360.0 pm, and K(2)··· Cl(2) = 316 pm. The highly symmetrical coordination of K(1) is also reflected in the I(2)··· K(1)···I(3) contact angle of 172.4°, while K(2) is displaced about 81 pm out of the plane of the crown ether and toward Cl(2). The most important bond distances and angles are given in **Fig. 92** [18].

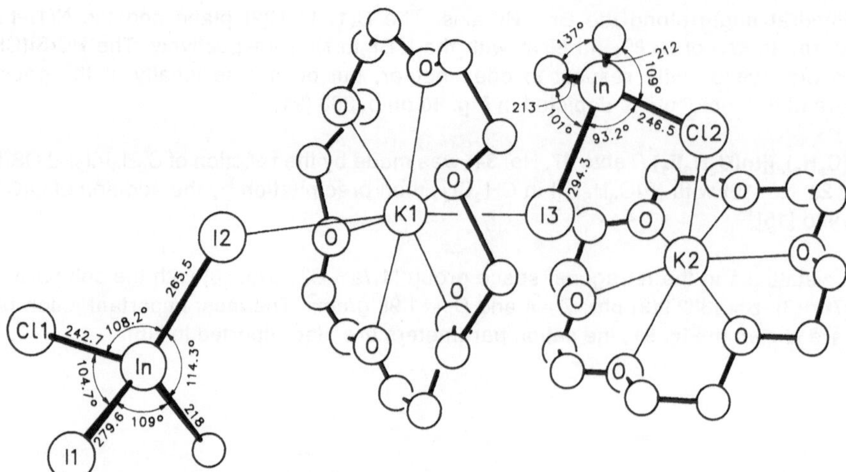

Fig. 92. Molecular structure of $[K(18\text{-crown-}6)]_2[In(CH_3)_2ClI][In(CH_3)ClI_2]$ [18].

References:

[1] Dötzer, R. (Ger. 1200817 [1965]; C.A. **63** [1965] No. 15896).
[2] Siemens-Schuckertwerke A.-G. (Fr. 1461819 [1966]; C.A. **67** [1967] No. 17386).
[3] Clark, H. C.; Pickard, A. L. (J. Organometal. Chem. **13** [1968] 61/71).
[4] Contreras, J. G.; Tuck, D. G. (J. Organometal. Chem. **66** [1974] 405/12).
[5] Waterworth, L. G.; Worrall, I. J. (J. Organometal. Chem. **81** [1974] 23/6).
[6] Widler, H. J.; Hausen, H.-D.; Weidlein, J. (Z. Naturforsch. **30b** [1975] 645/7).
[7] Guder, H.-J.; Schwarz, W.; Weidlein, J.; Widler, H.-J.; Hausen, H.-D. (Z. Naturforsch. **31b** [1976] 1185/9).
[8] Schwarz, W.; Guder, H.-J.; Prewo, R.; Hausen, H.-D. (Z. Naturforsch. **31b** [1976] 1427/9).
[9] Widler, H.-J.; Schwarz, W.; Hausen, H.-D.; Weidlein, J. (Z. Anorg. Allgem. Chem. **435** [1977] 179/90).
[10] Widler, H.-J.; Weidlein, J. (Z. Naturforsch. **34b** [1979] 18/22).

[11] Habeeb, J. J.; Said, F. F.; Tuck, D. G. (J. Organometal. Chem. **190** [1980] 325/34).
[12] Srivastava, T. N.; Singhal, K. (J. Indian Chem. Soc. **57** [1980] 225/6).
[13] Taylor, M. J.; Tuck, D. G.; Victoriano, L. (J. Chem. Soc. Dalton Trans. **1981** 928/32).
[14] Haaland, A.; Weidlein, J. (Acta Chem. Scand. **A36** [1982] 805/11).
[15] Peppe, C.; Tuck, D. G.; Vicotriano, L. (J. Chem. Soc. Dalton Trans. **1982** 2165/8).
[16] Khan, M. A.; Peppe, C.; Tuck, D. G. (J. Organometal. Chem. **280** [1985] 17/25).
[17] Atwood, J. L.; Bott, S. G.; Hitchcock, P. B.; Eaborn, C.; Shariffudin, R. S.; Smith, J. D.; Sullivan, A. C. (J. Chem. Soc. Dalton Trans. **1987** 747/55).
[18] Babaian, E. A.; Barden, L. M.; Hrncir, D. C.; Hunter, W. E.; Atwood, J. L. (J. Inclusion Phenom. **5** [1987] 605/10).
[19] Fagan, P. J.; Nugent, W. A. (J. Am. Chem. Soc. **110** [1988] 2310/2).
[20] Leman, J. T.; Barron, A. R. (Organometallics **8** [1989] 2214/9).

[21] Uhl, W.; Layh, M.; Hiller, W. (J. Organometal. Chem. **368** [1989] 139/54).

11.2.3.2 Compounds of the $M[InR_2XX']$ Type with X = Halogen and X' = Pseudohalogen

Simple $[InR_nX_{4-n}]^-$ anions containing pseudohalogens are unknown.

The mixed-substituent title compounds were obtained by treating a suspension of $(C_4H_9\text{-n})_2InX'$ ($X' = N_3$ or SeCN) in dry CH_3OH with an equimolar amount of $[N(CH_3)_4]I$ in the same solvent. After refluxing for 6 h, the salts were precipitated by reducing the volume of the reaction mixture and purified by recrystallization from CH_3OH. The compounds are only slightly soluble in nonpolar organic solvents, but dissolve well in CH_3OH or $C_6H_5NO_2$. In nitrobenzene the conductivity is 24 to 29 $\Omega^{-1} \cdot cm^2 \cdot mol^{-1}$, corresponding to a 1:1 electrolyte [1].

$[N(CH_3)_4][(C_4H_9)_2InN_3I]$

This salt forms pale yellow crystals, m.p. 136 °C. In the IR spectrum only the vibrations of the N_3 group were evaluated: ca. 2050 $\nu_{as}(N_3)$, 1340 $\nu_s(N_3)$, 665 $\delta(N_3)$; in addition 570 $\nu(InC_2)$ cm^{-1} [1].

$[N(C_2H_5)_4][(C_4H_9)_2InN_3I]$

The compound is a crystalline solid which melts at 260 °C. the IR data for the N_3 moiety are the same as above [1].

$[N(CH_3)_4][(C_4H_9)_2In(SeCN)I]$

This compound is a crystalline solid, m.p. >240 °C. The IR spectrum was evaluated only for the SeCN vibrations: ca. 2050 $\nu(CN)$, 660 $\nu(CSe)$, 400 $\delta(SeCN)$; in addition 570 $\nu(InC_2)$ cm^{-1} [1].

Reference:

[1] Srivastava, T. N.; Singhal, K. (J. Indian Chem. Soc. **57** [1980] 225/6).

11.2.3.3 Compounds of the $M[In_2R_6X]$ and $M[In_3R_8X_2]$ Types

These types of anions are not stable; none of the following compounds have been isolated in pure form nor have they been completely characterized, and their existence is speculative.

$[Sb(C_2H_5)_3C_3H_7\text{-n}][(C_2H_5)_3InBrIn(C_2H_5)_3]$

This salt was obtained by adding an equivalent of $In(C_2H_5)_3$ to molten $[Sb(C_2H_5)_3C_3H_7\text{-n}]$-$[In(C_2H_5)_3Br]$ (see p. 353), followed by 1 h warming at 80 °C. The resulting clear and almost colorless liquid froze below −90 °C. It decomposed above 135 °C with evolution of gas. Irradiation with light produces no changes. The specific conductivity is 0.96×10^{-2} $\Omega^{-1} \cdot cm^2$; analytical and spectroscopic data were not reported [1].

$[Sb(CH_3)_4][(CH_3)_2InCl_2(In(CH_3)_3)_2]$

The compound was obtained by the reaction of $(CH_3)_3SbCl_2$ with 3 mol $In(CH_3)_3$ in CH_2Cl_2. Even at room temperature and slightly reduced pressure $In(CH_3)_3$ split off, so that no pure compound could be isolated. Characterization in solution was carried out by ^1H NMR, IR, and Raman spectroscopies, but no details were given [2].

References:

[1] Dötzer, R. (Ger. Offen. 1200817 [1965]; C.A. **63** [1965] 15896).

[2] Widler, H.-J.; Schwarz, W.; Hausen, H.-D.; Weidlein, J. (Z. Anorg. Allgem. Chem. **435** [1977] 179/90).

11.2.3.4 Compounds of the Type $M[RInX_3]$

$[N(C_2H_5)_4][X_3InCH_2X]$ (X = Br, I)

Both compounds were obtained by electrochemical oxidation of In metal in CH_2X_2/CH_3CN (conditions: indium as anode, platinum wire as cathode, 30 mA) until the metal had dissolved. After the initially formed InX had dissolved in the solution (intermediate formation of X_2InCH_2X, see on p. 171), one equivalent of $[N(C_2H_5)_4]X$ was added to the mixture. Filtration and concentration of the filtrate gave colorless crystals on addition of ether and cooling; 84 and 79% yield for X = Br and I, respectively [1].

The conductivity in CH_3CN was 152 (X = Br) and 157 (X = I) $\Omega^{-1} \cdot cm^2 \cdot mol^{-1}$ [1].

The NMR spectra of the Br compound were run in CD_2Cl_2 (values in ppm). ^1H NMR: 1.35 (t, CH_3), 2.75 (s, CH_2In), 3.25 (q, CH_2N); ^{13}C NMR: 8.2 (CH_3), 18.3 (br, CH_2In), 53.3 (CH_2N) [1].

Reference:

[1] Annan, T. A.; Tuck, D. G.; Khan, M. A.; Peppe, C. (Organometallics **10** [1991] 2159/66).

11.3 Anions with In–O, In–S, and In–Se Bonds

$Na[In(CH_3)_2(OSi(CH_3)_3)_2]$

The compound was obtained by the reaction of dimeric $(CH_3)_2InOSi(CH_3)_3$ (see p. 188) with $NaOSi(CH_3)_3$; m.p. 87 to 89 °C. It is very soluble in water; hydrolysis with methane evolution, however, occurred only after it was acidified with dilute mineral acids. The compound is also very soluble in C_6H_6, and exhibits a doubled molecular weight in this solvent; from this it was concluded that the anions associate through sodium-oxygen bridges. Further structural details were not reported [1].

$[Sb(CH_3)_4][In(CH_3)_2(OSi(CH_3)_3)_2]$

This salt was postulated from ^1H NMR studies (no details) to have formed in a 1:1 mixture of $(CH_3)_2InOSi(CH_3)_3$ and $(CH_3)_4SbOSi(CH_3)_3$ [1].

$K_2[In_4(CH_3)_{12}S]$

The compound was obtained by condensing excess $In(CH_3)_3$ into a slurry of K_2S in $O(C_2H_5)_2$. After a 4 h reflux, the hygroscopic white microcrystals were filtered, washed with pentane, and dried in vacuum. The yield is practically quantitative. At 128 °C decomposition begins before the melting point [2].

The IR and Raman frequencies of the solid are summarized in Table 68. From the data a highly symmetrical structure was deduced as shown in Formula I, in which each In atom is bonded to two terminal CH_3 groups and one CH_3 group is bridging [2].

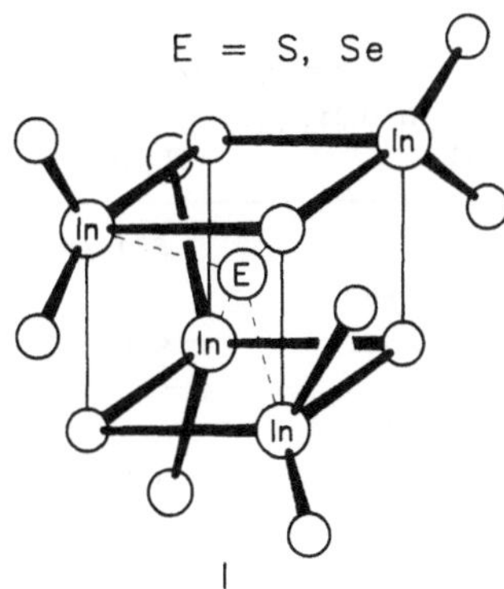

I

$K_2[In_4(CH_3)_{12}Se]$

The preparation and properties correspond to those of the above compound, except that toluene (instead of ether) at room temperature was used as solvent. Thermal decomposition begins at 161 °C.

The vibrational spectral data are compared with those of the corresponding sulfur derivative in Table 68 [2].

Table 68
Vibrational Spectra of $K_2[In_4(CH_3)_{12}X]$ Compounds with X=S, Se [2].
Wavenumbers in cm^{-1}.

$K_2[In_4(CH_3)_{12}S]$ IR	Raman	$K_2[In_4(CH_3)_{12}Se]$ IR	Raman	assignment
–	2968 m	–	2982 w	ν_{as}, $\nu_s(CH_3)$
2938 s	2930 m	2938 m	2930 w	
2890 s	–	2895 m	2870 w	
–	2858 m	–	–	
2825 s	–	2830 m	–	
2165 w	–	2165 w	–	$\delta_{as}+\delta_s(CH_3)$
1440 w, br	1450 vw	1440 w, br	1450 w, br	$\delta_{as}(CH_3)$
–	–	1168 w	1178 w	$\delta_s(CH_3)$ bridge
1150 sh	1152 w	1150 w	1165 sh	
–	1142 m	1140 sh	–	
1135 m	–	1132 m	1155 m	
1118 vw	1115 s	1120 w	1128 s	$\delta_s(CH_3)$ terminal
1101 s	1104 m	1102 m	1114 m	
1098 w	–	1098 m	–	
–	–	1010 vw	–	
735 sh	–	752 sh	–	$\varrho(CH_3)$ terminal
718 s	–	718 s	720 vw, br	
–	–	695 sh	–	
665 s	–	665 s	650 w, br	
625 sh	–	625 sh	–	$\varrho(CH_3)$ bridge
562 w	–	560 sh	–	

Table 68 (continued)

$K_2[In_4(CH_3)_{12}S]$ IR	Raman	$K_2[In_4(CH_3)_{12}Se]$ IR	Raman	assignment
486 vs	–	482 vs	481 vs	ν(InC) terminal
480 vs	479 vs	476 vs	–	
450 sh	460 s	460 vs	470 vs	ν(InC) bridge
400 sh	–	450 vs	–	
–	–	396 ?	–	
–	340 m, br	–	348 w	δ(InC)
312 w	–	315 vw	309 vw	
273 s	270 sh	–	290 vw	
255 sh	–	–	270 vw	$ν_{as}(XIn_4)$
–	222 w	–	224 w	δ(InC)
–	150 m	–	–	
–	115 m	–	–	

References:

[1] Schmidbaur, H. (Z. Chem. [Leipzig] **8** [1968] 254).
[2] van Dahlen, K. H.; Dehnicke, K. (Chem. Ber. **110** [1977] 383/94).

11.4 Anions with In–N Bonds

$Na[In(CH_3)_2(N_2C_3H_3)_2]$

The salt was obtained from a two-stage reaction. The first step involved equimolar amounts of $In(CH_3)_3$ and $Na[N_2C_3H_3]$ (sodium pyrazolide) in boiling THF. The $Na[In(CH_3)_3$-$N_2C_3H_3]$ formed was not isolated, but was treated with an equivalent of pyrazole and refluxed another 8 to 10 h. After removal of the solvent a very hygroscopic white solid remained; no yield or m.p. was reported [1].

The ^1H NMR spectrum in $(CD_3)_2CO$ shows the following signals (in ppm): −0.32 (s, CH_3In), 6.09 (br, H-4), 7.48 (br, H-3, H-5) [1].

Unlike the behavior of the analogous gallium compound, the indium compound appears to form no transition metal complexes; all attempts to prepare them have been unsuccessful [1].

$[N(C_2H_5)_4][In(CH_3)_2(N_2C_3H_3)_2]$

This salt was prepared from the above Na salt in THF and an equivalent of $[N(C_2H_5)_4]Cl$ in a little CH_3OH. After centrifuging off the NaCl formed, the solution was dried in vacuum, the hygroscopic white solid was dissolved in acetone, and reprecipitated by adding ether. No yield or m.p. was reported [1].

^1H NMR in $(CD_3)_2CO$ (δ values in ppm): −0.20 (s, CH_3In), 1.13 (m, CH_3 of C_2H_5), 3.85 (q, CH_2N), 4.03 (t, H-4, J = 1.7 Hz), 7.47 (d, H-3,5, J = 1.7 Hz) [1].

Reference:

[1] Breakell, K. R.; Patmore, D. J.; Storr, A. (J. Chem. Soc. Dalton Trans. **1975** 749/54).

11.5 Anions with In-Sn Bonds

$Li[In(CH_3)_3Sn(CH_3)_3]$

This compound has not been isolated. Its formation from $In(CH_3)_3$ and $Li[Sn(CH_3)_3]$ in dimethoxymethane at $-60\,°C$ was indicated by ^1H NMR (δ values in ppm): 0.23 (CH_3Sn, $^2J(H,Sn) = 22$ Hz); 0.82 (CH_3In, $^3J(H,Sn) = 21$ Hz). The values are referred to $Si(CH_3)_4$, recalculated from those reported versus $c\text{-}C_5H_{10}$ ($\delta = 1.513$ ppm) [2].

The compound decomposes at room temperature within 2 days, darkening and forming $Li[In(CH_3)_4]$ and $Li[Sn(Sn(CH_3)_3)_3]$ [1]. A comparison of the compounds $Li[M(CH_3)_3Sn(CH_3)_3]$ for M = Al, Ga, and In is given in [2].

References:

[1] Weibel, A. T.; Oliver, J. P. (J. Am. Chem. Soc. **94** [1972] 8590/2).
[2] Weibel, A. T.; Oliver, J. P. (J. Organometal. Chem. **74** [1974] 155/66).

11.6 Ionic Derivatives of Ylides

In this chapter various derivatives or adducts of the ylides $CH_2{=}P(CH_3)_3$, $CH_2{=}P(CH_3)_2\text{-}N{=}P(CH_3)_3$, $CH_2{=}S(O)(CH_3)_2$, $CH_2{=}P(C_6H_5)_3$, and $CH_2 \leftarrow N(CH_3)_2C_2H_4N(CH_3)_2$ are presented. Neutral addition compounds with ylides of the type $X_3InCH_2{=}PR_3$ and similar amino derivatives are summarized in Section 2.5.2.

$[(CH_3)_2In(CH_2)_2P(CH_3)_2]_2$

The compound was prepared by adding a benzene solution of $(CH_3)_2InCl$ at 6 °C to 2 equivalents of $(CH_3)_3P{=}CH_2$ in benzene and stirring for 15 h at 20 °C. After filtering off the $[P(CH_3)_4]Cl$ formed and removing the solvent, the residue was recrystallized from 1:1 $C_6H_5CH_3/C_5H_{12}$ at low temperature; colorless crystals, 98% yield, m.p. 114 °C, sublimes under vacuum. When the reaction was carried out with $(CH_3)_2InBr$ in toluene, a complicated mixture of oligomers resulted; the title compound was extracted with hot benzene in a yield of 91%. The compound is a dimer (cryoscopy in benzene) and has an eight-membered ring (Formula I) [1].

The ^1H NMR spectrum shows the following signals in C_6H_6 at 30 °C (δ values in ppm): -0.17 (s, CH_3In), 0.03 (d, CH_2P, $J(H,P) = 14.5$ Hz), 0.85 (d, CH_3P, $J(H,P) = 12.25$ Hz). The $\{^1H\}$ ^{31}P NMR spectrum (C_6H_6, 30 °C) is a singlet at $\delta = -30.0$ ppm [1].

The vibrational spectrum shows the characteristic deformation vibrations of the PCH_3, PCH_2, $InCH_3$, and $InCH_2$ groups, but data were not reported [1].

$[M_2 - CH_3]^+$ was the highest mass peak in the mass spectrum (no details of the measurement), from which it was concluded that dimers exist in the gas phase [1].

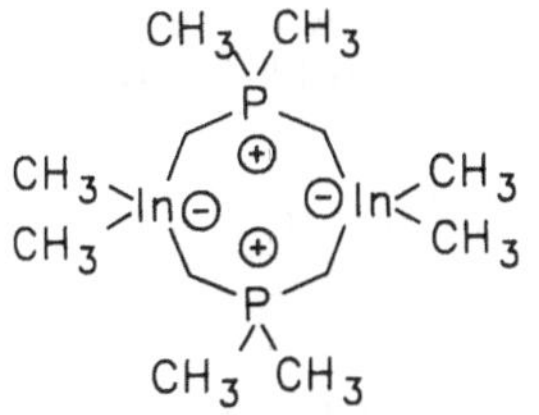

CH3 CH3 P ⊕ CH3>In⊖ ⊖In<CH3 CH3 ⊕ CH3 P CH3 CH3

I

H3C CH3 P CH3 ⊖In ⊕N CH3 P CH3 H3C

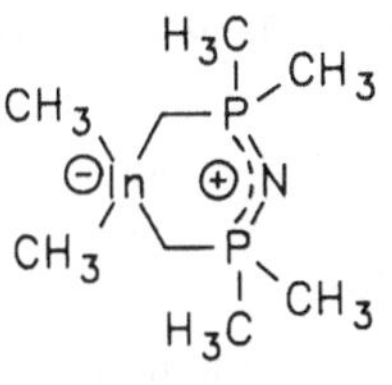

II

$(CH_3)_2In(CH_2P(CH_3)_2)_2N$

This compound was prepared analogously to the Ga derivative [2] from $In(CH_3)_3$ (or its etherate?) and $CH_2{=}P(CH_3)_2{-}N{=}P(CH_3)_3$ with elimination of CH_4, but was not characterized. A cyclic structure as shown in Formula II was postulated [3].

$\{In(CH_2S(O)(CH_3)_2)_3Cl_3\}_2$

The compound was prepared by mixing $InCl_3$ in THF with four equivalents of the ylide $(CH_3)_2S(O){=}CH_2$ (prepared in THF from $(CH_3)_3S(O)Br$ and NaH) and stirring the mixture for 3 d at room temperature. The white precipitate was filtered, washed with C_5H_{12}, and dried; 72.5% yield, dec. 160 °C. The compound is moderately soluble in DMF, DMSO, CH_3OH, and water; insoluble in $CHCl_3$, CH_2Cl_2, and $O(C_2H_5)_2$. It decomposes in water at 5 °C to form $[(CH_3)_3S{=}O]Cl$ [6].

The ^1H NMR in CD_3OD at −50 °C (δ values in ppm): 2.63 (s, CH_2In), 3.28 (s, CH_3S); overlaid by the solvent signals [6].

The conductivity in methanol at −50 °C is 100 $\Omega^{-1} \cdot cm^2 \cdot mol^{-1}$; an ionic structure such as $[\{(CH_3)_2S(O)CH_2\}_3In(\mu\text{-}Cl)In\{CH_2S(O)(CH_3)_2\}_3]Cl_5$ was derived [6].

$\{In(CH_2P(C_6H_5)_3)_2Cl_3\}_2$

The compound was obtained by adding $InCl_3$ to 2 equivalents of a THF solution of $(C_6H_5)_3$-$P{=}CH_2$. After the disappearance of the yellow ylide color (10 h), the title compound separated as a white precipitate, which was washed with anhydrous ether and dried; 64.3% yield, dec. 197 °C. It dissolves in alcohols, $CHCl_3$, CH_2Cl_2, DMF, DMSO, and water; however, these solutions are not stable at room temperature and decompose to form $[P(C_6H_5)_3CH_3]Cl$ [4].

^1H NMR (CD_2Cl_2, −80 °C, δ values in ppm): 2.2 (d, CH_2P, $^2J(H,P) = 15.2$ Hz), 7.2 to 7.9 (m, C_6H_5). The ionic structure, $[\{(C_6H_5)_3ECH_2\}_2InCl]_2Cl_4$ (Formula III, E = P) with a dimeric cation was suggested [4].

$$\left[\begin{array}{ccc} (C_6H_5)_3ECH_2 & & CH_2E(C_6H_5)_3 \\ & In(\mu\text{-}Cl)_2In & \\ (C_6H_5)_3ECH_2 & & CH_2E(C_6H_5)_3 \end{array}\right] Cl_4$$

III

The cytotoxic activity of this compound against mouse leukemic cells was studied [7].

$\{In(CH_2As(C_6H_5)_3)_2Cl_3\}_2$

This compound was prepared in the same way as the preceding compound from $InCl_3$ and $(C_6H_5)_3As{=}CH_2$; white solid, 70.5% yield, m.p. 122 to 125 °C. The structure corresponds to Formula III with E = As [5].

^1H NMR (CD_2Cl_2, −80 °C, δ values in ppm): 2.88 (s, CH_2In), 7.3 to 8.3 (m, C_6H_5) [5].

$[In\{CH_2N(CH_3)_2C_2H_4N(CH_3)_2\}_2I]I_2$

This cationic complex was obtained by stirring together InI and CH_2I_2 (1:6 mole ratio) in CH_3CN until a clear solution had formed. All volatile materials were removed in vacuum

(24 h at 50 °C), the residue dissolved in CH_3CN and treated with TMEDA (TMEDA = $(CH_3)_2$-$NC_2H_4N(CH_3)_2$), and the mixture stirred for 6 h at room temperature. The resulting precipitate (the solution contained $I_3InCH_2 \cdot$ TMEDA; see Section 2.5.2.1) was collected, dissolved in boiling water, and the solution cooled; colorless crystals [8].

The conductivity in CH_3CN solution (concentration ca. 1 mmol/L) was 230 $\Omega^{-1} \cdot cm^2 \cdot mol^{-1}$ [8].

The NMR spectra were run in D_2O (δ values in ppm): ^1H NMR: 2.58 (s, CH_3NIn), 3.14 (br, CH_2 of C_2H_4), 3.31 (s, CH_2In), 3.40 (CH_3NC), 3.66 (br, CH_2 of C_2H_4). ^{13}C NMR: 45.9 (CH_3NIn), 54.8 (CH_2In), 58.6 (br, CH_3NC_2), 59.8 (br, CH_2 of C_2H_4), 62.5 (CH_2 of C_2H_4) [8].

The compound crystallizes in the monoclinic space group $P2_1c-C^5_{2h}$ (No. 14) with the lattice constants a = 1141.4(4), b = 1459.2(6), c = 1614.4(7) pm, β = 110.1(4)°; Z = 4 gives D_c = 1.99, while D_m = 2.05 g/cm³. The In atom is in a distorted trigonal bipyramidal environment with an equatorial InC_2I plane. Important bond lengths and distances of the cation are shown in **Fig. 93** [8].

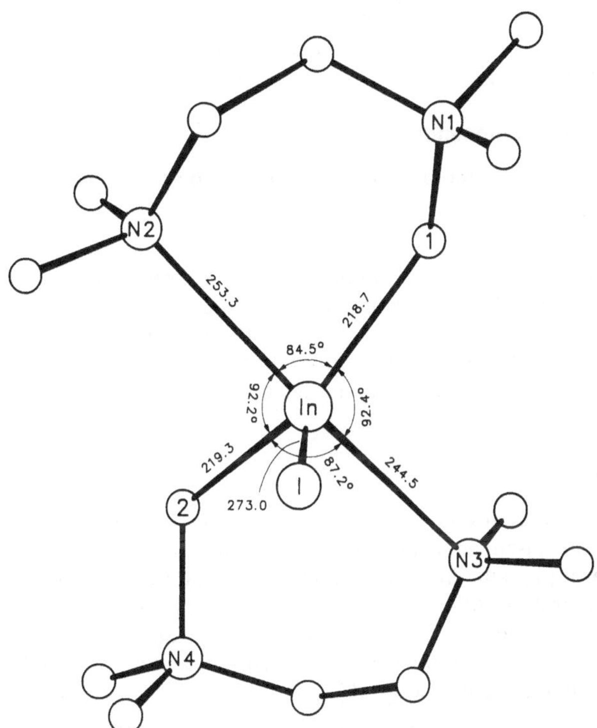

Fig. 93. Molecular structure of the cation $[In\{CH_2N(CH_3)_2C_2H_4N(CH_3)_2\}_2I]^{2+}$ [8].

Other bond angles (°)

I-In-C(1)	116.4(3)	I-In-C(2)	117.1(3)
I-In-N(2)	90.4(2)	I-In-N(3)	93.8(2)
N(2)-In-N(3)	175.4(4)		

References:

[1] Schmidbaur, H.; Füller, H.-J. (Chem Ber. **107** [1974] 3674/9).
[2] Schmidbaur, H.; Füller, H.-J. (Chem Ber. **110** [1977] 3528/35).
[3] Schmidbaur, H.; Füller, H.-J. (Ger. 2701143 [1978]; C.A. **90** [1979] No. 23243).
[4] Yamamoto, Y. (Bull. Chem. Soc. Japan **56** [1983] 1772/4).
[5] Yamamoto, Y. (Bull. Chem. Soc. Japan **57** [1984] 2835/8).
[6] Yamamoto, Y. (Bull. Chem. Soc. Japan **60** [1987] 1189/91).
[7] Yamamoto, Y.; Numasaki, Y.; Murakami, M. (Nippon Kaguku Kaishi **1985** 625/8 from C.A. **103** [1985] No. 47866).
[8] Annan, T. A.; Tuck, D. G.; Khan, M. A.; Peppe, C. (Organometallics **10** [1991] 2159/66).

12 Organoindium Compounds of Lower Oxidation States

The product obtained from the reaction of $In(CH_2Si(CH_3)_3)_3$ with NaH was first considered to be an organoindium(I) derivative with the formula $Na[In(CH_2Si(CH_3)_3)_2]$ [2]. Later investigations by the same researchers [4] demonstrated that they had synthesized a mixture of $Na[In(CH_2Si(CH_3)_3)_4]$ and $Na[In(CH_2Si(CH_3)_3)_3H]$ (thus, In^{III}); see also Table 64, No. 12. Also mentioned is one In^{I} compound of the InR type, **$InC_9H_7O_3$** (Formula I on p. 376) and its adduct **$Cl_3B \cdot InC_9H_7O_3$** (dec. at 145 °C); they are insoluble, and the structures of these compounds are unknown [1].

12.1 Organoindium(II) Compounds

In this chapter the only isolable compound containing an In-In bond is described.

CH_3InBr

Indium vapor and CH_3Br were co-condensed in an argon matrix at both 12 and 77 K. It was postulated that In had inserted into the C-Br bond, as evidenced by $\varrho(CH_3)$ absorption bands at 744 cm^{-1} (12 K) and 738 cm^{-1} (77 K) in the IR reflectance spectrum. Above 77 K the In and CH_3Br reactants were reformed. When the reaction was conducted with CD_3Br, the result was complicated but not comparable; the differences among Al, Ga, and In were discussed [3].

$[In\{CH(Si(CH_3)_3)_2\}_2]_2$

This compound was prepared from a pentane suspension of $In_2Br_4 \cdot 2\,(CH_3)_2NCH_2CH_2N(CH_3)_2$ with 4 equivalents of $Li[CH(Si(CH_3)_3)_2]$ in $O(C_2H_5)_2$ at −60 °C. The mixture was allowed to warm (metallic In separates at about −20 °C) and was stirred an additional 45 min at room temperature. After removing the solvent, the residue was taken up in pentane and precipitated by cooling to −30 °C; orange-red crystals, monomeric in benzene, dec. 154 °C, 54% yield. A second product of the disproportionation, $In\{CH(Si(CH_3)_3)_2\}_3$ (see p. 80), was isolated from the mother liquor [5].

The NMR spectra were run in C_6D_6 (δ values in ppm). ^{1}H NMR: 0.27 (s, CH_3Si), 0.90 (CHIn). ^{13}C NMR: 5.0 (CH_3Si), 28.2 (CHIn) [5].

The vibration frequencies are given in cm^{-1}. IR (solid): 1255 sh, 1250 s $\delta(CH_3)$, 1022 s δ(CH), 916 m, 898 m, 845 vs, 783 w, 768 s, 750 s, 725 sh, all $\varrho(CH_3Si)$, 682 m, 660 s, 595 m, all $\nu(SiC_3)$, 479 s $\nu_{as}(InC_2)$, 460 m $\nu_s(InC_2)$, 410 w, 331 w $\delta(SiC_3)$. Raman (C_6H_6): 798 w, 765 w, 748 vw, 729 vw, all $\varrho(CH_3Si)$; 681 m, 665 m $\nu(SiC_3)$, 485 sh $\nu_{as}(InC_2)$, 463 s $\nu_s(InC_2)$, 342 w $\delta(SiC_3)$, 246 m cm^{-1} unassigned; ν(In-In) not observed [5].

UV/VIS ($n-C_5H_{12}$, $c=3.9\times10^{-5}$ mol/L): $\lambda_{max}(\varepsilon)=240(24000)$, 270(9500), 380(800) nm ($L\cdot mol^{-1}\cdot cm^{-1}$). The band at 380 nm was interpreted as characteristic of the metal-metal bond, while the other two represented transitions from the ligands to empty metal orbitals [5].

The X-ray structure was determined at −80 °C. The compound crystallizes in the triclinic space group $P\bar{1}-C_i^1$ (No. 2) with the unit cell parameters a = 966.4(6), b = 1237.3(4), c = 2154.8(8) pm, $\alpha=74.93(3)°$, $\beta=82.54(4)°$, $\gamma=68.16(4)°$; Z = 2. The most important distances and angles of the diindane are shown in **Fig. 94**. The In(1)-In(2)-C(1 to 4) framework is nearly planar (the twist of the molecular halves is about 5°), and the indium atoms lie 3.9 and 4.2 pm above the plane formed by the three (one In and two C) surrounding atoms. The central C-H bonds are oriented in the same way around the In-In bond (not shown in Fig. 94). Because the $CH(Si(CH_3)_3)_2$ groups are twisted with respect to one another, the molecule possesses only C_1 symmetry. Comparisons were made to the analogous Al and Ga compounds [5].

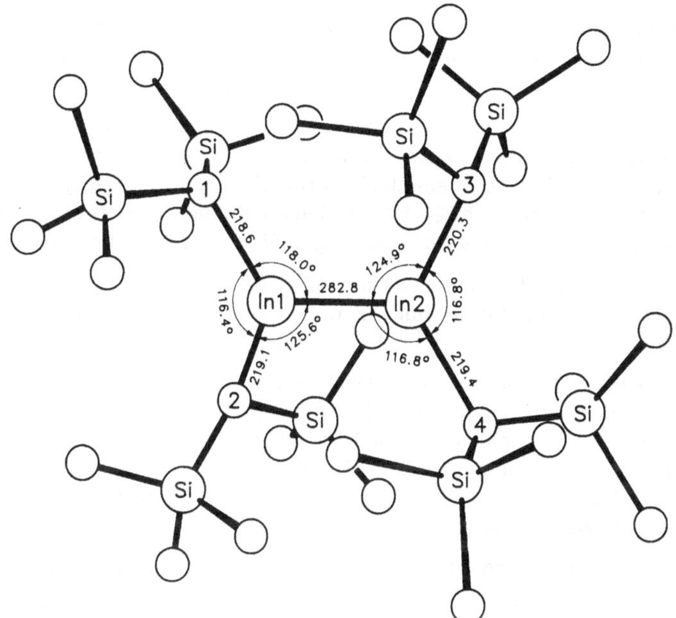

Fig. 94. Molecular structure of $[In\{CH(Si(CH_3)_3)_2\}_2]_2$ [5].

The following important fragments were found in the mass spectrum (70 eV, 365 K): $[M-CH(Si(CH_3)_3)_2]^+$ (1%), $[1/2\ M]^+$ (100%) [5].

References:

[1] Contreras, J. G.; Tuck, D. G. (Inorg. Chem. **12** [1973] 2596/9).
[2] Beachley, O. T., Jr.; Rusinko, R. N. (Inorg. Chem. **20** [1981] 1367/70).
[3] Tanaka, Y.; Davis, S. C.; Klabunde, K. J. (J. Am. Chem. Soc. **104** [1982] 1013/6).
[4] Hallock, R. B.; Beachley, O. T., Jr.; Li, Y.-J.; Sanders, W. M.; Churchill, M. R.; Hunter, W. E.; Atwood, J. L. (Inorg. Chem. **22** [1983] 3683/91).
[5] Uhl, W.; Layh, M.; Hiller, W. (J. Organometal. Chem. **368** [1989] 139/54).

12.2 Organoindium(I) Compounds

This section deals with compounds in which In^+ is bonded to $C_5H_5^-$ or C_6H_6 in an η^5 or η^6 coordination, respectively.

12.2.1 Cyclopentadiene Derivatives and Their Adducts

12.2.1.1 $In(C_5H_5\text{-}c)$

Preparation. The compound was first found in attempting to sublime $In(C_5H_5)_3$ from the reaction of $InCl_3$ with excess $Na[C_5H_5]$ in ether (see also p. 87) [1]. High-vacuum sublimation between 100 and 150 °C afforded the product in 65% yield [1, 3]. Other procedures differ only in the choice of reactants for the $In(C_5H_5)_3$ precursor ($InBr_3$ instead of $InCl_3$ and $Li(C_5H_5)$ or $K(C_5H_5)$ in lieu of $Na(C_5H_5)$ [3, 7]), and in the selection of the solvent (THF, dioxane, or ethylene glycol dimethyl ether instead of $O(C_2H_5)_2$). The yields varied from 55 to 69% [3].

According to [8], not only does simple pyrolysis of $In(C_5H_5)_3$ lead to $In(C_5H_5)$, but the accompanying redox reaction is much more significant: $In(C_5H_5)_3 + C_5H_5^- \rightarrow In(C_5H_5) + 3/2\ (C_5H_5)_2$.

The title compound was obtained in a yield of 72% by stirring InCl with a fivefold excess of $Li(C_5H_5)$ in $O(C_2H_5)_2$ for 12 h at room temperature followed by vacuum sublimation of the residue at about 100 °C. The postulated intermediate, **$Li[In(C_5H_5)Cl]$**, was not isolated [19]. This procedure was modified in [24] by reacting equimolar amounts of InCl and $Li(C_5H_5)$; 81.9% yield was obtained after vacuum subliming at 55 °C and resubliming at 45 °C.

$In(C_5H_5)$ was formed by the condensation of In vapor into a C_5H_6-c matrix at 77 K. The compound was not isolated pure, but was only identified from its typical IR absorptions [16].

Physical Properties and Spectra. $In(C_5H_5)$ is a pale yellow solid, crystallizing in characteristic clusters of needles up to 2 cm long [1], melting at 169.3 to 171.0 °C [24], 169.3 to 170.7 °C [26], and subliming in vacuum at 35 °C [26], 45 °C [24], 50 °C [1, 8], or 100 °C [19]. The crystals are slightly light-sensitive. The compound dissolves readily in THF, dioxane, and ethylene glycol dimethyl ether; it is moderately soluble in $O(C_2H_5)_2$, C_6H_6, and petroleum ether, and very slightly soluble in C_6H_{12} [1, 26].

The X-ray density is $D_c = 2.30$ [4] and 2.25 g/cm^3 [24]. The vapor pressure (by weighing the substance in a stream of dry N_2) is 12 Pa at 20 °C and 36 Pa at 40 °C [28].

The compound is diamagnetic with χ_{mol} (cm^3/mol) $= -61 \times 10^{-6}$ at 290 K, -80×10^{-6} at 77 K [1], or $-(64 \pm 17) \times 10^{-6}$ at 293, 200, and 88 K [2]. The dipole moment (ca. 0.1 M in $C_6H_5CH_3$) is 2.2 ± 0.2 D at 40 °C, versus a theoretical value (CNDO calculation) of 4.75 D [20].

Electron diffraction (GED) measurements of $In(C_5H_5)$ vapor were interpreted as an "open half-sandwich" with C_{5v} symmetry. The key bond distances are: In-C 262.1(5), C-C 142.6, C-H 110(6), and In···H 334.0(25) pm. The indium atom is situated over the center of the C_5 ring and 232.2 pm away. The five H atoms are 4.5(2)° out of the C_5 ring plane, opposite the In atom [5]. The vibration amplitude of the C-C bond, μ, has been used with the corresponding GED values of 56 other organic compounds to derive the empirical formula $\mu(CC) = -7.1856 + 0.124162 \cdot r - 2.8974 \times 10^{-4} \cdot r^2$, where r = C-C distance in pm. For r between 120.86 and 154.9 pm this equation gives the vibrational amplitude of the C-C bond in good agreement with the experimental values: for $In(C_5H_5)$, $r_m = 4.0 \pm 0.9$ and $r_c = 4.628$ [14]. Estimates of the overlap integral of the "centrally σ-bonded" $In(C_5H_5)$ structure were carried out and gave 5.8 eV for the energy of the d orbitals [6].

The photoelectron spectra (excitation by 21.22 eV He^I and 40.81 eV He^{II}) are discussed in [17] and [18]; each of the recorded curves are illustrated. Based on these measurements a variety of molecular orbital energies, ionization energies, eigenvalues, and charge densities have been calculated [20, 22, 23, 26]. Table 69 contains the measured data and important calculated values (ionization energies and eigenvalues in eV). For theoretical studies concerning the In-In interaction, see [31, 32].

Table 69
Photoelectron Spectra (PE) and Calculated Ionization Energies (IE) and Eigenvalues (η) for $In(C_5H_5)$.

observed energies				calculated energies				
PE [17] (eV)	PE [18] (eV)	He^I (intensities)	He^{II} [18] (intensities)	IE [23] (eV)	IE [20] (eV)	η [23] (eV)	assignment	
8.3	8.28	1.00	1.00	7.22	–	−4.41	$3e_1$ *)	(π)
9.3	9.23		0.91	8.30	9.34	−5.26	$4a_1$ *)	("lone pair")
–	–	–	–	11.79	–	−8.91	$3a_1$ *)	π(In-ring) + σ(C-H) + σ(C-C)
12.7	12.89	4.49	3.78	12.89	12.54	−9.56	$2e_2$ *)	
–	–	–	–	13.37	–	−10.08	$2e_1$ *)	
16.3	16.26	0.83	2.25	16.69	16.31	−13.47	$2a_1$	
–	16.95			17.13	16.76	−13.85	$1e_2$	
–	21.09	–	0.91	21.06	–	−17.72	$1e_1$	
–	–	–	–	24.32	–	−18.61	$In(4D)^2D_{5/2}$	
–	23.69	–	2.38	25.15	–	−18.66	$In(4D)^2D_{3/2}$	
–	24.54	–	2.44	25.25	–	−18.77		
–	–	–	–	25.76	–	−22.27	$1a_1$	

*) Contour plots of the $3e_1$, $4a_1$, and $3a_1$ MO are depicted in [23].

The proton resonance spectrum is a sharp singlet at δ = 5.93 ppm in C_6D_6 and in THF [24] (6.10 [24] or 6.13 ppm [8, 9] in $CDCl_3$), which remains unsplit and unbroadened even at −70 °C.

The IR spectra of solid $In(C_5H_5)$ have been measured by different researchers and interpreted as representing a C_5H_5 ion, even though the data disagree in detail [8, 11, 16, 24].

In the gas state $In(C_5H_5)$ appears to be monomolecular with C_{5v} symmetry; here the decisive $In-C_5H_5$ vibration was observed at 230 cm^{-1} and shows a force constant (calculated from a simple two-mass model) of 1.41 N/cm. This distinctive vibration is missing from the spectrum of solid samples, leading to the conclusion that the molecule in the crystal has a polymeric, largely ionic structure with planar C_5H_5 rings of D_{5h} symmetry [25].

Complete IR spectra (70 to 3100 cm^{-1}) of the solid at 12 and 200 K, in an argon matrix at 12 K, and of $In(C_5H_5)$ gas at 300 K, as well as the Raman spectrum at 195 K, are reported, assigned, and listed in Table 70. The recorded curves are illustrated along with those for $Tl(C_5H_5)$ [25].

Table 70
Vibrational Spectra of $In(C_5H_5)$ [25].
Wavenumbers in cm^{-1}.

IR (solid 200 K)	IR (gas 300 K)	Raman (solid 195 K)	assignment D_{5h}/C_{5v}	symmetry
3104 sh*)	3103 m	3094 m	ν(CH)	A_1'/A_1
3076 m	–	–	ν(CH)	E_1'/E_1
3062 sh	3058 w	3068 w	ν(CH)	E_2'/E_2
1610 w, br	–	–	overtones	
1515 vw, br	–	–		
1427 m	–	–	ν(CC)	E_1'/E_1
1422 m	–	–		
1417 sh	–	–		
1360 vvw*)	–	1369 sh	ν(CC)	E_2'/E_2
1338 w*)	–	1344 m, br		
1185 vw, br	–	1181 m	2×γ(CCC)	
–	1125 R	–		
1115 m*)	1117 Q, w	1116 vs	ν(CC)	A_1'/A_1
1108 sh	1109 P	1111 sh		
1090 vw	–	–		
1057 m*)	–	1060 w, br	δ(CH)	E_2'/E_2
–	1005 sh	–		
1009 m	1000 s	–	δ(CH)	E_1'/E_1
838 vvw	–	840 w, br	γ(CCC)	E_2'/E_2
804 vvw	–	–	ϱ(CH)	E_2''/E_2
772 s*)	–	775 sh	ϱ(CH)	E_1''/E_1
768 sh	779 R	761 m		
741 vs	773 Q, vs	–	γ(CH)	A_2''/E_2
–	767 P			
612 vvw	–	–	γ(CCC)	E_2''/E_2
–	–	230 m	$\nu(In{-}C_5)$	$-/A_1$
–	–	198 m, br	C_5-ring vibrations, lattice vibrations	
178 w	–	–		
147 m	–	–		
78 m	–	73 s		
–	–	38 vs		

*) Taken from Argon-matrix spectrum at 12 K.

The vacuum-UV spectrum of $In(C_5H_5)$ (no pressure or temperature was reported) is a broad band with $\lambda_{max}=220$ nm, but with no structure that could be evaluated. The cell used and the resulting spectrum are illustrated in [21].

X-Ray Diffraction Structure. Earlier, semiquantitative, structural analysis of crystalline $In(C_5H_5)$ have found $-C_5H_5-In-C_5H_5-In-$ polymers consisting of zigzag chains with planar C_5H_5 rings. Adjacent rings within each chain eclipse one another, the In–Z distance (Z= center of the C_5 ring) is 319 pm, the Z–In–Z angle is 137(5)°, and the shortest In···In distance between neighboring chains is 399 pm [4].

According to more recent measurements [24] the compound crystallizes in the monoclinic space group $Cc-C_s^4$ (No. 9) with the constants a=914.80(2), b=1008.75(19), c=591.16(25) pm, β = 102.795(26)°; Z = 2. The crystal structure is reproduced in **Fig. 95**. As depicted in the unit cell (A), each In atom is in contact with two other In atoms, with an In···In distance of 398.6 pm.

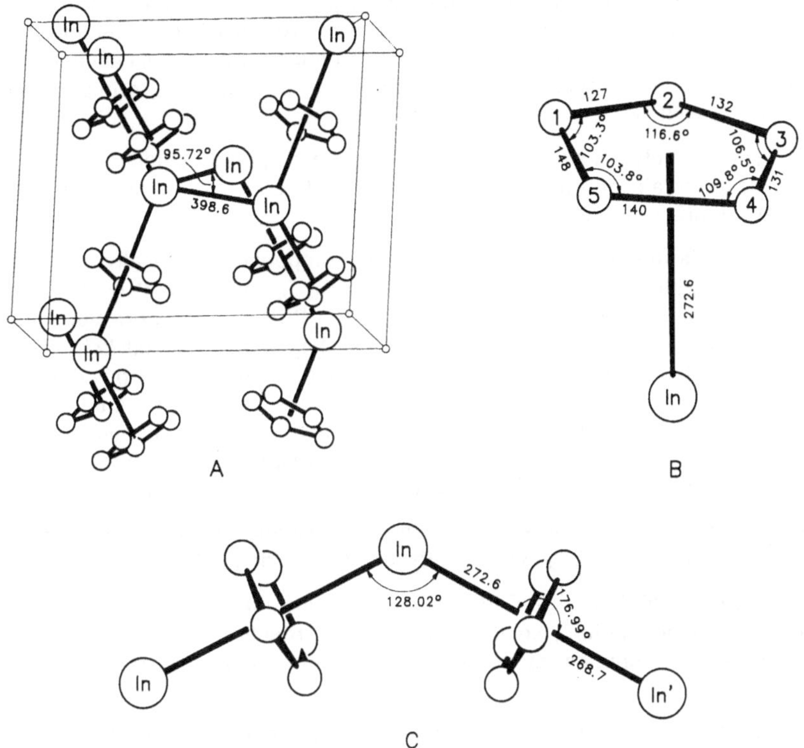

Fig. 95. Unit cell (A), crystallographic asymmetric unit (B), and environment of the In atom projected onto the centroid-In-centroid plane (C) of $(InC_5H_5)_n$ [24].

Other distances (pm)

In-C(1)	309.1(21)	In'-C(1)	286.3(20)
In-C(2)	300.4(18)	In'-C(2)	287.4(22)
In-C(3)	289.6(19)	In'-C(3)	298.3(17)
In-C(4)	285.3(22)	In'-C(4)	297.4(27)
In-C(5)	295.3(19)	In'-C(5)	292.7(18)

Overlap populations for an In···In interaction of 370 pm at different In-In-Cp(centroid) angles are given in [31].

Thermal Decomposition and Mass Spectrum. According to [1] thermal decomposition of $In(C_5H_5)$ has already begun at 110 °C, but no details were given. The mass spectrum (100 °C source) shows, along with the molecular peak, M^+, the fragments $[(C_5H_5)_2]^+$, $^{115}In^+$, $^{113}In^+$, and $[C_5H_5]^+$. The intensities of these peaks as well as the energy data for the fragmentation were also reported [7, 8].

Chemical Reactions and Applications. $In(C_5H_5)$ can react with retention or with cleavage of the In-ring bond. The first mode is associated with an increase in the oxidation state of the metal from +1 to +3; the second usually leads to other In^I compounds by losing C_5H_6-c.

Reactions with elementary iodine to give $(C_5H_5)InI_2$ [11] and with $CF_3C(\overline{S)=C(S})CF_3$ to produce $(C_5H_5)InS_2C_2(CF_3)_2$ [9] can be considered as oxidative insertions into the I-I and S-S bonds, respectively. The same procedure with bromine produces an unstable, unisolable product. Not only is the indium oxidized in this case, but the C_5H_5 group may possibly be brominated as well [11].

$In(C_5H_5)$ is extremely sensitive to oxygen. From mass spectral evidence (250 °C source) a sample exposed to air for 1 min forms In_2O_3 and $(C_5H_5OH)_2$ as end products; $(C_5H_5In)_2O$ was proposed as precursor [8].

In inert solvents $In(C_5H_5)$ reacts with protonic compounds practically quantitatively to form InX and C_5H_6. In the presence of X^- (X=Cl, Br, I) and N- or P-containing cations, ionic compounds like $[N(C_2H_5)_4][InX_2]$ or $[(CH_3(C_6H_5)_2PCH_2)_2][InI_3]$ were obtained [12, 13, 15]. With $HOC_6H_2(CF_3)_3$-2,4,6 dimeric $(InOC_6H_2(CF_3)_3)_2$ was produced [27]. Oxygen-free water attacks $In(C_5H_5)$ only slowly; dilute HCl immediately induces vigorous decomposition, in which an "indium ball" forms, then slowly dissolves with the evolution of H_2 [1]. A careful study of the hydrolysis by dilute hydrochloric acid [24] gave the equation $In(C_5H_5) + 3\,H^+(aq) \rightarrow In^{3+}(aq) + H_2 + C_5H_6$. The reaction involves disproportionation of initially formed $In^+(aq)$ to give $In^0 + In^{3+}(aq)$, followed by oxidation of In^0 to produce $In^{3+} + H_2$.

Retention of both the In-ring bond and the +1 oxidation state occurs when $In(C_5H_5)$ acts as a Lewis base toward BX_3 (X=F, Cl, Br, and CH_3); see the following section [10].

Reactions with $Na(C_5H_4D)$, $Na(C_5H_4CH_3)$, or metallic thallium result in partial ligand exchange (or transmetallation), so that after a short reaction period at 50 to 90 °C the reaction mixture also contains, along with $In(C_5H_5)$, $In(C_5H_4D)$, $In(C_5H_4CH_3)$, or $Tl(C_5H_5)$, respectively [7].

Maleic anhydride reacts with $In(C_5H_5)$ by Diels-Alder addition to form an insoluble compound with the suggested structure I [10].

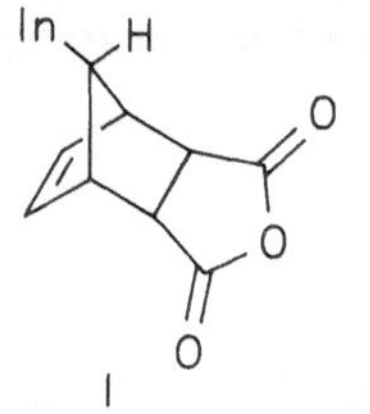

I

The use of $In(C_5H_5)$ as an MOVPE source material to prepare InP layers is described in [28 to 30].

References:

[1] Fischer, E. O.; Hofmann, H. P. (Angew. Chem. **69** [1957] 639/40).
[2] Fischer, E. O.; Joos, G.; Meer, W. (Z. Naturforsch. **13b** [1958] 456/7).
[3] Fischer, E. O.; Hofmann, H. P. (U.S. 2971017 [1961]; C.A. **1961** 17537).
[4] Frasson, E.; Menegus, F.; Panattoni, C. (Nature **199** [1963] 1087/9).
[5] Shibata, S.; Bartell, L. S.; Gavin, R. M., Jr. (J. Chem. Phys. **41** [1964] 717/22).
[6] Shchembelov, G. A.; Ustynyuk, Yu. A. (Dokl. Akad. Nauk SSSR **173** [1967] 1364/6; Dokl. Chem. Proc. Acad. Sci. USSR **172/177** [1967] 417/9).

[7] Lalancette, J. M.; Lachance, A. (Can. J. Chem. **49** [1971] 2996/9).
[8] Poland, J. S.; Tuck, D. G. (J. Organometal. Chem. **42** [1972] 307/14).
[9] Berniaz, A. F.; Tuck, D. G. (J. Organometal. Chem. **51** [1973] 113/8).
[10] Contreras, J. G.; Tuck, D. G. (Inorg. Chem. **12** [1973] 2596/9).

[11] Contreras, J. G.; Tuck, D. G. (J. Organometal. Chem. **66** [1974] 405/12).
[12] Habeeb, J. J.; Tuck, D. G. (J. Chem. Soc. Dalton Trans. **1975** 1815/6).
[13] Habeeb, J. J.; Tuck, D. G. (J. Chem. Soc. Chem. Commun. **1975** 600/1).
[14] Cyvin, S. J.; Mastryukov, V. S. (J. Mol. Struct. **30** [1976] 333/7).
[15] Habeeb, J. J.; Tuck, D. G. (J. Chem. Soc. Dalton Trans. **1976** 866/9).
[16] Kuz'yants, G. M. (Izv. Akad. Nauk SSSR Ser. Khim. **1976** 1895/6; Bull. Acad. Sci. USSR Div. Chem. Sci. **25** [1976] 1785/6).
[17] Cradock, S.; Duncan, W. (J. Chem. Soc. Faraday Trans. II **74** [1978] 194/202).
[18] Egdell, R. G.; Fragala, I.; Orchard, A. F. (J. Electron. Spectrosc. Relat. Phenom. **14** [1978] 467/75).
[19] Peppe, C.; Tuck, D. G.; Victoriano, L. (J. Chem. Soc. Dalton Trans. **1981** 2592/5).
[20] Lin, C. S.; Tuck, D. G. (Can. J. Chem. **60** [1982] 699/702).

[21] Haigh, J. (J. Mater. Sci. **18** [1983] 1072/6).
[22] Canadell, E.; Einstein, O.; Rubio, J. (Organometallics **3** [1984] 759/64).
[23] Lattman, M.; Cowley, A. H. (Inorg. Chem. **23** [1984] 241/7).
[24] Beachley, O. T.; Pazik, J. C.; Glassmann, T. E.; Churchill, M. R.; Fettinger, J. C.; Blom, R. (Organometallics **7** [1988] 1051/9).
[25] Garkusha, O. G.; Lokshin, B. V.; Materikova, R. B.; Golubinskaya, L. M.; Bregadze, V. I.; Kurbakova, A. P. (J. Organometal. Chem. **342** [1988] 281/90).
[26] Beachley, O. T., Jr.; Blom, R.; Churchill, M. R.; Faegri, K., Jr.; Fettinger, J. C.; Pazik, J. C.; Vicotriano, L. (Organometallics **8** [1989] 346/56).
[27] Scholz, M.; Noltemeyer, M.; Roesky, H. W. (Angew. Chem. **101** [1989] 1419/20).
[28] Staring, E. G. J.; Meekes, G. J. B. M. (J. Am. Chem. Soc. **111** [1989] 7648/50).
[29] Udagawa, T. (Japan. 01-094613 [1989] from C.A. **111** [1989] No. 164752).
[30] Staring, A. G. J. (EP 362952 [1990] from C.A. **113** [1990] No. 88768).

[31] Janiak, C.; Hoffmann, R. (J. Am. Chem. Soc. **112** [1990] 5924/46).
[32] Budzelaar, P. H. M.; Boersma, J. (Recl. Trav. Chim. Pays-Bas **109** [1990] 187/9).

12.2.1.2 BX_3 Adducts of $In(C_5H_5)$

The adducts were prepared by condensing a slight excess of BX_3 (X = F, Cl, Br, and CH_3) into a cooled solution of $In(C_5H_5)$ in $CHCl_3$. The mixture was warmed to room temperature and stirring continued for 30 min; the solid product was freed of solvent and unreacted BX_3 in vacuum. Yields were practically quantitative.

The 1:1 adducts, $(C_5H_5)In \cdot BX_3$, are insoluble in C_6H_6, CCl_4, CH_2Cl_2, $CHCl_3$, CH_3CN, CH_3NO_2, $(CH_3)_2SO$ and are very sensitive to moisture. They were characterized from their IR and mass spectra.

$C_5H_5In \cdot BF_3$

This adduct is a pale yellow solid, dec. 270 °C. IR (solid), see Table 71. The mass spectrum (160 °C source) shows the fragments $[InC_5H_5]^+$ (33%), $[InB]^+$ (21%), $^{115}In^+$ (100%) [1].

$C_5H_5In \cdot BCl_3$

The BCl_3 adduct is an orange-yellow solid, dec. 190 °C; IR (solid), see Table 71. The mass spectrum (200 °C): $[BCl_3]^+$ (20%), $^{115}In^+$ (100%), $[BCl_2]^+$ (17%), unassigned fragments with m/e = 117 (21%), 116 (25%), 113 (100%), 112 (75%), and 81 (27%) [1].

The reaction of the compound with maleic anhydride in $CHCl_3$ (1:1 ratio) yields, after stirring at room temperature for 24 h, the BCl_3 adduct of 7-indiobicyclo[2.2.1]-hept-5-ene-1,2-dicarboxylic acid anhydride, $InC_9H_7O_3$ (Formula I on p. 376, with BCl_3 coordinated to In) [1].

IR (solid): 3045 mw, 2930 ms, 2880 ms, 1855 ms, 1775 s, 1662 ms, 1631 s, 1375 s, br, 1262 m, 1002 w, 970 mw, 930 ms, 912 s, 835 w, 762 s, 721 mw, 675 w, 665 w, 622 mw, 475 mw, 425 w, 325 w cm^{-1} [1].

$C_5H_5In \cdot BBr_3$

The compound is a dark yellow solid, m.p. 286 °C; IR (solid), see Table 71. The mass spectrum (250 °C) shows the following fragments: $[InBr_2]^+$ (100%), $[InBBr]^+$ (52%), $[BBr_2]^+$ (33%), ^{115}In (100%) and an unassigned fragment with m/e = 139(48%) [1].

$C_5H_5In \cdot B(CH_3)_3$

The compound is a pale yellow solid, dec. 164 °C; IR (solid), see Table 71. The mass spectrum (250 °C) shows the following fragments: $[InC_5H_5]^+$ (99%), $^{115}In^+$ (100%), $[B(CH_3)_3]^+$ (20%). The compound apparently undergoes dissociation into its parent molecules on heating [1].

Table 71
IR Spectra for Solid $C_5H_5In \cdot BX_3$ (X = F, Cl, Br, CH_3) Compounds [1].
Wavenumbers in cm^{-1}.

X=F	X=Cl	X=Br	$X=CH_3$ [1)]	assignment
3042 w	3040 w	3035 w	3060 m	
2958 m	–	2930 s	2925 m	ν(CH)[2)]
2855 mw	2880 m, br	2870 s	2855 m	
1578 vw	1565 m	1545 ms	1555 s	ν(CC)[2)]
1305 vw	1305 mw	1297 w	1305 w	and
1275 m	1263 mw	1252 s	1255 m	δ(CH)[2)]
1101, 1065?	–	–	–	$\nu_{as}(BF_3)$
1009 s	1005 mw	1018 s	1005 s	δ(CH)[2)]
–	940 ms	935 m	–	and
–	911 ms	–	–	γ(CCC)[2)]
–	860 m	–	888 m	
808 m	807 ms	–	802 ms	ϱ(CH)[2)]
776?	–	–	–	$\nu_s(BF_3)$
758 w	760 mw	786 s, br	758 m	γ(CH)[2)]
721 ms	718 s	715 m	725 s	
–	692, 675?	–	–	$\nu_{as}(BCl_3)$
521?	–	–	–	$\delta(BF_3)$
–	–	580?	–	$\nu_{as}(BBr_3)$
515?	515?	515?	505?	ν(In–B)
–	375?	431?	–	$\nu_s(BX_3)$
–	260?	355?	–	$\delta(BX_3)$
–	285?	285?	285?	ν(In–ring)

1) No data reported for the $B(CH_3)_3$ moiety. (?) No intensity given. – 2) assignment was made by the author by analogy to assignment in Table 70.

Reference:

[1] Contreras, J. G.; Tuck, D. G. (Inorg. Chem. **12** [1973] 2596/9).

12.2.2 Cyclopentadiene Derivatives of the $In(C_5H_{5-n}R_n)$ Type

The compounds in Table 72 were prepared by the following procedures.

Method I: From InX and $Li[C_5H_{5-n}R_n]$.
A stoichiometric amount of InCl was introduced into an ether suspension of $Li[C_5H_{5-n}R_n]$. After stirring for several hours at room temperature, the LiCl and metallic In formed were separated, the filtrate evaporated to dryness, and the residue purified by repeated sublimation [5, 9]. For Nos. 4, 6, 7, and 9 the reagents were first combined at −80 °C and then worked up at room temperature as described above [4, 7]. For No. 2 recrystallization from C_5H_{12} at −5 °C was the most suitable purification technique [5].

Method II: Ligand exchange between InC_5H_5 and $Na[C_5H_{5-n}R_n]$ or $HC_5H_{5-n}R_n$.
Nos. 1 and 2 were obtained by mixing freshly sublimed $In(C_5H_5)$ and $Na[C_5H_4R]$ (R = D, CH_3) and then warming to 50 °C [1]. No. 5 was formed (with elimination of C_5H_6) when $In(C_5H_5)$ and $C_5H_5Si(CH_3)_3$ were allowed to react in benzene at 70 °C for 6 h [4]. In no case was a yield reported; Nos. 1 and 2 were identified by mass spectroscopy only as a mixture with $In(C_5H_5)$.

Nos. 1, 2, and 4 to 9 are colorless or yellow solids that are very soluble in the usual aprotic solvents and are extremely sensitive to oxidation and moisture. Hydrolytic decomposition with dilute HCl is strongly exothermic and forms H_2, In^{3+}, RH, or R_2; In^+ first disproportionates, as in the hydrolysis of $In(C_5H_5)$ (see p. 376) [5, 9].

Table 72
Compounds of the $In(C_5H_{5-n}R_n)$ Type.
Further information on numbers preceded by an asterisk is given at the end of the table.
Explanations, abbreviations, and units on p. X.

No.	compound method of preparation (yield)	properties and remarks
compounds with n = 1		
1	$In(C_5H_4D)$ II	no pure compound isolated, shown by mass spectrum in admixture with $In(C_5H_5)$ [1]
*2	$In(C_5H_4CH_3)$ I (87.8%)	colorless solid, m.p. 48.5 to 51.0 °C (yellow liquid), subl. under vacuum [5] vapor pressure: 53 Pa at 20 °C, 236 Pa at 40 °C [8] ^1H NMR (C_6D_6): 2.04 (CH_3), 5.82 (C_5H_4) [5] IR (solid): 1310 vw, 1039 vw, 1025 w, 997 w, 972 vw, 928 vw, 813 s, 781 s, 759 s, 722 s, 668 m, 614 m, 605 m [5] mass spectrum (source 75 °C): M^+, $^{115}In^+$, $[C_5H_4CH_3]^+$, $[C_5H_5]^+$ [2]

Table 72 (continued)

No.	compound method of preparation (yield)	properties and remarks
3	$In(C_5H_4C_2H_5)$ I (>80%)	mobile liquid, m.p. ca. 10 °C, purified by vacuum distillation [8] vapor pressure: 28 Pa at 20 °C, 52 Pa at 40 °C [8] used in MOVPE experiments [8]
4	$In(C_5H_4C_4H_9\text{-}t)$ I (91)	colorless crystals, m.p. 47.5 to 48.5 °C [11] molecular structure similar to No. 5 [11]
*5	$In\{C_5H_4Si(CH_3)_3\}$ I (91.3%) [9] I, II [4]	colorless crystals, m.p. 51.4 to 51.8 °C, subl. 27 °C/0.001 Torr [9] ^{1}H NMR (C_6D_6): 0.22 (s, CH_3Si), 6.12 (AA′BB′-m, C_5H_4) [4, 9] ^{13}C NMR (C_6D_6): 1.1 (CH_3Si), 111.0, 113.0 (C-2,3,4,5) [4] ^{29}Si NMR (C_6D_6): −10.6 [4] IR (solid): 1300 w, 1246 s, 1182 sh, 1172 s, 1035 s, 901 s, 834 s, 771 s, 752 sh, 690 w, 629 m, 412 m, 317 m, 300 sh [4, 9] mass spectrum: M^+ (14.8%), $[M-CH_3]^+$ (8.5%), $[C_5H_4Si(CH_3)_3]^+$ (1.8%), $^{115}In^+$ (100%), $[Si(CH_3)_3]^+$ (13.8%) [4]
6	$In\{C_5H_4Ge(CH_3)_3\}$ I (89.4%)	colorless solid, m.p. 50 to 52 °C, subl. 38 °C/0.001 Torr, dec. starts at ca. 120 °C [9] ^{1}H NMR: 0.30 (CH_3Ge), 6.05, 6.13 (2 t, C_5H_4) in C_6D_6, 0.21, 5.85 s in THF-d_8 [9] ^{13}C NMR(THF-d_8): 0.02 (CH_3), 107 (C_5H_4) [9] IR (solid): 1304 w, 1233 m, 1159 m, 1029 m, 823 vs, 768 s, 599 m, 570 m [9] mass spectrum (principal peaks only): M^+ (45%), $[M-CH_3]^+$ (60%), $^{115}In^+$ (100%) [9]
compounds with n = 2		
7	$In\{C_5H_3(Si(CH_3)_3)_2\}$ I	colorless crystals, subl. under vacuum [4] ^{1}H NMR (C_6D_6): 0.24 (s, CH_3Si), 6.29 ("d", H-4,5), 6.39 ("t", H-2) [4] ^{13}C NMR (C_6D_6): 1.2 (CH_3Si), 116.2 (C-4,5), 119.4 (C-2), 121.1 (C-1,3) [4] ^{29}Si NMR (C_6D_6): −10.8 [4] mass spectrum: M^+ (17.3%), $[M-CH_3]^+$ (23.5%), $[C_5H_3(Si(CH_3)_3)_2]^+$ (4.1%), $^{115}In^+$ (100%) $[Si(CH_3)_3]^+$ (62.3%) [4]
compounds with n = 3		
8	$In\{C_5H_2(Si(CH_3)_3)_3\}$ I	colorless crystals, subl. under vacuum [4] ^{1}H NMR (C_6D_6): 0.23 (1 CH_3Si), 0.32 (2 CH_3Si), 6.65 (s, H-3,5) [4]

Table 72 (continued)

No. compound method of preparation (yield)	properties and remarks
8 (continued)	^{13}C NMR ($CDCl_3$): 1.3, 2.5 (CH_3Si), 122.5 (C-4), 126.1 (C-3,5), 126.3 (C-1,2) [4] $^{29}Si(CDCl_3)$: −10.7, −10.0 [4] mass spectrum: M^+ (30%), $[M-CH_3]^+$ (36.6%), $[C_5H_2(Si(CH_3)_3)]^+$ (8.0%), $^{115}In^+$ (100%), $[Si(CH_3)_3]^+$ (85.2%)
compounds with n = 5	
*9 $In\{C_5(CH_3)_5\}$ I (62.1%)	golden yellow crystals, m.p. 92 to 93 °C [6], subl. 55 °C/10^{-3} Torr [3] monomeric in C_6H_{12}-c (degree of association 1.04 to 1.09) [6] ^1H NMR: 2.02 (s, CH_3) in C_6D_6, 2.05 in C_6D_{12} and C_4D_8O, and 2.06 in NC_5H_5 [6] IR (solid): 2720 vw, 1727 w, 1412 w, 1152 vw, 1014 w, 790 m, 583 m, 466 vw, 345 w, 285 sh, 280 vs, 268 vs [6]
*10 $In\{C_5(CH_2C_6H_5)_5\}$ I (51%)	white crystalline solid, m.p. 110 to 113 °C (dec.) [7] ^1H NMR (C_6D_6, c = 2 g/L): 3.88 (s, CH_2), 7.19 ("s", C_6H_5) [7] ^{13}C NMR (C_6D_6, c = 2 g/L): 31.78 (CH_2), 120.33 (C_5-ring), 125.96 (C-4), 128.57 (C-3,5), 128.99 (C-2,6), 144.44 (C-1) [7] IR (CsI-pellet): 3100 vw, 3080 w, 3055 m, 3020 m, 2995 vw, 2920 m, 2840 w, 1950 w, br, 1870 w, br, 1820 w, br, 1600 m, 1580 w, 1490 s, 1451 sh, 1420 w, 1325 w, 1290 w, br, 1275 w, 1260 w, 1195 vw, 1182 w, 1150 w, 1120 vw, 1075 m, 1030 m, 1000 to 800 w/vw, 750 m, 745 m, 730 m, 720 m, 700 s, 665 w, 650 vw, 633 w, 620 vw, 590 vw, 570 w, 470 w, 460 m, 320 m [7] mass spectrum (220 °C): M^+ (65%), $[M-In+H]^+$ (20%), $[M-In-H]^+$ (33%), $[M-In-2\,C_6H_5CH_3]^+$ (7%), $[M-In-3\,C_6H_5CH_3]^+$ (7%), $[C_{13}H_{11}]^+$ (10%), $^{115}In^+$ (44%), $[C_6H_5CH_2]^+$ (100%) [7]

* Further information:

$In(C_5H_4CH_3)$ (Table **72**, No. **2**) was prepared similarly to $In(C_5H_5)$ from $InCl_3$ and excess $Li(C_5H_4CH_3)$ in ether. While recrystallization of the reaction residue gave $In(C_5H_4CH_3)_3$ (Table 12, No. 3), sublimation from 60 to 80 °C at 0.1 Torr produced the title compound; [2] describes it as a pale yellow solid.

Cryoscopic measurements in C_6H_{12}-c indicate a concentration-dependent monomer-dimer equilibrium. The degrees of association (with molalities) are: 1.73 (0.0722), 1.59 (0.0626), and 1.49 to 1.47 (0.0337) [5].

Electron diffraction on the gas shows a "half-sandwich" structure, like that of the unsubstituted InC_5H_5 with the distances In-C = 260.7, In-centroid = 230.9, CH_3-C = 151.2 pm. In addition to the bond distances the following contact distances were reported: In···C (of CH_3) = 369, In···H(C_5) = 334, In···H(of CH_3) = 369 to 460, C(1)···C(3) = 230, C(2)···C (of CH_3) = 261, and C(3)···C (of CH_3) = 376 pm. The H atoms and CH_3 groups are bent 5° ± 4° and 7° ± 3°, respectively, out of the ring plane, away from the In [5].

The compound crystallizes in the monoclinic space group $P2_1/n-C^5_{2h}$ (No. 14) with the unit cell constants a = 616.36(14), b = 1156.84(17), c = 917.23(15) pm, β = 103.561(15)°; Z = 4 gives D_c = 2.03 g/cm³. The crystal structure is illustrated in **Fig. 96**. Each In atom of the infinite zigzag strand is in contact with one In atom (In···In = 398.6 pm) of a neighboring strand as shown in (B). The In-centroid-In angle is 179.74°, and the In-centroid and In′-centroid distances are 260.9 and 277.1 pm, respectively [5].

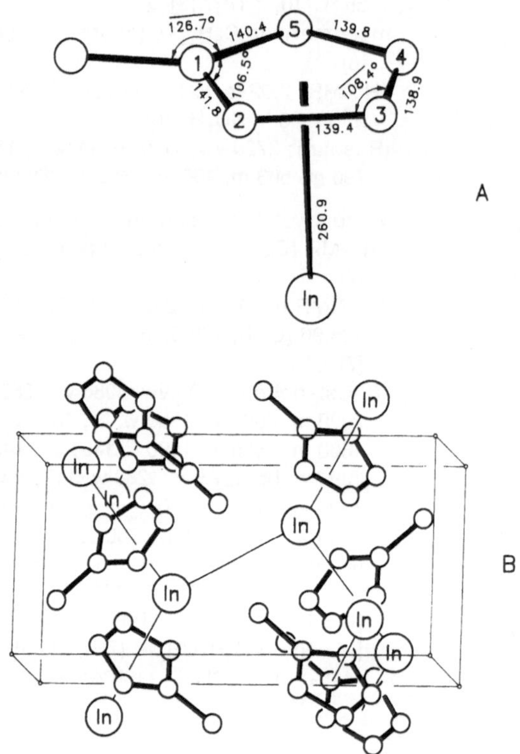

Fig. 96. Crystallographic asymmetric unit (A) and unit cell (B) of $In(C_5H_4CH_3)$ [5].

Other bond distances (°)

In-C(1)	283.8(4)	In′-C(1)	305.8(4)
In-C(2)	290.3(5)	In′-C(2)	297.3(4)
In-C(3)	292.4(5)	In′-C(3)	295.2(5)
In-C(4)	287.1(6)	In′-C(4)	301.4(6)
In-C(5)	280.0(5)	In′-C(5)	308.3(5)

The compound was tested as a source material for MOVPE [8].

$In\{C_5H_4Si(CH_3)_3\}$ (Table **72**, No. **5**) crystallizes in the monoclinic space group $P2_1/c-C_{2h}^5$ (No. 14) with the lattice constants a = 917.1(5), b = 991.0(6), c = 1167.7(7) pm, β = 97.30(5)°; Z = 4 gives D_c = 1.59 g/cm³. The compound forms zigzag chains in the crystal in which the In atoms are connected sandwich-like with two $C_5H_4Si(CH_3)_3$ groups. The most important angles in this chain are In···Z···In = 175.94, and Z···In···Z' = 131.78° (Z is the centroid of the C_5 ring); the distances are different with In–Z = 260.9 and In–Z' = 282.2 pm. In contrast to the other compounds, it is meaningless to talk of In···In interactions because the shortest In–In distances are between 543 and 591 pm. The structure of the asymmetric unit is depicted in **Fig. 97** [9].

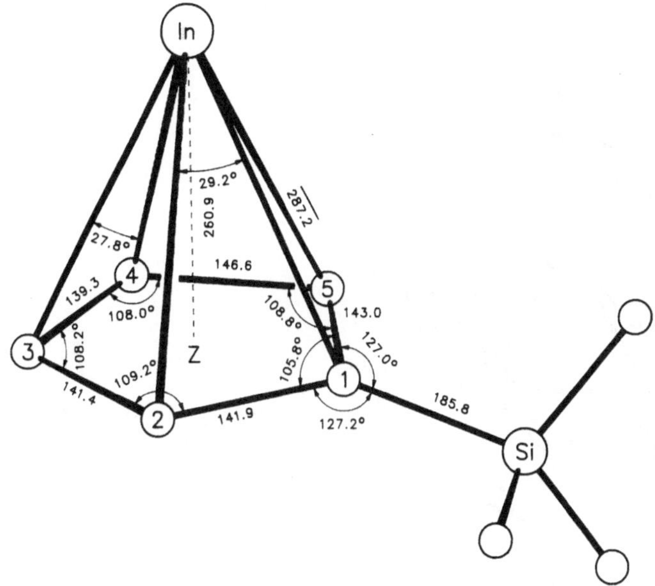

Fig. 97. Asymmetric unit of $In\{C_5H_4Si(CH_3)_3\}$ [9].

Other bond distances (°)

In-C(1)	285.4(4)	In-C(1')	302.9(4)
In-C(2)	276.9(4)	In-C(2')	316.1(4)
In-C(3)	283.7(5)	In-C(3')	317.5(5)
In-C(4)	294.7(5)	In-C(4')	303.9(5)
In-C(5)	295.3(4)	In-C(5')	293.0(4)

$In\{C_5(CH_3)_5\}$ (Table **72**, No. **9**). When prepared by Method I, 2 to 5% $\{C_5(CH_3)_5\}_2InCl$ is also formed (see p. 130) [3, 6].

Electron diffraction of the monomeric gas shows the molecule to be a half-sandwich like $In(C_5H_5)$ and $In(C_5H_4CH_3)$: In–C = 259(4), In–(ring center) = 228.8(4), C–C(ring) = 143.2(4), C–CH_3 = 150.5(5), and C–H = 110.3(6) pm; contact distances: In···C(of CH_3) = 362, C(1)···C(3) = 282, C(of CH_3)···C(of CH_3) = 320 and 517 pm; the methyl groups are bent out of the 5-ring plane by 4.1(3)°, away from the In atom. The data were used to calculate the orbital energies [3, 6].

The compound crystallizes in the rhombohedral space group $R\bar{3}-C_{3i}^2$ (No. 148) with the lattice constants a=2018.2(4), c=1343.6(3) pm; Z=18 (3 hexamers) gives D_c=1.58 g/cm^3. Six monomeric units are loosely associated in an octahedral cluster of In atoms. However, the In-(ring center) vectors do not point to the center of this cluster [3, 6]; an O_h(centroid)-In-Z (Z=centroid of the $C_5(CH_3)_5$ ring) angle of 147.5° was calculated [10]. The bond distances and angles within the monomer units are given in **Fig. 98** (B); the parameters are not significantly different from those in the gas state. Furthermore, the In-In distances of the octahedral In_6 skeleton (A) agree with the In···In contact distances between neighboring In atoms in the chain homologues, $In(C_5H_5)$ and $In(C_5H_4CH_3)$. The angle with which the CH_3 groups are bent from the C_5 plane (3.84° away from In) is almost the same as in the monomeric species in the gas phase [3, 6].

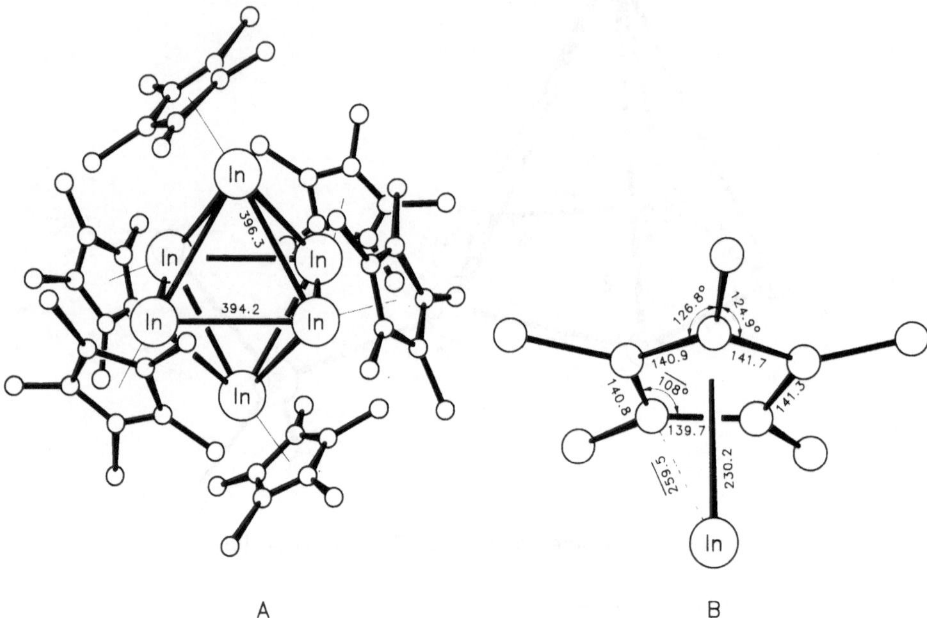

Fig. 98. Asymmetric unit of $In\{C_5(CH_3)_5\}$ (B) and the geometry of the hexamer unit (A) [6].

The compound is very soluble in C_6H_6, THF, and pyridine, but it decomposes slowly in solution to form metallic In and $C_5(CH_3)_5H$ and/or $\{C_5(CH_3)_5\}_2$. The pseudo-first-order decomposition was followed by ^{1}H NMR with k_0=0.0024 in C_6H_6, 0.0041 in THF, 0.0070 h^{-1} in NC_5H_5. Solutions in C_6H_{12} showed no decomposition after 5 d [6].

$In\{C_5(CH_2C_6H_5)_5\}$ (Table **72**, No. **10**) occurs in two allotropic forms. Colorless needles formed from concentrated hexane solutions and are converted to parallelepipeds on addition of toluene. This form is also obtained on slow crystallization from dilute C_6H_{14} solutions. The crystals are not sensitive to light and are stable for a few hours in air, but darken within a few weeks, even in an inert atmosphere at 5 °C. Repeated recrystallization (attempted with THF and $C_6H_5CH_3/C_6H_{14}$) was not successful, because on dissolution immediate decomposition to metallic In (?) began [7].

The compound crystallizes in the monoclinic space group $P2_1/c-C_{2h}^5$ (No. 14) with the unit cell constants a = 1022.8(1), b = 1575.3(2), c = 1871.3(3) pm, β = 91.01(1)°; the data were recorded at 140 ± 5 K. In the lattice two monomeric units are joined by an In···In′ interaction of 363.1(2) pm to form a quasi-dimer with an In···In′···Z (Z = ring center) angle of 136.46(5)°; In–Z = 238.2(2) pm. The shortest C···In contact distance in this dimer is that between C(23) and In′ at 347.5(2) pm. In addition to the distances and angles indicated in **Fig. 99**, the following torsion angles (±0.2°) of the $C_6H_5CH_2$ ligands were reported: C(5)-C(1)-C(6)-C(7) = −107.2°, C(4)-C(5)-C(34)-C(35) = 78.0°, C(2)-C(3)-C(20)-C(21) = −84.9°, C(1)-C(2)-C(13)-C(14) = 73.3°, and C(3)-C(4)-C(27)-C(28) = −97.1° [7].

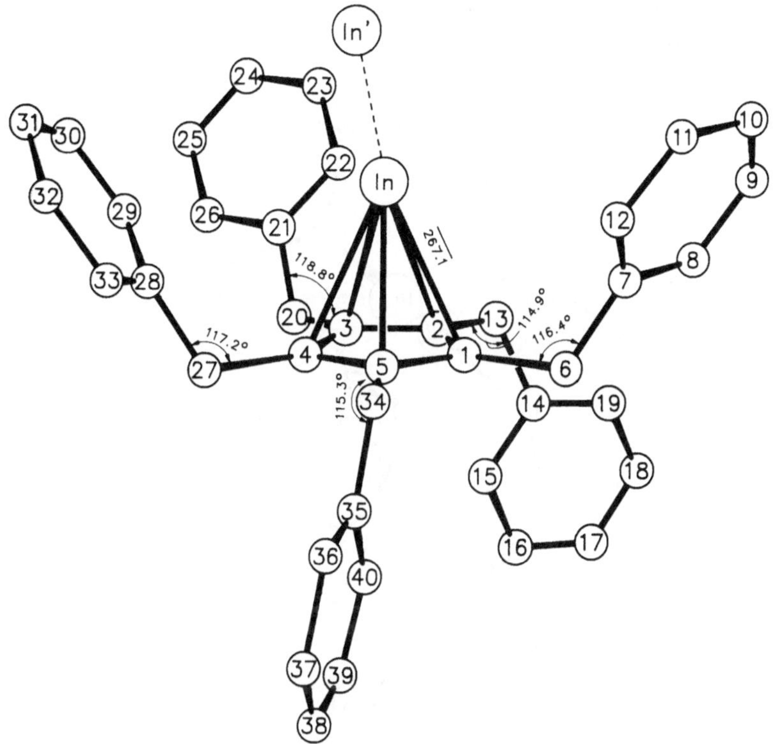

Fig. 99. Monomeric unit of In{$C_5(CH_2C_6H_5)_5$} [7].

References:

[1] Lalancette, J. M.; Lachance, A. (Can. J. Chem. **49** [1971] 2996/9).

[2] Poland, J. S.; Tuck, D. G. (J. Organometal. Chem. **42** [1972] 307/14).

[3] Beachley, O. T., Jr.; Churchill, M. R.; Fettinger, J. C.; Pazik, J. C.; Victoriano, L. (J. Am. Chem. Soc. **108** [1986] 4666/8).

[4] Jutzi, P.; Leffer, W.; Müller, G. (J. Organometal. Chem. **334** [1987] C 24/C 26).

[5] Beachley, O. T., Jr.; Pazik, J. C.; Glassmann, T. E.; Churchill, M. R.; Fettinger, J. C.; Blom, R. (Organometallics **7** [1988] 1051/9).

[6] Beachley, O. T., Jr.; Blom, R.; Churchill, M. R.; Faegri, K., Jr.; Fettinger, J. C.; Pazik, J. C.; Victoriano, L. (Organometallics **8** [1989] 346/56).

[7] Schumann, H.; Janiak, C.; Görlitz, F.; Loebel, J.; Dietrich, A. (J. Organometal. Chem. **363** [1989] 243/51).

[8] Staring, E. G. I.; Meekes, G. J. B. M. (J. Am. Chem. Soc. **111** [1989] 7648/50).
[9] Beachley, O. T., Jr.; Lees, J. F.; Glassmann, T. E.; Churchill, M. R.; Buttrey, L. A. (Organometallics **9** [1990] 2488/92).
[10] Janiak, C.; Hoffmann, R. (J. Am. Chem. Soc. **112** [1990] 5924/46).
[11] Beachley, O. T., Jr.; Lees, J. F. (J. Organometal. Chem. **418** [1991] 165/71).

12.3 Indium(I) Arene Complexes

The two thoroughly investigated examples in this section were made from $In^{I}[In^{III}Br_4]$ and are largely ionic. Other compounds of this type, $[In(C_6H_3(CH_3)_3\text{-}1,3,5)_2][AlBr_4]$, $[In(C_6(CH_3)_6)][InBr_4]$, and the very unstable, unisolated complexes of In^{I} with unsubstituted benzene, were reported in [1] but were not characterized.

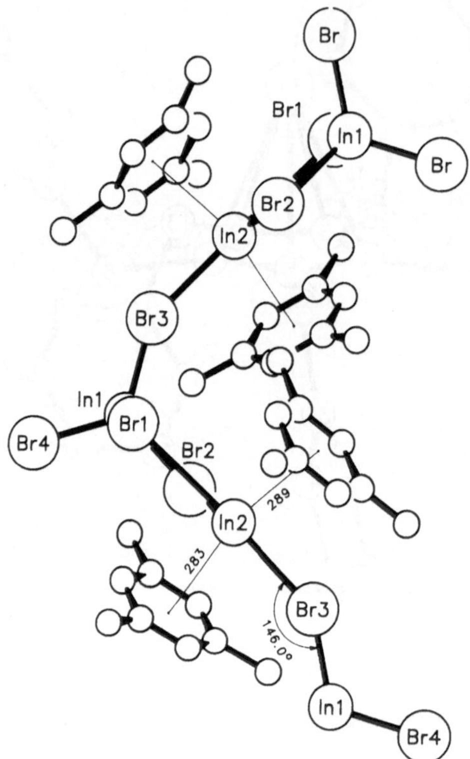

Fig. 100. Structure of $[In(C_6H_3(CH_3)_3\text{-}1,3,5)_2][InBr_4]$; projection of the chain onto the x, y plane [1, 2].

Other distances (pm) and angles (°)

Br(1)-In(1)	249.4(1)	Br(1)-In(2)-Br(2)	70.2(1)
Br(2)-In(1)	250.0(1)	Br(1)-In(1)-Br(2)	106.3(1)
Br(3)-In(1)	248.2(1)	In(1)-Br(2)-In(2)	91.0(1)
Br(4)-In(1)	246.6(2)	In(1)-Br(1)-In(2)	92.4(1)
Br(1)-In(2)	344.6(1)		
Br(2)-In(2)	350.3(1)		

[In($C_6H_3(CH_3)_3$-1,3,5)$_2$][$InBr_4$]

This salt was prepared from In[$InBr_4$] dissolved in warm anhydrous nitrogen-purged mesitylene. Cooling produced colorless crystals of a 2:1 ratio of reactants. The product is air- and moisture-sensitive and soon loses a portion of its arene content, especially at reduced pressure [1].

The compound crystallizes in the monoclinic space group $P2_1/n-C_{2h}^5$ (No. 14) with the lattice constants a = 1062.4(2), b = 1338.4(3), c = 1769.7(4) pm, β = 94.56(2)°; Z = 4 gives D_c = 2.091 g/cm³. The data were collected at −40 °C. The crystal consists of helical chain polymers running along the y axis. The two mesityl rings coordinated to each In^I form an angle of 47.3° with one another. In(2) (= In^I) is located almost directly over each of the ring centers. Br(1), Br(2), and Br(3) complete the pentacoordination of the In^I ion. The remaining parameters are shown in **Fig. 100** on p. 386 [1].

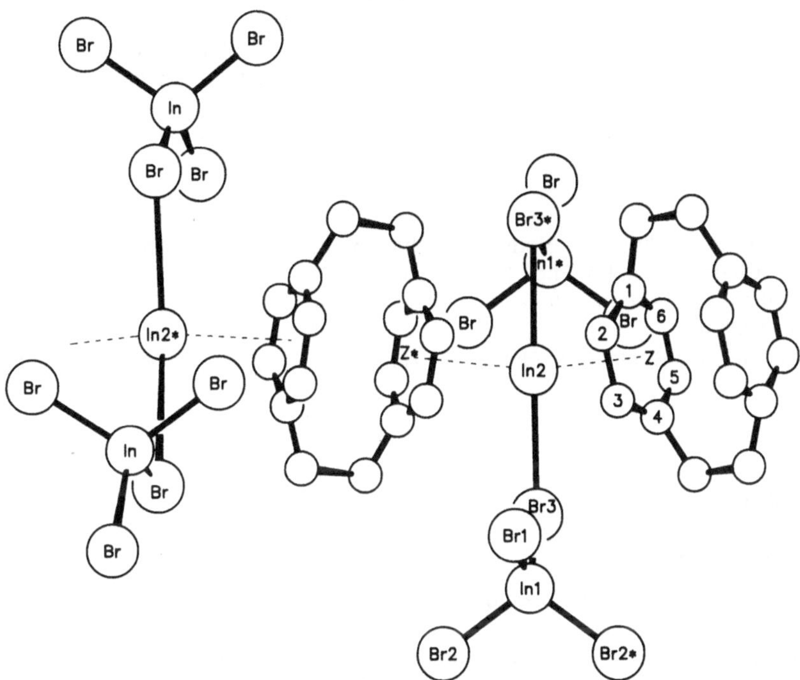

Fig. 101. Section of the crystal structure of [In(p-$C_6H_4CH_2CH_2$-)$_2$][$InBr_4$] [3]

Other distances (pm) and angles (°)

In(2)-C(1)	328.5(7)	Z-In(2)-Z*	124.4
In(2)-C(2)	314.0(6)	Br(3)-In(2)-Br(3*)	162.2(1)
In(2)-C(3)	321.6(7)	Br(1)-In(1)-Br(3)	108.3(1)
In(2)-C(4)	343.9(7)	In(1)-Br(3)-In(2)	125.1(1)
In(2)-C(5)	327.6(7)	In(1)-Br(3)-In(2*)	86.6(1)
In(2)-C(6)	320.3(7)	In(2)-Br(3)-In(2*)	150.3(1)
In(1)···In(2)	661.1(5)		
In(2)···Z*	295		

$[In(p-C_6H_4CH_2CH_2-)_2][InBr_4]$

A toluene solution of [2.2] paracyclophane was added at room temperature to $In[InBr_4]$ (ca. 1:1.5 mole ratio) in mesitylene. A slight cloudiness was removed by warming. On standing for several days colorless single crystals of the title compound formed in 37% yield; m.p. 220 °C (dec.) [3].

The IR spectrum (cm^{-1}) in Nujol shows absorption bands (in cm^{-1}) at 1170 w, 942 m, 895 s, 810 s, 710 s, 625 s, 505 s, and 245 vs [3].

The compound crystallizes in the orthorhombic space group $Pnma-D_{2h}^{16}$ (No. 62) with the lattice constants a = 1305.5(2), b = 1546.1(1), c = 1017.5(1) pm; Z = 4 and D_c = 2.450 g/cm³. The lattice contains polymers of paracyclophane rings that are stacked parallel to the crystallographic b axis and linked by the univalent metal ions. Each In(2) is coordinated over the center of phenyl rings (denoted as Z) from adjacent cyclophanes; the In(2)-C(1 to 6) distances, however, differ greatly since the bond angle strain in paracyclophane prevents planarity of the phenyl rings. The C-1,4 atoms of the phenyl rings are therefore farther from the In than are the other C atoms. The best planes of the two phenyl rings coordinated to the same In are tilted 61.2° to one another. The multilayer chains are networked together by $[InBr_4]^-$ anion bridges in which the Br(3) atom coordinates at In(2). **Fig. 101**, p. 387 shows a section of the twisting In-paracyclophane-In-chain [3]. This compound is compared with the corresponding arene complexes of univalent Ga and Tl in [2].

References:

[1] Ebenhöch, J.; Müller, G.; Riede, J.; Schmidbaur, H. (Angew. Chem. **96** [1984] 367/8; Angew. Chem. Intern. Ed. Engl. **23** [1984] 386/7).

[2] Schmidbaur, H. (Angew. Chem. **97** [1985] 893/904; Angew. Chem. Intern. Ed. Engl. **24** [1985] 893/904).

[3] Schmidbaur, H.; Bublak, W.; Huber, B.; Hofmann, J.; Müller, G. (Chem. Ber. **122** [1989] 265/70).

12.4 Indium(0) π-Complexes

$In(C_2H_4)_n$ (n = 1 to 3)

The title compounds were formed upon reactions of In atoms and C_2H_4 molecules in an Ar matrix. The In/Ar mole ratio was varied between 1:2000 and 1:300 and the C_2H_4/Ar ratio between 1:1000 to 1:50. The main reaction product was $In(C_2H_4)$; $In(C_2H_4)_2$ and $In(C_2H_4)_3$ were formed only in minor concentrations. Species with more than one In atom, presumably $In_2(C_2H_4)$ appeared, when the In:Ar ratio was 1:1200. Experiments were also run with $^{13}C_2H_4$, C_2D_4, and CH_2CD_2. All compounds were identified by IR spectroscopy, and the most important spectra were depicted. According to these measurements $In(C_2H_4)$ has a π-complex structure of C_{2v} symmetry as shown in Formula I.

In
H H
H H

I

The following IR absorptions were observed and assigned by comparison with the corresponding $Al(C_2H_4)$; frequencies in cm^{-1}:

$In(C_2H_4)$	$In(^{13}C_2H_4)$	$In(C_2D_4)$	$In(CH_2CD_2)$	assignment
–	–	–	3078	$\nu_{as}(CH_2)$
3060	3050	2329	–	$\nu(CH_2, CD_2)$
–	–	–	2310	$\nu_{as}(CD_2)$
1488	1473.5	1313.7	1441	$\nu(C{=}C)$ and $\delta_s(CH_2, CD_2)$
1403.5	1397	1054	1256	$\nu(C{=}C)$ and $\delta_{as}(CH_2, CD_2)$
–	–	–	1115	$\varrho(CH_2)$
1201	1171.5	944	993	$\nu(C{=}C) - \delta_s(CH_2, CD_2)$
792.5	791.5	567.5	631	$\varrho(CH_2, CD_2)$
763	759	584	–	wagging (CH_2, CD_2)
238	–	215	–	$\nu_s(In \cdots C)$

For normal-coordinate analysis to check the validity of the proposed assignments, a simplified harmonic force field was used. The two most important force constants are $f(C{=}C) = 6.4(2)$ N/cm (9.1 N/cm for free C_2H_4), and (with a large uncertainty) $f(In \cdots C) = 0.43$ N/cm [3].

$In(C_3H_6)$

Indium propene ions were generated by laser desorption Fourier transform mass spectrometry (FTMS) and were excellent reagents for selective chemical ionization mass spectrometry of mixtures. Abundant indium-containing molecule ions are obtained for olefins, aromatics, and compounds containing oxygen. Because this reagent is unreactive towards alkanes, it is possible to analyze unsaturated hydrocarbons in mixtures with alkanes without interference. No other details about the title compound itself were reported [1].

$In(CO)_2$

In atoms generated at ca. 1300 °C and CO were co-condensed in Ar matrices at ca. 4 K. IR studies of the bright red matrices (deposition on CsI disks cooled to 30 K) revealed the formation of the title compound. No evidence was obtained for the formation of In(CO). It was deduced that the dicarbonyl has a bent planar structure as shown in Formula II. In the proposed structure the In atom is sp^2 hybridized with its lone pair of electrons in the orbital pointing away from the ligands and with the unpaired electron in a p_z orbital perpendicular to the molecular plane. The compound is formally stabilized by two 2-electron donor bonds from the carbonyl ligands into vacant sp^2 orbitals and back-donation of the unpaired electron into π^* orbitals of the CO groups [2].

The IR shows $\nu_s(CO)/\nu_{as}(CO)$ vibrations for the following matrices (the spectra are depicted): $In/^{12}CO(10\%)/Ar$, 2037/1997; $In/^{13}CO(10\%)/Ar$, 1992/1953; $In/^{13}CO(5\%)/^{12}CO(5\%)/Ar$, 2025/1965. Similar spectra were observed in Kr and Xe matrices [2].

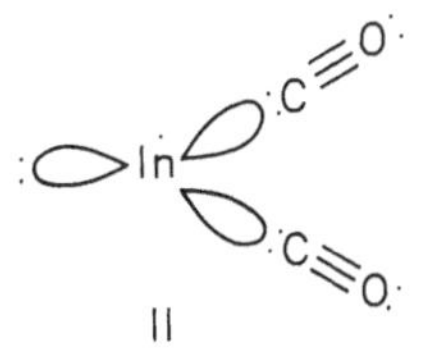

:⊂⊃In(:C≡O:)₂

II

The ESR spectrum from the In/CO(20%)/Ar system is illustrated and shows a decet. The g tensor and the ^{115}In hyperfine and quadrupole coupling tensors were determined to be as follows: $g_z = 1.987(3)$, $g_y = 1.980(3)$, $g_x = 1.969(3)$; $A_z = +418(10)$, $A_y = -399(10)$, $A_x = -440(10)$, and $p_z = +7.2(10)$, $p_y = p_x = -3.6(5)$ MHz [2].

References:

[1] Chowdhury, A. K.; Cooper, J. R.; Wilkins, C. L. (Anal. Chem. **61** [1989] 86/8).
[2] Hatton, W. G.; Hacker, N. P.; Kasai, P. H. (J. Phys. Chem. **93** [1989] 1328/32).
[3] Manceron, L.; Andrews, L. (J. Phys. Chem. **94** [1990] 3513/8).

Empirical Formula Index

In the following index the compounds are listed by their empirical formula in the order of increasing carbon content. Lewis bases of adducts are regarded as ordinary σ-bonded ligands and are thus included in the empirical formula, e.g., $In(CH_3)_3 \cdot N(CH_3)_3$ is listed as $C_6H_{18}InN$. Formulas of ionic compounds are given in brackets; ions as well as components of solvates are separated by a period.

In the second column, page references are printed in ordinary type, table numbers in bold face, and compound numbers within the table in italics.

Ligand Formula Index

The ligands containing carbon atoms (except CO, CN, CNO, and CNS) can be used to locate a compound. These ligands are listed in the Ligand Formula Index by number of carbon atoms in the empirical ligand formula. They are generally not further characterized by linearized formulas or names unless this is necessary to distinguish between isomers. The number of identical ligands in a compound and the nature of bonding are not taken into consideration. Thus several compounds may be listed at one position.

Compounds having two or more different carbon-containing ligands occur at more than one position. The variable organic ligands are placed in the first three columns, while nonorganic ligands such as H, halogen, chalcogen, NH_3, CN, etc., appear in the fifth column or in the third and fifth columns if necessary CO is given in the forth column.

Page references are printed in ordinary types, table numbers in boldface, and compound numbers in the tables in italics. The following examples illustrate the arrangement.

For $(CH_3)_2InI \cdot H_2NCH_2CH_2NH_2$ p. 158, Tab. 22, Nr. 14:

CH_3	$C_2H_8N_2$	–	–	I	158, **22**, *14*
$C_2H_8N_2$	CH_3	–	–	I	158, **22**, *14*

CH_3	–	–	–	H	347, **66**, *1*
					347, **66**, *2*
					347, **66**, *3*
CH_3	–	–	–	HO	180
CH_3	–	–	–	H_2N	255, **50**, *1*
CH_3	–	–	–	H_2O_2P	214, **41**, *10*
CH_3	–	–	–	H_2P	313, **61**, *1*
CH_3	–	–	–	NH_3	33, **6**, *1*
CH_3	–	I	–	NH_3	158, **22**, *12*
CH_3	–	–	–	PH_3	38, **6**, *35*
CH_3	–	–	–	SbH_3	40, **6**, *48*
CH_3	–	–	–	I	157, **22**, *1*
					161, **23**, *1*
					169/70, **24**, *13*
					170, **24**, *19*
					354, **67**, *28*
					355, **67**, *29*
					355, **67**, *30*
					355, **67**, *31*
					355, **67**, *32*
CH_3	–	–	–	N_3	177, **26**, *6*
CH_3	–	–	–	O	230
CH_3	–	–	–	S	245
					364/6
CH_3	–	–	–	Se	365/6
CH_3	CHO_2	–	–	–	197, **35**, *1*
CH_3	CH_2NO_2	–	–	–	213, **41**, *6*
CH_3	CH_3O	–	–	–	181, **29**, *1*
					183
					232, **45**, *1*
CH_3	CH_3O_2	–	–	–	209
CH_3	CH_3O_2S	–	–	–	212, **41**, *1*
CH_3	CH_3O_3S	–	–	–	212/3, **41**, *2*
					213, **41**, *5*
CH_3	CH_3S	–	–	–	240/1, **46**, *1*
CH_3	CH_5As	–	–	–	40, **6**, *44*
CH_3	CH_5N	–	–	–	33, **6**, *2*
CH_3	C_2D_6N	–	–	–	259
CH_3	$C_2F_3O_2$	–	–	–	198, **35**, *6*
CH_3	$C_2H_2N_2O_2$	–	–	–	229
CH_3	$C_2H_2N_2S_2$	–	–	–	293
CH_3	$C_2H_2N_3$	–	–	–	280, **56**, *15*
CH_3	C_2H_3	–	–	–	103, **17**, *4*
CH_3	$C_2H_3N_2$	–	–	–	51, **8**, *9*
CH_3	C_2H_3OS	–	–	–	204
CH_3	$C_2H_3O_2$	–	–	–	198, **35**, *2*
CH_3	$C_2H_3O_2$	$C_2H_5S_2$	–	–	239
CH_3	$C_2H_3O_2$	C_2H_6OS	–	–	224/5, **43**, *4*
CH_3	$C_2H_3O_2$	$C_2H_8N_2$	–	–	225, **43**, *6*
CH_3	$C_2H_3O_2$	$C_3H_7S_2$	–	–	239
CH_3	$C_2H_3O_2$	C_5H_5N	–	–	225, **43**, *5*

CH_3	$C_2H_3O_2$	$C_7H_7S_2$	–	–	239
CH_3	$C_2H_3O_2$	$C_{10}H_8N_2$	–	–	225, **43**, *7*
CH_3	$C_2H_3O_2$	$C_{12}H_8N_2$	–	–	225, **43**, *8*
CH_3	$C_2H_3O_2$	$C_{26}H_{24}P_2$	–	–	225, **43**, *9*
CH_3	C_2H_4N	–	–	–	273, **55**, *1*
CH_3	$C_2H_4N_2$	–	–	–	36, **6**, *18*
CH_3	C_2H_5	–	–	–	102, **17**, *1* 102, **17**, *2*
CH_3	C_2H_5O	–	–	–	181, **29**, *2*
CH_3	$C_2H_5O_3S$	–	–	–	213, **41**, *3*
CH_3	C_2H_6As	–	–	–	316, **61**, *14*
CH_3	$C_2H_6AsO_2$	–	–	–	215/6, **41**, *17*
CH_3	$C_2H_6B_4$	–	–	–	327/8
CH_3	C_2H_6N	–	–	–	255, **50**, *2* 296
CH_3	C_2H_6NOS	–	–	–	283, **57**, *1*
CH_3	$C_2H_6NO_4S_2$	–	–	–	256, **50**, *8*
CH_3	C_2H_6O	–	–	–	28, **5**, *1*
CH_3	C_2H_6OPS	–	–	–	215, **41**, *15*
CH_3	C_2H_6OS	–	–	–	29, **5**, *4*
CH_3	C_2H_6OS	$C_7H_6S_2$	–	–	249, **48**, *17*
CH_3	$C_2H_6O_2P$	–	–	–	215, **41**, *13*
CH_3	C_2H_6P	–	–	–	313, **61**, *2*
CH_3	$C_2H_6PS_2$	–	–	–	241, **46**, *3*
CH_3	C_2H_6S	–	–	–	29, **5**, *7*
CH_3	C_2H_7N	–	–	–	33, **6**, *3*
CH_3	C_2H_7N	C_5H_5	–	–	106, **17**, *19*
CH_3	C_2H_7P	–	–	–	38, **6**, *37*
CH_3	$C_2H_8N_2$	–	–	Cl	121, **18**, *12*
CH_3	$C_2H_8N_2$	–	–	I	158, **22**, *14*
CH_3	$C_2H_{11}B_{10}$	$C_4H_{10}O_2$	–	–	106, **17**, *21*
CH_3	C_2O_4	–	–	–	206/7
CH_3	C_3H_3	–	–	–	103/4, **17**, *6*
CH_3	$C_3H_3N_2$	–	–	–	278, **56**, *4* 279, **56**, *11* 366
CH_3	C_3H_5N	–	–	–	36, **6**, *17*
CH_3	$C_3H_5O_2$	–	–	–	198, **35**, *3*
CH_3	C_3H_6N	–	–	–	265/6 273, **55**, *2*
CH_3	C_3H_6NO	–	–	–	214, **41**, *7* 267
CH_3	$C_3H_6NO_2$	–	–	–	214, **41**, *9*
CH_3	$C_3H_6NS_2$	–	–	–	241, **46**, *2* 247, **48**, *2*
CH_3	C_3H_9As	–	–	–	40, **6**, *47*
CH_3	C_3H_9GeO	–	–	–	188, **31**, *3*
CH_3	C_3H_9N	–	–	–	34, **6**, *4*
CH_3	C_3H_9N	C_5H_{11}	–	–	105, **17**, *16*
CH_3	C_3H_9N	$C_7H_6S_2$	–	–	249, **48**, *18*
CH_3	C_3H_9NO	–	–	–	29, **5**, *5*

CH_3	C_3H_9NOP	–	–	–	215, **41**, *16*
CH_3	C_3H_9NP	–	–	–	283, **57**, *3*
CH_3	C_3H_9NPS	–	–	–	241, **46**, *4*
CH_3	C_3H_9OP	–	–	–	29, **5**, *6*
CH_3	C_3H_9OSi	–	–	–	188, **31**, *1*
					364
CH_3	C_3H_9P	–	–	–	39, **6**, *39*
CH_3	C_3H_9SSb	–	–	Br	148, **20**, *14*
CH_3	C_3H_9SSb	–	–	Cl	120, **18**, *10*
					138/9, **19**, *10*
CH_3	C_3H_9SSb	–	–	I	158, **22**, *11*
CH_3	C_3H_9Sb	–	–	–	40, **6**, *49*
CH_3	C_3H_9Sn	–	–	–	367
CH_3	$C_4H_3F_6O$	–	–	–	182, **29**, *5*
CH_3	C_4H_4N	–	–	–	277, **56**, *1*
CH_3	$C_4H_4N_2$	–	–	–	294
CH_3	$C_4H_5N_2$	–	–	–	278, **56**, *5*
					278, **56**, *6*
					279, **56**, *12*
CH_3	C_4H_6NO	–	–	–	267
CH_3	$C_4H_6N_2O_2$	–	–	–	229
					289
					290/1
CH_3	$C_4H_6N_2S_2$	–	–	–	293/4
CH_3	$C_4H_7O_2$	–	–	–	198, **35**, *4*
CH_3	C_4H_8N	–	–	–	273, **55**, *3*
CH_3	C_4H_8O	–	–	–	29, **5**, *3*
CH_3	C_4H_8O	C_5H_{11}	–	–	105, **17**, *15*
CH_3	$C_4H_9N_2$	–	–	–	268/9
CH_3	$C_4H_9N_2O$	–	–	–	271
CH_3	C_4H_9O	–	–	–	181/2, **29**, *3*
CH_3	$C_4H_9O_2$	–	–	–	209
CH_3	$C_4H_{10}N$	–	–	–	255, **50**, *3*
CH_3	$C_4H_{10}NO$	–	–	–	191, **33**, *1*
CH_3	$C_4H_{10}N_2$	–	–	–	295/6
CH_3	$C_4H_{10}O$	–	–	–	29, **5**, *2*
CH_3	$C_4H_{10}P$	–	–	–	313, **61**, *3*
					367
CH_3	$C_4H_{10}Sn$	–	–	–	104, **17**, *9*
CH_3	$C_4H_{11}N$	–	–	Cl	139, **19**, *11*
CH_3	$C_4H_{11}P$	–	–	–	38, **6**, *38*
					55/6
CH_3	$C_4H_{11}Sn$	–	–	–	104, **17**, *8*
					105, **17**, *12*
CH_3	$C_4H_{12}N_2$	–	–	–	49, **8**, *2*
CH_3	$C_4H_{12}P_2$	–	–	–	51/2, **8**, *13*
CH_3	C_4O_4	–	–	–	207/9
					208/9
CH_3	$C_5H_4F_3O_2$	–	–	–	191, **33**, *3*
CH_3	$C_5H_4F_3O_2$	$C_{10}H_8N_2$	–	–	224, **43**, *1*
CH_3	$C_5H_4F_3O_2$	$C_{12}H_8N_2$	–	–	224, **43**, *2*

CH_3	$C_5H_4F_3O_2$	$C_{26}H_{24}P_2$	–	–	224, **43**, *3*
CH_3	C_5H_4N	$C_7H_6S_2$	–	–	249, **48**, *19*
CH_3	C_5H_4NO	–	–	–	278, **56**, *2*
CH_3	C_5H_5	–	–	–	103, **17**, *5*
CH_3	C_5H_5N	–	–	Cl	121, **18**, *14*
					139, **19**, *12*
CH_3	C_5H_5N	–	–	I	158, **22**, *13*
CH_3	C_5H_5N	$C_{12}H_{10}O_2P$	–	HO	238
CH_3	$C_5H_7N_2$	–	–	–	278/9, **56**, *7*
CH_3	$C_5H_7O_2$	–	–	–	191, **33**, *2*
CH_3	$C_5H_9O_2$	–	–	–	198, **35**, *5*
CH_3	$C_5H_{10}N$	–	–	–	273, **55**, *4*
CH_3	C_5H_{11}	–	–	–	102, **17**, *3*
CH_3	C_5H_{11}	$C_6H_{16}N_2$	–	–	106, **17**, *17*
					106, **17**, *18*
CH_3	$C_5H_{11}N_2$	–	–	–	273/4, **55**, *6*
CH_3	$C_5H_{12}N$	–	–	–	106, **17**, *22*
					110, **17**, *38*
CH_3	$C_5H_{12}N_3$	–	–	–	271
CH_3	$C_5H_{13}NO$	–	–	–	192, **33**, *10*
CH_3	$C_5H_{14}N_2$	–	–	–	34, **6**, *8*
					49, **8**, *1*
CH_3	$C_6H_4NO_3$	–	–	–	192, **33**, *7*
CH_3	$C_6H_5N_2O$	–	–	–	214, **41**, *8*
CH_3	C_6H_5O	–	–	–	182, **29**, *4*
CH_3	$C_6H_5O_3S$	–	–	–	213, **41**, *4*
CH_3	C_6H_6NO	–	–	–	278, **56**, *3*
CH_3	C_6H_7As	–	–	–	40, **6**, *45*
CH_3	$C_6H_7N_2$	–	–	–	302/3
CH_3	C_6H_7P	–	–	–	38, **6**, *36*
CH_3	$C_6H_8BN_4$	–	–	–	279, **56**, *10*
CH_3	$C_6H_8BN_4$	–	–	Cl	296/7
CH_3	$C_6H_8N_2$	–	–	Cl	121, **18**, *16*
CH_3	$C_6H_{12}N_4$	–	–	–	34, **6**, *9*
					50, **8**, *7*
					51, **8**, *8*
					291/3
CH_3	$C_6H_{14}N$	–	–	–	255/6, **50**, *4*
CH_3	$C_6H_{14}P$	–	–	–	313, **61**, *4*
CH_3	$C_6H_{15}N$	–	–	–	34, **6**, *5*
CH_3	$C_6H_{15}NP$	–	–	–	283, **57**, *4*
CH_3	$C_6H_{15}OSn$	–	–	–	188, **31**, *4*
CH_3	$C_6H_{15}P$	–	–	–	39, **6**, *40*
CH_3	$C_6H_{15}Sb$	–	–	–	40, **6**, *50*
CH_3	$C_6H_{16}NP$	–	–	–	37, **6**, *25*
CH_3	$C_6H_{16}NP_2$	–	–	–	368
CH_3	$C_6H_{16}N_2$	–	–	–	50, **8**, *3*
CH_3	$C_6H_{16}N_2$	$C_8H_{15}B_{10}$	–	–	106, **17**, *20*
					329
CH_3	$C_6H_{17}NP_2$	–	–	–	56
CH_3	$C_6H_{18}GeNP$	–	–	–	37, **6**, *28*

CH_3	$C_6H_{18}NPSi$	–	–	–	37, **6**, *26*
CH_3	$C_6H_{18}NPSn$	–	–	–	38, **6**, *30*
CH_3	$C_6H_{18}NSi_2$	–	–	–	256, **50**, *7*
CH_3	$C_6H_{18}N_3P$	–	–	–	39, **6**, *43*
CH_3	$C_7H_5N_2$	–	–	–	279, **56**, *9*
					279, **56**, *13*
CH_3	$C_7H_5O_2$	–	–	–	191, **33**, *6*
					192, **33**, *9*
CH_3	$C_7H_6S_2$	–	–	–	247, **48**, *1*
CH_3	$C_7H_6S_2$	$C_{10}H_8N_2$	–	–	249/50, **48**, *20*
CH_3	$C_7H_6S_2$	$C_{12}H_8N_2$	–	–	250, **48**, *21*
CH_3	C_7H_8N	–	–	–	256, **50**, *6*
CH_3	$C_7H_{10}N_2$	–	–	–	36, **6**, *20*
CH_3	$C_7H_{14}N$	–	–	–	273, **55**, *5*
CH_3	$C_7H_{15}N$	–	–	–	35, **6**, *15*
					109/10, **17**, *37*
CH_3	$C_7H_{16}N$	–	–	–	106/7, **17**, *23*
					110, **17**, *39*
CH_3	$C_7H_{18}N_2$	–	–	–	50, **8**, *4*
CH_3	$C_7H_{18}N_2Si_2$	–	–	–	37, **6**, *24*
CH_3	$C_7H_{21}N_2SSi_2$	–	–	–	283, **57**, *2*
CH_3	C_8H_5	–	–	–	104, **17**, *7*
CH_3	$C_8H_5MoO_3$	–	–	–	332, **62**, *9*
CH_3	$C_8H_5O_3$	–	–	–	198, **35**, *7*
CH_3	$C_8H_6MoO_3$	–	–	–	56
CH_3	$C_8H_6O_3W$	–	–	–	56
CH_3	$C_8H_{10}F_3O_2$	–	–	–	191, **33**, *4*
CH_3	$C_8H_{17}N_2$	–	–	–	269
CH_3	$C_8H_{18}As$	–	–	–	316, **61**, *15*
CH_3	$C_8H_{18}P$	–	–	–	313/4, **61**, *5*
CH_3	$C_8H_{18}Sb$	–	–	–	316, **61**, *18*
CH_3	$C_8H_{21}N_2Si_2$	–	–	–	270/1
CH_3	$C_8H_{24}NOPSi_2$	–	–	–	38, **6**, *32*
CH_3	$C_8H_{24}N_2P_2Si$	–	–	–	38, **6**, *33*
					51, **8**, *12*
CH_3	C_9H_6NO	–	–	–	192, **33**, *8*
CH_3	$C_9H_9N_3$	–	–	–	226, **43**, *13*
CH_3	$C_9H_{12}N$	–	–	–	107, **17**, *25*
					107, **17**, *26*
					110, **17**, *40*
CH_3	$C_9H_{12}N_4O$	–	–	–	226, **43**, *16*
CH_3	$C_9H_{19}N$	–	–	–	35, **6**, *16*
CH_3	$C_9H_{21}N_2Si$	–	–	–	310/2
CH_3	$C_9H_{21}N_3$	–	–	–	36, **6**, *19*
CH_3	$C_9H_{24}GeNP$	–	–	–	38, **6**, *29*
CH_3	$C_9H_{24}NPSi$	–	–	–	37, **6**, *27*
CH_3	$C_9H_{24}NPSn$	–	–	–	38, **6**, *31*
CH_3	$C_{10}H_6F_3O_2$	–	–	–	191, **33**, *5*
CH_3	$C_{10}H_8N_2$	–	–	–	36, **6**, *21*
					36, **6**, *22*
					36, **6**, *23*

CH_3	$C_{10}H_8N_2$	–	–	–	51, **8**, *10* 51, **8**, *11* 225/6, **43**, *10* 294
CH_3	$C_{10}H_8N_2$	–	–	Cl	121, **18**, *15* 139, **19**, *13*
CH_3	$C_{10}H_8N_2$	–	–	I	158, **22**, *15* 163, **23**, *9*
CH_3	$C_{10}H_{14}N$	–	–	–	107, **17**, *24* 108, **17**, *28*
CH_3	$C_{10}H_{16}N_2$	–	–	–	34, **6**, *10* 50, **8**, *5*
CH_3	$C_{10}H_{22}N_2$	–	–	–	110, **17**, *41*
CH_3	$C_{10}H_{24}DN_2Si$	–	–	–	284, **57**, *9*
CH_3	$C_{10}H_{24}NOSi$	–	–	–	283, **57**, *7*
CH_3	$C_{10}H_{24}NOSi$	–	–	Cl	297
CH_3	$C_{10}H_{24}N_2Si$	–	–	–	284, **57**, *11* 310/1
CH_3	$C_{10}H_{25}N_2Si$	–	–	–	284, **57**, *8*
CH_3	$C_{11}H_{10}N_2$	–	–	–	226, **43**, *17*
CH_3	$C_{11}H_{16}N$	–	–	–	108, **17**, *27*
CH_3	$C_{11}H_{33}GaN_2P_2Si$	–	–	–	38, **6**, *34*
CH_3	$C_{11}H_{33}IrP_3$	–	–	–	331, **62**, *5* 332, **62**, *6*
CH_3	$C_{11}H_{33}IrP_3$	–	–	Br	332, **62**, *14* 333, **62**, *15*
CH_3	$C_{12}H_8N_2$	–	–	–	226, **43**, *14*
CH_3	$C_{12}H_8N_2$	–	–	Cl	121, **18**, *13*
CH_3	$C_{12}H_8N_3$	–	–	–	279, **56**, *14*
CH_3	$C_{12}H_{10}As$	–	–	–	316, **61**, *16*
CH_3	$C_{12}H_{10}O_2P$	–	–	–	215, **41**, *14*
CH_3	$C_{12}H_{10}P$	–	–	–	314, **61**, *6*
CH_3	$C_{12}H_{11}As$	–	–	–	40, **6**, *46*
CH_3	$C_{12}H_{12}N_2$	–	–	–	226, **43**, *11*
CH_3	$C_{12}H_{19}N_2$	–	–	–	108, **17**, *29* 111, **17**, *42*
CH_3	$C_{12}H_{22}N$	–	–	–	34, **6**, *7* 256, **50**, *5*
CH_3	$C_{12}H_{27}N$	–	–	–	34, **6**, *6*
CH_3	$C_{13}H_8N_4O_4$	–	–	–	35, **6**, *13*
CH_3	$C_{13}H_{15}N_6O$	–	–	–	233, **45**, *7*
CH_3	$C_{13}H_{15}N_6O$	NO_3	–	H_2O	233, **45**, *7*
CH_3	$C_{13}H_{17}N_2Pt$	–	–	–	332, **62**, *7*
CH_3	$C_{14}H_{12}N_2$	–	–	–	35, **6**, *12*
CH_3	$C_{14}H_{13}N_2$	–	–	–	269
CH_3	$C_{14}H_{14}N_5O$	–	–	–	233, **45**, *8*
CH_3	$C_{14}H_{14}N_5O$	NO_3	–	H_2O	233, **45**, *8*
CH_3	$C_{14}H_{16}N_2$	–	–	–	226, **43**, *12*
CH_3	$C_{15}H_{11}N_2$	–	–	–	279, **56**, *8*
CH_3	$C_{15}H_{11}N_3$	–	–	–	227, **43**, *18*
CH_3	$C_{15}H_{11}N_3$	–	–	Cl	139, **19**, *14*

CH_3	$C_{15}H_{13}N_4O$	–	–	–	234, **45**, *9*
CH_3	$C_{15}H_{13}N_4O$	NO_3	–	H_2O	234, **45**, *9*
CH_3	$C_{16}H_{12}N_3O$	–	–	–	234, **45**, *10*
CH_3	$C_{16}H_{12}N_3O$	NO_3	–	H_2O	234, **45**, *10*
CH_3	$C_{16}H_{14}N_2$	–	–	–	35, **6**, *14*
CH_3	$C_{16}H_{27}N_2$	–	–	Cl	134/5
CH_3	$C_{17}H_{13}N_2O$	–	–	–	234, **45**, *11*
CH_3	$C_{17}H_{13}N_2O$	NO_3	–	H_2O	234, **45**, *11*
CH_3	$C_{17}H_{22}N_2$	–	–	–	35, **6**, *11* 50, **8**, *6*
CH_3	$C_{18}H_{15}AsO$	–	–	Cl	120, **18**, *9*
CH_3	$C_{18}H_{15}NP$	–	–	–	283, **57**, *5*
CH_3	$C_{18}H_{15}OP$	–	–	Cl	120, **18**, *8*
CH_3	$C_{18}H_{15}OP$	–	–	I	158, **22**, *10*
CH_3	$C_{18}H_{15}OSi$	–	–	–	188, **31**, *2*
CH_3	$C_{18}H_{15}P$	–	–	–	39, **6**, *41*
CH_3	$C_{18}H_{15}P$	–	–	Cl	121, **18**, *17*
CH_3	$C_{18}H_{15}P$	–	–	I	158, **22**, *16*
CH_3	$C_{18}H_{28}N_2PSi_2$	–	–	–	283, **57**, *6*
CH_3	$C_{18}H_{42}LiN_4Si_2$	–	–	–	284, **57**, *10*
CH_3	$C_{21}H_{21}P$	–	–	–	39, **6**, *42*
CH_3	$C_{21}H_{23}N_3$	–	–	–	226, **43**, *15*
CH_3	$C_{26}H_{24}P_2$	–	–	–	52, **8**, *14*
CH_3	$C_{34}H_{33}P_3$	–	–	–	52, **8**, *15*
CH_3	$C_{36}H_{44}N_4$	–	–	–	305, **60**, *1*
CH_3	$C_{42}H_{42}P_4$	–	–	–	52, **8**, *16*
CH_3	$C_{44}H_{28}N_4$	–	–	–	306, **60**, *11*
CH_3O	CD_3	–	–	–	183
CH_3O	CH_3	–	–	–	181, **29**, *1* 183 232, **45**, *1*
CH_3O	C_2H_5	–	–	–	182, **29**, *6*
CH_3O	$C_4H_{11}Si$	–	–	–	183, **29**, *15*
CH_3O	C_5H_{11}	–	–	–	183, **29**, *14*
CH_3O_2	CH_3	–	–	–	209
CH_3O_2S	CH_3	–	–	–	212, **41**, *1*
CH_3O_3S					
OOS(O)CH_3	CH_3	–	–	–	213, **41**, *5*
OOS(O)CH_3	C_2H_5	–	–	–	216, **41**, *18*
OOSOCH_3	CH_3	–	–	–	212/3, **41**, *2*
CH_3S	CH_3	–	–	–	240/1, **46**, *1*
CH_5As	CH_3	–	–	–	40, **6**, *44*
CH_5N	CH_3	–	–	–	33, **6**, *2*
C_2 D_3N	C_7H_7	–	–	F	117
C_2D_6N	CH_3	–	–	–	259
$C_2F_3O_2$	CH_3	–	–	–	198, **35**, *6*
$C_2H_2N_2O_2$	CH_3	–	–	–	229
$C_2H_2N_2S_2$	CH_3	–	–	–	293
$C_2H_2N_3$	CH_3	–	–	–	280, **56**, *15*

C_2H_5	–	–	–	I	157, **22**, *2*
					162, **23**, *2*
					170, **24**, *14*
					355, **67**, *33*
					355, **67**, *34*
C_2H_5	–	–	–	N_3	177, **26**, *7*
C_2H_5	–	–	–	O	231
C_2H_5	–	–	–	S	245
C_2H_5	–	–	–	Se	252
C_2H_5	CH_2I	–	–	–	104, **17**, *10*
C_2H_5	CH_2NO_2	–	–	–	216, **41**, *20*
C_2H_5	CH_3	–	–	–	102, **17**, *1*
					102, **17**, *2*
C_2H_5	CH_3O	–	–	–	182, **29**, *6*
C_2H_5	CH_3O_3S	–	–	–	216, **41**, *18*
C_2H_5	C_2H_3OS	–	–	–	204/6
C_2H_5	$C_2H_3O_2$	–	–	–	198, **35**, *8*
C_2H_5	$C_2H_3S_2$	–	–	–	241, **46**, *6*
C_2H_5	C_2H_5O	–	–	–	182, **29**, *7*
					232, **45**, *2*
C_2H_5	$C_2H_5O_2$	–	–	–	210
C_2H_5	$C_2H_5O_3S$	–	–	–	216, **41**, *19*
C_2H_5	C_2H_6NO	–	–	–	192, **33**, *11*
C_2H_5	C_2H_6OS	–	–	I	163, **23**, *10*
C_2H_5	C_2O_4	–	–	–	207
C_2H_5	$C_3H_3N_2$	–	–	–	280, **56**, *16*
C_2H_5	$C_3H_5O_2$	–	–	–	199, **35**, *9*
C_2H_5	C_3H_6NO	–	–	–	267
C_2H_5	$C_3H_6NO_2$	–	–	–	216, **41**, *21*
C_2H_5	$C_3H_6NS_2$	–	–	–	247, **48**, *3*
C_2H_5	C_3H_9As	–	–	–	71
C_2H_5	C_3H_9SSb	–	–	Br	148, **20**, *15*
C_2H_5	C_3H_9SSb	–	–	Cl	121, **18**, *18*
C_2H_5	C_3H_9SSb	–	–	I	158, **22**, *17*
C_2H_5	$C_4H_5N_2$	–	–	–	280, **56**, *17*
C_2H_5	$C_4H_6N_2O_2$	–	–	–	290/1
C_2H_5	C_4H_9O	–	–	–	182, **29**, *8*
					233, **45**, *3*
C_2H_5	$C_4H_{10}N$	–	–	–	257, **50**, *9*
					296
C_2H_5	$C_4H_{10}NO$	–	–	–	192, **33**, *12*
C_2H_5	$C_4H_{10}O$	–	–	–	70/1
C_2H_5	$C_4H_{10}P$	–	–	–	314, **61**, *7*
C_2H_5	$C_4H_{12}P_2$	–	–	–	72
C_2H_5	C_4O_4	–	–	–	209
C_2H_5	C_5H_5	–	–	–	105, **17**, *13*
C_2H_5	$C_5H_7O_4$	–	–	–	192, **33**, *16*
C_2H_5	$C_5H_{10}N$	–	–	–	274, **55**, *7*
C_2H_5	$C_5H_{11}N$	–	–	–	71
C_2H_5	$C_5H_{12}N$	–	–	–	108, **17**, *30*
C_2H_5	$C_5H_{13}NO^+$	–	–	–	193, **33**, *18*

C_2H_5	$C_6H_{15}SeSi$	–	–	–	252
C_2H_5	$C_6H_{16}N_2$	–	–	Br	154, **21**, *9*
C_2H_5	$C_6H_{16}N_2$	–	–	I	163, **23**, *11*
C_2H_5	C_7H_5N	–	–	–	71
C_2H_5	C_7H_5OS	–	–	–	205
C_2H_5	$C_7H_{16}N$	–	–	–	108, **17**, *31*
C_2H_5	$C_8H_5MoO_3$	–	–	–	332, **62**, *10* 333, **62**, *17*
C_2H_5	$C_8H_{18}P$	–	–	–	314, **61**, *8*
C_2H_5	C_9H_6NO	–	–	–	193, **33**, *17*
C_2H_5	$C_9H_{10}NO$	–	–	–	267/8
C_2H_5	$C_9H_{11}O$	–	–	–	182, **29**, *9*
C_2H_5	$C_9H_{12}N$	–	–	–	108/9, **17**, *32*
C_2H_5	$C_{10}H_8N_2$	–	–	Br	154, **21**, *10*
C_2H_5	$C_{10}H_8N_2$	–	–	I	158, **22**, *18*
C_2H_5	$C_{11}H_{33}IrP_3$	–	–	–	332, **62**, *8*
C_2H_5	$C_{12}H_{10}O_2P$	–	–	–	217, **41**, *22*
C_2H_5	$C_{12}H_{19}N_2$	–	–	–	111, **17**, *43*
C_2H_5	$C_{13}H_{11}O$	–	–	–	182, **29**, *10*
C_2H_5	$C_{13}H_{11}S$	–	–	–	241, **46**, *5*
C_2H_5	$C_{14}H_{20}O_2$	–	–	–	230
C_2H_5	$C_{16}H_{25}O_2$	–	–	–	192, **33**, *13* 192, **33**, *14* 192, **33**, *15* 230
C_2H_5	$C_{16}H_{27}N_2$	–	–	–	111, **17**, *44*
C_2H_5	$C_{17}H_{22}N_2$	–	–	–	71/2
C_2H_5	$C_{26}H_{24}P_2$	–	–	–	72
C_2H_5	$C_{36}H_{44}N_4$	–	–	–	305, **60**, *2*
C_2H_5	$C_{44}H_{28}N_4$	–	–	–	306, **60**, *12*
C_2H_5O	CH_3	–	–	–	181, **29**, *2*
C_2H_5O	C_2H_5	–	–	–	182, **29**, *7* 232, **45**, *2*
C_2H_5O	C_4H_9	–	–	–	182, **29**, *11*
$C_2H_5O_2$	C_2H_5	–	–	–	210
$C_2H_5O_3S$					
$OOS(O)C_2H_5$	C_2H_5	–	–	–	216, **41**, *19*
$OOSOC_2H_5$	CH_3	–	–	–	213, **41**, *3*
C_2H_5S	C_6H_5	–	–	–	242, **46**, *14*
$C_2H_5S_2$	CH_3	$C_2H_3O_2$	–	–	239
C_2H_6As	CH_3	–	–	–	316, **61**, *14*
$C_2H_6AsO_2$	CH_3	–	–	–	215/6, **41**, *17*
$C_2H_6B_4$	CH_3	–	–	–	327/8
C_2H_6N	CH_3	–	–	–	255, **50**, *2* 296
C_2H_6NO	C_2H_5	–	–	–	192, **33**, *11*
C_2H_6NOS	CH_3	–	–	–	283, **57**, *1*
$C_2H_6NO_4S_2$	CH_3	–	–	–	256, **50**, *8*
$C_2H_6NO_4S_2$	$C_4H_{11}Si$	–	–	–	257/8, **50**, *12*
C_2H_6O	CH_3	–	–	–	28, **5**, *1*
C_2H_6O	C_2H_3	–	–	–	85, **12**, *6*

C_2H_6OPS	CH_3	–	–	–	215, **41**, *15*
C_2H_6OS	CF_3	–	–	Cl	122, **18**, *20*
C_2H_6OS	CH_3	–	–	–	29, **5**, *4*
C_2H_6OS	CH_3	$C_2H_3O_2$	–	–	224/5, **43**, *4*
C_2H_6OS	CH_3	$C_7H_6S_2$	–	–	249, **48**, *17*
C_2H_6OS	C_2H_5	–	–	I	163, **23**, *10*
C_2H_6OS	C_6F_5	–	–	–	92, **14**, *22*
$C_2H_6O_2P$	CH_3	–	–	–	215, **41**, *13*
C_2H_6P	CH_3	–	–	–	313, **61**, *2*
$C_2H_6PS_2$	CH_3	–	–	–	241, **46**, *3*
C_2H_6S	CH_3	–	–	–	29, **5**, *7*
C_2H_7N	CH_3	–	–	–	33, **6**, *3*
C_2H_7N	CH_3	C_5H_5	–	–	106, **17**, *19*
C_2H_7P	CH_3	–	–	–	38, **6**, *37*
$C_2H_8N_2$	CH_3	–	–	Cl	121, **18**, *12*
$C_2H_8N_2$	CH_3	–	–	I	158, **22**, *14*
$C_2H_8N_2$	CH_3	$C_2H_3O_2$	–	–	225, **43**, *6*
$C_2H_{11}B_{10}$	CH_3	$C_4H_{10}O_2$	–	–	106, **17**, *21*
C_2O_4	CH_3	–	–	–	206/7
C_2O_4	C_2H_5	–	–	–	207
C_3H_3	CH_3	–	–	–	103/4, **17**, *6*
C_3H_3	C_2H_3	–	–	–	105, **17**, *14*
$C_3H_3N_2$					
1*H*-Imidazol-1-yl	CH_3	–	–	–	279, **56**, *11*
1*H*-Pyrazol-1-yl	CH_3	–	–	–	278, **56**, *4*
					366
1*H*-Pyrazol-1-yl	C_2H_5	–	–	–	280, **56**, *16*
C_3H_5					
CH_2=$CHCH_2$	–	–	–	I	170, **24**, *18*
Cyclopropyl	–	–	–	–	77, **11**, *19*
C_3H_5N	CH_3	–	–	–	36, **6**, *17*
$C_3H_5O_2$	CH_3	–	–	–	198, **35**, *3*
$C_3H_5O_2$	C_2H_5	–	–	–	199, **35**, *9*
$C_3H_5O_2$	C_4H_9	–	–	–	199, **35**, *10*
$C_3H_5O_2$	C_4H_9	$C_{12}H_{27}OSn$	–	–	233, **45**, *4*
$C_3H_5O_2$	C_6H_5	–	–	–	199, **35**, *15*
					233, **45**, *6*
C_3H_6	–	–	–	–	389
C_3H_6N					
Azetidinyl	CH_3	–	–	–	273, **55**, *2*
N=C$(CH_3)_2$	CH_3	–	–	–	265/6
C_3H_6NO					
N$(CH_3)COCH_3$	CH_3	–	–	–	267
N$(CH_3)COCH_3$	C_2H_5	–	–	–	267
ON=C$(CH_3)_2$	CH_3	–	–	–	214, **41**, *7*
$C_3H_6NO_2$					
ON$(CH_3)C(O)CH_3$	CH_3	–	–	–	214, **41**, *9*
OON=C$(CH_3)_2$	C_2H_5	–	–	–	216, **41**, *21*
$C_3H_6NS_2$	CH_3	–	–	–	241, **46**, *2*
					247, **48**, *2*

$C_3H_6NS_2$	C_2H_5	–	–	–	247, **48**, *3*
$C_3H_6S_2$	C_6H_5	–	–	–	248, **48**, *13*
C_3H_7					
$(CH_3)_2CH$	–	–	–	–	75, **11**, *2*
					336/7
$(CH_3)_2CH$	–	–	–	Br	147, **20**, *4*
$(CH_3)_2CH$	–	–	–	Cl	119, **18**, *3*
					137, **19**, *3*
$(CH_3)_2CH$	–	–	–	I	157, **22**, *3*
					162, **23**, *4*
$(CH_3)_2CH$	C_4H_8O	–	–	–	336/7
$(CH_3)_2CH$	$C_4H_{10}N$	–	–	–	257, **50**, *10*
$(CH_3)_2CH$	$C_4H_{10}N$	–	–	Cl	298/9
					301/2
$(CH_3)_2CH$	$C_5H_{12}N$	–	–	–	109, **17**, *35*
$(CH_3)_2CH$	$C_7H_{16}N$	–	–	–	109, **17**, *36*
$(CH_3)_2CH$	$C_7H_{19}Si_2$	–	–	Cl	127/8
$(CH_3)_2CH$	$C_8H_{22}B_4Si_2$	–	–	–	328/9
$(CH_3)_2CH$	$C_{36}H_{44}N_4$	–	–	–	305, **60**, *3*
$(CH_3)_2CH$	$C_{44}H_{28}N_4$	–	–	–	306/7, **60**, *13*
C_3H_7	–	–	–	–	75, **11**, *1*
C_3H_7	–	–	–	Br	147, **20**, *3*
					153, **21**, *3*
					169, **24**, *3*
C_3H_7	–	–	–	I	162, **23**, *3*
					170, **24**, *16*
C_3H_7	$C_5H_{12}N$	–	–	–	109, **17**, *33*
C_3H_7	$C_7H_{16}N$	–	–	–	109, **17**, *34*
C_3H_7	$C_{12}H_{19}N_2$	–	–	–	111, **17**, *45*
C_3H_7NO	CF_3	–	–	Cl	122, **18**, *23*
C_3H_7S					
$SCH(CH_3)_2$	C_4H_9	–	–	–	242, **46**, *8*
					242, **46**, *12*
					247, **48**, *5*
SC_3H_7	C_4H_9	–	–	–	241, **46**, *7*
					242, **46**, *11*
					247, **48**, *4*
					248, **48**, *9*
SC_3H_7	C_6H_5	–	–	–	243, **46**, *15*
$C_3H_7S_2$	CH_3	$C_2H_3O_2$	–	–	239
C_3H_8OS	–	–	–	Cl	368
C_3H_9As	CH_3	–	–	–	40, **6**, *47*
C_3H_9As	C_2H_5	–	–	–	71
C_3H_9GeO	CH_3	–	–	–	188, **31**, *3*
C_3H_9N	CH_3	–	–	–	34, **6**, *4*
C_3H_9N	CH_3	C_5H_{11}	–	–	105, **17**, *16*
C_3H_9N	CH_3	$C_7H_6S_2$	–	–	249, **48**, *18*
C_3H_9N	C_2H_3	–	–	–	85, **12**, *7*
C_3H_9N	$C_4H_{11}Si$	–	–	–	78, **11**, *27*
C_3H_9N	$C_4H_{11}Si$	–	–	Cl	121, **18**, *19*
					139, **19**, *20*

C_3H_9N	C_5H_{11}	–	–	–	77, **11**, *21*
C_3H_9N	C_5H_{11}	–	–	Cl	139, **19**, *15*
C_3H_9N	C_6H_{11}	–	–	–	78, **11**, *24*
C_3H_9N	C_9H_{11}	–	–	Cl	140, **19**, *25*
C_3H_9NO	CH_3	–	–	–	29, **5**, *5*
C_3H_9NOP	CH_3	–	–	–	215, **41**, *16*
C_3H_9NP	CH_3	–	–	–	283, **57**, *3*
C_3H_9NPS	CH_3	–	–	–	241, **46**, *4*
C_3H_9OP	CH_3	–	–	–	29, **5**, *6*
C_3H_9OSi	CH_3	–	–	–	188, **31**, *1*
					364
C_3H_9P	CF_3	–	–	–	78, **11**, *29*
C_3H_9P	CH_3	–	–	–	39, **6**, *39*
C_3H_9SSb	CH_3	–	–	Br	148, **20**, *14*
C_3H_9SSb	CH_3	–	–	Cl	120, **18**, *10*
					138/9, **19**, *10*
C_3H_9SSb	CH_3	–	–	I	158, **22**, *11*
C_3H_9SSb	C_2H_5	–	–	Br	148, **20**, *15*
C_3H_9SSb	C_2H_5	–	–	Cl	121, **18**, *18*
C_3H_9SSb	C_2H_5	–	–	I	158, **22**, *17*
C_3H_9Sb	CH_3	–	–	–	40, **6**, *49*
C_3H_9Sn	CH_3	–	–	–	367
C$_4CoO_4$	C_6H_5	–	–	–	331, **62**, *4*
$C_4F_6S_2$	C_5H_5	–	–	–	249, **48**, *16*
$C_4F_6S_2$	C_5H_5	$C_{10}H_8N_4$	–	–	250, **48**, *22*
$C_4F_6S_2$	C_5H_5	$C_{12}H_8N_2$	–	–	250, **48**, *23*
C_4FeO_4	$C_{10}H_{27}Si_3$	–	–	Cl	333, **62**, *16*
$C_4H_3F_6O$	CH_3	–	–	–	182, **29**, *5*
C_4H_3S	–	–	–	–	91, **14**, *15*
C_4H_3S	$C_4H_8O_2$	–	–	–	94, **14**, *34*
C_4H_4	–	–	–	Cl	352, **67**, *11*
C_4H_4N	CH_3	–	–	–	277, **56**, *1*
$C_4H_4N_2$	CH_3	–	–	–	294
$C_4H_5N_2$					
2-Methyl-1*H*-Imidazol-1-yl	CH_3	–	–	–	279, **56**, *12*
3-Methyl-1*H*-pyrazol-1-yl	CH_3	–	–	–	278, **56**, *5*
3-Methyl-1*H*-pyrazol-1-yl	C_2H_5	–	–	–	280, **56**, *17*
4-Methyl-1*H*-pyrazol-1-yl	CH_3	–	–	–	278, **56**, *6*
C_4H_6NO	CH_3	–	–	–	267
$C_4H_6N_2O_2$					
$(NCH_3)_2C_2O_2$	CH_3	–	–	–	290/1
$(NCH_3)_2C_2O_2$	C_2H_5	–	–	–	290/1
$N_2(CCH_3)_2O_2$	CH_3	–	–	–	289
$O_2N_2(CCH_3)_2$	CH_3	–	–	–	229
$C_4H_6N_2S_2$	CH_3	–	–	–	293/4
C_4H_7	–	–	–	Br	169, **24**, *5*
$C_4H_7O_2$					
$C_2H_5OC(O)CH_2$	–	–	–	Br	147, **20**, *7*
$C_2H_5OC(O)CH_2$	–	–	–	I	170, **24**, *15*
$OOCCH(CH_3)_2$	CH_3	–	–	–	198, **35**, *4*

C_4H_8N	CH_3	–	–	–	273, **55**, *3*
C_4H_8O	CF_3	–	–	Cl	122, **18**, *21*
C_4H_8O	CH_3	–	–	–	29, **5**, *3*
C_4H_8O	CH_3	C_5H_{11}	–	–	105, **17**, *15*
C_4H_8O	C_3H_7	–	–	–	336/7
C_4H_8O	$C_4H_{11}Si$	–	–	Cl	139, **19**, *18*
C_4H_8O	C_6F_5	–	–	–	92, **14**, *21*
C_4H_8O	C_7H_7	–	–	–	78, **11**, *25*
$C_4H_8O_2$	C_4H_3S	–	–	–	94, **14**, *34*
$C_4H_8O_2$	C_6F_5	–	–	–	92, **14**, *20*
$C_4H_8O_2$	C_6F_5	–	–	Cl	140, **19**, *26*
$C_4H_8O_2$	C_6H_4Br	–	–	–	93, **14**, *29*
$C_4H_8O_2$	C_6H_5	–	–	–	92, **14**, *17*
$C_4H_8O_2$	C_6H_5	–	–	Cl	131
					139, **19**, *23*
$C_4H_8O_2$	C_7H_7	–	–	–	78, **11**, *26*
					93, **14**, *30*
					93, **14**, *31*
					93, **14**, *32*
$C_4H_8O_2$	C_7H_7	–	–	Br	154, **21**, *11*
$C_4H_8O_2$	C_7H_7	–	–	I	163, **23**, *12*
$C_4H_8O_2$	$C_{10}H_7$	–	–	–	94, **14**, *33*
$C_4H_8S_2$	C_6H_5	–	–	–	248, **48**, *14*
C_4H_9	$C_6H_{11}S$	–	–	–	247/8, **48**, *6*
C_4H_9					
$(CH_3)_2CH_2CH$	–	–	–	–	75, **11**, *4*
$(CH_3)_2CH_2CH$	$C_2H_3O_2$	–	–	–	199, **35**, *13*
$(CH_3)_2CH_2CH$	C_3H_7S	–	–	–	242, **46**, *11*
					242, **46**, *12*
					248, **48**, *9*
$(CH_3)_2CH_2CH$	$C_5H_{11}N$	–	–	–	77, **11**, *20*
$(CH_3)_2CH_2CH$	C_6H_5S	–	–	–	242, **46**, *13*
					248, **48**, *10*
$(CH_3)_2CH_2CH$	C_9H_6NO	–	–	–	193, **33**, *20*
$(CH_3)_2CH_2CH$	$C_{12}H_{27}OSn$	–	–	–	189, **31**, *6*
					189, **31**, *7*
$(CH_3)_3C$	–	–	–	–	75, **11**, *6*
$(CH_3)_3C$	–	–	–	Cl	119, **18**, *4*
$(CH_3)_3C$	C_2H_5O	–	–	–	182, **29**, *11*
$(CH_3)_3C$	$C_4H_9O_2$	–	–	–	210
$(CH_3)_3C$	$C_4H_{11}Sn$	–	–	–	104, **17**, *11*
$(CH_3)_3C$	$C_8H_5MoO_3$	–	–	–	332, **62**, *11*
$(CH_3)_3C$	$C_9H_{18}N$	–	–	–	274, **55**, *8*
$(CH_3)_3C$	$C_{36}H_{44}N_4$	–	–	–	305, **60**, *5*
$(CH_3)_3C$	$C_{44}H_{28}N_4$	–	–	–	307, **60**, *15*
$C_2H_5(CH_3)CH$	–	–	–	–	75, **11**, *5*
C_4H_9	–	–	–	–	75, **11**, *3*
C_4H_9	–	I	–	Cl	356, **67**, *40*
C_4H_9	–	–	–	I	355, **67**, *35*
C_4H_9	–	–	–	NCS	176, **26**, *4*
C_4H_9	–	–	–	NCSe	176, **26**, *5*

C_4H_9	–	–	–	N_3	177, **26**, *8*
C_4H_9	–	–	–	OCN	176, **26**, *2*
C_4H_9	$C_7H_5N_4$	–	–	–	280, **56**, *18*
C_4H_9	$C_7H_5N_4O$	–	–	–	280, **56**, *19*
C_4H_9	$C_7H_5N_4S$	–	–	–	280, **56**, *20*
C_4H_9-n	–	–	–	Br	147, **20**, *5*
					153, **21**, *4*
					169, **24**, *4*
C_4H_9-n	–	–	–	I	162, **23**, *5*
					170, **24**, *17*
C_4H_9-n	–	N_3	–	I	363
C_4H_9-n	–	SeCN	–	I	363
C_4H_9-n	$C_2H_2S_2$	–	–	–	248, **48**, *8*
C_4H_9-n	$C_3H_5O_2$	–	–	–	199, **35**, *10*
C_4H_9-n	$C_3H_5O_2$	$C_{12}H_{27}OSn$	–	–	233, **45**, *4*
C_4H_9-n	C_3H_7S	–	–	–	241, **46**, *7*
					242, **46**, *8*
					247, **48**, *4*
					247, **48**, *5*
C_4H_9-n	C_4H_9O	–	–	–	183, **29**, *12*
C_4H_9-n	$C_5H_7O_2$	–	–	–	193, **33**, *19*
C_4H_9-n	C_6H_5O	–	–	–	183, **29**, *13*
C_4H_9-n	C_6H_5S	–	–	–	242, **46**, *10*
					248, **48**, *7*
C_4H_9-n	$C_6H_{11}S$	–	–	–	242, **46**, *9*
C_4H_9-n	$C_7H_5O_2$	–	–	–	199, **35**, *12*
C_4H_9-n	$C_8H_{15}O_2$	–	–	–	199, **35**, *11*
C_4H_9-n	$C_{12}H_{27}OSn$	–	–	–	188, **31**, *5*
C_4H_9-n	$C_{36}H_{44}N_4$	–	–	–	305, **60**, *4*
C_4H_9-n	$C_{44}H_{28}N_4$	–	–	–	307, **60**, *14*
$C_4H_9N_2$	CH_3	–	–	–	268/9
$C_4H_9N_2O$	CH_3	–	–	–	271
C_4H_9O	CH_3	–	–	–	181/2, **29**, *3*
C_4H_9O	C_2H_5	–	–	–	182, **29**, *8*
					233, **45**, *3*
C_4H_9O	C_4H_9	–	–	–	183, **29**, *12*
C_4H_9O	$C_4H_{11}Si$	–	–	–	183, **29**, *16*
$C_4H_9O_2$	CH_3	–	–	–	209
$C_4H_9O_2$	C_4H_9	–	–	–	210
C_4H_9S	C_6H_5	–	–	–	243, **46**, *16*
$C_4H_{10}N$					
$HNC(CH_3)_3$	C_3H_7	–	–	–	257, **50**, *10*
$HNC(CH_3)_3$	C_3H_7	–	–	Cl	298/9
					301/2
$HNC(CH_3)_3$	C_7H_7	–	–	–	257, **50**, *11*
$N(C_2H_5)_2$	CH_3	–	–	–	255, **50**, *3*
$N(C_2H_5)_2$	C_2H_5	–	–	–	257, **50**, *9*
					296
$C_4H_{10}NO$	CH_3	–	–	–	191, **33**, *1*
$C_4H_{10}NO$	C_2H_5	–	–	–	192, **33**, *12*
$C_4H_{10}NO$	$C_4H_{10}Sn$	–	–	–	234, **45**, *12*

$C_4H_{10}N_2$	CH_3	–	–	–	295/6
$C_4H_{10}O$	CH_2Br	–	–	Br	172, **25**, *1*
$C_4H_{10}O$	CH_3	–	–	–	29, **5**, *2*
$C_4H_{10}O$	C_2H_5	–	–	–	70/1
$C_4H_{10}O$	$C_4H_{11}Si$	–	–	Cl	139, **19**, *17*
$C_4H_{10}O$	C_6F_5	–	–	–	92, **14**, *19*
$C_4H_{10}O$	C_6H_5	–	–	–	92, **14**, *16*
$C_4H_{10}O$	C_9H_7	–	–	–	86, **12**, *11*
					344/5
$C_4H_{10}O_2$	CH_3	$C_2H_{11}B_{10}$	–	–	106, **17**, *21*
$C_4H_{10}O_2$	$C_4H_{11}Si$	–	–	Cl	139, **19**, *19*
$C_4H_{10}P$					
$CH_2P(CH_3)_2CH_2$	CH_3	–	–	–	367
$HPC(CH_3)_3$	$C_4H_{11}Si$	–	–	–	315, **61**, *11*
$P(C_2H_5)_2$	CH_3	–	–	–	313, **61**, *3*
$P(C_2H_5)_2$	C_2H_5	–	–	–	314, **61**, *7*
$C_4H_{10}Sn$	CH_3	–	–	–	104, **17**, *9*
$C_4H_{10}Sn$	$C_4H_{10}NO$	–	–	–	234, **45**, *12*
$C_4H_{11}N$	CH_3	–	–	Cl	139, **19**, *11*
$C_4H_{11}P$					
$CH_2{=}P(CH_3)_3$	CH_3	–	–	–	55/6
$HP(C_2H_5)_2$	CH_3	–	–	–	38, **6**, *38*
$C_4H_{11}Si$	–	–	–	–	76/7, **11**, *15*
					339/40, **64**, *12*
					340, **64**, *13*
$C_4H_{11}Si$	–	–	–	Br	147, **20**, *6*
$C_4H_{11}Si$	–	–	–	Cl	120, **18**, *7*
					138, **19**, *6*
$C_4H_{11}Si$	–	–	–	D	347, **66**, *8*
$C_4H_{11}Si$	–	–	–	H	347, **66**, *6*
					347, **66**, *7*
$C_4H_{11}Si$	–	–	–	I	157, **22**, *4*
$C_4H_{11}Si$	CH_3O	–	–	–	183, **29**, *15*
$C_4H_{11}Si$	C_2H_3N	–	–	Cl	139, **19**, *21*
$C_4H_{11}Si$	$C_2H_6NO_4S_2$	–	–	–	257/8, **50**, *12*
$C_4H_{11}Si$	C_3H_9N	–	–	–	78, **11**, *27*
$C_4H_{11}Si$	C_3H_9N	–	–	Cl	121, **18**, *19*
					139, **19**, *20*
$C_4H_{11}Si$	C_4H_8O	–	–	Cl	139, **19**, *18*
$C_4H_{11}Si$	C_4H_9O	–	–	–	183, **29**, *16*
$C_4H_{11}Si$	$C_4H_{10}O$	–	–	Cl	139, **19**, *17*
$C_4H_{11}Si$	$C_4H_{10}O_2$	–	–	Cl	139, **19**, *19*
$C_4H_{11}Si$	$C_4H_{10}P$	–	–	–	315, **61**, *11*
$C_4H_{11}Si$	$C_6H_{18}AsSi_2$	–	–	–	316, **61**, *17*
$C_4H_{11}Si$	$C_6H_{18}AsSi_2$	–	–	Cl	325/6
$C_4H_{11}Si$	$C_6H_{18}NSi_2$	–	–	–	258, **50**, *13*
$C_4H_{11}Si$	$C_6H_{18}PSi_2$	–	–	–	316, **61**, *13*
$C_4H_{11}Si$	$C_6H_{18}PSi_2$	–	–	Cl	325
$C_4H_{11}Si$	$C_{12}H_{10}P$	–	–	–	315, **61**, *12*
$C_4H_{11}Si$	$C_{56}H_{107}P_4PtSi$	–	–	–	332, **62**, *13*
$C_4H_{11}Sn$	–	–	–	–	77, **11**, *16*

$C_4H_{11}Sn$	CH_3	–	–	–	104, **17**, *8*
					105, **17**, *12*
$C_4H_{11}Sn$	C_4H_9	–	–	–	104, **17**, *11*
$C_4H_{12}N_2$	CH_3	–	–	–	49, **8**, *2*
$C_4H_{12}P_2$	CH_3	–	–	–	51/2, **8**, *13*
$C_4H_{12}P_2$	C_2H_5	–	–	–	72
C_4O_4	CH_3	–	–	–	207/9
					208/9
C_4O_4	C_2H_5	–	–	–	209
C_5H_4D	–	–	–	–	379, **72**, *1*
$C_5H_4F_3O_2$	CH_3	–	–	–	191, **33**, *3*
$C_5H_4F_3O_2$	CH_3	$C_{10}H_8N_2$	–	–	224, **43**, *1*
$C_5H_4F_3O_2$	CH_3	$C_{12}H_8N_2$	–	–	224, **43**, *2*
$C_5H_4F_3O_2$	CH_3	$C_{26}H_{24}P_2$	–	–	224, **43**, *3*
C_5H_4N	CH_3	$C_7H_6S_2$	–	–	249, **48**, *19*
C_5H_4NO	CH_3	–	–	–	278, **56**, *2*
C_5H_5					
Cyclopenta-2,4-dien-1-yl	–	–	–	–	84/5, **12**, *2*
Cyclopenta-2,4-dien-1-yl	–	–	–	I	157, **22**, *5*
					162/3, **23**, *6*
					355, **67**, *36*
Cyclopenta-2,4-dien-1-yl	CH_3	–	–	–	103, **17**, *5*
Cyclopenta-2,4-dien-1-yl	CH_3	C_2H_7N	–	–	106, **17**, *19*
Cyclopenta-2,4-dien-1-yl	C_2H_5	–	–	–	105, **17**, *13*
Cyclopenta-2,4-dien-1-yl	$C_4F_6S_2$	–	–	–	249, **48**, *16*
Cyclopenta-2,4-dien-1-yl	$C_4F_6S_2$	$C_{10}H_8N_4$	–	–	250, **48**, *22*
Cyclopenta-2,4-dien-1-yl	$C_4F_6S_2$	$C_{12}H_8N_2$	–	–	250, **48**, *23*
Cyclopenta-2,4-dien-1-yl	$C_{10}H_8N_2$	–	–	–	85, **12**, *8*
Cyclopenta-2,4-dien-1-yl	$C_{10}H_8N_2$	–	–	I	163/4, **23**, *14*
Cyclopenta-2,4-dien-1-yl	$C_{12}H_8N_2$	–	–	–	86, **12**, *9*
Cyclopenta-2,4-dien-1-yl	$C_{12}H_8N_2$	–	–	I	164, **23**, *15*
Cyclopenta-2,4-dien-1-yl	$C_{18}H_{15}P$	–	–	–	86, **12**, *10*
Cyclopentadienyl	–	–	–	–	372/6
					377/8
					378
Cyclopentadienyl	–	–	–	Cl	372
C_5H_5N	CF_3	–	–	Cl	122, **18**, *24*
C_5H_5N	CH_3	–	–	Cl	121, **18**, *14*
					139, **19**, *12*
C_5H_5N	CH_3	–	–	I	158, **22**, *13*
C_5H_5N	CH_3	$C_2H_3O_2$	–	–	225, **43**, *5*
C_5H_5N	CH_3	$C_{12}H_{10}O_2P$	–	HO	238
C_5H_5N	C_6F_5	–	–	–	93, **14**, *26*
C_5H_5N	C_6H_5	–	–	–	92, **14**, *18*
C_5H_5N	$C_{28}H_{20}$	–	–	Cl	134/5
$C_5H_7N_2$	CH_3	–	–	–	278/9, **56**, *7*
$C_5H_7O_2$	CH_3	–	–	–	191, **33**, *2*
$C_5H_7O_2$	C_4H_9	–	–	–	193, **33**, *19*
$C_5H_7O_4$	C_2H_5	–	–	–	192, **33**, *16*
C_5H_9	–	–	–	Br	169, **24**, *6*

$C_5H_9O_2$	CH_3	–	–	–	198, **35**, *5*
$C_5H_{10}N$	CH_3	–	–	–	273, **55**, *4*
$C_5H_{10}N$	C_2H_5	–	–	–	274, **55**, *7*
$C_5H_{10}NS_2$	C_6H_5	–	–	–	243, **46**, *20*
C_5H_{11}					
$(CH_3)_3CCH_2$	–	–	–	–	76, **11**, *8*
$(CH_3)_3CCH_2$	–	–	–	Cl	119/20, **18**, *5*
					138, **19**, *4*
$(CH_3)_3CCH_2$	CH_3	–	–	–	102, **17**, *3*
$(CH_3)_3CCH_2$	CH_3	C_3H_9N	–	–	105, **17**, *16*
$(CH_3)_3CCH_2$	CH_3	C_4H_8O	–	–	105, **17**, *15*
$(CH_3)_3CCH_2$	CH_3	$C_6H_{16}N_2$	–	–	106, **17**, *17*
					106, **17**, *18*
$(CH_3)_3CCH_2$	CH_3O	–	–	–	183, **29**, *14*
$(CH_3)_3CCH_2$	C_3H_9N	–	–	–	77, **11**, *21*
$(CH_3)_3CCH_2$	C_3H_9N	–	–	Cl	139, **19**, *15*
$(CH_3)_3CCH_2$	$C_6H_{16}N_2$	–	–	–	77, **11**, *22*
$(CH_3)_3CCH_2$	$C_6H_{16}N_2$	–	–	Cl	139, **19**, *16*
$(CH_3)_3CCH_2$	$C_6H_{18}PSi_2$	–	–	–	315, **61**, *10*
$(CH_3)_3CCH_2$	$C_8H_5MoO_3$	–	–	–	332, **62**, *12*
$(CH_3)_3CCH_2$	$C_{12}H_{10}P$	–	–	–	315, **61**, *9*
$(CH_3)_3CCH_2$	$C_{12}H_{11}P$	–	–	–	78, **11**, *23*
$C_2H_5(CH_3)CHCH_2$	–	–	–	–	76, **11**, *9*
C_5H_{11}	–	–	–	–	76, **11**, *7*
$C_5H_{11}N$	C_2H_5	–	–	–	71
$C_5H_{11}N$	C_4H_9	–	–	–	77, **11**, *20*
$C_5H_{11}N_2$	CH_3	–	–	–	273/4, **55**, *6*
$C_5H_{12}N$	–	–	–	Cl	122, **18**, *25*
$C_5H_{12}N$	CH_3	–	–	–	106, **17**, *22*
					110, **17**, *38*
$C_5H_{12}N$	C_2H_5	–	–	–	108, **17**, *30*
$C_5H_{12}N$	C_3H_7	–	–	–	109, **17**, *33*
					109, **17**, *35*
$C_5H_{12}N_3$	CH_3	–	–	–	271
$C_5H_{13}NO$	CH_3	–	–	–	192, **33**, *10*
$C_5H_{13}NO$	C_2H_5	–	–	–	193, **33**, *18*
$C_5H_{14}N_2$	CH_3	–	–	–	34, **6**, *8*
					49, **8**, *1*
C_5MnO_5	C_6H_5	–	–	–	331, **62**, *2*
C_6 F_5	–	–	–	–	90, **14**, *4*
C_6F_5	–	–	–	Br	148, **20**, *12*
C_6F_5	–	–	–	I	157, **22**, *8*
C_6F_5	C_2H_6OS	–	–	–	92, **14**, *22*
C_6F_5	C_4H_8O	–	–	–	92, **14**, *21*
C_6F_5	$C_4H_8O_2$	–	–	–	92, **14**, *20*
C_6F_5	$C_4H_8O_2$	–	–	Cl	140, **19**, *26*
C_6F_5	$C_4H_{10}O$	–	–	–	92, **14**, *19*
C_6F_5	C_5H_5N	–	–	–	93, **14**, *26*
C_6F_5	$C_6H_{16}N_2$	–	–	–	93, **14**, *25*
C_6F_5	$C_6H_{16}N_2$	–	–	Cl	140, **19**, *27*

C_6F_5	$C_{10}H_8N_2$	–	–	–	94, **14**, *35*
C_6F_5	$C_{10}H_8N_2$	–	–	Cl	140, **19**, *28*
C_6F_5	$C_{12}H_8N_2$	–	–	Cl	140/1, **19**, *29*
					141, **19**, *30*
C_6F_5	$C_{15}H_{11}N_3$	–	–	Cl	141, **19**, *31*
C_6F_5	$C_{18}H_{15}As$	–	–	–	93, **14**, *28*
C_6F_5	$C_{18}H_{15}AsO$	–	–	–	92/3, **14**, *24*
C_6F_5	$C_{18}H_{15}OP$	–	–	–	92, **14**, *23*
C_6F_5	$C_{18}H_{15}P$	–	–	–	93, **14**, *27*
C_6F_5	$C_{26}H_{24}P_2$	–	–	–	94, **14**, *36*
C_6F_5	$C_{26}H_{24}P_2$	–	–	Cl	141, **19**, *32*
C_6F_5	$C_{36}H_{44}N_4$	–	–	–	306, **60**, *10*
C_6F_5	$C_{44}H_{28}N_4$	–	–	–	307/8, **60**, *20*
C_6F_5	$C_{48}H_{36}N_4$	–	–	–	308, **60**, *22*
					308, **60**, *24*
C_6HF_4	$C_{36}H_{44}N_4$	–	–	–	306, **60**, *9*
C_6HF_4	$C_{44}H_{28}N_4$	–	–	–	307, **60**, *19*
C_6HF_4	$C_{48}H_{36}N_4$	–	–	–	308, **60**, *21*
					308, **60**, *23*
C_6H_4Br	–	–	–	–	90, **14**, *6*
C_6H_4Br	$C_4H_8O_2$	–	–	–	93, **14**, *29*
C_6H_4Cl	–	–	–	–	90, **14**, *5*
C_6H_4F					
3-FC_6H_4	–	–	–	–	90, **14**, *2*
4-FC_6H_4	–	–	–	–	90, **14**, *3*
4-FC_6H_4	–	–	–	Br	147, **20**, *9*
					153, **21**, *6*
$C_6H_4NO_3$	CH_3	–	–	–	192, **33**, *7*
C_6H_5	–	–	–	–	90, **14**, *1*
					340, **64**, *14*
					340, **64**, *15*
					340, **64**, *16*
					340, **64**, *17*
					340, **64**, *18*
C_6H_5	–	–	–	Br	147, **20**, *8*
					153, **21**, *5*
					354, **67**, *27*
C_6H_5	–	–	–	Cl	130/1
					138, **19**, *8*
					353, **67**, *19*
C_6H_5	–	–	–	I	157, **22**, *6*
					163, **23**, *7*
					356, **67**, *37*
					356, **67**, *38*
C_6H_5	$C_2H_3O_2$	–	–	–	199, **35**, *14*
					233, **45**, *5*
C_6H_5	$C_2H_4S_2$	–	–	–	248, **48**, *12*
C_6H_5	$C_2H_4S_3$	–	–	–	248, **48**, *15*
C_6H_5	C_2H_5S	–	–	–	242, **46**, *14*
C_6H_5	$C_3H_5O_2$	–	–	–	199, **35**, *15*
					233, **45**, *6*

C_6H_5	$C_3H_6S_2$	–	–	–	248, **48**, *13*
C_6H_5	C_3H_7S	–	–	–	243, **46**, *15*
C_6H_5	C_4CoO_4	–	–	–	331, **62**, *4*
C_6H_5	$C_4H_8O_2$	–	–	–	92, **14**, *17*
C_6H_5	$C_4H_8O_2$	–	–	Cl	131 139, **19**, *23*
C_6H_5	$C_4H_8S_2$	–	–	–	248, **48**, *14*
C_6H_5	C_4H_9S	–	–	–	243, **46**, *16*
C_6H_5	$C_4H_{10}O$	–	–	–	92, **14**, *16*
C_6H_5	C_5H_5N	–	–	–	92, **14**, *18*
C_6H_5	$C_5H_{10}NS_2$	–	–	–	243, **46**, *20*
C_6H_5	C_5MnO_5	–	–	–	331, **62**, *2*
C_6H_5	$C_6H_5O_2S$	–	–	–	217, **41**, *23*
C_6H_5	C_6H_5S	–	–	–	243, **46**, *19*
C_6H_5	$C_7H_5FeO_2$	–	–	–	331, **62**, *3*
C_6H_5	$C_7H_5O_2$	–	–	–	199, **35**, *16*
C_6H_5	C_7H_7S	–	–	–	243, **46**, *17*
C_6H_5	$C_8H_5O_3W$	–	–	–	331, **62**, *1*
C_6H_5	$C_{10}H_8N_2$	–	–	Br	154, **21**, *13*
C_6H_5	$C_{10}H_8N_2$	–	–	I	164, **23**, *16*
C_6H_5	$C_{12}H_{25}S$	–	–	–	243, **46**, *18* 248, **48**, *11*
C_6H_5	$C_{36}H_{44}N_4$	–	–	–	306, **60**, *8*
C_6H_5	$C_{44}H_{28}N_4$	–	–	–	307, **60**, *18*
$C_6H_5N_2O$	CH_3	–	–	–	214, **41**, *8*
C_6H_5O	CH_3	–	–	–	182, **29**, *4*
C_6H_5O	C_4H_9	–	–	–	183, **29**, *13*
$C_6H_5O_2S$	C_6H_5	–	–	–	217, **41**, *23*
$C_6H_5O_3S$	CH_3	–	–	–	213, **41**, *4*
C_6H_5S	C_4H_9	–	–	–	242, **46**, *10* 242, **46**, *13* 248, **48**, *7* 248, **48**, *10*
C_6H_5S	C_6H_5	–	–	–	243, **46**, *19*
C_6H_6NO	CH_3	–	–	–	278, **56**, *3*
C_6H_7	–	–	–	–	85, **12**, *3* 379, **72**, *2*
C_6H_7As	CH_3	–	–	–	40, **6**, *45*
$C_6H_7N_2$	CH_3	–	–	–	302/3
C_6H_7P	CH_3	–	–	–	38, **6**, *36*
$C_6H_8BN_4$	CH_3	–	–	–	279, **56**, *10*
$C_6H_8BN_4$	CH_3	–	–	Cl	296/7
$C_6H_8N_2$	CH_3	–	–	Cl	121, **18**, *16*
C_6H_{11}					
$C_3H_7CH{=}CHCH_2$	–	–	–	Br	169, **24**, *7*
Cyclopentylmethyl	–	–	–	–	76, **11**, *13*
Cyclopentylmethyl	C_3H_9N	–	–	–	78, **11**, *24*
$C_6H_{11}S$	C_4H_9	–	–	–	242, **46**, *9* 247/8, **48**, *6*
$C_6H_{12}N_4$					
$(NCH_3)_4C_2$	CH_3	–	–	–	291/3

$C_6H_{12}N_4$					
1,3,5,7-Tetraazatricyclo-[3.1.1^{3,7}]decane	CH_3	–	–	–	34, **6**, *9* 50, **8**, *7* 51, **8**, *8*
C_6H_{13}					
$C_2H_5(CH_3)CH_2CH_2$	–	–	–	–	76, **11**, *11*
C_6H_{13}	–	–	–	–	76, **11**, *10*
$C_6H_{14}N$	CH_3	–	–	–	255/6, **50**, *4*
$C_6H_{14}P$	CH_3	–	–	–	313, **61**, *4*
$C_6H_{15}N$	CH_3	–	–	–	34, **6**, *5*
$C_6H_{15}NP$	CH_3	–	–	–	283, **57**, *4*
$C_6H_{15}OSn$	CH_3	–	–	–	188, **31**, *4*
$C_6H_{15}P$	CH_3	–	–	–	39, **6**, *40*
$C_6H_{15}Sb$	CH_3	–	–	–	40, **6**, *50*
$C_6H_{15}SeSi$	C_2H_5	–	–	–	252
$C_6H_{16}NP$	CH_3	–	–	–	37, **6**, *25*
$C_6H_{16}NP_2$	CH_3	–	–	–	368
$C_6H_{16}N_2$	CH_2	Cl	–	Br	175
$C_6H_{16}N_2$	CH_2	–	–	Cl	174/5
$C_6H_{16}N_2$	CH_2	I	–	Cl	176
$C_6H_{16}N_2$	CH_2Br	–	–	Br	172, **25**, *2*
$C_6H_{16}N_2$	CH_2I	–	–	I	172, **25**, *5*
$C_6H_{16}N_2$	CH_3	–	–	–	50, **8**, *3*
$C_6H_{16}N_2$	CH_3	C_5H_{11}	–	–	106, **17**, *17* 106, **17**, *18*
$C_6H_{16}N_2$	CH_3	$C_8H_{15}B_{10}$	–	–	106, **17**, *20* 329
$C_6H_{16}N_2$	C_2H_5	–	–	Br	154, **21**, *9*
$C_6H_{16}N_2$	C_2H_5	–	–	I	163, **23**, *11*
$C_6H_{16}N_2$	C_5H_{11}	–	–	–	77, **11**, *22*
$C_6H_{16}N_2$	C_5H_{11}	–	–	Cl	139, **19**, *16*
$C_6H_{16}N_2$	C_6F_5	–	–	–	93, **14**, *25*
$C_6H_{16}N_2$	C_6F_5	–	–	Cl	140, **19**, *27*
$C_6H_{17}NP_2$	CH_3	–	–	–	56
$C_6H_{18}AsSi_2$	$C_4H_{11}Si$	–	–	–	316, **61**, *17*
$C_6H_{18}AsSi_2$	$C_4H_{11}Si$	–	–	Cl	325/6
$C_6H_{18}GeNP$	CH_3	–	–	–	37, **6**, *28*
$C_6H_{18}NPSi$	CH_3	–	–	–	37, **6**, *26*
$C_6H_{18}NPSn$	CH_3	–	–	–	38, **6**, *30*
$C_6H_{18}NSi_2$	CH_3	–	–	–	256, **50**, *7*
$C_6H_{18}NSi_2$	$C_4H_{11}Si$	–	–	–	258, **50**, *13*
$C_6H_{18}NSi_2$	C_7H_7	–	–	Cl	299/300
$C_6H_{18}NSi_2$	C_9H_{11}	–	–	Cl	300/1
$C_6H_{18}N_3P$	CH_3	–	–	–	39, **6**, *43*
$C_6H_{18}PSi_2$	$C_4H_{11}Si$	–	–	–	316, **61**, *13*
$C_6H_{18}PSi_2$	$C_4H_{11}Si$	–	–	Cl	325
$C_6H_{18}PSi_2$	C_5H_{11}	–	–	–	315, **61**, *10*
$C_6H_{18}PSi_2$	$C_{10}H_{15}$	–	–	Cl	324/5
C$_7H_5FeO_2$	C_6H_5	–	–	–	331, **62**, *3*
C_7H_5N	C_2H_5	–	–	–	71

$C_7H_5N_2$					
1*H*-Benzimidazol-1-yl	CH_3	–	–	–	279, **56**, *13*
1*H*-Indazol-1-yl	CH_3	–	–	–	279, **56**, *9*
$C_7H_5N_4$	C_4H_9	–	–	–	280, **56**, *18*
$C_7H_5N_4O$	C_4H_9	–	–	–	280, **56**, *19*
$C_7H_5N_4S$	C_4H_9	–	–	–	280, **56**, *20*
C_7H_5OS	C_2H_5	–	–	–	205
$C_7H_5O_2$					
2-Hydroxycyclohepta-2,4,6-trien-1-onato	CH_3	–	–	–	191, **33**, *6*
2-$OCHC_6H_4O$	CH_3	–	–	–	192, **33**, *9*
$OOCC_6H_5$	C_4H_9	–	–	–	199, **35**, *12*
$OOCC_6H_5$	C_6H_5	–	–	–	199, **35**, *16*
$C_7H_6S_2$	CH_3	–	–	–	247, **48**, *1*
$C_7H_6S_2$	CH_3	C_2H_6OS	–	–	249, **48**, *17*
$C_7H_6S_2$	CH_3	C_3H_9N	–	–	249, **48**, *18*
$C_7H_6S_2$	CH_3	C_5H_4N	–	–	249, **48**, *19*
$C_7H_6S_2$	CH_3	$C_{10}H_8N_2$	–	–	249/50, **48**, *20*
$C_7H_6S_2$	CH_3	$C_{12}H_8N_2$	–	–	250, **48**, *21*
C_7H_7					
2-$CH_3C_6H_4$	–	–	–	–	91, **14**, *8*
2-$CH_3C_6H_4$	$C_4H_8O_2$	–	–	–	93, **14**, *30*
3-$CH_3C_6H_4$	–	–	–	–	91, **14**, *9*
3-$CH_3C_6H_4$	$C_4H_8O_2$	–	–	–	93, **14**, *31*
4-$CH_3C_6H_4$	–	–	–	–	91, **14**, *10*
4-$CH_3C_6H_4$	–	–	–	Br	147, **20**, *10* 154, **21**, *7*
4-$CH_3C_6H_4$	–	–	–	Cl	131
4-$CH_3C_6H_4$	–	–	–	I	157, **22**, *7*
4-$CH_3C_6H_4$	$C_4H_8O_2$	–	–	–	93, **14**, *32*
$C_6H_5CH_2$	–	–	–	–	76, **11**, *14*
$C_6H_5CH_2$	–	–	–	Cl	120, **18**, *6* 138, **19**, *5* 353, **67**, *16*
$C_6H_5CH_2$	–	–	–	F	117
$C_6H_5CH_2$	C_2D_3N	–	–	F	117
$C_6H_5CH_2$	C_4H_8O	–	–	–	78, **11**, *25*
$C_6H_5CH_2$	$C_4H_8O_2$	–	–	–	78, **11**, *26*
$C_6H_5CH_2$	$C_4H_8O_2$	–	–	Br	154, **21**, *11*
$C_6H_5CH_2$	$C_4H_8O_2$	–	–	I	163, **23**, *12*
$C_6H_5CH_2$	$C_4H_{10}N$	–	–	–	257, **50**, *11*
$C_6H_5CH_2$	$C_6H_{18}NSi_2$	–	–	Cl	299/300
$C_6H_5CH_2$	$C_{10}H_8N_2$	–	–	Br	154, **21**, *12*
$C_6H_5CH_2$	$C_{10}H_8N_2$	–	–	Cl	139, **19**, *22*
$C_6H_5CH_2$	$C_{10}H_8N_2$	–	–	I	163, **23**, *13*
C_7H_7O	–	–	–	–	90, **14**, *7*
C_7H_7S	C_6H_5	–	–	–	243, **46**, *17*
$C_7H_7S_2$	CH_3	$C_2H_3O_2$	–	–	239
C_7H_8N	CH_3	–	–	–	256, **50**, *6*
C_7H_9	–	–	–	–	380, **72**, *3*
$C_7H_{10}N_2$	CH_3	–	–	–	36, **6**, *20*

$C_7H_{14}N$	CH_3	–	–	–	273, **55**, *5*
$C_7H_{15}N$					
2,6-Dimethylpiperidine	CH_3	–	–	–	35, **6**, *15*
$CH_2CH_2CH_2N(CH_3)CH_2CH_2CH_2$	CH_3	–	–	–	109/10, **17**, *37*
$C_7H_{16}N$	–	–	–	Cl	122, **18**, *26*
$C_7H_{16}N$	CH_3	–	–	–	106/7, **17**, *23*
					110, **17**, *39*
$C_7H_{16}N$	C_2H_5	–	–	–	108, **17**, *31*
$C_7H_{16}N$	C_3H_7	–	–	–	109, **17**, *34*
					109, **17**, *36*
$C_7H_{18}N_2$					
$(CH_3)_2NCH_2CH_2N(CH_3)_2CH_2$	–	–	–	Br	172, **25**, *7*
$(CH_3)_2NCH_2CH_2N(CH_3)_2CH_2$	–	–	–	I	172, **25**, *9*
					368/9
$N(CH_3)_2CH_2CH_2CH_2N(CH_3)_2$	CH_3	–	–	–	50, **8**, *4*
$C_7H_{18}N_2Si_2$	CH_3	–	–	–	37, **6**, *24*
$C_7H_{19}Si_2$	–	–	–	–	77, **11**, *17*
					370/1
$C_7H_{19}Si_2$	–	–	–	Br	354, **67**, *23*
$C_7H_{19}Si_2$	C_3H_7	–	–	Cl	127/8
$C_7H_{21}N_2SSi_2$	CH_3	–	–	–	283, **57**, *2*
C$_8H_5$	CH_3	–	–	–	104, **17**, *7*
C_8H_5	$C_{36}H_{44}N_4$	–	–	–	306, **60**, *7*
C_8H_5	$C_{44}H_{28}N_4$	–	–	–	307, **60**, *17*
$C_8H_5MoO_3$	CH_3	–	–	–	332, **62**, *9*
$C_8H_5MoO_3$	C_2H_5	–	–	–	332, **62**, *10*
					333, **62**, *17*
$C_8H_5MoO_3$	C_4H_9	–	–	–	332, **62**, *11*
$C_8H_5MoO_3$	C_5H_{11}	–	–	–	332, **62**, *12*
$C_8H_5O_3$	CH_3	–	–	–	198, **35**, *7*
$C_8H_5O_3W$	C_6H_5	–	–	–	331, **62**, *1*
$C_8H_6MoO_3$	CH_3	–	–	–	56
$C_8H_6O_3W$	CH_3	–	–	–	56
C_8H_7	$C_{36}H_{44}N_4$	–	–	–	306, **60**, *6*
C_8H_7	$C_{44}H_{28}N_4$	–	–	–	307, **60**, *16*
C_8H_9	–	–	–	–	91, **14**, *11*
$C_8H_{10}F_3O_2$	CH_3	–	–	–	191, **33**, *4*
$C_8H_{13}Ge$	–	–	–	–	380, **72**, *6*
$C_8H_{13}Si$	–	–	–	–	380, **72**, *5*
$C_8H_{15}B_{10}$	CH_3	$C_6H_{16}N_2$	–	–	106, **17**, *20*
					329
$C_8H_{15}O_2$	C_4H_9	–	–	–	199, **35**, *11*
$C_8H_{17}N_2$	CH_3	–	–	–	269
$C_8H_{18}As$	CH_3	–	–	–	316, **61**, *15*
$C_8H_{18}P$	CH_3	–	–	–	313/4, **61**, *5*
$C_8H_{18}P$	C_2H_5	–	–	–	314, **61**, *8*
$C_8H_{18}Sb$	CH_3	–	–	–	316, **61**, *18*
$C_8H_{21}N_2Si_2$	CH_3	–	–	–	270/1
$C_8H_{22}B_4Si_2$	C_3H_7	–	–	–	328/9
$C_8H_{24}NOPSi_2$	CH_3	–	–	–	38, **6**, *32*

$C_8H_{24}N_2P_2Si$	CH_3	–	–	–	38, **6**, *33*
					51, **8**, *12*
$\mathbf{C}_9H_6NO$	CH_3	–	–	–	192, **33**, *8*
C_9H_6NO	C_2H_5	–	–	–	193, **33**, *17*
C_9H_6NO	C_4H_9	–	–	–	193, **33**, *20*
C_9H_7	–	–	–	–	85, **12**, *5*
					340, **64**, *19*
C_9H_7	$C_4H_{10}O$	–	–	–	86, **12**, *11*
					344/5
$C_9H_7O_3$	–	–	–	–	370
C_9H_9	–	–	–	Br	169, **24**, *8*
$C_9H_9N_3$	CH_3	–	–	–	226, **43**, *13*
$C_9H_{10}NO$	C_2H_5	–	–	–	267/8
C_9H_{11}	–	–	–	–	91, **14**, *13*
C_9H_{11}	–	–	–	Br	148, **20**, *11*
					154, **21**, *8*
C_9H_{11}	–	–	–	Cl	131/2
					138, **19**, *9*
					351, **67**, *5*
C_9H_{11}	–	–	–	F	117/8
C_9H_{11}	–	–	–	I	157, **22**, *9*
					163, **23**, *8*
C_9H_{11}	C_3H_9N	–	–	Cl	140, **19**, *25*
C_9H_{11}	$C_6H_{18}NSi_2$	–	–	Cl	300/1
$C_9H_{11}O$	C_2H_5	–	–	–	182, **29**, *9*
C_9H_{12}	–	–	–	–	387
$C_9H_{12}N$					
2-$(CH_3)_2NCH_2C_6H_4$	–	–	–	Cl	131/3
2-$(CH_3)_2NCH_2C_6H_4$	CH_3	–	–	–	107, **17**, *26*
					110, **17**, *40*
2-$(CH_3)_2NCH_2C_6H_4$	C_2H_5	–	–	–	108/9, **17**, *32*
2-$(CH_3)_2NC_6H_4CH_2$	CH_3	–	–	–	107, **17**, *25*
$C_9H_{12}N_4O$	CH_3	–	–	–	226, **43**, *16*
C_9H_{13}	–	–	–	–	380, **72**, *4*
$C_9H_{18}N$	C_4H_9	–	–	–	274, **55**, *8*
C_9H_{19}	–	–	–	–	76, **11**, *12*
$C_9H_{19}N$	CH_3	–	–	–	35, **6**, *16*
$C_9H_{21}N_2Si$	CH_3	–	–	–	310/2
$C_9H_{21}N_3$	CH_3	–	–	–	36, **6**, *19*
$C_9H_{24}GeNP$	CH_3	–	–	–	38, **6**, *29*
$C_9H_{24}NPSi$	CH_3	–	–	–	37, **6**, *27*
$C_9H_{24}NPSn$	CH_3	–	–	–	38, **6**, *31*
$\mathbf{C}_{10}H_6F_3O_2$	CH_3	–	–	–	191, **33**, *5*
$C_{10}H_7$	–	–	–	–	91, **14**, *14*
$C_{10}H_7$	–	–	–	Br	148, **20**, *13*
$C_{10}H_7$	$C_4H_8O_2$	–	–	–	94, **14**, *33*
$C_{10}H_8N_2$					
1,4-Dihydro-4-(1*H*-pyridin-4-ylidene)pyridin-1-ato	CH_3	–	–	–	294

$C_{10}H_8N_2$					
2,2′-Bipyridine	CH_3	–	–	–	36, **6**, *21*
					225/6, **43**, *10*
2,2′-Bipyridine	CH_3	–	–	Cl	121, **18**, *15*
					139, **19**, *13*
2,2′-Bipyridine	CH_3	–	–	I	158, **22**, *15*
					163, **23**, *9*
2,2′-Bipyridine	CH_3	$C_2H_3O_2$	–	–	225, **43**, *7*
2,2′-Bipyridine	CH_3	$C_5H_4F_3O_2$	–	–	224, **43**, *1*
2,2′-Bipyridine	CH_3	$C_7H_6S_2$	–	–	249/50, **48**, *20*
2,2′-Bipyridine	C_2H_5	–	–	Br	154, **21**, *10*
2,2′-Bipyridine	C_2H_5	–	–	I	158, **22**, *18*
2,2′-Bipyridine	C_5H_5	–	–	–	85, **12**, *8*
2,2′-Bipyridine	C_5H_5	–	–	I	163/4, **23**, *14*
2,2′-Bipyridine	C_6F_5	–	–	–	94, **14**, *35*
2,2′-Bipyridine	C_6F_5	–	–	Cl	140, **19**, *28*
2,2′-Bipyridine	C_6H_5	–	–	Br	154, **21**, *13*
2,2′-Bipyridine	C_6H_5	–	–	I	164, **23**, *16*
2,2′-Bipyridine	C_7H_7	–	–	Br	154, **21**, *12*
2,2′-Bipyridine	C_7H_7	–	–	Cl	139, **19**, *22*
2,2′-Bipyridine	C_7H_7	–	–	I	163, **23**, *13*
3,3′-Bipyridine	CH_3	–	–	–	36, **6**, *22*
					51, **8**, *10*
4,4′-Bipyridine	CH_3	–	–	–	36, **6**, *23*
					51, **8**, *11*
$C_{10}H_8N_4$	$C_4F_6S_2$	C_5H_5	–	–	250, **48**, *22*
$C_{10}H_{13}$	–	–	–	–	91, **14**, *12*
$C_{10}H_{14}N$					
2-$(CH_3)_2N(CH_3)CHC_6H_4$	–	–	–	Cl	133/4
2-$(CH_3)_2N(CH_3)CHC_6H_4$	CH_3	–	–	–	108, **17**, *28*
2-$(CH_3)_2NCH_2C_6H_4CH_2$	CH_3	–	–	–	107, **17**, *24*
$C_{10}H_{15}$					
(7,7-Dimethylbicyclo[3.1.1]hept-2-en-2-yl)methyl	–	–	–	Br	169, **24**, *12*
Pentamethylcyclopenta-2,4-dien-1-yl	–	–	–	Cl	130
					138, **19**, *7*
Pentamethylcyclopenta-2,4-dien-1-yl	$C_6H_{18}PSi_2$	–	–	Cl	324/5
Pentamethylcyclopentadienyl	–	–	–	–	381, **72**, *9*
$C_{10}H_{16}N_2$	CH_3	–	–	–	34, **6**, *10*
					50, **8**, *5*
$C_{10}H_{17}$	–	–	–	Br	169, **24**, *10*
					169, **24**, *11*
$C_{10}H_{22}N_2$	CH_3	–	–	–	110, **17**, *41*
$C_{10}H_{24}DN_2Si$	CH_3	–	–	–	284, **57**, *9*
$C_{10}H_{24}NOSi$	CH_3	–	–	–	283, **57**, *7*
$C_{10}H_{24}NOSi$	CH_3	–	–	Cl	297
$C_{10}H_{24}N_2Si$	CH_3	–	–	–	284, **57**, *11*
					310/1
$C_{10}H_{25}N_2Si$	CH_3	–	–	–	284, **57**, *8*

$C_{10}H_{27}Si_3$	–	–	–	Cl	353, **67**, *17*
$C_{10}H_{27}Si_3$	–	–	–	H	348/9
$C_{10}H_{27}Si_3$	–	O	–	HO	237
$C_{10}H_{27}Si_3$	C_4FeO_4	–	–	Cl	333, **62**, *16*
C$_{11}H_{10}N_2$	CH_3	–	–	–	226, **43**, *17*
$C_{11}H_{16}N$	CH_3	–	–	–	108, **17**, *27*
$C_{11}H_{21}Si_2$	–	–	–	–	380, **72**, *7*
$C_{11}H_{33}GaN_2P_2Si$	CH_3	–	–	–	38, **6**, *34*
$C_{11}H_{33}IrP_3$					
$Ir(CH_3)_2(P(CH_3)_3)_3$	CH_3	–	–	–	331, **62**, *5* 332, **62**, *6*
$Ir(CH_3)_2(P(CH_3)_3)_3$	CH_3	–	–	Br	332, **62**, *14* 333, **62**, *15*
$Ir(H)C_2H_5(P(CH_3)_3)_3$	C_2H_5	–	–	–	332, **62**, *8*
C$_{12}H_8N_2$	CH_3	–	–	–	226, **43**, *14*
$C_{12}H_8N_2$	CH_3	–	–	Cl	121, **18**, *13*
$C_{12}H_8N_2$	CH_3	$C_2H_3O_2$	–	–	225, **43**, *8*
$C_{12}H_8N_2$	CH_3	$C_5H_4F_3O_2$	–	–	224, **43**, *2*
$C_{12}H_8N_2$	CH_3	$C_7H_6S_2$	–	–	250, **48**, *21*
$C_{12}H_8N_2$	$C_4F_6S_2$	C_5H_5	–	–	250, **48**, *23*
$C_{12}H_8N_2$	C_5H_5	–	–	–	86, **12**, *9*
$C_{12}H_8N_2$	C_5H_5	–	–	I	164, **23**, *15*
$C_{12}H_8N_2$	C_6F_5	–	–	Cl	140/1, **19**, *29* 141, **19**, *30*
$C_{12}H_8N_3$	CH_3	–	–	–	279, **56**, *14*
$C_{12}H_{10}As$	CH_3	–	–	–	316, **61**, *16*
$C_{12}H_{10}O_2P$	CH_3	–	–	–	215, **41**, *14*
$C_{12}H_{10}O_2P$	CH_3	C_5H_5N	–	HO	238
$C_{12}H_{10}O_2P$	C_2H_5	–	–	–	217, **41**, *22*
$C_{12}H_{10}P$	CH_3	–	–	–	314, **61**, *6*
$C_{12}H_{10}P$	$C_4H_{11}Si$	–	–	–	315, **61**, *12*
$C_{12}H_{10}P$	C_5H_{11}	–	–	–	315, **61**, *9*
$C_{12}H_{11}As$	CH_3	–	–	–	40, **6**, *46*
$C_{12}H_{11}P$	C_5H_{11}	–	–	–	78, **11**, *23*
$C_{12}H_{12}N_2$	CH_3	–	–	–	226, **43**, *11*
$C_{12}H_{19}N_2$	–	–	–	Cl	140, **19**, *24*
$C_{12}H_{19}N_2$	CH_3	–	–	–	108, **17**, *29* 111, **17**, *42*
$C_{12}H_{19}N_2$	C_2H_5	–	–	–	111, **17**, *43*
$C_{12}H_{19}N_2$	C_3H_7	–	–	–	111, **17**, *45*
$C_{12}H_{22}N$	CH_3	–	–	–	34, **6**, *7* 256, **50**, *5*
$C_{12}H_{25}S$	C_6H_5	–	–	–	243, **46**, *18* 248, **48**, *11*
$C_{12}H_{27}N$	CH_3	–	–	–	34, **6**, *6*
$C_{12}H_{27}OSn$					
$OSn(C_4H_9)_2CH_2CH(CH_3)_2$	C_4H_9	–	–	–	189, **31**, *6*
$OSn(C_4H_9)_3$	$C_3H_5O_2$	C_4H_9	–	–	233, **45**, *4*
$OSn(C_4H_9)_3$	C_4H_9	–	–	–	188, **31**, *5* 189, **31**, *7*

C$_{13}$H$_8$N$_4$O$_4$	CH$_3$	–	–	–	35, **6**, *13*
$C_{13}H_{11}O$	C_2H_5	–	–	–	182, **29**, *10*
$C_{13}H_{11}S$	C_2H_5	–	–	–	241, **46**, *5*
$C_{13}H_{15}N_6O$	CH_3	–	–	–	233, **45**, *7*
$C_{13}H_{15}N_6O$	CH_3	NO_3	–	H_2O	233, **45**, *7*
$C_{13}H_{17}N_2Pt$	CH_3	–	–	–	332, **62**, *7*
$C_{13}H_{21}$	–	–	–	–	85, **12**, *4*
$C_{13}H_{23}Si_2$	–	–	–	Cl	353, **67**, *18*
C$_{14}$H$_{12}$N$_2$	CH_3	–	–	–	35, **6**, *12*
$C_{14}H_{13}N_2$	CH_3	–	–	–	269
$C_{14}H_{14}N_5O$	CH_3	–	–	–	233, **45**, *8*
$C_{14}H_{14}N_5O$	CH_3	NO_3	–	H_2O	233, **45**, *8*
$C_{14}H_{16}N_2$	CH_3	–	–	–	226, **43**, *12*
$C_{14}H_{20}O_2$					
2-O-3,5-$((CH_3)_3C)_2C_6H_3O$	C_2H_5	–	–	–	230
2-O-3,6-$((CH_3)_3C)_2C_6H_3O$	C_2H_5	–	–	–	230
$C_{14}H_{29}Si_3$	–	–	–	–	380/1, **72**, *8*
C$_{15}$H$_{11}$N$_2$	CH_3	–	–	–	279, **56**, *8*
$C_{15}H_{11}N_3$	CH_3	–	–	–	227, **43**, *18*
$C_{15}H_{11}N_3$	CH_3	–	–	Cl	139, **19**, *14*
$C_{15}H_{11}N_3$	C_6F_5	–	–	Cl	141, **19**, *31*
$C_{15}H_{13}$	–	–	–	Br	169, **24**, *9*
$C_{15}H_{13}N_4O$	CH_3	–	–	–	234, **45**, *9*
$C_{15}H_{13}N_4O$	CH_3	NO_3	–	H_2O	234, **45**, *9*
C$_{16}$H$_{12}$N$_3$O	CH_3	–	–	–	234, **45**, *10*
$C_{16}H_{12}N_3O$	CH_3	NO_3	–	H_2O	234, **45**, *10*
$C_{16}H_{14}N_2$	CH_3	–	–	–	35, **6**, *14*
$C_{16}H_{16}$	–	–	–	–	388
$C_{16}H_{25}O_2$					
2-$C_2H_5O((CH_3)_3C)_2C_6H_2O$	C_2H_5	–	–	–	230
2-C_2H_5O-3,5-$((CH_3)_3C)_2C_6H_2O$	C_2H_5	–	–	–	192, **33**, *13*
2-C_2H_5O-3,6-$((CH_3)_3C)_2C_6H_2O$	C_2H_5	–	–	–	192, **33**, *14*
2-C_2H_5O-4,6-$((CH_3)_3C)_2C_6H_2O$	C_2H_5	–	–	–	192, **33**, *15*
$C_{16}H_{27}N_2$	CH_3	–	–	Cl	134/5
$C_{16}H_{27}N_2$	C_2H_5	–	–	–	111, **17**, *44*
C$_{17}$H$_{13}$N$_2$O	CH_3	–	–	–	234, **45**, *11*
$C_{17}H_{13}N_2O$	CH_3	NO_3	–	H_2O	234, **45**, *11*
$C_{17}H_{22}N_2$	CH_3	–	–	–	35, **6**, *11* 50, **8**, *6*
$C_{17}H_{22}N_2$	C_2H_5	–	–	–	71/2
C$_{18}$H$_{15}$As	C_6F_5	–	–	–	93, **14**, *28*
$C_{18}H_{15}AsO$	CH_3	–	–	Cl	120, **18**, *9*
$C_{18}H_{15}AsO$	C_6F_5	–	–	–	92/3, **14**, *24*
$C_{18}H_{15}NP$	CH_3	–	–	–	283, **57**, *5*
$C_{18}H_{15}OP$	CH_3	–	–	Cl	120, **18**, *8*
$C_{18}H_{15}OP$	CH_3	–	–	I	158, **22**, *10*

$C_{18}H_{15}OP$	C_6F_5	–	–	–	92, **14**, *23*
$C_{18}H_{15}OSi$	CH_3	–	–	–	188, **31**, *2*
$C_{18}H_{15}P$	CH_2Br	–	–	Br	172, **25**, *3*
$C_{18}H_{15}P$	CH_2I	–	–	I	172, **25**, *6*
$C_{18}H_{15}P$	CH_3	–	–	–	39, **6**, *41*
$C_{18}H_{15}P$	CH_3	–	–	Cl	121, **18**, *17*
$C_{18}H_{15}P$	CH_3	–	–	I	158, **22**, *16*
$C_{18}H_{15}P$	C_5H_5	–	–	–	86, **12**, *10*
$C_{18}H_{15}P$	C_6F_5	–	–	–	93, **14**, *27*
$C_{18}H_{28}N_2PSi_2$	CH_3	–	–	–	283, **57**, *6*
$C_{18}H_{42}LiN_4Si_2$	CH_3	–	–	–	284, **57**, *10*
C$_{19}H_{17}As$	–	–	–	Cl	368
$C_{19}H_{17}P$	–	–	–	Br	172, **25**, *8*
$C_{19}H_{17}P$	–	–	–	Cl	368
$C_{19}H_{17}P$	–	–	–	I	172, **25**, *10*
C$_{21}H_{21}P$	CH_3	–	–	–	39, **6**, *42*
$C_{21}H_{23}N_3$	CH_3	–	–	–	226, **43**, *15*
C$_{26}H_{24}P_2$	CH_3	–	–	–	52, **8**, *14*
$C_{26}H_{24}P_2$	CH_3	$C_2H_3O_2$	–	–	225, **43**, *9*
$C_{26}H_{24}P_2$	CH_3	$C_5H_4F_3O_2$	–	–	224, **43**, *3*
$C_{26}H_{24}P_2$	C_2H_5	–	–	–	72
$C_{26}H_{24}P_2$	C_6F_5	–	–	–	94, **14**, *36*
$C_{26}H_{24}P_2$	C_6F_5	–	–	Cl	141, **19**, *32*
C$_{28}H_{20}$	–	–	–	–	340, **64**, *20*
$C_{28}H_{20}$	C_5H_5N	–	–	Cl	134/5
C$_{34}H_{33}P_3$	CH_3	–	–	–	52, **8**, *15*
C$_{36}H_{44}N_4$	CH_3	–	–	–	305, **60**, *1*
$C_{36}H_{44}N_4$	C_2H_5	–	–	–	305, **60**, *2*
$C_{36}H_{44}N_4$	C_3H_7	–	–	–	305, **60**, *3*
$C_{36}H_{44}N_4$	C_4H_9	–	–	–	305, **60**, *4*
					305, **60**, *5*
$C_{36}H_{44}N_4$	C_6F_5	–	–	–	306, **60**, *10*
$C_{36}H_{44}N_4$	C_6HF_4	–	–	–	306, **60**, *9*
$C_{36}H_{44}N_4$	C_6H_5	–	–	–	306, **60**, *8*
$C_{36}H_{44}N_4$	C_8H_5	–	–	–	306, **60**, *7*
$C_{36}H_{44}N_4$	C_8H_7	–	–	–	306, **60**, *6*
C$_{40}H_{35}$	–	–	–	–	381, **72**, *10*
C$_{42}H_{42}P_4$	CH_3	–	–	–	52, **8**, *16*
C$_{44}H_{28}N_4$	CH_3	–	–	–	306, **60**, *11*
$C_{44}H_{28}N_4$	C_2H_5	–	–	–	306, **60**, *12*
$C_{44}H_{28}N_4$	C_3H_7	–	–	–	306/7, **60**, *13*

$C_{44}H_{28}N_4$	C_4H_9	–	–	–	307, **60**, *14*
					307, **60**, *15*
$C_{44}H_{28}N_4$	C_6F_5	–	–	–	307/8, **60**, *20*
$C_{44}H_{28}N_4$	C_6HF_4	–	–	–	307, **60**, *19*
$C_{44}H_{28}N_4$	C_6H_5	–	–	–	307, **60**, *18*
$C_{44}H_{28}N_4$	C_8H_5	–	–	–	307, **60**, *17*
$C_{44}H_{28}N_4$	C_8H_7	–	–	–	307, **60**, *16*
C$_{48}H_{36}N_4$					
5,10,15,20-Tetrakis(3-methyl-phenyl)porphyrin	C_6F_5	–	–	–	308, **60**, *22*
5,10,15,20-Tetrakis(3-methyl-phenyl)porphyrin	C_6HF_4	–	–	–	308, **60**, *21*
5,10,15,20-Tetrakis(4-methyl-phenyl)porphyrin	C_6F_5	–	–	–	308, **60**, *24*
5,10,15,20-Tetrakis(4-methyl-phenyl)porphyrin	C_6HF_4	–	–	–	308, **60**, *23*
C$_{56}H_{107}P_4PtSi$	$C_4H_{11}Si$	–	–	–	332, **62**, *13*

Physical Constants and Conversion Factors

Avogadro constant N_A (or L) $= 6.02214 \times 10^{23}\ mol^{-1}$
Faraday constant $F = 9.64853 \times 10^{4}$ C/mol
molar gas constant $R = 8.31451\ J \cdot mol^{-1} \cdot K^{-1}$
molar volume (ideal gas) $V_m = 2.24141 \times 10^{1}$ L/mol
(273.15 K, 101325 Pa)

Planck constant $h = 6.62608 \times 10^{-34}\ J \cdot s$
elementary charge $e = 1.60218 \times 10^{-19}$ C
electron mass $m_e = 9.10939 \times 10^{-31}$ kg
proton mass $m_p = 1.67262 \times 10^{-27}$ kg

1 kg $= 2.205$ pounds
1 m $= 3.937 \times 10^{1}$ inches $= 3.281$ feet
1 m^3 $= 2.642 \times 10^{2}$ gallons (U.S.)
1 m^3 $= 2.200 \times 10^{2}$ gallons (Imperial)

Force	N	dyn	kp
1 N	1	10^{5}	1.019716×10^{-1}
1 dyn	10^{-5}	1	1.019716×10^{-6}
1 kp	9.80665	9.80665×10^{5}	1

Pressure	Pa	bar	kp/m^2	at	atm	Torr	lb/in^2
1 Pa $= 1 N/m^2$	1	10^{-5}	1.019716×10^{-1}	1.019716×10^{-5}	9.86923×10^{-6}	7.50062×10^{-3}	1.450378×10^{-4}
1 bar $= 10^6\ dyn/cm^2$	10^{5}	1	1.019716×10^{4}	1.019716	9.86923×10^{-1}	7.50062×10^{2}	1.450378×10^{1}
1 kp/m^2 = 1 mm H_2O	9.80665	9.80665×10^{-5}	1	10^{-4}	9.67841×10^{-5}	7.35559×10^{-2}	1.422335×10^{-3}
1 at (technical)	9.80665×10^{4}	9.80665×10^{-1}	10^{4}	1	9.67841×10^{-1}	7.35559×10^{2}	1.422335×10^{1}
1 atm = 760 Torr	1.01325×10^{5}	1.01325	1.033227×10^{4}	1.033227	1	7.60×10^{2}	1.469595×10^{1}
1 Torr = 1 mmHg	1.333224×10^{2}	1.333224×10^{-3}	1.359510×10^{1}	1.359510×10^{-3}	1.315789×10^{-3}	1	1.933678×10^{-2}
1 lb/in^2 = 1 psi	6.89476×10^{3}	6.89476×10^{-2}	7.03069×10^{2}	7.03069×10^{-2}	6.80460×10^{-2}	5.17149×10^{1}	1

Work, Energy, Heat	J	kW·h	kcal	Btu	eV
1 J = 1 W·s = 1 N·m = 10^7 erg	1	2.778×10^{-7}	2.39006×10^{-4}	9.4781×10^{-4}	6.242×10^{18}
1 kW·h	3.6×10^6	1	8.604×10^2	3.41214×10^3	2.247×10^{25}
1 kcal	4.1840×10^3	1.1622×10^{-3}	1	3.96566	2.6117×10^{22}
1 Btu (British thermal unit)	1.05506×10^3	2.93071×10^{-4}	2.5164×10^{-1}	1	6.5858×10^{21}
1 eV	1.602×10^{-19}	4.450×10^{-26}	3.8289×10^{-23}	1.51840×10^{-22}	1

1 cm^{-1} = 1.239842×10^{-4} eV
1 hartree = 27.2114 eV

1 Hz = 4.135669×10^{-15} eV
1 eV ≙ 23.0578 kcal/mol

Power	kW	hp	kp·m·s^{-1}	kcal/s
1 kW = 10^3 J/s	1	1.35962	1.01972×10^2	2.39006×10^{-1}
1 hp (horsepower, metric)	7.3550×10^{-1}	1	7.5×10^1	1.7579×10^{-1}
1 kp·m·s^{-1}	9.80665×10^{-3}	1.333×10^{-2}	1	2.34384×10^{-3}
1 kcal/s	4.1840	5.6886	4.26650×10^2	1

References:

Mills, I. (Ed.), International Union of Pure and Applied Chemistry, Quantities, Units and Symbols in Physical Chemistry, Blackwell Scientific Publications, Oxford 1988.

The International System of Units (SI), National Bureau of Standards Spec. Publ. 330 [1972].

Landolt-Börnstein, 6th Ed., Vol. II, Pt. 1, 1971, pp. 1/14.

ISO Standards Handbook 2, Units of Measurement, 2nd Ed., Geneva 1982.

Cohen, E. R., Taylor, B. N., Codata Bulletin No. 63, Pergamon, Oxford 1986.

Key to the Gmelin System of Elements and Compounds

System Number	Symbol	Element
1		Noble Gases
2	H	Hydrogen
3	O	Oxygen
4	N	Nitrogen
5	F	Fluorine
6	**Cl**	**Chlorine**
7	Br	Bromine
8	I	Iodine
8a	At	Astatine
9	S	Sulfur
10	Se	Selenium
11	Te	Tellurium
12	Po	Polonium
13	B	Boron
14	C	Carbon
15	Si	Silicon
16	P	Phosphorus
17	As	Arsenic
18	Sb	Antimony
19	Bi	Bismuth
20	Li	Lithium
21	Na	Sodium
22	K	Potassium
23	NH_4	Ammonium
24	Rb	Rubidium
25	Cs	Caesium
25a	Fr	Francium
26	Be	Beryllium
27	Mg	Magnesium
28	Ca	Calcium
29	Sr	Strontium
30	Ba	Barium
31	Ra	Radium
32	**Zn**	**Zinc**
33	Cd	Cadmium
34	Hg	Mercury
35	Al	Aluminium
36	Ga	Gallium

HCl → 6; $ZnCl_2$ → 32; $CrCl_2$ → 52; $ZnCrO_4$ → 52

System Number	Symbol	Element
37	In	Indium
38	Tl	Thallium
39	Sc, Y La—Lu	Rare Earth Elements
40	Ac	Actinium
41	Ti	Titanium
42	Zr	Zirconium
43	Hf	Hafnium
44	Th	Thorium
45	Ge	Germanium
46	Sn	Tin
47	Pb	Lead
48	V	Vanadium
49	Nb	Niobium
50	Ta	Tantalum
51	Pa	Protactinium
52	**Cr**	**Chromium**
53	Mo	Molybdenum
54	W	Tungsten
55	U	Uranium
56	Mn	Manganese
57	Ni	Nickel
58	Co	Cobalt
59	Fe	Iron
60	Cu	Copper
61	Ag	Silver
62	Au	Gold
63	Ru	Ruthenium
64	Rh	Rhodium
65	Pd	Palladium
66	Os	Osmium
67	Ir	Iridium
68	Pt	Platinum
69	Tc	Technetium[1]
70	Re	Rhenium
71	Np,Pu...	Transuranium Elements

Material presented under each Gmelin System Number includes all information concerning the element(s) listed for that number plus the compounds with elements of lower System Number.

For example, zinc (System Number 32) as well as all zinc compounds with elements numbered from 1 to 31 are classified under number 32.

[1] A Gmelin volume titled "Masurium" was published with this System Number in 1941.

A Periodic Table of the Elements with the Gmelin System Numbers is given on the Inside Front Cover